AF528599

CASHEW
Improvement, Production and Processing

The Editor

Prof. (Dr.) P.L. Saroj is an ARS Scientist (Horticulture) of 1990 batch. He has served in different organizations on various capacities such as Assistant Professor, CCS University, Meerut (UP) from 1989-90, Scientist, IIHR, Bangalore, (Karnataka) from 1991-92, Scientist IISWC, Dehradun (Uttarkhand) from 1992-1999, Senior Scientist and Principal Scientist, CIAH, Bikaner (Rajasthan) from 1999-2005, Prof & Head, Department of Horticulture/Dean (SW), SVPUAT, Meerut (UP) from 2005-2008 and Principal Scientist as well as Assistant Director General (Hort.-I) from 2009-2012 at ICAR HQ, New Delhi. In 2012, he joined as Director, ICAR- Directorate of Cashew Research, Puttur (Karnataka) and also shouldering the responsibility of Project Coordinator, ACIRP on Cashew. Prof. Saroj, posses versatile experiences while serving in different agro-ecosystems, including hill and mountain, arid, tropical, subtropical and coastal ecosystems. He is involved in research, extension, teaching, coordination, management and administration for the last 27 years. His major field of research is production system management in horticulture. He has standardized various production technologies, developed varieties and value added products for different fruit crops. He has published his first report on peach phytoplasma and developed jumbo nut size cashew variety in India. He was instrumental in institution building activities such as establishment of new Horticulture Block, IISWC, Dehradun. Fruit culture Lab & Nursery Unit, CIAH, Bikaner, NRC on Pomegranate, Solapur, MAP Centre and Hitech Floriculture Centre, SVPUAT, Meerut, KVK at Godhra, and Plant Protection Lab. & Silver Jubilee Building, DCR, Puttur. Besides founder Secretary of Indian Society for Arid Horticulture, he is associated with 12 professional societies as Advisor, Executive Councilor, Editor and Life Member, Over 125 research publication of national and international repute are in his credit. His book 'Advance in Arid Horticulture' has been widely appreciated by the scientist community. Prof. Saroj Is recipient of several awards and recognition including fellow of four professional societies, Dr. B.P Pal Scientist of the year-2009 and Scientist of the year 2011 Awards. Prof. Saroj also visited Kenya, Uganda, Tanzania, Nepal, Sri Lanka and USA under various official assignments.

CASHEW
Improvement, Production and Processing

— *Editor* —

Prof. P.L. Saroj

Director
ICAR-Directorate of Cashew Research,
Puttur – 574 202 (Karnataka)

2017

Daya Publishing House®

A Division of

Astral International Pvt. Ltd.

New Delhi – 110 002

Publisher's Note:

Every possible effort has been made to ensure that the information contained in this book is accurate at the time of going to press, and the publisher and author cannot accept responsibility for any errors or omissions, however caused. No responsibility for loss or damage occasioned to any person acting, or refraining from action, as a result of the material in this publication can be accepted by the editor, the publisher or the author. The Publisher is not associated with any product or vendor mentioned in the book. The contents of this work are intended to further general scientific research, understanding and discussion only. Readers should consult with a specialist where appropriate.

Every effort has been made to trace the owners of copyright material used in this book, if any. The author and the publisher will be grateful for any omission brought to their notice for acknowledgement in the future editions of the book.

Cataloging in Publication Data--DK
Courtesy: D.K. Agencies (P) Ltd. <docinfo@dkagencies.com>

Cashew : improvement, production and processing / editor, Prof. P.L. Saroj.
pages cm
Contributed articles.
Includes index.
ISBN 978-93-5124-825-5 (Hardbound)
ISBN 978-93-86071-55-2 (International Edition)

1. Cashew--India. I. Saroj, P. L., editor.

SB401.C3C37 2017 DDC 634.5730954 23

Published by : **Daya Publishing House®**
A Division of
Astral International Pvt. Ltd.
– ISO 9001:2015 Certified Company –
4736/23, Ansari Road, Darya Ganj
New Delhi-110 002
Ph. 011-43549197, 23278134
E-mail: info@astralint.com
Website: www.astralint.com

Digitally Printed at : **Replika Press Pvt. Ltd.**

This copy is for sale in Indian Sub-Continent.

प्रो. रमेश चन्द
सदस्य
Prof. Ramesh Chand
MEMBER
Tele. : 011-23096774
Telefax : 011-23096756
E-mail : rc.niti@gov.in

सत्यमेव जयते

भारत सरकार
नीति आयोग, संसद मार्ग
नई दिल्ली-110 001
Government of India
NATIONAL INSTITUTION FOR TRANSFORMING IN
NITI Aayog, Parliament Street
New Delhi-110 001

FOREWORD

With sustained research and development efforts by different research organizations in the country, cashew (*Anacardium occidentale* L.) has emerged as one of the most important export oriented horticulture crop in India. At present, India is major player in cashew trade and earning a substantial amount of foreign exchange by export of cashew kernel and cashew nut shell liquid. Besides, cashew processing industries in India is generating job opportunities especially to the womenfolk. However, productivity of cashew in the country has remained stagnant though the area and production have been increasing since last several years.

Research and development programmes at ICAR-Directorate of Cashew Research at Puttur and also at different Centres of All India Coordinated Research Project on Cashew under State Agricultural Universities have resulted in the development of a number of high yielding varieties and location specific production technologies. These technologies need to be transferred and adopted at farmers' field to break yield stagnation.

I feel happy to find that valuable information on technology and other aspects of cashew nut has been compiled by the ICAR-Directorate of Cashew Research in the form of a book titled "Advances in Cashew Research". The work done in the area of germplasm conservation, crop improvement, crop management, crop protection and post harvest technology done so far has also been included in this publication. I am hopeful the book will serve as useful guide in formulating future research and development strategies for improving production and productivity of cashew in the country beside providing reference material to the researchers, teachers, students, and extension workers. I take this opportunity to congratulate Dr. P. L. Saroj, Director, ICAR-Directorate of Cashew Research, Puttur for his efforts in compiling and presenting valuable information on cashew in one place in the form of book.

[Ramesh Chand]

Preface

Cashew (*Anacardiun occidentale* L.) is a precious gift of nature to the mankind. Cashew is produced commercially in as many as 32 countries in the tropical regions of Asia, Africa and Latin America. Asiatic zone comprising of India, Indonesia, Philippines, Sri Lanka, Thailand, Vietnam and Malaysia accounts for major global production of cashew nut. In India, cashew was introduced by Portuguese travelers in 16th century for afforestation of degraded lands in coastal region but now cashew has moved from forest confine to commercial horticulture crop. At present, cashew is grown in an area of about 10.27 lakh ha with a production of 7.25 lakh tonnes of raw cashew nuts and a productivity of 706 kg/ha (2014-15). Though, India is largest producer, processor, exporter and consumer of cashew in the world and earning foreign exchange of Rs. 5488.66 crores per annum (2014-15) by the export of cashew kernel and cashew nut shell liquid but still India is far behind in cashewnut productivity than potential productivity. This necessitates reorienting our research and development strategies in order to address the growing demand of raw cashewnut by the processing industries.

After establishment of ICAR-Directorate of Cashew Research, Puttur, Karnataka (Formerly National Research Centre of Cashew) in 1986 and also separate All India Coordinated Research Project on Cashew; large number of varieties and production as well as processing technologies have been developed for adoption of end users. Therefore, it was thought to have stock of research and development information available in the country in form of a book *"Cashew: Improvement, Production and Processing"* which will serve as bench mark to decide way forward in cashew sector.

I hope that the publication will serve as a valuable reference to the academicians, policy planners, development departments and extension workers engaged in cashew research and development in the country.

I take this opportunity to express our sincere gratitude to Dr. N. K. Krishna Kumar, Deputy Director General (Hort. Sci.), ICAR, New Delhi for his guidance to bring out this publication. I am grateful to former Directors of ICAR-Directorate of Cashew Research, Puttur; all ICAR-DCR staff and all those who are involved in R and D of cashew in the country including farmers. The support of my wife Smt. Neelam Saroj and Mr. O.G. Varghese, Private Secretary to the Director, ICAR-DCR in bringing out this publication is also acknowledged.

Prof. P.L. Saroj

Contents

Section III: Production System Management

Section IV: Plant Health Management

Section V: Postharvest Technology

Section VI: Technology Dissemination and Economics

Section VII: Impact Analysis

List of Contributors

Dr. N.K. Krishna Kumar
Deputy Director General (Hort. Science)
Indian Council of Agricultural Research
Krishi Anusandhan Bhavan-II
Pusa Campus, New Delhi – 110 012
e-mail: nkkrishna.icar@nic.in

Prof. P.L. Saroj
Director
ICAR – Directorate of Cashew Research
Puttur – 574 202, Dakshina Kannada, Karnataka
e-mail: plsaroj@yahoo.co.in

Dr. Abraham Verghese
Director
ICAR-National Bureau of Agricultural Insect Resources,
PB No.2491, H.A.Farm Post,
Bellary Road,
Bengaluru – 560 024, Karnataka
e-mail: directornbaii@gmail.com

Shri Venkatesh N. Hubballi
Director
Directorate of Cashewnut & Cocoa Development,
Kera Bhavan, Kochi-682 001, Kerala
e-mail: dccd@nic.in

Dr. P.B. Pushpalatha
Director (Extension)
CRS, Madakkathara, Kerala Agricultural University,
Thrissur, Kerala – 680 656
e-mail : pushpavellanikkara@gmail.com

Dr. D. Sundararaju
Former Pr. Scientist (Agril. Ento.),
NBAIR, Bengaluru
#G5, B-Block, JS Harmony Apartments,
3rd Main, Hoysala Nagar
RM Nagar Post,
Bengaluru – 560 016, Karnataka
e-mail: dsraju21@yahoo.com

Dr. P.S. Bhat
Principal Scientist (Agril. Entomology)
ICAR-Directorate of Cashew Research
Puttur -574 202, DK, Karnataka
e-mail: pshivarama59@gmail.com

Dr. T.N. Raviprasad
Principal Scientist (Agril. Entomology)
ICAR-Directorate of Cashew Research
Puttur – 574 202, DK, Karnataka
e-mail: tnrprasaad@gmail.com

Dr. V. Pandey
Principal Scientist (Hort.)
Indian Council of Agricultural Research
Krishi Anusandhan Bhavan – II
Pusa Campus, New Delhi – 110 012
e-mail: pandey_vs@rediffmail.com

Dr. M.G. Nayak
Principal Scientist (Hort.)
ICAR-Directorate of Cashew Research
Puttur – 574 202, DK, Karnataka
e-mail: mgnayak20@yahoo.com

Dr. T.R. Rupa
Principal Scientist (Soil Science)
ICAR – Directorate of Cashew Research
Puttur-574 202, DK, Karnataka
e-mail : sairupa.tr@gmail.com

Dr. D. Balasubramanian
Principal Scientist (AS&PE)
ICAR – Directorate of Cashew Research
Puttur – 574 202, DK, Karnataka
e-mail : bavika13@email.com

Dr. S. Jeeva
Professor
Horticultural College and Research Institute for Women,
Navalur Kuttappattu,
Tiruchirappalli – 620 009 (TN)
e-mail: jeeva_62@yahoo.com,
jeedharan@gmail.com

Shri R. Jnanadevan
Deputy Director (Marketing)
Directorate of Cashewnut & Cocoa Development,
Kera Bhavan
Kochi – 680 617, Kerala

Dr. V. V. Ramamurthy
Former Principal Scientist & National Coordinator,
NPIB, Division of Entomology
ICAR-Indian Agricultural Research Institute
New Delhi – 110 012

Dr. R. Asokan
Principal Scientist (Agrl. Ento.),
Division of Biotechnology
ICAR – Indian Institute of Horticultural Research,
Hessaraghatta Lake PO,
Bengaluru – 560 089, Karnataka
e-mail: asokan@iihr.ernet.in

Dr. A. Sobhana
Professor (Horticulture),
CRS, Madakkathara,
Kerala Agricultural University,
Thrissur, Kerala – 680656
e-mail : cohvka@kau.in

Dr. P. Indira Devi
Professor (Agri. Economics),
College of Horticulture
Kerala Agricultural University,
Thrissur, Kerala – 680 656
e-mail : cohvka@kau.in

Dr. J.D. Adiga
Professor (Hort.)
University of Horticultural Sciences,
College of Horticulture,
Bangalore, Karnataka
e-mail: dinakara.adiga@gmail.com

Dr. G.S. Mohana
Sr. Scientist (Genetics)
ICAR – Directorate of Cashew Research
Puttur – 574 202, DK, Karnataka
e-mail: mohangs2007@gmail.com

Dr. R. Rejani
Sr. Scientist (S&WCE)
ICAR – Central Research Institute for Dryland Agriculture,
Santoshnagar,
Hyderabad – 500 059, Andhra Pradesh
e-mail: rrejani10@gmail.com

Dr. K. Vanitha
Scientist (Agril Entomology)
ICAR – Directorate of Cashew Research
Puttur – 574 202, DK, Karnataka
e-mail: vanis102@gmail.com

Dr. Ramkesh Meena
Scientist (Hort)
ICAR – Directorate of Cashew Research
Puttur – 574 202, DK, Karnataka
e-mail: rkmeena8119@yahoo.com

Dr. Mini Poduval
Horticulturist
Regional Research Station (BCKV),
Jhargram, Midnapore West Dist.,
West Bengal-721 507
e-mail: poduvalmini@gmail.com

Dr. P.D. Sreekanth
Senior Scientist (Computer Science)
ICAR- National Academy of Agricultural Research Management,
Rajendranagar
Hyderabad – 500 030, Andhra Pradesh
e-mail: pd_sreekanth@yahoo.com

Dr. D. Kalaivanan
Scientist (Soil Science)
ICAR – Indian Institute of Horticultural Research,
Hessarghatta Lake Post,
Bengaluru – 560 089, Karnataka
e-mail : kalainith2003@gmail.com

Shri E. Eradasappa
Scientist (Plant Breeding)
ICAR – Directorate of Cashew Research
Puttur – 574 202, DK, Karnataka
e-mail: era.dasappa@gmail.com

Dr. M.V. Sajeev
Scientist (Agril Extn)
ICAR – Directorate of Cashew Research
Puttur – 574 202, DK, Karnataka
e-mail: sajeevmv@yahoo.co.in

Dr. K.K. Srikumar
Entomologist
UPASI, TRF, Tea Research Institute
Valparai – 642 127,
Coimbatore, T.N.
e-mail: sreeku08@gmail.com

Dr. M. Lognathan
Sr. Scientist (Plant Pathology)
ICAR – Directorate of Cashew Research
Puttur – 574 202, DK, Karnataka
e-mail: logumuruga@gmail.com

Dr. K. Subaharan
Senior Scientist (Agrl. Entomology)
ICAR-National Bureau of Agricultural Insect Resources,
PB No. 2491, H.A. Farm Post,
Bellary Road,
Bengaluru – 560 024, Karnataka
e-mail: subaharan_70@yahoo.com

Dr. K. M. Yuvaraj
Sr. Scientist (Hort.) & Head
Cashew Research Station,
Bapatla-522 101
(Dr. YSR Horticulture University),
Guntur Ditt., Andhra Pradesh
e-mail: cashewresebapt@gmail.com

Mr. K.B. Rebijith
Research Associate
Division of Biotechnology
ICAR-Indian Institute of Horticultural Research,
Hessaraghatta Lake PO,
Bengaluru – 560 089, Karnataka

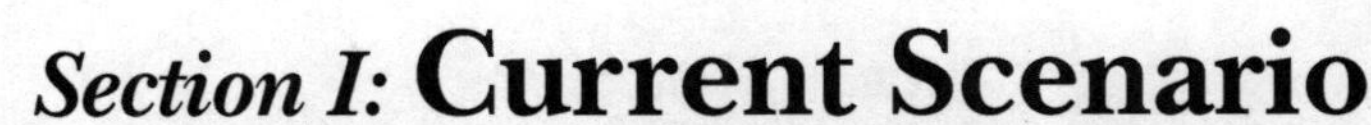

Section I: Current Scenario

Cashew: Improvement, Production and Processing (2017) *Pages* **1–29**
Editor: **Prof. P.L. Saroj**
Published by: **ASTRAL INTERNATIONAL PVT. LTD., NEW DELHI**

1

Cashew in India

P.L. Saroj

1. Introduction

The Cashew (*Anacardium occidentale* L.), is derived from the Portuguese word 'Caju'. The cashew tree belongs to the genus *Anacardium,* a member of the family *Anacardiaceae,* which comprises about 60 genera and 400 species. The chromosome number of *Anacardium occidentale was* reported to be 2n=42 (Darlington and Janaki Ammal, 1945). About 21 species of *Anacardium* are reported to exist, *viz.*, *A. amilcarianum, A. brasiliense, A. ciratellaefolium, A. corymlosum, A. encardium, A. excelsum, A. gigantum, A. humile, A. kuhlanannianum, A. mediterraneum, A. microcarpum, A. microsepalum, A. nanum, A. negrense, A. orthonianum, A. parvifolium, A. pumilum, A. rhinocarpus, A. rondonianum, A. spruceanum* and A. *tenuifolium.* Cashew is primarily a cross-pollinated crop and grows up to as much as 15 m height and canopy spread up to 12 m under favourable conditions. Vegetatively propagated cashew trees start bearing at the age of 2-3 years whereas seedling-raised trees may take 4 to 5 years for flowering. Cashew is an andromonoecious plant with staminate and hermaphrodite flowers, appearing on the same inflorescence or panicle (Rao and Hassan, 1957; Damodaran *et al.*, 1965). Flowering lasts for 2 to 3 months. It takes 50 to 70 days from pollination to maturity of the fruit. The cashew fruit has two distinct parts, *viz.* the swollen and pear shaped peduncle (cashew apple) and a kidney-shaped nut attached to the lower end of the apple. The nut contains the outer hard shell and the inside seed; the seed consists of the outer testa and the edible kernel.

Cashew is native of northern Brazil, where the fruit was first seen and Thevet (1958) was the first to describe the cashew tree by referring to its occurrence in the above region. From where, it has been spread to South and Central America. It reached West Indies in pre-Columbian times. The Portuguese also introduced it to East Africa. In the later part of 16th century and to be precise between 1563 and 1578, the Portuguese colonizing in the Goa region introduced cashew into India (Costa, 1578). Thereafter it was spread all along lateritic hill slopes of Konkan region. It has spread to Sri Lanka, Malaysia and Indonesia also. The Spaniards took it to the Philippines during the 17th century. Initially, cashew was considered as a forest crop for afforestation of wastelands and for soil conservation purpose as the root system of a mature tree consists of a very prominent taproot and a well-developed and extensive network of laterals. Realizing its importance as a hardy plant, precocious bearing habit, delicious snack food and good foreign exchange earning crop, it has been adopted as one of the important horticultural crops in India. At present, cashew is grown along the coastal regions mainly in Maharashtra, Goa, Karnataka and Kerala in the West coast and Tamil Nadu, Andhra Pradesh, Odisha and West Bengal in the East coast. It is spreading in non-traditional areas such as Bastar region of Chhattisgarh and Kolar (Plains) region of Karnataka, Gujarat, Jharkhand and in North East hill (NEH) region. In India, cashew nut production is 7.25 lakh tonnes from 10.27 lakh ha with average annual productivity of 706 kg/ha (2014-15).

Cashew kernel is a delicious snack food and rich source of protein (21 per cent), carbohydrate (22 per cent) and fat (47 per cent). Cashew nut shell liquid (CNSL) obtained while processing is of high industrial value such as manufacture of paints, varnishes, brake linings for automobiles, friction dust *etc* (Nayar, 1998). The thin-skinned edible cashew apple (pseudo fruit) has a light yellow spongy flesh, which is very juicy, pleasantly acidic and highly astringent when green. The cashew apple which is slightly astringent contains sugars, tannins, phenols, amino acids, ascorbic acid, minerals and fiber. Besides these, it also contains riboflavin and fairly good amount of ascorbic acid (240 mg/100 g). 'Feni' is a popular beverage from cashew apple but several other products like juice, jam, jelly, pickle, candy, toffee *etc.* can be prepared. There is also possibility of utilizing apple pomace as animal, fish and poultry feed. The cashew apple can also be utilized to some extent in biogas plant and for production of methane gas. The sap, which oxidizes black upon exposure to the air, is employed as an indelible ink marker for cottons and linens The wood is employed locally for construction, storage cases, boats, wheel hubs, and as a fuel wood (Mitchell and Mori, 1987). The cashew tree is mainly used as fire-wood but also for certain commercial purposes like manufacture of plywood, particle boards *etc.* (Nayar, 1998).

Besides above, some facts about Indian cashew sector are given below:

- ☆ Though, cashew is native of Brazil but India is the first country to commercialize it as one of the important horticultural crops.
- ☆ Cashew is cultivated in 32 countries of Latin America, Asia, Africa and Australia covering an area of 53.13 lakh ha with the production of 44.41 lakh tones (FAO, 2013).
- ☆ In India, cashew is grown in an area of about 10.27 lakh ha with a production of 7.25 lakh tonnes of raw cashew nuts and a productivity of 706 kg/ha (2014-15).

- ☆ As per statistics, India is third largest producer of raw cashew nut (16.4 per cent) after Vietnam (28.9 per cent) and Nigeria (19.5 per cent) but it is pertinent to mention that India is still on top position both in area and production of raw cashew nut as production data of some countries include imported quantity of raw nut also. As per Cashew Hand Book: 2014; the share of India in raw cashew nut production is highest (22 per cent) in the world, followed by Cote d Ivoire (19 per cent), Brazil (16 per cent), Indonesia (13 per cent), Vietnam (7 per cent), Tanzania (2 per cent), Mozambique (1 per cent) and others (20 per cent).
- ☆ The share of cashew kernel in global nut production is about 16 per cent.
- ☆ There are more than 10 lakh people involved in cashew sector, out of which about 90 per cent are women.
- ☆ India is the largest processor of raw cashew nut with processing capacity of 20 lakh tones, largest exporter of cashew kernel (about 1.5 lakh tones/annum) and earning of foreign exchange of Rs. 5489/annum through export of cashew kernel and cashew nut shell liquid (CSNL).
- ☆ There are about 3799 registered cashew nut processing units spread in different states as reported but total processing units are 6210 including large, medium and small.
- ☆ Cashew apple production is about 8 times more than that of nut production but 'Feni' is the only commercial fermented product prepared from cashew apple in Goa state.
- ☆ Production of quality planting materials through soft wood grafting has changed the scenario of cashew plantations in India.
- ☆ In India, 54 varieties of cashew including 33 selections and 21 hybrids have been developed so far.

2. Importance of Cashew

Culinary Value

Cashew kernels are eaten either as its own, salted or sweetened. The cashew kernels are widely used in confectionery, as an addition to biscuits, sweets and cakes. Ground and mixed with cocoa the kernels make a good chocolate. Lower grade kernels are processed in to cashew flour which has high protein content and is easily digested. The kernels contain 35-40 per cent edible oil comparable to olive oil. Kernel residue after extraction of kernel oil is used to produce cashew kernel butter which is almost similar to peanut butter. There is a possibility of developing sugar, honey and salt coated baby bits, which are organoleptically acceptable. Baby bits are the lowest grade kernels marketed commercially. Sweetened and flavored milk could be prepared from cashew kernel baby bits. Cashew spread can be prepared from kernel baby bits. Sweetened and Vanillin flavored cashew spread is most preferred to salted spread (Bhaskara Rao and Swamy, 2002).

Nutritional Value

The cashew kernel is a unique combination of fats, proteins, carbohydrates, minerals and vitamins. Cashew kernel contains 47 per cent fat, but 82 per cent of this fat is unsaturated fatty acid such as oleic acid (73.7 per cent), linoleic acid (7.67 per cent) and stearic acid (11.2 per cent). The unsaturated fat content of cashew kernel not only eliminates the possibility of increasing cholesterol, but also balances or reduces the cholesterol level in the blood. Besides these, it contains minerals such as Ca, P, Na, K, Mg, Fe, Cu, Zn and Mn. Cashew kernel contains few vitamins like thiamine, niacin and vitamin E (Ohler, 1979). Cashew kernel also contains 21 per cent proteins and 22 per cent carbohydrates and the right combination of amino acids, minerals and vitamins and therefore nutritionally, it stands on par with milk, eggs and meat. As cashew kernel has as low as 1 per cent soluble sugar, the consumer of cashew kernel is privileged to get a sweet taste without having to worry about excess calories. Consumption of cashew kernels does not lead to obesity and helps to control diabetes. In short, it is a good appetizer, an excellent nerve tonic, a stimulant and a body builder (Nayar, 1998).

Medicinal Value

Cashew kernel is packed with energy, fat, protein, antioxidants, minerals, vitamins and soluble dietary fiber that are essential for good health. Kernels are high in calories, as 100 g of kernels provide 553 calories. Eating cashew is considered "heart-friendly" diet as monounsaturated fatty acids like oleic and palmitoleic acids help to lower LDL (low-density lipoprotein) cholesterol or bad LDL cholesterol and increase HDL (high-density lipoprotein) cholesterol or good HDL cholesterol. Minerals especially manganese, potassium, copper, iron, magnesium, zinc and selenium are concentrated in the kernels. A handful of cashew kernels a day in the diet would provide enough of these minerals and prevent deficiency diseases. Selenium is an important micronutrient, which functions as co-factor for antioxidant enzymes such as Glutathione peroxidases, one of the most powerful antioxidant in the body. Copper is a cofactor for many vital enzymes. Zinc is a co-factor in many enzymes that regulate growth and development, sperm generation, digestion and nucleic acid synthesis. Cashews are also rich in many essential vitamins such as pantothenic acid (vitamin B_5), pyridoxine (vitamin B_6), riboflavin, and thiamin (vitamin B_1). These vitamins are essential in the sense that our body requires them from external sources to replenish and essential for metabolism of protein, fat, and carbohydrates at cellular levels. In addition, the kernels are also containing small amount of zea-xanthin, an important flavonoid antioxidant, which selectively absorbed into the retinal macula lutea in the eyes. It is thought to provide antioxidant and protective UV ray filtering functions and helps prevent age related macular degeneration in the elderly (Anon., 2012). The young leaves are edible and older leaves in combination with other plants are used to treat skin diseases and burns. The bark has a variety of medicinal uses such as the treatment of diarrhoea and constipation; mouth ulcers; tooth ache and sore gums. The bark and inflorescences are used in traditional Indian remedies for snake bite.

3. Research Accomplishment

The first *ad-hoc* scheme on cashew research was initiated during 1950's by the Indian Council of Agricultural Research (ICAR), New Delhi but cashew research got needed impetus with the establishment of Central Plantation Crops Research Institute (CPCRI) at Kasaragod in 1970 and the All India Coordinated Spices and Cashewnut Improvement Project in 1971. Moreover, implementation of the World Bank aided multi-state cashew project (MSCP) during 1982 to 1986 in Kerala, Karnataka, Andhra Pradesh and Odisha states had given cashew research a new direction. The Quinquennial Review Team (QRT) constituted by ICAR in 1982, after reviewing the entire research work on cashew, recommended delinking of cashew research from CPCRI and establishing an independent National Research Centre on Cashew (NRCC), thereby NRCC was established on 18th June 1986 at Puttur (Karnataka). The All India Coordinated Research Project on Cashew (AICRP on Cashew) was also delinked from AICRP on Spices and Cashewnut Improvement. During XI Plan, National Research Centre for Cashew was upgraded and renamed as Directorate of Cashew Research (DCR) by ICAR in 2009. AICRP on Cashew has now 14 centres to undertake location specific research programmes in the twelve cashew growing states (Andhra Pradesh, Odisha, Karnataka, Chhattisgarh, West Bengal, Gujarat, Goa, Kerala, Maharashtra, Tamil Nadu, Jharkhand and Meghalaya). Several R and D organizations are also engaged in the development of cashew. The Directorate of Cashewnut and Cocoa Development (DCCD), Kochi is the major agency involved in technology dissemination through Central Sector Schemes including National Horticulture Mission (NHM), besides Cashew Development/Plantation Corporations and Forest Departments in the states of Kerala, Karnataka, Goa, Tamil Nadu, Andhra Pradesh and Odisha are also involved in the developmental activities. Whereas, export is being looked after by Cashew Export Promotion Council of India (CEPCI), Kerala. Therefore, the scenario of cashew sector is changing in our country.

Cashew is a plant of tropical climate and can be grown in variety of soils like loamy red and lateritic soils, mixed red and black soils, coastal and deltaic alluvium derived soils. However, frost is main limiting factor of cashew plantation. The elevation of the cashew growing areas ranged from 0 to 1000 m above mean sea level (MSL) and the productivity of cashew is higher in regions up to 750 m above MSL. The mean annual rainfall distribution in cashew growing areas ranges from low rainfall (300-600 mm in Gujarat) to high rainfall (2700 to 3000 mm in West coast and NEH region) and the mean annual temperature ranges from 20.0 to even more than 27.5 °C. The productivity is highest in regions with a mean annual rainfall distribution of 600 to 1500 mm and a mean annual temperature of 22.5 to 27.5 °C. Similarly, the productivity is higher in regions where the minimum temperature ranged from 10 to 22 °C and was lower in regions where the minimum temperature drops below 10 °C. Cashew requires relatively dry weather and mild winter (15-20°C minimum temperature) coupled with moderate dew during night for profuse flowering. High temperature (>34.4° C) and low RH < 20 per cent) during afternoon results in drying of flowers and subsequent yield reduction. However, best production was noticed up to the altitude of 700 m, mean annual temperature range of 22.5-27.5°C with at least 8-9 hrs of sunlight/day from December-May. Prolonged and unseasonal rainfall along with high wind velocity, high temperature and moisture stress during flowering, fruit setting and nut development stages results in heavy yield loss and adversely affects the nut quality even under better management conditions.

Cashew is commonly grown in degraded slopy lands where poor fertility and soil erosion are the main limiting factors. In general, cashew experiences moisture stress from January to May, adversely affects its flowering and fruit set. To overcome this problem, *in-situ* rainwater harvesting and making it available to the cashew plant during critical periods is very important. For conserving soil moisture, preparing terraces around the plant and opening of catch pits have been found beneficial. Before the onset of south west monsoon (May-June), terraces of 1.5m radius should be made during the second year of planting and this should be widened up to 2.0m during the third year. A catch pit (200cm long x 30cm wide x 45cm deep) across the slope at the peripheral end of the terrace is made for withholding water during pre-monsoon and post-monsoon showers in slopy areas. A small channel connecting catch pit sideways is made to drain out the excess water during rainy season. Further, modified crescent bund (at 2 m radius having a crescent shaped bund of 6 m length, 1 m width and 0.5 m height on the upstream of the plant) was also found very useful. Formation of trenches (5 m length, 1m width and 0.5 m depth in the middle of 4 plants) and filling with coconut husk burial is also suggested for improving yield of cashew grown on degraded lands. Mulching by dry leaves, weed biomass, polythene etc. is also helpful in conservation of moisture. Using coconut coir pith as soil mulch in cashew plantations resulted in 14.15 per cent more water retention and suppression of weeds to an extent of 73.52 per cent (Kumar *et al.*, 1989).

Genetic Resource Management and Biotechnology

Though, there are 21 species reported in cashew but only 4 species are found in India *i.e. A. occidentale, A. pumilum, A. orthonianum, A. microcarpum*. However, *Semecarpus prainii* a related wild species of cashew with small fruit size is also collected from Andaman and planted at ICAR-DCR Farm Shantigodu. The variability that exists in natural population of cashew is from *A. occidentale* only and all released varieties and hybrids are from *Anacardium occidentale* L. which were selected either from natural population and maintained by vegetative means or selected with desirable traits after hybridization. In fact, the initial plantations of cashew in the country are of seedling origin and being a cross pollinated crop, a lot of variability in seedling plantations were observed and collected from different cashew growing areas. The Directorate of Cashew Research, Puttur, has country's largest cashew germplasm collection of 539 accessions maintained at its National Cashew Field Gene Bank (NCFGB). A total of 433 accessions have been assigned with National Collection numbers (IC Nos). A total of 478 accessions have been characterised as per IPGRI descriptors. Besides morphological characterization, these accessions were also characterized for nut and kernel quality attributes, cashew apple parameters, cashew nut cell liquid (CNSL) content and relative tolerance to tea mosquito bug. Five germplasm catalogues for 478 accessions have been brought out (Swamy *et al.*, 1997, 1998, 2000 and Nayak *et al.*, 2014 and 2015). Besides these collections, 1124 cashew accessions have been conserved in Regional Cashew Field Gene Banks (RCFGBs) in AICRP on Cashew centres located in different SAUs but still there is a scope for collection and conservation of cashew biodiversity for various useful traits.

Very little efforts have been made in cashew on application of biotechnology. Attempts were made to fingerprint cashew using RAPD markers, *i.e.*, 20 Tanzanian cashew accessions (Mneney *et al.*, 2001), 90 cashew germplasm accessions at DCR, Puttur (Dhanaraj *et al.*, 2002), 19 cashew germplasm accessions at NBPGR, New Delhi (Archak *et al.*, 2002). About 239 accessions at DCR have been fingerprinted by Thimmappaiah *et al.* (2009a). Fingerprinting of 40 varieties/lines was carried out with RAPD, ISSR and SSR markers at DCR and observed low diversity. Ten primers each of RAPD and ISSR markers and 15 pairs of primers of SSR were employed to generate polymorphic bands. The similarity index between the pairs of varieties varied from 0.70 to 0.84 with highest similarity observed between Bhaskara (Goa 11/6) and VRI-3 and Ullal-4 and H 32-4. In addition to DNA markers, Isozyme markers from 10 enzymes were used to characterize 30 varieties of cashew. DNA bulks were constituted for contrasting phenotype (characters) and screened with different markers. From Bulk Segregant Analysis of germplasm, four RAPD markers linked to economic characters like nut weight and plant stature were identified.

Crop Improvement

In crop improvement programme, major emphasis was given to the characters like bigger nut size and high production potential. With the concerted efforts of Directorate of Cashew Research, Puttur and Centres of AICRP on Cashew located in different SAUs/ICAR institutes, a total of 54 varieties has been identified in the country and 43 varieties (29 selections and 14 hybrids) have been developed and released through DCR/AICRP on Cashew for cultivation in varied agro-ecological situations. Out of 43 varieties, 21 having export grade kernels (W 180 to W 240). Cashew varieties recommended for different states are given in Table 1.1. The varieties released by DCR *viz.*, NRCC Selection-2 and Bhaskara are medium nut types with high yield and recommended for cultivation in west coast region. These varieties exhibited better yield performance in other cashew growing regions of the country as well. Among the hybrids under evaluation, hybrids namely H-43, H- 66, H-68 (cross combinations of NRCC Sel-2 × Bhutnath-II), H-125 and H-126 (cross combinations of NRCC Sel-2 × Bhedasi) have been found promising. Cashew varieties suitable for drought and high density planting (Salam, 1999) and nutrient deficient soils have been reported by Latha, 2000; Latha and Salam, 2001. In fact, precocious, less vigorous and pruning responsive varieties such as NRCC Sel-2, K 22-1, VRI 3, V-4 and Ullal-3 have shown good response to high density planting (Saroj and Nayak, 2013).

Propagation

Different methods of vegetative propagation, *viz.*, layering, stem cuttings, grafting and budding were tried but 'softwood grafting' was found to be the best technique for vegetative propagation of cashew (Sawke *et al.*, 1985; Swamy *et al.*,

Table 1.1: Cashew Varieties Recommended for different States

State	Recommended Varieties
Andhra Pradesh	BPP-4, BPP-6 and BPP-8
Chhattisgarh	BPP-4, BPP-6, BPP-8 and Indira Caju
Goa	Goa-1, Goa-2, Vengurla-4 and Vengurla-7.
Gujarat	Vengurla-4 and Vengurla-7
Karnataka	NRCC Sel-2, Bhaskara, Ullal-I, Ullal-3, Ullal-4, Vengurla-4, Vengurla-7, Madakkathara-2,VRI-3, Chintamani-1 and Chintamani-2.
Kerala	Madakkathara-1, Madakkathara-2, K-22-1, Kanaka, Dhana, Priyanka, Amrutha and VRI-3.
Maharashtra	Vengurla-I, Vengurla-3, Vengurla-4, Vengurla-7 and Vengurla-8.
Odisha	Bhubaneswar-1, Jagannath, Balabhadra, BPP-8 and Dhana.
Tamil Nadu	VRI-1, VRI-3 and VRI (Cw) H-1
West Bengal	Jhargram-1, Bidhan Jhargram-2 and BPP-8
North Eastern States	Ullal-3, Ullal-4, NRCC Sel. 2, Vengurla-1, BPP-8, Jhargram-1

1993). The percentage of success with softwood grafting technique varied in different agroclimatic regions (Ghosh, 1990). Softwood grafting of cashew under Northern dry zone of Karnataka showed higher rate of success (87 per cent) in the month of January, followed by October (60 per cent) (Mahesha *et al.*, 2005). Konhar and Das (1985) obtained 90 to 100 per cent success during January and February under Bhubaneswar conditions. Ghosh (1990) obtained 70 per cent success in the month of February under Jhargram conditions, while Radhakrishna *et al.* (1992) reported 79.80 per cent success in the month of December under Bapatala conditions. According to Kumar *et al.* (1989), the best time for softwood grafting of cashew is May followed by April. Moreover, cashew rootstocks are still being raised by seed nuts only and neither root stocks have been identified nor its clonal multiplication technique has been standardized. For raising root stocks, it has been reported that biofertilizers (*Azospirillum, Azotobacter* and VAM) increased the germination percentage of nuts, plant growth, number of graftable seedlings and reduced the incidence of fungal diseases in the nursery (Kumar *et al.*, 1998; Sinish *et al.*, 2005). In fact, production of quality planting materials of high yielding varieties through softwood grafting technique has made tremendous impact in improving cashew productivity. There are over 100 Regional Cashew Nurseries, including both public and private sector and over 150 lakh cashew grafts are produced annually.

Efforts were also made to develop tissue culture protocol from cashew explants. However, the regeneration protocol from matured explants are yet to be developed although reports are available regarding the multiplication and field establishment of cashew regenerated from young cashew nodal cuttings (Thimmappaiah and Shirly, 1999). Somatic embryogenesis was induced from nucellar tissue cultured from 2-3 week old immature nuts. So far nucellar tissues from 14 varieties were tried for embryogenesis. Embryogenesis and germination of somatic embryos was achieved in two varieties (Thimmappiah, 1997; Shirly and Thimmappaiah, 2005). Embryogenesis was induced in immature cotyledonary segments (Thimmappiah,

1997) and nucellar tissue (Nair and Mohankumar, 1993; Cardoza and D'Souza, 2002). Regeneration from mature tree source has been met with difficulties due to high rate of contamination and poor response (Thimmappaiah and Shirly, 2000; Thimmappaiah *et al.*, 2002a and b). Thimmappiah *et al.* (2002a) standardized micrografting technique in cashew.

Canopy Management

During initial 3-4 years, proper training of young plant is essential for providing desired shape and to promote better fruiting in subsequent years. Therefore, shoots coming below the graft union should be removed and staking of plants is required to avoid lodging due to high wind velocity. The water shoots coming from the base should also be removed. The plants should be allowed to grow straight by maintaining a single stem up to 0.75–1.00 m from the ground level. Thereafter, the main stem may be de-topped at a height of 2-2.5m from the ground level and allow the branches in different directions. Proper shape of plants facilitates terrace making, weeding, fertilizer application, nut collection and plant protection operations. The flower panicles emerging from the graft during the same year of planting may be removed. Though, cashew plants mostly grow in irregular shape but in general, modified system of training is adopted.

Cashew require proper sun light for better flowering and fruiting, thus in bearing trees, removal of dried/dead wood, crisscross branches and water shoots, should be done at least alternate years. This allows proper growth of the canopy and adequate exposure to sunlight. Pruning of cashew plants should be done after harvesting of nuts. The cut surfaces are smeared with Bordeaux paste after pruning. Sometimes, after beheading of thicker branches; there is attack of cashew stem and root borer (CSRB) and also attack of foliage pest on new flushes, that may be taken care of. Pruning intensity depends on variety, plant vigour, growing sites and planting systems. It has been shown that nut yield of cashew increased from 1.44 kg/tree before pruning to 2.36, 4.39, 5.53 and 6.08 kg/tree during the 1st, 2nd, 3rd and 4th years, respectively (Khan *et al.*, 1987). Pruning of leader shoots in July enhanced the production of productive laterals and number of bisexual flowers per panicle (Chattopadhyay and Ghosh, 1994). It has been shown that considerable improvement in nut yield in pruned plants of selected cashew varieties *viz.*, VTH 30/4, NRCC Sel-1, VRI-1 and Ullal-l from third year onwards after planting (Nayak *et al.*, 2008).

High Density Planting

The high density planting in cashew has not been exploited well, though, cashew plants respond to need based pruning. In fact, high density planting system not only provides higher yield and economic returns per unit area in the initial years, but also input use efficient if managed properly. Therefore, plants are required to be trained in initial years for desired and strong framework. High density planting system is more suitable for soils with low fertility. Further, high density planting reduces the weed growth by early canopy coverage, reduces soil temperature thereby increasing the soil moisture content especially during peak summer season and provides mulching effect. Besides, nutrition and irrigation management of high density planting for a particular variety and resource situation, the aspects like; canopy cover, photosysnthetic efficiency, root development pattern, pest and disease interaction, longevity of orchard, operational feasibility etc. need to be worked out. In general, recommended spacing for cashew is 7.5 m × 7.5 m or 8 m x 8 m. Maintaining a tree density of 625 plants/ha (4 m × 4 m) for the first 11 years and diagonal thinning thereafter to reduce the population to 50 per cent doubled the nut yield during the first 10 years of planting (Yadukumar *et al.*, 2001). Salam (1999) reported that the varieties 44/3, Anakkayam-1 and H-1608 were suitable for high density planting system. The precocious, less vigorous and pruning responsive varieties such as NRCC Sel-2, K 22-1, VRI 3, V-4 and Ullal-3 have shown good response to high density planting (Saroj and Nayak, 2013). Also, high density planting system with spacing of 6.5 m × 4 m (384 plants/ha), 5 m × 5 m (400 plants/ha) and 5 m × 4 m (500 plants/ha) were found superior over normal spacing.

Integrated Nutrient Management

Even today, integrated nutrient management in cashew has not been taken up seriously by the farmers. There is a perception among the farmers that cashew does not require nutrition like other crops. However, researches clearly shown that cashew respond to nutrition. Application of fertilizer dosage and time schedule under different agroclimatic zones has been standardized (Veeraraghavan *et al.*, 1985; Grundon, 2001; Salam *et al.*, 2008; Yadukumar *et al.*, 2008; Aikpokpodion *et al.*, 2009; O'Farrell *et al.*, 2010), though it varies in different states (Saroj *et al.*, 2014). Studies conducted at ICAR-Directorate of Cashew Research, Puttur indicated that application of 500 g N and 125 g each of P and K and 10 kg poultry manure/tree/year under normal density planting system (200 trees/ha) and 250 g N and 50 g each of P and K and 10 kg poultry manure per tree per year under high density planting system (625 trees/ha) is found superior in terms of higher nut yield for rainfed cashew. Based on the initial fertility status of soil, nutrient dose may vary from location to location. The recommended dose of fertilizers during first year after planting is 1/5th of the full dose, 2nd year after planting is 2/5th, 3rd year after planting is 3/5th, 4th year after planting is 4/5th and 5th year onwards is full dose should be applied. The fertilizers should be applied in 2 split doses during pre-monsoon (May-June) and post-monsoon (September-October) periods. However, if single application is to be done, it should be during the post-monsoon season (September-October). In red loamy soils and in low rainfall areas (east coast), the fertilizers should be applied in circular bands at a distance of 0.5m, 0.75m, 1.0m and 1.5m away from the tree trunk during first, second, third and fourth year onwards of planting respectively. In laterite soils and soils of slopy land in heavy rainfall areas (west coast), the fertilizers should be applied in circular trenches of 25cm width and 15cm depth at above distances. The trenches should be closed immediately after the fertilizer application.

Foliar sprays of nutrients (urea 2 to 4 per cent; DAP 1 per cent; orthophosphoric acid; $ZnSO_4$ 4 per cent; Cu 0.3 to 0.6 per cent) at the emergence of the flush, panicle initiation and fruit set stages ensure better fruit set and also enhance nut yield in cashew (Ankaiah and Rao, 1987; Sapkal, 2000). Yellow leaf spot in low soil pH (4.5-5.0) could be corrected by foliar sprays of Mo salts (Subbaiah *et al.*, 1986). Foliar spray of growth regulators Planofix, Nutron, IAA, IBA, NAA, 2,4-D and ethrel were effective in increasing the total number of flowers, hermaphrodite flowers, sex ratio, fruit and nut yield per panicle, besides improving physico-chemical composition of apples and nuts (Ghosh, 1988; Singh *et al.*, 1992; Kumar *et al.*, 1994). There is also need to standardize organic production of cashew but it is only possible after developing non-chemical based pest management strategy. Cashew plantations have large amount of organic biomass available for recycling. The availability of leaf litter from cashew plantations of 10 to 40 years old varied from 1.38 to 5.20 t/ha (Guruprasad *et al.*, 2009). About 5.5 t of available cashew biomass waste per ha can be converted into 3.5 t of compost or vermicompost and helps in meeting nutrient requirement to cashew by 50 per cent (Yadukumar and Nandan, 2005; Yadukumar, 2007). *Ex-situ* green manuring by using biomass of *Gliricidia sepium* is also beneficial.

Water Management and Fertigation

Cashew is generally grown in coastal regions as a rainfed crop but it responds well to soil moisture management. Moreover, stagnation of water near cashew tree hampers the plant growth. Therefore, a proper drainage in cashew orchards is essential. Even in high rainfall areas of west coast, cashew experiences severe moisture stress from January to May, which adversely affects its flowering and fruit set, resulting in immature nut drop and lower productivity of cashew. During fruiting season of cashew (February to May), a mean rainfall of around 67 to 415 mm is received. The water deficit is highest during March to May (112 to 183 mm). Cashew yields can be enhanced by providing protective irrigation with 200 L of water per tree once in 15 days from January to March. Irrigation can be started after the commencement of flowering for better nut set, filling and yield. The frequency and quantity of irrigation varies with soil conditions and age of trees.

Fertigation of water soluble fertilizers such as urea, DAP, and MOP can be done once in a week from December to March. With fertigation, the quantity of nutrients to be applied can be reduced to a half of the quantity of recommended dose. Richards, 1993; Yadukumar and Mandal 1994 and Mishra *et al.*, 2008 reported that fertigation can save 50 per cent fertilizer requirement and doubled the cashew yield. The experiments conducted at Directorate of Cashew Research (DCR), Puttur on fertigation in cashew indicated that under normal density planting system (7 m x 7 m), the effective canopy coverage per tree is 12.56 m^2. The quantity of irrigation water calculated based on the effective canopy area was 12.56 L/tree/day from December to January (daily open pan water evaporation is 5 mm) and 20 L/tree/day from February to March (daily open pan water evaporation is 6.5 mm) to meet 20 per cent of the cumulative pan evaporation (CPE). Similarly, for 40 per cent CPE and 60 per cent CPE, the irrigation rate was 24 L/tree/day and 38 L/tree/day from December to January and 36 L/tree/day and 58 L/tree/day from February to March, respectively. In order to meet 20 per cent CPE, four drippers of 2 L/hour discharge rate can be fitted at two equidistant points 1 m away from the base of the tree. Similarly, to meet 40 per cent and 60 per cent CPE, four drippers and six drippers of 4 L/h discharge rate can be fixed. Drip irrigation can be given for 1 h 30 min during December and January and 2 h in February and March (Yadukumar and Rejani, 2008).

Weed Management

Weeds compete for moisture, nutrients and light in cashew plantations, owing to heavy yield loss, if not managed properly. Sometimes, perennial weeds cover the cashew plants, if not removed in early stage. Often, field preparation for cashew planting is not taken seriously and planting is done without leveling of field and removing of perennial bushes. Often cashew orchards are neglected, thus weeds act as host for the pests also. The major weeds of cashew plantations are *Eupatorium odoratum, Tridex procumbent, Mimosa pudica, Lantana camara, Brachiaria* spp., *Ischaemum indicum, Chromolaena odorata* L, *Syndrella nodiflora* Gaertn, *Elephantopus scabber* L. *Desmodium triflorum* DC, *Glycosmis arborea* DC, *Hemedesmus indicus* R. Br.,*Ichnocarpus frutescens, Passiflora foetida* L.,*Cyclea peltata* Cooke, *Tragia involucrate L., Triumfetta rhomboidea* Jacq, *Ziziphus* spp., *Narengamia elata* W., *Macaranga peltata etc.* In most of the cases, weeding is done manually. Moreover, weeds must be slashed or uprooted before seed setting in order to minimize the weed intensity. The slashed biomass can be used as mulch material or after decomposition may be applied to the plants. Application of glyphosate at 0.8 kg/ha (single spray) or glyphosate at 0.4 kg per ha + 2, 4-D 1.0 kg per ha (single spray) or paraquat 0.4 kg per ha + 2, 4-D 1.0 kg per ha (single spray) during August can effectively control the weed growth in cashew plantations (Abdul Salam *et al.*, 1993).

Intercropping Systems

Intercropping is the best option to minimize erosion, nutrient losses and conserve soil and moisture and to realize higher returns from an unit area. Cashew takes 8 to 10 years for the canopy to cover the entire area under normal spacing; thereby there is good scope for growing intercrops. Trailing pepper on the stem and branches of grown up cashew trees (more than 6 years) is also adopted. Maize and groundnut can be grown successfully as intercrops in newly planted and two-year-old cashew orchards (Abeysinghe *et al.*, 2003). Other crops such as tapioca, turmeric, ginger, cucurbits, colocasia, elephant foot yam have been found to be suitable intercrops in West coast region while colocasia, brinjal, groundnut, blackgram in East coast region. Pineapple can be grown as a suitable intercrop between two rows of cashew for the first seven years. The spacing to be maintained for cashew is 8m x 8m (156 trees/ha) or 7.5 m x 7.5 m (175 trees/ha) or 10 m x 5 m (200 trees/ha). Growing pineapple in trenches across the slope between two rows of cashew helped to conserve the soil moisture, which in turn increased the yield of cashew (main crop) by 1.5 times compared to cashew alone. Therefore, pineapple can be grown as a profitable intercrop under irrigated as well as rainfed conditions in west coast region (Yadukumar *et al.*, 2003), though there is a little problem in nut collection. Popular cover crops in cashew plantations are *Peuraria javanica, Calapagonium mucunoides* and *Centrosema pubescens* which improve soil fertility and moisture balance and help in conserving orchard soils. Even, growing of medicinal plants like; *Aloe vera* in Tamil Nadu, vegetable like; okra and pulses like; cowpea and pigeon pea in Gujarat and seed spices like; fenugreek and dill in eastern humid region are highly profitable. In Thane and Nasik districts of Maharastra, *Mogra is* grown by the farmers, whereas in Baptala areas of Andhra Pradesh, inter cropping of marigold with cashew recorded the highest net profit of Rs.65, 967/- ha.

Plant Growth Regulators

Meagre work has been done on use of growth regulators in cashew. Mariappan *et al.* (1995) reported that best production of hermaphrodite flowers and highest nut count and nut weight were found with application of NAA (200 ppm) followed by ethephon (100 ppm). The study conducted in Nigeria by Aliyu *et al.* (2011) indicated that exogenous application of GA_3 (50 mg L^{-1}) at pre-blooming stage of cashew enhanced flowering precocity, shortened flowering duration, increased production of hermaphrodite flowers and fruit set and resultant nuts development optimally with higher percentage of retention. The application of GA_3 at 50-100 mg L^{-1} gave five-fold improvement in flowering (precocity and number of hermaphrodite flowers) and fruit setting and about 60 per cent increase in fruit retention ability and 25 per cent in nut size compared to untreated. Lakshmipathi *et al.* (2014) reported that application of 50 ppm Ethrel and 25 ppm NAA were found to be beneficial for increasing the nut yield through improvement in sex ratio, fruit set and fruit retention in cashew var. Bhaskara. For increasing yield of cashewnut, two sprays of 10 ppm ethrel *i.e.* first at flushing and second at flowering stage are recommended from AICRP on Cashew centre, Vengurla (Maharashtra). Under field plantations of cashew variety Ullal-3 at spacing of 3 m apart, soil application of Paclobutrazol (2.0 g alternate year) at pre-flushing stage is effective in reducing plant height, canopy spread, intermodal length but increased nut yield to the tune of 51.78 per cent over control (Meena *et al.*, 2014). A vivid account on application of growth regulators in cashew flowering has been given by Saroj *et al.* (2014).

Plant Protection

Tea mosquito bug (TMB) and cashew stem and root borer (CSRB) are the major pests of cashew plantations which limit the cashew production. There are three species of TMB *Helopeltis* (*H. antonii* S., *H. theivora* W. and *H. bradyi* W.) infesting cashew plantations and yield losses caused by TMB is to the tune of 40 per cent. Both nymphs and adults of *Helopeltis* suck sap from tender shoots and leaves, floral branches and from developing nuts and apples by making a number of feeding lesions. In severe cases complete burning appearance of cashew plantation can be seen. Mid to late season flowering cashew variety such as Bhaskara, is able to escape from the severity of the pest to a certain extent. Even though, all groups of insecticides and plant products (botanicals) were evaluated against this pest, none of them exhibited any ovicidal action (Raviprasad *et al.*, 2005). However, λ-cyhalothrin (0.003 per cent) and carbaryl (0.1 per cent) had shown longest residual action against nymphs and adults (Sundararaju *et al.*, 1993). Need based sprays are recommended during most vulnerable periods of crops such as flushing, flowering and fruiting stages. Though, cashew is an insect pollinated crop, spraying of these insecticides during flowering season did not influence the fruit set (Pillai *et al.*, 1984; Rai, 1984, Sundararaju *et al.*, 1993a). Development of pheromone technology, use of egg parasitoids like; *Erythmelus helopeltidis* Gahan. (Hymenoptera: Mymaridae), *Telenomus* sp. (Laricis group) (Scelionidae), *Chaetostricha* sp. (Trichogrammatidae), *Gonatocerus* sp. nr. *bialbifuniculatus* and *Ufens* sp. and nymphs predator like; *Crematogaster wroughtonii* Forel (Formicidae) can be utilized in biological control strategy. Several species of spiders, *Hyllus* sp., *Oxyopes sehireta*, *Phidippus* patch and

Matidia sp. five species of reduviid bugs (*Sycanus collaris* (Fab), *Sphedanolestes signatus* Dist. and *Endochus inornatus* Stal., *Irantha armipes* Stal. and *Occamus typicus* Dist. have also been recorded as predators of TMB. *Aspergillus flavus* and *A. tamarii* were confirmed to be pathogenic to *H. antoni* (Ambika and Abraham, 1979; Devasahayam and Nair, 1986; Sundararaju, 1984).

Cashew stem and root borer (CSRB) (*Plocaederus ferrugineus* L.) (Coleoptera: Cerambycidae) is another important pest infesting cashew in all parts of India. Once plant is affected by this pest it is difficult to save. Thus, pest management approaches include careful examination of trunk portion for any symptoms, collection of grubs by careful removal of the bark and swabbing of the affected portion with either carbaryl or chlorpyripos (0.2 per cent). Swabbing the pest infested portion after removal and drenching the root zone with chlorpyriphos (0.2 per cent) as post extraction prophylaxis (PEP) was found effective in reducing reinfestation. There exists large population of CSRB grubs in dead cashew trees and also in CSRB infested cashew trees. The trees having more than 50 per cent of bark circumference damaged/having yellow canopy are considered as trees beyond recovery. It is essential to adopt phytosanitation by removing the dead trees and trees beyond recovery by uprooting during monsoon season, prior to emergence of adult beetles and dispose off suitably without delay to reduce build-up of pest inoculum in a given locality. There is need to evaluate healer-cum-sealer for recovery of borer infected plants. Also, research work must be intensified on development of karomone technology. In addition to these major pests, leaf and blossom webbers, shoot tip caterpillar, apple and nut borers, *Ambrosia* beetle *etc.*, are also causing damage to cashew plantation. Though, they are not so serious now, but their management strategy must be worked out in order to avoid any future threat.

Cashew plants are also affected by various diseases. In nursery stage; seedling blight (*Cylindrocladium scoparium*), damping-off (*Phytopthora palmivora* Butler) and root-rot disease (*Pythium ultimum* Tron); foliage diseases such as anthracnose (*Phytophthora nicotinae* and *Asterina carbonacea*), diplodia rot (*Diplodium anacardiacearim*), leaf blight (*Pestalotia paeoniae*); inflorescence diseases such as anthracnose (*Colletotrichum gloesporioides*) and inflorescence blight (*Gloeosporium mangiferae* and *Phomopsis anacardii*); nut disease (*Aspergillus tamari, Penicillium citrium, Lasiodiplodia theobromae* and *Gloeosporium* spp.); besides dieback (*Pellicularia salmonicolor*), gummosis (*Diplodia natalensis*) and powdery mildew (*Odium* spp.) are common diseases. Under heavy rainfall situations, occurrence of collar rot and root rot due to several fungal pathogens may be noticed in nursery. For their control, it is recommended to remove the poly bags having infected seedlings/grafts and spray the remaining seedlings/grafts with 1 per cent Bordeaux mixture when there is a short dry spell of at least 4-6 h in order to avoid spread of the infestation to healthy seedling/grafts. Dieback disease occurs due to severe infestation by tea mosquito bug is more pronounced under cloudy weather. For management of this disease, the infected twigs need to be pruned off. The cut ends need to be smeared with Bordeaux paste. For management of blossom blight, spraying with Dithane M-45 (2 g/L) is recommended.

Orchard Rejuvenation

There is a sizeable area under seedling plantation which are more than 20-30 years old with poor productivity. Such plantations are either replaced by fresh planting or may be top worked with desired varieties. The technique for rejuvenation of unproductive trees has been standardized by beheading the healthy trees at 1 m height after harvesting. The cut end should also be pasted with Bordeaux paste in order to avoid any secondary infections. Thereafter, large number of new sprouts will emerge but retain only 4-6 healthy branches, so as to develop a good canopy structure and rest should be thinned out. These new sprouts after attaining appropriate size should be top worked by softwood grafting method. In case of 10-15 years old cashew plantations, highest success rate (80 per cent) was achieved in the month of April followed by June grafting (77.08 per cent). In case of 5-10 years old plantations, grafting in the month of June (70 per cent) was better followed by grafting in May (55.3 per cent). The uneconomical cashew trees top worked with high yielding Ullal-1 variety resulted in four-fold increase in nut yield per tree within five years (Kumar, 1990). Cashew growing regions in which the population of CSRB is high, the top working technology may not be suitable unless regular follow-up action is taken to manage the incidence of CSRB (Swamy, 1995). Besides CSRB, attack of TMB and other pests are also common on new sprouts which should be managed properly. Proper nutrition and moisture management of rejuvenated trees are also essential.

4. Safety Concern

The health consciousness is increasing world over, thereby demand for safe food/organic products are increasing at faster rate. Cashew is generally grown in India with minimum use of plant protection chemicals, however sometimes cashew yield is affected significantly due to infestation of tea mosquito bug (TMB) and cashew stem and root borer (CSRB). For their management, need based spray of lambda cyhalothrin (0.003 per cent) and chlorpyriphos (0.2 per cent), respectively is recommended. As per EU pesticide database, the maximum residue limit (MRL) permissible for lambda cyhalothrin and chlorpyriphos is 0.05 mg/kg respectively. The study conducted at ICAR-DCR, Puttur (Karnataka) to analyze cashew kernel obtained from treated plots with lambda cyhalothrin (0.003 per cent) in collaboration with Quality Control Laboratory, CEPCI, Kollam indicated that no residue was detected at detection limit of 100 ppb (DCR Annual Report: 2005-06). However, in cashew apple samples collected after spray of lambda cyhalothrin (0.003 per cent), trace of residue was recorded within 15 days of treatment but residue level fell below the MRL (DCR Annual Report: 2010-11). The cashew kernel obtained from the trees treated with chlorpyriphos (0.2 per cent) in Tamil Nadu, Odisha, Maharashtra and Karnataka for the management of CSRB at recommended concentration and double the recommended concentration were analyzed but found free from the pesticide residue (Bhat and Raviprasad, 2008). Thus, consumption of cashew kernel is safe.

5. Post-Harvest Management

Cashew nut is either collected with fallen cashew apple or harvested at ripe stage of apple and then the nuts are to be separated from the cashew apple and sun dried for 2 to 3 days. Raw nuts at the time of harvest will have moisture content ranging between 16 to 25 per cent. Drying is important in order to prevent spoilage of nuts during storage due to fungal attack. Safe permissible moisture level of raw nuts is 8 per cent and well dried nuts can be stored at ambient temperature for 12 months without any quality deterioration. In India, cashew processing is mostly manual and it consists of steaming or roasting, shelling, kernel drying, peeling, grading and packing. Although grading raw cashew nuts before processing reduces broken kernel, it is not followed in India. Conditioning makes the raw cashew nuts hard and loosens the kernel within the shell. Various methods such as Drum roasting, Steam boiling and Oil bath roasting have been employed for commercial processing of raw cashewnuts. In recent times commercial processing by steam roasting involves drying of raw nuts, steam roasting for 18-20 min, shelling of the raw nuts, peeling, grading and packing the kernels under vacuum (Ohler, 1979). The brief flow chart of cashew nut processing in given in Figure 1.1.

Cashew apple which is about 8 times that of nut is not used except in Goa where it is fermented to an alcoholic beverage called 'Feni". Cashew apple is highly nutritious and rich in vitamins and minerals. Various products such as juice, jam, jelly, pickle etc. could be prepared from cashew apple. In case the cashew apple is processed either for producing syrup/liquor or for bio-fuel or for extracting nutraceuticals, the income of farmers can be increased. Cashew apple pomace (the residue left after removal of juice), which is rich in mineral and fibre could be utilized as animal feed. Another by-product known as cashew nut shell liquid (CNSL) finds extensive industrial use like furnace lining, textile industry, preparation of varnish, paints, resins, timber protection *etc.*

6. Impact of Cashew Research and Development

Area and Production

Cashew, after its introduction has naturalized well to Indian climatic conditions and country has emerged as third largest producer of raw cashew nut (15.7 per cent) after Viet Nam (28.9 per cent) and Nigeria (19.5 per cent). Few years back, India was credited with the fame of largest producer, processor and exporter of cashew in the world. Since other cashew producing countries acquired the know-how of processing and entered in international market, India's share in the International market has shrunk to 30 per cent. On the other hand, the domestic consumption in India has increased considerably, owing to largest consumption of cashew kernel. It is pertinent to mention that India is still on top position both in area and production of raw cashew nut but production data of some countries include imported quantity of raw cashew nut also. As per Cashew Hand Book: 2014; the share of India in raw cashew nut production is highest (22 per cent) in the world, followed by Cote d Ivoire (19 per cent), Brazil (16 per cent), Indonesia ((13 per cent), Viet Nam (7 per cent), Tanzania (2 per cent), Mozambique (1 per cent) and others (20 per cent).

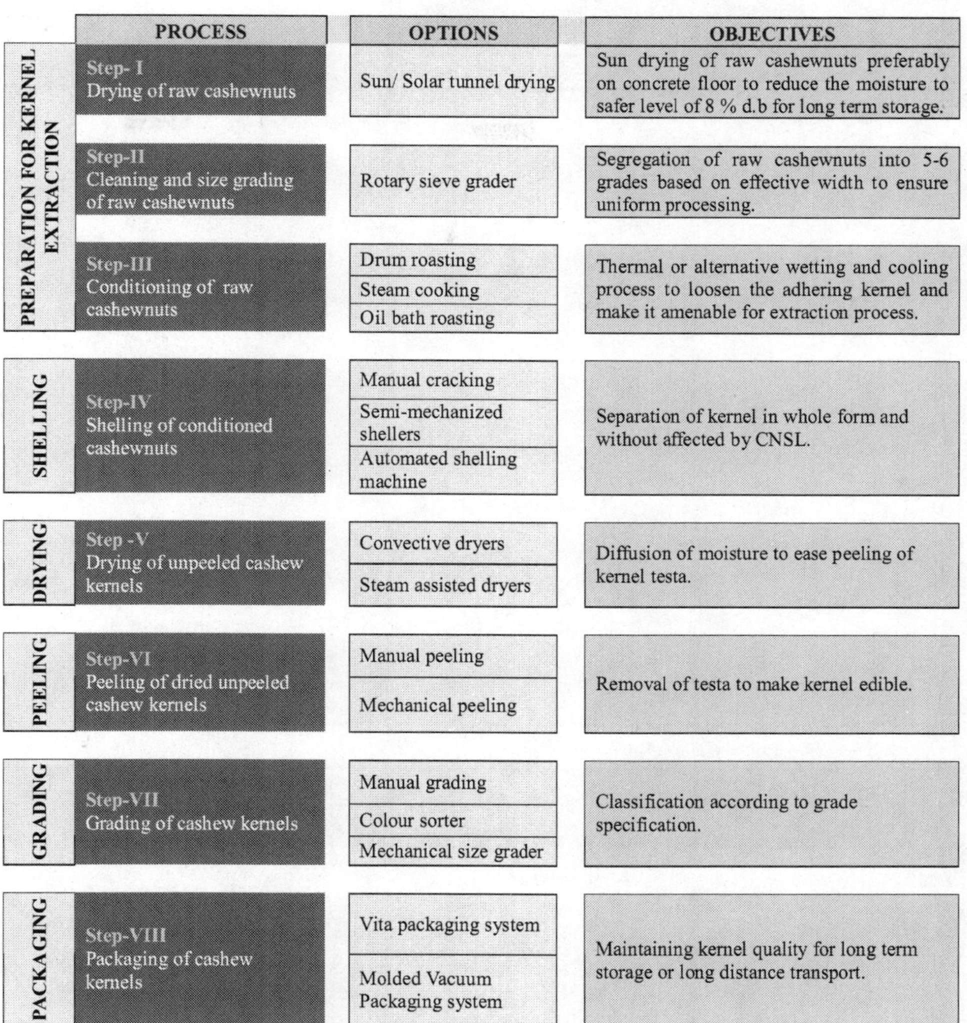

Figure 1.1: Various Operational Stages in Cashew Nut Processing.

At present, cashew is grown in an area of 10.27 lakh ha with a production of 7.25 lakh tonnes of raw cashew nuts and a productivity of 706 kg/ha (Table 1.2). Though, area and production of cashew in India is increasing continuously but growth in productivity is at very slow rate. From the last 14 years, productivity of raw cashew nut is in between 600-800 kg/ha. The trend in area, production and productivity of cashew for the last 14 years is shown in Figure 1.2. However, there is great regional imbalance in productivity of raw cashewnut in the country which varies from as low as 303 kg/ha in Jharkhand to 1262 kg/ha in Maharashtra. Andhra Pradesh and Maharashtra both are having individually more than 18 per cent area under cashew but Maharashtra alone contributing 32.42 per cent of raw cashewnut production while Andhra Pradesh contributes only 13.78 per cent. This

Table 1.2: State-wise Area, Production and Productivity of Cashew in India: 2014-15

State	*Area ('000 ha)*	*Production ('000 tonnes)*	*Productivity (kg/ha)*
Kerala	84.53	80.00	946
Karnataka	124.71	80.50	645
Goa	58.17	32.00	550
Maharashtra	186.20	235.20	1262
Tamil Nadu	140.42	67.00	478
Andhra Pradesh	185.45	100.00	539
Odisha	180.41	85.50	474
West Bengal	11.36	13.00	1096
Chhattisgarh	13.70	8.50	620
Jharkhand	14.83	4.50	303
Gujarat	7.22	8.50	1177
Puducherry	5.00	3.00	600
Assam	1.05	0.57	543
Tripura	4.25	2.50	588
Meghalaya	8.50	4.50	529
Manipur	0.90	0.15	167
Nagaland	0.50	0.20	400
Total	1027.20	725.45	706

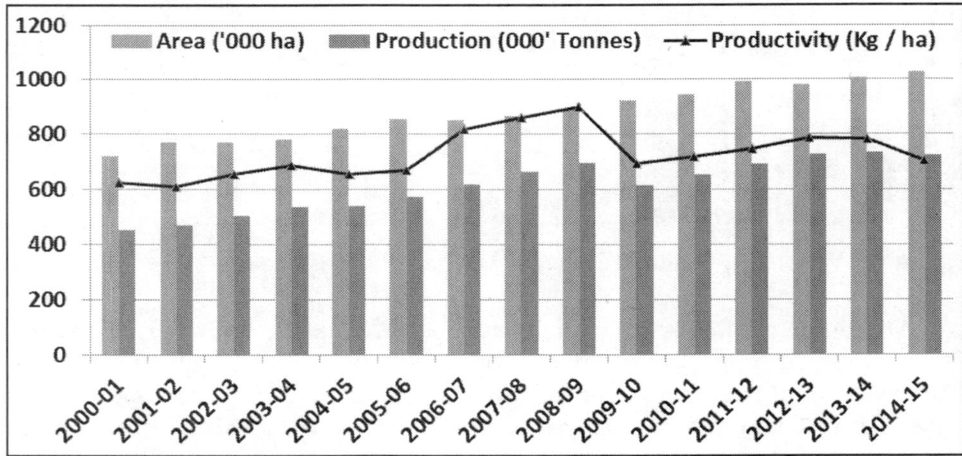

Figure 1.2: Trends in Area, Production and Productivity of Cashew in India.

is only because of the adoption of new varieties and production technologies in Maharashtra. The productivity level of traditional cashew growing states like Tamil Nadu, Andhra Pradesh and Karnataka is very low. There is a good scope of area expansion under cashew in non-traditional states like, Chhattisgarh, Jharkhand, Gujarat and North-Eastern states.

Import and Export

India is a key player in the global cashew industry and has developed better capability in raw cashew nut processing. There are more than 3799 processing units in the country, which require about 15-20 lakh tonnes of raw cashewnuts per annum. However, our annual raw nut production is 7.25 lakh tonnes only, thus India imports annually over 7-8 lakh tonnes of raw cashewnuts from African and other countries worth of Rs.4000-5000 crores (Table 1.3). Since many of these cashew producing African countries have started strengthening their processing capacities, the availability of raw cashewnuts for import will be reduced drastically in future. Hence, there is an immediate need to double the production level in order to meet the requirement of the cashew industries. Now, Indian Government has imposed

Table 1.3: Import of Raw Cashew

Year	*Quantity* *M.T*	*Value* *Rs. Crore*
1990-91	82639	134.00
1991-92	106080	266.68
1992-93	134985	376.33
1993-94	191322	482.70
1994-95	228109	690.94
1995-96	222819	760.08
1996-97	192285	640.60
1997-98	224968	744.00
1998-99	181009	680.24
1999-00	200584	953.52
2000-01	249318	960.80
2001-02	356566	960.01
2002-03	400659	1236.57
2003-04	452898	1400.93
2004-05	578884	2183.24
2005-06	565400	2162.95
2006-07	592604	1811.62
2007-08	605970	1746.80
2008-09	605850	2632.41
2009-10	752894	3037.35
2010-11	529370	2649.56
2011-12	809371	5337.76
2012-13	892000	5331.10
2013-14	771000	4564.00

Source: DGCI and S, Kolkata (2014) and CEPC Kochin.

import duty on raw nut import (5 per cent), thus there is every possibility of increase on unit cost of cashew kernel in future.

Cashew is one of the most valuable processed nuts in global commodity markets and has the potential to generate employment and revenue for developing countries. After Vietnam, India is the second largest exporter of cashew kernels, accounting for 34 per cent of the world's export of cashew kernels and earns a sizeable amount of foreign exchange. India has a comparative advantage in the production and processing of cashew nuts on account of its cheap and skilled labour force. Indian cashew kernel has global repute and every year about 1.2 lakh tones of cashew kernels are exported to various countries. The USA, UAE, Netherlands, Saudi Arabia, Japan, Germany, Spain, UK, France, Belgium, Korea, Singapore etc. are the major export destination. During 2014-15, total earning of foreign exchange was Rs. 5488.66 crores through the export of cashew kernels (Rs. 5432.85 crores) and cashew nut shell liquid (Rs. 55.81crores). The export of cashew kernel and CNSL from last 24 years is given in Table 1.4. The export quantity of both kernel and CNSL are increasing year after year indicating the bright future of Indian cashew sector in global scenario. However, to sustain our position in global trade, there is a tough competition in future with major cashew growing countries. Therefore, India has to intensify its domestic production and automation in processing to face the future challenges in cashew sector.

Socio-economic and Ecological Impact

There is a considerable amount of human resources involved in cashew cultivation and processing in India employing over 10 lakh persons with more than 90 per cent of them being women. The processing of cashew is a highly labour intensive activity. Employment generation in cashew processing units was 11 man days in Odisha and 7 man days in Tamil Nadu and Kerala for processing 1 bag of raw nuts (80 kg). Since the share of women in this sector is large, the growth of this sector will contribute towards women's employment (Shalini Yadav, 2010). In some areas cashew is grown, where tribal population are in predominant stage and if they are trained properly for cashew cultivation and processing, it can improve the livelihood of the inhabitants.

Cashew plantations are also important in ecological restoration. In general, cashew is grown in resource poor situations of tropical region on slopy and undulating lands with poor soil fertility. Soil erosion is the another limiting factor of farming in these areas. In such areas, cashew being low input demanding crop and having extensive deep root system with better soil binding capacity is playing a vital role in creating ground cover and minimizing soil erosion. Cashew plantations have vast potential of organic biomass available for recycling. The two years mean value for total dry weight of canopy biomass fall out was 55 per cent as leaves, 27.3 per cent as apples, 5.6 per cent as kernels and 12.1 per cent as shells. The major portion of N, P, K, Ca, and Mg of canopy biomass came from leaves, branches, kernels and apples. The amounts of nutrient elements recycled in canopy fallout may partially meet the nutrient requirements of cashew. Yadukumar and Nandan (2005) determined the nutrient composition of organically recyclable biomass

Table 1.4: Export of Cashew Kernels and CNSL

Year	*Cashew Kernel Export*		*CNSL Export*		*Total Value Rs. Crore*
	Quantity	*Value*	*Quantity*	*Value*	
	MT	*Rs. Crore*	*MT*	*Rs. Crore*	
1990-91	49874	4422.4	5658	5.56	447.96
1991-92	47738	6690.9	4542	4.02	6694.92
1992-93	53436	7454.9	4258	3.81	7458.71
1993-94	69884	1046.02	3525	2.90	1048.42
1994-95	77000	1246.02	3807	2.4	1248.42
1995-96	70334	1240.50	760	1.45	1241.65
1996-97	68663	1285.50	1735	2.77	1288.27
1997-98	76593	1396.10	4446	7.17	1403.87
1998-99	15026	1609.90	1572	3.26	1613.16
1999-00	92461	2451.45	764	1.84	2453.29
2000-01	89155	2049.75	2246	38.94	2088.69
2001-02	97550	1776.80	1814	4.19	1780.99
2002-03	104137	1933.02	7215	9.25	1942.27
2003-04	100828	1804.42	6926	7.03	1811.45
2004-05	126667	2709.24	7474	7.91	2717.15
2005-06	114143	2514.86	6405	7.09	2521.95
2006-07	118540	2455.15	5589	10.29	2165.44
2007-08	114340	2288.90	7813	11.97	2300.37
2008-09	109522	2988.40	9099	26.06	3014.46
2009-10	108120	2801.60	11227	27.62	2829.22
2010-11	105755	2819.39	12051	33.77	2853.16
2011-12	131760	4390.68	13575	59.46	4450.14
2012-13	104015	4067.20	9192	29.83	4097.03
2013-14	114791	5095.73	9480	38.61	5134.34
2014-15	118952	5432.85	10938	55.81	5488.66

Source: DGCI and S, Kolkata (2014) and CEPCI, Kochin.

compost with 20 per cent cowdung slurry as N as 0.91 to 1.5 per cent, available P as 0.34 to 0.6 per cent, K as 0.39 to 0.46 per cent. Whereas, in the decomposed leaf litter, organic carbon content ranged from 0.7 to 1.61 per cent, total N from 0.18 to 0.25 per cent, available P_2O_5 from 0.13 to 0.28 per cent and K_2O from 0.29 to 0.40 per cent (Kumar and Hegde, 1999). About 15.5–37.7 per cent of tree total requirements of macro-nutrients are recycled from canopy biomass fallout of leaves, cashew apples and flowers from six-year-old cashew trees in Australia (Richards, 1993). The micronutrient concentrations (Zn, Mn, Cu and Fe) were higher in the litter fall compared to the green leaves in cashew (Isaac and Nair, 2002).

Carbon sequestration from plant biomass into soil organic carbon is a key strategy for controlling gaseous carbon emission from agriculture. The cashew plantations also have good carbon sequestering ability. It was recorded that cashew plantation produces biomass of about 4.27 t/ha/yr. with CO_2 sequestration of 6.26t/ha/yr. Carbon storage by cashew has been estimated as 32.25, and 59.22 t CO_2/ha at 5th and 7th year of growth respectively under high density planting system. Seven-year-old tree of cashew (VTH-174) sequestered about 2.2-fold higher C under high density planting system (625 plants/ha) as compared to normal density planting system (156 plants/ha).

REFERENCES

Abdul Salam, M. and Peter, K.V. (2010). Cashew - A Monograph Stadium Press (India) Pvt. Ltd., pp. 26-51.

Abdul Salam, M., Pushpalatha, P. B., Suma, A. and Abraham, C. T. (1993). Efficacy of chemical weed control in cashew plantations. *J. Plantation Crops*, 21 (1): 54-56.

Abeysinghe, D. C., Sangakkara, U. R. and Jayasekera, S. J. B. A. (2003). Intercropping of young cashew (*Anacardium occidentale* L.) and its effects on crop productivity and land utilization. *Tropical Agricultural Research*, 15: 10-19.

Aikpokpodion, P.E., Uloko, B. and Edibo, G. (2009). Nutrient dynamics in soil and cashew (*Anacardium occidentale* L.) leaf and kernel in Kogi State, Nigeria *Journal of Applied Biosciences*, 25: 1573-1578.

Aliyu, O.M., Adeigbe, O.O. and Awopetu, J. A. (2011). Foliar application of exogenous plant hormones at pre-blooming stage improves flowering and fruiting in cashew (*Anacardium occidentale* L.). *J.Crop Sci. Biotech.*, 14(2): 143-150.

Ambika, B. and Abraham, C.C. (1979). Bioecology of *Helopletis antonii* Sign. (*Miridae Hemiptera*) infesting cashew trees. *Entomon.*, 4: 335-342.

Ankaiah, S. and Rao, P.V. (1987). Studies on the age of leaf, time of absorption and concentration of urea spray on cashew. *Indian Cashew J.*, 15(3): 9-11.

Anonymous. (2012). Cashew nut nutrition facts, medicinal properties and health benefits. http: //www.nutrition-and-you.com/cashew_nut.html

Archak, S., Gaikwad, A. B., Gautum, D., Rao, E.V.V.B., Swamy, K.R.M. and Karihaloo, J. L. (2002). Comparative assessment of DNA fingerprinting techniques (RAPD, ISSR and AFLP) for genetic analysis of cashew (*Anacardium occidentale* L.) accessions of India. *Genome*, 46: 362-369.

Bhaskar Rao, E.V.V. and Swamy, K.R.M. (2002). In: Cashew Research Scenario in India. (Eds. H.P. Singh, P. P. Balasubramanian and Venkatesh N. Hubballi), The Cashew industry in India. Directorate of Cashewnut and Cocoa Development, Kochi-11. pp. 30-41.

Bhat, P. S. and Raviprasad, T. N. (2008). Studies on lindane and chlorpyriphos residues in cashew kernels. *Journal of Plantation Crops*, 36(2): 142-144.

Cardoza, V. and D'Souza, (2002). Induction, development, and germination of somatic embryos from nucellar tissue of cashew. *Anacardium occidentale* L. *Sci. Hort.*, 93: 367-372.

Chattopadhyay, N. and Ghosh, S. N. (1994). Studies on the effect of time and extent of pruning in increasing the yield of cashew. *Journal of Plantation Crops*, 22(2): 111-114.

Costa, C.Da. (1578). Tratado das drogas e medicinas das Indias Orientalis. Junta de Investigacoes do Ultramar, Lisoba (Portugal). 1964.

Damodaran, V. K., Abraham, J. and Alexander, K. M. (1965). The morphology and biology of the cashew flower I- Flowering habit, flowering season, morphology of the flower and sex ratio. *Agric. Res. J. Kerala*, **3**(1 and 2): 23-28.

Darlington, C. D. and Janaki Ammal, E. K. (1945). Chromosome Atlas of Cultivated Plants. George Allen and Unwin, London, p.397.

Devasahayam, S. and Nair, C. P. R. (1986). The mosquito bug, *Helopeltis antonii* Sign. on cashew in India. *Journal of Plantation Crops*, 14: 1-10.

Dhanaraj, A. L., Rao, E. V. V. B., Swamy, K. R. M., Bhat, M.G., Prasad, D.T. and Sondur, S. N. (2002). Using RAPDs to assess the diversity in Indian Cashew (*Anacardium occidentale* L.) germplasm. *J. Hortic. Sci. Biotechnol.*, 77(1): 41-47.

Ghosh, S. N. (1988). Effect of nitrogen, phosphorus and potassium on flowering duration, yield and shelling percentage of cashew (*Anacardium occidentale* L.). *Indian Cashew J.*, 19(1): 19-23.

Ghosh, S. N. (1990). Effect of season on success of softwood grafting in cashew in semiarid region in West Bengal. *Cashew Bulletin*, 27 (12): 15-19.

Grundon, N. J. (2001). A desktop study to predict the fertilizer requirements of cashew trees in Northern Australia. CSIRO Land and Water, Atherton, Technical Report 32/01.

Guruprasad, T.R., Ramaswamy, G.R. and Hanumanthappa (2007). Recent techniques and strategies for improving productivity of senile cashew plantations. In: 6th National Seminar on Indian Cashew in the Next Decade-Challenges and Opportunities held during 18-19 May 2007 at Raipur, Karnataka, India. p. 35.

Guruprasad, T. R., Hanumanthappa, M., Sudhir, K. V. and Dhananjaya, B. (2009). Technologies for improvement of senile cashew gardens. *The Cashew and Cocoa Journal*, 1(3): 18-22.

Isaac, S. R. and Nair, M. A. (2002). Nutrient and biochemical composition of common tree leaf biomass. *Indian J. Agroforestry*, 4(2): 145-147.

Khan, M. M., Kumar, D. P. and Hiremath, I. G. (1987). Pruning studies in cashew. *The Cashew*, 1(1): 11-12.

Konhar, T. and Das, G. C. (1985). Epicotyl and softwood grafting - new techniques of vegetative propagation of cashew in Odisha. *Cashew Causerie*, 7(3): 13-13.

Kumar, D. P., Hegde, M. and Guruprasad, T. R. (1998). Fertigation for higher nut production in cashew (*Anacardium occidentale* L.). *Cashew Bulletin*, 35(4): 2-4.

Kumar, D. P., Hegde, M. (1999). Recycling of plant nutrient through cashew leaf litter. *The Cashew*, 132): 2-8.

Kumar, D. P., Subbarayappa, A., Hiremath, I. G., Khan, M. M. and Sadashivaiah (1989). Use of coconut coir-pith: a biowaste as soil mulch in cashew plantations. *The Cashew*, 3(3): 23-24.

Kumar, D.P. (1990). The growth and yield performance of top-worked cashew trees. *Cashew Bulletin*, 27(9): 7-10.

Kumar, D.P., Khan, M. M. and Melanta, K. R. (1994). Effect of growth regulators on sex expression, fruit set, fruit retention and yield of cashew. Proc. of the XI Symposium on Plantation Crops. 30th November-3 December, 1994, Calicut, Kerala, India, pp. 610-627.

Latha, A. (2000). Cashew varieties tolerant to N deficient soils. *The Cashew*, 14(2): 29-39.

Latha, A. and Salam, M. A. (2001). Screening of cashew varieties for drought tolerance. *Journal of Plantation Crops*, 29(2): 27-30.

Lakshmipathi, Adiga, J. D. and Kalaivannan, D. (2014). Influence of growth regulators on certain reproductive parameters of cashew (*Anacardium occidentale* L.) variety Bhaskara, *Journal of Plantation Crops*, 2014, 42(1): 113-116

Mahesha, U.S., Hanamashetti, S. I. and Hubballi, V. N. (2005). Effect of season on success of softwood grafting in cashew (*Anacardium occidentale* L.) under Northern dry zone of Karnataka conditions. *The Cashew*, 19 (3): 17-19.

Mariappan, S., Prabakaran, J. and Sambandamoorthy, S. (1995). Effect of growth regulators on sex expression and fruit set in cashew (*Anacardium occidentale* L.). Cashew, 9(1): 11-13.

Meena, R.K., Adiga, J. D., Nayak, M.G., Saroj, P. L. and Kalaivanan, D. (2014). Effect of Paclobutrazol on growth and yield of cashew (*Anacardium occidentale* L.), *VEGETOS*, 27(1): 11-16.

Mishra, J. N., Paul, J.C. and Pradhan, P.C. (2008). Response of cashew to drip irrigation and mulching in coastal Odisha. *J. of Soil and Water Conservation*, 7(3): 36-40.

Mneney, E. E., Mantell, S. H., Bennett, M. (2001). Use of random amplified polymorphic DNA (RAPD) markers to reveal genetic diversity within and between populations of cashew (*Anacardium occidentale* L.). *J. Hortic. Sci. Biotechnol.*, 76: 375-383.

Nair, S. R. and Mohankumar, N. (1993). Standardization of tissue/apical meristem culture techniques on horticultural crops of Kerala. Final Technical Report, USDA, IN-AES-34 Thiruvananthapuram, Kerala, pp. 3-15.

Nayak, M. G., Swamy, K. R. M., Yadukumar, N. and Raviprasad, T. N. (2008). Effect of limb pruning on growth and yield of four cashew (*Anacardium occidentale* L.) varieties. *Journal of Plantation Crops*, 36(3): 300-303.

Nayar, K.G. (1998). Cashew: A versatile nut for health. In C.T.Topper, P.D.S. Caligari, A.K.Kulaya, S.H.Shomari, L.J. Kasuga, P.A.L. Masawe, and A.A.Mpunami (Eds.), Proceedings of International Cashew and Coconut Conference. Trees for Life – the Key to Development. BioHybrids International Ltd., UK. pp. 195–199.

Nayak, M.G., Mohana, G.S., Bhat, P.S., Saroj, P.L. and Swamy, K.R.M. (2014). Minimum descriptors of cashew germplasm accessions- Catalogue IV. Pub. ICAR-DCR, Puttur, pp. 1-32.

Nayak, M.G., Mohana, G.S., Bhat, P.S., Saroj, P.L., Swamy, K.R.M. and Bhat, M.G. (2015). Minimum descriptors of cashew germplasm accessions- Catalogue V. Pub. ICAR-DCR, Puttur, pp. 1-44.

O'Farrell, P.J., Armour, J. D. and Reid, D. J. (2010). Nitrogen use for high productivity and sustainability in cashew. *Scientia Horticulturae,* 124(1): 19-28.

Ohler, J.G. (1979). Cashew processing. *Tropical Abstracts.,* 21(9): 1792-2007.

Pillai, G. B., Singh Vijay, Dubey, O.P. and Abraham, V. A. (1984). Seasonal abundance of tea mosquito bug, *Helopeltis antonii* on cashew in relation to meteorological factors. In: Cashew Research and Development (Eds: Rao, E.V.V.B. and Khan, H.H.), Indian Society for Plantation Crops, CPCRI, Kasaragod, India, pp. 103-110.

Radhakrishna, Y., Subbarao, N., Sathyanarayanareddy and Ramadevi, M. (1992). Propagation of cashew (*Anacardium occidentale* L.) by softwood grafting under Bapatla conditions. *The Cashew,* 6(2): 3-5.

Rai, P.S. (1984). Handbook of Cashew Pests. Pub: Rsearchco Publications, Delhi.

Rao, V. N. M., and Hassan, M. V. (1957). Preliminary studies on the floral biology of cashew (*Anacardium occidentale* L.). *Indian Journal of Agricultural Sciences,* 27(3): 277-288.

Raviprasad, T. N., Sundararaju, D. and Bhat P. S. (2005). Efficacy of botanicals against *Helopeltis antonii* Sig. infesting cashew. *The Cashew,* 19(4): 9-14.

Richards, N. K. (1993). Cashew Research in Northern Territory, Australia, 1987-1991. *Tech. Bull.* No. 202.

Salam, M. A. (1999). Cashew varieties suitable for high density planting. *Cashew Bulletin,* 37(6): 3-5.

Salam, M. A., John, P. J., Joseph, M., Poduwal, M., Kumar, P., Yadukumar, N. and Bhat, M. G. (2008). Quantitative estimation of soil fertility and fertilizer recommendations (QUEFC) for Cashew (*Anacardium occidentale* L.). J. *of Plantation Crops,* 36(2): 86-94.

Sapkal, B. B. (2000). Integration of foliar application of urea and insecticidal spray for yield maximization in cashew. *The Cashew,* 14(1): 41-44.

Saroj, P. L. and Nayak, M. G. (2013). Potential of high density plantation for enhancing quality production of cashew. In: Canopy Management and High Density Planting in Subtropical Fruit Crops. Pub. CISH, Lucknow, pp. 55-63.

Saroj, P.L., Nayak, M.G. and Meena, R.K. (2014). Physiology of flowering, fruit and nut development in cashew. In: Souvenir of International Symposium on Plantation Crops at IISR, Calicut from 10-12 December, 2014 pp. 49-71.

Saroj, P.L., Krishna Kumar, N.K. and Jankiram, T. (2014). Converting wasteland into goldmine by cashew cultivation. *Indian Horticulture*, 59(6): 49-56.

Sawke, D. P., Salvi, M. J. and Patil, M. M. (1985). Softwood grafting - A sure technique of clonal propagation in cashew. *Indian Cashew Journal*, 17(1): 17-18.

Shalini Yadav (2010). Economics of Cashew in India. Occasional Paper- 50. Department of Economic Analysis and Research, National Bank for Agriculture and Rural Development, 4th Floor, 'C' Wing, Plot No. C-24, G-Block, PB No. 8121, Bandra-Kurla Complex, Bandra (East) Mumbai - 400 051.

Shirly, R.A. and Thimmappaiah, 2005. Somatic embryogenesis in nucellar callus of cashew (*Anacardium occidentale* L.). *J. Hort. Sci. and Biotechnol.*, 80(3): 327-331.

Singh, S. K., Syamal, M. M. and Maurya, A. N. (1992). Effect of NAA and Ethrel on vegetative growth, flowering, fruiting and yield of cashewnut. *The Cashew*, 6(2): 11-12.

Singh, S.N. (1997). Effect of bio-fertilizer (*Azotobacter chroococcum*) in cashew plantation. *Environment and Ecology*, 15(2): 482-484.

Sinish, M.S., Mercy, G. and John, P.S. (2005). Organic methods for cashew root stock production. *The Cashew*, 19(1): 8-15.

Subbaiah, C.C., Manikandan, P. and Joshi, Y. (1986). Yellow leaf spot of cashew: a case of molybdenum deficiency. *Plant and Soil* (Netherlands), 94(1): 35-42.

Sundararaju, D. (1984). Studies on cashew pests and their natural enemies in Goa. *Journal of Plantation Crops*, 12: 38-46.

Sundararaju, D. (1993a). Studies on the parasitoids of the mosquito bug, *Helopeltis antonii* Sign. (Hymenoptera: mymaridae) on cashew with special reference to *Telenomous* sp. (Hymenoptera: Scelionidae). *J. Biol. Control.*, 7: 68.

Sundararaju, D., Bakthavatsalam, N. and Bhat, P. S. (1993). Ovicidal activity and residual toxicity of certain insecticides of tea mosquito bug *Helopeltis antonii* Sign. (Hymenoptera: mymaridae) infesting cashew. *Pestology*, 17: 14-17.

Swamy, K. R. M. (1995). Top working of cashew in Goa and Maharashtra States - a case study. *The Cashew*, 9(2): 12-17.

Swamy, K. R. M., Bhaskara Rao, E. V. V. and Bhat, M. G. (1997). Catalogue of minimum descriptors of cashew germplasm accessions - I, pp. 1-41.

Swamy, K. R. M., Bhaskara Rao, E. V. V. and Bhat, M. G. (1998). Catalogue of minimum descriptors of cashew germplasm accessions - II, pp. 1-54.

Swamy, K. R. M., Bhaskara Rao, E. V. V. and Bhat, M. G. (2000). Catalogue of minimum descriptors of cashew germplasm accessions - III, pp. 1-54.

Swamy, K. R. M., Rao, E. V. V. B., Nagaraja, B., and Nayak, M. G. (1993). Commercial propagation of cashew varieties. Proc. of the Golden Jubliee Symp. on Horticultural Research – Changing Scenario, Bangalore.

Thevet, A. (1958). Singularidades da Franca Antartica. Companhia Editora Nacional Sao Pavlo (Brazil).

Thimmappaiah (1997). "*In vitro* studies on cashew (*Anacardium occidentale* L.)" Ph.D thesis in Biosciences submitted to Mangalore University.

Thimmappaiah and Shirly, R. A. (2000). Micropropagation in mature tree of cashew (*Anacardium Occidentale* L.) In: Recent Advances in Plantation Crops (Eds: Muralidharan, N. and Raj Kumar, R.) Allied Publishers Ltd., New Delhi, pp. 45-48.

Thimmappaiah and Shirly, R.A. (1999). *In vitro* regeneration of cashew (*Anacardium occidentale* L.). *Indian J. Exp. Biol.*, 37: 384-390.

Thimmappaiah, Santhosh, W.G., Shobha, D., Melwyn, G. S. (2009a). Assessment of genetic diversity in cashew germplasm using RAPD and ISSR markers. *Sci. Hortic.*, 120: 411-417.

Thimmappaiah, Shirly, R.A., and Sadhana, P. H. (2002a). *In vitro* propagation of cashew (*Anacardium occidentale* L) from shoot explants of young trees. *In Vitro Cellular and Develop. Biol. Plant*, 38(2): 152-156.

Thimmappaiah., Melwyn, G. S., Shobha, D., Shirly, R. A. (2009b). Assessment of cashew species for molecular diversity. *Journal of Plantation Crops*, 37: 146-151.

Veeraraghavan, P. G., Celine, V. A., Balakrishnan, S. (1985). Study on the fertilizer requirements of Cashew (*Anacardium occidentale* L.). *Cashew Causerie*, 7(2): 6-8.

Yadukumar, N. (2007). Organic farming in cashew. Proceedings of the National Seminar on Research, Development and Marketing of Cashew, 20-21st November, Goa, pp. 36-46.

Yadukumar, N. and Mandal, R. C. (1994). Effect of supplementary irrigation on cashewnut yield. Water Management for Plantation Crops - Problems and Prospects. Centre for Water Resource Development and Management, Calicut, Kerala, pp. 79-84.

Yadukumar, N. and Nandan, S. L. (2005). Recycling organic wastes of cashew plantations by aerobic composting. *Journal of Plantation Crops,* 33(2): 99-102.

Yadukumar, N. and Rejani, R. (2006). Temporal and spatial variation of soil moisture content and development of a yield model based on soil moisture content. *Journal of Plantation Crops,* 34(3): 364-367.

Yadukumar, N. and Rejani, R. (2008). Soil and water management in cashew plantations - Technical Bulletin (10). NRC - Cashew, Puttur, Karnataka, pp. 20.

Yadukumar, N., Rao, E. V. V. B. and Mohan, E. (2001). High density planting of cashew. *Tropical Agriculture*, 78: 19-28.

Yadukumar, N., Raviprasad, T. N., Nagaraja, K.V., Haldankar, P. M., Godase, S. K., Susanamma, K., Gajendran, G., Mahalingam, T., Lenka, P.C., Mohapatra, R.N. and Bandyopadhyay, B. (2003). National Agricultural Technology Project. Final Report on developing integrated production packages for enhancing productivity of cashew. National Research Centre for Cashew, Puttur, D.K., Karnataka, pp. 1-95.

Cashew: Improvement, Production and Processing (2017) *Pages* **31–43**
Editor: **Prof. P.L. Saroj**
Published by: **ASTRAL INTERNATIONAL PVT. LTD., NEW DELHI**

Cashew Development in India: Past, Present and Future

Venkatesh N. Hubballi and R. Jnanadevan

1. Introduction

Cashew (*Anacardium occidentale* L.), a native of Brazil was introduced to India in Goa by the Portuguese in the 16th century. India continues to be the largest producer of raw nut in the world producing 7.28 lakh metric tones (MT) of cashew from an area of 9.82 lakh hectares (2012-13). The dietary value of cashew kernels is making the product attractive for consumption throughout the world. Cashew kernel is a rich source of protein, carbohydrate and fat, and is an ideal supplement in the diet of children, pregnant women and lactating mothers. Cashew kernel compares well with almond with respect to protein, carbohydrate, fibre and mineral matter contents. Apart from cashew kernels, the important nutritious nut, Cashewnut Shell Liquid (CNSL) expelled during processing is a valuable industrial raw material, which finds use in a number of chemical industries such as friction linings, paints, varnishes etc. Other valuable products from cashew industry are cashew testa and tannin, which have also industrial applications. Cashew apple is a rich natural source of vitamin C and fibre. Cashew apples also contain free soluble sugars, most of which are reducing sugars. Consumption of cashew apple could help in overcoming the vitamin C deficiency and constipation.

Cashew plays an important role in the Indian economy. It is commercially important because of its substantial export earnings that India gaining by its international trade. India is the largest exporter of cashew kernel, exported 1,04,015 MT of cashew kernel valued of Rs. 4067.21 crores and 9,192 MT of cashew nut shell liquid (CNSL) valued at Rs.29.84crores during 2012-13. Ever since its introduction, besides significant contribution to national economy, it also provides substantial employment generation, provides employment to1.5 million people in cashew farming and industry sector, out of which 90 per cent are women labour in the processing sector. Today, India is the largest producer and processor and second in exporter of cashew in the world followed by Vietnam. Nearly 50 percent of the world export of cashew kernel accounted for by India. Among the various agri-horticultural commodities involved in the global trade, cashew has attained a prominent place by providing significant contribution to Indian exchequer ranking among the ten top agriculture export commodities contributing nearly 0.25 per cent of the total export earnings of the country.

Cashew is well adapted to every part of tropical and sub tropical regions of India. It spread all along the laterite hill slopes in the Western region from Bombay to Cape Comorin and to the sandy soils on the Eastern coast as well as over inland areas in the Southern parts. Its potential to transform wastelands to economically productive lands is widely exploited for commercial plantation development. The Directorate of Cashew and Cocoa Development, Kochi established under the Ministry of Agriculture, Government of India, is providing support for production and productivity oriented programmes since its inception. The Directorate of Cashewnut and Cocoa Development implements the programmes through the State Departments of Agriculture/Horticulture, ICAR-Directorate of Cashew Research, Puttur and State Agricultural Universities. The support provided by these agencies for cashew has amply proved its merits and their efforts have brought about substantial changes in development, processing and marketing, besides helping substantially to reduce dependence on imports.

2. Historical Perspectives

Cashew development programme started in the First Five Year Plan with distribution of non-descript seedlings of the crop to farmers. Establishment of Directorate of Cashewnut Development in 1966 by bi-furcating the erstwhile Indian Central Committee on Spices and Cashew under Union Ministry of Agriculture paved the way for well-designed strategic approach for various cashew development programmes in India and equally supported by research advances. The Cashew Development programmes have been executed through Directorate of Cashewnut Development from third Plan onwards *i.e.,* its inception of in 1966. A real integrated and coordinated approach in development of cashew became possible only after inception of Directorate of Cashewnut and Cocoa Development (DCCD), Kochi; till then some stray attempts were made by certain states towards the promotion of cashew. However, concerted effort for development of cashew started only from IV Five Year Plan onwards with focus on area expansion, production of quality planting material and development of production technology, which included rejuvenation and plant protection. Research was strengthened by establishing All India Coordinated Research Project on Cashew. At the same time, State Forest Departments started systematic plantation of this crop using seedlings from high yielding plants. The formation of Cashew Development Corporations and Forest

Development Corporations was a significant step towards the promotion of cashew in the public sector during V Plan period. The following Centrally Sponsored Integrated Development Programmes were implemented with financial assistance of Government of India until the end of VII Plan:

1. Area Expansion.
2. Production and distribution of planting materials.
3. Laying out demonstration plots.
4. Adoption of prophylactic plant protection measures.
5. Establishment of progeny orchards/scion banks.

The salient achievement of this period was the introduction of World Bank Aided Cashew Project - Multi State Cashew Project (MSCP) during VI Plan. The project was implemented from 1980-81 to 31.08.1987, both under the public and private sector in four states *i.e.* Kerala, Karnataka, Andhra Pradesh and Odisha. Area expansion was resorted through seedlings under this massive project due to absence of any specified variety. An area of 61,729 hectares was planted in the above states under this project. A study was also conducted to examine the need for possible institutional reorganization of the cashew industry with particular emphasis on improving efficiency and competitiveness of the Industry. Cashew research and training was another component of the project. The technological advances supported the strategy of cashew development of VIII Plan in a sound manner. As a result, the objective of VIII Plan was to increase the production and productivity of cashew by two pronged approach–extensive and intensive cultivation. The extensive cultivation includes area expansion and replanting programmes. Adoption of comprehensive production technology and adoption of intensive pest control measures forms part of intensive cultivation. Supporting the programmes like infrastructural development for the generation of planting material and transfer of scientific technology to the farming community was contemplated. To achieve the above objectives, the following programmes were drawn and executed in a project manner;

- ☆ Development of new plantations with clones conforming to export qualities and maintenances–to expand the area with high yielding clones of recommended varieties.
- ☆ Replanting/rejuvenation by top working of old senile and unproductive cashew gardens–to reduce the senility as it was posing a serious problem towards augmenting the production.
- ☆ Adoption of comprehensive production technology in grower's orchards–to improve the productivity status of the yielding plantations by way of adoption of recommended package of practices in a comprehensive manner.
- ☆ Adoption of intensive pest control measures–to control the incidence of major pests attacking cashew by way of adoption of prophylactic plant protection measures to prevent the loss in yield.

- ✰ Establishment of Regional Nurseries for generation of planting materials– to eliminate the use of seeds and seedlings for plantation purposes by producing required quantity of clones of high yielding nature.
- ✰ Transfer of scientific technology to the farming community

The salient achievements of the VIII Plan in the field of cashew promotion were: A greater impetus to cashew development could be achieved in the VIII Plan when a high financial support became available. The use of seeds and seedlings for plantation purpose are now not recommended and wider adoption of clonal planting materials was made. Nearly 63,000 ha were planted with clones of high yielding varieties of cashew in this Plan period. The replanting programme even though was introduced to replace the highly uneconomical old and senile plantations, instantaneous removal of older trees did not take place in many of the states as the farmer was not willing to loose the revenue obtains continuously and incentive provided was unattractive. Due to improper selection of the plantations, the productivity level could not be increased much as most of the areas selected were not responding to the technology due to inherent negligence and senility. The area treated under the intensive pest control measures had showed less affected by major pests, which could help restoration of the high yielding ability of such plantations. The regional nursery concept was a novel approach of the VIII Plan, where in nurseries to generate clones of high yielding varieties to distribute the beneficiaries. Involvement of private sector in the nursery activity besides the public sector organizations was another significant turn in the promotion of development programmes of cashew.

With this, an awareness on the use of clones and adoption of proper nutrition and crop protection measures besides the conservation measures on soil and water could be brought out on a wider range whereby the concept of scientifically economical plantation development is percolating among the farming sector. The farmers training programme was another salient approach towards the transfer of scientific technologies which gained a substantial momentum by way of imparting training to 6227 farmers in the vicinity of the farmers fields at village/ block level and with the association of scientists and extension workers. Publicity on crop promotion was taken up to disseminate the scientific information through literatures developed on regional vernacular languages. We have seen that the VIII Five Year Plan programmes have consolidated the technologies, needs of the industries and the aspiration of the farming sector on a sound footing by way of infrastructural development. Awareness has been created on the use of clones. Improved management system has started percolating in a modest way. The challenges ahead were to achieve the industrial requirement by making use of the available resources, technology and infrastructure build up.

Potentiality for new plantation development as such was limited in traditional area *viz.*, Kerala and Karnataka. States of Maharashtra, Andhra Pradesh, Tamil Nadu and Odisha had scope for new plantation development. Chhattisgarh and North Eastern regions, where huge potential exists, cashew has been intruded only recently. A compromising technological support was needed for tapping the areas.

The infrastructural build up needed expansion in terms of administrative machinery and resource personnel. Apart from new plantation development, senile areas needed replacement on a war footing basis. Public sector corporations having large compact senile areas had to play a major role in this regard. Transfer of technology from the research to farmer has been the responsibility of the extension department and through this mechanism transfer of technology has not been effective resulting in gaps between adopted and developed technology. This has been perhaps due to inadequate knowledge of extension workers. Thus, there was a need for developing a scientific mechanism for the transfer of technology involving self-help group. Developing model gardens particularly with the technology on local needs was one of the better approach to convince farmers about the advantages of technologies.

The objective of the IX Plan was to increase domestic production of raw cashewnut to meet the needs of the industry by reducing the dependence on import to augment the export trade in addition to the demand for domestic consumption. Specific emphasis was given in the clonal plantation development, its maintenance and technology dissemination centering the various practices on clonal plantation development. Replanting programme carried out in private sector areas during VIII Plan had its own difficulties.

For this purpose, it was planned to tackle the available large chunk of compact areas under public sector like corporations. To avoid intermediaries in the disposal of raw nuts produced by the growers and obtaining remunerative price for their produce, development of infrastructure either for marketing or if necessary for processing of raw cashewnuts was also programmed for IX Plan. More or less the similar approach of VIII Plan has been adopted. During IX Plan an outlay of Rs.70.00 crores had been allocated for the implementation of cashew development programmes in the country. Even though the IX Plan started from 1997-98, the actual execution of the approved programmes commenced in 2000-01 only, till then VIII Plan programmes were continued. Further, the development programmes comes under the purview of State Work Plan (Macro Management System) from 2000-01 onwards, as per new policy of the Government of India. Under this, the State has its own priority to choose the crop for its development. Because of this policy, some of the approved programmes of IX Plan have been included under the State Work Plan.

3. National Horticulture Mission

Considering the vast potential for development of this highly economical crop, government of India selected this crop for promotion under the Centrally Sponsored Scheme on 'National Horticulture Mission' (NHM) which was launched during 2005-06 throughout the Country except in North Eastern and three Himalayan States, wherein Technology Mission for Integrated Development of Horticulture (TMNE) is being implemented. The Horticulture Schemes including Cashew which were subsumed under Macro Management Scheme during 2001 were taken out and subsumed under NHM for their implementation and since then more thrust has been given for cashew development in the country. To oversee the activities of NHM and TMNE and for approving the Action Plan/Projects of the States, Central Steering Committee headed by Secretary (A and C) was constituted in 2005. At State level, State Level Steering Committee headed by Chief Secretary/APC has been constituted to monitor the activities carried out by the States under the Schemes. Priority with greater importance for holistic development of cashew by adopting an integrated approach concentrating on production and productivity improvement programmes, human resource development through training and field visits along with creation of infrastructure for postharvest management and marketing, are being accorded in the schemes. With the launching of National Horticulture Mission (NHM) in 2006, development of Cashew was covered under this Mission. The crop has been included for holistic development in the States of Andhra Pradesh, Chhattisgarh, Goa, Gujarat, Jharkhand, Karnataka, Maharashtra, Odisha, Tamil Nadu and West Bengal. Under NHM, programmes relating to improvement of production and productivity, postharvest management and marketing are being implemented by the respective State Horticulture Mission (SHM) whereas activities relating to technology dissemination through front line demonstration (FLD), HRD, pest/disease surveillance and publicity campaigns are taken up by DCCD with involvement of SAUs/ICAR institutes. Expansion of area under cashew in potential areas, replanting old and senile plantations, technology demonstrations, intensive transfer of technology progrmmes, production and distribution of quality planting materials and human resource development are the promotional activities carried out under NHM.

The cultivation and promotion of cashew in major cashew growing states are progressing well and increase in production from non-traditional areas of Chattisgarh and North Eastern States is really significant. Among the North Eastern states, Meghalaya, Tripura, Manipur and Nagaland have good potential for cultivation and promotion of cashew. Some parts of Mizoram also have potential to grow this crop on a commercial scale. It is important to note that cashew plantations under the control of forest departments and state cashew corporations, which were planted before 1966, have become senile and their productivity has become uneconomical. Nearly 36 per cent of senile plantations, including public sector plantations, contribute to only 15 per cent of total production of the country. Removal of these plantations and replanting with high yielding varieties will be the better option to boost the production and to attain sustainability. The area existed prior to VIII Plan constituted by seeds and seedling progenies amounts to 5.34 lakh ha and the production obtained from these plantations were 3.05 lakh MT with a productivity level of 650 kg/ha. The developmental measures so adopted have helped in developing over 4.00 lakh ha. of cashew plantation with clones of high yielding varieties during VIII Plan and thereafter till 2012-13.

4. Impact of Development Programmes

Past developmental efforts have helped us both in increasing the area and production of cashew in India. The area and production of cashew in 1965-66 was 2.41 lakh ha and 1.04 lakh MT with a productivity of 400 kg/ha. Efforts made in the II and III Plan and with the continued efforts in subsequent plan periods, the area and production of cashew increased to 3.75 lakh ha and 1.62 lakh MT in 1975-76 to 9.82 lakh ha and 7.28 lakh MT in 20012-13, (Table 2.1).

Table 2.1: Year-wise Area, Production and Productivity of Cashew

Year/Period	Area (ha)	Production (MT)	Productivity (Kg/ha)
1965-66	2,41,134	1,03,818	400
1975-76	3,74,653	1,61,536	500
1985-86	5,18,379	2,34,480	626
1995-96	6,34,970	4,17,830	720
2005-06	8,37,000	5,73,000	684
2006-07	8,54,000	6,20,000	726
2007-08	8,68,000	6,65,000	766
2008-09	8,93,000	6,95,000	778
2009-10	9,23,000	6,13,000	695
2010-11	9,45,000	6,53,000	720
2011-12	9,91,000	6,92,000	749
2012-13	9,82,210	7,28,474	772

Here, it can be seen that during the period of 1965-66 to 1985-86, both area and production of cashew increased on an average of 6 per cent per annum. During the period of 1985-86 to 2011-12, even though area under cashew showed an increase of 3.50 per cent per annum, there was a tremendous increase of 7.50 per cent per annum in production. It was during this period research findings of high yielding varieties of grafts suitable to different agro climatic conditions and standardization of softwood grafting techniques were gained popularity in all the cashew growing states with the help of development departments of the concerned states. Now cashew occupies an area of 9.82 lakh ha with a production of 7.28 lakh MT of rawnuts as on 2012-13 with average productivity of 772 kg/ha.

Cashew plantations in India are mostly distributed under small and marginal sector with the holding size of <1 ha, 1-2 ha, 2-5 ha and >5 ha. Out of this, senile cashew plantations of 3.20 lakh hectares are available in the country which is the bottleneck for enhancement of production. The production and productivity is highest in the state of Maharashtra, since majority of plantations are developed primarily by grafts of high yielding varieties and also farmers are adopting better management practices and provides protective irrigation. These promotional efforts helped in creating awareness among growers on the importance of cashew to gain remunerative income from marginal lands resulted in increasing area, production and productivity of cashew as shown above. Of the total area of 9.82 lakhs ha of cashew plantations existed in India, nearly 4.00 lakhs ha area is being developed from the beginning of VIII Plan onwards with the elite clones of high yielding varieties. Most of these plantations have already reached economical bearing and the benefits are being reflected now in the field. Whereas one of the major limiting factor contributing towards low level of productivity is due to increasing senile status of existing plantations. Cashew plantations having generated from indiscrete seeds and seedlings progenies in the earlier years have become senile and uneconomical in production and the total area under senile is about 3.19 lakh ha.

Besides the contribution of Indian Cashew industry of contributing valuable foreign exchange to our national exchequer and it is also one of the potential industries which provide sustainable employment to nearly 1.5 million people in the agrarian and industrial sector.

5. Cashew Processing

The Indian Cashew industry is one of the few industries in the country, which is almost export oriented. Apart from earning valuable foreign exchange from the industry, the major substantial contribution is in the field of providing sustainable employment opportunities to about 1.5 million people and especially for women folk. Cashew is processed and marketed on an industrial basis. Even though, organized marketing structures are not available but cashew does not face any serious marketing problem in India. Private entrepreneurs to a great extent are engaged in this industry. At present 3797 processing units are functioning in India under organized and unorganized sector with a processing capacity of 16.23 lakh MT. The state-wise distribution of processing units, along with their capacities is given in Table 2.2. At 16.23 lakh MT, the aggregate processing capacity in the country exceeds the total production of raw cashew. Therefore, raw nuts are imported to feed the processing units. The produced output is used both for domestic consumption and export.

Table 2.2: Distribution of Cashew Processing Industries

State	*Processing Units (Nos)*	*Processing Capacity (000 MT)*
Kerala	432	600
Karnataka	266	300
Goa	45	50
Maharashtra	2200	50
Tamil Nadu	417	400
Andhra Pradesh	175	100
Odisha	207	100
West Bengal	30	8
Chattisgarh	3	5
North Eastern States	22	10
Total		**3797 1623**

6. Export of Cashew Kernels and Import of Raw Nuts

India continue to be the premier exporter of cashew kernels and earns a sizable amount of foreign exchange (Rs. 4390 crores during 2011-12 was the highest record). Whereas raw nuts valued at Rs.5337 crores were imported during last year to meet the demand of the processing units. There was 24.5 per cent increase in the quantity of cashew kernel exported during 2011-12 as compared to that of 2010-11. However, the quantity of cashew kernel exported has decreased by 23.51 per cent in 2012-13 compared to the previous year. The detail of export of cashew during the last nine years is given Table 2.3.

Table 2.3: Export of Cashew Kernels and CNSL from India

Year	*Cashew Kernels*		*Cashew Nut Shell Liquid*		*Total Value*
	Qnty.	*Value*	*Qnty.*	*Value*	
	M.T	*Rs. Crore*	*M.T*	*Rs. Crore*	*Rs. Crore*
2003-04	100828	1804.42	6926	7.03	1811.45
2004-05	126667	2709.24	7474	7.91	2717.15
2005-06	114143	2514.86	6405	7.09	2521.95
2006-07	118540	2455.15	6139	10.29	2465.44
2007-08	114340	2288.90	7813	11.97	2300.87
2008-09	109523	2988.40	9099	26.06	3014.46
2009-10	117991	2801.60	11227	27.62	2829.22
2010-11	105755	2819.39	12051	33.77	2853.16
2011-12	131760	4390.68	13575	59.46	4450.14
2012-13	104015	4067.21	9192	29.84	4097.05

Source: DGCI and S Kolkata (2011).

The local production of raw cashewnut in India is almost 50 per cent of the domestic demand and hence India depends heavily on import of raw nuts from Asian and African countries like Tanzania, Guinea Bissau, Ivory Coast, Indonesia, Mozambique, Kenya, Nigeria, Senegal, Ghana, Gambia and Madagascar. In the year 2012-13, India imported nearly 8.92 lakh MT of raw cashew nuts at a cost of Rs. 5331.12 crores. The details of import of raw cashewnut for the past ten years are given in Table 2.4.

Table 2.4: Import of Raw Cashewnuts by India

Year	*Quantity (MT)*	*Value (Rs. Crore)*
2003-04	452399	1400.93
2004-05	578884	2183.24
2005-06	565400	2162.95
2006-07	592604	1811.62
2007-08	605970	1746.80
2008-09	605850	2632.41
2009-10	755959	3047.50
2010-11	529370	2649.56
2011-12	809371	5337.76
2012-13	892160	5331.12

Source: DGCI and S, Kolkatta (2011).

7. Future Development Strategies

The development of cashew in India has gained momentum since VIII Five Year Plan. The objective of VIII Plan and IX Plan was to increase the domestic production of raw cashew nut to meet the needs of the industry by reducing the dependence on import and to augment the export trade, in addition to the demand for domestic consumption. Special impetus was given in the clonal plantation development and productive technology focussing centering the various practices needed for clonal plantation development. It was also planned and given, priority for replanting programme under private sector and corporation sector, but this programme was not yielded the desired results due to its own demerits under private sector.

From 2000 onwards, the development of cashew in India, especially programmes like New Plantation Development and Replantation programme were kept under the purview of State Work Plan by the Government of India. The main aim was to give more flexibility to State Development Departments in order to prioritize developmental strategies for important horticultural crops. However, the indication arrived by analyzing the developmental strategies adopted by various State Development departments and Directorate of Cashewnut and Cocoa Development has not yet given the desired result. As on today, India is the lead player in the global cashew scenario, but its position is not sustainable for the years to come. Small countries like Vietnam and Brazil are developing their own industries by manifold and now they are the tough competitors to India's position in the cashew world market. In this context, it is utmost necessary to select and prioritize the important components for the development of cashew in India, in order to meet the demands of domestic industry as well as to maintain India's prime position in the global cashew scenario.

Area expansion under new area is the limiting factor in the State of Kerala, whereas states of Maharashtra, Andhra Pradesh, Tamil Nadu, Karnataka and Odisha have scope for new plantation development. Chattisgarh, Jharkhand, Gujrat and North Eastern regions, where huge potentiality exists and cashew has just introduced there. The intensive technological support is needed for tapping these areas. Therefore, specific emphasis was given in the expansion of area in traditional and non traditional tracts with elite clonal planting materials, apart from new plantation development, senile areas needs replacement for which public sector corporations having large compact senile areas is being chosen along with senile plantations of the private sector. However, the replacement of old and senile cashew plantations is being replaced to the tune of more than 60,000 ha, under private and corporate sectors. Transfer of technology from the research to farmer was the responsibility of the extension department. However, transfer of technology could not be made effective in the past plan periods which resulted in wide gaps between developed and adopted technologies. Therefore, a scientific mechanism for the transfer of technology involving self help group, Non Governmental Organizations and other development departments by providing adequate training and other support have been developed, so as to make the farmers well aware of modern scientific technologies. High-density concept was also included as a measure for popularization with different production technologies and in most of the states, this concept was not accepted by the farming community and normal density planting was generally adopted.

The infrastructure build-up also needs improvement and expansion. Altogether 42 cashew nurseries are functioning well with the financial assistance of Government of India. These nurseries have capacity to produce 10 million grafts which can cater the area coverage of 50,000 ha per annum. Therefore, there will not be any scarcity in planting material production, if the strategy of the development department is moving in the right track.

In order to popularize the usage of cashew apple, for the preparation of cashew apple products such as cashew apple juice, jam, candy, syrup *etc.*, a processing unit for cashew apple on demonstration scale is being established in various states with the technologies available at Kerala Agriculture University. To make full awareness on modern production technologies on cashew among the farming community literatures were distributed and seminars, workshops and field days were organized to have greater impact among farmers.

8. Projected Growth and Demand

The extent of processing of raw cashew nuts during 1994-95 was 5.53 lakh metric tonnes (MT) of which 3.22 lakh MT was indigenously produced and 2.31 MT tons was imported. During 2012-13, the raw cashew nuts processed in India has reached 16 lakh tons of which the indigenous raw nuts production is to the tune of 7.28 lakh MT and the balance requirement is met through import. Thus in the last 10 years, the quantity of raw nuts processed in India has increased three times. If the same trend continues for next 10 years, the requirement shall be at least 22 lakh tons by 2020. Since area expansion under cashew is picking up in many states, especially in non-traditional areas and also the low yielding cashew plants of senile and non-descript origin are being replaced by cashew grafts of high yielding varieties and improved package of practices is being adopted by farmers; interest in cashew cultivation is expected to increase substantially. Hence, it is expected that India will be able to keep pace with the increasing demand of raw cashew nuts from cashew processing units and will become self-sufficient by 2020. The estimated demand and anticipated production of raw cashew nuts during next decade is presented in Table 2.5.

9. Constraints in Cashew Production

Even though, cashew is blessed with research technologies, vast net work of processing industries and elite private processing sector and most congenial climatic parameters to cultivate the cashew in 15 states but main constraint is because of large scale plantation under seedlings, owing to poor productivity. The lack of intensive farmer's friendly transfer of technologies programmes by all the sectors involved in the cashew sector. There are still many lacunas noticed at present system of development strategies adopted by the various state governments as well as farmers. To introduce cashew in the non-traditional districts of cashew growing states of Odisha, Maharashtra, Andhra Pradesh and Karnataka and removal of senile cashew plantation and replanting with regionally recommended clones of

Table 2.5: Projected Growth and Demand of Cashew Production

Year	*Estimated Demand of Raw Cashew Nut (lakh MT)*	*Anticipated Production of Raw Cashew Nut (lakh MT)*
2013	15.50	7.28
2014	16.00	8.00
2015	16.50	8.80
2016	17.00	9.60
2017	18.00	10.20
2018	19.00	11.40
2019	20.00	12.40
2020	21.00	13,80
2021	22.00	15.00

high yielding varieties to be considered as a thrust area. More emphasis should be given to introduce technology driven programme in nontraditional areas like Chhattisgarh, Gujarat, Jharkhand and North Estern states like Meghalaya, Tripura, Assam, Manipur and Nagaland. Identify and establish cashew export zones in the state of Maharashtra and Odisha, these states are having huge potentiality to bring more area under cashew and the desired focus need to be given to establish cashew export zones in these states.

10. Conclusion

The development departments of cashew growing states have to give more priority on the use of quality planting materials. With due technical evaluation of planting materials of desired quality should be selected by foregoing tender system. Identify and establish cottage industries which is necessary for the processing of rawnuts and cashew apple. Nearly, 70.00 lakhs MT of cashew apple goes as a waste every year. If the cashew apple used properly, it will help the farming community to gain more income per unit area. Effective transfer of technology programmes, the interaction between the Directorate of Cashewnut and Cocoa Development, State Development Departments and Research Organizations are very much required on war foot basis to create clear awareness among cashew growers. The options also need to be sorted out to include non-governmental organizations to disseminate productive transfer of technology programmes. The thematic training programme on regular basis, exposure visit and need based campaigns at innovative farmer's field are to be given top priority. Contract farming system in cashew need to be introduced with the active involvement of processing industries. Since major chunk of the senile area is under various state level corporations and the senile area under these corporations must be brought under lease of about 20-25 years with government support.

The challenges ahead would involve measures to enhance the production and productivity particularly from the senile cashew plantations which are more than 30 years old, promoting cashew in nontraditional tracts, strengthening cashew nurseries for production of quality planting materials, creation of infrastructure facilities for postharvest management of cashew apple and other products like CNSL and testa derived during processing of raw cashew nut. Organic farming of cashew is another area, which requires attention to meet the increasing demand for hygienic foods in the international market. Over viewing the threats and opportunities of cashew development in India, it is satisfying to note that there has been a substantial increase in production and productivity. However, cashew nut has to face the threats from other countries, as well as from other edible nuts. Thus, there is need for efficiency in production, processing and utilization, so that cashew can continue to be competitive. If all the areas mentioned above are given prime importance and implemented with quality approach in cashew development under National Horticulture Mission, it will be possible to become self sufficient by the near of this decade.

Cashew: Improvement, Production and Processing (2017) *Pages* **45–90**
Editor: **Prof. P.L. Saroj**
Published by: **ASTRAL INTERNATIONAL PVT. LTD., NEW DELHI**

3

Processing of Cashew: International and National Scenario

D. Balasubramanian and P.L. Saroj

1. Introduction

Commercial production of raw cashewnuts is confined to the tropical coastal regions of the world especially Asia, North America and African zones while the temperate regions of the world account for processing and the export demand for consumption. Raw cashewnut is made up of three different portions *viz.*, the shell, the kernel and the adhering testa. The primary product of cashewnut is the kernel which is the edible portion of the nut and is consumed directly by the consumer as plain cashew or roasted and salted nuts, flavoured and spice coated as value added cashew or as an ingredient in confectionery and bakery products. Cashew processing is a tedious and arduous task because of the irregular shape of the cashewnut, the presence of tough outer shell and the presence of corrosive cashewnut shell liquid (CNSL) within the shell.

Although cashew is native to Brazil, cashew processing evolved in India. In the beginning, only India was able to profitably perform the elaborate and strenuous manual processing required to extract the cashew kernel. This resulted in India's domination of the supply of kernels in the world market until the last century. The introduction of mechanical processing of raw cashewnuts in Africa on a commercial scale in the mid 1960's transformed the cashew trade. The cashewnut processing developed in the last century was labour oriented, employing largely rural women who contributed 90 per cent of the work force and with minimal mechanization. The industry has made strides in twentieth century by developing an efficient processing technology over a period of time. Value chain in cashew generates employment and fetch significant turnover for agri-business dependent countries.

2. Nutritional and Medicinal Values

Cashewnuts are an up-market snack food and the demand is set to increase in view of its health benefits and the rising purchasing power of emerging economies. It is a complete nourishing diet and suitable food medicine for ailments. The analysis of the nut shows that on an average it contains 21.2 per cent protein, 22.3 per cent carbohydrate, 46.9 per cent fat, 2.4 per cent mineral matter, 0.05 per cent calcium, 0.45 phosphorous, 5.0 mg per 100 gm iron and 1.3 per cent fibre. It also contains carotene calculated as vitamin A 100 IU, thiamine 630 mg, niacin 2.1 mg and riboflavin 190 mg per 100 g of kernel (USDA, 2002).

Cashew is an aphrodisiac that can boast of many curative properties. Laden with 21 per cent protein and an equally high percentage of poly unsaturated fatty acids, cashew helps in reducing the blood cholesterol level considerably preventing possibilities of heart attacks. As the ratio of Saturated: Monounsaturated: Polyunsaturated is 1:2:1, it can be considered as an ideal food for human consumption. The most prominent vitamins in cashew are vitamin A, D and E. These vitamins help in assimilating the fats and increase the immunity level. The presence of vitamin E in cashew takes care of all reproductive problems and prevents the development of oxidative rancidity in fats. High content of linoleic acid in cashew kernel makes it ideal digestive assimilative stimuli. Cashew kernel is rich in calcium, phosphorous and iron which can prevent anaemia and nervous system ailments. In fact, the vegetable proteins contained in cashew kernels stand at par with milk, eggs and meat with an added advantage of not producing uric acid in the body. It also contains a high concentration of much needed acids in right proportions, generally very rare in nuts. Recently discovered vitamin-PP also develops while roasting cashew and it can exert a sparing action on the B group vitamins and assist in metabolism of lactose and thiamine.

Cashew kernel oil is also considered a good mechanical and chemical antidote for irritant poisons and it is a vehicle for ligaments and other external applications. The kernel is also used as a substitute for almond mixture and helps the patients suffering from incessant and chronic vomiting. Raw cashew fruit is used as an anesthetic in leprosy and also for curing warts, corns and ulcers. The juice of the nut is used as a substitute for iodine while the oil obtained from the shell is good for cracks in feet. The cashew apple contains 10.44 per cent of fermentable sugars and 261.5 mg per 100 g of vitamins C, giving antiscorbutic properties to both the fruit and the wine made out of it. The liquor is also valued as a diuretic with healthy effect on kidneys and advanced cases of cholera. Cashew feni is said to have high medicinal value and as a cure for ailments ranging from worm sickness in children to diarrhoea and even cholera.

3. Historical Timeline of Processing

Indigenous stands of cashew trees along the North-eastern coast of Brazil were used by natives for food and medicinal purposes. After colonization of the area in the mid-15th century, the trees were cultivated by European settlers and the edible kernels used by navigators as a reserve food source on long voyages (Leckie, 1989) and through this process the cashew was disseminated to other tropical areas in the world, specifically East Africa and India. Finally, its present diffusion can be geographically located between 31° North latitude and 31° South latitude, both as the wild species and under cultivation. Despite this diffusion, cultivation and processing techniques remained primitive and commercialization limited, until the 1940s. During World War II, the cashew tree became highly prized as the source of valuable oil drawn from the shell. Moreover, as a return for technical assistance in aviation and general industrialization, the Brazilian government allowed the United States to run a series of expeditions through Brazil in search of raw materials that could be used in the war effort. During this time modern uses for cashew nut shell liquid were discovered and the world became more familiar with the taste of the cashew kernel. After the war, demand for CNSL fell, but as the world economy recovered, the demand for the kernel as a delicacy food rose.

India was the first country to build up a processing industry, but domestic production has been insufficient to meet the requirements of the country's small and medium-scale processing outlets. The imports from East Africa generally took place from December to May, which complemented the national harvest which commences from March and extend up to mid of July. Thus the Indian processors have been able to operate over a prolonged period without having to maintain large stores of raw nuts. India also has a long tradition and good reputation as a high quality processor of cashew. A number of countries therefore prefer to export their raw nuts to India for processing rather than to process themselves and produce lower quality kernels. The Cashew industry has grown from cottage industry status in 1924 into an organized factory set up due to bulk export to overseas markets in developed countries.

4. Production Vs Processing

Significant increase in the production of cashew at global level registered over the past 50 years (Table 3.1). Africa was the largest producer of cashew till the 70s, accounting for about 68 per cent of total world output. During this period, Mozambique and Tanzania were the dominant players in the market. In the last three decades, however, there has been a transformation in the global production scenario with Asia overtaking Africa. India till recently led the world production, but the production data available from Food and Agricultural Organization (FAO, 2011) indicated that Vietnam and Nigeria have outstripped India and Brazil in cashew production during the year 2011 (Figure 3.1).

In India, cashew is largely grown in the coastal regions *viz.*, Kerala, Karnataka, Goa, Maharashtra, Tamil Nadu, Andhra Pradesh, Odisha, West Bengal and North East Hilly region (NEH). The Directorate of Cashew and Cocoa Development, the Directorate of Cashew Research (erstwhile National Research Centre for Cashew), the Cashew Export Promotion Council and the Agricultural Universities across the states have been continuously intervening for the development of high yielding varieties of seeds, grafts and planting systems for augmenting the production

Table 3.1: Cashew Production in Major Cashew Growing Zones (2000-2012)

Zones	*2000*	*2001*	*2002*	*2003*	*2004*	*2005*	*2006*	*2007*	*2008*	*2009*	*2010*	*2011*	*2012*
Asia													
India	520000	450000	470000	500000	535000	544000	573000	620000	665000	695000	613000	674600	680000
Indonesia	69927	91586	110232	106931	131020	135070	149226	146148	156652	147403	115000	114600	117400
Philippines	4610	5310	5340	5490	6130	6200	6810	7050	7120	6770	6970	6890	6510
Sri Lanka	111019	111153	111223	111291	116910	116533	113071	112558	112334	111993	134681	133388	132541
Thailand	44789	39071	35345	43492	42126	60506	46377	43048	40612	38184	37857	29060	31500
Vietnam	270400	292800	515200	657600	818800	960800	1092400	1249600	1234000	1165600	1242000	1237300	1190900
Sub Total Asia	**1037901**	**1009083**	**1272023**	**1453294**	**1682613**	**1858585**	**2019810**	**2191895**	**2229153**	**2178370**	**2163691**	**2211676**	**2175601**
Africa West													
Benin	40000	43264	46771	46002	45000	52500	55000	60000	86000	117000	102137	162986	170000
Ghana	7697	8325	9000	16500	25000	29000	34000	40511	22000	27000	30000	35736	36500
Guinea Bissau	72725	85000	86000	84586	96649	89000	95000	98000	99000	99500	108029	128684	130000
Ivory Coat	63380	87573	104985	84811	140636	185000	235000	280000	330000	350000	380000	393000	450000
Nigeria	466000	485000	514000	524000	555000	594000	636000	660000	727603	800000	830000	835000	836500
Senegal	7000	4000	4500	4426	5057	5663	6332	5000	5050	5500	6000	6500	6650
Burkina Faso	3732	4037	4364	4292	4904	5492	6141	3600	3969	4544	4933	5876	5750
El Salvador	2530	2600	3484	3787	3910	3235	4397	2700	3687	3380	3380	3400	3450
Sub Total Africa West	**663804**	**720979**	**775250**	**770709**	**881729**	**971875**	**1079597**	**1160878**	**1289622**	**1421214**	**1476399**	**1589290**	**1657100**

Contd...

Table 3.1–*Contd...*

Zones	*2000*	*2001*	*2002*	*2003*	*2004*	*2005*	*2006*	*2007*	*2008*	*2009*	*2010*	*2011*	*2012*
Africa East													
Kenya	12500	12540	10031	10000	9332	10924	11349	13520	17000	17683	17568	20927	29026
Madagascar	6500	6543	6349	6613	7289	8533	6700	6900	7607	5913	5605	6677	7000
Mozambique	57894	58000	50177	63818	42988	104337	62821	74395	85000	64000	96558	112796	64731
Tanzania	121600	122300	67380	91340	92810	89980	90382	92600	99100	79100	74170	122274	122274
Sub Total Africa East	**198494**	**199383**	**133937**	**171771**	**152419**	**213774**	**171252**	**187415**	**208707**	**166696**	**193901**	**262674**	**223031**
Latin America													
Brazil	138608	124073	164539	183094	187839	152751	243770	140675	243253	220505	104342	230785	80630
Peru	2717	2465	2142	2173	2408	2213	2250	2350	2350	2535	2124	2588	2700
Sub Total Latin America	**149514**	**135230**	**180174**	**195536**	**202101**	**167835**	**259522**	**154461**	**259858**	**236663**	**119221**	**246387**	**96583**
World Total	**2049713**	**2064675**	**2361384**	**2591310**	**2918862**	**3212069**	**3530181**	**3694649**	**3987340**	**4002943**	**3953212**	**4310027**	**4152315**

Source: FAO (2011).

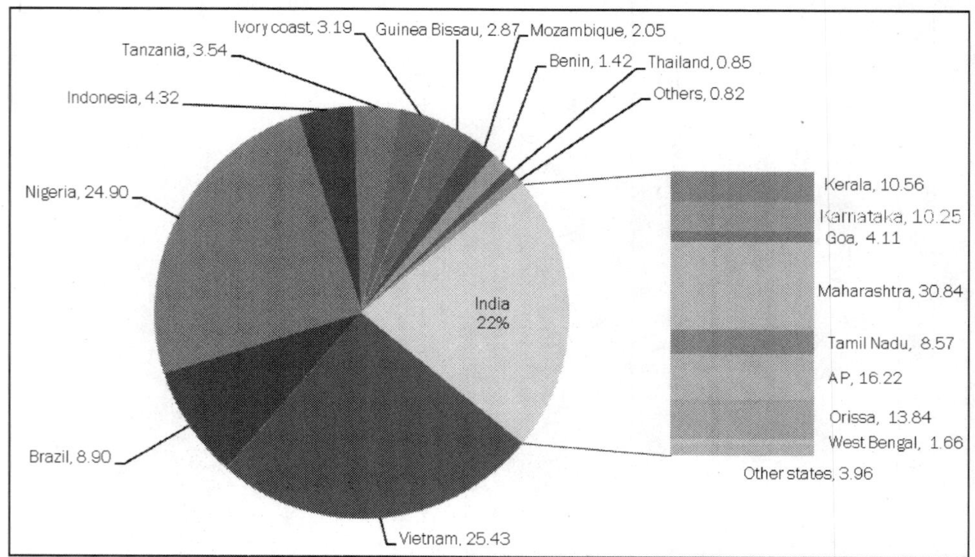

Figure 3.1: Share of Cashewnut Production in the Domestic and International Level.

of quality cashew nuts in this country. Current production data reveals that Maharashtra had emerged as the dominant player in the market contributing almost one third of the national production but accounting for only one fifth of the land area under cashew cultivation, showing a higher efficiency in production (Saroj and Balasubramanian, 2013). Productivity is found to be nearly one and half times that of national average.

5. Maturity and Harvesting of Raw Cashewnuts

Maturity is defined as the stage of development giving minimum acceptable quality to the ultimate consumer. Any harvest either too early or too late in the season is more susceptible to physiological disorders and has a shorter storage life than those harvested at proper maturity. Objective maturity indices enable growers to know whether their commodity can be harvested when the market is buoyant. The maturity of raw material is important in controlling both quality of the final product and effectiveness of processing. Immaturity reduces yield and the final product are liable to have substandard colour and texture. Flowering process in cashew tree starts approximately 5 days after pollination and continue to flower and set fruit and nut in cycles based on the environment conditions and fruit load throughout season (Mandal, 1992). Due to this indiscriminate fruiting pattern, nuts of different maturities are harvested during season supplied to postharvest processing posing the problem in attaining uniform quality in final product.

Investigations carried out on linear dimensions and kernel yield of raw cashewnut at different levels of maturity for the selected cultivar VRI 2 and VRI 3 at Directorate of Cashew Research, India revealed that tri-axial dimensions of raw cashewnut *viz.*, length, width and thickness decreased by 19.48 per cent, 17.73 per cent and 24.07 per cent in VRI 2 and 13.26 per cent, 11.77 per cent and 21.65 per cent for VRI 3 respectively from the stage of nut formation (Stage I) to full maturity (Stage IV) (Figure 3.2). Comparing shrinkage in all the three dimensions, width of the nut recorded the lowest changes with maturity. Changes in nut dimensions in measured axes were observed to be non-uniform resulting in irregular shape of the nut.

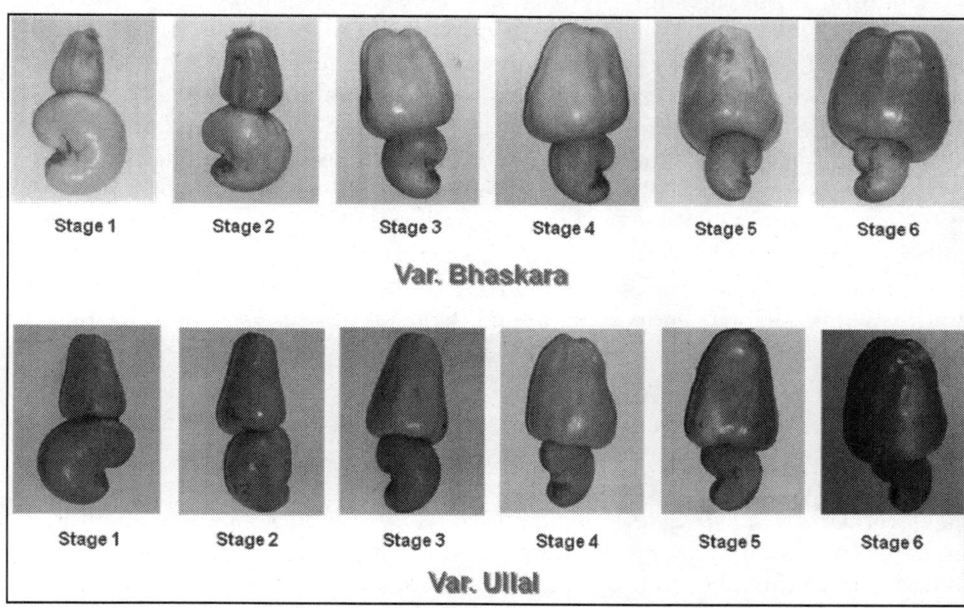

Figure 3.2: Changes in the Size and Surface Colour of Raw Cashewnuts at Various Stages of Maturity.

Bulk density of raw cashewnuts increased from 534.21 to 611.99 kg m^{-3} (14.56 per cent) and from 559.09 to 647.09 kg m^{-3} (15.74 per cent) in VRI 2 and VRI 3 respectively during nut maturation. Dimensional shrinkage from stage I to IV is the due reason for the increase in bulk density.

Moisture content of whole nut, shell and kernel showed a declining trend from 17.58 to 10.52 per cent d.b, 18.61 to 11.67 per cent d.b and 12.29 to 9.07 per cent d.b respectively for VRI 2 from the nut formation to full maturity. Similar trend was also observed for VRI 3 during different stages of maturity with the moisture content changing from 17.91 to 10.96 per cent d.b, 18.31 to 11.78 per cent d.b and 12.45 to 9.14 per cent d.b for whole cashewnut, shell and kernel respectively (Anonymous, 2013). This phenomenal change in moisture content coupled with non-uniform changes in tri-axial dimensions indicating the size of the nuts lead to changes in shape and bulk density of raw cashewnuts during maturation.

Surface colour of the nuts changes from brownish green at fruit set to light green at 50 per cent growth of adhering pseudo fruit and thereafter turns grey irrespective of varieties. Diminishing chlorophyll pigment present in the shell and the formation of cashewnut shell liquid in endocarp region of raw cashewnut as the nuts attain maturity are the possible reasons for the changes in surface colour of the shell. In addition, the cutting force generally increased from nut formation stage to full maturity and this increasing tendency of failure force was mainly due to moisture loss and nut hardening towards full maturity (Anonymous, 2013).

Raw cashewnuts are normally harvested in four different ways *viz.* collection after natural drop; thrashing by stick and collection; climbing on the tree and thrashing and shaking the tree. But in majority cases, fallen fruits with attached nuts are collected from the ground manually or harvested using bamboo pole having hook at the bottom end. In exceptional cases, cultivators have the tendency to pluck the nuts before the pseudo fruit develops completely in order to prevent pilferage. Fully mature apples are plucked from the tree with the help of a long pole with a sickle tied to it at one end. When the cashew apple is fully ripe, with a simple tap on the fruit, it drops off. Shaking the branches for harvesting the nuts is also practiced. Although this results in dropping of fully mature nuts, leads to dropping of flowers and immature nuts too. The best method of harvesting the nuts is to wait till it falls, but this damages the pseudo fruit. If one intends to utilize cashew apple, it has to be plucked to avoid damage to the fruit.

Studies conducted on harvesting of raw cashewnuts in India indicated that only 16 per cent of farm follows the collection of nuts by natural drop which is recommended for obtaining quality produce with full maturity. Majority of the farm up to 72 per cent follow harvesting the nuts by thrashing with stick and collection method (Dalvi *et al.*, 1992) and this mode of harvesting increased the quantum of immature nuts supplied for processing. This is mainly due to cultivation of cashew in underutilized land at a far distance from farm and to overcome the problem of pilferage. Besides climbing on the tree and thrashing and shaking the tree to harvest the nuts amounts to 10 per cent and 2 per cent respectively. These two methods obviously increase the immature nuts and leads to inferior quality in the end product after processing. Harvesting by thrashing or climbing also results in dropping of young flowers resulting in lower production. Besides, the nuts have to be harvested at right stage. The colour of the nuts changes from brownish green at fruit set to light green at ¼th growth of apple and thereafter turns grey irrespective of varieties.

Development of cashew apple and nut has been studied by number of researchers in different cashew varieties (Antarkar and Joshi, 1987; Mohan Kumar *et al.*, 1984; Sumrit *et al.*, 1989; Augustine and Unnithan, 1981; Rao and Joshi, 1987). According to report, specific gravity could be considered as physical index of maturity of nut in cashew varieties (Antarkar and Joshi, 1987; Mohan Kumar *et al.*, 1984). According to a report from Thailand the nut reached maximum size, approximately four weeks after fruit set, hardened in the subsequent week and declined in size by 10 per cent at harvest (Sumrit *et al.*, 1989). The duration of harvest extends from 45 to 70 days depending upon the cashew growing regions in the world. Different harvesting period of cashew in the world ensures the supply of nuts around the year. This helps to have control on imported nut price and assisting processors of Indian origin to not to carry long term stock. Harvesting of raw cashewnuts in major cashew growing countries is represented in the Figure 3.3.

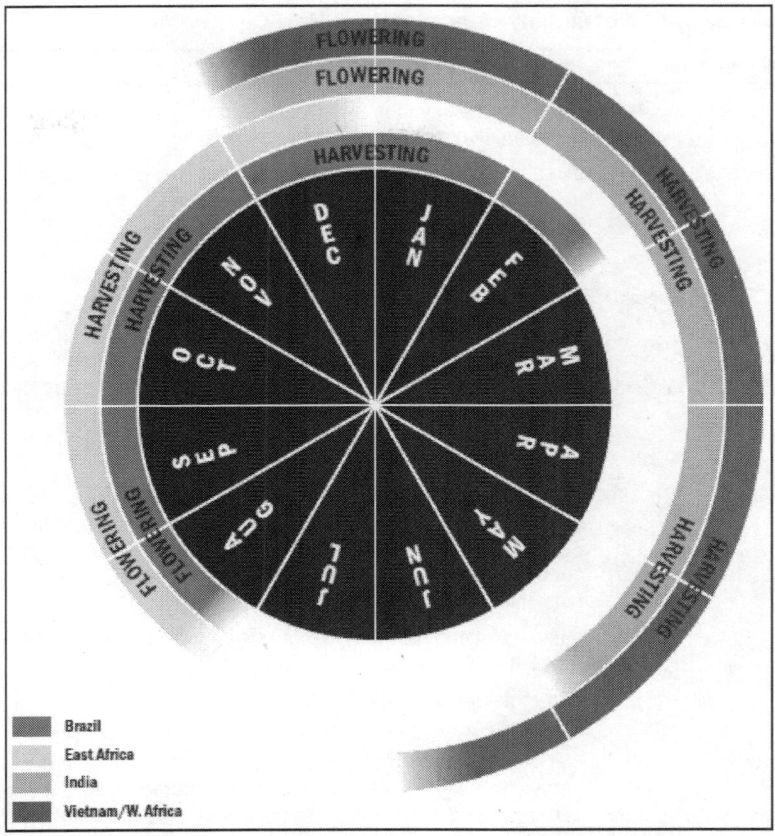

Figure 3.3: Harvesting Period of Cashew in Major Cashew Growing Countries.

6. Quality Standards for Raw Cashewnuts

Quality starts in the field with the time of harvest and depends on properly following subsequent steps after harvest as it has the potential to either maintain or reduce quality. Few postharvest procedures can improve the quality of the commodity (Cavalieri, 1999). Grade standards identify the degree of quality in a commodity that is basis of its usability and value. Cashew processing is relatively labour and energy intensive industry wherein raw material accounts for a large proportion of the processing cost (70 per cent). The process suitability of raw material is determined by a balanced assessment of its properties such as its geometric, physical, functional, growth, mechanical, thermal and electrical properties. The relative importance of these quality factors and their contribution to the suitability vary depending on the material, the process applied to it and the end product made from it. Grading of raw cashewnut can assist in size and weight control. As raw materials with over size and under size materials present processor with economic problem, sampling and testing before acceptance must be carried out to ensure conformity with processing requirements (Table 3.2).

Table 3.2: Description of Quality Parameters for Raw Cashewnuts

Quality Parameter	*Description*
Number of Nuts	Counting the number of nuts per kilogram and higher the number of nuts per kilogram, poorer is their quality.
Extraneous Matter	Dust, dirt, stones, earth, chaff, straw and any other impurity mixed with the nuts are termed as extraneous matter. It can be regarded as a discounting quality factor of grade and price.
Void Nuts	Nuts in which there are no kernels (Number/kg).
Broken and Damaged Nuts	Broken/internally-damaged or discoloured nuts affecting the quality (Number/kg).
Cutting Test	A random sample of a kilogram of raw nuts is taken from the bulk and cut open using hand-cutting tool. Based on the kernel in terms of surface colour *i.e.* white, shriveled, dotted or reject (spoiled), the percentage of good kernel can be calculated and it has direct influence on the quality of nuts.
Floating Test	One kilogram of sample is transferred to a vessel containing water, after continuous stirring floaters are collected and counted. Immature nuts due to low density than water, improperly filled nuts and deteriorated nuts floats. Higher the percentage of floaters, poorer is the quality of cashew nuts. It is a discounting factor for both quality and price.
Size	It is determined with the help of length, width, and thickness of the nut. Bigger the size, better is the quality.
Moisture	The moisture (per cent) in cashew nuts inversely affects the price and grade.
Kernel recovery	Out of 1-kg of raw nuts with shell and without shell are weighed and percentage of nuts without shell is calculated. Higher the percentage better is the quality of cashew nuts. It is a factor contributing to higher price and better grade for cashew nuts.

Raw nut grading is the first step in mechanising the cashew processing system (Ohler, 1968). The kernel yield is primarily a function of raw material, control over which is not much exerted prior to processing. Quality indicators of raw nut are moisture content, fractions of defective nuts, size and extraneous matter etc. Providing graded nuts can improve operator efficiency during manual shelling process. Grading nuts at preliminary level can ensure control over kernel drying cycle for better white whole kernel recovery and minimize workload at kernel grading stage. Certain physical parameters related to quality of raw cashewnuts were identified and its description is given in Table 3.2 (Potty *et al.*, 2013).

7. Cashewnut Processing

Structure and Properties of Raw Cashewnuts

Raw cashewnut is attached to the exterior end of the swollen peduncle *i.e.* false fruit and is positioned at the distil end of the fruit, the point furthest away from its attachment to the stem giving unique appearance to the fruit. It is a kidney shaped one with 2-3 mm thick soft leathery outer skin (Endocarp) and thin hard inner skin (Exocarp). Between these two walls of the shell is a honeycomb structure (Pericarp), which contains the phenolic material, commercially, known as cashew nut shell liquid (CNSL) (Figure 3.4). The kernel is enclosed by a leathery testa which is the embryo comprised of two large cotyledons, a radical and a plemule inside the shell.

Certain physical properties of raw cashewnuts are essential for optimizing processing and useful in the design of machinery. It is reported that the average length (L), width (W), thickness (T) of raw cashewnuts at 8.46 per cent was 31.0, 22.9 and 16.9 mm respectively, while the corresponding values of kernels were 24.7 (l), 12.9 (w) and 12.1 mm (t) (Figure 3.5). About 52 per cent of the nuts had length of between 30 and 35 mm whereas 11.5 per cent and 36.5 per cent of nuts categorised under >35mm and <30mm respectively. The average unit mass of the nut and kernel was 5.9 and 1.9 g respectively. The average sphericity of raw cashew nut and kernel at moisture content 8.46 per cent db were worked out to be 0.74 and 0.63 respectively (Balasubramanian[1], 2000).

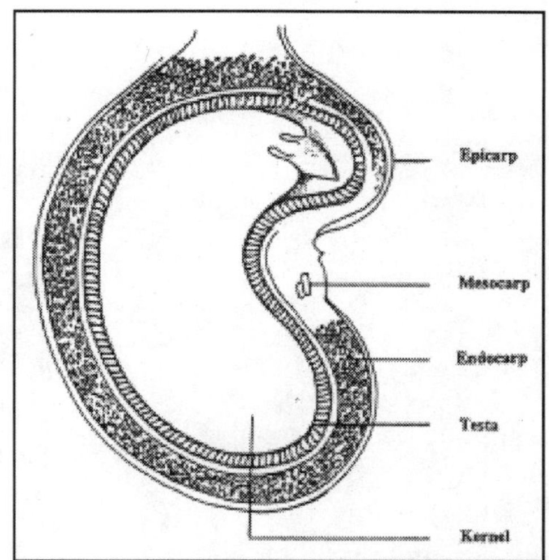

Figure 3.4: Cross Sectional View of Raw Cashewnuts.

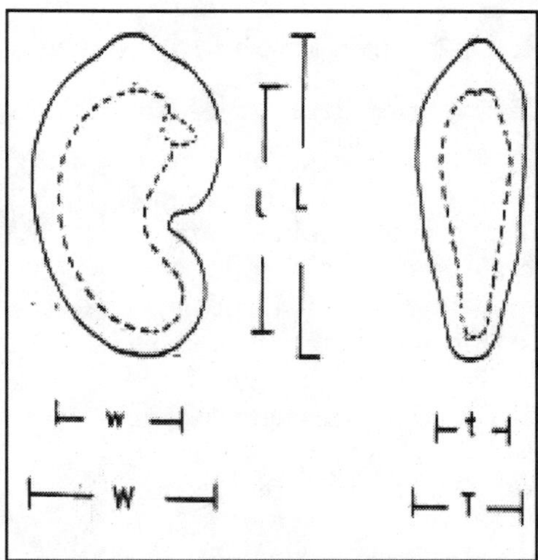

Figure 3.5: Characteristic Dimension of Cashewnut and Kernel.

One hundred nuts mass of raw nut varied between 610.6 to 735.1 g and it increased linearly with increased moisture content. The porosity increased from 48.0 to 52.3 per cent, while the bulk density decreased from 624.2 to 591.9 kg m^{-3} as the moisture content increased from 3.15 per cent to 20.06 per cent db. The true density of nuts followed a linear correlation with moisture content and increased from 1.20 to 1.24 kg m^{-3} in the moisture range investigated. The static co-efficient of friction for nuts increased linearly with moisture content irrespective of surface employed *viz.*, Glass (0.19 to 0.25), Aluminium (0.25 to 0.30), Galvanised iron (0.25 to 0.30) and Cardboard (0.25-0.31).

Processing

It is defined as an operation of materials separation designed to recover the edible kernel and the by-product commercially known as Cashewnut Shell Liquid

(CNSL), contained in the tissues of the middle shell layer of the raw nuts. In its natural state, the shell is pliable and unsuitable for any type of manual or mechanical opening. Heating, however, will harden and make the nut brittle, susceptible for cracking or splitting. Oven drying and subsequent cooling either at ambient or controlled conditions cause the testa to loosen from the cotyledon and become amenable to mechanical or hand removal. Quality sorting as per the standards is the last operation before packing the kernels into cardboard box containing either tin containers or multilayered polythene bags. At each of these steps, quantity and quality losses can occur. The avoidance or control of which decisively influences revenue per unit raw material. Quantitative efficiency is captured in the yield of packaged kernels per unit weight of raw nuts. Qualitative efficiency is harder to measure as it has several criteria like size, shape, colour and wholesomeness. The main distinctions in kernel grading are related to wholesomeness, surface texture and its colour.

8. Unit Operations in Cashewnut Processing

Drying and Storage of Raw Cashewnuts

Presumably, moisture content of raw cashewnuts just after harvest ranges from 16 to 25 per cent d.b depending upon the variety, maturity level and nut to kernel ratio. Once the cashew fruits are either harvested or collected from the field, the nuts are to be detached from the cashew apple and sun dried for 2 to 3 days (Figure 3.6). This is very important in view of the quality of kernel so obtained during subsequent processing. Sun dried nuts expected to have a moisture content of ≤ 8 per cent d.b. Undried nuts lead to fungal spoilage due to adhering fleshy portion of cashew apple, resulting in poor quality of kernels. Fungi such as *Gonatobotryum, Helminthosporium* sp., *Corynespora* sp., *Alternaria* sp., *Verticillium* sp. and many species of *Aspergillus* have been isolated from stored cashewnuts (Joseph, 1981). In view of the occurrence of fungi, drying of the raw nuts before storage becomes absolutely essential.

Figure 3.6: Sun Drying of Raw Cashewnuts.

Raw cashewnuts could be exposed to sun by uniformly spreading in thin layer on either semi-finished floor or over polythene sheets and the quantity of nuts to be dried shall not exceed 20 kg m^{-2}. Uniform drying can be achieved by constantly turning nuts over to prevent partial drying or non-uniform heating. Alternatively, raw nuts can be dried in electrically or bio fuel operated cross-flow dryers. Time required to dry raw cashewnuts at air temperature of 65°C inside commercial dryer is in the range of 7 to 10 h whereas drying time under sun is found to be in the range of 33 to 40 h (Balasubramanian and Sandeep, 2011). Irrespective of drying methods, moisture content of raw nuts should be brought down to ≤ 8 per cent d.b for safe storage.

As per the standards, permissible moisture content for oil seeds for safe storage is around 13 per cent of moisture calculated on the non-oleaginous portion of the seed. Raw cashewnut is considered as oily seed, due to presence of oil in shell structure (CNSL) and kernel (fat). Chemical composition of cashew kernel indicates that it has 47 per cent of fat and 1/3 rd of the whole nut is cashew kernels. Besides, one third of shell is non-fatty oil liquid *i.e.* CNSL. Considering the above fact, the total non-oil portion of the whole nut would be 62 per cent and the calculated permissible moisture content for raw cashewnuts would be around 8.07 per cent (Ohler, 1979). Apart from this, moisture sorption isotherm studies clearly indicated that 50.4 per cent of total moisture is free moisture *i.e.* 8.54 per cent and removal of this free moisture ensures protection of raw cashewnuts against microorganisms. Therefore it is strongly recommended to bring down the moisture to safe level of 8 per cent for long term storage (Anonymous, 2013). Cashew being a seasonal crop available only from March to May every year, it is essential to decrease the moisture to prolong its shelf life so as to operate processing units throughout the year. Therefore, a moisture loss (driage) of 5-6 per cent takes place depending on the size of the nut, maturity and drying time.

Dried nuts should be stored in jute or sisal bags preferably of 80 kg nut per bag and not heaped or stored in silos to avoid warm humid condition in storage chamber. Storing in woven high density polyethylene (HDPE) bags are not recommended for raw nuts. While storing, it is necessary to provide proper wooden dunnage to prevent damage to the cashew nut in bags to avoid direct contact of bags to the warehouse floor (Figure 3.7). This can be economically achieved by keeping bamboos, criss-crossed and tied and placed on the warehouse floor. Dried nuts should be stored in well-ventilated, low humidity environment, free of stored product pests and well insulated to prevent water seepage during rains.

Figure 3.7: Storage of Raw Cashewnuts.

Okwelogu and Mackay (1969) have reported that moisture content of whole nut and kernel could be predicted from that of shell with more than 99 per cent accuracy. A higher kernel-shell weight ratio has been shown to be inversely proportional to moisture content. Effect of storage temperature and relative humidity has been studied by Okwelogu and Mackay (1969) and reported that nut stored at 27°C at a relative humidity (RH) of 70 per cent had a moisture content of 9.2 per cent d.b. Nuts exposed to higher RH (>75 per cent) lead to mould infection within few weeks.

Therefore, RH of the storage godown plays an important role. Under high RH, nuts pick up moisture from the atmosphere till they attain equilibrium.

Gunny bag storage of 80 kg capacity is the most common practice all over the world and it is shown that storage of raw nuts with a moisture content of 5 to 6 per cent for a period of 12 months at ambient temperature (25 to 30°C) did not affect either processing or biochemical quality. Moreover, kernel rejects of 2 to 3 per cent obtained during processing of cashew by steam method has been shown to be influenced by immature nuts and floaters (Nagaraja and Prabhu, 1996). Recent studies on effect of maturity on the processing quality have clearly indicated that shelling percentage, peeling outturn, whole kernel recovery increases while kernel rejects decreases with maturity. Application of insecticides or pesticides is prohibited to improve shelf life of raw cashewnuts in warehouses.

Nut Conditioning

Understanding the export potential after the first shipment during 1920, process to extract edible kernel was refined. Owing to its caustic properties, the CNSL in the pericarp, it is first burnt out to eliminate the chances of contaminating the kernels and minimize the risk of injury to the fingers of shelling workers. Although the shelling process remains essentially the same until today in the Indian cashew processing industry located in certain regions, the discovery of commercial value and the versatility of CNSL for industrial uses encouraged adoption of methods to salvage the oil.

Open Pan Method

This is the oldest method that is still prevalent in the rural areas in many cashew growing countries in one or other form. The nuts are roasted while being constantly stirred in shallow pans over earthen ovens. The oil exuding from the pericarp catches fire and at appropriate time the pan is lifted and the contents emptied on the ground. The flame is extinguished by splashing a little water over them and the nuts are rubbed with a little ash to absorb the oil. This method do not ensure uniform roasting and CNSL is not recovered.

Drum Roasting Method

This is an improvised method over traditional method and it was first introduced in the year 1931 in India. It is still in practice in majority of the organized sector of the industry located in India due its enhancement of flavour to edible kernels during processing. It consists of a galvanized iron cylinder or a hollow drum open on both ends suspended at an angle over a furnace equipped with a chimney to carry off objectionable fumes. The nuts are fed through a hopper connected to a bin wherein its feed rate is regulated manually. Rotational motion of the drum makes the nuts to slide down across the red hot drum. During its travel, it catches fire due to the presence of corrosive oil and reaches the lower end by gravity. Water is sprinkled to quench off the fire and mixed with ash to absorb oil. Later it is transferred to shelling section. The degree of roasting depends on size, feed rate, rotational speed and moisture content of the raw cashewnuts. It is essential to shell the nuts in brittle condition to avoid breakage of kernels, otherwise it absorb moisture, if prolonged for

Figure 3.8: Drum Roasting Process during Operation.

Figure 3.9: Smoke Release through Chimney.

shelling and lead to higher breakage (Anonymous, 2013). Presently, this method is discouraged due to scorching of tips of cashew kernels, loss of valuable by-product *i.e.* CNSL and the emission of obnoxious smoke polluting the atmosphere. Research work is underway to develop an efficient system to either minimize air pollution within the permissible limit or total prevention.

Oil Bath Method

In this method, raw nuts are soaked in water for 3 h and stored in silos after draining excess moisture in order to attain an optimum moisture level of 15 to 25 per cent d.b (Ratnam, 1969). Oil bath tank consists of rectangular tank, conveyor assembly, feeding chute, roasted nut outlet and CNSL over flow system. Soaked raw nuts from silo is passed into the oil bath tank just above belt conveyor and allowed to immerse completely in hot CNSL which is maintained at 190 to 195°C. The total length of tank is approximately 3.6 m and the belt conveyor rotate at the speed of 35 rpm. When nut travels from inlet to outlet through hot CNSL, it gets cooked and 70 per cent of total CNSL available in shell structure oozes out. The roasted nuts are passed on to centrifuge to remove adhering oil and through bucket elevator conveyed to shelling section. It requires 25 bags of cashew shell cake as fuel for processing a quantity of 6 TPD. The temperature of CNSL, rpm of conveyor inside tank and feed rate depends on the size, moisture and origin of raw nuts.

Steam Boiling Method

Steam boiling mode of conditioning raw cashewnuts is considered to be cost effective system and gaining importance all over the world. As CNSL is preserved in the shell after conditioning, it can be extracted adding revenue to the processor, it is quiet advantageous. During steaming process, nut expands and become soft due to infusion of moisture. Rate of moisture infusion should be controlled in order to prevent the transfer of pigment from testa to kernel surface. Therefore, steam pressure and duration of exposure are the decisive factors during steaming operation which are influenced by size, dryness and storage time (Balasubramanian[2], 2000). After steaming, the nuts are air-cured by spreading out on the floor in the ambient environment. These ultimately harden the shell and make it amenable for subsequent kernel extraction process.

Shelling

The nuts after conditioning are shelled manually or mechanically using hand cum pedal operated shelling gadget. The manual shelling is an operation which requires high dexterity and knocked 2 to 3 times on each of the long edge by wooden mallets or light hammers ensuring the release of whole kernels without damage or breakage. Workers acquire the skill through practice and outturn is expected to be in the range of 92-95 per cent of the whole kernels with operational capacity of 1.75 to 2.5 kg h^{-1} of kernels. It is a standard practice to smear ash or clay or oil to prevent the contact of CNSL on their hands during operation (Balasubramanian[3], 2000).

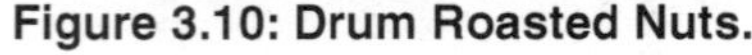

Figure 3.10: Drum Roasted Nuts.

Figure 3.11: Manual Shelling of Drum Roasted Nuts.

The mechanical shelling gadget is used to extract steam conditioned nuts and it consists of two blades between which a raw nut is inserted in such a way that it's notched portion facing the twin blade assembly. Gap between the blades is adjustable so as to accommodate graded nuts based on its minor axis dimension for improving efficiency. When the pedal is pressed, the linked cam move the single blade towards twin blade making slit on the contour of the inserted nut.

At this moment, the hand lever is lifted to actuate the twin blade to split open the shell. The nut along with shells is allowed to pass through the spout downwards, once the applied force on the pedal is released. Recently modified shellers are operated by single action to hold nut and split the shell by turning action. If the kernels are found to be intact with an half of the shell, it can be scooped out by means of a sharp needle (Balasubramanian, 2010). The manual shelling is known to be an undesirable job which causes painful sores on the hands and fingers from the irritating liquid from the shell. With the proper attention, this can be avoided through the constant application of castor oil before and during the shelling.

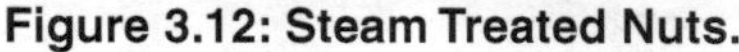

Figure 3.12: Steam Treated Nuts.

Figure 3.13. Mechanical Shelling of Steam Treated Nuts.

Kernel Drying

Shelled kernels are exposed to controlled environment to facilitate the removal of skin adhering to edible kernel. Due to relatively faster diffusion of moisture from testa layer than kernel, it shrinks and become loose to ease the process of manual or mechanical removal of skin. Care should be taken to regulate the air temperature and limit the drying period so as to avoid scorching of kernels. In the commonly used 'Borma', drying chamber is indirectly heated by flue gases from the furnace at the bottom wherein spent cashew shells or cashew shell cake are used as fuel. In order to overcome differential rate of drying at top and bottom location, it is a standard practice to change the position of the trays to ensure uniform drying (Balasubramanian, 2001). The conventional "Borma" is presently replaced by more efficient electrical/diesel/de-oiled shell cake or wood operated tray driers. This system ensures uniform heating of kernels for a pre-set temperature [80±2°C] and time [4-6 h] by means of reliable electronic digital controller and timer. This system reduces the heating time considerably and maintains uniform heat throughout the drier by forced air circulation. Steam assisted dryer deployed in processing line, utilises dry air converted from super heated steam through radiator assembly. This type of dryer, preserves the original color of the cashew kernel and quality consistency is achieved within a batch and in every batch dried.

Humidification

Thermally treated cashew kernels are humidified *i.e.* moisturized by water-mist depending on the environment primarily to avoid breakage during peeling process. Trolleys with wire mesh trays having unpeeled cashew kernels spread on its surface are loaded in humidifying chamber. Duration of humidification depends on the atmospheric conditions and moisture level of cashew kernels. The brown skin *i.e.* testa of kernels, being hygroscopic in nature, absorb moisture during humidification and helps to dislodge easily the adhering skin from the kernel during manual or mechanical process without damaging or breakage.

Peeling

Peeling is the process of removing outer skin called testa. The kernel shrinkage away from testa during unpeeled kernel drying facilitates easy removal of skin manually. Specially designed stainless steel knife is used to scrap adhering testa without damaging the edible kernel. Average peeling capacity ranges between 4-12 kg $head^{-1}$ day^{-1} and peeling efficiency or whole kernel recovery observed to be 65-72 per cent. Preliminary grading of peeled kernels *i.e.* 8-10 grades is segregated at this stage itself. Wages are fixed on the basis of whole kernel recovery as it serves as a control for careful work (Balasubramanian[2], 2000). Mechanical graders working on the principle of impact or abrasion or both are employed to overcome the problem of scanty labourer.

Figure 3.14: Manual Peeling of Unpeeled Cashew Kernels.

Grading

Kernel grading is primarily based on wholesomeness, size and surface colour and manually segregated in processing units located in many countries. Mechanical graders with fluted rollers and colour sorters are also available for grading kernels based on size and surface colour respectively. Mechanical graders are utilized depending on availability of labour force and volume of production. At this stage rigorous hygienic standards are applied and provision of apron and hair net for workers is more prominent. White whole kernels of larger size fetches premium price in the International market, therefore ultimate aim in cashewnut processing is to extract whole kernel with unchanged surface colour. Grading standards developed in India refer to number of kernels in a pound weight, for example, WW320 indicates 320 numbers of white whole kernels in a pound (lbs) weight. Number of kernels in the largest whole kernel grade 'W180' is in the range of 265 to 395 per kg and the smallest whole kernel grade 'W500' in the range of 1000 to 1100 per kg. Thus, lower the number, bigger the size and higher the price per unit weight. Broken kernels are either butts or splits depending on whether they are broken across or along the cleavage and size grading is based on sieve through which it passes through. System of grading standards are developed by various organization *viz.*, Cashew Export Promotion Council of India (CEPCI), United Nations Economic Commission for Europe (UNECE), Brazilian Cashew Industry (BCI), Vietnamese Cashew Industry,

Bureau of Indian Standards (BIS) and Association of Food Industries (AFI, New York) (Annexures I,II and III).

Figure 3.15: Grading of Cashew Kernels in a Clean Environment.

Conditioning of Cashew Kernels

Permissible moisture in the graded cashew kernel is 5 per cent d.b by weight at the port of importation (CEPCI, 2013). While a moisture level above 5 per cent at the time of final packing is not preferred as the same may cause fungi and aid infestation, low moisture level below 2 per cent is undesirable as it causes breakage of kernel during processing and shipment. Processors by and large control the final moisture level of kernel before final packing up to 3-5 per cent by weight. Cashew kernel are humidified to increase moisture percentage to the required level or hot air treated to reduce the excess moisture to maintain desired final moisture level before final packing.

Packaging

It has been recorded that cashew processing on a commercial scale was first started in mid 1920's by Roch Victoria, a Sri Lankan who migrated to Quilon, Kerala state, India. During the first export of cashew kernels under the name Indian Nut Company in Quilon, cashew kernels were packed in wooden tea chests lined with newspaper before shipping to the USA. Later on, metal tin containers replaced tea chests. Tin containers were vacuumed using hand operated vacuum pumps and were sealed. It was only in 1954 that 'Vita pack' machines were used to infuse tin containers with carbon dioxide gas.

Vita packing is the process of vacuumising and injecting inert gas *viz.* Carbon-di-oxide or Nitrogen into the cashew kernel filled tins. The gas infused tins are hand-soldered hermetically using lead free solder. Such tins are packed in a carton (2 nos.) and strapped and stencil marked for shipment. The consignment is shipped after per-shipment inspection formalities. Flushing with CO_2 avoids microbial spoilage during shipment or distant transport. Tin containers of 25 lbs equivalent to 11.34 kg, are used for packing and tested for leakage by immersing in a pool of water.

After securing ISO 9000 and introducing HACCP (Hazard Analysis and Critical Control Points), importing countries are following stringent quality standards for cashew kernels and the present system of tin packaging is bulky and it is difficult to dispose off these tin containers. Moulded vacuum packaging (MVP) developed by M/s Vanleer private limited, South Africa, replaced vita packing as it produces consistent rectangular blocks ranging in size from 500g to 25kg of cashew kernels. Besides, MVP system has advantages during transport, handling, display, stock count etc. This also helps to save the total cost incurred on secondary packaging considerably. The vacuum barrier bag and the cardboard box are totally recyclable. The rectangular shape of primary packs ensures minimum movement during transport and handling ensuring maximum protection to kernels. The removal of air and gas flush reduce the incidence of rancidity and insect infestation.

Figure 3.16: Vita Packaging of Cashew Kernels.

Figure 3.17: Moulded Vacuum Packaging of Cashew Kernels.

9. Processing of Cashew: International Scenario

Brazil

In contrast to other cashew producing countries, most of the processing units in Brazil are mechanized. Highly mechanized system following oil bathe roasting mode of operation results in fewer whole kernel recovery and darker in color. Despite these shortcomings, it is essential for Brazil to develop increasingly sophisticated technology for mechanization as relatively higher labor costs preclude more labor intensive manual processing. These characteristics limit the access of the Brazilian cashew industry to value-added markets and the low price earned for the final product prevents processors from providing higher remuneration to producers (Leckie, 1989). This low yield of whole nuts *i.e.* 50-55 per cent for large processing firms in Brazil compared to over 70 per cent in India further inhibits competitiveness in production capacity. However, despite low yields of whole nuts, Brazil's cashews are larger, whiter and considered to be higher quality than its competitors.

Small processing facilities are limited by access to finance which prevents from adopting new technology as well as meeting international standards such as HACCP, ISO 9000, American Food Institute etc., and market access due to disadvantages of scale. The competitiveness of mini-mills is also threatened by a tendency to have limited access to information, lower management capacity and less financial records which makes access to credit more difficult (Marcio and

Francisco, 2001). The increased cost is compensated by the higher production of whole cashews (75-85 per cent for the mini-mills compared with 50-55 per cent for the large processors), the better appearance and better taste of the nuts that implies a higher price and a higher margin for the producer. However, as only 2 per cent of the export volume comes from mini-mills, the other 98 per cent are processed by a few large firms.

Vietnam

The cashew processing industry has undergone a fast development. Until, 1994, Vietnam had insufficient capacity to process all internally produced cashew nuts. Around 20 per cent of the raw cashew nuts production was exported to India and other countries for processing. This situation has changed dramatically and as of today, Vietnamese processing facilities exceed the present production and imports by around 50,000 to 100,000 tons of raw cashews for final processing to meet domestic capacities. This change from a nuts-in-shell (NIS) exporter to an importer of NIS was important for Vietnam to increase value adding of cashew in Vietnam and to become a direct trading partner to the international cashew buyers rather than only providing semi-processed products (Kihn, 2004). Vietnam is currently the second largest exporter of processed cashew nuts in the world and it achieved this feat within a span of last twenty years.

These reforms focused on securing agricultural prices, linking production and industries and increasing off farm activities (processing capacity) to reduce under employment in agriculture and rural areas. In the early 1990s, when international cashew prices were rising, Vietnam recognized the potential use of cashews as a poverty reduction or cash crop. Both public and private sectors began to invest in cashew research and development, agricultural inputs, and processing facilities. The combined efforts of farmers, the government, and increasingly, the private sector led to rapid growth in the quantity and quality of cashews produced and exported from Vietnam. The government eliminated export taxes, which further contributed to rising exports.

Approximately 98 per cent of Vietnam's processed cashews are exported and only 2 per cent sold in the domestic market. The primary market is the United States (capturing 41 per cent of Vietnamese exports) followed by China (20 per cent), the UK and the Netherlands (12 per cent) and Australia (10 per cent). Vietnam's cashew industry has greatly benefited from China's export market due to proximity and less strict quality standards. As Vietnam's processing capability exceeds its production capacity, the country must import raw cashewnuts to be processed. Currently, processors import approximately 300,000 MT of raw cashewnuts. Primarily it is imported from Africa and Indonesia with little or no import tax at the end of the year when domestically harvested cashews are exhausted. Competition for domestically produced cashews has driven many inefficient processors out of business, and has coincided with an increasing trend towards privatizing state-owned processing facilities.

Vietnam's current efforts at cashew diversification centre on producing the Cashew Nut Shell Liquid (CNSL) for use in paints. The CNSL is mainly exported to China but is also commercialized domestically. Currently, ten companies are exporting CNSL but only in very minimal amounts. The quantity of CNSL currently produced is not sufficient to diversify and increase incomes of farmers and small and medium sized processing plants. Larger processors which are currently exporting CSNL can achieve a slight augment in profit, but the amount is fairly insignificant and therefore benefits do not trickle down to actors further down the value chain. If processors increase production and exportation of CNSL, this could potentially increase prices received by farmers in the future, thus increasing farmer income.

African Countries

West Africa's cashew processing sector which includes Benin, Burkino faso, Ivory Coast, Ghana, The Gambia, Guinea Bissau, Mali, Nigeria, Senegal and Togo is quite small given the region's status as the world's largest producer of raw cashew nuts (Nicholas and Brook, 2007). These cashew growing countries follow either cottage level or semi industrial processors (<250 TPA). Certain countries like Benin, Burkina Faso, Ivory Coast, Guinea-Bissau and Nigeria reportedly started large cashew processing recently (1000-15000 TPA) following modern Indian semi mechanical system. Due to demand of larger volume of cashew kernels of specific grade and higher price fluctuations in the international market, processed kernels are targeted for local market only. It is estimated that only 10 per cent of the total production of raw cashewnuts are utilized for processing facilities developed at domestic level, otherwise exported to India and Vietnam. But, institutional changes presently planned in Africa in order to promote processing to drastically restrict the export of raw cashewnuts to India.

In East African countries like Tanzania and Mozambique, less than 40 per cent of the cashewnuts produced is utilized for primary processing and the rest is mainly shipped to India where it gets processed by large companies with labour intensive technology (Nazneen *et al.*, 2004, Waithaka, 2002). Cashewnut processing in this region is still in its infant stage, despite all the efforts that have been made in the past to introduce various technologies. There are three different processing technologies adopted *viz.* drum roasting at cottage level, medium scale with Indian steam boiling and Italian or Japanese oil bath roasting suiting to large scale. Mozambique is the first African country to install large scale cashewnut processing following oil bathe roasting. Although processing facilities developed have installed capacity ranged between 300 and 10,000 TPA, only 29 per cent is utilized, due to further investment for rehabilitation and lack of working capital. As the large scale units following Italian or Japanese oil bath roasting has the advantage of food quality and safety, becomes profitable only if operated at full capacity. Presently, large scale units are non-operational and in the processors are changing over to semi mechanized system. Cashew by products *viz.*, shell and testa are not consistently utilized loosing additional revenue to the industry. Given the current situation of most processors, substantial skills development and training is required. Capacity strengthening should embrace all levels of processing, be it first or second level processing of cashew nuts, CNSL extraction or adding value on apples.

10. Processing of Cashew: National Scenario

India is the world leader in cashew processing and exportation. However, India is at an important juncture in the development of its cashew industry due to increased international competition. At the present time only half of the cashews processed in India are produced domestically, reflecting that production has not kept pace with its growing processing capacity. While Kerala has been the centre of cashew production and processing for over fifty years, both activities are now moving out of the state due to high labour costs and land shortages. Production is now more fragmented and the states of Tamil Nadu, Maharashtra, Karnataka, Andhra Pradesh and Odisha have emerged as leading producers of raw cashewnuts.

Dexterity of labour is one of India's comparative advantages due to its three-generation history of processing. It is estimated that manual processing technique perfected in India results in only 20 per cent broken kernels, as compared to about 47 per cent broken kernels in North American and African regions (Figure 3.18).

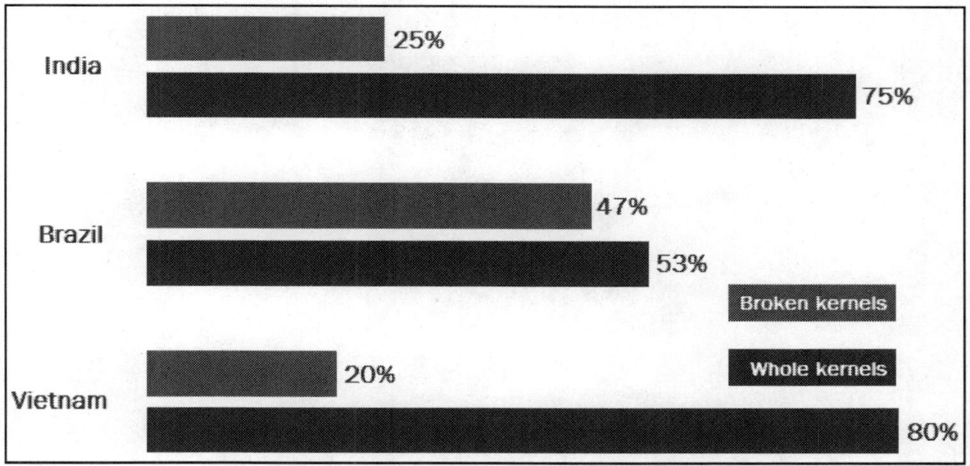

Figure 3.18: Estimated Whole and Broken Kernel Recovery in Major Cashew Processing Countries

However, there are significant differences in the way each state within India has supported its cashew industry. The potential benefits to small producers and workers are highly dependent upon the type of cultivation (individual homestead farms vs. plantations) and processing system (cooperative, cluster, private and government factories). Technological innovation has been concentrated in the development of new varieties and planting techniques and in the final packaging phase, but there appears to be little interest in the mechanization of processing. Due to the long history of the sector, India has the strength of highly skilled labor in all stages of manual processing and therefore is committed to the continuation of this "hand-crafted" approach. This involves the use of mallets to remove cashews from their shells, as opposed to a more mechanized technique that results in a higher percentage of broken nuts. India's low labor costs have enabled it to sustain this method and maintain a quality advantage in the industry.

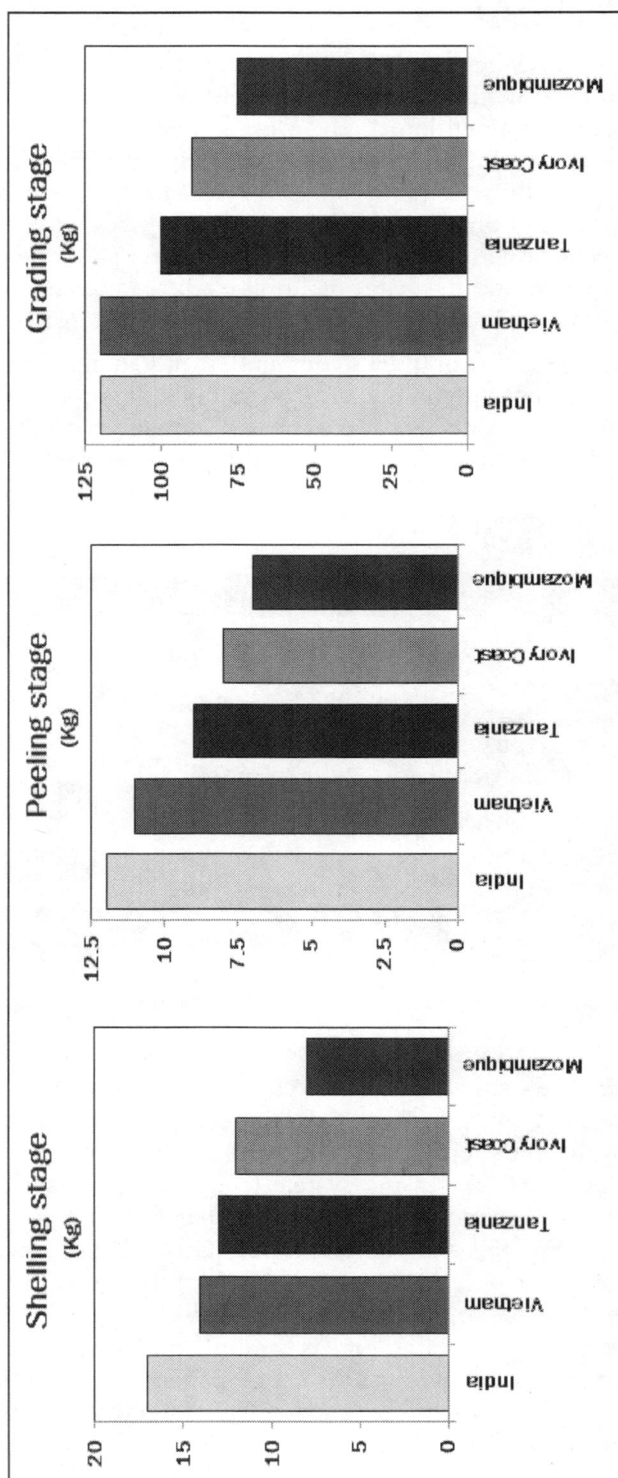

Figure 3.19: Comparative Performance of Manual Processing in different Countries.

India is the global leader in cashew processing, and Kerala is the most productive processing state in the country. Kerala processed nearly 50 per cent of the country's total capacity of one million metric MT of raw nuts in 2000 and exports the highest share of cashew kernels. However, rising labour costs and strengthening labour unions have resulted in the development of processing units elsewhere. It is estimated that there are 3799 processing units nation-wide (Table 3.3) (DCCD, 2013).

Table 3.3: Utilization of Processing Capacity in India

States	*Processing Units (Nos.)*	*Capacity (000' MT)*	*Utilization (000' MT)*		
			Indigenous	*Import*	*Total*
Kerala	432	600	67	320	387
Karnataka	266	300	45	20	65
Goa	45	50	21	—	21
Maharashtra*	2200	50	20	—	20
Tamil Nadu	417	400	294	225	519
Andhra Pradesh	175	100	92	—	92
Odisha	209	100	11	—	11
West Bengal	30	8	8	—	8
Chattisgarh	3	5	—	—	—
NE States	22	10	15	—	15
Total	**3799**	**1623**	**573**	**565**	**1138**

* Includes 1850 small scale cashewnut processing having processing capacity up to 40 TPA.

Source: Directorate of Cashew Cocoa Development, Cochin, India.

11. Cashew Processing in Indian: States Scenario

Kerala

Cashew processing industries in Kerala are mostly concentrated in Quilon district. Besides, processing industries are operating in Trivandrum, Allepy, Calicut, Pathanamthitta, Thrissur and Kasaragod districts also. Kerala is the birth place of drum roasting process and system of functioning is somewhat unique due to the influence of government. As a result of pressure generated by differences in organisations of labour, legislation and wage rates many factories have been shifted elsewhere. Only 25 per cent of the total industries have the capacity to run the factory throughout the year and their utilised capacity falls under the category of 1000-5000 TPA. Labour wages are fixed by the state Government and workers are motivated by share in the profit. In view of employment potential and poverty elevation, the government incorporated the Kerala State Cashew Development Corporation (KSCDC) in 1969. The Kerala State Cashew Workers Apex Industrial Co-operative Society (CAPEX) also operates worker-owned factories. Mostly raw nuts of indigenous origin are procured from local market situated in Anchal, Paracode, Thamarakulam, Kadaikan, Kallara, Milimanoor and Kadambanadu.

But, 60-75 per cent of total requirement of raw nuts are derived through imports (Balasubramanian[3], 2000). Most of the processors are following drum roasting as preliminary process, but steam boiling technique is also practiced in this region.

The Government of India established the Cashew Export Promotion Council in the year 1955, with the active cooperation of the cashew processing industry and with the object of prompting exports of cashew kernels, Cashewnut shell liquid and allied products from India. This council provides the necessary institutional framework for performing the different functions that serve to intensify and promote exports of cashew kernels, cashewnut shell liquid and allied products. In addition, detailed information on trade and market potential of cashew are gathered time to time and provided to the exporters. High quality and superiority of Indian Cashews are given publicity abroad as promotional activity by the council.

Karnataka

Cashewnut processing in Karnataka state plays a prominent role in the economy of the state and towards employment generation. Although steam treatment is the commonly followed prior to extraction of edible kernels, large scale processing is practiced in coastal Karnataka *i.e.* Dakshina Kannada and Udupi district and home level processing in plain region *i.e.* Kolar district. Cashew processing in Karnataka had its beginning in Mangalore and spread to other coastal and cashew growing regions afterwards. Dakshina Kannada serves as the nucleus centre for the cashew trade and home of cashew processing in Karnataka.

Cashew processing as an industry started in Mangalore district of Karnataka state in 1920, but, it was shifted to Quilon district of Kerala state later. About 275 processing units having installation capacity up to 325 TPA constitutes cashewnut processing sector in Karnataka state. Only 26 per cent of raw cashewnuts are derived from this state itself and processors depend on either neighbouring states like Kerala, Maharashtra and Goa state (34 per cent) or import from African countries (40 per cent) to operate the units throughout the year. Processing facility developed in Karnataka is utilized to the extent of 88 per cent indicating the growth of this agro processing and profitability of business.

Well organized system of cashewnut processing had its beginning during 80's in Dakshina Kannda and Udupi districts and these units are primarily classified as proprietorship and partnership companies. Most of the industries are categorized under partnership due to high capital investment on machinery and inadequate finance as working capital. About 53 per cent of the industries supplied plain cashew kernels to domestic market whereas 39 per cent exported the end product and balance for secondary processing. None of the processors involved in value addition in bulk quantity, confirming strong demand of plain cashew kernels in domestic market. Cashew nut shell liquid, one of the by-products of processing was extracted in 8.5 per cent of the total industry and crude oil supplied to allied industries. Majority of the processing plants had installation capacity in the range of 100 to 5000 TPA and most of the operations in the line of processing were carried out employing women force. In order to assess the quality of raw cashewnuts, processors followed three different tests *viz.*, visual, floating and cutting tests. Total employee's strength of a factory varied between 100 and 400 and men labourers are mostly involved in drying, stacking, steam boiling raw nuts, kernel drying and tin packing of kernels (Balasubramanian[2], 2000).

In Kolar region, harvested raw cashewnuts are systematically marketed through Agricultural Produce Market Committee (APMC) situated in Chintamani. Although edible kernels are extracted by cracking dried raw nuts using mallet in the beginning, steam boiling technique is widely adopted now. Down sized cashewnut processing system is followed in this region (Balasubramanian, 2003). Operational capacity varied from 10 to 80 TPA depending upon the raw cashewnut availability and labour force. Presently, large scale industries located away from production catchments and in urban areas are facing the problem of labour scarcity. At this juncture, automated machines are finding their way in the line of processing. Although, there was reluctance in the past to invest in capital intensive cashew processing in this region, i) Changes in the labor cost and demand for better conditions; ii) Decline in the operational efficiency of the labor with age; iii) Compensatory wages during non-working days; v) Non-preference of traditional industry by the younger generation; vi) Improvement in the country's economy providing opportunities in newer areas; vii) Wages insufficient to lead a urbanized life; viii) Monotonous job in the factory, now mechanization has become feasible solution.

Tamil Nadu

In Panruti, Tamil Nadu, cluster based cashewnut processing is followed. Cashew farmers supplement their own incomes by operating small units wherein own harvests as well as nuts purchased from export houses that source raw materials from other parts of India or abroad are being processed. These processing units are not obligated under contract to sell the finished product back to the raw nuts suppliers or export houses. But a stable network has already been developed either to link to specific export houses or domestic market. Besides, small processors in Panruti are able to rent machinery and other infrastructure within the cluster, resulting in low barriers to entry. The Panruti case is unique because it is one of the few examples in which rural people are involved in both the production and processing phases of the cashew industry and participation in both activities generates employment for considerable period in a yaer. Cluster members have also devised risk sharing schemes that protect them from the daily prices fluctuations. Because it typically takes 10-15 days for a batch of raw nuts to be processed into kernels, processors in Panruti avoid these price changes that would affect their margins by carrying out limited quantity and then selling the semi-finished product immediately. This cluster experience demonstrates benefit of the producers with limited resources for primary production and processing when the relationships within the value chain are renegotiated. The involvement of farmers in processing in Panruti is very innovative and distinct (Shalini, 2010). Overall, privately-owned firms dominate processing in India and traders serve as important middlemen between farmers and factories. The government plays a crucial role in trying to enhance the capacity of private processors and exporters. Cashew processing factories located in Kuzhithurai and Marthandam of Tamil Nadu are more or less simulated processing in Kerala as many of them are migrated from the Kerala state.

Andhra Pradesh

In Andhra Pradesh state, raw nuts are obtained from Srikakulam, Guntur and East Godavari district. In order to fulfil the demand, it is also procured from Karnataka and Tamil Nadu state either domestically produced nuts or imported from African countries. Floating and cutting test are performed to check the quality of nuts during winter and summer respectively. Apart from regular practice of storing the nuts in gunny bags 'Gadher' structure' which is nothing but cow dung coated bamboo bins, cylindrical in shape with 5 feet height and 3 feet diameter at farm level. Processors believe that nuts harvested from hilly region could be stored up to 2 years without spoilage in warehouse, whereas, the nuts from coastal area can be stored for 4-5 months only. Cashew processing units in Andhara Pradesh are concentrated in Srikakulam (Figure 3.20), Rajahmundry and Vetapalem regions.

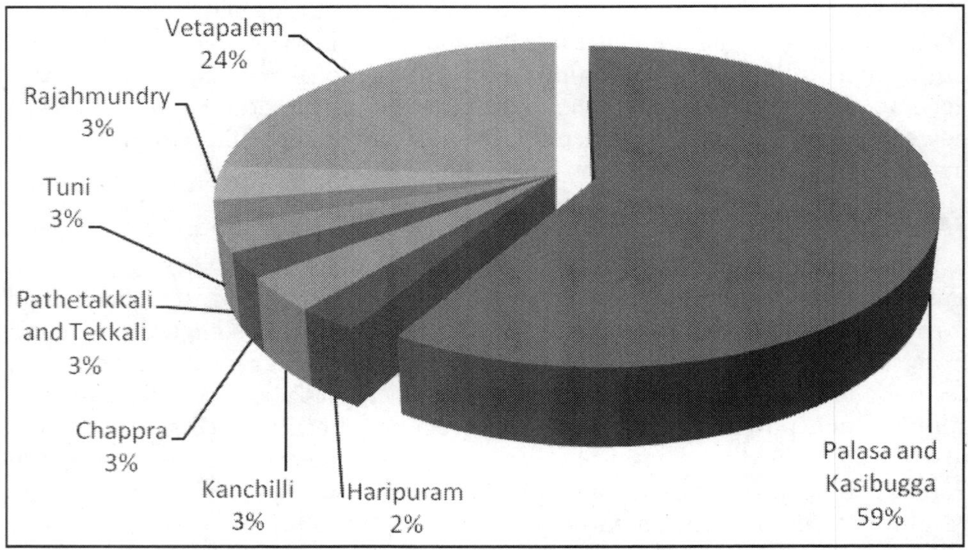

Figure 3.20: Distribution of Cashew Processing Units in Srikakulam District, Andhra Pradesh.

Processing of cashewnuts started in drum roasting mode, locally called as 'Bhatti' with an estimated capacity of 28,000 TPA. But after 1996, most of the processors' switched over to steam boiling system as these regions had been declared as pollution affected area due to emission of smoke during roasting process. Capacity utilization of developed facility in the processing units' ranges from 0.96 - 1.44 TPD. Most of the process like shelling, peeling and grading are done manually and now semi-mechanized or fully automatic machines are gaining importance due to labour scarcity and need for bulk production (Balasubramanian[4], 2000). As rural economy is in close association with development of cashew processing industries in this region, commercial banks and financial agencies are supporting the change in the system and growth of cashew sector in this region.

Maharashtra

Cluster based cashewnut processing is predominant in Mahrashtra state especially in Sidhudurg and Ratnagiri districts. Small-scale cashew nut processing developed in this region is in consideration of resources and need. It is a simple and low cost technology for processing and the technological design maintains a balance between human and machine energy to strengthen work opportunities. These small scale units having operational capacity < 1.0 TPD, has outnumbered large scale factories. Financial assistance provided by Government of Maharashtra, demand for cashew kernels in the domestic market, interest among the self help group (SHG) and generation of rural employment are the major reasons for the success of the small scale processing sector in this region. Raw nuts are either sourced from own/neighbouring farm and/or from local traders. The nuts dried under sun in single layer for a day or two to reduce the moisture level. Steam boiling technique is predominantly followed and four types of boilers *viz.*, direct steaming vertical type, Indirect steaming horizontal type, Indirect steaming vertical type and compact type are being used for nut conditioning. After nut conditioning, it is dried in ambient environment to make nut hard to suit for hand cum pedal operated type shelling. The shelling capacity is around 20 kg/head/day and the whole kernel recovery ranges between 80 to 90 per cent. Trained personals are engaged with conventional type driers to avoid kernel scorching. Electrically powered tray driers with capacity up to 25 kg per batch are also used for the purpose. Peeling is done manually and the operational capacity ranges between 4 to 6 kg/head/day depending on their skill and experience. After skin removal, the kernels are graded in to wholes and broken and packed in polythene bags.

The processing technology developed is socially appropriate, economically accessible and can be used and maintained in the local environment without causing damage. It is a household technology package tailored to escalate the economic status of shelf help group. The feasibility of small-scale cashew processing units operating in this region reveals that the net value addition per unit at farm level is worked out to be Rs 23,700 and the value addition per unit weight of raw cashew nut is around Rs. 38. The economics for the unit having utilization capacity of 3 TPA is found out as Rs. 32,983 per month (Balasubramanian, 2008). Besides, about 300 man-days per unit have been generated during season.

Goa

Raw cashewnuts produced from Goa fetches better price due to superior quality, regulated market and minimum support price. Harvesting of fully ripened cashew apple for the preparation of alcoholic beverage called 'Feni' is the prime factor for the superior quality of Goan cashewnuts. Cashewnut processing units in Goa are mostly concentrated in Ponda, Valpoi and Pernem districts. Although processing started with 'Drum roasting' method of conditioning, presently, majority of the processing factories employ 'Steam boiling' method due to inherent advantages. Majority of the small scale cashewnut processing units having capacity <1 TPD utilizes man power for various operations *viz.*, shelling, peeling and grading. Recently, large scale industries *i.e.* capacity utilization is in the range of 1-4 TPD, mechanized shelling and peeling machines are introduced to counter labour shortage (Potty *et al.*, 2013). Few industries have incorporated steam assisted unpeeled kernel dryer recently for better utilization of thermal power and uniform drying of kernels to ensure higher whole kernel recovery in the peeling process. Only in Goa, unpeeled cashew kernels after 'Borma' drying is directly sold in the market with a tag of 'Natural Cashew kernels'. This particular grade, not only reduces the work load in the industry and but also the price of kernels as well. It was informed that this type of grade is very much preferred in the local market.

Odisha

Cashewnut processing in Odisha started during early fifties in Rambha and Ganjam districts following the cashewnut processing methodology from Palasa and Kasibugga of Srikakulam district, Andhra Pradesh. Later, it spread to cashew growing districts and adjoining regions. Most of the processing units follow 'Drum Roasting' process due to availability of skilled man power suitable for the process and demand of cashew kernels obtained from this mode of conditioning in the domestic market. At present, cashewnut processing units are functioning in five clusters *viz.*, Cluster I (Keonjhar district), Cluster II (Dhenkanal and Athagarh district); Cluster III (Puri and Brahmagiri district); Cluster IV (Ganjam and Khurda district) and Cluster V (Koraput and Nuvrangapet district).

A total of 240 cashewnut processing units are under operation in Odisha state in an organized way. Few processing units in remote locations classified under cottage to small scale *i.e.* capacity utilization is below 500 kg day^{-1} functions in an unorganized manner. About 80 per cent of the industries under organized sector process around 1.2 to 1.6 TPD whereas 20 per cent of the units utilizes 50 per cent of the installed capacity *i.e.* 3.0 to 4.0 TPD. Cashewnut processing in Odisha is self sufficient with raw material production in the state and during lean season, required quantity is imported from African countries. After drum roasting, nuts are cracked to extract unpeeled cashew kernels. Either fully electrically operated or hybrid of electrical and biofuel based kernel dryers are prevalent in majority of the processing units. Operational capacity of the labourers engaged in shelling or peeling or grading are on par with other parts of the country.

Steam boiling process is gaining momentum in this region. About 75 per cent of the cashewnut processors in Ganjam district switched over to latest technology, steam boiling, in view of additional revenue through its byproducts and appealing quality of end product. Few units in Koraput have facility to extract cashewnut shell liquid from steam boiled nuts. Manually operated hand cum pedal shelling units is used to extract cashew kernels from steamed nuts. Recently, mechanized shelling units are introduced in this region and most of the units distribute unpeeled cashew kernels for home peeling. Few processing units have already installed mechanized peeling machine either imported from Vietnam or indigenously developed (Abrasion and pneumatic type) to overcome the problem of shortage of labourer and its performance is stated to be better. WW180, WW210, WW240, WW320, WW450, JH, K, Butts, LWP, BB (Rice/many pieces) Scorched JH, Red JH,

Scorched K, Red K, Red Pisces, PKW (Spotted kernels) are certain grades sorted in the grading section for domestic market.

West Bengal

Cashewnut processing is concentrated in Contai and Purba Medinipur in the state of West Bengal as more than 80 per cent of raw cashewnut is produced in this district. Although raw cashewnut is cultivated in the forest areas in the adjacent districts of Paschim Medinipur, Bankura and Purulia, there is no processing facility developed in those districts. In West Bengal, cashew processing started at home level wherein raw nuts were roasted using earthen pots along with sand, a similar process generally followed for making puffed rice commonly known as 'Muri'. This crude method of roasting led to more breakage of edible kernels. Further, the process exuded cashewnut shell liquid which is highly corrosive in nature. The first commercial cashew processing unit is started in the year 1975 and gradually cashew processing units grew to the extent that around 250 units constitute cashew processing sector in this state (Figure 3.21). Increase in the demand of cashew kernels by the turn of twentieth century is the motivation factor for the growth of processing units. Most of the units are proprietary in nature; however, there exists few partnership firms which are estimated to be around 10 per cent of the total units. Generally growers sell the product in the village market where the small traders procure and supply the same in the urban markets. While the demand of raw cashewnuts in Purba Medinapur district is estimated to be 20,000 TPA, only 8 per cent of the total requirement is derived from local production and the balance quantity from neighbouring states and imported from African countries.

Drum roasting is predominant in this region and majority of the units are categorized under small scale sector having processing capacity ranging from 0.8 to 3.2 TPD. Most of the processing units operate between 120 to 150 days in a year due to non-availability of raw cashewnuts at a reasonable price. Large scale units operate up to 250 days depending upon the domestic market potential (Balasubramanian and

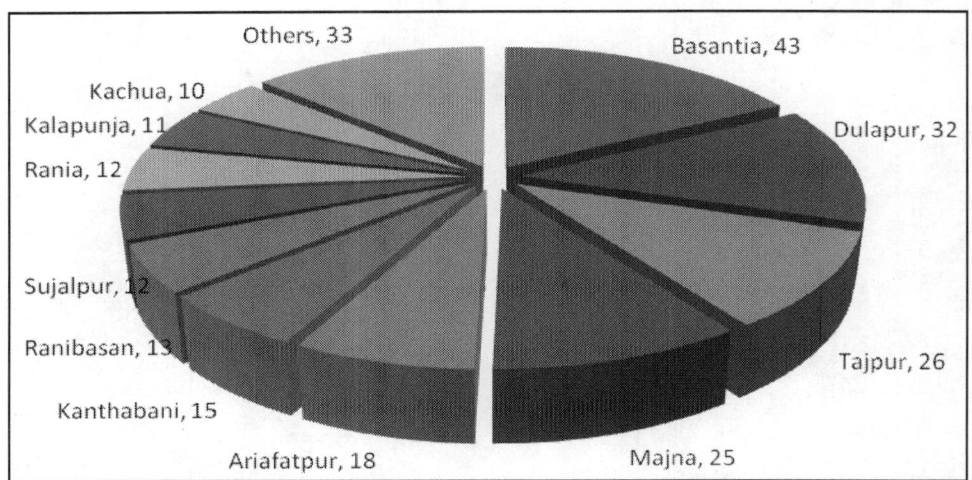

Figure 3.21: Distribution of Cashewnut Processing Units in West Bengal.

Joycy, 2012). Women labourers are engaged in shelling, peeling and grading processes and production potential comparatively lower than neighbouring state, Odisha. Lack of knowledge on cashew processing and degree of skill are the contributing factors for inferior production efficiency. Financial support is extended by banking sector and certain local institutions play vital role for the co-operation among the processors. SWOT analysis suggests that an integrated effort is required by Government, processors and farmers for the improvement of the cashew industry in this region.

12. SWOT Analysis on Cashew Processing

India

Strenghts	*Weakness*	*Opportunities*	*Threats*
☆ Support for the cashew sector through a range of government institutions at the state and national levels. Institutions such as the Cashew Research Station, Directorate of Cashew Research (DCR), and relevant ministries have prioritized plantation development and organic production. The Cashew Export Promotion Council (CEPC), an organization funded by the federal government of India,	☆ Limited use of by-products for the preparation of food and beverages. The general perception is that this is wasteful and a missed opportunity to diversify farmers' incomes, but attempts to adjust consumers' tastes to products from the cashew apple have not been successful. ☆ In the case of the Kerala State government, employment creation is more of a priority than	☆ Fair trade and organic production. ☆ The Cashew Export Promotion Council has proposed the branding of a "Made in India" cashew which could pave way to differentiate their product. ☆ Improved market for the by-products especially CNSL.	☆ As the economic life a cashew tree is 30-35 years many are becoming less productive in this well established sector. Especially a threat in Kerala where production was historically concentrated, but there are promising higher yields in other states such as Maharashtra where trees have been planted more recently. ☆ Although financing is widely available, smallholder farmers assert

Strenghts	Weakness	Opportunities	Threats
coordinates nearly all of the exporters of cashews and allied products in its membership. ☆ Some of India's largest exporters are leaders in the use of sorting and packaging technology. ☆ Growing domestic market with India's emerging middle class. ☆ Strong financial infrastructure due to India's social banking regulations. ☆ Highly skilled labourers in all steps of manual processing due to long history of sector. ☆ Availability of land for new cultivation. ☆ Improved physical infrastructure (good roads, high population density, accessible ports)	the competitiveness of the sector. ☆ Dependence on imported cashews mainly from African countries where processing facilities are strengthened by Government. ☆ Certain cashew growing regions, homestead farming is predominant and cashew production is just one part of farmers' income generation activities. This method of land use has also limited organic certification. ☆ Inconsistent acceptance of the varieties, methods and use of chemicals promoted by extension agencies and the role of NGOs in the provision of extension services is limited.		that the credit products do not meet their needs. ☆ Price fluctuations, vulnerability to weather changes and the short term yield are causing farmers to switch over to rubber cultivation. Cashew farming is considered to have a lower social status compared to rubber production, so there is less incentive to invest in upgrading. ☆ Increasing competitiveness in the International market especially from Vietnam where cost of processing is comparatively lower.

Vietnam

Strenghts	*Weakness*	*Opportunities*	*Threats*
☆ No export tax and no tax on land used for agricultural purposes. ☆ Government investment in R and D and extension towards increasing the yield. ☆ Quality cashewnuts are considered to be superior, resulting in better quality end products after processing. ☆ Easy access to agricultural inputs. ☆ Increased involvement of the private sector in processing. ☆ Improved market for lower quality cashews in neighbouring country *i.e.* China.	☆ Little organization among actors along the value chain. ☆ Lack of access to information *viz.*, price, production techniques, market information, quality standards etc., ☆ Increased competition for land from other cash crops like rubber. ☆ Inadequate support from the cashew association to increase growth and success of the sector. ☆ Lack of access to credit for machinery, agricultural inputs, and new tree varieties. ☆ Insufficient quality control, inconsistent quality and shortage of domestic cashew production to fulfil processing capabilities. ☆ Unorganised trading of raw cashewnuts leading to high fluctuation of price.	☆ Increased income in Vietnam and China may increase demand augmenting market potential. ☆ Branding of Binh Phuoc cashews will differentiate Vietnamese cashews and increase prices. ☆ WTO accession with cashew leaders like Brazil and India ensures the industry in terms of obtaining new technology, controlling price, increased access to information. ☆ Cashew kernel price is highly volatile in the International trade.	☆ Labour constraints for both processors and producers. ☆ Unstable and unpredictable weather.

Mozambique

Strenghts	*Weakness*	*Opportunities*	*Threats*
✰ Tremendous growth of processing capacity in a matter of very few years. ✰ Interest created among new entrepreneurs to start up cashewnut processing. ✰ Processed cashews or the kernel sells at a higher price in the international market than raw cashew, encouraging the stake holders in the cashew value chain. ✰ Processing plants are located in areas of concentrated production, which strengthens the links between processors and producers and minimizing involvement of intermediaries. ✰ Promotion of small and medium size plants which do not require highly sophisticated managerial	✰ Insufficient capital is available for processors to capitalize on profitable use of by-products. Neither CNSL is extracted nor cashew shell used as fuel. ✰ Poor public or private sector support for investment on cashewnut processing industry. ✰ Increases the cost of transferring the product to the port due to improper infrastructure. ✰ Low processing capacity limits the supply of high volume demand of cashew kernels of specific grade to International market. ✰ High level of pests and diseases inherent in Mozambique's climate leading to poor quality of raw cashewnuts.	✰ South Africa offers an opportunity for further expansion. ✰ Processing is gaining political leverage as it witnessed rapid growth, rural employment and alleviation of poverty in the last decade. ✰ The processing industry is relatively new, best practices in all areas of production can be more easily incorporated to the industry such as quality up gradation, technological developments, marketing and branding, etc.	✰ Financial crisis of existing processing units led to their production less competitive and in some cases withdrawing their investments or closing. ✰ The USAID fund that is currently acting as a loan guarantee for working capital for processors could be removed in the short-term making access to financing even more difficult.

Strenghts	*Weakness*	*Opportunities*	*Threats*
skills, substantially smaller initial capital requirements, use of manual shelling process resulting in better quality. ☆ Coordination among cashew associations ensures commercialization, marketing, political leverage, etc to be competitive in the kernel market.			

Brazil

Strenghts	*Weakness*	*Opportunities*	*Threats*
☆ Well developed infrastructure for processing and few large scale units could process the whole production of cashew and well connected to port infrastructure. ☆ Proximity to USA is a comparative advantage for marketing cashew kernels. ☆ Cashew apple is very much utilized providing more employment and encouraging cashew growers. CNSL used in the chemical industry for the production of dyes, lubricants and cosmetics and tanin, extracted from the tree trunk and nut shells used in chemical compost.	☆ Raw cashewnut production is stagnant for many years and neither expansion nor productivity increased for decades. ☆ Quality of cashew kernels manufactured by highly mechanized mode of processing *i.e.* oil bath roasting is considered lower than steam boiling method followed elsewhere in the world. ☆ Lack of access to credit for Micro and Small Enterprises (MSE) is one of the obstacles for the growth of cashew sector.	☆ Strategic public sector support to promote private sector engagements in cashew sector. ☆ Participation in cashew fair trade to promote products in International level. ☆ Senile cashew trees are replaced with high yielding dwarf varieties and assuring increase in quality and quantity of raw material to the processing industry. ☆ Growing domestic market for cashew kernels.	☆ Increased competition in the International market lowering the price of kernels. ☆ Quality standards are not strongly enforced through a price differential when purchased by an intermediary.

13. Future Thrusts

Spectrum of valuable cashew kernels obtained after processing by any mode, primarily depends on the quality of raw cashewnut. Therefore, procedural technique to assess its quality needs to be standardized to benefit both farmers and processors. It is imperative to set up an effective marketing information system that is able to monitor and analyze global production and market trends and relay such information to cashew stakeholders. It involves collection of price data at farm gates, factory gates, warehouse gates, at ports and at international trading places and setting up of a price information service with market intelligence.

Although different harvesting season ensures supply of raw cashewnuts throughout the year at global level, supply of this industrial raw material is inadequate due to stagnation in raw cashewnut production, imbalanced production and processing facility and increasing demand for cashew kernel either in the domestic market or International markets or both. This seasonal production of raw cashewnuts for a particular country leads to procuring huge quantity towards operating the processing units throughout year. In tropical countries, the problem of insects, pests and microbes in storage subsides considerably in winter and summer due to extreme temperatures. However, fluctuating temperatures are delirious during storage due to condensation of moisture in the ware houses at certain pockets and consequent microbial growth. This involves a revision of existing storage procedures, identification of best practices, and setting up of knowledge network on storage technology. Farm level technologies needs to be identified and modern technologies have to be promoted. Besides, such technologies have to be strengthened through research and development institutions ensuring good governance.

Various modes of cashew processing are followed in cashew growing countries with varied degree of mechanization for effective utilization of man, machine and materials. The economic efficiency of the processing depends on the production of white whole kernels which fetches premium price at consumer's level. Therefore, pertinent technical parameters at all stages have to be optimized to recover maximum white whole kernels irrespective of the processing method followed. Further, various unit operations in cashewnut processing should be energy efficient, environment friendly and highly suitable for production catchments in order to promote rural economy.

The quality of kernels has a direct and significant influence on price and hence profitability, it is therefore important to have clear and common standardized systems in place. This implies, enforcement of application of quality criteria at all levels in the value chain of cashew which includes Good Manufacturing Practices (GMP) and Good Hygienic Practices (GHP) as well as food safety standards such as Hazard Analysis Critical Control Points (HACCP) and ISO 22000 have to be introduced and promoted.

Finance is the basic fuel of cashew industry and the financial needs in the value chain should be understood and supported through Government policies and mobilizing increased investments through private sector. Financial agencies should gear up to support the short, medium and long-term financial needs of processors, setting up of guarantee to lower risks of lending, the development of a low-interest funding to enable processors to stock the raw material.

Interventions that target development constraints and opportunities at the various levels of the cashew value chain including the promotion of knowledge exchange, the use of effective, sustainable and affordable technologies and innovative partnerships among chain actors and development agencies should be strengthened. It is essential to build up internationally competitive business environments and promote enterprises to make the cashew processing sector an engine of the economic growth. It should particularly promote agricultural development-led industrialization to support successful implementation.

References

Anonymous (2013). DCR Annual Report. Directorate of Cashew Research, Dakshina Kannada, Karnataka, India. p. 86.

Antarkar, M.V. and Joshi, G.D. (1987). Studies on physical indices of maturity of some cashew nut varieties. *The Cashew* 1: 18-21.

Augustin, A. and Unnithan, V.K.G. (1981). An attempt on maturities of cashew apple. *Indian Cashew J.* 14: 9-11.

Balasubramanian[1], D. (2000). Physical properties of raw cashew nuts. *J. of Agricultural Engineering Research*, 78(3): 291-297.

Balasubramanian[2], D. (2000). Status report of cashew industries in Dakshina Kannada. *The Cashew*, 14 (2): 12-25.

Balasubramanian[3], D. (2000). Cashew processing industries in Kerala – Quilon and Calicut. *The Cashew*, 15(2): 28- 35.

Balasubramanian[4], D. (2000). Performance of cashew nut processing in Andhra Pradesh-A case study. *The Cashew*, 14(4): 7-16.

Balasubramanian, D. (2001). Cashew processing industries in India- An overall analysis. *The Cashew*, 15(2): 14-20.

Balasubramanian, D. (2003). Consumer processing and its feasibility in Chintamani district of Karnataka: A case study. *The Cashew*, 17(4): 6-11.

Balasubramanian, D. (2008). Farm level Cashewnut processing. *Cashew Bulletin*, 45(10): 3-4.

Balasubramanian, D. (2010). Design and development of radial arm type cashew kernel extracting machine. *Agricultural Mechanization in Asia, Africa and Latin America*, 42(2): 49-55.

Balasubramanian, D. and Sandeep, T. N. (2011). Performance evaluation of dual mode dryer for raw cashewnuts. *J. of Plantation Crops*, 40(1): 28-39.

Balasubramanian, D. and. Joycy, R. L. K. (2012). Cashewnut Processing in West Bengal – Current Status and Challenges, *The Journal of Cashew and Cocoa*, 42(2): 14-22.

Cavalieri, R. P. (1999). Fresh fruits and vegetable: quality and food safety. *Postharvest Biology and Technology*, 15: 195-340.

CEPCI (2013). Available at http: //www.cashewindia.org/php/cepc Contents. php?cat ID=29.

DCCD (2013). Available at http: //dccd.gov.in/stat.htm#stat.

Dalvi, V. D. and Thakare, C. G and Borude, S. G. (1992). Economics of production of cashew nut in Sindhudurg district in Maharashtra. *The Cashew*, 5(3): 15-17.

FAO (2011). Available at http: //www.faostat.fao.org/site/339/default.aspx.

Joseph, K.T. (1981). Seed-borne fungi on cashew (*Anacardium occidentale* L). *Cashew Causerie*, 3: 12.

Kihn, T B, 2004. Cashew Industry to Crack Targets. Vietnam Economic Times. Available at: http: //vietnamnews. vnagency.com.vn/2004-02/27/Stories/16. htm.

Leckie, A. M. (1989). The cashew industry of Ceara, Brazil: A development alternative. Proceedings –AAG Middle State Division. 11: 35-44.

Mandal, R.C. (1992). Cashew production and processing technology. Pub. Agrobotanical Publishers. pp. 132-142.

Marcio, C. M. P and Francisco, F.A. P. (2001). Cashew nut miniplants in North-Eastern Brazil: A successful partnership. In. Technical workshop on methodologies, organization and management of global partnership programmes. IFAD, Rome. pp. 11.

Mohan Kumar, G. N, Narayanaswamy, P. and Mukashi, A. N. (1984). Fruit growth and development of cashew. *J. Plantation Crops*, 12: 81-85.

Nagaraja, K.V. and Giridhar Prabhu, G. (1996). Investigations in to the factors influencing the kernel rejects during processing. Presented in the National Seminar on Development of Cashew Industry in India. 14-15 December, Bhubaneswar, India.

Nazneen, K., Carin, V., Luis, A. and Carla, B. (2004). Liberalization, Gender and Livelihoods: The Mozambican Nut Case. Viewed at www. Iied.org/ pdfs/9554IIED.pdf.

Nocholas, B. and Brook, A. (2007). Cashew processing, Marketing and consumption in West Africa. Current status and opportunities. *Technical Report* No. 22. p. 31.

Ohler, J.G. (1968). Cashew. Dept. of Agr. Res. Royal tropical Inst. Amsterdam, p. 260.

Ohler, J.G. (1979). Cashew. Communication 71. Department of Agricultural Research. Kaninklijk. Institute voor de Tropen, Amsterdam, pp. 72-78.

Okwelogu, T. N. and Mackay, P. J. (1969). Cashew moisture relations. *J. Sci. Food Agric.*, 20: 697-702.

Potty, V. P., Prabhakumari, C., Balasubramanian, D., Palanimuthu, K. A. and Ratheesh, K. A. (2013). A Value Chain in Cashew for Domestic and Export Markets.Cashew export Council of India, Kollam, Kerala, India. pp. 100.

Rao, C. and Joshi, G. D. F. (1987). Studies on physical indices of maturityof some cashew nut varieties. *The Cashew*, 1(4): 180-21.

Ratnam, N. V. (1969). Economic analysis of the competitive position of India in world cashew trade. Microfilm of Ph.D. thesis submitted to University of Hawaii. University Microfilms, Inc., *Ann Arbor*, Michigan.

Saroj, P. L. and Balasubramanian, D. (2013). Cashew Industry in India – a sustainable road map. *Indian Horticulture*, 58(1): 9-15.

Shalini Yadav. (2010). Economics of cashew in India. Report published by Department of Economic Analysis and Research. National Bank for Agriculture and Rural Development, Mumbai, India.

Sumrit Feungehan., Taweekiat Yimaswat, and Sorot Chindaprasert. (1989). Physiological development of cashewnut on canopy flower and fruit. *Khon Kaen Agric. J.*, 17: 239-253.

United States Department of Agriculture (USDA) (2002). National nutrient database for standard reference. Release 15. http: //ndb.nal.usda.gov/ndb.foods/show/3677.

Waithaka, J. H. G. (2002). Kenya: Assessment of the situation and development prospects for the cashewnut sector. Viewed at www.unctod.info/upload/infocomm/DOCS/Cashew/Keny.pdf.

Annexure I: Comparison of the Quality Standards and Specifications for Cashew Kernels

Country/Grade	*CEPCI*	*UNECE*	*BIS*	*AFI*	*BCI*
			Whole kernels		
Definition	Kernels with characteristic shape and there is no specific mentioning about the allowable limit for broken off part.	Kernels with the characteristic shape and the presence of a small hole at the proximal end of the kernel or a central split or crack are not considered as a defect. A kernel with no more than 1/8th of the kernel broken off is also considered as whole	There is no specific mentioning about the allowable limit for broken off part of the kernel	Kernels having the characteristic shape of the cashew kernels and not more than 1/8th of the kernel has been broken off	
Classification	Wholes kernels are further classified as "White wholes" (W), "Scorched wholes" (SW) and "Dessert wholes" (DW) based on colour and appearance. Sizing is given for "white wholes" and "scorched wholes" and not for "Dessert wholes"	Kernels are classified as "Extra" class, Class I and Class II based on the colour, appearance, uniformity, etc. compulsory for Extra class and optional for Class I and Class II.		Classified as First quality fancy, Second quality scorched/color variation, Third quality scorched/ color variation, Fourth quality and Dessert based on colour and appearance. Compulsory for first quality fancy and optional for other whole grades	Wholes kernels are designated as "wholes' (W1), 'scorched wholes" (W2), "second scorched wholes"(W3) and "Fourth quality wholes" (W4). sizing exists for "wholes" and "Scorched wholes" but not for "second scorched wholes" and "Fourth quality wholes"
Tolerance limits*	Specifications are given for allowable percentage of broken and maximum moisture content and it is 5 per cent for both	Specifically for Extra class, Class I and Class II.	Broken and next lower grade should not exceed by 5 per cent of the weight of the packing	Specifications are given clearly for serious damage and defects for all the whole grade classifications	
Size tolerance	The allowable limit for next lower grade is 5 per cent for white wholes and 7.5 per cent for Scorched and Dessert wholes.	NLG should not exceed by 5 per cent for "Extra" class and 7.5 per cent for Class I	The allowable limit for next lower grade for wholes is 5 per cent.	Next lower grade (NLG) should not exceed by 10 per cent of the weight of the wholes	

Contd...

Annexure I–*Contd...*

Country/Grade	*CEPCI*	*UNECE*	*BIS*	*AFI*	*BCI*
			Butts and Splits		
Definition	Kernels that are broken cross wise and naturally attached are described as butts and the kernels that are split naturally lengthwise are described as splits	Butts are defined as the kernels of not less than 3/8 of a whole kernel which have been broken cross wise but the coty-ledons are still naturally attached. Splits are the kernels that are split length wise. Pieces are the kernels that are broken more than two pieces	Kernels that are broken cross wise and naturally attached are described as butts and the kernels that are split naturally lengthwise are described as splits	Kernels that are less than a 7/8 but not less than 3/8 of a whole kernel and the cotyledons are still naturally attached. The splits are defined as the kernel one half of which has been split length wise provided not more than 1/8 of this cotyledon has been broken off	
Size tolerance	The grades, SSP, LWP, and SWP: NLG should not exceed by 5 per cent by weight. SB, SS, SP, SPS, DP: NLG should not exceed 7.5 per cent by weight.	NLG should not exceed 5 per cent by weight for all the three classes namely, "Extra", "Class I" and "Class II".		NLG (Next Lower Grade) should not exceed by 10 per cent by weight	
			Pieces		
Classification	Classified as "white pieces", "scorched pieces" and "Dessert pieces" based mainly on the colour and appearance	Classified as "Large pieces", "Small pieces", "Very small pieces", "Baby bits or granule" based on the size		Classified as "large pieces", "small pieces", "chips or small pieces", "grains, granules or baby bits", "fine grains", "fines", and "mixed pieces" based on the size	Classified as "large pieces", "small pieces" "small Brazilian pieces" and "Brazilian granules" based on size

Contd...

Annexure I–*Contd...*

Country/Grade	*CEPCI*	*UNECE*	*BIS*	*AFI*	*BCI*
Size tolerance	Baby bits: NLG should not exceed by 1 per cent by weight	NLG should not exceed 5 per cent by weight for "Extra" class and 7 per cent by weight for the other two classes	For white pieces and scorched pieces: NLG should not exceed 5 per cent by weight. For dessert pieces: NLG should not exceed 10 per cent by weight	NLG should not exceed by 5 per cent by weight	

* Other requirements such as free from moulds, insects, extraneous matters etc are mentioned and no specific tolerable limits are given. But it is specified that the kernels should be completely free from rancidity.

Annexure II: Sizes and Counts per Pound for Whole Cashew Kernels

Size Designation	*AFI Standards*	*BIS*	*Brazilian Standards*	*CEPC Standards*	*UNECE Standards*
150		–	–	–	120-147
180	120-180	120-180	–	170-180	148-180
210	180-210	200-210	–	200-210	180-210
240	220-240	220-240	220-240	220-240	220-240
280	–	260-280	–	–	261-281
320	300-320	300-320	300-320	300-320	300-320
400	–	350-400	–	–	321-400
450	400-450	400-450	400-450	400-450	400-450
500	–	450-500	–	450-500	450-500

Annexure III: Size Descriptions for Broken Cashew Kernels

Standards	*Major Classification*	*Grade Designations*	*Sieve Size in mm*	
			Passing through	*Retained on*
AFI	Large pieces	LWP, SP, SPS, DP, P1, P2, P3	8.0	4.75
	Small pieces	SWP, SSP, DSP, SP1, SP2, SP3	4.75	2.80
	Chips or special small pieces	CHIPS or SSP1, SSP2, SSP3	2.80	2.36
	Grains, granules or baby bits	BB or G1, G2, G3	2.36	1.70
	Fine grains	X	1.70	1.18
	Fines	FE	1.18	N/A
	Mixed pieces	P1M, P2M, P3M	To be determined by the buyer and seller	
UNECE	Large pieces	–		4.75
	Small pieces	–	4.75	2.80
	Very small pieces	–	2.80	2.36
	Baby bits or granules	–	2.80	1.70
Brazilian	Large pieces	Large white pieces (P1), Scorched pieces (P2), Special scorched pieces (P3)	–	4.75
	Small pieces	Small white pieces (SP1), Scorched small pieces (SP2), SP3	–	2.80
	Small Brazilian pieces	SSP1, SSP2, SSP3	–	2.36
	Brazilian Granules	G1, G2, G3	–	1.70
		X1, X2, X3,	–	1.19
BIS		LWP, SP, SPS, DP	–	4.75
		SWP, SSP, DSP	–	2.80
		BB		1.70
CEPC		LWP, SP, SPS, DP	–	4.75
		SWP, SSP	–	2.80
		BB	–	1.70

Cashew Export Promotion Council of India (CEPCI), United Nations Economic Commission for Europe (UNECE), Brazilian Cashew Industry (BCI), Vietnamese Cashew Industry, Bureau of Indian Standards (BIS) and Association of Food Industries (AFI, New York).

Section II: Crop Improvement

Cashew: Improvement, Production and Processing (2017) *Pages* **93–104**
Editor: **Prof. P.L. Saroj**
Published by: **ASTRAL INTERNATIONAL PVT. LTD., NEW DELHI**

Botany and Taxonomy

E. Eradasappa and P.L. Saroj

1. Introduction

Cashew (*Anacardium occidentale* L.) belongs to the Family Anacardiaceae under the Order Sapinadales. This family comprises 60 to 74 genera and 400 to 600 species (Barros, 1991). According to Susan (2009) Anacardiaceae, the cashew family includes approximately 800 species in 82 genera. Members of the family are cultivated throughout the world for their edible fruits and seeds, medicinal compounds, valuable timber, and landscape appeal. Some other important species of Anacardiaceae include mango (*Mangifera indica*), pistachio (*Pistacia vera*) and pink peppercorn (*Schinus terebinthifolia*) and products of these species are consumed worldwide. Other distinguished species such as the pantropical *Spondias* fruits, the marula of Africa (*Sclerocarya birrea*), and the neotropical fruits of the genus *Antrocaryon* are considered to be localized cultivation and consumption and are not generally transported to distant markets. Twenty-one species of the genus *Anacardium* were identified through classical taxonomy (Table 4.1) (Barros, 1995). Among these species, *Anacardium occidentale* L. is the only cultivated and widely distributed (Jhonson, 1973; Ohler, 1979; Mitchelle and Mori, 1987). Most species of the *Anacardium* genus are found all over Brazil (Nomisma, 1994). The genus *Anacardium* is a native to Latin America and has a primary centre of diversity in Amazonia and secondary one in the Planalto of Brazil. Behrens (1998) described the crop as a tropical tree species that originated from South America. The diversity of *Anacardium* is greatest in the Amazon basin and in the central uplands, the highest degree of diversification of the cultivated species is found in Northeast Brazil. For this reason Johnson (1973) considered that Ceará state was where the cashew originated and Barros (1995) suggested that the origin and speciation of cashew crop is linked to this region. Now about 98 per cent of cultivated cashew is grown in the Northeast Brazil (Paula Pessoa *et al.*, 1995). Natural occurrences of cashew have been reported from Mexico to Peru and in the West Indies. The crop was introduced into India, the East Indies and Africa by the Portuguese explorer in the 16th century. Thereafter, exploitation of cashew for its fruits (nut and apple) among local people appears to have been the pattern for more than 400 years in Asia and Africa (Mitchell and Mori, 1987).

Table 4.1: Species of *Anacardium* Described by Classic Systematic Botany (Johnson, 1973)

Species	*Country of Origin*
Anacardium amilcarianum Machado	Brazil
A.brasiliense Barb.Rodr.	Brazil
A.corymbosum Barb.Rodr.	Brazil
A.curatellaefolium St.Hil	Brazil
A.encardium Noronha	Malaysia
A.excelsum Skeels	Brazil
A.giganteum Hancock ex.Engl.	Brazil
A.humile St.Hil	Brazil
A.kuhlmannianum Machado	Brazil
A.mediterranneum Vell.Fl.Flum	Brazil
A.microcarpum Ducke.	Amazon region
A.microsepalum Loesn	Amazon region
A.nanum St.Hil	Brazil
A.negrense Pires and Fro'es	Brazil
A.occidentale Linn.	Brazil
A.othonianum Rizz.	Brazil
A.parvifolium Ducke	Amazon region
A.rhinonocarpus D.C.Prod.	Brazil
A.rondonianum Machado	Brazil
A.spruceanum Benth Ex.Engl.	Brazil
A.tenuifolium Ducke	Amazon region

2. Botanical Classification

Kingdom: Plantae

Sub-kingdom: Tracheobionta

Super division: Spermatophyte

Division: Magnoliophyta

Class: Magnoliopsida

Sub-class: Rosidae

Order: Sapindales

Family: Anacardiaceae

Genus: *Anacardium*

Species: *occidentale*

Among the 21 species known, three species *viz., A. pumilum* (*A. humile*), *A. othonianum and A. microcarpum* have been collected and conserved in our National Cashew Field Gene Bank (NCFGB) of Directorate of Cashew Research, Puttur. Among these, *A. pumilum* (also called monkey nut) is dwarf (shrub like) with weak shoots growing upto 3-4ft and big leaves, with profuse flowering and poor fruit-set. *A. othonianum* has short sturdy stem with compact canopy, small, thick leathery leaves and low fruit-set. In the above two species the nut and apple size is very small with bright pink apples. While *A. microcarpum* (also called miniature cashew) is similar to *A.occidentale* in both morphology and other characters and has apples with acidic sweet flavour. *A. gigantium* is a very large tropical tree up to 150ft with the biggest apple; tree may be resistant to anthracnose. *A. spruceanum* and *A. rhinocarpus* (wild cashew) are possessing hard wood; as such they may be useful as root stocks.

3. Cytology

The cytological studies in cashew are not many compared to other crops. The chromosome number of *A. occidentale* as 2n=42 was first reported by Darlington and Janaki Ammal (1945). Further the same chromosome number was corroborated by some workers in the same species (Hutchinson and Dalziel, 1954; Purseglove 1974; Aliyu and Awopetu 2007). Aliyu and Awopetu (2007) karyotyped the chromosomes of the cashew populations of Indian and Brazilian origin. The total length of the homologous chromosomes recorded for the Indian cashew population was found to be 51.10 µm, and were designated 1 - 21, according to decreasing lengths. The chromosome compliment gave a karyotypic formulae of 6Asm + 1Am + 4Bsm + 5Bm + 5Cm, while A represent chromosomes >3.00 µm, B = 1.50 - 2.99 µm and C < 1.49 µm. Meanwhile, the chromosome lengths ranged between 1.00 and 4.200 µm for the shortest and the longest respectively. Based on the morphology of the chromosomes, the complement comprises of 6 long submetacentric, 1 long metacentric, 4 intermediate submetacentric, 5 intermediate metacentric and 5 small metacentric chromosomes. The mitotic metaphase chromosome of Brazilian cashew population presented a length of 56.00 µm. Individual chromosome length ranged between 1.00 and 4.50 µm. The chromosome karyotype was very similar to that of Indian population comprising, 6Asm + 1Am + 1Ast + 9Bm+ 2Bsm + 2Cm. It, however, shows that the complement includes, 6 long submetacentric, 1 long metacentric, 1 long subtelocentric, 9 intermediate metacentric, 2 intermediate submetacentric and 2 small metacentric chromosomes with regular mitotic division.

4. Growth and Development

Morphology

Cashew is an evergreen dicotyledonous tropical tree with low branching and medium size canopy (Figure 4.1). On an average the plant attains 5- 8 m height and 10-12 m width, however some plants with 15 m height and 20 m width have been observed. The tree bears stout branches and thick, resinous, round and scaly bark. The wood of tree is yellow, moderately soft and light and has relative density of 0.5 (Lima, 1954; Tavares, 1959). The sapwood is pale brown when it is dry. The heartwood zone can easily be distinguished by the brown colour. The wood has an even texture with moderately close straight grains (Sebastine, 1955). The leaves are alternate, simple, glabrous, oblong, leathery, often notched at the apex, veins prominent, pinnately veined, lateral veins spreading with 10-20 pairs. The size of leaf varies from 6 -24 cm in length and 4-15 cm in width (Jhonson, 1973; Kumaran *et al.*, 1976a). Petioles are short with 1-2 cm length and the leaves are commonly crowded at the ends of branches.

Figure 4.1: A Cashew Plant in Flowering and Fruiting.

Root System

The root system of complete grown cashew tree consists of a taproot surrounded by a well developed and extensive network of lateral roots, 90 per cent of which lie on the 15-32 cm depth of soil layer and possess rootlets. The ratio between lateral and canopy width is close to 2:1 in plants up to six years.

Growth Pattern

The pattern of growth of cashew tree alternates with vegetative and reproductive phases. The initiation and duration of these phases vary among varieties as these phases are regulated by both genetic and environmental factors. There are two types branching in cashew; one is intensive and another extensive (Dasarathi, 1958).

The intensive shoot grows to a length of about 25-30 cm and ends in a panicle. Concurrently, three to eight lateral shoots come up below 10-15 cm of the apex and few of these laterals may also bear panicles. This kind of growth pattern is repeated in this type to give bushy appearance to the tree. On the other hand in the extensive type, the shoot grows to 20-30 cm length and rests. Thereafter a bud sprouting 5-8 cm below the apex gives rise to further growth. This type of growth process continues for two or three years without giving flowers and results in spreading tree habit. In high yielding trees more than 60 per cent of intensive branches are seen whereas low yielders possess less than 20 per cent of intensive branches.

Inflorescence and Sex Ratio

The cashew plant starts flowering in three to five years. Grafted plants come to flowering in 3 years whereas seedling plants may take 4 to 5 years for flowering. The inflorescence of cashew is called terminal panicle (Figure 4.2) which bears both male (staminate) (Figure 4.3) and hermaphrodite (perfect) (Figure 4.4) flowers in the same panicle (Figure 4.5). For this reason, cashew is considered as andromonoecious species. The inflorescence may be conical, pyramidal or irregular in shape. Number of panicles per plant, flowers per panicle and distribution of male and hermaphrodite flowers (sex ratio) in each panicle vary significantly. Morada (1941) counted 3 to 11 branches in each panicle, depending on the vigor of the tree, with 40 to 100 individual flowers on each panicle branch or 120 to 1100 flowers with 90-99 per cent staminate flowers in one panicle. Rao and Hassan (1957) reported that 96 per cent of the flowers are staminate in a panicle; Bigger (1960) observed a ratio of 6:1 staminate to perfect flowers; Damodaran *et al.* (1966) reported perfect flowers from as low as 0.45 per cent to 24.9 per cent.The variations in these traits are observed due to genetic and environmental factors. Flowers are produced in gradual manner and hence each panicle may stay up to three months giving continuous fruits. The duration of flowering phase depends on the genotype and environmental conditions.

Figure 4.2: A Flower Panicle.

Flowering Period and Phases

Flowering normally occurs following the growth flush at the end of the wet season, but its timing and duration are strongly influenced by temperature. Flowers are produced at the end of the new shoots. Thus flowers and fruits are borne on

Figure 4.3: Male Flower.

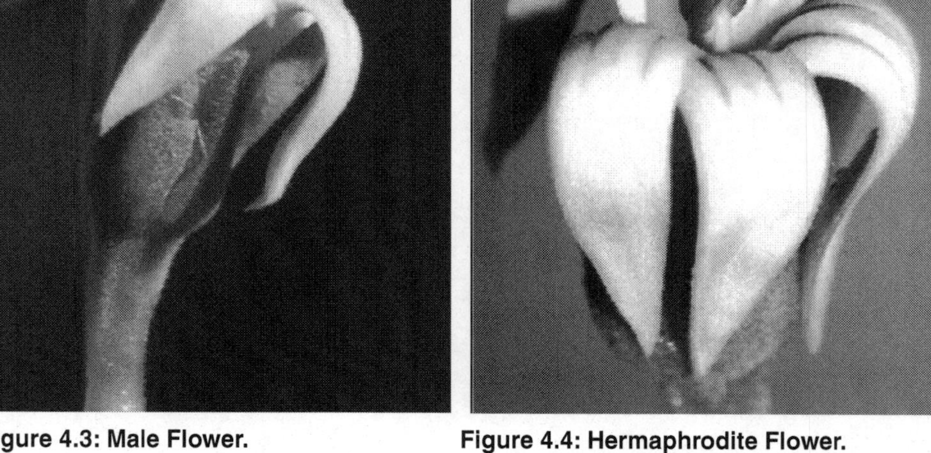

Figure 4.4: Hermaphrodite Flower.

the outer extremity of the canopy. Flower bud emergence in cashew initiates by the middle of September and continue until the end of February, the main season being October-November (Damodaran *et al.,* 1965). Nambiar (1977) reported that the flowering season in cashew varies with country depending upon its altitude. The flowering season is from June to November in Tanzania with peak in August-September (Northwood, 1966). In coastal region and transitional zones of Northeastern Brazil (around 6°S and up to 100m altitude), flowering lasts 4-6 months in the common type (from July/August to December/January) and 6-8 months (June/July to January/February) in the precocious dwarf type (Barros *et al.,* 1984; Barros 1988; Freitas 1994). The flowering period is from December to March with a peak in January-February in Central America (El Salvador) and West Africa. Two flowering periods were reported in Kenya, one from September to November and second from December to January (Agnoloni and Guiliani, 1977).

Figure 4.5: Cluster of Nuts.

Even within India the variation for flowering time was observed across different regions. In the West Coast of India, the peak flowering is in early January and the peak harvest is in early April (Rao, 1956) whereas in the East Coast, the peak flowering is from mid January to mid February and the crop is harvested in the end of April (Dasarathi, 1958). Cashew shows the trend of late flowering and fruiting at higher elevation irrespective of latitude (Nambiar, 1977). For instance, cashew flowering in Northeastern States gets delayed by 2-3 months and flowering is seen during April-June compare to other cashew growing regions of India where the flowering period is November/December to February/March. This is mainly due to reduction in temperature at higher altitudes. Generally flowering occurs in two or three typical phases and that appearing in the intermediate stage is considered as the productive phase. Three distinct phases of flowering are observed in cashew (Pavithran and Ravindranathan, 1974). They are (i) the first male phase with 19-100 per cent male flowers, (ii) the mixed phase with 0 to 60 per cent male and 0 to 20 per cent hermaphrodite flowers and (iii) the second male phase with 0 to 6.7 per cent male flowers. Many other cashew workers have also found that flowers produced early in a panicle are by and large male. The duration of these flowering phases varies with genotype. The mean duration of flowering observed was 85.2 days in which the duration of first male phase was 2.4 days, mixed phase 69.4 days and second male phase 13 days. Based on the season of flowering cashew genotypes are classified as early (Nov-Dec), mid (Dec-Jan) and late (Jan-Feb) flowering types.

Floral Biology

The flowers are small, white or light green at the time of opening, later turn to pink. Two kinds of flowers, *viz.*, hermaphrodite (bisexual/perfect) and male (staminate) are present in the same panicle. The perfect flowers are larger than staminate flowers (Damodaran *et al.*, 1965). The flowers are pentamerous. The calyx is green and oval with five free sepals. The corolla is linear to lanceolate in shape, white or creamy white at the time of opening with five free petals. The external surface of sepals and petals is pubescent with simple hairs. The androecium consists of one fully developed stamen and 7-9 staminodes. The developed stamen has pink anther. The anther is basifixed, bilobed, dehisces through a slit between the two pollen sacs of each lobe. The staminodes possess short filaments and are hidden in the lower half of the open flower. The developed stamen in the hermaphrodite flower has only short filament and its anther is far below the level of stigma. The pistil is dorsiventral; ovary is superior, reniform and monocarpellate. The style is long and slender, springs from distal margin of the ovary, tapering towards the end with slightly expanded stigma. The ovary is rudimentary in male flowers. It was suggested that the staminate flowers are derived from the ancestral hermaphrodite flowers by gradual reduction and loss of function of the gynoecium (Ascenso and Mota, 1972b).

Anthesis takes place between 9 AM and 2 PM in India and hermaphrodite flowers open mostly between 9 and 11 AM though some flowers open beyond this time. Staminate flowers were found to open very early in the morning and continue till about 2 PM. Over 80 percent of the perfect flowers open between 10 AM and 12 Noon. The peak period of dehiscence of anthers was from 9.30 to 11.30

AM and the rate of dehiscence was slightly higher on the sunny side of the tree as compared to that on the shady side. The viability of pollen is usually high and it was 94 per cent in types studied and the stigma becomes receptive one day prior to anthesis and its receptivity stays for two days (Damodaran *et al.*, 1966; Eradasappa *et al.*, 2012). *In vitro* germination of pollen was reported by Subbaiah *et al.* (1982) and also germination of pollen grains up to 50 per cent has been observed on high sucrose medium at Directorate of Cashew Research, Puttur.The majority of male flowers open between 7 and 9 AM while majority of bisexual flowers open between 8 AM and 12 noon in the West Coast region of India characterized by heavy rainfall and humid climate. Stigma is receptive throughout the day after anthesis of the flower (Rao and Hassan, 1957). The cashew produces scented flowers which attract pollinating insects mainly honey bees (*Apis mellifera*). The pollen grains of cashew are not easily carried by the wind (Paulino, 1992; Freitas, 1994; Freitas and Paxton, 1996). Reddi (1991) observed in the 20 inflorescences left open to insects bore 89 fruits, whereas those protected with butter paper bags (exclude insects and wind) or mosquito nets (exclude insects and allow wind) had no fruits formed. It was also confirmed the study that cashew pollen grains were sticky and were not released into the atmosphere as no pollen grains were trapped by the air sampler or by the stigmas of flowers left open to the wind. Hence, it was proved that the possibility of the wind acting as a pollinating agent was ruled out in cashew. It was reported that only pollens from large stamens of both staminate and perfect flowers are viable and play role in the reproduction. Staminodes (smaller stamens) do not produce viable pollen grains.

The cashew is chiefly allogamous tree species because of its reproductive system. However, due to coincident flowering of the two kinds of flowers on the same tree and same panicle also favours self-pollination. Self-incompatibility hitherto was not reported among the species of the genus.

Fruit Setting

As described earlier the cashew produces abundant flowers but only less than 10 per cent of which are hermaphrodite. Although close to 85 per cent of the perfect flowers are fertilized under standard conditions, only 4-6 per cent of them reach maturity to give fruits, the remaining shed away at different stages of development. The fruit drop in cashew during the early stages of development is attributed to physiological reasons (Nothwood, 1966). Insects attack also plays an important role in immature fruit drop (Pillay and Pillai, 1975). Reddi (1987) reported that cashew plants can permit about 27 per cent of their well pollinated flowers to develop into fruits. But in nature only 10.5 per cent yield is possible because of under-pollination and this was substantiated by stigmatic- pollen load analysis data. About 25-72 per cent of the stigmas were found unpollinated due to limitation of pollinators leading to lower than potential yields. A yield increase of 157.8 per cent is possible if the flowers receive adequate pollen. Earlier workers also suggested that pollination in nature is inadequate (Rao, 1974; Kumaran *et al.*, 1976). Fruit set through bagging of panicles in two varieties of cashew, Bhaskara and Ullal-3 was observed at the Directorate of Cashew Research, Puttur, Karanataka, India indicating the possibility of self-fertilization to a limited extent. However, with assisted pollination fruit-set up to 37 per cent was achieved (Eradasappa, 2012, unpublished).

After pollination, the fruit takes 6 to 8 weeks to develop. The nut develops first and the apple fills up during the last fortnight before nut drop occurs. The nut drop continues for 6 to 8 weeks (Figure 4.6).

Figure 4.6: Cashew Apple with Nut.

Apple and Nut Characters

The fruit of cashew is called nut. The nut is grey coloured, kidney shaped achene (a dry one-seeded indehiscent fruit with the seed distinct from the fruit wall) consisting of epicarp, mesocarp, endocarp and a kernel wrapped by a peel (testa). The epicarp is smooth, coriaceous and grey or grayish green and it forms the epidermis. The mesocarp is thickest of three layers, spongy tissue, containing a sticky, acrid and corrosive reddish liquid, the Cashew Nut Shell Liquid (CNSL). The CNSL is resinous oil rich in phenolic compounds. The endocarp is hard and is formed of a compact mass of schlerenchymatous cells. These three layers form the thick shell, *i.e.*, the pericarp which forms 45 to 50 per cent of the nut. The kernel, the edible part of nut is formed by two cotyledons forms 20 to 22 per cent of nut. The kernel is covered by a brown membranous testa (peel), which forms about 5 per cent of the weight of the nut. The peeling is one of the most important problems in the industrial processing of cashew nuts as it is very difficult to remove in as much as 20 per cent of the seeds. The nut size varies from 3 to 33g and the shelling percentage varies from 15-30 per cent. The nuts also vary in shape.

The cashew apple is a false fruit and it develops from the hypertrophied pedicel. The apple is fleshy and juicy and varies in size, weight, shape, texture, colour and taste. The nut grows slowly in the initial stages and then accelerates to attain its maximum size by 28 days. In subsequent days, the nut size reduces as it matures and becomes visibly smaller (Damodaran *et al.*, 1966). The cashew apple, in contrast, reaches its maximum size when it matures without any further change in size.

References

Agnoloni, M. and Giuliani, F. (1977). Cashew cultivation. Library of Tropical Agriculture, Ministry of Foreign Affairs. Institute Agronomic Per L' Olhemare. p.168.

Aliyu, O. M. and Awopetu, J. A. (2007). Chromosome studies in Cashew (*Anacardium occidentale* L.). *African Journal of Biotechnology*, 6(2): 131-6.

Ascenso, J. C. and Mota, M. I. (1972). Phylogenetic derivation of the cashew flower. *Boletimda Sociedade Broteriana*, 42: 253-7.

Barros, L. M. (1991). Caracterizacao morfologica e isoenzymatica do cajueiro (*Anacardium occidentale* L.) Tipos comum e anao precoce, por meio de tecnicas multivariadas. PhD. Thesis, USP-ESALQ. p.256.

Barros, L. M. (1988). Melhoramento. In: Lima VPMS (ed), A Cultura do Cajueiro no Nordeste do Brazil, pp.321-355. BNB/ETENE, Fortaleza.

Barros, L. M., Araujo, F. E., Almeida, J. I. L. and Teixeira, L. M. S. (1984). A cultura do cajueiro ana˜o (Dwarf cashew crop). EPACE Document 2. Fortaleza, Ceara´, Brazil.

Barros, L. M. (1995). Botanica, origem e distribuicao geografica. *In* Cajucultura: Modernas Technicas de Producao. (Araujo, J.P.P. and Silva,V.V.Eds.). EMBRAPA-CNPAT, Fortuleza, pp. 73-93.

Behrens, R. (1998). About the spacing of cashew nut tree. Proceeding of the International Cashew and Coconut conference, 17-21 February 1997, Dares Salam, Tanzania. BioHybrids International Ltd., Reading, UK. pp. 48-52.

Bigger, M. (1990). *Selenothrips rubrocinctus* (Grand) and the floral biology of cashew in Tanganyika. *East African Agriculture Journal*, 25: 229-234.

Damodaran, V. K., Abraham, J. and Alexander, K. M. (1965). The morphology and biology of the cashew flower I - Flowering habit, flowering season, morphology of the flower and sex ratio. *Agricultural Research Journal of Kerala*, 3(1 and 2): 23-8.

Damodaran, V. K., Abraham, J. and Alexander, K. M. (1966). The morphology and biology of the cashew flower II - Anthesis, dehiscence, receptivity of stigma, pollination, fruit set and fruit development. *Agricultural Research Journal of Kerala*, 4: 78-84.

Darlington, C. D. and Janaki Ammal, E. K. (1945). Chromosome Atlas of Cultivated Plants. George Allen and Unwin, London, p.397.

Dasarathi, T. B. (1958). A study of the blossom biology and growth features of cashew nut (*Anacardium occidentale* L.). M.Sc. Thesis. Andhra University, Waltair (unpublished).

Freitas, B. M. (1994). Beekeeping and cashew in North-eastern Brazil: the balance of honey and nut production. *Bee World* 75(4): 160-8.

Freitas, B. M. and Paxton, R. J. (1996). The role of wind and insects in cashew (*Anacardium occidentale* L.) pollination in NE Brazil. *Journal of Agricultural Science*, 126: 319-26.

Hutchinson, J. J. and Dalziel, M. (1954). Flora of West Africa. Crown Agent for Overseas Government and Administration, Millbank, London, S.W. I. Vol. 1: 428-429.

Jhonson, D. V. (1973). The botany, origin and spread of cashew (*Anacardium occidentale* L.). *Journal of Plantation Crops,* 1: 1-7.

Kumaran, P. M., Nayar, N. M., Nambiar, M. C., Mohan, E. and Vimala, B. (1976). Cashew-varietal improvement. Central Plantation Crops Research Institute, Annual Report for 1975, Kasaragod, India p.125.

Lima, D. de A. (1954). Contribution to the study of the flora of perhembuco, Brazil. *Mon.Univ.Rur.Pern.Recife* 1.

Mitchell, J. D. and Mori, S. A. (1987). The cashew and its relatives (*Anacardium: Anacardiaceae*). Memories on The New York Botanical Garden 42: 1-76.

Morada, S. K. (1941). Cashew Culture. *Philippines Journal of Agriculture.* 12: 89-103.

Nambiar, M. C. (1977). Cashew. In Ecophysiology of Tropical Crops. Academic Press, Inc., San Francisco, pp.461-478.

Nomisma, (1994). *The World Cashew Economy* Oltremare Sp A, via Piemonte 5, Zola Predosa, Bologna, Italy, pp. 218.

Nothwood, P. J. (1966). Some observations on the flowering and fruit-setting in the cashew (*Anacardium occidentale* L.). *Tropical Agriculture* (Trinidad)., 43: 35-42.

Ohler, J. G. (1979). Cashew. Kominklijk Institunt Voor de Tropen, Amsterdam, p.260. (Communication 71).

Paula Pessoa, P. F. A., Leite, L. A. de S. and Pimentel, C. R. M. (1995). Situação atual e perspectivas da agroindústria do caju. p. 23 – 41. *In*: J.P.P de Araújo, and V.V. da Silva (org.) Cajucultura: Modernas técnicas de produção. Embrapa-CNPAT. Fortaleza.

Paulino, F D S. 1992. Polinizacao entomologica em cajueiro (*Anacardium occidentale* L.) no litoral de Pacajus, CE. Thesis M.Sc.USP-ESALQ, p.70.

Pavithran, K. and Ravindranathan, P. P. (1974). Studies on the floral biology in cashew, *Anacardium occidentale* L. *Journal of Plantation Crops,* 1: 32-3.

Pell, S. K. (2009). Anacardiaceae. Cashew family. Version 03 January 2009 (under construction). http: //tolweb.org/Anacardiaceae/21262/2009.01.03 In: The tree of life web project, http: //tolweb.org.

Pilla, P. K. T. and Pilla, G. B. (1975). Note on the shedding of immature fruits in cashew. *Indian Journal of Agricultural Sciences,* 45: 233-234.

Purseglove, J. W. (1968). Tropical Crops: Dicotyledons. Longman. Harlow, UK.

Rao, V. N. M. and Hassan, M. V. 91957). Preliminary studies on the floral biology of cashew (*Anacardium occidentale L.*). *Indian Journal of Agricultural Sciences,* 27(3): 277-288.

Rao, V. N. M. (1956). Multiply better yielding cashewnut. *Indian Farming,* 6: 15-7.

Rao, V. N. M. (1974). Crop improvement of cashew. Mimeo. Report of the All India Summer Institute on Improvement and Management of Plantation Crops. Central Plantation Crops Research Institute, Kasaragod. pp. 128-134.

Reddi, E. U. B. (1987). Under-pollination: a major constraint of cashewnut production. *Proceedings of the Indian National Science Academy*, 53(3): 249-52.

Reddi, E. U. B. (1991). Pollinating agent of cashew-wind or insects? *Indian Cashew Journal*, 20(4): 13-8.

Sebastine, K. M. (1955). The emigrant economic plants of India (1) *Anacardium occidentale* L. *Proceedings of Indian Academy of Sciences*, 42: B.239-48.

Tavares, S. (1959). Medeiras do nordeste do Brazil. *Mon.Univ. Rur. Pern. Recife.*, 5: 9-171.

Cashew: Improvement, Production and Processing (2017) *Pages* **105–122**
Editor: **Prof. P.L. Saroj**
Published by: **ASTRAL INTERNATIONAL PVT. LTD., NEW DELHI**

Genetic Resources in India

G.S. Mohana, P.L. Saroj and E. Eradasappa

1. Introduction

Germplasm in any crop is the basic resource for a breeder to develop improved varieties and hybrids. Hence robust germplasm collection representing wide array of variability is essential for any meaningful improvement in terms of yield and quality parameters. In many crops, across the world, considerable efforts towards germplasm collection and conservation have been documented. However, comprehensive evaluation and characterization of germplasm needs greater attention and its subsequent utilization in crop improvement program is very crucial. This would also facilitate further systematic exploration for trait specific germplasm. With this in view, an account of conservation and utilization of cashew genetic resources in India is provided in this chapter.

Cashew (*Anacardium occidentale* L.) belongs to the family *Anacardiaceae* and is a native of Brazil. The family comprises of about 60 genera and 400 species of trees and shrubs with resinous bark, and grows most abundantly in the tropics in both eastern and western hemisphere (Ohler, 1979). Several other important fruits and nuts such as mango (*Mangifera indica* L.), the Pistachio nut (*Pistacia vera* L.) and various speices of *Spondias* such as Otaheitc apple (*S. cytherea*), Hog-Plum (*S. mombin*) and Spanish plum or red mombin (*S. purpurea* L.) belong to this family. According to Baily (1949), the genus *Anacardium* contains eight tropical American species. Parente (1972) names 10 species but Peixoto (1960) names twenty different species several of which had edible peduncles such as *A. nanum*, St. Halaire, a very early bearing small shrub, *A. subterranium* Liais, a small shrub with its trunk almost completely underground containing water reserves; *A. microcarpum* Ducke, a small tree from sandy savannas, *A. spruceanum* Benth. a large tree and the largest species of the genus *A. gigentium* Hanca which grows in Amazon forest. However, as per the latest working list of all plant species, the *Anacardium* genus comprises of 20 species (Table 5.1).

Table 5.1: Species of the *Anacardium* Genus (Anonymous, 2010)

1.	*Anacardium amapaense* J.D.Mitch.
2.	*Anacardium amilcarianum* Machado
3.	*Anacardium brasiliense* Barb.Rodr.
4.	*Anacardium caracolii* Mutis ex Alba
5.	*Anacardium corymbosum* Barb.Rodr.
6.	*Anacardium curatellifolium* A.St.-Hil.
7.	*Anacardium excelsum* (Bertero ex Kunth) Skeels
8.	*Anacardium fruticosum* J.Mitch. and S.A.Mori
9.	*Anacardium giganteum* Hancock ex Engl.
10	*Anacardium humile* A.St.-Hil.
11.	*Anacardium kuhlmannianum* Machado
12.	*Anacardium microsepalum* Loes.
13.	*Anacardium nanum* A.St.-Hil.
14.	*Anacardium negrense* Pires and Fróes
15.	*Anacardium occidentale* L.
16.	*Anacardium othonianum* Rizzini
17.	*Anacardium parvifolium* Ducke
18.	*Anacardium rondonianum* Machado
19.	*Anacardium spruceanum* Benth. ex Engl.
20.	*Anacardium tenuifolium* Ducke

The cultivated species *A. occidentale* L. is andromonoecious, with male and hermophrodite flowers in the same inflorescence and the phenomenon is almost similar in all the species of the genus *Anacardium* (Damodaran, 1977). Within the species *A. occidentale* also, there is a wide variation in colour, size and shape of the apple, as well as in size and shape of the nuts. The time of flushing, flowering varies among different types. There are also differences in leaf size and leaf shape and numerous other characters.

2. Origin and Distribution

Brazil is the original home of cashew. The earliest reports of cashew are from Brazil coming from French, Portuguese and Dutch observers (Johnson, 1973). The French naturalist and monk, Thevet was the first to describe, in 1558, a wild plant extremely common in Brazil, the cashew tree and its fruits. He recounted that cashew apple and its juice were consumed and that the nuts were roasted in fire and the kernels eaten. He provided the first drawing of the cashew showing the local people harvesting cashew fruits and squeezing juice from cashew apples into a large jar (Johnson 1973).

There are indications that local Tupi Indians had used cashew fruits for centuries. They probably played a major role in the dispersion in their temporary migrations towards the coast of north-eastern Brazil where a considerable interspecific variation has been recorded (Ascenso, 1986). The entire cashew fruit with pseudo apple when matures usually floats on the surface of water. This could account in Brazil for coastward dispersal of the species by river draining north and east. Fruit bats might also have been involved in seed movement. Within the Amazon forest, fruit bats were the most important agents of seed dispersal of tree species (Johnson, 1973). From its origin in north-eastern Brazil, cashew spread into South and Central America.

The presence of cashew in other continents is to be attributed to man's intervention. The Portuguese discovered cashew in Brazil and spread first to Mozambique (Africa) and later into India between sixteenth and seventeenth centuries (De Castro, 1994). According to Agnoloni and Giuliani (1977), it arrived in Africa during the second half of the 16th century, first on the east coast and then on the west and lastly in the islands. In Africa, although it can be guessed that the cashew was introduced at an early period by the Portuguese, there are no records which provide specific dates. Dispersal of the cashew in East Africa may in part be due to the elephant, whose fondness for fruits is well known (Johnson, 1973). Attracted by the colour of the false fruit, they swallowed this together with the nut which was too hard to be digested. This was then expelled with their droppings, a natural manure, and trodden far enough into the ground by the animals leaving behind to root and grow, into a seedling first and then a tree. This is how the cashew was spread along the East coast of Africa facing the Indian Ocean. The spread of cashew within the South American continent was gradual and spontaneous.

3. Cashew in India

Cashew was introduced to India by Portugese during 16th century. Molecular studies have shown the possibility of its introduction repeatedly over a period of time but at a single location *i.e.* west coast (Archak *et al.*, 2009). Presently, the cashew plants in wild state as well as in well managed orchards are seen in Maharashtra, Goa, Karnataka and Kerala along the west coast, Tamil Nadu, Andhra Pradesh, Odisha and West Bengal on the east coast. To a limited extent, the crop is also seen growing in Chattisgarh, Gujarath, Assam, Arunachal Pradesh, Meghalaya, Tripura, Manipura, Nagaland and Andaman and Nicobar Islands (Singh, 1998). After the establishment of National Research Centre for Cashew (NRCC) at Puttur, Karnataka in 1986 (Now the Directorate of Cashew Research–DCR), the germplasm collection through vegetatively propagated material started and since then, a coordinated approach was brought in the cashew gemplasm collection by organizing joint survey teams consisting of scientists of NRCC and the centres of All India Coordinated Research Project on Cashew (AICRP on Cashew) of the respective States (Bhaskara Rao and Swamy, 2000).

The germplasm survey and collection were carried out in cashew growing States namely, Karnataka, Kerala, Maharashtra, Goa, Tamil Nadu, Andhra Pradesh, Odisha and West Bengal. The non-traditional areas such as Garo Hills (Meghalaya), Bastar (Chhattisgarh), Gujarath, Dadra and Nagar Haveli and Andaman and Nicobar Islands were also surveyed for germplasm collection. So far, 528 accessions have been collected and conserved in the National Cashew Field Gene Bank (NCFGB) at the Directorate. Similarly, Regional Cashew Gene Banks (RCGBs) have been established at AICRP Centres which are maintaining a total of 1104 accessions (Table 5.2).

Table 5.2: Status of Cashew Germplasm Holding in India

State	*At NCFGB*	*At AICRP centers*	*Total*
Andaman and Nicobar Islands	10	—	10
Andhra Pradesh	103	48	151
Arunachal Pradesh	2	—	2
Assam	3	—	3
Chattisgarh	1	61	62
Goa	45	—	45
Karnataka	128	128	256
Kerala	72	181	253
Maharashtra	45	297	342
Manipur	1	—	1
Meghalaya	11	—	11
Mizoram	1	—	1
Odisha	21	97	118
Tamil Nadu	46	200	246
Tripura	3	—	3
West Bengal	14	92	106
Exotic	22	—	22
Total	528	1104	1632

NCFGB: National Cashew Field Gene Bank; AICRP: All India Coordinated Research Project.

4. Conservation and Utilization of Genetic Resources

The efforts are made at the Directorate of Cashew Research, Puttur, Karnataka for systematic characterisation of cashew germplasm. The collected scion material of the accession is grafted onto a suitable root stock and each grafted accession is then grown in the field gene bank. Recommended agronomic practices are adopted and observations are recorded on 3 selected plants in each accession after 10th year of planting and after obtaining 6 annual harvests for 68 characters following "Cashew Descriptors" (IBPGR 1986; Table 5.3). So far 478 clonal accessions out of 528 accessions have been evaluated and 444 are conserved in a conservation block by planting 4 plants per accessions at a closer spacing of 4 m x 4m.

Table 5.3: Descriptors Used for Characterization of Cashew Germplasm

General Characteristics	*Plant Characteristics*	*Flowering Characteristics*	*Nut Characteristics*	*Apple Characteristics*
Accession number	Age of tree	Season of flowering	Colour of mature nut shell	Mature cashew apple colour
Donor name	Tree habit	Inflorescence shape	Nut shape	Cashew apple shape
Donor identification No.	Internodal length of twig	Flower colour - Length	Nut weight	Size of cashew apple
Scientific name	Leaf shape	Colour of boot leaf	Shape of nut base	Size of cashew apple – Width
Type of maintenance	Tree height	Inflorescence size – Length	Suture of nut	Weight of cashew apple
	Tree spread	Inflorescence size – Width	Flanks of nut	Shape of cashew apple base
	Cracks on the trunk bark	Compactness of inflorescence	Stylar scar on nut	Ridges on cashew apple
	Crotch angle of main branches	Type of inflore-scence branching	Shape of nut apex	Cashew apple apex
	Ease of peeling bark from trees	Sex ratio	Relative position of suture and apex	Grooves on apex of cashew apple
	Extension growth of twigs	Secondary flowering	Shell thickness	Cavity at apex of cashew apple
	Branching pattern	Flowering duration (days)	Uniformity of shell thickness	Skin of cashew apple
	Twig diameter	Flowering intensity (per cent)	Cashew nut dime-nsion – Length	Attachment of nut to apple
	No. of leaves per twig		Cashew nut dime-nsion – Width	
	Colour of young leaves		Cashew nut dime-nsion - Thickness	
	Colour of mature leaves		Apple to nut ratio	
	Odour of leaves		Shelling percentage	
	Leaf margin		Kernel weight	
	Leaf apex shape		Attachment of peel to kernel	
	Leaf size		Kernel dimension – Length	
	Brittleness of leaf		Kernel dimension – Width	
	Angle of leaf petiole		Kernel dimension – Thickness	
	Leaf cross-section		Cotyledonary grooves	
	Cumulative yield per plant			

The information on first set of 56 accessions planted in 1986 has published in the "Catalogue of Minimum Descriptors of Cashew Germplasm Accessions-I", 1997. The second set of 97 accessions planted in 1987 and 1988 have been documented in the "Catalogue of Minimum Descriptors of Cashew Germplasm Accessions-II", 1998. The third set of 102 accessions planted in 1989 and 1990 have been included in the "Catalogue of Minimum Descriptors of Cashew Germplasm Accessions-III, 2000. These are the first efforts made in characterisation of clonal accessions of cashew in the world and so far, 255 accessions have been characterised and catalogued (Swamy *et al.*, 1997, 1999 and 2000). Recently, fourth catalogue containing information on 108 accessions planted during 1991-97 and fifth catalogue containing information on 115 accessions planted during 1198-2003 have been prepared and are in the process of publication. An attempt has been made to develop a core collection of cashew germplasm from the 255 accessions of cashew which have already been characterized. These accessions were grouped into 22 clusters following K-clustering analysis and from each group a 10 per cent sample size was randomly selected as core entries. The germplasm accessions which are unique and have potential (verified/verifiable) attributes of scientific/commercial value are registered in NBPGR, New Delhi (Table 5.4).

Table 5.4: Germplasm Accessions from DCR Registered with NBPGR, New Delhi

Accession	*Traits of Importance*
NRC-59 (VTH 196/18)	Big apple, bold nut size and high shelling percentage,
NRC-111 (Mardol-4)	Mid season flowering, big apple and big nut size
NRC-116 (CNSL Free)	Cashew Nut Shell Liquid (CNSL) free type
NRC-120 (Nairobi)	Early season flowering, bold apple and nut size
NRC-121 (Purple genotype)	Purple stem and leaves and high shelling percentage
NRC-140 (VTH 155 L)	Semi-tall, early season (Nov-Dec) and long flowering duration (120 days)
NRC 142 (VTH 578/1)	Wild relative with genetic diversity
NRC-152 (VTH 713/4)	Wild relative with genetic diversity
NRC-201 (Pl.No.1254)	Uright and compact habit, semi-tall type

DCR: Directorate of Cashew Research, NBPGR: National Bureau of Plant Genetic Resources.

5. Genetic Architecture of Cashew Germplasm

The variability and genetic architecture was assessed deploying 13 important quantitative characters of 478 cashew germplasm accessions evaluated and conserved in National Cashew Field Gene Bank, Directorate of Cashew Research, Puttur (research paper in press). Considerable variability was observed for all charâcters (Table 5.5) and the highest CV was observed for sex ratio followed by cumulative yield per plant and apple weight. The lowest CV was observed for shelling percentage followed by shell thickness. Frequency distribution patterns (Figure 5.2) showed highly positively skewed distribution for characters such as nut weight, sex ratio, apple weight and apple to nut ratio. Genetically, it is evident that decreasing alleles are in excess and dominant for these characters. Whereas tree spread, kernel weight and cumulative yield per plant showed moderately positively skewed distribution indicating decreasing alleles are in slight excess and dominant. Flowering intensity showed moderately negatively skewed distribution indicating the presence of increasing alleles in slight excess and their dominant nature. Tree height, shell thickness, flowering duration, shelling percentage and leaf area showed approximately symmetric distribution indicating increasing and decreasing alleles are in equal proportion and the dominance is ambi-directional.

Figure 5.1: Purple Accession of Cashew.

Significant positive correlations with cumulative yield per plant were observed for tree height, tree spread, sex ratio, flowering duration, apple to nut ratio, shelling percentage and leaf area and significant negative correlation for shell thickness (Table 5.6). The present germplasm collection represents sufficient number of accessions for both quantitative and qualitative characters in desired direction (Tables 5.7 and 5.8). However based on the frequency distribution patterns, it is imperative to collect germplasm with dwarfness, less tree spread, high nut weight, apple weight and high yield.

6. Unique Germplasm Accessions

The germplasm accessions conserved in the NCFGB at ICAR-DCR, Puttur include the diverse types such as high yield, bold nut, semi-tall, compact, Cashew Nut Shell Liquid (CNSL) free, purple pigmented, high shelling percentage, cluster bearing, big apple and early maturity types. Further biochemical profiling of cashew varieties has been accomplished and accession with high mineral composition, neutraceuticals, proteins, starch, lipids, sugars, tannins, phenols and ascorbic acid have been identified. Three wild species namely, *Anacardium pumilum, A. othonianum* and *A. microcarpum* are also conserved. The collection also has seedling accessions of 23 exotic collections of which nine were collected from Brazil, Nairobi, Mtwara, Lindi, Nacala, Mozambique, Ex Tanganyka, Singapore and Australia and 14 from Republic of Panama.

7. Utilisation of Germplasm

Of the 28 cashew varieties and 13 hybrids released in the country, the varieties are *per se* selections made from the germplasm material. These 28 varieties were identified and different centres. About 155 germplasm accessions have been effectively utilized for crossing programme at the Directorate of Cashew Reserach, Puttur and several of these were also supplied to other cashew research centres

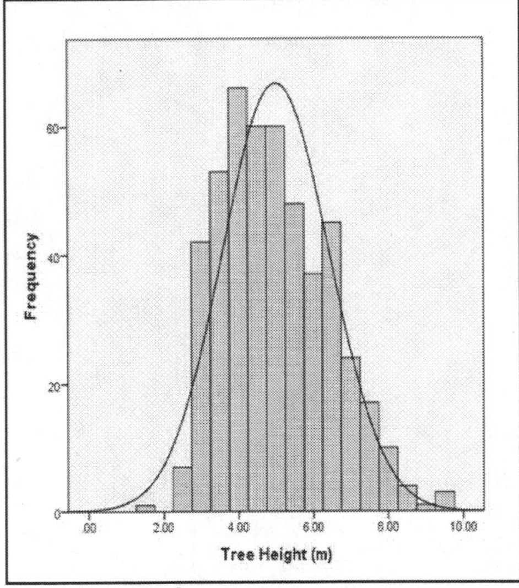

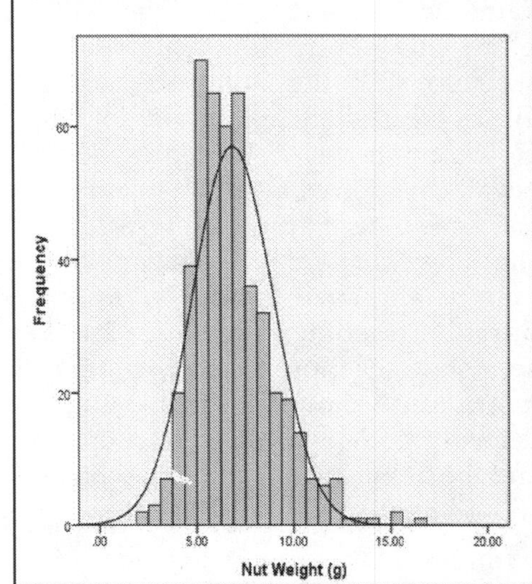

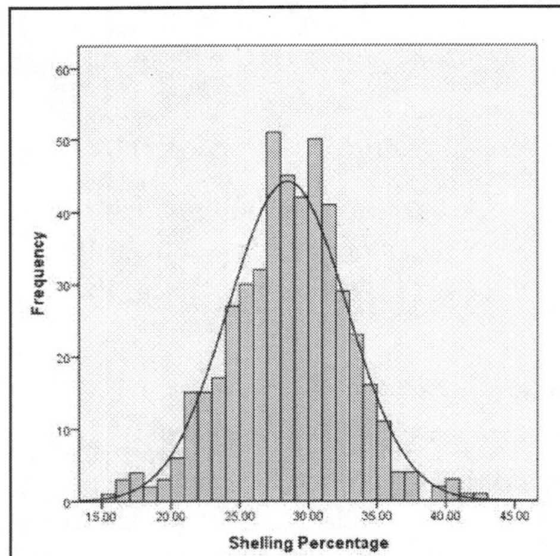

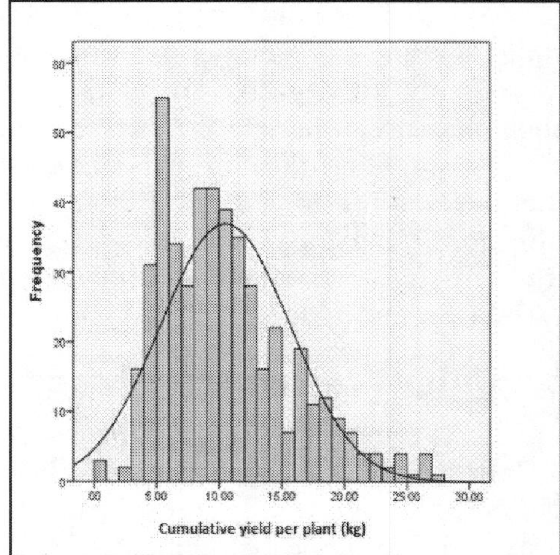

Figure 5.2: Frequency Distribution of some Quantitative Characters.

Figure 5.3: Compactness of Inflorescence.

Table 5.5: Descriptive Statistics of 478 Germplasm Accessions

Character	*Range*	*Minimum*	*Maximum*	*Mean*	*SE of Mean*	*SD*	*CV (per cent)*	*Skewness*	*SE of Skewness*	*Kurtosis*	*SE of Kurtosis*
Tree Height (m)	8.20	1.50	9.70	5.01	0.07	1.43	28.56	0.47	0.11	–0.22	0.22
Tree Spread (m)	9.60	1.50	11.10	6.20	0.07	1.57	25.30	0.72	0.11	0.36	0.22
Nut Weight (g)	14.78	2.00	16.78	6.88	0.10	2.09	30.41	1.02	0.11	1.91	0.22
Sex Ratio	0.29	0.01	0.30	0.09	0.00	0.05	52.21	1.08	0.11	1.76	0.22
Apple Weight (g)	170.00	10.00	180.00	61.82	1.07	23.37	37.81	1.10	0.11	2.45	0.22
Shell Thickness (mm)	3.20	1.50	4.70	3.06	0.02	0.52	16.83	0.36	0.11	0.68	0.22
Flowering Duration (days)	88.00	42.00	130.00	82.54	0.82	18.03	21.85	0.16	0.11	–0.78	0.22
Apple to Nut ratio	26.10	2.00	28.10	9.32	0.14	3.14	33.64	1.10	0.11	2.75	0.22
Shelling Percentage	27.30	15.30	42.60	28.50	0.20	4.32	15.15	–0.10	0.11	0.42	0.22
Kernel Weight (g)	4.00	0.40	4.40	1.94	0.03	0.55	28.30	0.79	0.11	1.49	0.22
Leaf Area (sq.m)	131.30	37.00	168.30	87.33	1.11	24.30	27.83	0.36	0.11	–0.36	0.22
Cumulative Yield per plant (kg)	27.24	0.29	27.53	10.49	0.24	5.16	49.19	0.88	0.11	0.49	0.22
Flowering Intensity (per cent)	82.50	14.30	96.80	65.78	0.73	16.06	24.42	–0.60	0.11	0.00	0.22

Table 5.6: Correlation among different Characters in 478 Germplams of Cahsew

Character	*TH*	*TS*	*NW*	*SR*	*AW*	*ST*	*FD*	*ANR*	*SP*	*KW*	*CYP*	*FI*	*LA*
Tree Height (m)	1												
Tree Spread (m)	0.753**	1											
Nut Weight (g)	0.106*	0.102*	1										
Sex Ratio	0.042	–.054	0.000	1									
Apple Weight (g)	0.176**	0.192**	0.547**	0.041	1								
Shell Thickness (mm)	–.060	–.124**	0.527**	0.104*	0.223**	1							
Flowering Duration (days)	0.001	–.159**	–.094*	0.324**	0.091*	0.089	1						
Apple to Nut ratio	0.141**	0.198**	–.173**	0.030	0.553**	–.169**	0.143**	1					
Shelling Percentage	0.103*	0.252**	–.326**	0.032	–.115*	–.373**	–.155**	0.079	1				
Kernel Weight (g)	0.166**	0.241**	0.804**	0.005	0.511**	0.325**	–.170**	–.099*	0.215**	1			
Cumulative Yield per plant (kg)	0.397**	0.500**	–.088	0.171**	0.031	–.107*	0.090*	0.108*	0.315**	0.075	1		
Flowering Intensity (per cent)	0.022	0.088	0.152**	–.279**	0.177**	0.069	0.062	0.098*	–.051	0.131**	–.067	1	
Leaf Area (sq.m)	0.068	0.133**	0.315**	0.117*	0.196**	0.085	–.132**	–.022	0.054	0.340**	0.162**	0.029	1

** Significant at the 1 per cent; *: Significant 5 per cent.

TH: Tree Height (m); FD: Flowering Duration (days); TS: Tree Spread (m); ANR: Apple to Nut Ratio; NW: Nut Weight (g); SP: Shelling Percentage; SR: Sex Ratio; KW: Kernel Weight (g); AW: Apple Weight (g); CYP: Cumulative Yield per Plant (kg); ST: Shell Thickness (mm); FI: Flowering Intensity; LA: Leaf Area (sq.m).

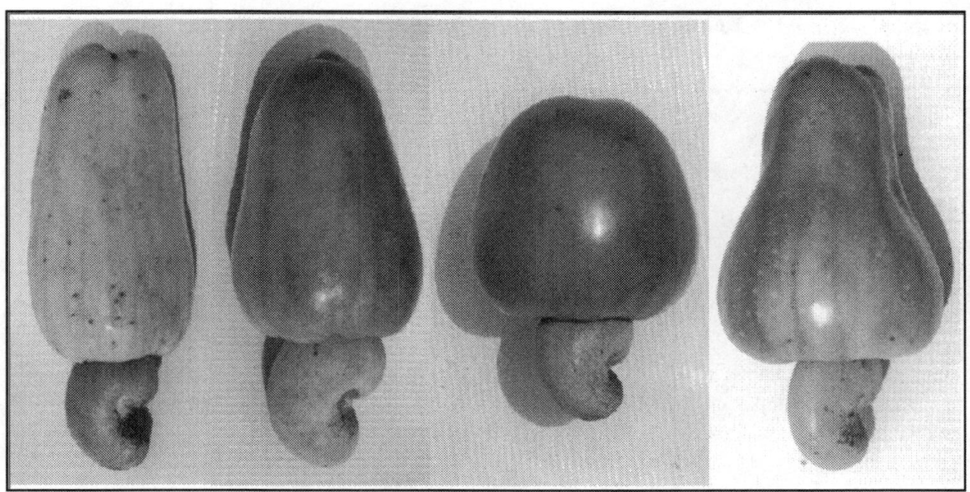

Figure 5.4: Different Shapes of Cashew Apple.

Figure 5.5: Different Tree Habits.

for hybridization programme and other studies. For instance, a total of 75 cashew accessions have been supplied to AICRP on Cashew Centers/ICAR Research Complex for Goa for evaluation and hybridization programme. A total of 107 accessions (65 during 2001 and 42 during 2002 fruiting season) in NCFGB have been utilized as parents under the adhoc research scheme "Network Programme on Hybridization in Cashew" which was in operation during 2000-2003. Leaf samples of 34 varieties and 153 gemrplasm accessions have been supplied to Division of Horticulture, UAS, Bangalore for DNA Finger Printing of varieties and germplasm under the DST funded project. Leaf samples of 142 accessions have also been supplied to NRC DNA Finger Printing, New Delhi.

Table 5.7: Number of Accessions for Desirable Quantitative Characters

Character	*#*	*Top Accessions (Character values in parenthesis)*
Tree Height (< 2.5 m)	5	NRC 153 (1.5), NRC 128(2.3), NRC 131 (2.4),NRC 100(2.5), NRC 239 (2.5)
Tree Spread (<3.0 m)	4	NRC 153 (1.5), NRC 121(2.5), NRC 131(2.7), NRC 190(3.0)
Leaf Area (> 120 sq.m)	50	NRC 270(168.3), NRC 291(159.0), NRC 279(156.0), NRC 277(152.0), NRC 278(150.3)
Nut Weight (> 7.0 g)	190	NRC 269(16.78), NRC 183 (15.4), NRC 161(15.0),NRC 402(14.2), NRC 383(13.4)
Sex Ratio (> 0.13)	74	NRC 60 (0.30), NRC 63 (0.29), NRC 68(0.27), NRC 279 (0.27), NRC 278(0.25)
Weight of Cashew Apple (> 52 g)	306	NRC 385(180.0), NRC 301(169.8),NRC 140(142.8), NRC 164 (141.0), NRC 333(135.0)
Weight of Cashew Apple (>100 g)	29	NRC 385(180.0), NRC 301(169.8),NRC 140(142.8), NRC 164 (141.0), NRC 333(135.0)
Shell Thickness (<2.5 mm)	40	NRC 152(1.5), NRC 153(1.5), NRC 285(1.5),NRC 87(1.7), NRC 281(1.8)
Shell Thickness (>4.0 mm)	16	NRC 160(4.7), NRC 278(4.6), NRC 166 (4.5), NRC 180(4.5), NRC 270(4.5)
Flowering Duration (<60 days)	50	NRC 266(42), NRC 246(47), NRC 238(48), NRC 265(48), NRC 221(50)
Flowering Duration (>90 days)	161	NRC 24(130), NRC 12(128), NRC 11(128), NRC 03(121), NRC 20(121)
Flowering Intensity (> 70 per cent)	205	NRC 126(96.8), NRC 175(95.5), NRC 141(95.2),NRC 148 (95.0), NRC 385 (93.7)
Apple to Nut Ratio (<6.0)	51	NRC 298(2.0), NRC 156(3.2), NRC 255(3.3), NRC 238(3.4), NRC 460(3.6)
Apple to Nut Ratio (>12)	74	NRC 41(28.1), NRC 385(18.9), NRC 370(18.6), NRC 115(18.5), NRC 327(18.1)
Shelling Percentage (>28 per cent)	265	NRC 406(42.6), NRC 343(41.0), NRC 393(40.5),NRC 405(40.5),NRC 327(40.3)
Kernel Weight (>2.5 g)	62	NRC 183(4.4), NRC 323(4.0), NRC 333(4.0)NRC 160(3.7), NRC 409(3.5)
Cumulative yield per plant (> 18 kg) –6 years	48	NRC 352 (Ullal- 1;27.53), NRC 457 (Estamol-1;26.82), NRC 349 (NDR-2-1; 26.21), NRC 356 (Chintamani-1;26.08), NRC 354 (Ullal-3; 25.95), NRC 465 (Banjha Kusum-1;24.96), NRC 346 (Vengurla-4;24.65), NRC 475 (Amritha; 24.42), NRC 452 (Anakkayam-1; 24.06), NRC 434 (Petamalapalli-1; 23.70)

Presently, the hybridization programme is going on at DCR, Puttur and cashew research stations at Bapatla, Bhubaneswar, Vridhachalam, Madakkatahra and Vengurle. The review of performance of varieties and hybrids indicated that in the States where both selections and hybrids were released for cultivation, the performance of hybrids has been better than the selections. Hybrid vigour can easily be exploited in cashew because of the amenability of this crop for vegetative propagation. Recently, a program for development of varieties for cashew apple has been initiated at the Directorate and this is expected to provide impetus for cashew apple utilisation.

Table 5.8: Number of Accessions for Desirable Qualitative Characters

Qualitative Characters	#
Tree Habit (Upright and compact)	62
Tree Habit (Upright and open)	378
Season of flowering (Early- November to-December)	131
Season of flowering (Late- January to February)	45
Cashew Apple Shape (Cylindrical)	137
Secondary Flowering (Nil)	449
Attachment of Nut to Apple (Loose)	167
Attachment of Peel to Kernel (loose)	412

At ICAR-Directorate of Cashew Research, Puttur; it was observed that when tall accessions are crossed with dwarf accessions, the majority of the resulting progenies have tall stature indicating that tall is dominant over dwarf character. In Madakkathara centre, when three parents with prolific bearing and three bold nut type parents were used for hybridisation, It was evident that wherever Brazil-18, an exotic bold nut accession was used in hybridisation, the percentage of high yielding progenies was more compared to other accessions within the country.

8. Diversity Analysis through Molecular Markers

Moderate to high genetic diversity has been observed in germplasm collections in studies with RAPD markers (Anik *et al.*, 2002). Further it was found that among RAPD, ISSR and AFLP markers, AFLP was found to have superior marker efficiency in differentiating germplasm accessions. A total of 172 accessions comprising collections from 9 states of India and exotic sources were fingerprinted using both RAPD and ISSR markers (Thimmappaiah *et al.*, 2009). Based on Shannon's information index and percentage of polymorphic loci, it is evident that high genetic variation was observed in the collections of Karnataka, Kerala and Andhra Pradesh. There was more diversity (96 per cent variation) existed within the groups than between the collections (4 per cent variation) from different states. Among the accessions NRC-432 and NRC-119 were highly divergent and NRC-235 and NRC-216 were highly similar. The cluster analysis performed to create dendrogram distinguished 17 clusters in all (Figure 5.6). Although there was no correspondence between the centre of collections and clusters, there were some exceptions as species from Brazil like *A. othonianum* and *A. pumilum* were found to cluster together in the same sub-group and some sub-clusters were in agreement with morphological clusters. From 17 clusters, 63 accessions were identified to form a 'core collection'.

Similarly genetic diversity and species relationship in 10 diverse types of cashew including three species (*Anacardium pumilum* St. Hillarie, *A. microcarpum* Ducke, *A. othonianum*, three inter-specific hybrids *i.e.* V-5 (*A. occidentale*) x *A. pumilum*, *A. pumilum* X V-5 (*A. occidentale*) and *A. orthonianum* X V-5 (*A. occidentale*)

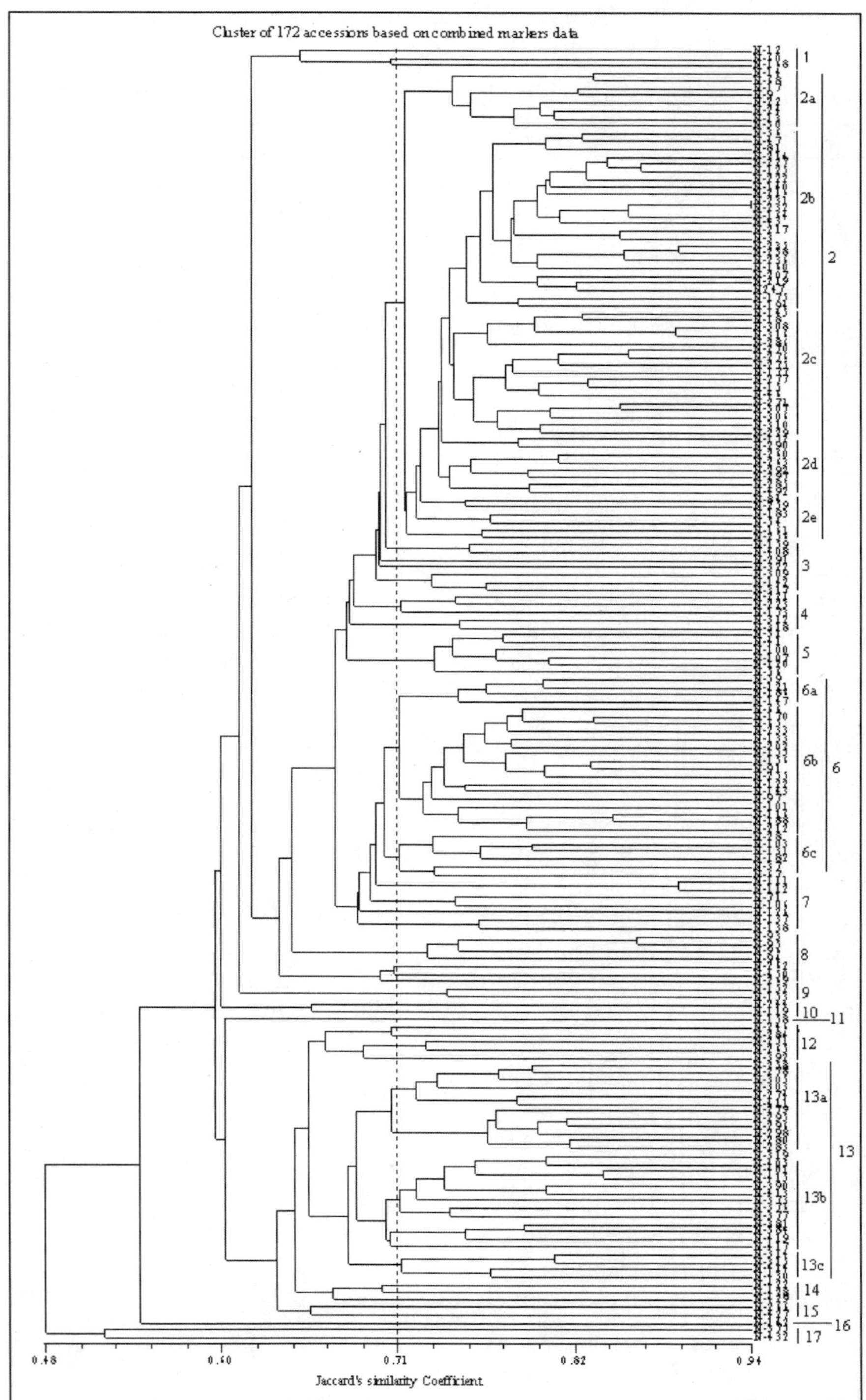

Figure 5.6: Cluster of 172 Germplasm Accessions of Cashew Based on RAPD and ISSR Markers.

and four genotypes of *A.occidentale* was assessed using RAPD, Isozymes and SSR markers. In the cluster analysis three broad groupings were distinguished: In first group *Anacardium pumilum* was found clustering with two of its inter-specific hybrids, in the second group *Anacardium othonianum* clustered with one of its inter-specific hybrid and a dwarf accession Kodippady and while in the third group contained most accessions of *Anacardium occidentale* clustering with *Anacardium microcarpum* thus indicating close affinity between *A.occidentale* and of wild species *A. microcarpum*.

9. Fingerprinting of Varieties

Fingerprinting of 40 varieties released in the country was carried out at ICAR-DCR with RAPD, ISSR and SSR markers. Polymorphic markers generated with a combination of 10 primers each of RAPD and ISSR markers and 15 primer pairs of SSR of cashew were used in the analysis. Marker analysis was carried individually and by combining all the three markers. Based on the combined markers, Jaccard's coefficient of genetic similarity between the different pairs of varieties varied from 0.54 to 0.81 with an average similarity of 0.68 indicated low diversity existing among the varieties studied. Highest similarity was observed between Goa 11/6 and VRI-3 and lowest similarity was observed between Kanaka and V-2. UPGMA dendrogram grouped 40 varieties in to 8 to 10 clusters at 75 per cent similarity (Figure 5.7). Among the varieties, Kanaka, Jhargram and V-6 were highly divergent. Varieties clustered together irrespective of their geographic origin indicating no relation between the clusters and the origin of varieties.

Apart from DNA markers, Isozyme markers which are co-dominant were used for characterization of varieties. Isozyme extraction from young cashew leaves was standardized using Arulsekhar and Parfitt buffer. Extraction and staining protocols for 14 enzymes have been standardized so far. Fingerprinting of 30 varieties of cashew was carried out with Isozyme polymorphism of 10 enzymes. By assaying 10 enzymes, 33 Isozyme bands (loci) were observed (1-4 bands per enzyme) with an average of 3.3 bands per enzyme; of which 23 bands were polymorphic (69.7 per cent) and 10 were monomorphic. Among the different enzymes, Isozymes of Shikimate dehydrogenase were found to be highly informative. Cluster diagram made with Isozyme markers indicated that Ullal-1 and V1 as most divergent. Groupings based on Isozyme and RAPD were not identical nor with any marker type.

10. Future Line of Action

Since there is a moderate amount of diversity available in Indian cashew germplasm due to limited introduction episodes, it is essential to introduce and enhance genetic variability from countries of Central and South America. The introduction of dwarf accessions from Brazil needs and subsequent development of dwarf hybrids needs special mention since dwarf types are very much required for high density planting systems to improve productivity (Table 5.9). Introduction of *A. gigantium* from Surinam which with biggest apple (200 g) will be advantageous, especially in states like Goa where the cashew apple utilization contributes substantially to the economy of the state.

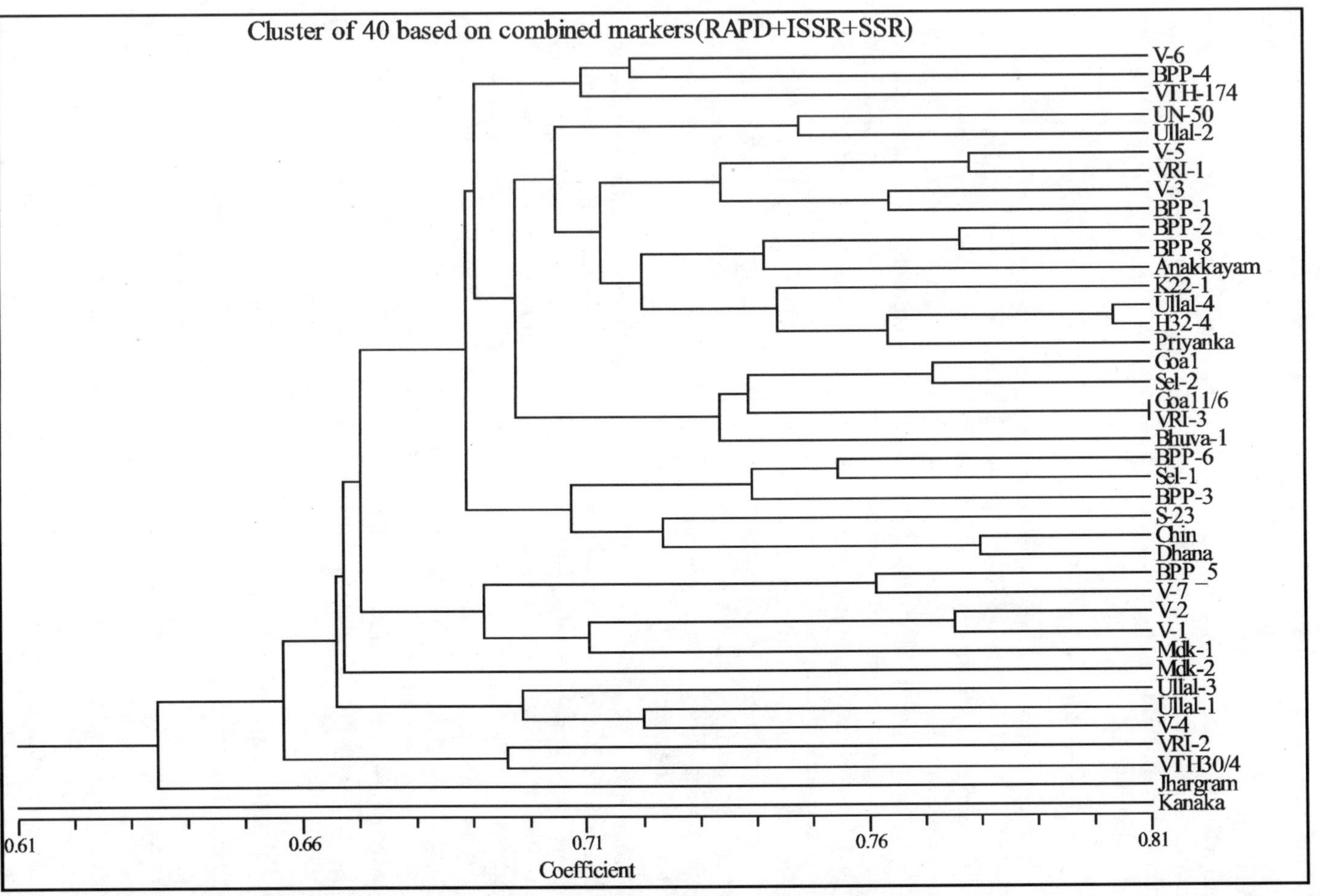

Figure 5.7: Dendrogram of 40 Varieties of Cashew Based on RAPD, ISSR and SSR Markers.

Table 5.9: List of Cashew Germplasm to be Introduced from Outside India

Sl.No.	Germplasm	Source
1	High yielding clones (*Anacardium occidentale* L.)	Federal University of the State of CEARA, Fortaleza, CE, Brazil
2	High yielding dwarf clones (*Anacardium occidentale* L.)	National Centre for Tropical Agro-Industry Research (CNPAT), Fortaleza, Brazil
3	Tea mosquito resistant types (*Anacardium occidentale* L.)	
4	Other diverse types (*Anacardium occidentale* L.)	
5	Related species such as *Anacardium giganteum* (Big sized apple), *A. excelsum, A. rhinocarpus* (Hardwood with smooth, shiny and non-scaly bark), *A. spruceanum* (Hardwood with smooth, shiny and non-scaly bark), *A. microsepalum* (Hypocarp absent), *A. microcarpum* (Slow growing species), *A. humile, A.nanum, A. Corymbosum, A.parvifolium* and *A.fruiticosum*	

One of the main problems in cashew is that all the existing germplasm are susceptible to CSRB which kills the trees. There is a need to screen the allied species (for their suitability as root stocks) which have relatively hard wood and also posses smooth bark. It is suggested that introduction of species like *A. rhinocarpus* and *A. spruceanum* from Brazil which are reported to possess hard wood will be useful for testing their suitability as root stocks. However, more comprehensive exploration is required for target specific traits such as resistance to Tea Mosquito Bug (TMB) and Cashew Stem and Root Borer (CSRB), high yield, dwarf, bold nut with cluster bearing, tolerance to drought, frost, salt and other problematic soils etc.

Generation of core collections, utilization of germplasm accessions in hybridization programs, exploitation of unique types such as CNSL free and rich types are some of the areas that need to be attempted in the ensuing days. Further, germplasm conservation through in vitro/cryopreservation needs attention as conserving in field gene bank requires considerable space and time. However, this requires in first place, the standardization of regeneration protocols for cashew which hitherto has not been successful. Pollen cryopreservation to conserve nuclear genetic diversity also needs to be tried in cashew as it is successful in some horticultural crops like mango, citrus, grape and papaya.

References

Agnoloni, M. and Giuliani, F. (1977). *Cashew cultivation*. Library of Tropical Agriculture, Ministry of Foreign Affairs. Instituto Agronomico Per L' oltremare, Florence. p.168.

Anik Luke Dhanaraj, Bhaskara Rao, E.V.V., Swamy, K.R.M., Bhat, M.G., Theertha Prasad, D. and Suresh N. Sondur, (2002). Using RAPDs to assess the diversity in

Indian cashew (*Anacardium occidentale* L.) germplasm. *The Journal of Horticultural Science and Biotechnology*, 77(1): 41-47.

Anonymous (2010). The Plant List- A working list of all plant species. http: / /www. theplantlist.org/browse/A/Anacardiaceae/Anacardium.

Archak, S., Gaikwad, A. B., Swamy, K. R.M. and Karihaloo, J. L. (2009). Genetic analysis and historical perspective of cashew (*Anacardium occidentale* L.) introduction into India. *Genome*, 52(3): 222-230.

Ascenso, J. C. (1986). Potential of the cashew crop-1. *Agric. Intl.*, 38(11): *324-327.*

Baily, L. H. (1949). Manual of cultivated plants. Mac Milan Co. New York (USA).

Bhaskara Rao, E.V.V. and Swamy, K.R.M. (2000). Cashew Research Scenario in India. Indian Cashew Industry. (Eds H.P. Singh *et al.*) Directorate of Cashewnut and Cocoa Development, Kohi-11. 30-41.

Damodaran, V. K. (1977). F1 population variation in cashew. *Journal of Plantation Crops*, 5(2): 89-91.

De Castro, P. (1994). Summary of the study. In: *The World Cashew Economy* (Eds. A M Delogu and G Haeuster). NOMISMA, L'Inchiostroblu, Bologna, Italy. pp. 11-12.

IBPGR (1986). Cashew Descriptors. International Board for Plant Genetic Resources (Presently, Bioversity International), Rome. p.33.

Johnson, D. (1973). The botany, origin, and spread of the cashew *Anacardium occidentale* L. *J. Plant. Crops*, 1: 1-7.

Singh, H.P. (1998). Keynote address. National Seminar on Cashew Devlopment in India challenges and opportunities. Held at Goa on 1st and 2nd March 1998.

Swamy, K.R.M., Bhaskara Rao, E.V.V. and Bhat, M.G. (1997). Catalogue of Minimum Descriptors of Cashew – Germplasm Accession-I. National Research Centre for Cashew, Puttur. p.40.

Swamy, K.R.M., Bhaskara Rao, E.V.V. and Bhat, M. G. (1999). Catalogue of Minimum Descriptors of Cashew – Germplasm Accession-II. National Research Centre for Cashew, Puttur. pp.54

Swamy, K.R.M., Bhaskara Rao, E.V.V. and Bhat, M.G. (2000). Catalogue of Minimum Descriptors of Cashew – Germplasm Accession-III. National Research Centre for Cashew, Puttur. pp.1-54.

Ohler, J.G. (1979). Cashew. Kohinklijk Instuut voor de Tropeu; Amsterdam. 50-78.

Peixoto, A. (1960). Caju. Servvico de In formacao Agaricola, Ministerio de Agricultura Rio de Joneiro (Brazil). 61p.

Parente, J. I.G. (1972). Cajueiro. Aspectos economicos e agronomicos. Circulara, instituto de Pesquisa Agropecuaria do Nordeste (Brazil) No. 19.

Thimmappaiah., Santhosh, W.G., Shobha, D. and Melwyn, G. S. (2009). Assessment of genetic diversity in cashew germplasm using RAPD and ISSR markers. *Sci. Hortic.*, 120, 411-417.

Cashew: Improvement, Production and Processing (2017) *Pages* **123–145**
Editor: **Prof. P.L. Saroj**
Published by: **ASTRAL INTERNATIONAL PVT. LTD., NEW DELHI**

Crop Improvement and Varietal Wealth

P.L. Saroj, J.D. Adiga and G.S. Mohana

1. Introduction

Cashew (*Anacardium occidentale* L.) is an evergreen tropical tree nut belongs to family Anacardiaceae under the Order Sapinadales and 21 species of the genus *Anacardium* were identified through classical taxonomy (Barros and Crisostomo, 1995) but only *Anacardium occidentale* L. is exploited for commercial cultivation. The *Anacardiaceae* family comprises of 60 to 74 genera and 400 to 600 species (Barros, 1991). Cytology of *Anacardium occidentale* L. has not been studied in detail. The chromosome number is reported in the literature only for *A. occidentale* ranging from 2n=24 (Khosla *et al.*, 1973; Goldblatt, 1984), 2n=30 (Machado, 1944), 2n=40, (Simmonds, 1954; Goldblatt, 1984) to 2n=42 (Darlington and Janaki Ammal, 1945; Khosla *et al.*, 1973; Goldblatt, 1984; Purseglove, 1988). The cultivated species *A. occidentale* L. is andromonoecious, with male and hermophrodite flowers in the same inflorescence and the phenomenon is almost similar in all the species of the genus *Anacardium* (Damodaran *et al.*, 1979).

Cashew is primarily grown in Asia, Africa, Latin America and Australia. Asiatic region mainly includes India, Vietnam, Indonesia Philippines, Malaysia, Thailand and Sri Lanka. African countries producing cashew are Côte d'Ivoire, Nigeria, Tanzania, Mozambique, Kenya, Benin, Guinea-Bissau, Mozambique, Ghana, Senegal and Madagascar. Whereas, Latin American countries producing cashew consist of Brazil, Columbia, Costa Rica, Honduras and Salvador. Brazil is said to be the home of cashew but now distributed over most of the tropical areas of the world from

27° N in southern Florida, to 28° S in South Africa (Frota and Parente, 1995). In the later part of 16th century and to be precise between 1563 and 1578, the Portuguese colonizing in the Goa region introduced cashew into India (Costa, 1578). Kerala served as a dispersal point for South-East Asia as well (Johnson, 1973). India was the first country to realize its importance and adopted as commercial plantation crop. At present cashew is grown in an area of 10.27 lakh hectares, with production of 7.25 lakh metric tons and productivity of 706 kg/ha (2014-15). However, there is a wide gap between production of raw cashew nut and demand of processing industries in India. The low level of productivity is the main cause for this gap between demand and supply. Cashew improvement programme through development of high yielding varieties is one the main strategies to address this issue. Under the scenario of climate change, development of high yielding varieties needs high priority as the risk of production is ever increasing which has association with biotic and abiotic factors. It is, therefore imperative to give due attention in crop improvement efforts for improving production and productivity of cashew.

ICAR-Directorate of Cashew Research, Puttur (Karnataka) is the nodal agency for cashew research in the country. The cashew research was initiated as early as in 1951 with the sanction of an *ad-hoc* scheme for crop improvement on cashew at Ullal station (Karnataka) and regional stations at Kottarakkara (Kerala) and Vengurla (Maharashtra) by Indian Council of Agricultural Research (ICAR). Another station was started at Bapatla (Andra Pradesh) and Jorhat (Assam) in 1955. Subsequently, in 1962, the Kottarakkara station was terminated when the state government started a research station at Anakkayam in Malappuram District (Kerala). In 1970, the All India Coordinated Spices and Cashew Nut Improvement Project was sanctioned by the ICAR under which a co-ordinating centre was established at the Central Plantation Crops Research Institute, Kasaragod with participating centres at Anakkayam subsequently shifted at Madakkathara (Kerala), Vridhachalam (Tamil Nadu), Vengurla (Maharashtra) and Bapatla (Andra Pradesh). The hybridization in cashew was started at Kottarakkara, Kerala (1963) and later continued at Cashew Research Station (CRS), Anakkayam and currently at CRS, Madakkathara. The research efforts were intensified with the establishment of National Research Center for Cashew (NRCC), Puttur in 1986 along with a separate AICRP on Cashew. The NRCC was upgraded to ICAR-Directorate of Cashew Research, Puttur in 2009 and thereafter massive hybridization programme was initiated.

2. Genetic Resource Management

Since, cashew was initially adopted as a plant for soil conservation and afforestation in the coastal regions, thus large scale plantation was done by seedlings. Therefore, considerable variability was created because of cross pollination. As first step of crop improvement programme, the germplasm survey and collection was carried out in cashew growing States namely, Karnataka, Kerala, Maharashtra, Goa, Tamil Nadu, Andhra Pradesh, Odisha and West Bengal. The non-traditional areas such as Garo Hills (Meghalaya), Bastar (Chhattisgarh), Gujarath, Dadra and Nagar Haveli and Andaman and Nicobar Islands were also surveyed for germplasm collection. At present, National Cashew Field Gene Bank (NCFGB) of ICAR-DCR has the largest collection of cashew germplasm (539 accessions) and maintained as clonal progenies. Similarly, Regional Cashew Gene Banks (RCGBs) have been established at AICRP Centres which are maintaining a total of 1104 accessions. Besides *Anacardiun occidentale; A. pumilum (A.humile), A. orthonianum* and *A. microcarpum* have also been collected and conserved in NCFGB. About 155 germplasm accessions have been effectively utilized for crossing programme at the Directorate of Cashew Reserach (DCR), Puttur and also supplied to other cashew research centres for hybridization programme and other studies.

A total of 433 accessions have been assigned with National Collection numbers (IC Nos) and 478 accessions have been characterised as per IPGRI descriptors and five germplasm catalogues have been brought out. The germplasm accessions include the diverse types such as high yield, bold nut, semi-tall, compact, Cashew Nut Shell Liquid (CNSL) free, purple pigmented, high shelling percentage, cluster bearing, big apple and early maturity types. Further biochemical profiling of cashew varieties has been done and accession with high mineral composition, neutraceuticals, proteins, starch, lipids, sugars, tannins, phenols, ascorbic acid, Cashew Nut Shell Liquid (CNSL) and relative tolerance to tea mosquito bug have been identified.

Molecular characterization of germplasm using RAPD was done at AICRP centre at Goa, where 57 genotypes of cashew were differentiated into two broad groups. The first group comprised of 35 and other comprised of 22 genotypes which in turn grouped the genotypes into eight and three sub-clusters. It was inferred that the comparative analysis of clustering pattern based on morphometric and molecular diversity data reflected rather a partial consensus (Desai, 2008). At ICAR-DCR, 172 accessions comprising collection from 9 states of India and exotic sources were fingerprinted using both RAPD and ISSR markers (Thimmappaiah *et al.*, 2009). Archak *et al.* (2009) attempted to construct historical events related to cashew introduction by investigating the level of genetic variation and genetic structure of cashew populations collected from different geographical regions of India. A total of 91 individuals from four populations were analyzed using AFLP markers and morphometric data. AFLP analysis based on 354 polymorphic loci revealed Indian cashew to have low but relatively substantial genetic diversity for an introduced species (HE=0.262 and IS=0.404). Twenty-seven qualitative and quantitative traits also revealed the existence of considerable morphometric variation. Bayesian cluster analysis based on AFLP data did not indicate the existence of definite population differentiation. The results supported the possibility of cashew having been introduced into India repeatedly over a period of time but at a single location (west coast).

At ICAR-DCR fingerprinting of 40 released varieties was also carried out with RAPD, ISSR and SSR markers. Polymorphic markers generated with a combination of 10 primers each of RAPD and ISSR markers and 15 primer pairs of SSR of cashew were used in the analysis. Marker analysis was carried out individually and by combining all the three markers. Based on the combined markers, Jaccard's coefficient of genetic similarity between the different pairs of varieties varied from 0.52 to 0.81 with an average similarity of 0.68 indicated low diversity existing among the varieties studied. Highest similarity was observed between Goa 11/6 and VRI-3 and lowest similarity (52 per cent) was observed between Kanaka and V-2. UPGMA dendrogram grouped 40 varieties in to 8 to 10 clusters at 75 per cent similarity. Among the varieties, Kanaka, Jhargram and V-6 were highly divergent. Varieties clustered together irrespective of their geographic origin indicating no relation between the clusters and the origin of varieties. Archak *et al.* (2003) analyzed diversity in 24 selections and 11 hybrids of cashew using a combination of both RAPD and ISSR markers and observed narrow range of similarity values (low diversity) among the major cashew breeding centres and also between selections and hybrids. Similarly, Samal *et al.* (2003) analyzed genetic relationship in 20 varieties of cashew using RAPD markers and observed no coincidence in the similarities on comparison of morphological with molecular data.

3. Biometric Correlations

Within the species *A. occidentale*, there is a wide variation in colour, size and shape of the apple, as well as in size and shape of the nuts. The time of flushing, flowering varies among different types. There are also differences in leaf size and leaf shape and numerous other characters. Least information is available on the genetics of different characters' association with cashew nut yield, though attempts were made to study the genetic variability and also to correlate yield with ratio of perfect flowers, short and synchronized flowering, branching and flowering intensity, fruits per panicle etc. In general, the variability observed with respect to plant canopy, leaves, flowering period, proportion of male and perfect flowers, percentage of fruit set, size, shape, colour, taste and astringency etc. may be due to the segregation of inherent heterozygosity (Rao and Hassan, 1957; Albuquerque *et al.*, 1960; Cordoba, 1967; Northwood, 1967; Morton, 1970; Murthy and Yadava, 1972).

It has been observed that the yield of nuts from a cashew tree is strongly correlated with flowering shoots per unit area, total canopy area, number of perfect flowers and percentage fruit drop (Pugalendhi *et al.*, 1990). Whereas, Manoj *et al.* (1994a) observed that weight of kernel, mean canopy spread, number of nuts/ panicle, girth of tree, leaf area, duration of flowering and height of tree are the important biometric characters which contribute towards nut yield/tree in cashew. The positive correlation between yield and percentage of perfect flowers was reported by Rao (1974) and Damodaran *et al.* (1979) and also suggested that it is highly desirable to select types with high percentage of perfect flowers for increasing production of cashew nut yield.

Northwood (1966) observed that the trees which produced large number of small size which are unsuitable for cashew trade but smaller nuts gave higher germination percentage (Pugalendhi *et al.*, 1990). Work done at Cashew Research Station, Ullal (Karnataka) indicted that high yielding tress were more likely to produce medium sized nuts and hence medium-sized nuts should be preferred in selection. Medium sized nuts also had higher percentage of germination than either heavy or light nuts (Rao and Hassan, 1956). On the other hand, Albuquerque *et al.* (1960) reported that larger the nuts, larger will be the apples. Similarly, Swamy *et al.* (1993) obtained strong positive correlation between nut weight and apple weight.

Yellow apples are often less astringent, heavier and softer than the red apples and also heavier and larger the apple, the higher will be the juice content (Albuquerque, *et al.*, 1960).

The genetic architecture studies showed highest coefficient of variation (CV) for sex ratio followed by cumulative yield per plant and apple weight, whereas lowest CV was observed for shelling percentage followed by shell thickness. Frequency distribution patterns showed highly positively skewed distribution for characters such as nut weight, sex ratio, apple weight and apple to nut ratio. Genetically, it is evident that decreasing alleles are in excess and dominant for these characters. Whereas tree spread, kernel weight and cumulative yield per plant showed moderately positively skewed distribution indicating decreasing alleles are in slight excess and dominant. Flowering intensity showed moderately negatively skewed distribution indicating the presence of increasing alleles in slight excess and their dominant nature. Tree height, shell thickness, flowering duration, shelling percentage and leaf area showed approximately symmetric distribution indicating increasing and decreasing alleles are in equal proportion and the dominance is ambi-directional. Significant positive correlations with cumulative yield per plant were observed for tree height, tree spread, sex ratio, flowering duration, apple to nut ratio, shelling percentage and leaf area and significant negative correlation for shell thickness. The present germplasm collection represents sufficient number of accessions for both quantitative and qualitative characters but based on the frequency distribution patterns, it is imperative to collect germplasm with dwarfness, less tree spread, high nut weight, apple weight and high yield.

In cashew, seed and seedling characters with yield were studies. There were significant correlations between juvenile plant and mature tree characters, especially with yield. The positive and significant correlations of juvenile plant characters like; height of the plant, girth of the trunk, number of primary and secondary branches, leaf length and spread of the canopy with yield. The path analysis revealed that four characters, namely, seed weight, seed length, length from cotyledon to first leaf and mature tree canopy had direct effect on nut yield. There was significant negative correlation between nut yield and days taken for germination. This association suggests that seedlings that are late in germination should be discarded. Similarly, correlation and regression analysis of bearing cashew trees showed that number of flowering laterals, fruiting intensity and yield per square meter of canopy are the important yield component characters. Both fruiting intensity and yield per square meter of canopy had high positive correlation with estimated yield per tree, thus same can be used for evaluating breeding material for improvement of nut yield per tree.

4. Crop Improvement Approaches

a) Cashew Ideotype

Though, it is very difficult to have all desirable attributes in one genotype but combination of major attributes is essentially required in an ideal plant type. For cashew, following attributes are envisaged:

i) Dwarf tree stature and compact canopy suitable for high density planting.
ii) Bearing in cluster of > 8 nuts/bunch.
iii) Synchronize flowering with high sex-ratio (hermaphrodite: staminate)
iv) Higher nut yield of about 2 t/ha.
v) Medium to bold nuts (7.5 – 10.0 g/nut) with high shelling percentage (>28 per cent) and export grade kernel (W150 to W240).
vi) Big size apple (>75 g), high juice recovery (>75 per cent), TSS (>12° Brix) and high Vitamin C and low astringency.
vii) Both high and low CNSL types for industrial value and reduction of women drudgery, respectively.
viii) Field tolerance to TMB and CSRB.

b) Breeding Methods

In India, genetic base of the cashew is narrow as most of the available germplasm are from the single genus *i.e. Anacardium occidentale* L. Therefore, there is a need to broaden the genetic base of cashew by targeted introgression of new desirable alleles (Cavalcanti *et al.*, 2003). Cashew is a highly cross pollinated crop and hence it is highly heterozygous in nature, which was exploited in identification of superior types. Cashew is propagated through vegetative means but seed propagation is commonly adopted for raising root stocks and evolving hybrids. According to Barros *et al.* (2002) cashew breeding programs comprises of plant introduction, progeny testing, individual selection and hybrid breeding. Emphasis was given earlier to exploit heterosis in cashew improvement as the hybrids performed better than selections. Hence, breeding was focused on generation and selection of superior hybrids and subsequent perpetuation of elite clones by vegetative techniques. Recently molecular Marker Assisted Selection (MAS) is also being employed in cashew breeding programmes to reduce the breeding time. The breeding methods applicable in cashew are as follows;

Introduction

Plant introduction refers 'transfer of a genotype or a group of genotype of crop plants from the place of their cultivation into a new area where they were not grown previously'. Movement of crop varieties from one environment into another within the country is also known as introduction. The main sources of plant introduction are; Centre of its diversity, gene bank, gene centauries, seed companies and farmer' field. It is the easiest way of collecting genetic resources from the country of origin or from the countries or regions where variability already exists. The fruit like; Kiwi (*Actinidia chinensis*)-New Zealand, Simarouba (*Simarouba glauca*)-USA, Hops (*Humulus lupulus*)-USA, Oil palm (*Elaesis guineensis*)-Malayasia, South-East Asia and Africa, Pecan nut (*Carya illinoensis*)-USA, Hazel nut (*Corylus avellana*), Macadamia nut (*Macadamia tetraphylla*)-USA, Japanese Persimon (*Diospyrus* spp.)-USA, Sea-buck-thorn (*Hippophae rhamnoides*)-China and Nepal etc. are the main examples of plant introduction in India. Cashew belongs to Brazil but it has now been introduced to other parts of the world and adapted to environments prevailing in those regions.

By this way other countries are exploiting the benefits of cashew. Besides, related species and other genotypes of cashew can also be introduced particularly from Brazil through NBPGR, New Delhi to enrich genetic diversity.

Selection

Based on the phenotype expressions, the cashew plant is selected for particular trait of interest but there is influence of environmental factors on characters under consideration. In order to reduce the time lag, there is a scope to select the phenotype in the desired environment only. Also there is no control on the male parent and selected plants are pollinated with a random mixture of pollen which ensures cross pollination with related genotypes. The disadvantage of this methodology is the possibility of loss of genotypes with potentialities not expressed in the environment in which the selection is made because of adverse environmental conditions as well as inadequate evaluation. Parent control occurs only on the female side. Lack of control on environmental effect and male parent effect can be minimized when clonal selection is done through progeny test or hybridization followed by individual selection. Most of the cashew varieties have been evolved earlier are by selection process.

Hybridization

Before taking up hybridization activity in cashew, proper knowledge of floral biology including flower anthesis is essential. Cashew is an andromonoecious plant having hermaphrodite or bisexual and staminate or male flowers in the same panicle or inflorescence. Cashew is a cross pollinated crop and pollination occurs in nature through insects (entomophily). The pollen grains of cashew are sticky and heavy; thus least possibility of wind pollination. The hybridization technique in cashew consists of clipping off all the open flowers in a panicle on female parent tree on the day of pollination, opening of bisexual flower bud, removal of anthers from bisexual flower bud, followed by dusting of pollen from selected male parent on stigmatic surface of emasculated bisexual flower and finally covering the pollinated flower (stigma along with style) with butter paper bag or butter paper roll, though covering by butter paper roll is better (NRCC Annual Report: 2003-04). Now, emphasis is given on development of cashew hybrids for desired traits and genetic purity can be maintained by clonal perpetuation.

Hybrid Vigour

The hybridisation work carried out at Anakkayam (Kerala) indicated that whenever an exotic parent was involved, the progeny showed better performance than crosses between local types. Comparison of the performance of hybrids with open pollinated and selfed progenies also showed the superiority of the hybrids over others (Nambiar *et al.*, 1990). Hybridization work carried out at various cashew research centres in India (Puttur, Vengurla, Bapatla, Madakkathara and Bhubaneswar) also confirmed that hybrids performed better than selections. Evidence of hybrid vigour with an increase in the nut yield was also reported by Damodaran (1975). Manoj and George (1993) reported heterosis for nut yield/ tree in 10-year-old F_1 hybrids and the standard heterosis (standard variety-

Madakkathara-1) ranged from 11.0–33.4 per cent, mean nut weight (11.4–54.8 per cent) and weight of kernel (29.8–84.7 per cent).

Mutation Breeding

So far, no success has been obtained in mutation breeding of cashew. Moreover, Occurrence of natural mutants and bud sports in woody trees like cashew is very rare. However, the work initiated at Regional Fruit Research Station, Vengurle (Maharashtra) gave some indications about mutation breeding, where irradiation of cashew bud sticks with one, two and three Kr gamma rays at the Bhaba Atomic Research Centre, Bombay and soft wood grafting led to mortality of all the bud sticks subjected to 3 Kr dose. Whereas those subjected to 1 and 2 Kr dose led to 100 per cent sprouting and variation in phenotypic characters (Anon., 1985). The variations included changes in leaf shape, leaf thickness and leaf venation. On long-term evaluation, most of them were reverted to original characters in subsequent growth. At Cashew Research Station, Madakkathara (Kerala), dwarfism was observed in seedlings produced from the nuts irradiated at 40 Kr to 60 Kr using Cobalt 60. The LD 50 value for cashew nut is between 40-50 Kr. The dwarf seedlings have been planted in the field and are being observed. Beyond 60 Kr, the seed nuts did not germinate (Abdul Salam *et al.*, 1992).

c) Specific Breeding Programmes

Dwarf Tree Stature

Dwarf genotypes are essentially required for obtaining higher productivity through accommodation of more number of plants per unit area, but in India availability of dwarf and fruitful genotypes are meagre. In Brazil, dwarf root stock seedlings have been used for several years. More recently, research workers were able to succeed by using seedlings from *A. microcarpum* Ducke and *A. Pumilum,* the slow growing species. These two root stocks, more markedly the latter, exert a dwarfing effect and induce earlier bearing upon the grafted trees. No symptoms of root stock–scion incompatibility have been observed so far (Ascenso, 1986). In India, neither rootstocks have been identified nor its clonal multiplication has been standardized. Efforts are also on at Directorate of Cashew Research to develop dwarf and compact hybrids with high yield and better nut characters.

Bold Nut Size

Often bold nut size produces export grade kernel (W150-W240) which fetch premium price in the international markets. Therefore, in breeding programme always emphasis was given to develop varieties with bold nut size and high yield potential. More than 70 bold nut types having a mean nut weight of 12-21 g have been collected by Regional Fruit Research Station, Vengurla (Maharashtra). Some of these bold nut types have been used in the hybridisation with existing high yielding varieties like Vengurla-2 and Vengurla-5. Some of the hybrids, *viz.*, H-610, H-613 had given high yield and bold nut size (10 g) (Gunjate and Deshpande, 1994). Under an ICAR ad-hoc scheme "Network programme on hybridization in cashew", hybrids namely H-43, H-66, H-68 (cross combinations of NRCC Selection-2 × Bhutnath-II),

H-125 and H-126 (cross combinations of NRCC Selection-2 × Bhedasi) have been identified as promising and found to be consistently performing good for both annual and cumulative yield. In a field evaluation under replicated trial, the hybrids, namely H-125 and H-126 have produced jumbo nut of 11-12g with moderate yield. The kernel grade is better than W150 and suitable for export purpose.

Quality Kernels

Absolutely, no attempts were made to improve quality of cashew kernel through breeding approaches, though the demand for superior quality kernel is increasing in international market. It is also important to mention that quality of cashew kernel not only refers to size grade, colour, brittleness, free from residues etc. but also rich in nutritional and nutraceutical properties. Data on defatted cashew kernel flour indicated that there is considerable variability for protein ranging from 32-44 per cent, lysine 35-75 µg/mg protein, vitamin-C 144-274 mg/100g and kernel sugar 10-19 per cent. It is desirable to identify the varieties with the protein level of over 35 per cent, lysine level of over 50 µg/mg protein and sugar content of not more than 14 per cent (Bhagavan, 1986).

Drought Tolerance

Cashew often experience moisture stress as it is predominantly grown in rainfed conditions on marginal soils. The yield is drastically influenced by moisture stress and drought if it occurs during reproductive phase of plant life. At a given edaphoclimatic situations, high water use efficiency (WUE) of a genotype has paramount significance in drought tolerance. Moreover, it is difficult to determine the WUE of field grown cashew trees. Therefore, an investigation was carried out by Udaya Kumar *et al.* (2000) with the objective to assess the 'carbon isotope discrimination', which is an indicator of WUE. The result indicated that the considerable genetic variability in WUE exist in cashew which can be exploited in crop improvement programme. Latha (1998) also reported that the genotypes *viz.*, H-1591, M-26/2 and M-44/3 were drought tolerant while K22-1 was drought sensitive. Therefore, screening of germplasm/varieties under moisture stress and drought prone areas may be one of the researchable issue in cashew.

Resistance/Tolerance to Pest and Diseases

Tea mosquito bug, cashew stem and root borer, apple and nut borer and thrips are the major pests of cashew for which resistant/tolerant types need to be developed. Laboratory screening of 27 accessions at DCR to tea mosquito but tolerance has indicated that two accessions/types, namely Bhaskara and VTH 153/1, are relatively tolerant than the susceptible check and some of the test accessions (Nagaraja *et al.*, 1990). There are various diseases of cashew also like, anthracnose, die back, gummosis, powdery mildew etc. However, no resistant source is reported for pest and diseases in cashew.

An attempt of interspecific hybridization between four varieties (Ullal-1, Ullal-3, Vengurla-4 and Bhaskara) of cultivated cashew (*Anacardium occidentale* L.) and two wild species viz, *A. microcarpum* and *A. othonianum* was made during 2010 with the objective of introgressing characters related to biotic and abiotic stress tolerance. The average seed set of all the successful cross combinations put together was 53 per cent with highest (71 per cent) in Ullal-3 x *A. microcarpum* and lowest (25 per cent) in Vengurla-4 x *A. microcarpum*. The varied success could be attributed to extent of cross compatibility between different combinations involving wild species of cashew. The frequency distribution of girth, height, spread (in both directions) showed approximately symmetrical distribution. The number of flowering laterals in East, West, North and South directions showed highly positively skewed distribution indicating the absence of individuals having higher number of flowering laterals. However, non-flowering laterals in East, West and South directions showed approximately symmetric distribution and in North direction, it showed moderately positively skewed distribution. The yield of interspecific progenies (N=189) ranged from 0 - 0.83 kg/tree in the fourth year after planting with a mean yield of 0.34 kg/tree. The frequency distribution pattern of yield per tree revealed that it was approximately symmetrically distributed (Skewness = 0.23). It can be observed that yield in interspecific progenies in very low due to introgression of wild alleles. Further backcrosses may help in improving the yield levels along with tolerance to biotic and abiotic stress in these progenies (Adiga *et al.*, 2015)

5. Varietal Wealth of Cashew

As many as 54 varieties of cashew have been identified countrywide. Among them, 33 are selections and 21 are hybrids. Most of the varieties popular are in medium nut category (6-8 g nut weight). Twenty-four varieties having kernels with export grade (W 180 to W 240). The breeding efforts in cashew so far concentrated on development of varieties for cashew nut only. Now, efforts are initiated towards development of varieties for cashew apple (fruit) and Cashew Nut Shell Liquid (CNSL) which have high industrial application. Recently, attempts are being made to change the genetic architecture of the plant from tall to dwarf, to cater to the needs of high density planting systems. However, to initiate the work on resistant breeding, the source of resistance for major pests of cashew *i.e.* Tea Mosquito Bug (TMB) and Cashew Stem and Root Borer (CSRB) are not available.

The classification of cashew varieties developed so far is as under:

i) Organization-wise Varieties Developed

At present ICAR-Directorate of Cashew Research, Puttur, Karnataka along with its 14 AICRP centers located in 08 State Agricultural Universities and 2 ICAR Institutes are directly working on cashew research and development. The varieties developed by different organizations are given in Table 6.1, which includes all 54 varieties developed so far.

ii) Varieties Based on Flowering and Fruiting Time

Some commercial varieties of cashew were evaluated at ICAR-Directorate of Cashew Research, Puttur. Based on their flowering time, flowering duration and fruiting time under west coast region they were grouped as early, mid and late season varieties (Table 6.2).

Table 6.1: Cashew Varieties Developed by different Organizations

Name of the Organization	*Experimental Station/Farm*	*Varieties Developed*
State Agricultural Universities		
Dr. YSR Horticultural University, Venkataramannagudem, AP (Formerly. Acharya NG Ranga Agricultural University)	AICRP on Cashew Centre: Cashew Research Station, Bapatla	BPP-1, BPP-2, BPP-3, BPP-4, BPP-5, BPP-6, BPP-8
Tamil Nadu Agricultural University, Coimbatore, Tamil Nadu	AICRP on Cashew Centre: Regional Research Station, Vridhachalam	VRI-1, VRI-2, VRI-3, VRI-4, VRI-5
Odisha Univiersity of Agriculture and Technology, Bhubaneswar, Odisha	AICRP on Cashew Centre: Cashew Research Station, Bhubaneswar	Bhubaneswar-1, Jagannath, Balabhadra
Bidhan Chandra Krishi Vishwavidyalaya, Kalyani, West Bengal	AICRP on Cashew Centre: Regional Research Station, Jhargram	Jhargram-1, Bidhan Jhargram-2
Indiara Gandhi Krishi Vishwavidyalaya, Raipur, Chhatishgarh	AICRP on Cashew Centre: SG College of Agriculture and Research Station, Jadgalpur	Indira Kaju -1
Dr. Balasaheb Sawant Konkan Krishi Vidyapeeth, Dapoli, Maharashtra	AICRP on Cashew Centre: Regional Fruit Research Station, Vengurle	Vengurla-1, Vengurla-2, Vengurla-3, Vengurla-4, Vengurla-5, Vengurla-6, Vengurla-7, Vengurla-8, Vengurla-9
Kerala Agricultural University, Thrissur, Kerala	AICRP on Cashew Centre: Cashew Research Station, Madakkathara	Madakkathara-1, Madakkathara-2, K 22-1, Kanaka, Dhana, Priyanka, Amrutha, Sulabha, Damodar, Poornima, Raghav
	Cashew Research Station, Anakkayam	Anakkayam-1, Dharashree, Akshaya, Anagha
University of Agricultural Sciences, Bangalore, Karnataka	AICRP on Cashew Centre: Agricultural Research Station, Chintamani	Chintamani-1, Chintamani-2
	Agricultural Research Station, Ullal (Presently under UAHS, Shimoga)	Ullal-1, Ullal-2, Ullal-3, Ullal-4, UN 50.
ICAR Institutes		
ICAR-Directorate of Cashew Research, Puttur (Karnataka)	Experimental Farm, Puttur	NRCC Selection-1, NRCC Selection-2, Bhaskara
ICAR-Central Coastal Agriculture Research Institute, Goa	AICRP on Cashew Centre: Experimental Farm, Ela, Old Goa	Goa-1, Goa-2

Table 6.2: Cashew Varieties Based on Flowering and Fruiting Season under West Coast Region

Names of the Varieties	*Period of Peak Flowering*	*Flowering Duration (months)*	*Peak Fruiting*
	Early-season varieties		
NRCC Sel-2	October – November	2-3	February – March
Vengurla-1	October – November	2	January –February
Madakkathara-1	October – November	2	January –February
Ullal-4	October – November	2	January –February
Kanaka	October – November	2-3	January – March
VRI-3	October – November	4-5	January – May
	Mid-season varieties		
Vengurla-4	November – December	2-3	January – March
Vengurla-3	November – December	3-4	January – March
Vengurla-7	November – December	3-5	February – May
Ullal-3	November – December	3-4	January – April
K-22-1	November – December	3-4	January – April
Priyanka	November – December	3-4	January – April
	Late-season varieties		
Ullal-2	December – January	2-3	February – April
Dhana	December – January	3-5	February – May
Ullal-1	December – February	3-4	March – June
Bhaskara	December – February	3-4	February – May
Madakkathara-2	January – February	2-3	March - May

iii) Varieties Based on Nut Weight

There is no authentic and standard reference on classification of cashew varieties based on nut weight. However, in Nigeria six different sizes of cashew nuts are discernible *i.e.* Extra Small or Madras Nut (<2g), Small Nut (2-5 g), Medium Nut (6-7 g), Large Nut (8-11 g), Extra Large Nut (12-15 g) and Jumbo Nut (> 16 g). These categories were further refined as Extra Small or Madras Nut (<2g), Small Nut (2-3.9 g), Medium Nut (4-7.9 g), Large Nut (8-11.9 g), Extra Large Nut (12-15.9 g) and Jumbo Nut (> 716 g). As per IBPGR Cashew Descriptor (1986), there are only thee category of raw nut *i.e.* Low (< 5 g), Intermediate (5-7 g) and High (>7 g), which is popular as Small, Medium and Bold Nut. In fact, these classifications are based on relative weight of raw nut and availability of genetic diversity in the nature, but commercial cultivation point of view, the nut weight less than 6 g require more processing cost as shelling of cashew is done nut by nut basis. Therefore, varieties with smaller nut size neither growers are interested nor processors are preferring. On the other hand, the accessions available with jumbo nut category are in general low yielding. Also, no variety is available in India for commercial cultivation in jumbo nut category (> 16 g), as per earlier classification. Most of the commercial varieties available in the country are in the category of medium to bold nut size, thereby it is imperative to reclassify the varieties developed in India based on their relative nut weight. The cashew varieties with special reference to Indian conditions are grouped into four major categories as given in Table 6.3.

Table 6.3: Cashew Barieties Based Raw Nut Weight

Category	*Nut Weight (g)*	*Varieties*
Small Nut	<6.0	BPP-1, BPP-2, BPP-3, BPP-4, BPP-5, BPP-6, VRI-2, Bhubaneswar-1, Jhargram-1, Vengurla-2, Vengurla-5, Ullal-2
Medium Nut	6.0 – 8.0	VRI-1, VRI-3, VRI-5, Balabhadra, Vengurla-1, Vengurla-4, Vengurla-6, Goa-1, Madakkathara-1, Madakkathara-2, K 22-1, Kanaka, Amrutha, Poornima, Anakkayam-1, Dharashree, Ullal-1, Ullal-3, Ullal-4, Bhaskara, Chintamani-1, Chintamani-2.
Bold Nut	8.1 – 10.0	BPP-8, Jagannath, Bidhan Jhargram-2, Vengurla-3, Vengurla-7, Goa-2, Sulabha, Damodar, Raghav, Anagha, UN-50, NRCC Selection-2.
Jumbo Nut	>10.0	Priyanka, Akshaya, Indira Kaju -1, H-126, H-130

iv) Varieties with Special Traits

Though, genetic diversity in cashew is narrow but there are some accessions and varieties are available with special traits (Table 6.4). These accessions and varieties are of high importance in future breeding programmes.

Table 6.4: Cashew Accessions/Varieties for Special Traits

Accession/Variety	*Special Traits*
NRC-121	Plant with purple stem, leaves, apple and high shelling percentage.
NRC-301	Big apple type (Av. weight 160 g)
NRC-406	High shelling percentage (42.6 per cent)
NRC-269	High nut weight (16.78 g)
NRC-183	High nut weight (15.4 g) and high kernel weight (4.4 g)
NRC-298	High apple to nut ratio (28.1)
NRC-59	Big apple, bold nut size and high shelling percentage.
NRC-116	Cashew Nut Shell Liquid (CNSL) free type
H-125, H-126, H-130	Jumbo nut size (>11.5 g)
Indira Kaju-1	Moderately tolerant to low temperature and TMB and CSRB
Vengurala-8	For nut and apple both with high juice recovery (85 per cent)
Priyanka, Akshaya, UN-50, Vegurala-7, Bidhan Jhargram-2	Varieties for export grade kernel (W 180)
Bhaskara	Potential to escape from the attack of TMB at low to moderate out break situation.

v) Varieties Recommended for different States

A National Group meeting was convened in 1988 to finalise production strategy of different plantation crops. This group suggested cultivation of several varieties of cashew in different states based on the varietal performance in different regions and their availability. Subsequently several cashew varieties have been released for cultivation in different states. Some varieties developed in one state/region were found to perform well in other states/regions as well. The state-wise varieties recommended are given in Table 6.5.

Table 6.5: Cashew Varieties Suitable for different States

State	*Varieties*
Andhra Pradesh	BPP-4, BPP-6 and BPP-8
Chhattisgarh	BPP-4, BPP-6, BPP-8 and Indira Caju
Goa	Goa-1, Goa-2, Vengurla-4 and Vengurla-7.
Gujarat	Vengurla-4 and Vengurla-7
Karnataka	NRCC Sel-2, Bhaskara, Ullal-I, Ullal-3, Ullal-4, Vengurla-4, Vengurla-7, Madakkathara-2,VRI-3, Chintamani-1 and Chintamani-2.
Kerala	Madakkathara-1, Madakkathara-2, K-22-1, Kanaka, Dhana, Priyanka, Amrutha and VRI-3.
Maharashtra	Vengurla-I, Vengurla-3, Vengurla-4, Vengurla-7 and Vengurla-8.
Odisha	Bhubaneswar-1, Jagannath, Balabhadra, BPP-8 and Dhana.
Tamil Nadu	VRI-1, VRI-3 and VRI (Cw) H-1
West Bengal	Jhargram-1, Bidhan Jhargram-2 and BPP-8
North Eastern States	Ullal-3, Ullal-4, NRCC Sel. 2, Vengurla-1, BPP-8, Jhargram-1

vi) Main Features of Cashew Varieties

The cashew varieties identified by different organizations, their year of release and mode of development, parent, nut size, shelling percentage, yield, kernel grade, apple size and colour etc. are given in Table 6.6.

vii) Varieties Based on Kernel Grade

In case of cashew, unit price is based on the grade of kernels. Standard specifications for Indian cashew kernels for export have been laid down by the Government of India under the Quality Control and Inspection Act 1963. Cashew kernels are classified into as many as 33 grades of whole, pieces, splits and butts depending upon shape, size, colour, quality *etc*. However, there are only 23 grades commonly used in trade. In classification of cashew kernels, three alphabets *i.e.*, W, S and B for whole, split and broken kernels respectively are often used. For white whole kernels, based on their numbers per pound and quality, different trade symbols like; King of Cashew, Super Premium, Premium, Jumbo nut, Regular *etc.* are popular. In the market, there is a high demand for white whole kernels used primarily as snacks while in confectionary, splits and broken pieces are in good demand. The cashew varieties developed in India are classified based on their whole kernels grade as given in Table 6.7.

Table 6.6: Salient Features of Main Cashew Varieties

State	*Variety*	*Year of Release*	*Description*
Andhra Pradesh	**BPP 6**	1980	A selection of Tree No.56 from germplasm collection. The mean bisexual flowers are 8 per cent and mean yield of nuts is about 10.5kg/tree. The nut weight is 5.2g with shelling percentage 24 and kernel protein 20.3 per cent. Juice content of apple is very high (74 per cent).
	BPP 8	1993	A hybrid (H2/16) between Tree No.1 x Tree No.39. Performing well in Andhra Pradesh, Odisha and West Bengal. The variety has mean yield of 14 kg/tree with better nut size (8.2g) and shelling percentage (29 per cent) Kernel grade is W 210.
Tamil Nadu	**VRI-3**	1991	A selection from seedling progeny of a high yielding tree collected from Edayanchavadi village in South Arcot District of Tamil Nadu. It has 12.1 per cent perfect flowers. The average yield is 10 kg/tree. The nut weight is 7.2g and shelling percentage of 29.1. The kernel grade is W 210. This variety is picking up fast among farmers of not only Tamil Nadu but also of other states.
	VRI-5	2007	A hybrid between M 26/2 (VRI-3) x M 26/1 with compact canopy and spreading branches. The average nut yield 13.2 kg/tree. The nut weight 7.2g nut and shelling percentage of 30.5 per cent. The kernel grade is W 210. The apple colour is pink with yellow tinge and the shape is round and the apple weight is ranging from 50.0 to 53.5 g. This variety is recommended for all the cashew growing districts of Tamil Nadu.
Odisha	**Bhubaneswar-1**	1989	A selection from seedling progeny of Vengurla 36/3, collected from RFRS, Vengurla. Flowering season is from January to March with medium duration of 70 days. It has cluster bearing habit with about 12 fruits/bunch. Average nut yield of 10 kg/tree with 4.6g nut weight. The shelling percentage is 32.0 with kernel grade of W 320. It has been found suitable for cultivation in the sandy and laterite soils of the East Coast.
	Jagannath	2008	A mid-season flowering (Jan-March) variety having bold nuts with 8.6 g nut weight. The average nut yield of 10.5 kg/tree (2.1 t/ha) with high shelling percentage (32.5 per cent). Kernel grade W-180.
	Balabhadra	2008	An early flowering (Dec-Feb) variety having nut weight of 7.4 g. Average nut yield of 10.0 kg/tree (2.0 t/ha) with high shelling percentage (30.0 per cent). Kernel grade W-210.
West Bengal	**Jhargram-1**	1989	A selection from T.No.16, collected from Bapatla with medium compact canopy and intensive branching habit. On an average, 6 fruits/bunch and yield of 8.5 kg/tree with nut weight of 5g. Shelling percentage is 30 and kernel grade is W 320.

Contd...

Table 6.6–*Contd...*

State	*Variety*	*Year of Release*	*Description*
	Bidhan Jhargram-2	2013	A selection from seedling population of H-2/15. The variety has mid-season flowering habit with 3-4 fruits/ panicle. Apple is golden yellow of 63g weight and juice content of 68.9 per cent. The average nut weight is 9.2g with a kernel weight of 2.85g and high shelling (32 per cent). The kernel grade is W 180. The variety can yield 13.5 kg/tree in 7th harvest.
Chhattisgarh	**Indira Kaju-1**	2010	A clonal selection from natural pollinated seedling and moderately cold tolerant. Nut weight is 10.50g. It has field tolerance to tea mosquito bug and cashew stem and root borer. Shelling percent is 28.65 . Average yield is 15.53 kg/tree. This is recommended for Bastar Plateau Zone, Jashpur and Raigarh (Zone VII of Indian Agroclimatic Zone).
Maharashtra	**Vengurla-1**	1974	A selection from the germplasm (Ansur-1), collected from Ansur village in Vengurla Tehsil of Maharashtra. The average percentage of perfect flowers is about 8 per cent. The average yield is 19 kg/tree. The nut weight is 6.2g with the shelling percentage of 31. It is early flowering variety. Colour of the apple is yellow. Kernel grade is W 240.
	Vengurla-4	1981	A hybrid between Midnapore Red x Vetore 56. It is a cluster bearing type with percentage of perfect flowers of 35 and fruit set of 6 per cent. The yield of nuts is 17.2 kg/tree. The nut weight is 7.7g and shelling percentage is 31. The colour of the apple is red. Kernel grade is 210.
	Vengurla-6	1991	A hybrid between Vetore 56 x Ansur-1. The variety has average yield of 13.8 kg/tree with nut weight of 8g. The kernel grade is W 210. This variety is popular among farmers.
	Vengurla-7	1997	A hybrid between Vengurla-3 x M 10/4 (VRI-1). The percentage of perfect flowers is very high (40 per cent). Average yield is 18.5 kg/tree with nut weight of 10g and shelling percentage of 30.5. Kernel grade is W 180. The colour of apple is yellow with apple weight of 60g and with juice content of 75 per cent. Average weight of kernel is 2.9g. Variety is recommended for Konkan region of Maharashtra and adjoining cashew growing regions of Goa and Karnataka
	Vengurla-8	2001	A hybrid between Vengurla - 4 x M10/4. It yields 15-20 kg/tree. It has medium bold nuts and red apples with juice recovery is 85 per cent.
	Vengurla-9	2015	A hybrid between Vengurla 4 x VRI 1. The average nut weight 8.9 g, shelling percentage-29.21, average yield-7.24 kg/tree. Kernel weght-2.2 g of W 210 grade. Apple colour is reddish yellow, apple weight-69.71 g, TSS-13 ºB and juice content-76 per cent. Recommended for cultivation in Konkan region.

Contd...

Table 6.6–*Contd...*

State	Variety	Year of Release	Description
Goa	**Goa-1**	1999	A selection from accession Balli-2 which is originated from a tree located in Balli village of Quepem taluk of Goa. The average yield is 7.0 kg/tree with nut weight of 7.6 g (range: 7.3 - 7.9 g) and the shelling percentage of 30.0 (range: 28.9 - 31.0 per cent). Kernel weight is 2.2 g. The kernel grade is W 210. Apple colour is yellow and with average weight of 66.7 g and with average juice content of 68.0 per cent. It is recommended to the state of Goa.
	Goa-2	2007	A selection from Ela village of Tiswadi taluk of North Goa District. The average yield is 5.5 kg/tree with nut weight of 9.4 g (range: 9.2 – 9.6 g) and the shelling percentage of 29.25. Kernel weight is 2.3 g. The kernel grade is W 210. Apple colour is yellowish orange with cylindrical shape and with average weight of 105 g. Juice content ranges from 68.0-72.0 per cent. Variety is recommended to the state of Goa.
Kerala	**Madakkathara-1**	1987	A selection from seedling progeny of Tree No. 39 of Bapatla. The flowering season is from November to January. The mean yield is 13.8 kg/tree with average nut weight of 6.2g. Shelling percentage is 26.8. The kernel weight is 1.6g and kernel grade is W 280. Apple colour is yellow with a weight of 52g. Reducing sugar content is 10.5 per cent.
	Madakkathara-2	1987	A selection from germplasm collected from Neduvellur in Kerala maintained at CRS, Anakkayam. The mean yield is 17 kg/tree and nut weight of 7.3 g with shelling percentage of 26.2 per cent. Kernel weight is 2g with kernel grade W 240. Apple colour is red and 63.3g weight. Reducing sugar content is 7.8 per cent.
	K-22-1	1987	A selection from clonal progeny of Kottarakkara-22 (Layer 23) maintained at CRS, Kottarakkara. Mean yield of 13.2 kg/tree and nut weight is 6.2g with shelling percentage of 26.5. The kernel weight is 1.6g with kernel count of W 280. The apple colour is red and 74g weight. Reducing sugar is 7.2 per cent.
	Kanaka	1993	A hybrid between BLA 139-1 X H 3-13 released from CRS, Madakkathara. An early variety with average yield of 19 kg/tree and a mean nut weight of 6.8g. Shelling percentage is 31 per cent. Kernel weight is 2.1g and quality of kernels is W 210. Colour of apple is yellow.
	Dhana	1993	A hybrid between ALGD-1 x K 30-1 released from CRS, Madakkathara. It has cluster bearing habit. The mean yield is 17.5 kg/tree with a shelling percentage of 28. Kernel weight is 2.2g conforming to export grade of W 210. Apple colour is yellow.
	Priyanka	1995	A hybrid between BLA 139-1 x K 30-1 with jumbo nut size released from CRS, Madakkathara. The yield of nuts is 16.9 kg/tree. The nut weight is 10.8g with kernel weight of 2.87g. Shelling percentage is 26.5 and kernel grade is W 180. Colour of apple is yellowish red. Apple weight is 135g with 57.4 per cent of juice.

Contd...

Table 6.6–*Contd...*

State	*Variety*	*Year of Release*	*Description*
	Amrutha	1999	A hybrid between BLA 139-1 x H 3-13 released from CRS, Madakkathara. It has yield potential of 18.4 Kg/tree with nut weight of 7.2 g. Shelling percentage is 31.6 and with kernel weight of 2.2 g and kernel grade W 210. Colour of apple is yellow and apple weight is 76.0 g. Apple has 57.4 per cent juice content. Recommended to the state of Kerala.
	Sulabha	1996	A selection released with compact canopy and intensive branching. It is bold nut type with 9.8 g nut weight. The tree yields 21.9 kg of nuts with high shelling percentage (29.4 per cent). The kernel weight is 2.88 g and grade is W 210. It bears light orange apples.
	Damodar	2002	A hybrid between BLA 139-1 X H 3-13. It has yield potential of 13.7 kg/tree with nut weight of 8.2 g. Shelling percentage is 27.3. The kernel weight is 2.0 g and kernel grade is W 240. It bears yellow red apples.
	Poornima	2006	A hybrid between BLA 139-1 X K -30-1. It has yield potential of 14.1 kg/tree with nut weight of 7.8 g. Shelling percentage is 31.0. The kernel weight is 2.6 g and kernel grade is W 210. It bears yellow apples.
	Raghav	2002	A hybrid between ALGD -1 X K -30-1. It has yield potential of 14.7 kg/tree with nut weight of 9.2 g. Shelling percentage is 26.6. The kernel weight is 2.27 g and kernel grade is W 210. It bears yellow apples.
	Anakkayam-1	1985	A selection from the seedling progeny of Tree No 139 of germplasm collection of Agricultural College, Bapatla, Andhra Pradesh. The variety has a short flowering duration and yield is 12 kg/tree. The nut weight is 6g and shelling percentage 28. Colour of apple is pink with average weight of 67.5g. Reducing sugar content is 10 per cent. Kernel grade is W 280.
	Dharasree	1996	A hybrid between T 30 X Brazil-18 with yield potential of 15.0 kg/tree with nut weight of 7.8 g. Shelling percentage is 30.5. The kernel weight is 2.4 g and kernel grade is W 240. It bears yellowish pink apples.
	Akshaya	1998	A hybrid between H-4-7 X K -30-1. It has yield potential of 11.0 kg/tree with nut weight of 11.0 g. Shelling percentage is 28.4. The kernel weight is 3.12 g and kernel grade is W 180. It bears yellow apples.
	Anagha	1998	A hybrid between T 20 X K -30-1. It has yield potential of 13.7 kg/tree with nut weight of 10.0 g. Shelling percentage is 29.0. The kernel weight is 2.9 g and kernel grade is W 180. It bears orange red apples.
Karnataka	**Ullal-1**	1984	A selection from the germplasm collected from Taliparamba in Kerala (8/46 Taliparamba) and released by ARS, Ullal of UAS. The variety has 2-3 per cent of bisexual flowers. The average yield is 16 kg/tree. The duration of harvest is long (about 110 days). The nut weight is 6.7g with shelling percentage of 30.7 per cent. The colour of apple is yellow. Kernel grade is W210.

Contd...

Table 6.6–*Contd...*

State	*Variety*	*Year of Release*	*Description*
	Ullal-3	1993	A selection from 5/37 Manjeri. It is early in flowering (November - January) and fruiting period is very short (50-60 days). The fruiting is from January to March and sometimes starts from last week of December. It is a high yielding variety with average yield of 14.7 kg/tree. The nut weight is 7g with shelling percentage of 30.0 and the kernel grade W 210. The colour of apple is red.
	Ullal-4	1994	A selection from 2/77 Tuni Andhra. The average yield is 9.5 kg/tree. The nut weight is 7.2g with shelling percentage of 31 per cent. Kernel grade W 210. The colour of apple is yellow and apple weight is 75g.
	UN-50	1995	A selection from Nileshwar 2/27 (T.No.25). This is a medium duration variety and average nut yield is 10.5 kg/tree. The nut weight is 9g and shelling percentage is 32.8. More than 85 per cent of kernels are W 180 grade. Apple colour is yellow.
	NRCC Selection-2	1989	A selection from the segregating seedling progeny of 2/9 Dicherla (VTH 40/1) collected from Andhra Pradesh. A mid-season flowering habit (November - January) with a flowering duration of 74 days. The number of fruits per bunch is 3. The average yield is 9 kg/tree. The nut weight is 9.2g. The shelling percentage is 28.6 and kernel grade is W 210. Colour of apple is orange red.
	Bhaskara	2006	Variety with late season flowering habit (Dec-Mar) and flowering duration of 60 days, potential to escape from the attack of the tea mosquito bug (TMB) under low to moderate outbreak situation. The number of fruits per panicle (bunch) ranged from 4-13. The average yield on 13th year was 10.7 kg/tree with highest yield of 19 kg/tree. The nut and kernel weight are 7.4 g and 2.2 g respectively. The shelling percentage is 30.6 and kernel grade is W240. The apple colour is pinkish orange and juice content is 67.5 per cent. This variety is very popular among the farmers of Dakshina Kannada District of Karnataka and also in neighbouring districts of Karnataka and Kerala.
	Chintamani-1	1993	A selection from 8/46 Taliparamba, collected from Taliparamba in Kerala and released from ARS, Chintamani. Its flowering period is from January to April with 2-4 nuts per panicle. The average yield is 7.2 kg/tree. The nut weight is 6.9g with shelling percentage of 31 per cent. The kernel grade is W 210. Recommended for plain region of Karnataka.
	Chintamani-2	2007	A seedling selection from ME 4/4 of ARS, Ullal and released from ARS, Chintamani. The canopy type is compact with intensive branching. Its flowering period is from December to January. The average yield is 12.4 kg/tree. The nut weight is 7.9g with shelling percentage of 30.0. The kernel weight is 2.35 g. The kernel grade is W 210. The colour of the apple is red purple with average weight of 70g. Juice content is 60 per cent. Recommended for plain region of Karnataka.

Table 6.7: Cashew Varieties based on their White Whole Kernel Grade

Grades	*No. of Kernels/ Pound (454 g)*	*No. of Kernels/kg*	*Varieties*
W180	170-180	375-396	Priyanka, Vengurla 7, Akshaya, Anagha, Bidhan Jhargram 2, UN 50, Jagannath
W210	200-10	440-465	BPP 8, VRI 3, VRI 5, Vengurla 4, Vengurla 6, Vengurla 9, Goa 1, Goa 2, Dhana, Kanaka, Amrutha, Sulabha, NRCC Sel. 2, Poornima, Raghava, Ullal 1, Ullal, 3, Ullal 4, Chintamani 1, Chintamani 2, Balabhadra
W240	220-240	485-530	Bhaskara, Vengurla 1, Madhakkathara 2, Damodar, Dharasree
W280	260-280	575-620	Madhakkathara 1, K 22-1, Anakkayam 1
W320	300-320	660-705	Bhubaneswar 1, Jhargram 1

6. Future Line of Action

Crop improvement in cashew was mainly confined to higher nut yield. Recently, efforts are on to breed dwarfing, compact canopy types to suit high density orchard, bold nut and higher yield, to introgress desirable genes from wild species of *Anacardium* to the recommended varieties, evolving varieties for apples, high CNSL content etc. Therefore, following future line of actions are suggested;

- ✰ Broadening of genetic base to address the challenges anticipated in the scenario of climate change.
- ✰ Initiation of work on standardization of *in-vitro* protocols for clonal multiplication and cryo-preservation.
- ✰ Molecular marker assisted selection/molecular breeding to speed up the process of evolving desired varieties
- ✰ Development of genotypes with high productivity under resource poor conditions.
- ✰ Rootstock breeding to address issues related to tree vigour (dwarfing), adaptability to lower soil fertility status, biotic and abiotic stresses.
- ✰ Breeding for resistance/tolerance to major insect pests like Cashew Stem and Root Borer and Tea Mosquito Bug.
- ✰ Breeding for resistance/tolerance to diseases like anthracnose, powdery mildew, panicle drying, gummosis *etc.*
- ✰ Breeding for big cashew apple with higher, shelf life, juice content, TSS, vitamin C, antioxidants *etc.*
- ✰ Breeding for varied Cashew Nut Shell Liquid (CNSL) content.

References

Abdul Salam, M., Pushpalatha, P.B. and Suma, A. (1992). Effect of gamma irradiation on cashew. *Indian Cashew J.*, 20(4): 19-22.

Adiga, J.D., Mohana, G.S. and Eradasappa, E. (2015). Interspecific hybridization in cashew. Proceedings of the National Meet on Distant Hybridization in Horticultural Crop Improvement, 22nd -23rd January 2015, IIHR, Bangalore. p. 269.

Albuquerque, S.D.S., Hassan, M.V. and Shetty, K. R. (1960). Studies on apple characters of cashew. *Mysore Agric. J.*, 35: 2-8.

Anomymous (1985). Annual. Report. Konkan Krishi Vidyapeeth, Dapoli. p. 57.

Archak, S., Gaikwad, A.B., Swamy, K.R.M. and Karihaloo, J. L. (2009). Genetic analysis and historical perspective of cashew (*Anacardium occidentale* L.) introduction to India. Genome, 52(3): 222-230.

Ascenso, J.C. (1986). Potential of the cashew crop-l. *Agric. Intl.*, 38(11): 324-327.

Barros, L. M. (1991). Caracterizacao morfologica e isoenzymatica do cajueiro (*Anacardium occidentale* L.) Tipos comum e anao precoce, por meio de tecnicas multivariadas. PhD. Thesis, USP-ESALQ. p.256.

Barros, L. M. and Crisostomo, J. R. (1995). Melhoramento genético do cajueiro. In: ARAÚJO, J.P.P.; SILVA, V.V. da (Orgs.) *Cajucultura*: modernastécnicas de produção. Fortaleza: Embrapa-CNPAT, pp. 73-93.

Barros, L.M., Cavalcanti, J. J.V., Paiva, J. R., Crisostomo, J. R.,Correa, M. P.F. and Lima, A.C. (2000). Seleção de clones de cajueiro-anão para o plantio comercial no Estado do Ceará. *Pesquisa Agropecuária Brasileira*, 35: 197-2204.

Barros, L.M., Paiva, J.R., Cavalcanti, J.J.V. and Araujo, J.P.P. (2002). Cajueiro. In: Bruckner, C.H. (Ed.) Melhoramento de Frutei-ras Tropicais. Editora UFV, Viçosa, pp. 159-176.

Bhagavan, S. (1986). Annual. Report. National Research Centre for Cashew, Puttur (Karnataka). pp. 24-25.

Cavalcanti, J. J.V., Crisostomo, J. R., Barros, L.M. and Paiva, J. R. (2003). *Heterose em cajueiro anao precoce*. Ciencia e Agrotecnologia, 27: 565-570.

Cardoba, J. A. (1967). El ramio en la alimentacifn animal, *Agricultura Trop.* (Bogota), 23: 390-403.

Costa, C.Da. (1578). Tratado das drogas e medicinas das Indias Orientalis. Junta de Investigacoes do Ultramar, Lisoba (Portugal). 1964.

Damodaran, V.K. (1975). Hybrid vigour in cashew (*Anacardium occidentale* L.). *Agric. Pes. Journal*, Kerala. 13: 195-196.

Damodaran, V.K., Vilaschandhran, T.and Valasakumari, P.K. (1979). Research on Cashew in India. Techn. Bull. Kerala Agriculture University, Trichur (Kerala).

Darlington, C. D. and JanakI Ammal, E. K. (1945). Chromosome Atlas of cultivated plants. Allen and Unwin., London. p. 397.

Deasi, A.R. (2008). Molecular diversity and phenotyping of selected cashew genotypes of Goa and physiological response of cv. Goa 1 to *in-situ* moisture conservation. Ph. D. Thesis submitted to University of Agriculture Sciences, Dharwad (Maharashtra).

Frota, P.C.E. and Parente, J.I.G. (1995). Clima e fenologia. In: J.P.P de Araujo, and V.V. da Silva (org.) Cajucultura: Modernas técnicas de produção. Embrapa-CNPAT, Fortaleza. p. 43-54.

Goldblatt, P (Eds.), (1984). Index to plant chromosome numbers 1979 – 1981. Missouri Bot. Garden.

Gunjate, R.T. and Deshpande, P.B. (1984). In: Advances in Horticulture Vol. 9 – Plantation and Spice Crops Part-I. (Chadha, K.L. and Rethinam, P. Eds). Malhotra Publishing House, New Delhi. pp. 281-292.

IBPGR (1986). Cashew Descriptor. International Board for Plant Genetic Resources (Presently Bioversity International), Rome, pp. 1-33.

Johnson, D. (1973). The Botany, Origin, and Spread of the Cashew *Anacardium occidentale* L. *J. Plant. Crops,* 1: 1-7.

Khosla, P.K., Sareen, T.S. and Mehra, P.N. (1973). Cytological studies on Himalayan Anacardiaceae. *Nucleus,* 16: 205-209.

Latha, A. (1998). Varietal reaction to nutrient and moisture stress in cashew (*Anacardium occidentale* L.). *Ph.D. Thesis,* KAU, Thrissur, Kerala.

Machado, O. (1944). Estudos novos sobre uma planta vehla cajueiro (*Anacardium occidentale* L.). *Rodriguesia,* 8: 19-48.

Manoj, P.S. and George, T.E. (1993). Heterosis in cashew (*Anacardium occidentale* L.). *Cashew,* 7: 7-9.

Manoj, P.S., George, T.E. and Krishnan, S. (1994a). Correlation studies and path coefficient analysis in cashew (*Anacardium occidentale* L.) hybrids. *The Cashew,* 8(2): 10-14.

Morton, J.F. (1970). Prospects for cashew advancement in Colombia and Venezuela. *Proc. Fla. State. Hort. Soc.,* 83: 405-410.

Murthy, K. N. and Yadava, B.B.R. (1972). *Indian J. Agric. Sci.,* 42(10): 960-961.

Nagaraja, K.V., Sundararaju, D., Prasanna Rai, K., and Joy N. John. (1990). In: Annual. Report (1990-91). National Research Centre for Cashew, Puttur (Karnataka). pp. 14-17.

Nalini, P.V., Pushpalatha, P.B. and Chandy, K.C. (1994). Hybrids for nut size. *The Cashew,* 8(4): 23-24.

Nambiar, M.C., Bhaskara Rao, E.V.V. and Thankamma Pillai, P.K. (1990). In: Fruits-Tropical and Sub-tropical (Bose, T.K. and Mitra S.K. eds.). Naya Prokash 206 Bidhan Sarani, Calcutta.

Northwood, P.J. (1966). Some observations on flowering and fruit-setting in cashew. *Trop. Agric.* (Trinidad), 43: 35-42.

Nortwood. P.J. (1967). The effect of specific gravity of seed and the growth and yield of cashew (*Anacardium occidentale* L.). *East Afr. Agric. For. J.,* 33: 159-162.

Pugalendhi, L., Shah, A.H., Manoharan, V. and Manivannan, K. (1990). Studies on seedling vigour in cashew. *South Indian Hort.,* 38(3): 129-132.

Purseglove, J.W. (1988). In: Tropical Crops - Dicotyledons. English Language Book Society, Longman, London. pp. 18-32.

Rao, V.N.M. and Hassan, M.V. (1956). A note on the variations in seed characters of cashew. (*Anacardium occidentale* L.), *Indian J. Agric. Sci.*, 26: 211-216.

Rao, V.N.M. and Hassan, M.V. (1957). Preliminary studies on the floral biology of cashew (*Anacardium occidentale* L.), *Indian J. Agric. Sci.*, 27: 277-288.

Rao, V.N.M. (1974). In: Mimeo. Rept. All India Summer Int. Improvement and Management of Plantn. Crops. Central Plantation Crops Research Insttitute, Kasaragod. pp. 128-134.

Simmonds, N.W. (1954). Chromosome behaviour in some tropical plants. *Heredity*, 8: 139.

Swamy, K.R.M., Bhaskara Rao, E.V.V., Nagaraja, B., and Nayak, M.G. (1993). In: Abstract of Golden Jubliee Symp. Hort. Res. – Changing Scenairo, held at Bangalore from May 24-28, 1993. *Hort. Soc. India*, pp. 448-449.

Thimmappaiah, Sequeira, M.G., Shobha, D. and Anil, S. R. (2009). Assessment of cashew species for molecular diversity. *J. Plantation Crops*, 37(2): 146-51.

Udaya Kumar, M., Janardhana, K.V. and Devendra, R. (2000). In: Annual Report 1999-2000 of the Ad-hoc Scheme "Genetic variability in water use efficiency (WUE) in cashew – identification based on carbon isotope discrimination (CID)". University of Agricultural Sciences, Bangalore.

Section III: Production System Management

Cashew: Improvement, Production and Processing (2017) *Pages* **149–169**
Editor: **Prof. P.L. Saroj**
Published by: **ASTRAL INTERNATIONAL PVT. LTD., NEW DELHI**

Delineation of Cashew Cultivation through GIS Application

R. Rejani and T.R. Rupa

1. Introduction

In India, cultivation of cashew is confined in tropical part of the country along West and East coasts due to its specific agroclimatic requirement. Moreover, we do not have proper data base to delineate cashew cultivation in India. Very little work has been done so far regarding the edaphoclimatic requirement and responses of cashew to water and nutrients. Even today cashew is cultivated mainly on poor soils with little care, owing to poor productivity. The existing area under cashew and barren areas suitable for cashew can be determined using remote sensing images. Tottrup *et al.* (2007) used multi-temporal MODIS 250 m data to map mature forest. Matsuoka *et al.* (2007) classified land cover over East Asia using 250 m MODIS land surface reflectance, MODIS snow cover and OLS human settlement data. Rao *et al.* (2003) used LISS III data of IRS-1D and classified rubber, paddy, mixed forests, coconut, shallow and deep water bodies using supervised classification and maximum likelihood data. Choudhury and Saha (2003) integrated remote sensing and GIS for cropping pattern analysis and optimal land use planning in Haryana. Many studies have been reported by number of workers in different parts of the world using GIS and Remote Sensing for determination of changes in cropping pattern, vegetation, forest cover, crop yield estimation, determination of runoff, soil loss, planning of suitable soil and water conservation structures (IMSD, 1995; Ramakrishnan *et al.*, 2009; Shanwad *et al.*, 2011). No studies have been conducted in this direction for cashew orchards. Studies conducted at Directorate of Cashew

Research (DCR), Puttur on suitability of cashew crop to different soil and climatic conditions determined by using GIS is presented in this chapter (Rejani *et al.*, 2013). The study focused on two objectives *i.e.*, 1) To standardize a methodology for identification and quantification of cashew area using remote sensing images 2) To prepare a soil moisture data base for irrigation scheduling and a nutrient data base for site specific nutrient management in cashew.

Description of Study Area

The study area lies in Puttur Taluk, (Dakshina Kannada District of Karnataka) with an elevation ranged from 40-150 m above the mean sea level and characterized by seasonally wet, hot humid with dry season (January to May) during the flowering and fruiting period of cashew (Figure 7.1). This region receives an average annual rainfall of 3500 mm with 120-140 rainy days and distributed from May to November. The soil is laterite and texturally sandy clay loam in the surface. It is acidic in nature with a pH of 5.25, low to medium in N content (150- 250 kg/ha) and low in P_2O_5 (2 -10 kg/ha) and K_2O (40 - 80 kg/ha) contents.

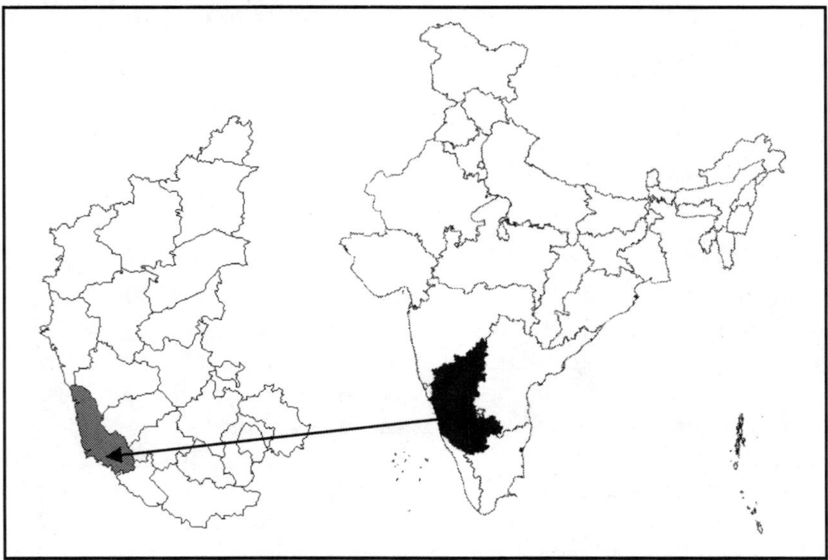

Figure 7.1: Location Map of the Study Area.

Determination of Cashew Area using Remote Sensing

The IRS P6 LISS IV MX data with a spatial resolution of 5.8 m and LISS III data with a spatial resolution of 23.5 m was procured from NRSC, Hyderabad and was classified (unsupervised and supervised) for determining the area under cashew, other vegetations, barren areas *etc.*, using ERDAS Imagine. The error matrix was prepared after collecting the ground truth and the classification efficiency was worked out.

LISS III and LISS IV Image Classification

The major cover types derived using LISS IV data in Figure 7.2 are waterbodies (W), forest (F), bushes (Bu), cashew (C) and barren land (BL). For each of the classified cover types, 10 sample points were established and collected the ground truth (for verifying the actual cover type). Then error matrix was prepared by comparing the classifications with the ground truth. The overall classification accuracy was 73 per cent and for cashew it was 70 per cent at farm level.

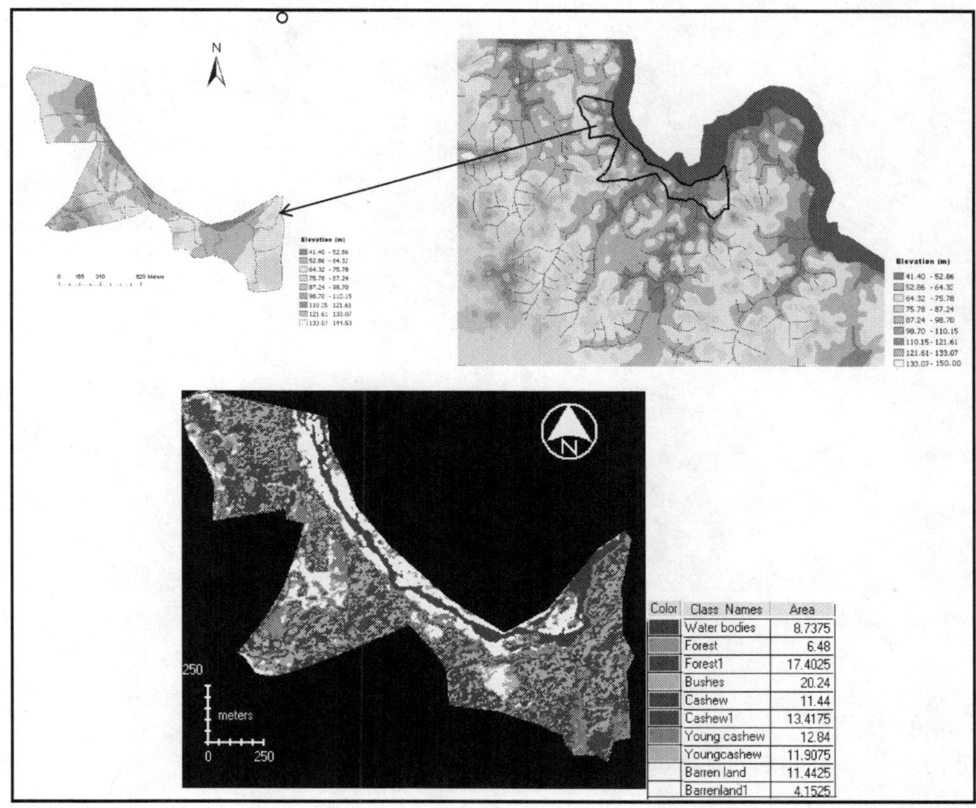

Color	Class Names	Area
	Water bodies	8.7375
	Forest	6.48
	Forest1	17.4025
	Bushes	20.24
	Cashew	11.44
	Cashew1	13.4175
	Young cashew	12.84
	Youngcashew	11.9075
	Barren land	11.4425
	Barrenland1	4.1525

Figure 7.2: LISS IV Image Classification at Farm Level (Area in ha).

The major cover types derived using LISS IV data in Figure 7.3 are waterbodies (W), forest (F), arecanut (A), cashew (C), bushes (Bu), barren land (BL) *etc*. For each of the classified cover types, 12 sample points were established and the overall classification accuracy was computed as 67 percent and for cashew it was 50 percent. The major cover types derived using LISS III data in Figure 7.4 are waterbodies (W), forest (F), Arecanut (A), cashew (C), Bushes (Bu), barren land (BL) *etc*,. For each of the classified cover types, 15 sample points were established and the overall classification accuracy was 73 percent and for cashew it was 80 percent. Among the LISS III and LISS IV images, the LISS III image was found to be suitable for classifying cashew vegetation since the error in classification of data was comparatively less.

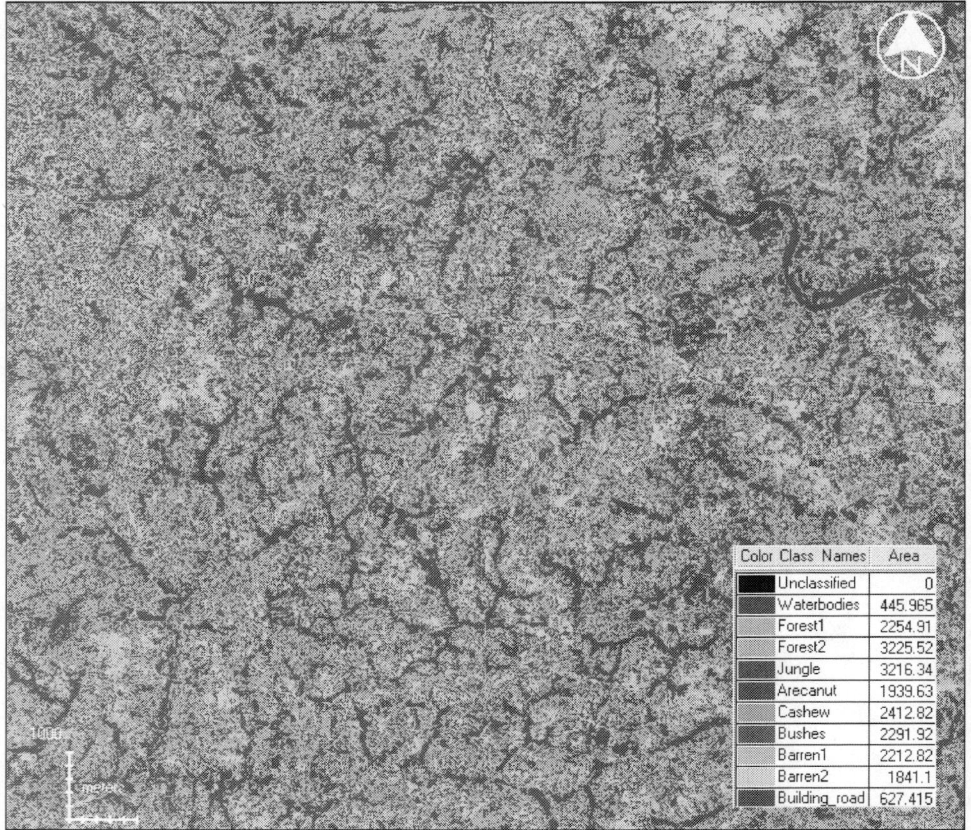

Figure 7.3: LISS IV Image Classification for Larger Area (Area in ha).

Classification of images for different months showed that images pertaining to the month of November were suitable for finding area under cashew since the colour of cashew leaves changes during that period.

GIS Based Spatial Database of Soil Moisture and Nutrient Availability for Soil and Water Management

The soil moisture content at three different depths (0-30, 30-60 and 60-90 cm) in different plots of DCR Farm was determined using TDR Soil Moisture Profiling System (January to March) and spatial database of soil moisture was prepared using Arc GIS for the irrigation scheduling in cashew. During January and February 2008, the soil moisture content in the irrigated plots at two depths ranged from 19.8 to 22.7 per cent and in the unirrigated plots, it ranged from 9 to 13.3 per cent at 0-30 cm and 10 to 15 per cent at 31-90 cm depth on dry basis. In March 2008, the soil samples were saturated after a rainfall of ~ 100 mm. During January and February 2010, the soil moisture ranged from 20 to 21 per cent (dry basis) at 0-30 cm and 16.2 to 18.8 per cent (dry basis) at 31-60 cm depth in the irrigated plots. In control

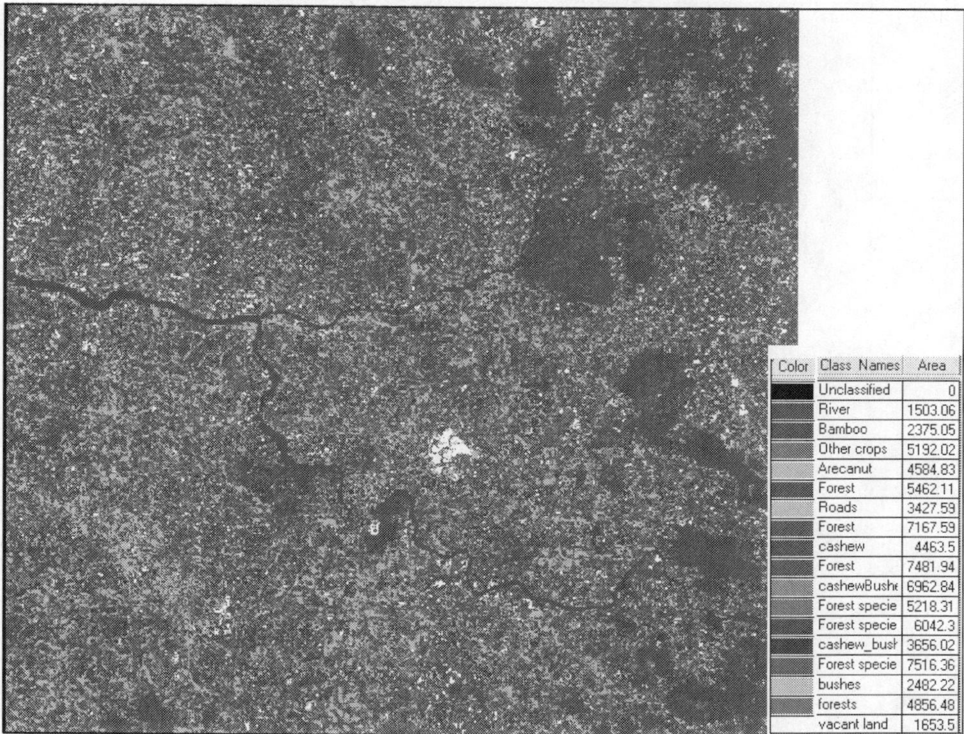

Figure 7.4: LISS III Image Classification for Larger Area (Area in ha).

plot without irrigation, it was 12 per cent (dry basis). During March 2010, the soil moisture content in the unirrigated plots reduced to 10.0 to 11.8 per cent on dry basis (below the available soil moisture range) whereas in irrigated plots, it was 19 to 21 per cent on dry basis. The spatial base of soil moisture content (Figures 7.5 and 7.6) will help in irrigation scheduling for cashew in different plots during the critical period of cashew (January to March). Earlier studies also showed that the irrigation during critical period can substantially increase the yield of cashew.

Soil samples at two different depths (0-30 cm and 31-60 cm) were collected from different plots of DCR Farm, analyzed for nutrients using the standard procedure (Yadukumar *et al.*, 2009) and the spatial database was prepared using Arc GIS. The organic carbon content of the soil at 0 -30 cm depth in DCR Shanthigodu Farm was highest in 11 plots (0.75 to 0.98 per cent) and medium in 5 plots (0.50 to 0.74 per cent) (Figure 7.7). For 31-60 cm depth, it was medium in 8 plots and low in 8 plots. The available nitrogen content was medium in all the plots (320 to 338 kg/ha for 0-30 cm depth and 315 to 323 kg/ha for 31-60 cm depth). Available P_2O_5 content of the soil (0-30 cm depth) was medium in 5 plots (12.5 to 17 kg/ha) and low in 11 plots (3.8 to 12.4 kg/ha) (Figure 7.8). At 31-60 cm depth, it was low in all 16 plots (1.9 to 8.4 kg/ha). Whereas, available K_2O content was high in 2 plots, medium in 4 plots and low in 10 plots.

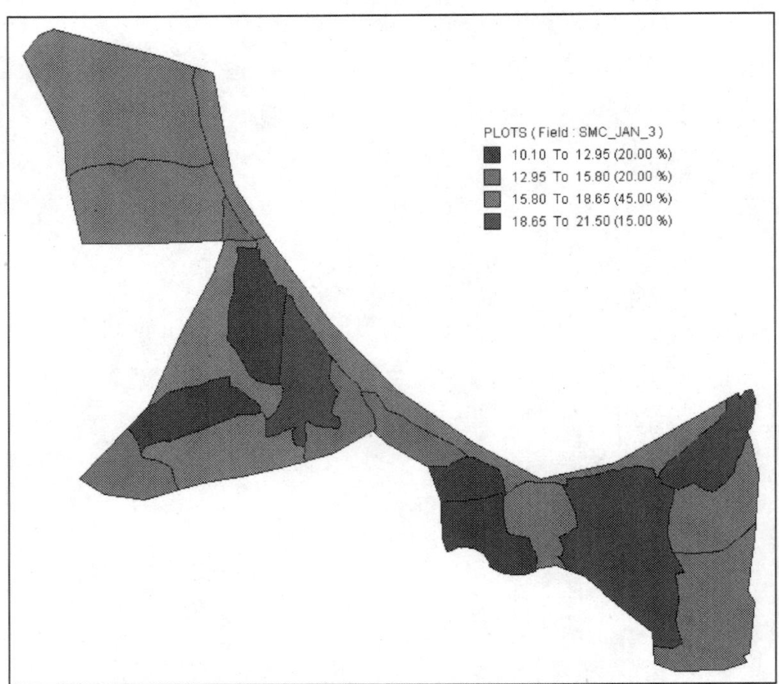

Figure 7.5: Soil Moisture Content (Per cent dry basis) at DCR Farm (0-30 and 31-60 cm depth) (January 2008).

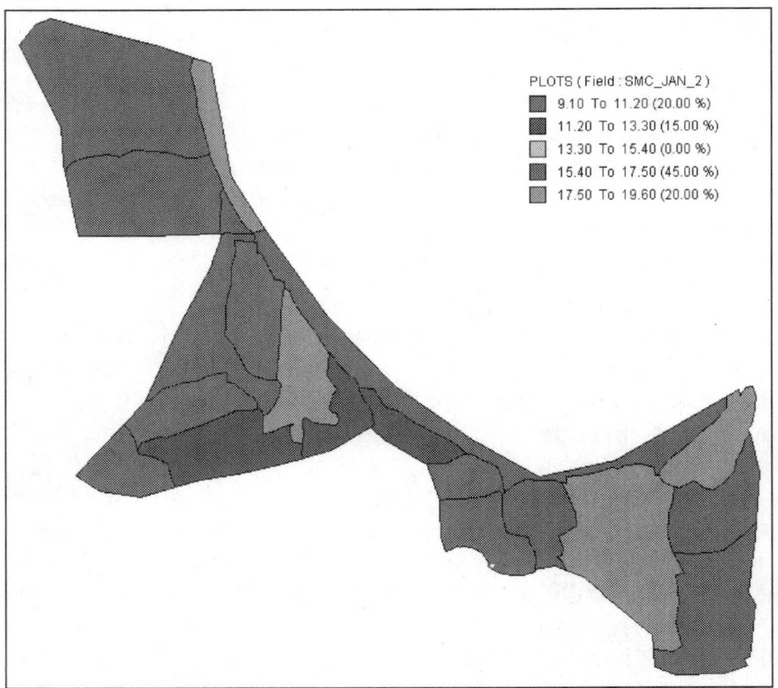

Figure 7.6: Soil Moisture Content (Per cent dry basis) at DCR Farm (0-30 and 31-60 cm depth) (January 2008).

Figure 7.7: Organic Carbon Content (Per cent) and Available P_2O_5 Content (kg/ha) of the Soil (0-30 cm depth).

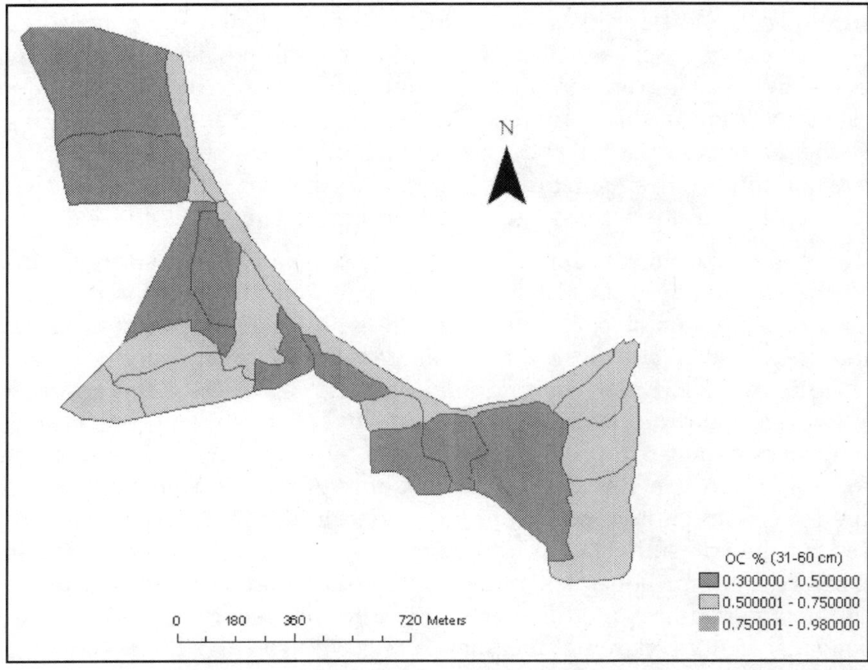

Figure 7.8: Organic Carbon Content (Per cent) and Available P_2O_5 Content (kg/ha) of the Soil (0-30 cm depth).

Hence, the nutrient inputs required for different plots are not uniform. The spatial data base of available N, available P_2O_5 content and available K_2O content with nutrient decision support system (NDSS) developed for cashew will help to estimate the site specific nutrient requirement (Rejani and Yadukumar, 2012). The application of optimal fertilizer and manure to the soil instead of recommended dose will help to reduce the input required, cost of cultivation and finally the ecological hazards.

Land Slope and Soil and Water Conservation

The SOI toposheet (48P5) was digitized; land slope, catchment area and stream orders were determined using GIS. As per the criteria for selection of structures suggested by Shanwad *et al.* (2011), IMSD (1995) and based on local experience, an attempt was made for the identification of suitable soil and water conservation interventions required for the DCR Farm. The study area consists of sandy clay loam soil with high permeability, the runoff and soil loss are quite high due to the undulated topography (steep slopes). Check dams are generally recommended in cashew plantations with slopes less than 15 per cent, upto 3rd order streams with a minimum catchment area of 25 ha and land with runoff potential moderate to high. Farm ponds are recommended for slopes less than 5 per cent, first order streams with a catchment area of 1 to 2 ha and land with runoff potential moderate to high. Soil and water conservation measures are essential for cashew gardens grown on medium as well as steep slopes. The land slope of DCR Farm was determined using GIS. The land slope ranged from low to high (up to 33 per cent) (Figure 7.9). Higher yield of cashew was observed even in steep slopes with soil and water conservation measures. The runoff observed from plots without soil and water conservation was 37 per cent of the mean annual rainfall whereas in plots with soil and water conservation (terracing and modified crescent bunds) and contour planting, it was only 20-27 per cent of mean annual rainfall (Rejani and Yadukumar, 2010). This shows the requirement of soil and water conservation structures for this area. The slope map (Figure 7.9), stream orders (Figure 7.10) and catchment area will help to plan the suitable soil and water conservation structures required for the selected area.

There are two, third order, two, second order and five, first order streams within the selected area (DCR Farm) of 200 acres. No treatment is required for the third order stream since it flows through the selected area only for a shorter distance (less than 100 m) close to the Kumaradhara River. In order to reduce the high velocity of runoff and resulting soil loss, two more check dams are required in the two, second order streams. Of the five, first order streams, one farm pond and two check dams are already constructed in three streams, of which one check dam was a failure structure due to the lack of required catchment area. There is a scope for construction of one more farm pond in the study area which can be utilized for the critical irrigation for cashew (Figure 7.10). In cashew plantations along low to medium slopes, soil and water conservation measures like terracing, trenches (staggered) and modified crescent bunds are recommended, whereas in steep slopes, contour planting of cashew along with soil and water conservation techniques was found suitable.

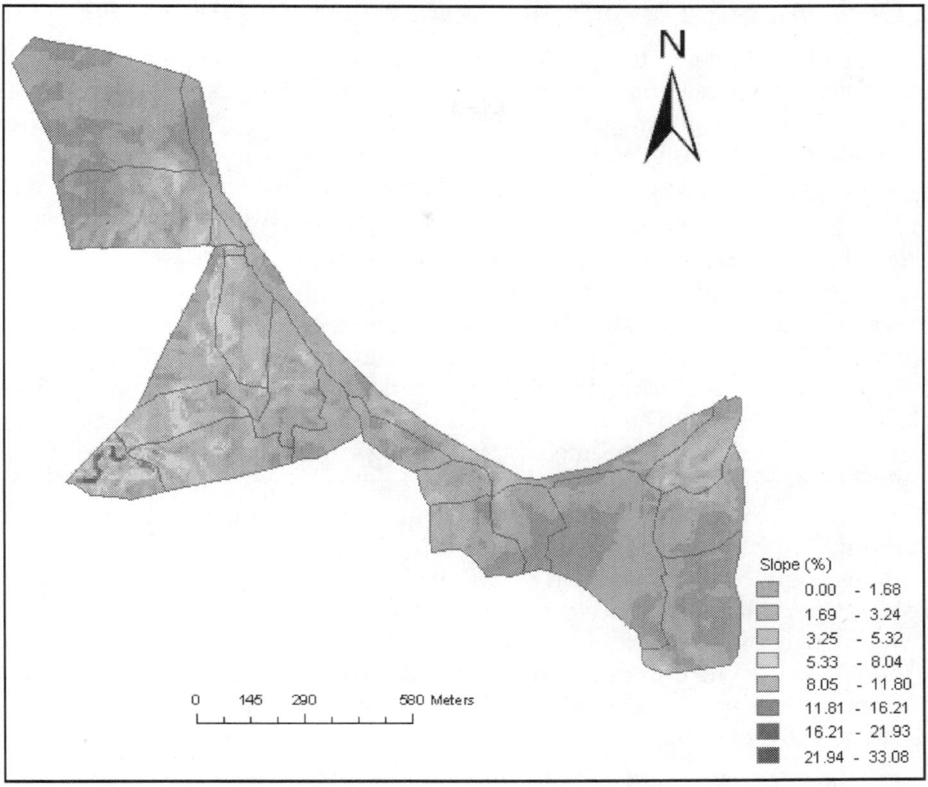

N

Slope (%)
0.00 - 1.68
1.69 - 3.24
3.25 - 5.32
5.33 - 8.04
8.05 - 11.80
11.81 - 16.21
16.21 - 21.93
21.94 - 33.08

0 145 290 580 Meters

Figure 7.9: Land Slope of DCR Farm.

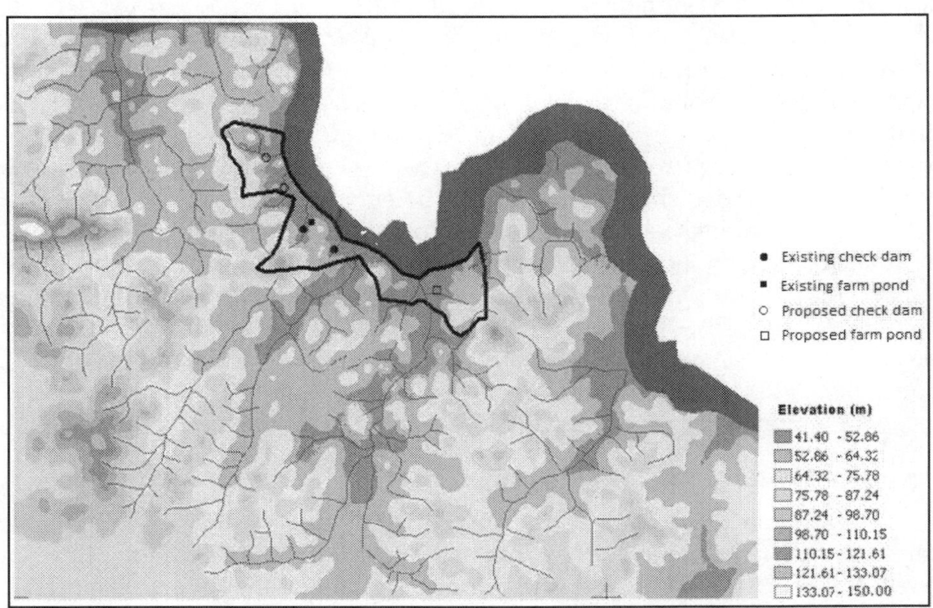

Figure 7.10: Structures Proposed for the DCR Farm.

2. Determination of Suitable Cashew Growing Areas Using GIS

While selecting any new area, land suitability and climate play an important role in achieving sustainable production. Cashew comes up well even in steep slopes with proper soil and water conservation followed by manure application and pest management. Cashew is generally grown as rainfed crop on marginal lands unsuitable for other crops. With climate change there will be variation in yields, flowering, fruit setting, nut development, kernel quality, diseases and water stress.

The mean annual temperature for the country as a whole has risen by 0.56 °C during 1901-2009. Since 1990, the minimum temperature is steadily rising and rate of its rise is slightly more than that of maximum temperature (IMD Annual Climate Summary, 2009). Cashew requires a mild humid tropical climate. The variation in relative humidity during pre-flowering period influences the yield of cashew (Haldankar *et al.*, 2003).The humidity of suitable region ranged from 60-80 per cent (Nair *et al.*, 1979). A well distributed rainfall during growing and pre-flowering phase (from September to November) favours higher productivity (Venugopal and Khader, 1991). Any unusual rains during November and December inordinately delay the reproductive phase of late-season varieties (Rao, 1994; Rao, 1999). Prolonged and unseasonal rainfall accompanied with high velocity wind during flushing and flowering period of cashew adversely affects the yield (Yadukumar *et al.*, 2010). The flowering of cashew requires low minimum surface air temperature ranging from 16 to 20 °C with more dew nights and moderate dew in humid climates. It requires bright sunshine (>9 h/day), moderate dry weather with relative humidity between 40 and 60 per cent in the afternoon.

Cashew responds well to water and nutrients (Rejani and Yadukumar, 2010). Drought conditions drastically reduce the cashew nut production. Earlier researchers reported the criteria/guide lines for selection of sites for cashew (Devanandam, 1983 and Mishra, 1984). In India, cashew is mainly grown along the lateritic soil, red soil and coastal sands along Kerala, Maharashtra, Goa, Karnataka, Tamil Nadu, Andhra Pradesh, Odisha, West Bengal, NEH regions and Gujarat. The most fertile soil lies along the Western Ghats. Mishra (1985) conducted land suitability studies for cashew in Sundargarh district of Odisha. The present study was focused to assess the land suitability of cashew for current and potential regions in India based on soil and climatic conditions using geographic information system (GIS). The influence of climate change on incidence of pests in cashew has been reported by Godse *et al.* (2005) and Rupa *et al.* (2012). Cashew is severely infested by a sucking pest; tea mosquito bug (TMB). The peak period of infestation coincided with minimum relative humidity levels of 52 per cent and a low minimum temperature of 19.4 °C (Rupa *et al.*, 2012). The production loss from TMB alone is 30 per cent and this pest has got potential to cause cent percent loss in yield in several cases.

Area and Productivity of Cashew

The variation in the area, production and productivity of cashew in India and Maharashtra for the period 1993-2010 is shown in Figure 7.11. Among different states, the highest production and productivity of cashew was observed in Maharashtra due to the expansion of area under cashew, adoption of improved

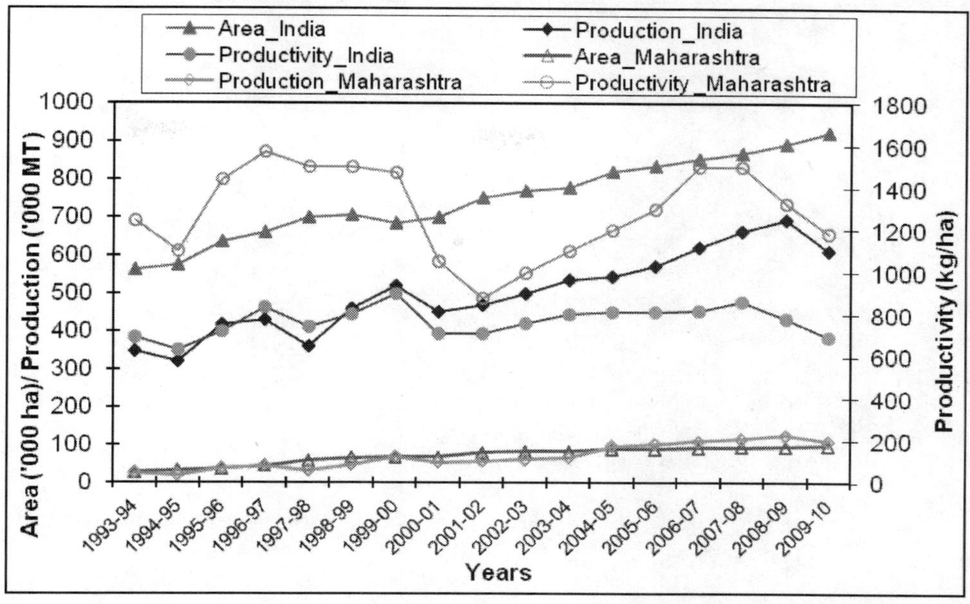

Figure 7.11: Area, Production and Productivity of Cashew in India and Maharashtra.`

varieties and management practices. The decrease in productivity of cashew in some years is mainly due to the change in weather parameters and pest infestation.

Suitability Analysis for Cashew

In the present study, the spatial data base of soil, surface elevation, mean annual rainfall, mean annual temperature, mean minimum temperature and mean maximum temperature at country level was prepared using Arc GIS for the crop suitability studies. The data source for soil, mean annual rainfall and mean annual temperature was obtained from Mandal *et al.* (1999) and for other climatic parameters data IMD was used. Cashew productivity data (DCCD, 2012) and cashew survey data were also used. Cashew area and productivity distribution map was prepared by digitization of the cashew growing locations obtained from the survey and interpolation of the productivity data (Figure 7.12). This map was overlayed with the soil, surface elevation, means annual rainfall, means annual temperature, means minimum temperature and mean maximum temperature maps and derived the suitability criteria pertaining to each parameter for high, medium and low productivity regions.

The results obtained for each parameter was verified using the criteria for suitability of cashew based on field studies reported by Nair (1979); Devenandam (1983) and Mishra (1984) (Table 7.1). The final suitability map was arrived by applying the suitability criteria derived for all the selected parameters after intercepting the different layers in Arc GIS. In this study, an attempt was also made to study the suitability of cashew crop with increase/decrease in minimum temperature, which can arise due to climate change.

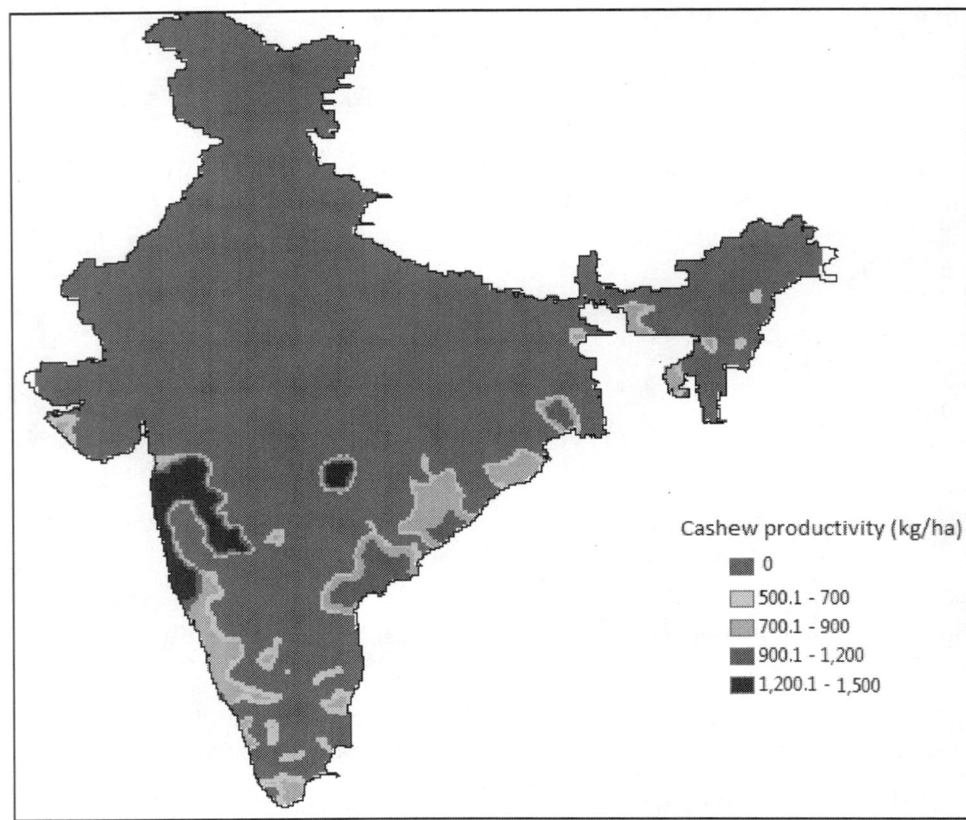

Figure 7.12: Distribution of Cashew Areas in India.

Suitability of Soil and Altitude Range for Cashew

Cashew growing area lies along loamy red and lateritic soil, mixed red and black soil, coastal and deltaic alluvium derived soil. Soil and climatic conditions of different cashew growing states in India with respect to production and productivity reveals that the well-drained laterite soil with high organic matter content is ideal for growing cashew. Productivity of cashew was found to be medium in eroded laterite soil and low in coastal sandy soil, saline/alkali soil and waterlogged soils. Soil salinity higher than 2 per cent was found to be unsuitable for the growth of cashew (Venugopal and Khader, 1991). The soil textures in high productivity area were found to be loam, sandy loam, silty loam and coastal sand. Similarly, in medium production areas, the texture was silty clay loam, sandy clay loam, clay loam and loamy skeletal. Similar results were reported by Nair *et al.* (1979). The depth of soil should be more than 90 cm and the soil can be neutral to medium acidic (5.6 to 7.3 pH). Venugopal and Khader (1991) reported that brown forest soil or a deep well drained laterite soil with high water holding capacity and rich in organic matter content appears to be an ideal soil type for better growth and higher productivity of cashew.

Table 7.1: The Guidelines for Selection of Site Suitable for Cashew Growth

Sl.No.		*Very Good*	*Good*	*Fair*	*Poor*	*Very Poor*
I.	**Soil physical characteristics**					
	1. Soil depth (cm)	>180	90-180	45-90	23-35	<23
	2. Soil texture	Sandy loams and loams	Coastal sands, loamy sands, silty loams	Light clay loams, light silty clay loams, heavy sandy clay loams	Heavy clay loams, heavy silty loams, heavy sandy clay loams	Clay, sandy clay, silty clay
	3. Drainage	Well drained	Well drained to somewhat well drained	Moderately well drained to excessively drained	Excessively drained	Poorly drained
II.	**Land features**					
	1. Slope (per cent)	<3	3-5	5-15	15-25	>25
	2. Erosion	None to slight	Slight	Moderate	Severe	Very severe
III.	**Climate and environmental factors**					
	1. Altitude (m above MSL)	<25	25-250	250-500	500-750	>750
	2. Rainfall (mm/year)	1200-2500	900-1200	700-900	500-700	<500
	3. Maximum temperature (summer) (ºC)	33-37	37-39	39-41	41-43	>43
	4. Minimum temperature (winter) (ºC)	16-18	13-16 18-20	10-13 20-22	8-10 22-24	<8 >24
	5. Max Relative Humidity (per cent)	70-75	75-80	80-85	85-90	>90
	6. Min Relative Humidity (per cent)	70-75	60-70	30-60	25-30	<25
IV.	**Agronomic aspects**					
	1. Expected yield (kg/ha)	>1000	600-1000	300-600	200-300	<200

Source: Devanandam, 1983; Mishra, 1984; Rejani *et al.*, 2013.

Devanadam (1983) reported that in Andhra Pradesh, cashew is raised on a wide variety of soils such as skeletal soils, eroded soils, undulating and slopy hilly terrains, coastal sands, red soils, lateritic soils, light clay loams, black soils and sandy soils along coast. The elevation of the cashew growing areas ranged from 0 to 1000 m above mean sea level (MSL) and the productivity of cashew was higher in regions up to 750 m above MSL (Figure 7.13). The cyclonic storms along the East Coast resulting in heavy rainfall, floods and waterlogging also seriously affect the productivity of cashew. Augmenting supplemental irrigation facilities in the rain deficit regions coupled with appropriate soil conservation and fertility management practices are essential in raising the cashew productivity.

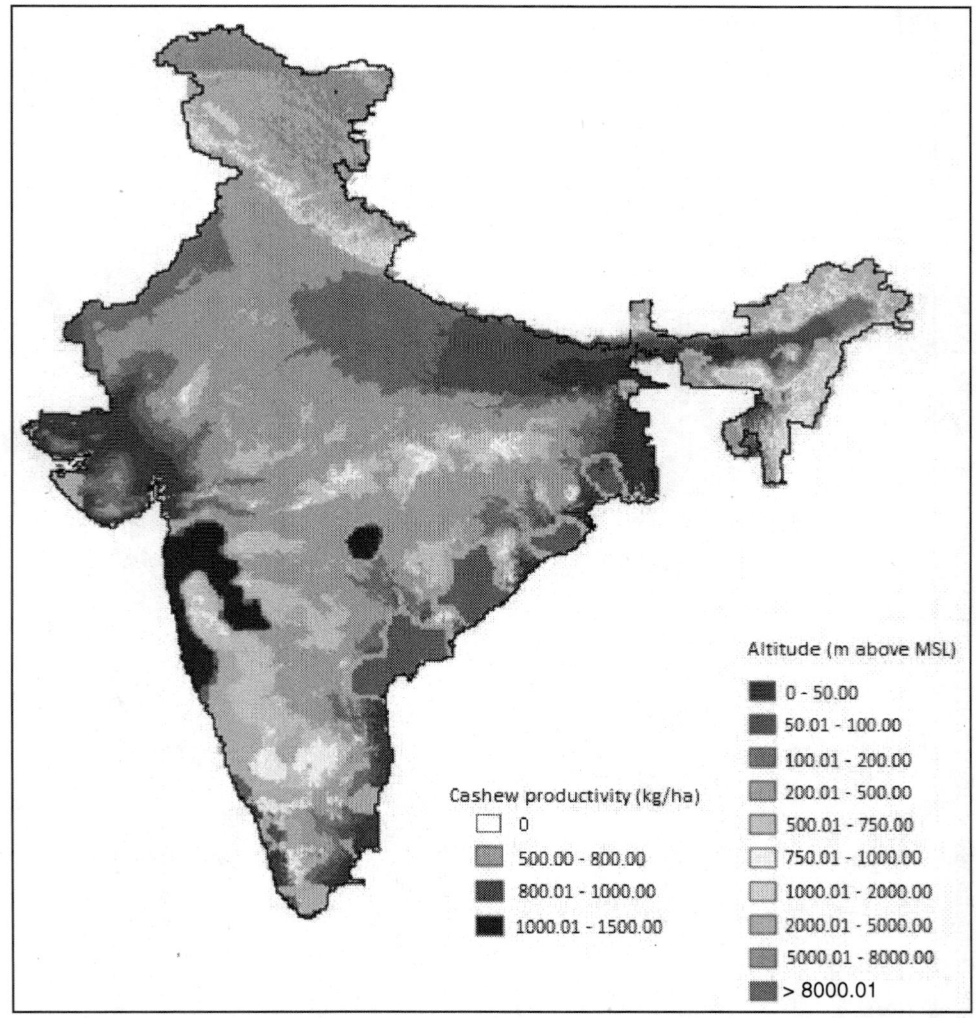

Figure 7.13: Cashew Area and Productivity (kg/ha) Overlay with Altitude (m above MSL).

Climatic Suitability for Cashew

Cashew is generally grown as a rainfed crop along steep slopes of hills or on neglected land unsuitable for any other crop. In India, cashew experiences severe moisture stress from January to May, adversely affects its flowering and fruit set. The results showed that the mean rainfall distribution in cashew area ranged from low rainfall (300-600 mm in Gujarat) to high rainfall (2700 to 3500 mm in West coast and NEH region). The productivity was highest in Maharashtra region with an annual rainfall range of 600 to 1500 mm and next to that portion of Andhra Pradesh, Odisha and West Bengal with a rainfall range of 600 to 1800 mm (Table 7.2) (Figure 7.14). Cashew yield increases with few light rains during January to March. But the yield will get severely affected with heavy rains resulting in immature nut drop and inferior kernel quality. The overlay maps showed that the mean annual temperature in cashew area ranged from 20 to even more than 27.5 °C whereas, the productivity was higher in regions with 22.5 to 27.5 °C (Figure 7.15). The minimum and maximum temperature in suitable area ranged from 10 to 22 °C and 32 to

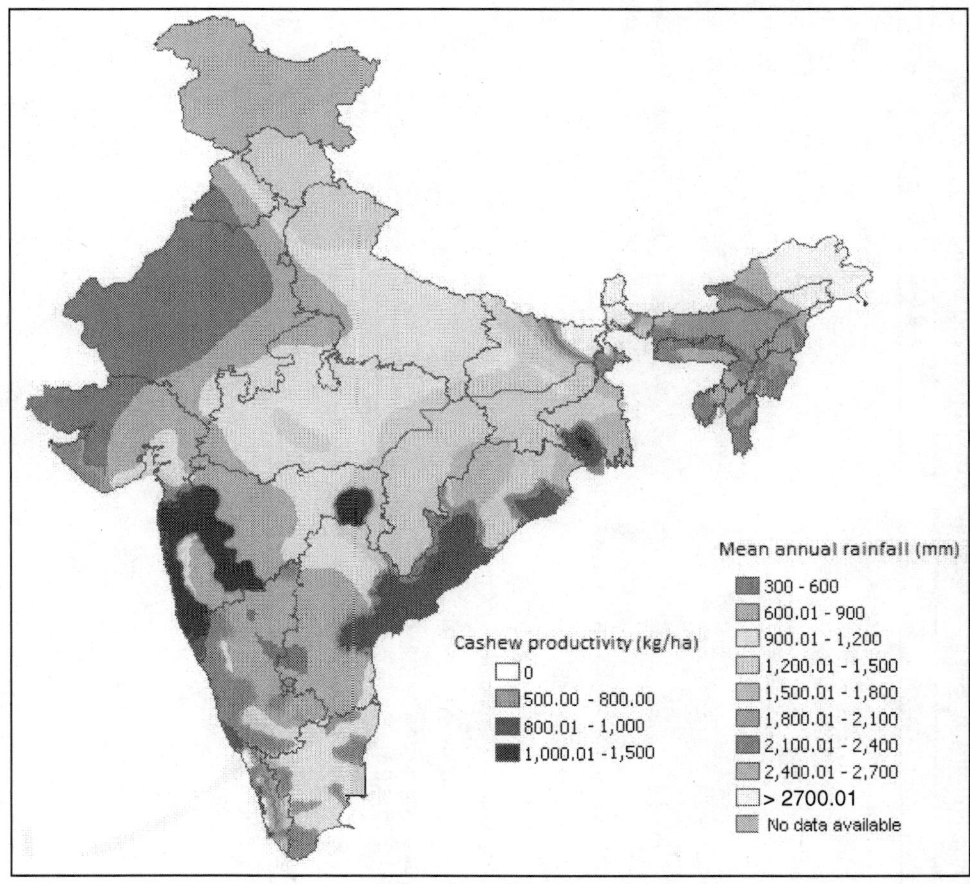

Figure 7.14: Cashew Area and Productivity (kg/ha) Overlay with Mean Annual Rainfall (mm).

Table 7.2: Suitability of Sites for Cashew growing Derived Using GIS

		Suitable	*Moderately Suitable*	*Not Suitable*
I.	**Soil physical characteristics**			
	Soil depth (cm)	>90	45-90	<45
	Soil texture	Loamy red and lateritic soil, mixed red and black soil, alluvium derived soils; Sandy loams and loams, coastal sands, loamy sands, silty loams	Light and heavy clay loams, silty clay loams, sandy clay loams	Clay, sandy clay, silty clay
	Drainage	Well drained	Moderately well drained to excessively drained	Poorly drained
II.	**Climate and environmental factors**			
	Altitude (m above MSL)	<750	750-1000	>1000
	Rainfall (mm/year)	>600	300-600	<300
	Mean annual temperature (ºC)	22.5 - 27.5	20-22.5	>27.5
	Max temperature (summer) (ºC)	32-39	39-40	>40
	Min temperature (winter) (ºC)	12-22	10-12	<10
III.	**Agronomic aspects**			
	Expected yield (kg/ha)	>1000	500-1000	<500

Source: Rejani *et al.*, 2013.

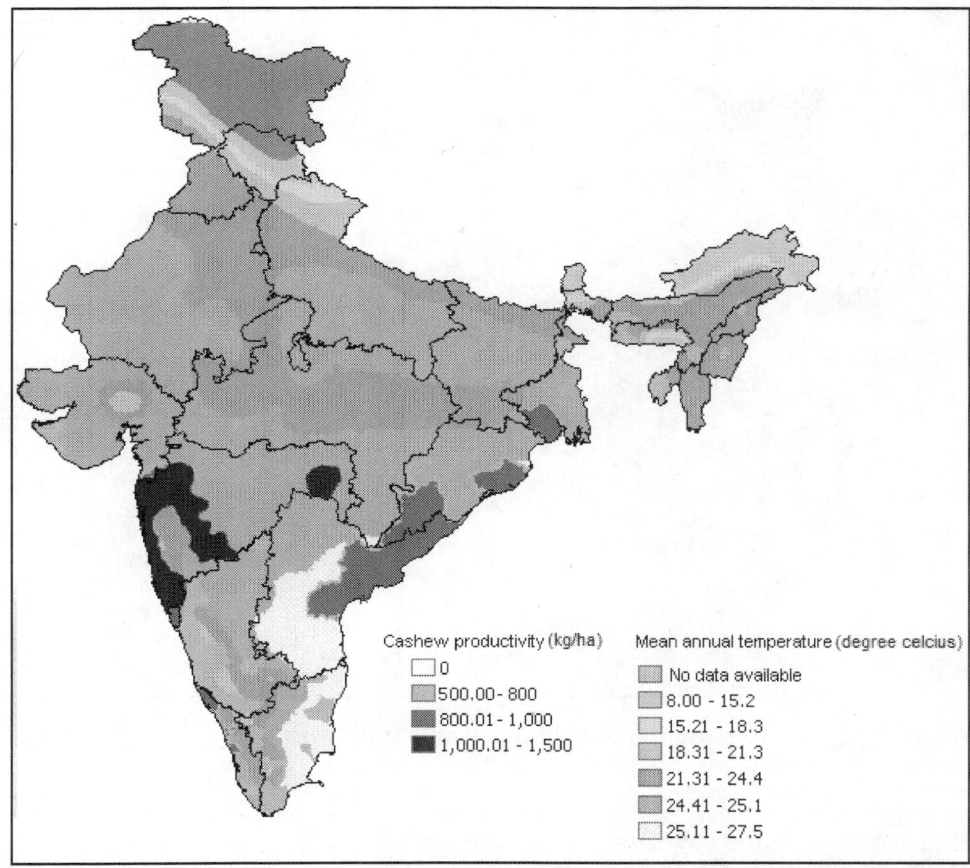

Figure 7.15: Cashew Area and Productivity (ka/ha) Overlay with Mean Annual Temperature (ºC).

40.1 °C respectively. The productivity of cashew was lower in regions where the minimum temperature drops below 10 °C. Similar results have also been reported by Devanandam (1983) and Mishra (1984).

Land Suitability for Cashew

The site suitability map indicated that suitability for cashew has direct impact on production. Parts of Maharashtra, Goa, Kerala, West Bengal, NEH regions and Odisha are highly suitable while parts of Andhra Pradesh, Tamil Nadu, Karnataka and Gujarat are moderately suitable (Figure 7.17). Parts of Jharkhand, North Eastern Hilly (NEH) Region, Andaman and Nicobar Islands and Bastar region of Chhattisgarh are also moderately suitable for cashew cultivation. The area under the cashew crop in India has increased from 565,420 ha in 1993-94 to 923,000 ha in 2010. Similarly, the production has also increased from 348,350 to 613,000 tonnes over the same period (DCCD, 2012; Figure 7.11). The area under the cashew crop in Maharashtra has increased from 51,220 ha in 1993-94 to 175,000 ha in 2010.

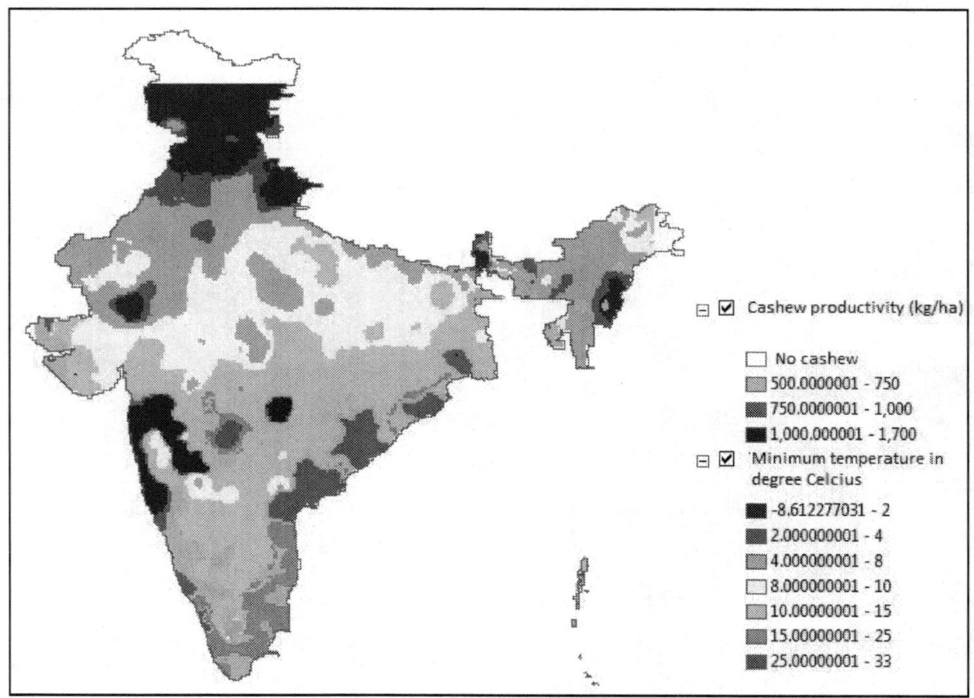

Figure 7.16: Cashew Area and Productivity Overlay with Decrease in Minimum Temperature (-2 ºC).

Similarly, the production has also increased from 46,860 to 198,000 tonnes over the same period (DCCD, 2012). As per the estimate of Directorate of Cashew and Cocoa Development, Maharashtra had the highest production of 225,000 tonnes of raw cashew nuts, followed by Andhra Pradesh with 112,000 tonnes, Odisha with 95,000 tonnes, Kerala with 75,000 tonnes and Tamil Nadu with 68,000 tonnes in 2008-09.

The Konkan region of Maharashtra comprising of Thane, Raigad, Ratnagiri and Sindhudurg districts is the major tract of cashew cultivation with more than 80 per cent in South Konkan region. Andhra Pradesh has about 183,000 hectares with an annual production of 99,000 tonnes (2009-10). The productivity of cashew in Andhra Pradesh is moderate and hence the productivity is not increasing considerably with the increase in area under cashew. It is mainly cultivated in Srikakulam, Visakhapatnam, East Godavari, West Godavari, Krishna, Guntur, Prakasam and Nellore districts in the State. Kerala is very suitable for cashew cultivation and now the rubber cultivation has replaced the area under cashew. Hence, the area under cashew is only 72,000 ha (2009-10) with a production of 66,000 tonnes. In Kerala, the cashew area was 155,810 ha with a production of 140,200 tonnes during 1993-94. In Odisha, the area under cashew was 143,000 ha with a production of 84,000 tonnes (2009-10). Four districts *viz.*, Koraput, Ganjam, Nawrangpur and Khurda occupy 60 per cent area under cashew while Denkanal and Gajpati have 10 per cent.

Karnataka is moderately suitable for cashew. The total area under cashew is 118,000 ha with a production of 53,000 tonnes (2009-10). The suitability of cashew in Tamil Nadu is also moderate. The area under cashew is 133,000 ha whereas the production is only 60,000 tonnes (2009-10). The area under cashew cultivation in Assam has increased over years and is grown in Dhubri, Suksor, Rangapani and Malatijhora areas. In Meghalaya, along West Garo Hills, the productivity is very high. Cashew is cultivated in few pockets in Nagaland and Tripura also.

3. Cashew and Climate Change

Climate change is one of the important alerts for present era. Several recent studies have indicated that annual rainfall, diurnal temperature, maximum and minimum temperatures are in a fluctuating trend. Future prediction shows that a slight increase or decrease in minimum temperature (±0.5 °C) may not considerably affect the yield of cashew. But a decrease in minimum temperature of about 2°C may tremendously reduce the yield of cashew in some regions of North Eastern states, Maharashtra and Andhra Pradesh, by changing the area highly suitable for cashew in Maharashtra to marginally suitable and moderately suitable area in NEH regions to unsuitable, indicating the effect of climate change on cashew

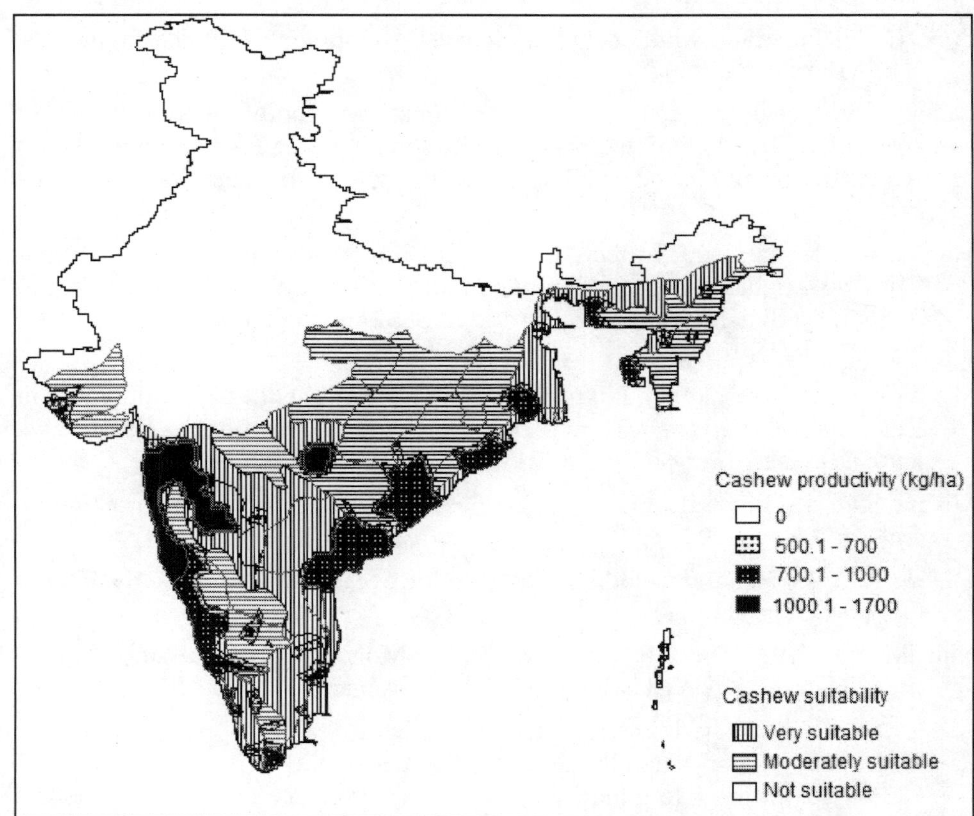

Figure 7.17: Crop Suitability Map for Cashew.

suitability (Figure 7.17). Lanel and Jarvis (2006) projected the future data for 2055 and predicted that climate change will cause shifts in areas suitable for cultivation of a wide range of crops.

References

Choudhury, S. and Saha, S.K. (2003). Cropping Pattern change analysis and optimal land use planning by integrated use of satellite Remote Sensing and GIS–A case study of Barwala C.D. Block, Panchkula Dist., Haryana Indian Cartographer: 111-123.

DCCD (2012). Directorate of Cashew and Cocoa Development. Available online. http: //dacnet.nic.in/cashewcocoa/stat.htm#stat.

Devanandam, M. (1983). Soils for cashew. *Cashew Causerie*, 5(1): 1-6.

Godse, S. K., Bhole, S. R. and Patil, B.P. (2005). Population fluctuation studies of tea mosquito on cashew and its relation with weather parameters. *Journal of Agrometeorology*, 7(1): 107-109.

Haldankar, P. M., Deshpande, S. B., Chavan, V.G. and Rao, E.V.V.B. (2003). Weather associated yield variability in cashewnut. *Journal of Agrometeorology*, 5(2): 73- 76.

IMD Annual Climate Summary (2009). Published by National Climate Centre, Pune.

IMSD (1995). Integrated Mission for Sustainable Development: Technical guidelines, NRSA, Hyderabad, India: 1-127.

Lanel, A. and Jarvis, A. (2006). Changes in climate will modify the geography of crop suitability: Agricultural biodiversity can help with adaptation. An Open Access Journal published by ICRISAT. http: //www.icrisat.org. Journal Special Project/sp2.pdf.

Mandal, C., Mandal, D. K., Srinivas, C.V., Sehgal, J. and Velayutham, M. (1999). Soil-Climatic Database for Crop Planning in India. NBSS Publication 53. Technical Bulletin. National Bureau of Soil Survey (NBSS), and Land Use Planning (LUP), Nagpur, India, 1014p.

Matsuoka, M., Hayasaka, T., Fukushima, Y. and Honda, Y. (2007). Land cover in East Asia classified using Terra MODIS and DMSP OLS products. *International Journal of Remote Sensing*, 28(2): 221 – 248.

Mishra, R.C. (1984). Land suitability classification of cashew I The Criteria. *Cashew Causerie*, 6(4): 12-20.

Mishra, R.C. (1985). Land suitability classification of cashew II The Criteria. *Cashew Causerie*, 7(1): 7-11.

Nair, M. K. (1979). Crop management, Cashew Monograph of Plantation Crops. (Eds.) Nair, M. K, Bhaskara Rao, E. V. V., Nambiar, K. K. N and Nambiar, M. C., CPCRI, Kasaragod pp. 43-45.

Ramakrishnan, D., Bandyopadhyay, A. and Kusuma, K. N. (2009). SCS-CN and GIS-based approach for identifying potential water harvesting sites in the Kali Watershed, Mahi River Basin, India. *J. Earth Syst. Sci.*, 118 (4): 355–368.

Rao, D. V. K. N., Meti, S., Nair, A. N. S., and Punnose, K. I. (2003). Report of a project on development of Rubber Information System: Current status, Agronomy/ Soils Division, Rubber Research Institute of India, Kottayam, Kerala, 14 pp.

Rao, G. S. L. H. V. P and Gopakumar, C. S. (1994). Climate and cashew. *The Cashew*, 8(4): 3-9.

Rao, G. S. L. H. V. P. (1999). Weather inflicted damage on cashew production- a remedy. *The Cashew*, 3(4): 36-43.

Rejani, R. and Yadukumar, N. (2010). Soil and water conservation techniques in cashew grown along steep hill slopes. *Scientia Horticulturae*, 126: 371-378.

Rejani, R. and Yadukumar, N. (2012). A nutrient decision support system for rainfed cashew. *Journal of Plantation Crops*, 40 (1): 40-49.

Rejani, R., Rupa, T.R. and Nayak, M.G. (2013). Suitability of cashew cultivation in India – an application using GIS. *Journal of Agrometorology*, 5: 123-128.

Rupa, T. R., Rejani, R. and Bhat, M. G. (2012). Climate change effects on cashew and adaptation strategies. Proceedings of the *National Dialogue on Climate Resilient Horticulture* held on 28-29 January at IIHR, Bangalore, pp. 200-211.

Shanwad, U. K., Patil, V. C., Gowda, H. H., Dasog, G. S. and Shashidhar, K. S. (2011). Generation of water resources action plan for Medak Nala Watershed in India using Remote sensing and GIS technologies. *Australian Journal of Basic and Applied Sciences*, 5(11): 2209-2218.

Tottrup, C., Rasmussen, M. S., Eklundh and Jönsson, P. (2007). Mapping fractional forest cover across the highlands of mainland Southeast Asia using MODIS data and regression tree modeling. *International Journal of Remote Sensing*, 28(1): 23-46.

Venugopal, K. and Khader, K. B. A. (1991). Effect of soil and climate on the productivity of cashew. *Indian Cashew Journal*, 20(3): 19-24.

Yadukumar, N., Raviprasad, T. N., and Bhat, M.G. (2010). Effect of climate change on yield and insect-pests incidence on cashew. Challenges of Climate Change - Indian Horticulture. (Eds.) Singh, H. P., Singh, J. P. and Lal, S. S. pp.224. Westville Publishing House, New Delhi.

Yadukumar, N., Rejani, R. and Prabhakar, B. (2009). Fertigation for efficient water and nutrient management in high density cashew plantation. *Journal of Plantation Crops*, 37(2): 108-121.

Cashew: Improvement, Production and Processing (2017) *Pages* **171–185**
Editor: **Prof. P.L. Saroj**
Published by: **ASTRAL INTERNATIONAL PVT. LTD., NEW DELHI**

Soil and Water Conservation Practices

R. Rejani and T.R. Rupa

1. Introduction

Soil erosion either by water or by wind is a serious issue in our country. The areas like hilly terrain, arid region, ravines, torrents, sandy soils, salt affected lands etc are more prone to soil erosion. The developmental activities and unscientific agricultural practices further aggravating the problems. Deforestation and shifting cultivation is another concern causing soil erosion. Nearly one-sixth of the land area has serious limitations for crop production such as erosion, aridity, water logging, acidity, salinity, and alkalinity. It is estimated that nearly 80 million ha of cultivated area requires soil conservation measures (Murthy *et al.*, 2013). In many parts of the world, the overdraft of groundwater is resulting in declining groundwater levels, freshwater scarcity, saltwater intrusion and land subsidence (Rejani *et al.*, 2003; Rejani *et al.*, 2008; Rejani *et al.*, 2009). It is estimated that if the current trend continues, about two-thirds of global population will face moderate to severe water stress by 2025 (Kuylenstierna *et al.*, 1997). According to the Ministry of Water Resources, Government of India, groundwater levels in the 16 states of India have dropped to more than 4 m during the 1981-2000 (GOI, 2001). In many parts of Gujarat and Rajasthan, groundwater level fell more than 16 m. A few districts in coastal Gujarat, central Rajasthan, Madhya Pradesh, Uttar Pradesh, West Bengal, Karnataka and Tamil Nadu showed a decline of 4 to 8 m. In addition, a 1 to 4 m decline was observed in many other states (Sekhri, 2012). Cashew is a suitable crop in this case since no intensive soil disturbing activities are required for this crop. In the beginning cashew was introduced for soil and water conservation purpose. Now, adoption of cashew in large scale will generate rural employment related with its nut collection and processing and aid in sequestering carbon in the soil. There is a possibility of increased production of cashew by adopting soil and water conservation measures with high yielding varieties of cashew. With proper soil and water conservation practices, the soil loss can be minimized; the runoff water from post-monsoon and pre-monsoon rainfall can be harvested and make available to the plant during the critical period (Yadukumar and Rejani, 2004; Rejani and Yadukumar, 2007; Rejani and Yadukumar, 2010; Rejani and Yadukumar, 2012).

2. Soil and Water Conservation Techniques

Soil and water conservation activities are an essential part of the cashew production technology followed in slopy areas where the top soil and surface runoff have to be conserved. Most of the cashew plantations in Myanmar was also established on degraded slopes with poor fertility. Although soil and water conservation measures are advocated but most of the farmers continue to grow the crop without adopting conservation measures such as bunds, terraces, inverted crescent basins, catch pits *etc* (FAO, 2013). Manivannan *et al.*, 2007 conducted a field experiment on lateritic soil of Western Ghat region in Goa State to study the effect of bioengineering measures *viz.*, Half Moon Terraces + *Vetiveria zizanioides* + *Stylosanthes scabra*, continuous contour trench + *Vetiveria zizanioides* + *Stylosanthes scabra*, graded trench + *Vetiveria zizanioides* + *Stylosanthes scabra*, staggered contour trench + *Vetiveria zizanioides* + *Stylosanthes scabra* and semi elliptical trench + *Vetiveria zizanioides* + *Stylosanthes scabra* on runoff, soil and nutrient loss and growth of cashew. The study revealed that the reduction of runoff varied from 3.2 to 7.6 per cent of the annual rainfall under different conservation treatments. Minimum runoff (3.5 per cent) and soil loss (3.7 t/ha/year) were recorded under the treatment *viz.*, continuous contour trench + *Vetiveria zizanioides* + *Stylosanthes scabra*. Overall, continuous contour trenches with *Stylosanthes scabra* and *Vetiveria zizanioides* was found to be the best soil and water conservation treatment among all other bioengineering measures for cashew plantations in Western Ghat region. Better growth and yield for mango and cashew was observed with platform bench terraces due to the increased moisture content in the root zone of plants (Deshmukh *et al.*, 1992). Using coconut coir pith as soil mulch in cashew plantations resulted in 14.15 per cent higher water retention and suppression of weeds up to 73.52 per cent (Kumar *et al.*, 1989). Mane *et al.* (2009) demonstrated that continuous contour trench (0.5 m x 0.6 m) was the best soil-conservation practice for cashew on hands with 7 to 8 per cent slope.

In order to evaluate the effect of *in-situ* soil and water conservation measures on cashew the following soil and water techniques for cashew have been tried at DCR.

1. Modified crescent bund (at 2 m radius having a crescent shaped bund of 6 m length, 1 m width and 0.5 m height) on the upstream of the plant so that a trench of 6 m length and 50-75 cm deep will be formed while making the bund (Figure 8.1).

Figure 8.1: Modified Crescent Bund.

2. Staggered trenches opened across the slope with coconut husk burial between 2 rows of cashew (Trenches of size 5 m length, 1 m width and 0.5 m depth in the middle of 4 plants with coconut husk buried) (Figure 8.2).
3. Reverse terraces (2 m length, 2 m width and 0.7 m depth of cut) (Figure 8.3).
4. Catch pits (3 m length, 0.5 m width and 0.5 m depth) (Figure 8.4).
5. Control plot without any soil and water conservation measure.

Among different soil and water conservation techniques tried, no significant difference in plant height was observed. Stem girth and canopy spread was higher in modified crescent bund treatment, coconut husk burial and reverse terrace treatments and lowest in control and catch pit treatments. The higher cashew nut yields were obtained in coconut husk burial treatment (2.04 t/ha), modified crescent bund (2.03 t/ha) and reverse terrace treatments (1.94 t/ha), compared to catch pit

Figure 8.2: Coconut Husk Burial Treatments.

Figure 8.3: Reverse Terrace.

Figure 8.4: Catch Pit.

(1.39 t/ha) and control (1.48 t/ha) (Table 8.1). The *in situ* soil and water conservation structures effectively harvested the pre monsoon rainfall and increased the cashew yield significantly. A second time flowering and fruit set was observed after the pre monsoon rain. The cumulative yield of five harvests were also high in coconut husk burial (6.60 t/ha), and modified crescent bund treatments (6.45 t/ha) compared to control (4.88 t/ha) and other treatments. *i.e.*, an increase in cashewnut yields by 32 to 35 per cent of control due to soil and water conservation. Deshmukh *et al.* (1992) found better growth and yield of mango and cashew with platform bench terraces due to increased moisture conserved in the root zone of plants. Yadukumar and Mandal (1994) and Ghosh (1995) reported that irrigation during fruit setting stage (February) increased the fruit set, its retention and increased the nut yield in cashew. Ghosh (1999) and Ghosh *et al.* (2000) found improved the nut production in cashew by the adoption of soil and water conservation structures such as circular trench around the cashew tree with dry cashew leaves mulching. The net profit from cashew orchard for the first seven years (5 harvests) with proper soil and water conservation ranged from Rs.2,02,445 to 2,02,304 per ha whereas in control it was Rs. 1,44,630/ha (Table 8.2).

Table 8.1: Effect of Soil and Water Conservation Techniques on Yield of Cashew

Treatments	*Yield (t/ha)*					
	2005-06	*2006-07*	*2007-08*	*2008-09*	*2009-10*	*Cum. yield*
Modified crescent bund	0.355	0.671	1.41	2.03	2.00	6.45
Coconut husk burial	0.390	0.702	1.50	2.04	1.97	6.60
Reverse terraces	0.336	0.654	1.24	1.94	1.71	5.88
Catch pits	0.316	0.561	1.21	1.39	1.57	5.05
Without soil and water conservation	0.262	0.451	1.14	1.48	1.55	4.88
Mean	0.332	0.608	1.27	1.77	1.76	5.77
CD (p = 0.05)	0.057	0.152	0.088	0.224	0.147	0.14

Table 8.2: Cost of Cultivation and Net Profit (Rs.) from Cashew Orchard with Soil and Water Conservation Measures for the First Seven Years

Labour Cost	*Soil and Water Conservation Techniques*				
	Modified Crescent Bund	*Coconut Husk Burial*	*Reverse Terrace*	*Catch Pits*	*Control*
Weeding once in a year	14010	14010	14010	14010	14010
Manuring	4483	4483	4483	4483	4483
Plant protection measures	2382	2382	2382	2382	2382
Digging pits and soil and water conservation	7986	14010	5978	5604	4016
Picking nuts	4296	4390	3923	3362	3269
Total Labour cost	33157	39275	30775	29841	28160
Cost of materials					
Fertilizers	17513	17513	17513	17513	17513
Plant protection Chemicals	4904	4904	4904	4904	4904
Total cost of cultivation	55573	61691	53191	52257	50529
Nut yield (t/ha)	6.45	6.60	5.88	5.05	4.88
Income	258018	263995	235181	201978	195206
Net profit	202445	202304	182037	149767	144630
Benefit cost ratio	3.64	3.28	3.42	2.87	2.86

The runoff was quite high in all the treatments during June and July due to the high intensity of rainfall during monsoon/rainy season (Table 8.3 and Figure 8.5). The mean annual runoff was less in staggered trenches with coconut husk burial (20 per cent of annual rainfall), modified crescent bund (22 per cent of annual rainfall), and reverse terrace treatments (27 per cent of annual rainfall) compared to control plot (37 per cent of annual rainfall). The soil and water conservation structures harvested a part of the rainfall available and reduced the runoff of water and its flow velocity. Hence the erosion and soil loss was relatively less in treatments with soil and water conservation structures. A mean annual rainfall of 3031 mm was recorded during this period (Table 8.4). The soil loss was also less in the case of modified crescent bund (4.6 t/ha/yr), coconut husk burial (4.8 t/ha/yr) and reverse terrace treatments (5.1 t/ha/yr) compared to control plot (9.7 t/ha/yr). Bruijnzeel *et al.* (1998) reported in similar type of land that the runoff from bench terraces in upland West Java, Indonesia as 28 per cent of rainfall. Tejwani *et al.* (1975) reported the runoff as 27.1 per cent of rainfall and soil loss as 1.0 t/ha/yr from a land covered with natural grass in silty loam soil on 9 per cent slope, having a rainfall of 1250 mm in Dehradun. The mean soil moisture content in modified crescent bund and coconut husk burial treatments were on par (15.6 per cent and 15.8 per cent dry basis respectively in March) and were high compared to control (11.6 per cent dry basis in March) and other treatments (reverse terrace - 14.6 per cent and catch pit - 12.7 per cent dry basis) (Table 8.5).

Table 8.3: Monthly Rainfall and Runoff as Affected by Soil and Water Conservation Techniques

Months	*June*	*July*	*Aug*	*Sept*	*Oct*	*Nov*
Rainfall (mm)	919	1025	425	388.5	164.5	0.0
Soil and Water Conservation Techniques			*Runoff (Per cent of rainfall)*			
Modified crescent bund	25.9	25.0	9.0	7.0	3.2	0.0
Coconut husk burial	29.8	29.0	10.0	9.0	2.8	0.0
Reverse terrace	31.3	31.2	14.0	11.2	3.8	0.0
Control	51.2	43.3	25.0	18.5	6.2	0.0

Table 8.4: Annual Rainfall and Runoff as Affected by Soil and Water Conservation Techniques

Year	*2004*	*2005*	*2006*	*Mean*
Rainfall (mm)	3297	2922	2873	3031
Soil and Water Conservation Techniques		*Runoff (Per cent of rainfall)*		
Modified crescent bund	27.6	19.3	15.7	22.3
Coconut husk burial	18.6	22.4	17.0	20.4
Reverse terrace	29.5	24.5	21.2	26.7
Catch pits	35.0	33.5	25.5	33.2
Control	36.3	37.7	30.8	36.9

Table 8.5: Variation of Mean Soil Moisture in Per cent Dry Basis with Depth

Soil and Water Conservation Techniques	*Depth (cm)*			*Mean*
	0-30	*30-60*	*60-90*	
Modified crescent bund	13.3	16.0	17.6	15.6
Coconut husk burial	13.4	16.0	18.2	15.8
Reverse terrace	12.9	14.7	16.4	14.6
Catch pits	11.3	12.5	14.2	12.7
Without any soil and water conservation	10.3	11.4	13.0	11.6
Mean	**12.0**	**14.1**	**15.9**	**14.0**
CD (p = 0.05) Treatment				**1.4**
CD (p = 0.05) Depth				**0.4**

Similar trend of soil moisture distribution was found during other months (Table 8.6). The available soil moisture ranges from 12 to 22 per cent dry basis. The soil moisture content in all the three layers (0-90 cm) in soil and water conservation treatments was in the available soil moisture range and ranged from 12.9 to 18.2 per cent dry basis. The soil moisture content in the first two layers of soil (0-60 cm)

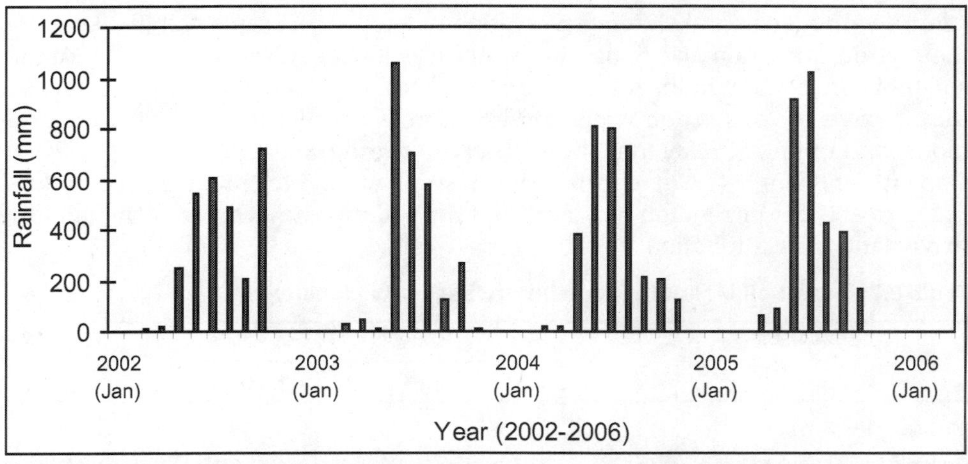

Figure 8.5: Monthly Rainfall Data during the Experimental Period (2002-2006).

in the control plot were 10.3 to 11.4 per cent dry basis respectively and was below the available soil moisture range in March itself (Table 8.6).

Table 8.6: Variation in Mean Soil Moisture in Per cent Dry Basis

Treatments	*Mean Soil Moisture (Per cent dry basis)*					
	Dec	*Jan*	*Feb*	*Mar*	*Apr*	*May*
Modified crescent bund	19.9	18.6	17.2	15.6	12.1	23.1
Coconut husk burial	20.5	18.6	17.1	15.8	12.1	22.5
Reverse terrace	19.0	17.5	16.3	14.6	11.4	22.0
Catch pits	17.5	16.5	15.3	12.7	10.8	18.7
Without any soil and water conservation	16.8	15.4	12.7	11.6	10.6	17.3
CD (p = 0.05) Treatment	**2.1**	**2.0**	**1.8**	**1.4**	**0.5**	**0.48**
CD (p = 0.05) Depth	**0.7**	**0.6**	**0.5**	**0.4**	**0.3**	**0.17**

A rainfall of 50, 49.4 and 26.9 mm were observed during April 2005, May 2006 and April 2009 respectively and no rainfall was observed before soil sample collection during other period. After the pre-monsoon rainfall of 50 mm in April 2005, almost same soil moisture content was observed in the first two layers of soil in all the treatments including control (20.0 to 21.4 per cent db in first layer and 18.0 to 20.3 per cent db in second layer). Soil moisture content in the third layer was higher in the coconut husk burial and crescent bund treatments (19.1 and 19.0 per cent db) and lowest in the control plot (16.6 per cent db) (Table 8.7). The soil and water conservation treatments have 5 to 6 m trenches, which harvested the pre-monsoon rainfall during April and hence more infiltration time was available for the harvested rainwater to get infiltrated into the soil and it increased the soil moisture content in the first three layers of soil (0-30, 31-60 and 61-90 cm depth). In case of control without soil and water conservation, the top soil became wet (0-30 cm) by the pre-monsoon rain and remaining water was lost as surface runoff. Hence the soil moisture content in the lower layers of control treatment was less. This shows the effectiveness of soil and water conservation techniques in conserving the pre-monsoon rainfall. Similar trend was observed during May 2006 and April 2009. Chovatia and Singh, (1996) reported that crescent bund, catch pits, trench planting, mulches and drip irrigation has increased the soil moisture content considerably in dry land fruit cultivation.

Table 8.7: Mean Soil Moisture (Per cent dry basis) Availability

Treatment	*Depth (cm)*			*Mean*
	0-30	*30-60*	*60-90*	
Modified crescent bund	21.00	20.30	19.05	20.10
Coconut husk burial	21.40	20.03	19.18	20.20
Reverse terrace	20.43	19.80	18.60	19.61
Catch pits	20.20	19.43	18.10	19.24
Without any soil and water conservation	20.03	18.00	16.63	18.22
Mean	**20.61**	**19.51**	**18.31**	
CD(p = 0.05) Treatment				**1.20**
CD (p = 0.05) Depth				**0.35**

The organic carbon content (OC per cent), P_2O_5 and K_2O content of the soil at a depth of 0 to 60 cm was high in coconut husk burial, modified crescent bund and reverse terrace treatments (organic carbon content - 0.57 to 0.66 per cent compared to 0.44 per cent in control, P_2O_5 content – 4.6 to 7.0 kg/ha compared to 2.0 kg/ha in control, K_2O content – 57.8 to 65 kg/ha compared to 40.5 kg/ha in control). An intensive growth of roots inside the coconut husk burial trenches was noticed. This may be due to the increased soil moisture content availability inside the husk burial trenches. Coconut husks also serves as a source of nutrients especially potassium. The major and micro nutrients were on par in different soil and water conservation treatments but significantly higher compared to control (Table 8.8). The major and micronutrients present in leaf were also high in case of soil and water conservation treatments (Table 8.9).

3. *In-situ* Soil and Water Conservation Techniques Coupled with Nutrient Application

Main Plot

S1: Individual tree base terrace with catch pit (Catch pits of size 4 m length, 0.45m depth and 0.3 m width at 2 m from the base of the plant).

S2: Individual tree base terrace with crescent bund (Terrace at 2 m radius around the plant with a crescent shaped bund of size 6 m length, 0.5 m height and 1 m width at the base).

S3: Staggered trenches between two rows of cashew (8 trenches of size 2 m length, 0.45 m width and 0.45 m depth in the middle of 4 plants).

S4: Control plot without any soil and water conservation technique.

Table 8.8: Nutrient Status of the Soil

Treatments	*OC Per cent*	*P (kg/ha)*	*K (kg/ha)*	*Ca (kg/ha)*	*Mg (kg/ha)*	*Cu (mg/kg)*	*Zn (mg/kg)*	*Mn (mg/kg)*
Modified crescent bund	0.60	7.0	61.0	6.0	1.60	1.24	5.90	11.5
Coconut husk burial	0.66	5.1	65.0	3.5	1.42	1.19	4.25	10.9
Reverse terrace	0.64	4.9	57.8	5.9	1.22	1.08	4.55	14.3
Catch pits	0.57	4.6	57.0	11.5	1.31	1.00	2.62	6.7
Control	0.44	2.0	40.5	4.5	1.25	0.95	1.41	3.7
CD (p = 0.05)	0.08	1.33	5.94	1.81	0.22	0.12	0.97	2.26

Table 8.9: Mineral Leaf Nutrient Concentration

Treatments	*N per cent*	*P per cent*	*K per cent*	*Zn ppm*	*Mn ppm*	*Cu ppm*
Modified crescent bund	1.80	0.51	0.32	1.8	4.5	0.9
Coconut husk burial	1.84	0.41	0.36	1.2	4.2	0.3
Reverse terrace	1.79	0.39	0.37	1.5	5.2	0.9
Catch pits	1.75	0.32	0.29	1.0	3.8	0.9
Control	1.63	0.18	0.23	0.8	2.8	0.2
CD (p = 0.05)	0.20	0.07	0.08	NS	1.2	0.1

Sub Plot

M1: Application of recommended doses of fertilizers only (750 g N: 150 g each of P_2O_5 and K_2O/tree/year).

M2: Application of recommended doses of fertilisers + organic manure (10 kg poultry manure).

M3: Application of organic manure only (20 kg poultry manure).

During 2002-06, it was found that individual tree base terracing with crescent bund has resulted a significant increase in the mean soil moisture content (SMC) over the control (13.5 per cent dry basis compared to 10.5 per cent dry basis in control in March). The catch pit and staggered trench treatments have increased the SMC to 12.5 per cent and 11.9 per cent dry basis respectively (Table 8.10 and Figure 8.6). The available soil moisture ranges from 12 to 23 per cent dry basis.

Annual runoff from terrace with catch pit, crescent bund, staggered trench and control plots was 756, 722, 782 and 850 mm (30 per cent of the annual rainfall) and the peak runoff from these plots was 4.6, 4.6, 5.0 and 5.1 l/s/ha respectively. The first two treatments have considerably reduced the annual runoff (15 per cent of

Table 8.10: Temporal Variation of Mean Soil Moisture Content (Per cent dry basis)

Treatment	*Soil Moisture Content (per cent dry basis)*			*Mean*
	January	*February*	*March*	
Individual tree base terrace with catch pit	15.5	13.5	12.5	13.8
Individual tree base terrace with crescent bund	16.5	14.2	13.5	14.7
Staggered trenches between two rows of cashew	14.6	13.0	11.9	13.2
Control plot without any soil conservation technique	12.6	11.3	10.5	11.5
Mean	14.9	13.1	12.2	
CD Treatment (p=0.05)				0.306
CD Month (p=0.05)				0.271
T x M				NS

Available soil moisture ranges from 12 to 23 per cent.

control), peak runoff (10 per cent of control) and soil loss (50 per cent of control). Soil conservation measures like tree base terrace with crescent bund resulted in higher production of nuts (707.3, 1157.7, 1192 and 1543 kg/ha/year respectively) than the rest of the treatments including control (439, 650, 668, 1030 kg/ha/yr) during the period 2003-2006 (Table 8.11). During the first year (2002) significant difference among the treatments was not observed. The mean yield data over five years indicated that tree base terrace with crescent bund resulted in significantly higher production of nuts (986 kg/ha/yr- 163 per cent of control) than rest of the treatments including control (602 kg/ha/yr). Crescent bund treatment coupled with

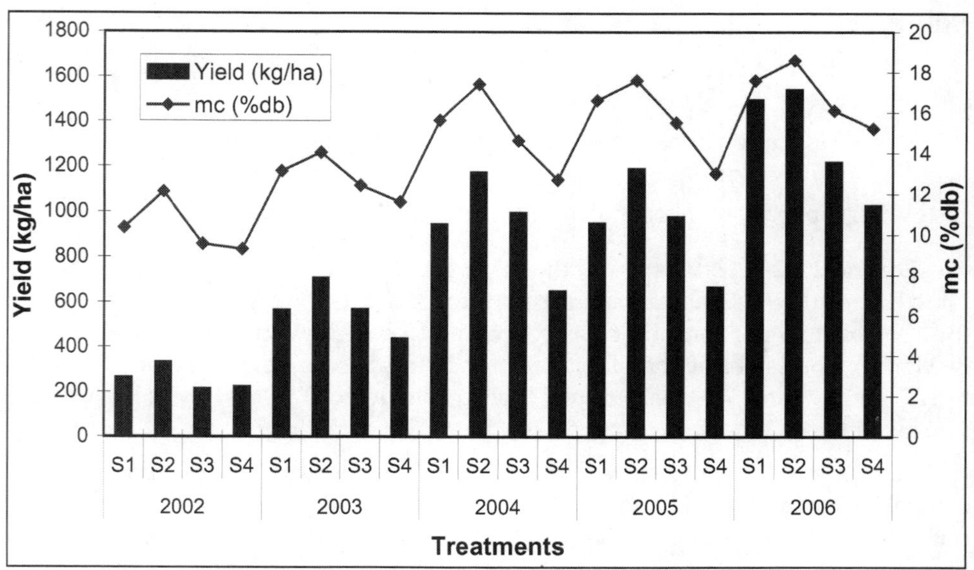

Figure 8.6: Temporal and Treatment-wise Variation of Soil Moisture Content and Yield.

Table 8.11: Effect of Soil and Water Conservation Coupled with Manuring on Nut Yield (kg/ha) for Five Years (2002 to 2006)

Treatment	*2002*				*2003*				*2004*			
	Yield (kg/ha)				*Yield (kg/ha)*				*Yield (kg/ha)*			
	M1	*M2*	*M3*	*Mean*	*M1*	*M2*	*M3*	*Mean*	*M1*	*M2*	*M3*	*Mean*
S1	202	295	294	263.7	538	553	604	565.0	869	947	1068	961.33
S2	301	290	402	331.0	674	734	714	707.3	1050	1212	1211	1157.67
S3	189	195	257	213.7	548	620	538	568.7	831	1129	1032	997.0
S4	151	202	315	222.7	376	438	502	438.7	560	614	776	650.0
Mean	210.7	245.5	317.0		534.0	586.2	589.5		687.5	822.0	827.7	
CD for Main plot (P=0.05)				NS				74.93				127
CD for Sub plot (P=0.05)				68.4				NS				112
Main x Sub				NS				NS				120

Treatment	*2005*				*2006*				*Mean*			
	Yield (kg/ha)				*Yield (kg/ha)*				*Yield (kg/ha)*			
	M1	*M2*	*M3*	*Mean*	*M1*	*M2*	*M3*	*Mean*	*M1*	*M2*	*M3*	*Mean*
S1	850	980	1020	950.0	1295	1547	1650	1497.4	751	864	927	847.5
S2	1131	1195	1250	1192.0	1340	1747	1541	1542.4	899	1036	1024	986.1
S3	890	970	1080	980.0	992	1209	1461	1220.8	690	825	874	796.1
S4	550	714	740	668.0	908	1042	1139	1029.8	509	602	694	601.8
Mean	855.3	964.8	1022.5		1133.5	1386.5	1447.7		949.7	1109	1172.9	
CD for Main plot (P=0.05)				140				154				117
CD for Sub plot (P=0.05)				118				121				105
Main x Sub				132				145				112

the application of recommended dose of fertilizer +10 kg poultry manure resulted in an yield of 1109 kg/ha/yr and 20 kg poultry manure alone resulted in maximum yield of 1173 kg/ha/yr compared to 950 kg/ha/yr in control. Soil moisture contents in soil and yield are very closely related and the corresponding increase in yield was observed with increase in soil moisture in five consecutive years (Figure 8.6). The rainfall distribution during this period is shown in Figure 8.5. In West Bengal, Ghosh (1999) reported that harvesting of rainwater by different soil and water conservation structures significantly improved the nut production in cashew.

Cost of cultivation and income from cashew for five years (2002 to 2006) is shown in Tables 8.12 and 8.13 respectively. It was found that that highest net profit of Rs. 124815/- per ha (173 per cent of control) was obtained from crescent bund treatment plot compared to Rs. 71,859/- per ha in control plot *i.e.* without soil and water conservation structure. Among the subplots, application of recommended doses through inorganic fertilizers + 10 kg poultry manure fetched highest profit of Rs. 129460/- per ha. The combination of crescent bund with 20 kg poultry manure resulted in a net profit of Rs. 128706/- per ha compared to Rs. 84,678/- per ha in control (Figure 8.7). From the point of view of organic farming, crescent bund with 20 kg poultry manure is the most recommended practice.

Table 8.12: Cost of Cultivation of Cashew for 5 Years (2002-2006) in Rs/ha

Treatments	*Cost of Cultivation (Rs.)*			*Mean*
	M1	*M2*	*M3*	
S1	43893	54649	54243	50928
S2	46737	57675	56674	53695
S3	51559	62153	62227	58646
S4	31625	38221	41767	37204
Mean	43454	53175	53728	

Table 8.13: Income from Cashew for 5 Years (2002-2006) in Rs/ha

Treatments	*Income (Rs.)*			*Mean*
	M1	*M2*	*M3*	
S1	135605	156170	167360	153045
S2	163015	187135	185380	178510
S3	125200	149155	158280	144211
S4	91825	108920	126445	109063
Mean	128912	150345	159366	

In situ soil and water conservation is one of the effective methods for harvesting the pre-monsoon and post-monsoon rainfall and make it available to the cashew plant during this critical period. Soil and water conservation measures such as modified crescent bund and coconut husk burial treatments increased the growth of cashew plants and cumulative cashew yield indicating 32 to 35 per cent increase in

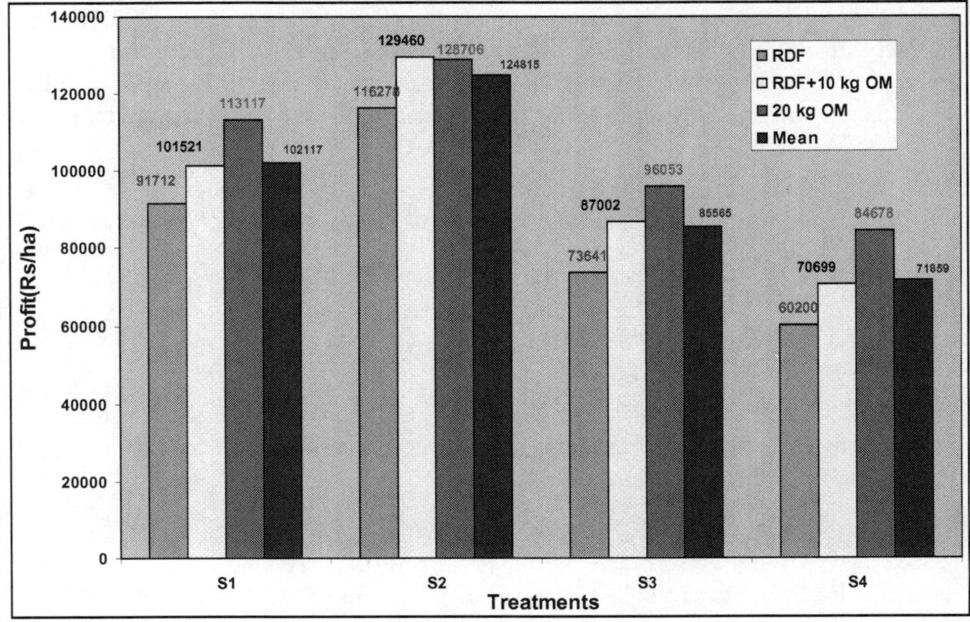

Figure 8.7: Net Profit for Five Years (2002-06).

yield. These treatments reduced the annual runoff (20 and 22 per cent of the annual rainfall compared to 37 per cent of the annual rainfall in control) and soil loss (47 and 49 per cent of control), increased the mean soil moisture content (15.6 per cent and 15.8 per cent dry basis compared to 11.6 per cent dry basis in control in March), nutrient content of the soil and leaf. The harvested rainwater increased the ground water level in nearby wells and ponds. The barren land even in steep slopes can be effectively utilized for cashew cultivation with proper soil and water conservation measures like modified crescent bund or coconut husk burial in staggered trenches opened across the slope.

References

Murthy, G.R.K., Soam, S.K., and Manoj P. Samuel. (2013). Reading Material. 97th Foundation Course for Agricultural Research Service, NAARM, Hyderabad. Available online. http: //www.naarm.ernet.in/images/stories/focars/97 per cent 20FOCARS per cent 20Reading per cent 20Material.pdf.

Bruijnzeel, L.A., Eijk-B-van, Purwanto, E., van-Eijk, B., Summer-W., Klaghofer, E. and Zhang, W. (1998). Runoff and soil loss from bench terraces in upland West Java, Indonesia and implications for process modelling. *Proc. International symposium*, 13 to 17 July. 1998. Vienna, Austria. IAHS Publication, 249: 211-220.

Chovatia, R.S., Singh, S.P. (1996). Management techniques in dry land fruit culture. *Scientific Horticulture*, 5: 51-55.

Deshmukh, M.T., Sawke, D.P., Borude, S.G., Hurni, H. and Tato, K. (1992). Effects of platform bench terraces on the growth and yield of mango and cashew grafts. Erosion-conservation and small scale farming report. Department of Agricultural Engineering, Konkan Agriculture University, Ratnagiri, India. pp. 477-482.

FAO (2013). Chapter 6. Integrated production practices of cashew in Myanmar - Maung Maung Lay. Available online. http: //www.fao.org/docrep/005/ac451e/ac451e06.htm

Ghosh, S.N. (1995). *The Cashew,* 9(3): 5-8.

Ghosh, S.N. 1999. Effect of rainwater harvesting on nut yield of cashew. Journal *of Plantation Crops,* 27(3): 225-229.

Ghosh, S.N., Mathew, B. and Manna, S. (2000). Effect of *in situ* moisture conservation on growth and yield of cashew. In: *Proceedings of International Conference on Plantation Crops,* Hyderabad, India, pp.61.

GOI (2001). General Overview of Drought - 2001. *Weekly Drought Report No. 9,* Government of India (GOI), New Delhi.

Kumar, D.P., Subbarayappa, A., Hiremath, I.G., Khan, M.M. and Sadashivaiah (1989). Use of coconut coir-pith: a biowaste as soil mulch in cashew plantations. *The Cashew,* 3: 23-24.

Kuylenstierna, J.L., Bjorklund, G. and Najlis, P. (1997). Sustainable water future with global implications: Everyone's responsibility. *Natural Resources Forum,* 21(3): 181-190.

Mane, M.S., Mahadkar, U.V., Ayare, B.L. and Thorat, T.N. (2009). Performance of mechanical soil conservation measures in cashew plantation grown on steep slopes of Konkan. *Indian Journal of Soil Conservation,* 37: 181-184.

Manivannan, S., Sikka, A.K., Manjunath, B.L., Priya Devi, S. and Ashok Kumar, J. (2007). Effect of bioengineering measures on runoff, soil and nutrient losses under cashew plantations in Western Ghat region. *Journal of Agricultural Engg.,* 44(4).

Rejani, R., Jha, M.K., Panda, S.N and Mull, R. (2003). Hydrologic and hydrogeologic analyses in a coastal groundwater basin, Odisha, India. *Applied Engineering in Agriculture* (ASABE), 19(2): 177–186.

Rejani, R., and Yadukumar, N. (2007). Response of cashew plantations grown in slope areas of West Coast region of India for soil and water conservation measures. *Proceedings of the National seminar on Research, Development and Marketing of Cashew,* 20- 21st November, Goa, pp. 63-64.

Rejani, R. and Yadukumar, N. (2010). Soil and water conservation techniques in cashew grown along steep hill slopes. *Scientia Horticulturae,* 126: 371-378.

Rejani, R. and Yadukumar, N. (2012). Chapter 13 - Response of cashew plantations grown in slope areas of West Coast region of India for soil and water conservation measures, pp. 95-104. Book chapter in "Research advances in cashew production utilization and management techniques", (Eds. A.R. Desai and N.P. Singh). New India Publishing Agency, Pitam Pura, New Delhi – 110 088, pp. 259.

Sekhri, S. (2012). Sustaining Groundwater: Role of Policy Reforms in Promoting Conservation in India. *Report of Indian Policy forum 2012*. India International Centre, New Delhi.

Tejwani, K.G., Gupta, S.K., and Mathur, H.N. (1975). *Soil and Water Conservation Research*, ICAR, NewDelhi, 1956-1971.

Veigus, G. (2009). Karnataka Cashew Development Corporation. National Group Meeting of Scientists on AICRP Cashew – 2009, TNAU, Coimbatore. November 13-15.

Yadukumar, N. and Mandal, R.C. (1994). Effect of supplementary irrigation on cashewnut yield. Water Management for Plantation Crops - Problems and Prospects. Centre for Water Resource Development and Management, Calicut, Kerala. pp. 79-84.

Yadukumar, N. and Rejani, R. (2004). Evaluation of soil and moisture conservation techniques coupled with manuring in cashew grown under medium slope. *Journal of Plantation Crops*, 32 (Suppl.): 190-195.

Cashew: Improvement, Production and Processing (2017) *Pages* **187–191**
Editor: **Prof. P.L. Saroj**
Published by: **ASTRAL INTERNATIONAL PVT. LTD., NEW DELHI**

Establishment of Orchard

K.M. Yuvaraj

1. Introduction

Cashew (*Anacardium occidentale* L.) is a tropical plant and acclimatized to wide rage range of soil and climatic conditions. In India, cashew is considered as a crop of marginal lands and grown mostly under rainfed conditions. In general, distribution of cashew is restricted to altitudes below 750m asl, where temperature does not fall <20 °C for prolonged periods but it can be grown up to 1000m asl. Cashew cannot tolerate prolonged periods of extremes of cold and frost. The average rainfall distribution in cashew growing areas ranged from 300-600 mm in Gujarat to 2700- 3000 mm in west coast and NEH region. However, higher productivity was recorded in regions with mean annual rainfall distribution of 600-1500 mm. The cashew plants respond well to basic inputs used for orchard management, thus more investment is directly proportion to higher yields but still cashew is grown where no other crops can grow. Though, cashew can be cultivated with little care but for sustainable yield, scientific establishment of orchard is very important.

2. Selection of Site and Land Preparation

Selection of land for large scale plantations of cashew poses an entirely different set of problems in India. While selecting the land for cashew, soils with water logging and excessive salinity or alkalinity should be avoided. Otherwise, soil depth, slope, stoniness, soil fertility and availability of water seem to impose little constraints, since cashew is a hardy plant. For raising new cashew plantations, land preparation should begin as soon as the pre-monsoon showers are received. The area should be cleared of bushes and other wild growth before digging for planting. The land should be ploughed thoroughly and leveled in case of agricultural lands. In case of forest lands, the jungle should be cleared well in advance and the debris should be burnt. After clearing the jungles, land is to be terraced or bunds constructed on slopy land. In order to ensure better moisture conservation, soil trenches are dug nut across the contours. Then the land is ploughed on both ways to uproot the stumps, stubbles, which are removed promptly. The land preparation work should be completed prior to the onset of monsoon season *i.e.* during May – June.

Cashew can be cultivated in almost all types of soils ranging from coastal sandy and red lateritic soils to red sandy loam soils in plains and soils with poor fertility status. It is largely believed that cashew is a wasteland crop and does not need much care, as a result of which it is given least attention. This mind-set needs to be changed. It grows well and performs better in well drained red sandy loams with pH in the range of 5.8 to 6.8. Acidic soils below this pH need to be ameliorated with proper soil amelioration practices. Sandy loam soils having depth of >180 cm, permeability 20-50 mm/h, salinity $<1dSm^{-1}$, pH 6.1-7.3, available water holding capacity >18cm and slope 3 per cent are considered as best suitable land for cashew cultivation (Mishra, 1984). Solis with poor drainage and water logging problems with excessive alkalinity and salinity are not suitable for cultivation of cashew and hence may be avoided for commercial orchards establishment.

3. Layout and Planting Systems

After preparation of land, systematic layout may be done for planting. Selection of appropriate planting system based on the topography of the land, fertility and type of soil is very important for establishing cashew plantations in different locations. Generally, Square, Rectangular and Triangular System of planting are recommended for cashew plantations.

i. **Square system:** Spacing between rows and columns of trees is same. Spacing of each tree in all the four directions is same. Recommended spacings are: 8m x 8m for highly fertile soils, 7m x 7m for medium and light soils and 5m X 5m (high density planting) for sandy or poor or very light soils.

ii. **Rectangular System:** Spacing between rows and columns of trees is different depending on the purpose and topography of the land.

iii. **Triangular System:** In this system, the spacing of all the trees will be equidistant in six directions and three trees are arranged in triangular manner. This accommodates 15 per cent more population compared to square or rectangular system.

iv. **Contour System**: In hilly terrain, slopy lands and undulating areas, planting should preferably be done along the contours, with cradle pits or trenches provided at requisite spacing in a staggered manner to arrest soil erosion and help moisture conservation. However, there is always better moisture buildup on lower portion of the slope than top portion, owing some variations in plant performance.

4. Plant Spacing

The plant spacing depends upon the vigour of the variety, soil types, fertility status, management practices and purpose of plantations. Cashew trees are generally planted with a normal spacing of 7.5 m X 7.5 m or 8 m x 8 m (175 plants/ha) for poor soils 10 m x 10 m for rich and deep soils and sandy coastal areas is recommended. High density planting of cashew at a closer spacing of 4 m X 4 m (625 plants/ha) in the beginning and thinning out in stages to maintain a final spacing of 8 m X 8 m in the tenth year is also recommended. This enables higher returns during initial years. In case of slopy lands, the triangular system of planting is recommended to accommodate 15 per cent more plants without affecting the growth and development of the trees. Now a day, ultra-high density system of planting is also started at a spacing of 4m x 3 m, 3m x 3 m, 2.5m x 2.5m. However, ultra-high density system of planting requires growing of dwarf stature varieties, canopy management responsive varieties, use of dwarfing root socks, use of growth regulators to promote flowering etc. Therefore, management aspect is crucial in ultra-high density system of planting.

5. Digging and Filling of Pits

The work of digging of pits has to be completed during summer (May – June). The size of pits for good soils is 60cm x 60cm x 60cm and for hard lateritic soils, 1m x 1m x 1m are recommended. The top soil and sub-soil are kept separately and allowed to wither under sun. It helps in migration of termites and ants. Burning of the debris and forest wastes inside the pits before planting is advantageous. The pits are then filled with topsoil mixed with farmyard manure or compost (10-15 kg/pit) or poultry manure (2 kg) and rock phosphate (200 g). Malathion @ 100g/pit is also added to the soil mixture, it helps in the prevention of termites and ants. A small drainage channel at the elevated side of the pit is made to divert the rain water to the sides of the pit in slopy lands. In slopy areas it is preferable to plant the grafts along with the contour. At the top of the pit should be opened across the slope to trap rain water and also to prevent soil erosion.

6. Planting Materials and Season of Planting

Selection of planting material is most important in perennial horticulture and that is true for cashew culture also. The planting materials must be procured from accredited nurseries or government nurseries. Seedlings raised from the raw nuts should not be used as planting material, as it gives lot of variation in growth and yield. Since, cashew is cross pollinated crop, vegetative propagation is recommend to produce planting materials true to type. Grafts are recommended for precocity, uniformity in quality and for higher yields. Soft wood grafting is the best technique for production of quality planting materials of cashew. The grafts should be 4-6 month old, healthy and vertical growing, 45-60 cm height, 5-7 functional leaves, 15-20 cm height from collar region to graft joint, perfect graft joint without any girdling or constriction, intact poly bag without damage, free from side sprouts from stock should be selected for planting. The most ideal time for planting of cashew grafts is second fortnight of June to July coinciding with the onset of monsoons. Planting during this period ensures better establishment of grafts in the field. If irrigation is assured, planting can be taken up at any time during the year.

7. Method of Planting

Proper planting ensures better field survival. Planting of grafts is done preferably during monsoon season. The soil in the center of the filled pit is scooped out. The polythene bag of the grafted plant is removed carefully without disturbing the ball of earth. Then the graft along with ball of earth is placed in the centre of the pit where the soil was scooped out and it is covered with soil and pressed gently. Care must be taken to see that the graft joint remains at least 10-15 cm above the ground level at the time of planting. The grafted plant should be supported with a stake and tied loosely with a gunny thread immediately after planting to avoid breakage at the graft joint due to strong winds. Staking should be continued during second year of planting also. Immediately after planting, the basin around the grafts should be mulched with green leaves or dry grasses. This suppresses weed growth and conserves soil moisture (Saroj *et al.*, 2014).

8. After Care

Initial care is essential for better establishment of cashew plantations. For the first two years, regular watering is necessary for the young grafts. The grafts are staked and protected from scorching sun in summer by providing shade with palmyra leaves or any other available thaching materials and from cattle damage by fencing. The basins of young plants may be mulched with dry leaves to conserve moisture in summer and also to suppress the weed growth (Nayak *et al.*, 2011). Weeds can also be controlled by spraying of glyphosate (5ml/litre of water).

The cashew is generally planted on the wastelands and hence availability of soil moisture is always low, hence, mulching is essential. Mulching with black polythene is beneficial to conserve soil moisture and subsequent better growth of plants. However, locally available materials like green or dry grass or weeds can be utilized for mulching the basins. Small pebbles or stones can also be used for mulching of the basin. The plastic or stone mulch does not improve soil health but ensures better moisture retention in the soil and also prevents attack of soil borne insects and pests.

Any sprouts coming from root stock *i.e.* below the graft union should be removed regularly. Also, if inflorescences are coming in some plants in first year of planting, they should be removed in order to provide better strength to the plant. Plants must be trained to develop desired frame work by promoting branches in all directions. No branch should be allowed up to 1m above the ground level. The need based disbudding/pinching of undesired shoots should be done. Recommended dose of manures and fertilizers may be applied to promote better growth, however it varies with the site conditions. After care of cashew plants in sandy soils is more difficult compared to coastal regions. The fertilizers should be applied immediately after cessation of rains. While application of fertilizers, adequate moisture in the soil should be ensured so as to absorb the applied nutrients by the plant. Sometimes, plants are also affected by foliar pests and fungal diseases; they should be controlled after proper ascertaining of causal organism.

References

Mishra, D.P. (1984). Land suitability classification III. A case study. *Cashew Causherie*, 7(2): 12-19.

Nayak, M.G., Bhat, P.S. and Sajeev, M.V. (2011). Cashew Production Technology. Pub. ICAR-Directorate of Cashew Research, Puttur, pp. 1-43.

Saroj, P.L., Krishna Kumar, N. K. and Jankiram, T. (2014). Converting wastelands into goldmine by cashew cultivation. *Indian Horticulture*, 59(6): 49-56.

Cashew: Improvement, Production and Processing (2017) *Pages* **193–206**
Editor: **Prof. P.L. Saroj**
Published by: **ASTRAL INTERNATIONAL PVT. LTD., NEW DELHI**

10

Production of Quality Planting Materials

P.L. Saroj

1. Introduction

Establishment of an ideal nursery and production and supply of quality planting materials are the vital components for development of good cashew plantations. The quality planting materials refer the plants stock *i.e.* i) true-to-type, ii) propagated by appropriate vegetative technique, iii) healthy and free from any biotic pressure and iv) should be as per standard. Before standardization of vegetative propagation technique, cashew was multiplied by seeds and being a cross pollinated crop a lot of variability in plant vigour, precocity, bearing habit, nut yield, fruit colour and its quality etc. were observed. For production of planting materials through vegetative means, researchers have tried various techniques of cutting, budding, layering and grafting with varying levels of success both under open and partially controlled conditions. After massive efforts, soft wood grafting has standardized. Now, cashew industry has been revolutionized with the development of soft wood grafting technique for the mass multiplication of quality planting materials. However, the clonal multiplication of rootstocks in cashew is yet to be standardized, thus the raising of rootstocks by seeds are still continue. Some efforts were also made for *in-vitro* multiplication of cashew but results were not very encouraging. The regeneration protocol from matured explants are yet to be developed although reports are available regarding the multiplication and field establishment of cashew regenerated from young cashew nodal cuttings (Thimmappaiah and Shirly, 1999). Somatic embryogenesis was induced from nucellar tissue cultured from 2-3 week old immature nuts. So far nucellar tissues from 14 varieties were tried for embryogenesis. Embryogenesis and germination of somatic embryos was achieved in two varieties (Thimmappiah, 1997; Shirly and Thimmappaiah, 2005). Thimmappiah *et al.* (2002a) also got success in micrografting of cashew.

2. Establishment of Cashew Nursery

There are different components of nursery required for production of quality planting materials, as discussed here under;

i) Establishment of Scion Bank

Establishment and maintenance of scion bank is prerequisite for collection of scion shoots and subsequent grafting operation. The Scion Bank should be established with the recommended cultivars of the region and planted at spacing of 4m × 4m. The area of scion bank depends upon the target of planting materials but in order to get continuous supply of scions, sufficient plants should be accommodated. The plants in the scion bank should be maintained by heading back to a height of 1.5 m from ground level and regular pruning of dead and dried branches should be done. Pruning of tees may be carried out annually during September–October. The flower panicles should be removed every year in order to get more number of scions. Proper nutrition, irrigation and aftercare of scion bank are very essential. From second year after planting, scion shoots can be obtained. The ideal scion shoots are non-flowered, 3-5month old current season's growth, about 10-12 cm long, straight, pencil thickness, brown coloured with dormant plumpy terminal bud. Clipping off leaf blades, leaving petiole should be done about a week before detaching the scion shoots from the mother plant. The scion sticks should be taken from the mother plants early in the morning to avoid desiccation. Soon after separation from the mother plant, scion sticks should be wrapped in polythene and brought to the nursery shed for grafting. For long period storage or transportation; the scion sticks should be wrapped in sphagnum moss/cotton cloth and placed in a polythene bags.

ii) Raising of Rootstocks

In cashew rootstocks are still raised by seeds. Various size and colour of polybags were tried for raising of rootstocks of cashew but more commonly used high density polybags are 25 cm x 15 cm size of 300 guage and black in colour. If the grafts are to be maintained only for about six months before planting in the field, the size of ploybags can be reduced to 20 x 15 cm. Make 15-30 holes of 0.5 cm diameter uniformly on the polythene bags by punching for ensuring drainage of excess water during heavy rains. In low rainfall areas the number of drainage holes can be reduced also. Now a day, various types of root trainers are also available which should be tried for raising of rootstocks. Proper filling media is required for satisfactory seed germination and subsequent seedling growth. For cashew, the filling mixture is prepared in the proportion of 1:1:1 (red soil: sand: compost) in heavy rainfall areas and 1:1(red soil: compost) in low rainfall areas and mixed with 5g of rock phosphate. However, vermi-compost, poultry manure, coco-pith etc. are also used as filling mixture. Fill the mixtures in the polybags up to the brim and arrange in nursery beds of 10 numbers in each row to a desired length. In between two polybag beds, there should be sufficient space for grafting operation.

The current season harvested nuts, sun dried for 3 days and about 6-8 g weight is used for sowing. Seed nuts should be soaked in water over night before sowing in polybags. Pre-soaked nuts are sown in the centre of the polybag with stalk end up at a depth of 2.0–2.5 cm. At the time of sowing, the soil in the polythene bags should be moist and loose. Immediately after sowing seed nuts, light watering should be done with sprinkler. Polythene bags should be covered with paddy straw/dry grass/palm leaves to facilitate early germination. The seed nuts usually germinate within 15-20 days after sowing during monsoon season and within 8-10 days during summer months. Daily watering with sprinkler is required for germination and growth of seedling. The mulch material should be removed once the emergence of seedlings observed. These seedlings will be ready for grafting in 45-60 days after germination. Moreover, use of biofertilizers (*Azospirillum*, *Azotobacter* and VAM) increased the germination percentage of nuts, plant growth, number of graftable seedlings and reduced the incidence of fungal diseases in the nursery (Sinish *et al.*, 2005). Oblisami *et al.*,1985 also reported that inoculation of *Azotobacter* resulted in higher root growth of cashew seedlings. Lakshmipathi *et al.* (2013) reported that osmopriming of seed nut by GA_3 (200 ppm) for 48 hours has increased seedling growth and vigour index; thereby obtained healthy seedlings in short period for grafting operation. Drenching of Bordeaux mixture (1 per cent) is recommended to avoid damping off young seedlings.

iii) Establishment of Growing Structures

There are various types of growing structures based on ecological situations, type of covering material used, weather control system, growing needs and cost involvement with or without misting system. Some structures are fully auto operated; some are partially controlled while some are manually managed. In hi-tech structures, there is fully controlled system of temperature, humidity, fogging and photoperiod. For cashew nursery, after grafting operation, partial shade is required. In general, green shade net or poly house are used for this purpose to create better microclimatic conditions inside the structure for quick union of rootstock and scion and subsequent sprouting of scion shoots. Once, shoots are sprouted, shade net can be removed for better exposure of spouts to sun light but optimum humidity must be maintained by regular watering with sprinkler. Moreover, proper sanitation and maintenance is required to maintain growing structures. Still most of the cashew nurseries are operated under partial shade net houses.

iv) Care of Nursery Plants

Raising of rootstocks, grafting operation and aftercare of grafted plants are very essential operations. Since, seedlings are tender in nature at initial stage, proper and timely irrigation is very important to obtain healthy growth. Also, avoid excess irrigation which otherwise cause of wilting of seedlings due to fungal attack. There should not be water stagnation in and around nursery beds. In case of cashew, the proper filling mixture contains sufficient nutrients for the growth of seedlings and no addition supplement is required. However, foliar feeding of micronutrients, plant growth regulators and use of bio-consortia improve healthy seedling. After grafting, any sprouts below the union should also be removed. In general, complete nursery should be weed free but any weed growth inside polybags should not be allowed and removed manually. Any symptom of pest and disease occurrence must be taken care immediately. Tall growing plants along the boundary of the nursery should be lopped regularly to avoid shade.

v) Hardening and Disposal Unit

Proper hardening is required before disposal of planting materials. In fact, hardening is done to make the plant tissues firm so as to withstand against possible adverse conditions during transportation and assure high field survival. Hardening should be done by withholding watering and followed by exposure to sun light which may stop the growth of plant. However, hardening process should be gradual in order to avoid severe setback. Thus, only after attaining about 5-6 months' age with 5-7 functional leaves and properly hardened plant should be sold or planted. Timely disposal of plants is also important to get good income. Keeping the plants for longer period in the nursery not only involves additional expenditure but also start coiling of roots. Timely lifting of polybags plants and proper packaging is also important so that plants should not be damaged during long distance transportation. As far as possible, there should be minimum time gap between lifting of the plants from the nursery and planting at desired site.

3. Propagation of Cashew

i) Seed Propagation

The concept of seed garden in cashew is not popular and mature seed nuts collected from the orchard after drying are sown for raising the seedling rootstocks. However, only vigorous seeds (6-8 g) and free from any damage should be used. The current season collected mature nuts can be sown for raising of seedlings, though the nuts are viable for one year. After collection of cashew fruits along with nuts from the field, nuts may be detached from the fruits and sun dried for about 3-4 days. The properly dried nuts can be stored in gunny bags on a cool and dry place. The seeds can be stored in baskets but the best method of storage of seeds is in air tight tins. The seed nuts floating after soaking in water should not be used for sowing. Overnight soaking of selected seed nuts in water before sowing improves germination. It is also advisable that up to seven months old seeds can be used for sowing and thereafter germination reduced drastically and after 12-14 months of storage, germination ceases fully. Moreover, seed propagation is done only for raising of rootstocks. It is not recommended for establishment of cashew orchard becuase of following limiatations:

1. The plants raised by seeds are not true-to type and possess a lot of variability with respect to plant vigour, yield and quality.
2. The orchards raised by seedlings have long gestation period.

3. Sometimes, there is possibility of transmission of seed born diseases.
4. Cultural operations are also difficult in orchards raised by seedlings due to vigorous and uneven growth.
5. In general, yield potential of seedling raised orchard is low.

ii) Vegetative Propagation

The vegetative propagation of cashew is important because; i) the plants produced are true-to-type and genetically uniform, ii) comparatively short stature, short gestation period and hence early bearing, iii) higher yield and quality produce and iv) easy in intercultural operations. Propagation of cashew by different methods of vegetative propagation techniques have been tried with varying levels of success are as follows:

Cutting

Propagation of cashew by the cuttings was tried by different research workers but very little success was obtained, though Peixoto (1960) reported that propagation of cashew by cuttings is possible. The rooting was obtained with lateral shoots and water shoots from lateral buds. He stated that large quantity of carbohydrates in relation to the available soluble nitrogen in the shoots favours rooting. The rooting can be improved by ringing the bark or by wrapping a thin wire around the twig, forty days before separating it from the twig. Milheiro (1969) reported that in Mozambique cuttings from mature twigs only rarely produced some roots and with green twigs, no rooting was obtained at all. The first successful trial on cutting was done by Argles (1969) in Mysore, where he used etiolated shoots (30 days before taking cuttings) in the month of November and got 40 per cent rooting. Retaining leaves on cutting improved rooting but cuttings with flowers did not root well. The response was increased by dipping the cutting base in sugar solution (15 per cent) The best treatment for callus formation was 15 °C and for rooting 24 °C. Coester and Ohler (1976) obtained satisfactory results with cuttings taken from cashew tree growing in green house, treated with 1 per cent IBA and planted in perlite+coarse sand mixture. They retained leaves of cuttings, sprinkled water daily, covered with plastic and maintained humidity between 90-95 per cent during day time and 100 per cent during night time. Day temperature was kept 25-28 °C and night temperature at 15 °C. Thus, shade, good aeration of rooting medium, high relative humidity and exogenous hormonal treatments are important factors determining the success of the cuttings. However, propagation of cashew by stem cutting was not successful at Baptala (Andhra Pradesh) and Ullal (Karnataka) stations (Rao, 1979 and Krishna Murthy *et al.*, 1985). Nageswara Rao *et al.* (1988a) reported that even juvenile shoot cuttings are difficult to root probably due to low level of endogenous rooting factors. Further, Nageswara Rao *et al.* (1988b) observed highest rooting percentage, number of roots and root length with post ringing of 180 days. The ringed cuttings were treated with IBA (10000 ppm) by quick dip and placed in vermiculite for rooting under mist. The hardening of rooted cuttings was better under restricted mist than continuous mist as far as the root dry weight and field establishment were concerned (Nageswara Rao *et al.*, 1989). Overall, propagation by cuttings gave less success, more expensive and plant raised by cuttings have poor tap root system thus, vulnerable to strong winds. Now, propagation of cashew by cutting is not a recommended practice.

Layering

In case of layering, shoots are detached from mother plant only after rooting; hence there is a better possibility of propagation by layering as compared to cutting. In older trees, lower branches tend to trail on ground, if covered with soil and keeping moist may produce roots. However, the ground layers cannot be transported to other places, thus it may not be efficient method of cashew propagation. Encouraging higher branches to root into pots placed on a support has been tried in Madras but this was not successful (Argles, 1969). Sometimes, broken, bend and damaged plant can be rejuvenated by this method. Rao (1979) reported that cashew branches of 2.5 cm thickness trailing near the ground were girdled and pressed in the soil were produced profuse rooting from June to October under Bapatla conditions. However, mound layering/stool layering tried at CPCRI, Regional Station, Vittal has revealed that it is possible to obtain profuse rooting when etiolated shoots are cinctured and treated with 10,000 ppm IBA. In this technique, heading back of main trunk and encouraging coppicing from the ground level followed by girding of shoots and covering with soil can give some success. The cinctured shoots induce roots within 30-45 days. Nagbhushanam and Menon (1980) reported good success through mound layering with treating the shoots after girdling by IBA 10, 000 ppm while Suryanarayana and Melanta (1989) observed significant decrease in rooting of stool layers with increase in age of shoots.

Though less efficient but air layering was one of the successful methods of vegetative propagation of cashew till 1980. Naik (1949) first reported that layering done during rainy season readily rooted within two months with high percentage of success at Fruit Research Station, Kodur (Andhra Pradesh). Abraham (1956) also attempted air layering in cashew successfully. Rao and Hassan (1957) reported that maximum rooting was obtained when air layering was done during flowering season. Even in less humid region of East-Coast and during flushing period in West-Coast, it was successful. The percentage success of establishment of layers is depends on seasonal variation in different regions (Rao, 1958). The influence of parent tree, age and type of shoots, position of cincturing, use of rooting media, use of plant growth regulators etc. were also studied by different workers. Chhokar and Singh (1967) obtained 85 per cent rooting by treating the shoots with 75 ppm IBA in lanolin paste. Rai (1970) also got good success in air layering of cashew shoots. Whereas, Acharya and Das (1972) reported that treating the shoots with 300 ppm IBA in petroleum jelly gave maximum successful layers (85 per cent). Similar observations made by Sadhu *et al.* (1972) by using *p*-hydroxybezoic acid alone and additive effect in combination with IBA. Krishna Murthy *et al.* (1985) narrated that use of IBA 300 ppm shortened the period of rooting emergence with maximum number of rooted layers. IAA at 250 ppm gave highest percentage of rooting with higher length of roots and more number of roots per shoot.

Nagabhushanam and Menon (1978) have assessed prospects of vegetative propagation of cashew by air layering and suggested that air layering can be done on non-flowered shoots of previous year's growth of pencil thickness and brown in colour after ringing and treating with IBA. Under east Coast conditions, the better success of air layers was found when trees are in resting phase (monsoon season) than during the period of growth and reproduction (October-April) (Rao, 1979 and Palaniswami *et al.*, 1979). July- August appears to be most ideal period for air layering in East Coast region, when rooting and field establishment both is good. Whereas, during summer season (January-April), cashew trees are in active phase in the West Cost and was found suitable for air layering with 73-100 per cent success (Damodaran, 1979; Nagabhushanam, 1979b; Nagabhushanam and Murthy, 1979; Valsakumari *et al.*, 1979 and Krishna Murthy *et al.*, 1985). Damodaran (1979) reported that the best time for air layering was February-March, which will enable detachment of layers in May-June and subsequent planting in field for better establishment.

Not only age of shoots but age of cashew trees and their health also play some role in success of air layering. The air layered shoots of trees below ten years rooted early than 20 years old trees (Rao and Hassan (1957 and Krishna Murthy *et al.*, 1985), whereas Damodaran (1979) reported that age of tree had no effect on extent of rooting of air layers. It was also found that the one year old shoots rooted better than current season growth (Rao and Rao, 1957). The longer shoots rooted better as having relatively more number of leaves which have been conducive for higher rooting due to increased availability of photosynthates (Krishna Murthy *et al.*, 1985). Besides length of shoots, Damodaran (1979) reported that non-flowered shoots gave significantly higher percentage of rooting than flowered shoots. He further reported that the thicker shoots (girth of 4 cm and above) were found more vigorous in growth during first year of orchard life as compared to medium shoots (3-4 cm) or thin shoots (< 3 cm). Ringing/cincturing of shots at nodal region gave better success (Rao and Hassan, 1957 and Krishna Murthy, 1985).

Regarding use of different kinds of rooting media, Rao and Hassan (1957) and Krishna Murthy *et al.* (1985) reported positive response while Damodaran (1979) and Palaniswami *et al.* (1979) did not find good impact of rooting media on rooting success. Among different types and thickness of wrapping materials, polythene film of 100 gauge proved to be the best for rooting of air layers (Rao and Hassan, 1957 and Krishna Murthy *et al.*, 1985). Survival of layers under field conditions is normally poor, hence post separation treatment and providing conducive conditions are very essential. Krishna Murthy *et al.* (1985) suggested that layers must be defoliated after separation to minimize transpiration loss. Whereas, Rajeevan and Srinivasan (1979) reported that for better survival layers must be planted in decomposable containers of coconut husk pot and straw pot. It was also suggested that air layers after separation may be dipped in cow dung-urea solution (100 g urea in 100 liters of cow dung slurry for about an hour) and thereafter planting in polybags (45x30 cm) of 500 gauge (Anon., 1979).

Budding

Propagation of cashew by budding was attempted successfully by different workers. In fact, budding is preferred over other vegetative methods because of better efficiency, as from single scion shoots more than 4-6 buds can be taken for budding operation. In India, Naik (1949) had first reported successful budding in case of cashew. Thereafter, Phadnis *et al.* (1974) successfully demonstrated patch budding in the month of October-November with 67 per cent success in Maharashtra. Whereas, Hameed (1976) reported 76 per cent success of patch budding in the month of July in Tamil Nadu. Bhatee (1977) claimed that better success can be obtained with bud graft on older rootstocks (2 year) raised in polybags. A success of 80 per cent in the month of March-April has been claimed by Gunjate and Limaye (1979) by forkert method of budding on two months old seedling rootstocks raised in polythene bags. In fact, the best time for budding is when the bark can easily be removed from the wood which is the indicator that the tree is in active growth phase. Under East Coast of Vridhachalam conditions, *in-situ* patch budding in July gave 71 per cent success (Palaniswami *et al.*, 1979), at Bapatla 32-41 per cent during June-September (Rao, 1979) and less than 16 per cent (T budding) under Odisha conditions during January-February, June-July and September (Das and Mishra, 1979). Now a day, this method is also not popular.

Grafting

The first systematic work on propagation of cashew by approach grafting was initiated at Central Cashewnut Research Station, Mangalore; wherein 11 months old seedlings with a girth of about 4.5 cm and a height of 50-60 cm were used. Nagabhushanam and Vekata Rao (1977) reported 49 per cent success when veneer grafting was performed in the month of September with 15-20 months old seedlings at cashew Research Station, Bapatla (Andhra Pradesh). The wedge grafting was successfully tried on cashew seedlings raised in polythene bags with pre-cured or fresh scion (Bhandary *et al.*, 1974). In case of epicotyls grafting, Bhandary *et al.* (1974) reported that 21 days old seedlings when grafted with thin scion (0.3 cm) gave 63 per cent success at Dharwar. Konhar and das (1985) reported that 5-7 days old seedlings are most suitable for grafting throughout the year except in March under Odisa conditions with 56-100 per cent success. Whereas, Nagbhushanam (1983) reported 10-15 days old seedlings are ideal for epicotyls grafting during monsoon period (June-November) at Vittal (Karnataka) with 86 per cent success. Sawke (1984) reported that with 6-10 days old rootstocks, the period from February to May was congenial for grafting under Vengurala conditions with 60-74 per cent success. Khrisna Murthy *et al.* (1985) observed that February to may were ideal time for epicotyl grafting with 72-80 per cent success at Mangalore. Konhar and Das (1985) stated that soft wood grafting on seedlings of 15-60 days can be done throughout year except October and November under odisha conditions with 53-100 per cent success. In order to avoid mortality of epicotyls grafts, modified method of epicotyl grafting was suggested by Seshadri and Rao (1985) wherein the of the rootstock is removed leaving bottom two leaves intact. Then a longitudinal cut of 4-5 cm from tip was made on the rootstock so that each half contains one leaf followed by inserting wedge of scion and tied with plastic strip. Flush grafting with 15-21 days old flushes as scion stick in case of epicotyl grafting or soft wood grafting has also been reported from AICRP centre Madakkathara, Kerala (Veeraraghavan, 1990 and Pushpalatha *et al.*, 1090).

Among various vegetative propagation methods tried, softwood grafting was found to be the best technique for mass multiplication of cashew (Sawke *et al.*, 1985; Swamy *et al.*, 1993). Under warm humid conditions of Konkan, Maharashtra, the technique was successful (65-86 per cent) throughout the year except in December. The percentage of success of softwood grafting technique varied in different agroclimatic regions. Under Ullal conditions, soft wood grafting gave 70-75 per cent success during March-May (Krishna Murthy *et al.*, 1985). While Swamy *et al.* (1988) incated that monsoon period (June-November) is ideal for commercial production of soft wood grafts with highest success in September (82.3 per cent), followed by October (78.5 per cent) and August (72.5 per cent). Konhar and Das (1985) obtained 90 to 100 per cent success during January and February months under Bhubaneswar conditions. According to Kumar *et al.* (1989), the best time for softwood grafting of cashew is May followed by April. Ghosh (1990) obtained 70 per cent success in the month of February under Jhargram conditions while Radhakrishna *et al.* (1992) reported 79.80 per cent success in the month of December under Bapatala conditions. Clonal propagation trial by softwood grafting in cashew under Northern dry zone of Karnataka showed higher rate of success (87 per cent) in the month of January followed by October (60 per cent) (Mahesha *et al.*, 2005). *In-situ* soft wood grafting by wedge method on one-year-old cashew plants under field conditions having bronze coloured leaves with 71.4 percent success was also standardized by Amin (1978). Now, this is most widely adopted method of cashew propagation in the country.

In soft wood grafting technique, about 45-60 days old rootstocks raised in polybags and current season scion shoots are used for grafting purpose. The grafting operations should be done either in shade net house or under shady place. There should be proper arrangement of irrigation by sprinkler system. Two pairs of bottom leaves are retained in the selected root stock and other leaves are removed using a sharp grafting knife. At a height of 15-20 cm from ground level where soft wood portion is available on the root stock, a transverse cut is made and the terminal shoot is removed. A cleft of 4-5 cm deep is made in the middle of the decapitated stem by giving a longitudinal cut. A matching thickness scion shoots of 10-20 cm length should be selected. The cut end of the scion is shaped to wedge of 4-5 cm long by chopping the bark and wood from two opposite sides. The cut surfaces of scion should not be soiled by touching with fingers. The wedge cut of the scion is inserted into the cleft of the rootstock in order to make a perfect contact of cambial layers of both root stock and scion. The graft joint is tied firmly with 1.5 cm wide and 30 cm long polythene strip of 100 gauge thickness followed by covering with inverted narrow white polytube/polycap of 12 cm x 4 cm size to protect the scion stick from drying up and to create conducive microclimate for union and subsequent to encourage sprouting of the terminal buds. After two weeks the polytubes/caps are removed gently. After sprouting, graft can be shifted or shade net can be removed. Within 3-4 weeks, most of the grafts will sprout, which are ready for sale/planting within 4-5 months after attaining height of about 30-45 cm. Shifting of grafts by holding the root stock should be done in order to prevent from striking roots into ground.

4. Plant Standard

Plant standard is defined as an ideal physical state of a plant based on certain quantitative and qualitative parameters. The standard for seedlings, vegetatively propagated plants, clonal root stocks, scion and component plants are different which depends upon type of plants, variety and method of propagation. In case of cashew, following standards have been suggested for soft wood grafting based on experimentation and experiences:

Parameters	*Standard*
Method of propagation	Soft wood grafting by wedge method
Type of root stock	Healthy (Green foliage, free from any insect problem, without any symptom of dieback and nutrient deficiency), 0.5-0.7 cm diameter
Size of polythene bag	25 x 15 cm (300 guage)
Age of root stock	45 – 60 days
Age of scion shoots	3-5 months
Diameter of scion	0.5 – 0.7 cm
Union matrix	15-20 cm above ground level
Length of scion sticks	10 –12 cm
Season for grafting	Throughout the year, preferably June –November
Grafted plant height	30-45 cm
Grafted plant stem girth	2 – 2.5 cm
Conditions of grafted plant	Healthy, graft joint should be perfect without any girdling, with 5-7 functional leaves, free from side sprouts from rootstock,

5. Entrepreneurship in Cashew Nursery

Accreditation of cashew nursery is essential for production of quality planting materials. For accreditation, the nursery owner must have to maintain all records related to production of planting materials. The accreditation of nursery is done by a group of experts through physical verification. Based on the status of nursery the accreditation agency will award certificate for production of planting materials. In general, certification of nursery is done respective state department based existing nursery act. However, in case of cashew such certification was not taken up earlier by the government department as most of the cashew nurseries were started in public departments like ICAR, SAUs, DCCD, Forest Department, Horticulture Department and Cashew Corporations. Now a lot of private nurseries have been started in private sector also. Initially, accreditation of cashew nursery was started by National Horticulture Board but now the responsibility of cashew nursery accreditation also entrusted to DCCD, Kochi both for public and private sector nurseries. Therefore, to ensure the quality and genuineness, growers are advised to purchase the planting materials either from government nurseries or from accredited nurseries.

Establishment of cashew nursery is a profitable enterprise if managed properly. However, it depends upon the availability of input, labour, skilled grafter and other nursery infrastructures. An estimate was made for producing 50,000 saleable cashew grafts at the rate of about 70 per cent success. To obtain 50,000 saleable cashew grafts about 72,000 grafting has to be done. Moreover, there are some components like, establishment of shade net, procurement of scion shoots and sillpauline; the cost can be minimized. As per this estimate, total cost involve for production of 50,000 saleable cashew grafts is Rs. 7,93,350/-. If it is sold at Rs.20/graft, one can earn Rs.2,06,650/-year as net profit. Moreover, this profit margin can be increased by utilizing family manpower, maintenance of own scion bank, reduction of miscellaneous cost as the materials can be used in subsequent years also etc.

6. Future Line of Action

- ☆ There is an urgent need for development of *in-vitro* protocol for mass multiplication of quality planting materials.
- ☆ Identification and multiplication of compatible dwarfing root stocks in cashew and related genera.
- ☆ Development of chain of accredited cashew nurseries.
- ☆ Skill development in cashew nursery management.

References

Abraham, E.V. (1959). Air layering is a success with cashew. *Indian Farming*, 9(3): 14-15.

Achraya, N. and Das, P.C. (1972). Effect of two plant growth substances on cashew air layers, *Current Science*, 41(14): 534-535.

Amin, R.S. (1978). Softwood Grafting – A new technique for hard wood plants. *Current Science*, 47: 468-69.

Anonymous, (1979). Success in cashew vegetative propagation. Proceedings of Group Discussion on Cashew Vegetative Propagation, held at CPCRI, Regional Station, Vittal, 26-27 November 1979. p. 9.

Argels, G.K. (1969). *Anacardium occidentale* (cashew). Ecology and Botany in relation to propagation. Paper presented in conference on 'Propagation of Tropical and Sub-tropical Fruits', London (UK), 22-24 September, 1969.

Bahttee, S.S. (1977). Vegetative propagation of cashew by bud grafting of young seedling in nursery. *Cashew Bulletin*, 14(2): 8-10.

Bhandary, K.R., Shetty, K.P.V. and Shet, M. (1974). Propagation of cashew by wedge grafting. *J. Plantation Crops*, 2: 37.

Chhokar, V.S. and Singh, R. (1967). Effect of plant growth regulators on air layering in cashewnut (*Anacardium occidentale* L.), *Indian J. Hort.*, 24: 26-29.

Coester, W.A. and Ohler, J.G. (1976). Cashew propagation by cuttings. *Tropical Agriculture* (Trinidad and Tobago), 53(4): 353-358.

Damodaran, BV.K. (1979). Vegetative propagation of Cashew – Review of work done in Kerala. *Acta Horticulture*, No. 108 (1985): 51-56.

Das, R.C. and Mishra, S.N. (1979). Vegetative propagation in Cashew. *Acta Horticulture*, 108 (1985): 205.

Ghosh, S. N. (1990). Effect of season on success of softwood grafting in cashew in semiarid region in West Bengal. *Cashew Bulletin*, 27 (12): 15-19.

Gunjate, R.T. and Limaye, V.P. (1979). Vegetative propagation in cashew. Pub. Cashew Research Station, Vengurla of Dr. B.S.K.K.V. Dapoli, Maharashtra.

Konhar, T. and Das, G. C. (1985). Epicotyl and softwood grafting - new techniques of vegetative propagation of cashew in Odisha. *Cashew Causerie*, 7(3): 13-13.

Krishna Murthy, K., Khan, M.M., Hiremath, I.G., Nagaraju, A.P., Mallik, B. ande Hegde, M. (1985). Three decades of cashew research at Agricultural Research Station, Ullal (1954-1984). *Tech. Bull.*, No. 3. p. 44.

Kumar, D. P., Subbarayappa, A., Hiremath, I. G., Khan, M. M. and Sadashivaiah (1989). Use of coconut coir-pith: a biowaste as soil mulch in cashew plantations. *The Cashew*, 3(3): 23-24.

Lakshimipathi; Adiga, J.D., Kalaivannan, D. and Saroj, P.L. (2013). Response of cashew (*Anacardium occidentale* L.) to osmopriming with gibberellic acid (GA_3). *Journal of Plantation Crops*, 41(3): 455-459.

Mahesha, U.S., Hanamashetti, S. I. and Hubballi, V. N. (2005). Effect of season on success of softwood grafting in cashew (*Anacardium occidentale* L.) under Northern dry zone of Karnataka conditions. *The Cashew*, 19 (3): 17-19.

Milheiro, A. Vaz. (1969). A propagacao vegetative do cajuerio por enxertia de gono terminal. *Agronomia Mocambicana* (Mozambique), 3(4): 179-186.

Nagabhushanam, S., Murthy, M.G. and Mohan, E. (1980). A note on field establishment of cashew (*Anacardium occidentale* L.) air layers, *Indian J. Hort.*, 37: 151-53.

Nagabhushanam, S. (1983). A study on epicotyl grafting in cashew (*Anacardium occidentale* L.). *Indian Cashew J.*, 15 (1: 13-16.

Nagabhushanam, S. and Venkat Rao, P. (1977). Propagational trial in cashewnut, *Indian Cashew J.*, 11(3): 7-11.

Nageswara Rao, M.B., Satyanarayana, G., Ranga Reddy, B. and Shivaraj, A. (1988a). Effect of duration of post ringing period on rooting of shoot cuttings of cashew (*Anacardium occidentale* L.). *Indian Cashew J.*, 18 (4): 13-15.

Nageswara Rao, M.B., Satyanarayana, G., Rameswar, A., Shivaraj, A. Gnana Kumari, N. and Padmanabhan, V. (1988b). Interaction effect of certain phenolic compounds with IBA on induction of rooting in stem cuttings of Cashew (*Anacardium occidentale* L). *Indian Cashew J.*, 18 (4): 9-10.

Nageswara Rao, M.B., Satyanarayana, G., Gnana Kumari, N. and Padmanabhan, V. (1989). Effect of method of potting and hardening on field establishment of rooted cuttings of cashew (*Anacardium occidentale* L.). *Indian Cashew J.*, 19 (2): 15 – 17.

Naik, K. C. (1949). *South Indian Fruits and their Culture*, Pub. P. Varadachari and Company, Madras.

Oblisami, G., Santhanakrishnan, P., Pappaiah, C. M. and Shanmugavelu, K. G. (1985). Effect of Azotobacter inoculant and growth regulators on the growth of cashew. *Acta Horticulture*, 108: 40-45.

Palaniswami, V., Hameed, A.S. and Vijaya Kumar, M. (1979). Vegetative propagation in cashew – Work done at Vridhachalam. *Acta Horticulturae*, No. 108 (1985): 67-70.

Peixoto, A. (1960). Caju.Servico de Informacao Agricola, Ministerio de Agricultura, Rio de Janeiro (Brazil), p.61.

Phadnis, N.A., Choudhary, K.G. and Banberkar, D.G. (1974). Studies in raising of cashew (*Anacardium occidentale* L.) clonal material *in situ*. *Indian Cashew J.*, 8(2): 7-13.

Pushpalatha, P.B., Veeraraghavan, P.G. and Sitarama Rao, D. (1990). Flush grafting in cashew. *Cashew Bull.*, 27 (12): 9-14.

Radhakrishna, Y., Subbarao, N., Sathyanarayanareddy and Ramadevi, M. (1992). Propagation of cashew (*Anacardium occidentale* L.) by softwood grafting under Bapatla conditions. *The Cashew*, 6(2): 3-5.

Rai, B.G.M. (1970). Cashew through vegetative propagation. *Intensive Agriculture*, 7: 17-19.

Rajeevan, M.S. and Srinivasan, K. (1979a). Studies on cashewnut air layers-1. Use of different decomposable containers and potting mixtures as aids in maximizing field survival and establishment. *Acta Horticlture*, No. 108 (1985): 287-88.

Rao, V.N.M. (1957). A note on side grafting of cashew (*Anacardium occidentale* L.). *Indian Journal of Agricultural Sciences*, 27(4): 451-452.

Rao, V.N.M. (1958). Studies on vegetative propagation of cashew (*Anacardium occidentale* L.). I Air layering by cincturing and etiolation of shoots. *Indian J. Agril. Sciences*, 28(2): 181-197.

Rao, V.N.M. and Hassan, M.V. (1957). Studies on vegetative propagation of cashew (*Anacardium occidentale* L.). III. Further studies on air layering. *Indian J. Agril. Sciences*, 27(4): 453-464.

Rao, S.N. (1979). Vegetative propagation of cashew-Review of work done at Baptala. *Acta Horticulture*, 108 (1985): 64-66.

Sadhu, M.K., Naskar, B.B. and Basu, R.N. (1972). Auxin synergist in rooting of air layers of tropical fruit trees, *Indian Agriculturist*, 16(3): 251-257.

Sawke, D.P. (1984). A new technique of vegetative propagation 'epicotyl grafting'. *Cashew Causerie*, 6 (1): 2.

Sawke, D. P., Salvi, M. J. and Patil, M. M. (1985). Softwood grafting - A sure technique of clonal propagation in cashew. *Indian Cashew J.*, 17(1): 17-18.

Seshadri, K.V. and Rao, R.R. (1985). Modified method of epicotyl grafting in cashew for commercial propagation. *Indian Cashew J.*, 17 (4): 11-13.

Shirly, R.A. and Thimmappaiah, 2005. Somatic embryogenesis in nucellar callus of cashew (Anacardium occidentale *L.*). J. Hort. Sci. and Biotechnol., *80(3): 327-331.*

Suryanarayana, M.A. and Melanta, K.R. (1989). Effect of age of shoots on rooting of cashew stool layers. *Indian Cashew J.*, 19 (2): 10 – 11.

Sinish, M.S., Mercy, G. and John, P.S. (2005). Organic methods for cashew root stock production. *The Cashew*, 19(1): 8-15.

Swamy, K. R. M., Rao, E. V. V. B., Nagaraja, B., and Nayak, M. G. (1993). Commercial propagation of cashew varieties. Proc. of the Golden Jubliee Symp. on Horticultural Research – Changing Scenario, Bangalore.

Thimmappaiah (1997). *In-vitro* studies on cashew (*Anacardium occidentale* L.). Ph. D. Thesis in Biosciences submitted to Mangalore University.

Thimmappaiah and Shirly, R.A. (1999). *In-vitro* regeneration of cashew (*Anacardium occidentale* L.). *Indian J. Exp. Biol.*, 37: 384-390.

Thimmappaiah, Shirly, R.A., and Sadhana, P. H. (2002a). *In-vitro* propagation of cashew (*Anacardium occidentale* L) from shoot explants of young trees. *In-Vitro Cellular and Develop. Biol. Plant*, 38(2): 152-156.

Valsakumari, P.K., Vidhyadharan, K.K. and Damodaran, V.K. (1979). A comparative study of different methods of vegetative propagation of cashew. *Acta Horticulture*, No. 108 (1985): 289.

Veeraraghaan, P.G. (1990). Highlights of research and extension activities of the Cashew Research Station, Madakkathara (Kerala). *The Cashew*, 4 (2): 5-7.

Cashew: Improvement, Production and Processing (2017) *Pages* **207–215**
Editor: **Prof. P.L. Saroj**
Published by: **ASTRAL INTERNATIONAL PVT. LTD., NEW DELHI**

11

Rootstocks

J.D. Adiga and D. Kalaivanan

1. Introduction

The most significant feature of majority of the perennial fruit crops is the heterozygous nature of the plant which necessitates its propagation by vegetative means to maintain uniformity in the progenies. The asexual propagation methods like budding and grafting involves the role of rootstocks which forms the root system of the newly propagated plant. A rootstock is part of a plant, often an underground part, from which new above-ground growth can be produced. It can refer to a rhizome or underground stem (Hickey and King, 2001). In grafting, it refers to a plant, sometimes just a stump, which already has an established, healthy root system, onto which a cutting or a bud from another plant is grafted. In some cases, such as vines of grapes and other berries, cuttings may be used for rootstocks, the roots being established in nursery conditions before planting them out. The plant part grafted onto the rootstock is usually called the scion. The rootstock is selected for its interaction with the soil, providing the roots and the stem to support the new plant, obtaining the necessary soil water and minerals, and resisting the relevant pests and diseases. After a few weeks the tissues of the two parts will have grown together, eventually forming a single plant. After some years it may be difficult to detect the site of the graft although the product always contains the components of two genetically different plants.

The use of rootstocks is most commonly associated with fruiting plants and trees, but is the only way to mass propagate many other types of plants that do not breed true from seed, or are particularly susceptible to biotic or abiotic stress when grown on their own roots. Although grafting has been practiced for many hundreds of years, even in Roman times, most orchard rootstocks in current use were developed in the 20th century (Mudge *et al.*, 2009). A variety of rootstocks may be used for a single species or cultivar of scion because different rootstocks impart different properties, such as vigour, fruit size and precocity. Rootstocks also may be selected for traits such as resistance to drought, root pests, and diseases. Grapevines for commercial planting are most often grafted onto rootstocks to avoid damage by phylloxera, salinity, drought, *etc.*

The rootstock may be a different species from the scion, but as a rule it should be closely related, for example, many commercial pears are grown on quince rootstock. Grafting can also be done in stages; a closely related scion is grafted to the rootstock, and a less closely related scion is grafted to the first scion. Serial grafting of several scions may also be used to produce a tree that bears several different fruit cultivars, with the same rootstock taking up and distributes water and minerals to the whole system. Those with more than three varieties are known as 'family trees'. When it is difficult to match a plant to the soil in a certain field or orchard, growers may graft a scion onto a rootstock that is compatible with the soil. It may then be convenient to plant a range of ungrafted rootstocks to see which suits the growing conditions best; the fruiting characteristics of the scion may be considered later, once the most successful rootstock has been identified. Rootstocks are studied extensively and often are sold with a complete guide to their ideal soil and climate. Growers determine the pH, mineral content, nematode population, salinity, water availability, pathogen load and sandiness of their particular soil, and select a rootstock which is matched to it. Genetic testing is increasingly common, and new cultivars of rootstock are always being developed.

Cashew (*Anacardium occidentale* L.) being a cross pollinated crop, its progenies raised through seeds cannot be expected to perform true to type due to heterozygous nature. Most of the cashew plantations in the country have been raised from open pollinated seeds of unknown pedigree. Hence, if high yields of superior quality are desired, then the best clones must be selected and propagated vegetatively (Swamy, 1997). Standardization of vegetative propagation techniques has been one of the important thrust areas under the research programme for more than three decades. Research efforts were made to develop suitable vegetative propagation methods in cashew and accordingly several methods *viz.*, air layering, approach grafting, veneer grafting, budding and epicotyl grafting have been tried in different areas with varying success and each technique having certain limitations. Of late, softwood grafting technique is being practiced commercially at almost all the research centres of the country. In view of these facts, standardization of softwood grafting for large scale multiplication of high yielding cashew varieties was done (Sawke, 1992). Since cashew is commercially propagated by soft wood grafting (wedge-cleft method), rootstocks are the essential component of the planting material. Rootstocks have been known to resist biotic and abiotic stresses apart from controlling scion vigour in many fruit crops. The ability of rootstocks to control tree size and vigour has been exploited for high density planting in many temperate fruit crops. However, work done on rootstocks in cashew is very limited and the same is highlighted in this chapter.

2. Influence of Rootstocks on Grafting Success and Seedling Growth

Age of Rootstocks

Softwood grafting is generally done on the newly emerged or emerging bronze coloured terminal shoots. Gowda and Melanta (1989) tried rootstocks of different age group (2, 3, 4, 5, 6, 7and 8 months) for softwood grafting in cashew and recorded highest success (60 per cent) with 4 month old seedlings for softwood grafting under Bangalore conditions. Satpathy (1989) used 50 to 60 day old cashew seedlings for softwood grafting and recorded a success of 43 to 100 per cent from June to May, except in March. The highest success of 90 to 100 per cent was noticed in January and February, respectively under Odisha conditions. Experiments conducted at various cashew research stations in the country have indicated the use of younger seedlings for softwood grafting. At Vengurla, rootstocks of different age groups (1, 2, 3, 4, 5 and 6 months) were tried and the highest success (84.63 per cent) was observed with three month old seedlings closely followed by two months old rootstocks (81.85 per cent). Gangadhara Nayak *et al.* (1995) made an attempt to screen the dwarfing rootstocks at nursery stage from among five tall and five semi dwarf cashew types. The screening of two month old seedlings revealed that morphological and anatomical characteristics of young seedlings do not serve as an index for identifying dwarf types. However, 8 month old seedlings raised from seeds of Brazilian origin and of the identified semi dwarf trees exhibited differences with respect to height of the seedling and internodal length. The height of seedlings and internodal length in case of semi dwarf types were about half that of tall types (Brazilian origin).

Similarly, at Jhargram, rootstocks of four different age groups were tried and the highest percentage of success (59 per cent) was observed with two month old rootstocks, while, 30 – 40 days old seedlings (after germination) were found to be ideal as rootstocks with a success of 70 per cent under Shantigodu (Dakshina Kannada) conditions, whereas, 28 day old seedlings gave 100 per cent sprouting and survival with seven day pre-cured scions at Madakkathara conditions. At Vridhachalam, it was found that the graft take was comparatively high (75 per cent) on 30 day old seedlings than 60 day old ones (68.0 per cent) with the scion of the same cultivar precured a week prior to grafting (Anon., 1989b). An investigation undertaken to assess the effect of season and age of the rootstock on success of grafting in cashew varieties V-4 and V-5 by Shingre *et al.* (2003) revealed that the most appropriate age of rootstock for cashew grafting was 45 days. There was a decreasing trend of success linearly from 15 days to 90 days after grafting with the increase of age. The percentage of success was in the range of 39.43-67.54 per cent at the end of 90 days after grafting. The highest percentage (67.54 per cent) of success was recorded in three months old rootstock followed by two and four months old rootstocks (55.73 per cent and 54.95 per cent) which were found on par to each other and the least (39.43 per cent) was observed in six months old rootstock. The physiological maturity of rootstocks is very necessary to obtain higher success and growth of softwood grafts in cashew. Too old or too young rootstock seems to be unsuitable for grafting. Hence, two and three month old rootstocks were found to be optimum. Favourable weather conditions for rootstocks growth prior to grafting and congenial conditions of high temperature coupled with higher relative humidity after grafting under mist house caused quick and more callus production, which may be the reason for higher success and growth rate of softwood grafts. (Krishnaraj Sagar, 2007).

Varietal Influence

An experiment was conducted with nursery trial in 2007 at the Experimental Station of Directorate of Cashew Research (DCR), Puttur, Dakshina Kannnada District, Karnataka, India (12.45°N latitude, 75.4°E longitude and 90 m above MSL). The scion varieties comprised of popular vigorous cultivars like Ullal-3, VRI-3, NRCC Selection-2 and Vengurla-4(V-4). The rootstocks comprised of dwarf types like NRC-492 and Taliparamba-1 along with control (check) rootstock V-4. All the stionic combinations, excepting the ones which involved Taliparamba-1 as rootstock, recorded nearly cent per cent grafting success. The stionic combinations involving Taliparamba-1 as rootstock recorded around 50 per cent grafting success. The stionic combination of V-4 on NRC 492, VRI-3 on V-4, NRCC Selection-2 on V-4 and V-4 on V-4 recorded 100 per cent grafting success. This clearly indicates the suitability of either V-4 or NRC 492 as rootstock in obtaining higher grafting success at nursery stage. This could be because the rootstocks with different genetic makeup may have influence on compatibility of scion materials (Hartmann and Kester, 1986).

Swamy and Mohan (1991) found that cashew types (*viz.*, VTH-174, V-1 and Selection-1) with medium sized nuts of 6.6 to 7.6 g produced vigorous seedlings as shown by their increased seedling height, stem girth, average number of leaves and internodal length. They conducted an experiment with 8 rootstock varieties and five scion varieties and found no stock-scion interaction at nursery stage for graft success. However, Dinakara Adiga *et al.* (2014) reported that dwarf types as rootstocks reduce the grafting success to the tune of 50 per cent compared to vigorous types. Further, Swamy and Nayak (2001) studied stionic interactions in cashew involving vigorous and semi vigorous rootstocks. The air layers of 4 semi tall types (VTH-762/2, VTH-762/4, S 11/1 and S 11/2) and one tall type (Selection-1) were *in situ* grafted with scions of tall type (Selection-1). The study indicated that growth of the tall scion variety (Selection-1) was found to be tall and vigorous on the rootstocks irrespective of whether a rootstock is a tall type(Selection-1) or semi tall type(VTH-762/2, VTH-762/4, S 11/1 and S 11/2). The study revealed that semi tall rootstocks may not have any influence on the growth of tall growing scion variety.

A study was conducted at the Department of Horticulture, KNUST Kumasi, Ghana to investigate the effects of three local rootstock accessions, scion wood types and its defoliation period on scions of a Brazilian dwarf cashew accession. The three rootstock accessions were from Ejura, Damongo and Obom and the two scion wood types namely, hardwood and semi-hardwood were used for the study. There were two defoliation periods of one week (prior to grafting) and same-day grafting of the scion wood types. Side grafting technique on 18-weeks old stock was used. The Ejura accession had the highest mean graft 'take' of 62 per cent with Damongo and

Obom recording 55 per cent and 50 per cent respectively at 45 days after grafting. The semi hardwood recorded 66 per cent "take" as compared to 46 per cent than scions taken from hardwood materials (Adjei and Mante, 2007).

3. Influence on Field Established Plants

Since availability of land area is limited in India, one of the ideal propositions would be to increase the plant density per unit area. However, the presently available cashew varieties are either vigorous or moderately vigorous in nature. This necessitates the need for reducing the tree size and vigour in order to accommodate more plants per unit area. Genetic control by the use of dwarfing rootstocks is one of the means to control tree vigour and size. Hence, an investigation was conducted at ICAR-DCR, Puttur, to screen the rootstocks for dwarfness and to find out the influence of different stionic combinations on growth and yield parameters of popular cultivars of cashew. The two month old seedlings of dwarf rootstocks (NRC 492 and Taliparamba-1) were screened for dwarfness based on visual observations for plant height and intermodal length, as the seedlings of dwarf types segregated for plant height. The two month old seedlings from variety V-4 and the dwarf types were used for soft wood grafting (wedge-cleft method) involving scion varieties like Ullal-3, VRI-3, NRCC Selection-2 and Vengurla-4(V-4). There were 12 different stionic combinations involving three types of rootstocks and 4 scion varieties. Grafting was done during July and August and success rate among different stionic combinations were recorded. The successful grafts were field planted at a closer spacing of 6m x 6m during 2008. Among the field planted stionic combinations, the parameters like plant height, number of leaves, leaf area, bark percentage, stomatal density and yield per plant varied among different stionic combinations. This clearly indicated the role of rootstocks in altering the vigour and yield of different scion cultivars in early stages of growth in cashew.

The field performance of different stionic combinations indicated the influence of rootstocks on growth and yield performance of popular scion cultivars of cashew. The stionic combinations had a significant influence on plant height during the first year, while on number of leaves and leaf area on both the years of study. The plant height was least in stionic combination NRCC Selection -2 grafted upon Taliparamba-1 rootstock during the first year of study in field. The plant height was highest in stionic combination VRI-3 grafted on NRC 492 rootstock during first year and Ullal-3 on NRC 492 rootstock during second year of study. The increment in plant height during the second year over the previous year was least in stionic combination VRI-3 grafted on NRC 492 rootstock and increment was highest in stionic combination NRCC Selection -2 grafted upon Taliparamba-1 rootstock. This indicated that vigour reduction by dwarf rootstocks is variable depending on stionic combination and is not uniform among different stionic combinations. Similar observations were reported in kiwi fruit (Moeen-ud-din *et al.*, 2001).

The number of leaves per shoot and leaf area were significantly influenced by different stionic combinations. The stionic combination VRI-3 grafted on NRC 492 rootstock recorded the highest number of leaves per shoot in the first year while, stionic combination VRI-3 grafted on V-4 rootstock recorded highest number of leaves in the second year. The highest leaf area was recorded in a stionic combination Ullal-3 grafted on V-4 rootstock during both the years of observation. The stionic combinations which involved Taliparamba-1(dwarf type) as rootstock recorded lesser leaf area compared to other rootstocks. This is in accordance with the findings of Clearwater *et al.* (2006) who reported in kiwi fruit that scions on low-vigor rootstocks had lesser leaf area than scions on the most vigorous rootstocks. However, the canopy spread was influenced by the stionic combinations. Though not significant, the stionic combination NRCC Selection -2 grafted upon Taliparamba-1 rootstock recorded least canopy spread during both the years of observation. Similar findings in Kinnow mandarin were reported by Waqar Ahmed *et al.* (2006) who found out that canopy increment proved to be non-significant as affected by various rootstocks.

Stomatal density is used as an index to correlate with plant vigour. The stomatal density was significantly influenced by stionic combinations. The various stionic combinations, rootstocks and scion cultivars were compared for stomatal density. Among the stionic combinations, NRCC Selection-2 as well as V-4 grafted on Taliparamba-1 rootstock recorded the highest stomatal density. The combination of Ullal-3 on grafted on NRC 492 rootstock recorded least stomatal density. Among the rootstocks, Taliparamba-1 had highest stomatal density and V-4 had least density of stomata. Among the scion varieties, VRI-3 had highest stomatal density compared to other varieties. In the present study stomatal density was negatively associated with plant height as seen in case of dwarf rootstocks. Similar negative correlations have been reported earlier in cashew hybrids (Aneesa Rani *et al.*, 2011) and avocado (Barriento Priego and Sanchez Colin, 1987). The bark percentage in shoots of field planted cashew varied among different stionic combinations.

The highest bark percentage was associated with the stionic combination Ullal-3 grafted on V-4. The graft percentage was reduced when Ullal-3 was grafted on either NRC 492 or Taliparamba-1 rootstock. Similarly, Taliparamba-1 as a rootstock reduced the bark percentage in scion variety V-4 as compared to V-4 grafted on its own root. This suggests the association between bark percentage and tree vigour as well as the role of rootstocks in altering the bark percentage of scion varieties in cashew. Similar relationships have been established with respect to avocado rootstocks (Alfredo Lopez Jimenez and Alejandro Barrientos Priego, 1987).

The stem girth and intermodal length were not significantly affected by different stionic combinations. However, stionic combination NRCC Selection -2 grafted upon Taliparamba-1 rootstock recorded least stem girth below and above the graft union during both the years of study. The increment in stem girth below the graft union over the previous year was least in stionic combination VRI-3 grafted on Taliparamba-1 rootstock. The increment in stem girth above the graft union over the previous year was least in stionic combination NRCC Selection -2 grafted upon Taliparamba-1 rootstock.

The intermodal length of shoots under different stionic combinations decreased in the second year of observation over the first year. The scion varieties on dwarf rootstock (NRC 492 and Taliparamba-1) recorded lesser intermodal length compared to scion varieties on vigorous rootstock(V-4).The highest decrease in intermodal length was seen in the stionic combination NRCC Selection -2 grafted upon Taliparamba-1 rootstock closely followed by NRCC Selection -2 grafted upon NRC 492 rootstock. The influences of rootstocks on intermodal length have been reported in apples (Mohammad Reza Dodangeh *et al.*, 2012) and grapes (Somkuwar *et al.*, 2006).

The different stionic combinations significantly influenced the yield of cashew in the first harvest. Among the different stionic combinations, the combination of VRI-3 grafted on NRC 492 rootstock recorded the highest nut yield in first harvest closely followed by Ullal-3 grafted on NRC 492 rootstock. The yield performance of popular varieties varied based on the rootstocks. This clearly indicates the role of rootstocks on yield performance in cashew. Similar findings in apple (El-Shammaa *et al.*, 2011) and mandarin (Demirkeser *et al.*, 2009) have been reported.

4. Rootstocks for Abiotic Stress Tolerance

The cashew crop is of great economic and social importance for Northeast Brazil, a region usually affected by water and soil salinity. A study was conducted in a greenhouse to evaluate the effects of four salinity levels established through electrical conductivity of irrigation water (ECw: 0.7, 1.4, 2.1 and 2.8 dSm^{-1}, at 25°C), on growth and physiological indices of five rootstocks of dwarf-precocious cashew varieties CCP06, CCP09, CCP1001, EMBRAPA50, and EMBRAPA51. Plant height, leaf area, dry weight of root, shoot and total water content of leaves, root/shoot ratio, leaf area ratio, absolute and relative growth rates and rate of net assimilation were evaluated. The majority of the evaluated variables were found to be affected by ECw and the effects varied among clones; however, no significant interactive effects were observed for factors. The value of ECw = 1.39 dS m-1 was considered as a threshold tolerance for the precocious cashew rootstocks used in this study. The dwarf-precocious cashew is moderately sensible to soil salinity during the formation phase of rootstock. Clones EMBRAPA 51 and EMBRAPA 50 presented, respectively, the least and the best development indices (Carneiro Paulo Torres *et al.*, 2004).

5. Future Line of Action

- ☆ Rootstock breeding has not been attempted hitherto in cashew.
- ☆ It is worthwhile to study the influence of wild species of cashew as rootstocks to screen them for biotic and abiotic stresses.
- ☆ Clonal propagation of rootstocks hold key for exploiting the beneficial effects of rootstocks for various aspects as seed propagation of rootstock leads to varied stionic interaction.

References

Adjei, P. Y. and Mante, D. (2007). Sources of rootstock, scion wood type and duration of scion defoliation on graft success in cashew (*Anacardium occidentale* L.). *Ghana Journal of Horticulture*, 6: 59-64.

Alfredo Lopez Jimenez and Alejandro F Barrientos Priego. (1987). Selection of Dwarfing Rootstocks of Avocado (*Persea americana* Mill.). *California Avocado Society 1987 Yearbook*, 71: 225-234.

Aneesa Rani, M. S., Kumar, N. and Marimuthu, R. (2011). Evolving cashew F1 hybrids suitable for high density planting system. *Indian Journal of Horticulture*, 68(2): 152-155.

Barrientos Priego, A. and Sanchez Colin, S. (1987). Stomatal density and its relationship to growth habit in avocado. *South African Avocado Growers Association Yearbook*, 10: 66-67.

Carneiro Paulo Torres, Fernandes Pedro Dantas, Gheyi Hans Raj, Soares Frederico Antônio Loureiro and Viana Sergio Batista Assis. (2004). Salt tolerance of precocious-dwarf cashew rootstocks: physiological and growth indexes. *Scientia Agricola*, 61 (1): 9.

Clearwater, M. J., Seleznyova, A. N., Thorp, T. G., Blattmann, P., Barnett, A. M., Lowe, R. G. and Austin, P. T. (2006). Vigor controlling rootstocks affect early shoot growth and leaf area development of kiwifruit. *Tree Physiology*, 26(4): 505-15.

Demirkeser, T. H., Kaplankýran, M., Toplu, C. and Yýldýz, E. (2009). Yield and fruit quality performance of Nova and Robinson mandarins on three rootstocks in Eastern Mediterranean. *African Journal of Agricultural Research*, 4 (4): 262-268.

Dinakara Adiga J, Kalaivanan D, Meena R K, Mohana G S and Lakshmipathi. (2014). Performance of vigorous cashew cultivars as influenced by dwarf rootstocks. *Vegetos*, 27 (2): 242-248.

El-Shammaa, M. S., Mansour, A. E. M., Cimpoies, G. P., Nageib, M. M. and Saleh Malaka. (2011). Effect of Rootstocks on Flowering, Yield and Fruit Quality of Anna Apple Trees (*Malus domestica*, Borkh). *Research Journal of Agriculture and Biological Sciences*, **7**(2): 190-195.

Gangadhara Nayak, M., Swamy K. R. M. and Palaniswamy, K. (1995). Studies on screening of cashew types for dwarfing character. *The Cashew*, 9(4): 18-24.

Gowda, B. J. and Melanta, K. R. (1989). Vegetative propagation of cashew (*Anacardium occidentale* L.) by grafting. *The Cashew*, 3 (2): 13-14.

Hartmann, H. T. and Kester, D. E. (1986). *Plant Propagation- Principles and Practices* (4th Edn), pp. 337- 338. Prentice Hall of India Pvt. Ltd, New Delhi, India.

Hickey, M. and King, C. (2001). *The Cambridge Illustrated Glossary of Botanical Terms*. Cambridge University Press.

Krishna Raja Sagar, M. K. (2007). Propagation Studies in Cashew Nut (*Anacardium occidentale* L.) under Mist House Conditions, M.Sc (Agri.), Thesis submitted to the University of Agricultural Sciences, Dharwad.

Mohammad Reza Dodangeh, Mohammad Javad Shakouri, Zeinab Hamzehei and Ahmad Dadashpour. (2012). *Indian Journal of Science and Technology*, 5 (1): 46-48.

Mudge, K., Janick, J., Scofield, S., Goldschmidt, E. E. (2009). *A history of grafting*. In Horticultural reviews 35, pp. 437–493. Janick J (editor). John Wiley and Sons, Hoboken, New Jersey.

Sawke, D. P. (1992). Standardization of soft wood grafting technique in cashewnut. *Ph.D. Thesis* submitted to University of Agricultural Sciences, Dharwad.

Somkuwa, R. G., Satisha, J. and Ramteke, S. D. (2006). Effect of Different Rootstocks on Fruitfulness in Thompson Seedless (*Vitis vinifera* L.) Grapes. *Asian Journal of Plant* Sciences, 5: *150-152*.

Swamy, K. R. M. (1997). Technology for planting material multiplication in cashew. *The Cashew*, 11 (1): 22-32.

Swamy, K. R. M. and Mohan, E, (1991). Screening of cashew (*Anacardium occidentale* L) varieties/types for rootstocks at nursery stage. *The Cashew,* 5 (2): 3-5.

Swamy, K. R. M. and Nayak, M.G. (2001). Rootstock scion interaction in cashew. RPF-III of project report. ICAR-DCR, Puttur.

Waqar Ahmed, M. A, Pervez, M., Amjad, M., Khalid, C., Ayyub, M. and Azher Nawaz, M. (2006). Effect of stionic combination on the growth and yield of kinnow mandarin. *Pakistan Journal of Botany*, 38(3): 603-612.

Cashew: Improvement, Production and Processing (2017) *Pages* **217–231**
Editor: **Prof. P.L. Saroj**
Published by: **ASTRAL INTERNATIONAL PVT. LTD., NEW DELHI**

12

High Density Planting

S. Jeeva and P.L. Saroj

1. Introduction

Cashew (*Anacardium occidentale* L.) is important plantation crop largely grown for its delicious kernel. Demand for cashew kernels is increasing day by day creating a gap between supply and demand both in domestic and international markets. Though, but India is the leading country in production of cashew nuts in the world but the productivity is low when compared to the newly emerging country like Vietnam. With a production of 7.28 lakh tons of raw nuts in the country, it is sufficient to meet only 50 per cent of the requirement by the processing industries which is 14 lakh tons. Prime reasons behind the low productivity of Cashew in India are existence of senile unproductive plantations of seedling origin, non-adoption of improved package of practices and poor plant population resulting in low productivity per unit area. The productivity of cashew could be increased by the use of elite planting material of improved varieties and adoption of scientific technologies like High Density Planting (HDP) technique. With shrinking land area for agriculture due to urbanization, the necessity of tapping the maximum potential of the unit land becomes essential now-a-days. High density planting technique is a modern method of cultivation involving planting of trees densely, allowing small or dwarf trees with modified canopy for better light interception and distribution and ease of field operations. Such system produces precocious cropping, high and regular yields of good quality produce and low labour requirement to meet ever rising production costs. High density planting is a very important and effective venture to attain higher productivity per unit urea even in early years. However, initial establishment cost of high density planting is comparatively more than normal density planting.

In cashew, during early years of planting, considerable inter space remains unutilized by the crop. This leads to wastage of large amount of horizontal and vertical space, underground soil resources like nutrients and moisture and above ground resources like solar radiation. The canopy expansion rate in cashew is slow during early years. The canopy development pattern in cashew offers considerable scope, either for inter cropping or for high density planting. If interspace is left unutilized, it will be occupied by unproductive vegetation (weeds). Therefore, it is essential to adopt a management strategy capable of exploiting the environmental resources both in time and space dimensions. High density planting is a technique aimed in this direction.

2. Merits of HDP over Normal Planting

- Increasing pressure on land owing to diversion of lands to various other obvious reasons as well as rising energy and land-costs have made it imperative to achieve higher productivity from limited space.
- High density planting facilitates better utilization of solar radiation and increase in bearing surface per unit land area.
- In HDP, plants are precocious, easily manageable and fetch higher returns per unit area.
- High density plantations have better amenabilities to modern input saving horticultural techniques such as drip irrigation.
- Managing excessive vegetative growth gives higher productivity and harvest index as well as early economic returns.
- Weed growth is minimum due to early ground coverage by the crop canopy causing shade over the ground surface.
- Due to less exposure of ground to sunlight the soil temperature during peak summer season is reduced resulting in reduced soil moisture loss through evaporation.
- The evaporation from ground surface is also reduced considerably due to heavy deposit of cashew leaves under HDP system.

3. High Density Planting

High density planting technique is useful for enhancing production and productivity of cashew in India (Salam, 1997). The technology has been standardized for cashew and also adopted by many growers in India. High density orchards not only have provided higher yield and net economic returns per unit area in the initial years, but also facilitated more efficient use of inputs. Closer spacing has given two times more yield than normal spacing. Adoption of high density planting system 6.5m x 4 m (384 plants/ha), 5m x 5m (400 plants/ha) and 5m x 4 m (500 plants/ha) is the best method to increase the productivity of cashew. However, this needs suitable dwarf and or compact cashew varieties so that overlapping of canopy is kept at minimum. There is lack of suitable dwarf and compact cashew varieties with high yield. Limb pruning and diagonal thinning also need to be evaluated for their effectiveness in high density planting systems (Rupa *et al.*, 2013). High density planting system is more suitable for soils with low fertility and it helps to utilize the space available between cashew plants during the initial years of orchard, thereby increase the yield per unit area. Further, high density planting reduces the weed growth by early canopy coverage, reduces soil temperature thereby increasing the soil moisture content especially during peak summer season and provides mulching effect.

This technique involves planting more number of grafts per unit area initially and thinning them out at later stages. In high density planting, grafts may be planted at a spacing of 4 m x 4 m or 5 m x 5 m or 8 m x 4 m. While adopting planting at a spacing of 4 m x 4 m in square system of planting, there will be 625 plants per hectare. This population can be retained for a period of seven to nine or ten years depending upon the canopy expansion rate. If the soil is very rich, the canopy development rate will be faster. High density planting technique would be more useful in poor soils where the rate of canopy expansion is slow. Considering the fertility status of the soil, the level of management in terms of fertilization, irrigation and initial plant population is to be decided carefully. Later, after monitoring the canopy pressure between adjacent plants, the alternate plants are to be removed. Finally, when the plants attain full growth, the spacing between the plants will be optimum.

If uniform management practices are adopted, during early years of yield the per tree nut yield will be more or less the same with all the trees, both in the normal density planting and in the high density planting. But per hectare yield will be more from high density plantations (due to higher plant population) compared to the normal density plantations. During later years when the plant population is equalized to that of normal density plantation, the productivity of both the plantations would be more or less the same. The bonus yield obtained during the early years would be substantial in high density plantations. In addition to higher yields, substantial quantities of firewood can be obtained during thinning which may fetch additional revenue to the farmer.

The initial spacing recommendation for cashew in Tanzania was 9 X 9 m, but a wider recommendation of 12 X 12 m (69 trees ha^{-1}) was adopted in the 1970s. With a wide-spreading tree like cashew, there is a large variation in the optimum tree spacing as the tree ages; without thinning, the optimum spacing will mean compromising between high initial yields (close spacing) and high yields later in the plantations' life (wide spacing). With the growth rates that occur in Tanzania and for materials that are currently being distributed, a 12 x 12 m spacing is suitable for most locations and allows intercropping for several years. On the Makonde Plateau, however, where trees grow more vigorously a wider spacing of 15 x 15 m is desirable. Although, high density planting followed by thinning may be appropriate for intensive cultivation, it is unlikely to be suitable for smallholders. Amongst the new clones being distributed there are a few available that have a small stature (*e.g.* AZA 2); disease control would be easier on these trees and some intensification of production would be possible by planting them closer than normal. (Martin *et al.*, 1997).

A 10-year long study by Suma *et al.* (1998) revealed that 625 or 312 plants/ha is optimum for cashew upto 9th year of planting under the agroclimatic conditions prevail in the state of Kerala. The canopy development pattern and thinning pattern of high-density plantation is given by Abdul Salam *et al.* (1998) (Figure 12.1). Ghosh *et al.* (2000) recorded highest yield from the plants spaced at 8m x 8m throughout the period of investigation followed by 10m x 5m rectangular planting. The unit area yield had markedly increased with the increase in plant density and it was highest at a plant density of 400 plants/ha (5m x 5m) followed by 312 plants/ha (8m x 6m) and minimum at 100 plants/ha (10m x 10m).

A high density trial of cashew was carried out at the ICAR-National Research Centre for Cashew, Experimental Station, Shantigodu, India, during 1982-94 with the objectives of studying growth behaviour, root distribution, photosynthesis,

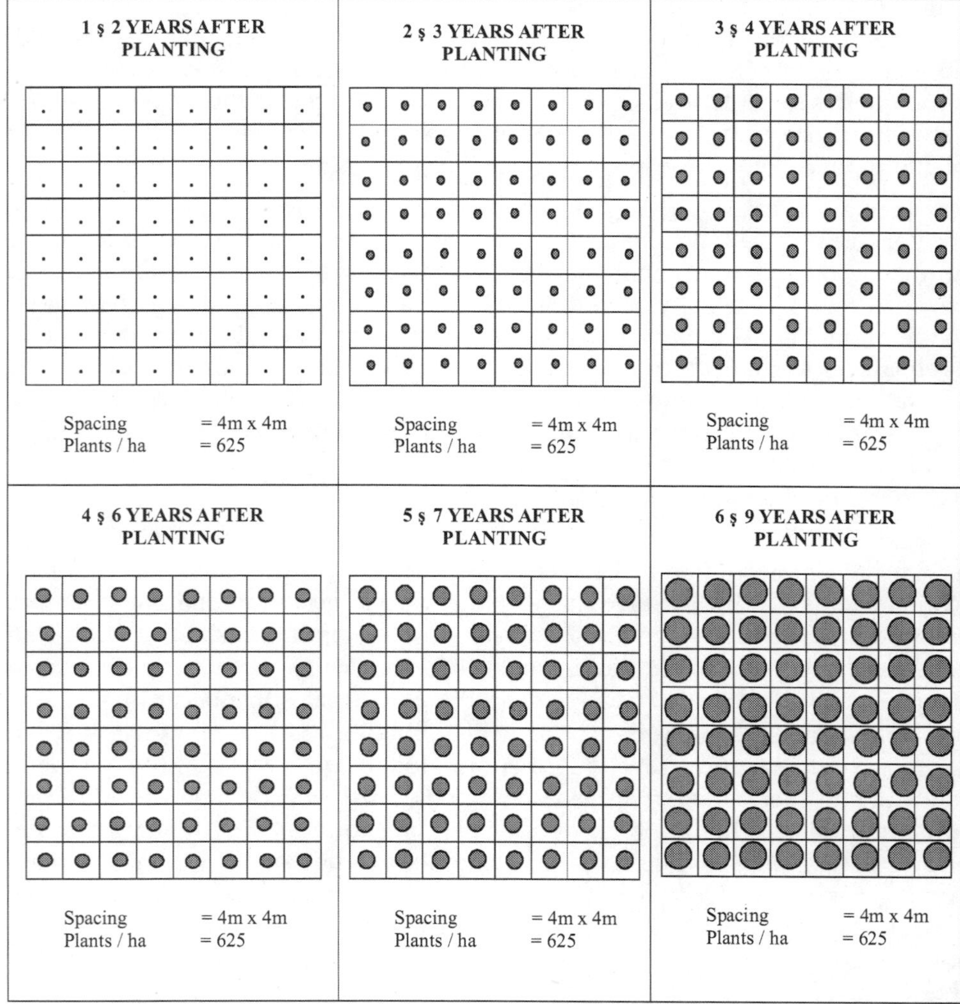

Figure 12.1: Canopy Development Pattern Over Years in High Density Planting (4mx4m.

weed biomass, dry branch accumulation, and yield under the varying densities. Densities adopted were 156 trees, 278 trees, 625 trees, 1111 trees, and 2500 trees ha^{-1}. The density was reduced to 50 per cent in the treatment with 1111 trees ha^{-1} and to 25 per cent in the treatment with 2500 trees ha^{-1} in the seventh year. Density was reduced to 50 per cent by the 11th year in the treatment with 625 trees ha^{-1}. Higher density treatment resulted in deeper root penetration, lower moisture at depths above 60 cm, greater light interception, and as a result, greater mutual shading induced drying of branches. Yield levels were maintained in higher density plots by pruning to 80 per cent light interception or at later stages by diagonal thinning. Maintaining a tree density of 625 ha^{-1} (4 m x 4 m) for the first 11 years and diagonal thinning thereafter to reduce the population to 50 per cent (8 m x 5.7 m x 5.7 m)

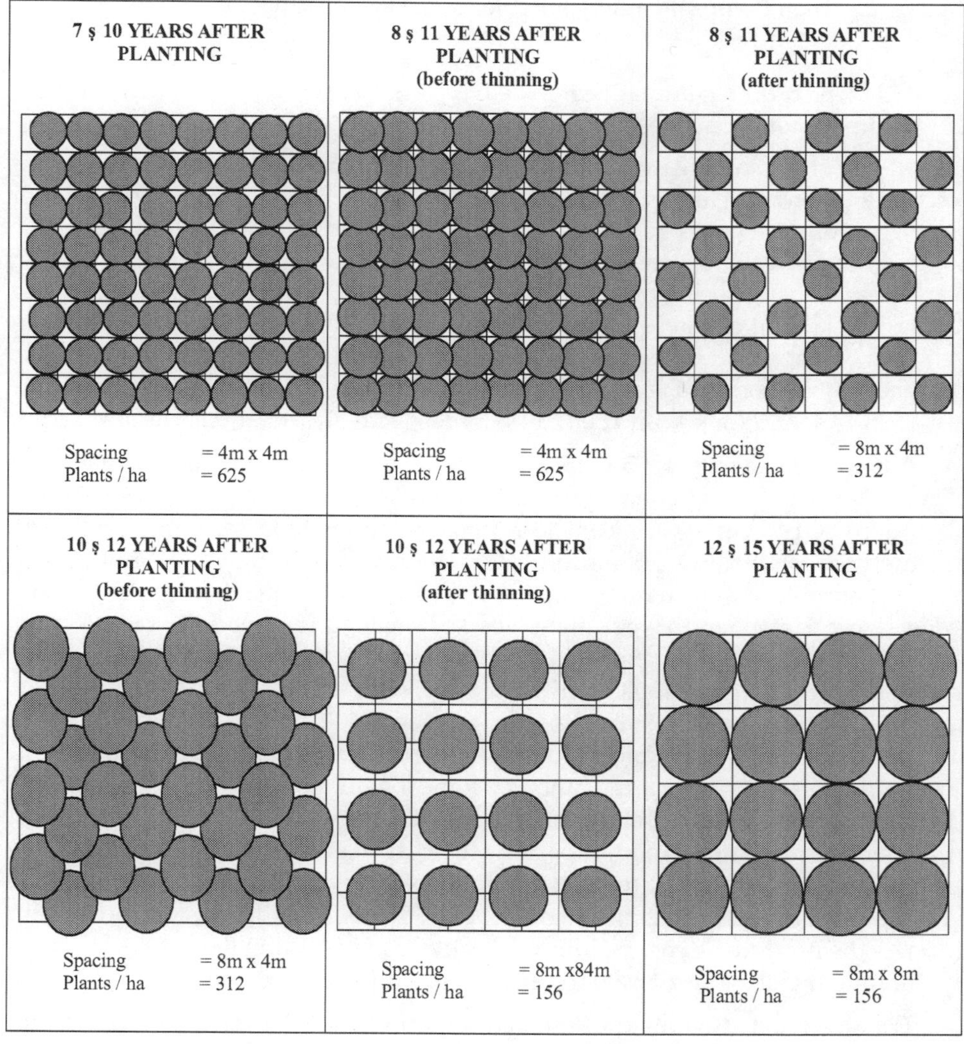

Figure 12.2: Canopy Development Pattern Over Years in High Density Planting (4mx4m.

Figure 12.3: High Density Planting at Regional Research Station, Vridhachalam.

resulted in maximum cumulative nut yield (4.94 ha^{-1}) and profit (U.S. \$2464.90 ha^{-1}) (Yadukumar *et al.*, 2001).

At Jhargram (West Bengal) under high density plantation 6m x 4m spacing is recommended with a fertilizer dose of 150kg N+ 50 Kg P_2O_5 + 50 Kg K_2O/ha/yr for getting maximum yield per unit area. (Mini Poduval, 2012). At Cashew Research Station, Madakkathara, studies on the effect of high density planting (1546-625 plants/ha) revealed that cumulative yield for the initial 9 years were best (5611-6981 kg/ha) at planting densities of 625 plants/ha (4m x 4m) and 312 plants/ha (4m x 8m). In a recent study conducted at Cashew Research Station, Madakkathara, the cumulative yield for eight years with 200 plants/ha was only 3956 kg while the same was 7598 and 9139 for 400 and 500 trees/ha respectively. (Pushpalatha *et al.*, 2013)

A planting spacing of 7.5 x 7.5, 8 x 8 or 9 x 9 m is recommended for cashew at Nigeria, which gives a tree density of 175, 156 and 123 plants per hectare respectively. High density planting is a recent technique recommended for enhancing early productivity of cashew plantations. This involves planting more number of plants at 4m x 4m or 8m x 4m giving a tree density of 625 or 312 trees per hectare, respectively. However, this population can be retained for a period of 6 to 7 years depending on the canopy expansion rate before selectively thinning them down in stages to reach a final spacing of 9m x 9 m. (Asogwa *et al.*, 2008). At Australia, suggested spacing is 12m x 12 m, giving 70 trees/ha. A closer spacing of 12 x 6 m (140 trees/ha) could be used to maximize early production, but the planting would need thinning to 12 x 12 m as the canopies come together (Wait and Jamieson,1986). Common spacing used in Tanzania is 12 m within rows and 12 m between rows. High density planting is good, especially in areas of good soils (Kasuga, 2010). In most locations in Tanzania, the recommended spacing is 12m x 12m (59 trees/ha) as this allows intercropping in the early years. Where trees grow vigorously 15m x 15m is the preferred density although high density planting (9m x 9m) followed by thinning may be appropriate for intensive cultivation.

Trials conducted at Cashew Research Station, Odisha University of Agriculture and Technology, Bhubaneswar, Odisha, revealed significant variations for vegetative as well as yield attributing traits among the different germplasm of Cashew evaluated. Cashew germplasm with dwarf plant habit, exhibiting their suitability for high density planting were H 68, H 303, H 367, M15/4, H 1598 and H 1600 (2.42 to 3.83m) (Kabita Sethi *et al.*, 2014).

4. Ultra High Density Planting

An advanced method of High Density Planting System called Ultra High Density Planting System is being experimented at Regional Research Station, Vridhachalam to accommodate a plant population of 1666 plants per hectare at a closer spacing of 3m x 2 m or 1250 plants at 4m x 2 m. Such ultra high density planting systems have already been established and proven successful in mango. Moreover, training the plants from the initial stage is necessary. Intensive pruning for canopy architecture makes the system successful to avoid intermingling of branches. It is also possible to take ultra high density planting in cashew at a very close planting of 2 X 2 m or 2.5 X 2.5m and thus a total of 2500 or 1600 cashew grafts can be accommodated in a hectare area. The grafts need to be forced to develop a frame work within a height of 1m by pruning and thereafter canopy management by annual pruning be done for containing it in within the allotted space. Use of clonal dwarfing rootstocks and application of growth retardants may be useful in this regard but it has not been standardized properly in case of cashew.

5. Varieties Suitable for HDP

Dwarf varieties with compact canopy are ideal for high density planting. If dwarf types are not available, relatively less statured and more compact varieties may be selected. Cashew varieties M 44/3, Anakkayam-1 and H1608 are relatively less statured and less spreading compared to other varieties and as such they are suitable for high density planting (Abdul Salam, 1999). Relatively short stature and compact varieties such as VRI 2, Anakkayam 1 and Dhana are more suitable for High Density Planting, since thinning operation can be delayed to more than 10 years. (Kurian and Peter, 2007). The precocious, less vigorous and pruning responsive varieties such as NRCC Sel-2, K 22-1, VRI 3, V-4 and Ullal-3 have shown good response to this technique (Saroj and Nayak, 2013).

At Nigeria, nine accessions were characterized by trees with small and compact canopy, though they are highly prolific in fruiting but the total output per tree were significantly influenced by smallness of its nuts and kernels. The genetic attributes of these nine cashew accessions include small sized trees with compact canopy. These plant materials could be useful for future breeding of cashew cultivars that would be adaptable to high density planting in an effort to improve outputs per unit area (Aliyu, 2012).

At Brazil, cashew has been grouped into two types, regarding its genetic variability, common and dwarf, named according its height. The common type is the most cultivated, presenting high canopy, height between 8 and 15 m, and canopy diameter reaching 20 m. The individual productive capacity varies widely, with some plants producing less than 1 kg and others up to 180 kg of raw cashew nut per harvest. The dwarf type is characterized by being less than 4 m, having a homogeneous canopy, with stem diameter and canopy diameter smaller than the

Cashew under UHDP – At planting.

Cashew under UHDP – One year after planting.

Cashew under UHDP – 1 ½ years after planting.

Cashew under UHDP – 2 years after planting.

Cashew under UHDP – At flowering.

Cashew under UHDP – At fruiting.

Figure 12.4: UHDP at RRS, Vridhachalam, Tamil Nadu.

Figure 12.5: UHDP trial by Jain Irrigation, Udumalaipet, Tamil Nadu.

common type, and by being more precocious, starting flowering between 6 and 18 months after planting. Cashew dwarf clones most cultivated in Brazil (CPC 06, CPC 09, CPC 76 and CPC 1001) were obtained from plant introductions at the Embrapa's Experimental Station, in Pacajus City, from natural populations of the

Brazilian Northeast coastal region, followed by phenotypic individual selection and annual production control of selected plants. Subsequently, new methodological procedures, as the policross method, selection between and within progenies and inter-and intra-specific hybridization, resulted in the development of dwarf cashew clones Embrapa 50, Embrapa 51, BRS 189, BRS 226 and BRS 265. Recently, Embrapa Tropical Agro industry launched two new clones: the BRS 274, the first clone of common cashew for commercial planting, and the BRS 275, the first clone of cashew hybrid, created by crossing the common cashew with the dwarf cashew clone (CPC 1001). The expected yield, for the dwarf cashew under rained conditions is around 1,000 kg/ha of raw cashew and 10,000 kg/ha of cashew apple. Under irrigated conditions it may reach 3,800 kg/ha of raw cashew nut and 30,000 kg/ha of cashew apple. With the introduction of the dwarf cashew clones, the production system based on the use of high-yield cashew varieties, high-density planting, fertilizer application and plant protection has evolved significantly (Oliveira, 2008).

6. Canopy Management in HDP

Under High Density system, canopy development within the manageable size is very essential. In the absence of dwarf and high yielding varieties, dwarfing rootstocks, chemicals to induce dwarfing, the only option remains to control the desired canopy by pruning (Nayak and Swamy, 2010). In HDP, plants tend to grow lanky and erect for want of sunlight resulting in drying of lower branches. Therefore, pruning and training needs to be attempted from the initial years of planting. The best time for pruning is just after harvesting of the crop *i.e.* normally in May-June under west coast region. Pruning should be done regularly without disturbing initial frame work. Plants can be pruned to bush shape at a height of 0.75-1.00 m. The yield of bush planted plants is superior even at a closer spacing of 2.5m x 2.5m during several years at the beginning. Thus a yield of more than 4 tons/ha can be achieved (Singh, 2010).

7. Use of Paclobutrazol in Cashew

Controlling excessive vegetative growth for increased or sustained productivity is the major principle of high density planting (Santra,1996). In cashew, due to non availability of dwarf clones, dwarfing root stocks or a pruning technology for the management of vigorous canopies, use of growth retarding chemicals assumes significance. An experiment was conducted on cashew cultivar, Ullal-3 to study the influence of Paclobutrazol (PBZ) on its vigour and yield. The paclobutrazol was applied on 1-year-old plant in 1.0, 2.0 and 3.0 g per plant as soil drench annually, biennially and once in three years before vegetative flushing. The results revealed that the plant height of PBZ treated plant ranged from 229.2 to 261.1 cm in different doses while in untreated plants, it varied from 262 to 275.7 cm. The canopy spread in E-W direction among PBZ treated plants ranged from 223.7 to 252.5 cm while the untreated plants recorded 269.4 to 274.5 cm in same direction. The canopy spread in N-S direction among PBZ treated plants ranged from 234.5 to 268.2 cm, while the untreated plants varied from 273.6 to 277.7 cm in the same direction. The inter-nodal length in PBZ treated plants ranged from 0.6 to 3.1 cm, while the untreated plants recorded 3.4 to 3.5 cm indicating the role of PBZ in reducing the plant vigour. The PBZ treated plants recorded nut weight of 6.63 to 7.56 g and nut yield of 756 to 1127 g/plant, while untreated plants recorded nut weight of 7.02 to 7.36 g and a nut yield of 668 to 728 g per plant (Meena *et al.*, 2014).

8. Nutrient Management

The study on nutrient budgeting and nutrient balance was conducted to determine the optimal fertilizer requirement for cashew under high-density planting system in relation to yield and net profit (Yadukumar *et al.*, 2013). The six-year-old plantation of cashew variety Bhaskara under high-density planting system (625 trees/ha) was used. A negative N, P and K balance of 113, 22 and 45 kg/ha/year was found in the control plot where no fertilizers were applied. A strong positive N balance ranged from 137 to 251, P balance from 34 to 75 and K balance from 89 to 164 kg/ha/year were found in trees with two-thirds and full doses of fertilizer treatments. In treatments with one-third dose of fertilizers, a positive N, P and K balance of 40, 19 and 40 kg/ha/year were found. This indicates that in high-density planting system, existing fertilizer schedule, *viz.* recommended dose of fertilizers (RDF) is on higher side and optimal dose of fertilizer required is only one-third of the RDF. The yield and net profit in full dose and one-third dose of fertilizer treatments were on par. The higher yield of 1.47 and 1.41 tonnes/ha/year and a net profit of Rs. 40,030 and Rs. 41,274/ha/year was obtained in these two treatments. The application of one-third dose of fertilizer to soil instead of recommended dose of fertilizer not only helps to reduce the input required and cost of cultivation but also the ecological hazards arising from excess inorganic fertilizer application. Hence, one-third dose of fertilizer is optimal for cashew plantation under high-density planting system. Cashew is an ideal crop for carbon sequestration. Based on research undertaken at Directorate of Cashew Research, it was found that cashew genotype (VTH- 174) trees of 7 years old sequestered about 2.2-fold higher carbon (C) under high density planting system (625 trees/ha). Carbon storage by cashew has been estimated as 32.25 and 59.22 t CO_{2}/ha at 5 th and 7 th year of growth respectively under high density planting (Rupa *et al.*, 2013).

9. Soil Moisture and Fertigation

In any close plantation, the loss of surface soil moisture is less because the surface canopy covers owing to less evaporation. In a study, under high density plantation (625 plants/ha) at DCR, Puttur; the soil moisture was higher up to 60 cm depth but decreased at lower depth (60-90 cm). The higher soil moisture at upper horizons was due to heavy shading effect over the ground and thick leaf litter deposit on the ground surface. The soil moisture at lower depth was less by 3.9 per cent in case of high density planting than low density planted plots. The lesser soil moisture at lower depth indicated that the soil moisture was extracted by the plants from the lower depths during summer months (Yadukumar *et al.*, 2001). The weed occurrence is also low in high density plantations. The lower biomass of weeds in high density plantations are due to low light interception.

The efficacy of fertigation for increasing the productivity of cashew under high density planting system (625 plants/ha) was reported by Yadukumar *et al.* (2009). The optimal rate of fertilizer application was found as 125 g N, 31.25 g P_2O_5 and 31.25 g K_2O/tree/year through fertigation and 2 kg castor cake/tree through soil application. The highest mean cashew nut yield of 2 t/ha/year and 1.96 t/ha/year was in treatments with 50 per cent of recommended doses of fertilizers (RDF) through fertigation and 4 kg castor cake/tree through soil application and the treatment with half the aforesaid dose. The protein content was high in fertigation treatment with 50 per cent RDF and 4 kg neem cake/tree (42.6 per cent). In soil application with drip irrigation separately, the nut yield ranged from 1.45 to 1.73 t/ha/year and protein content from 35.9 to 38.9 per cent, respectively. The lowest nut yield of 1.12 t/ha/year was in absolute control. Fertigation increased the nut weight to 7.0 g, apple weight to 76.9 g and the shelling percentage to 30.06 compared to soil application with a nut weight of 6.8 g, apple weight of 70.8 g and a shelling percentage of 29.5 and absolute control with a nut weight of 6.7 g, apple weight of 69.4 g and a shelling percentage of 28.8, respectively.

10. Photosynthetic Efficiency in HPD

In the HDP, the canopy cover, leaf area, light interception, soil and canopy temperature, evaporation and transpiration etc. are altogether different than normal density; thereby photosynthetic efficiency varied significantly. In a study at DCR, Puttur, it was observed that the net photosynthesis (Pn) had reached a minimum level in high tree density plot during 6th, 7th and 8th years. This was attributed to mutual shading. The differences of Pfd (Photon flux density) and Pn between the tree densities were significant, although differences between the tree years (6, 7 and 8 years) were not significant (Balasimha and Yadukumar, 1993). Subsequently, on the 9th, 10th and 11th year Pn at the top positions of the canopy of trees in high tree density and low tree density were similar. In low tree density plots (156 trees/ha), the Pn values for the middle height of the canopy were significantly more than high tree density plot. Pn increased in middle position of the canopy from 5.98 in 9th year to 7.48 µmol/m^2/s in the 11th year in high tree density plot. This was mainly due to pruning of the overlapping branches to maintain 80 percent light interception in the 9th and 10th year and due to thinning in the 11th year. One the other hand, overlapping branches were removed (8 kg dry weight per tree) each year (8, 9 and 10th year) to limit light interception to 80 per cent.

The yield/tree did not increase considerably particularly on the tenth year. The effect of pruning of overlapping branches on 10th year was not evident as the ground coverage and light interception increased to more than 90 and 80 percent respectively in the same season. Incidence of less light (Pfd) in the middle position of the canopy (927 and 998 µmol, resulting in reduced photosynthesis (5and 6.98 µmol CO_2/m^2/s) may be the reason or yield levels not improving even after pruning of overlapping branches on tenth year. When total tree population was reduced to 312 trees/ha in high density plot, the Pn in the remaining trees increased considerably (11th year) both at top and middle height positions of the canopy. This is reflected in the increased yield obtained one year after thinning in 12th year (Yadukumar *et al.*, 2001).

11. Plant Protection in HDP

In HDP, protection of plants against pest and fungal diseases is crucial. In fact, there is build up of humidity in side of HDP than widely spaced plantations, owing to more chances of pest and diseases. Sometimes, if HDP are left unmanaged, they become shelter for several pests and diseases. Therefore, soon after pruning of shoots, there should be one spray of Bordeaux (1 per cent) and on the cut surface of thicker branches pasting of Bordeaux paste takes care of possible fungal invasion. There is further attack of pests like Tea Mosquito Bug and thrips on new flushes arising after pruning, thus need based sprays of λ-cyhalothrin (0.003 per cent) and carbaryl (0.1 per cent) are recommended during most vulnerable periods of crops such as flushing, flowering and fruiting stages.

12. Economic of HDP

While emphasizing the benefit of high density planting, Despande (2002) reported that gross income, gross expenditure and net profit were considerably higher from the high density (5m x 5m) treatments. Comparative economics of cashew plantations under different plant densities (175-625) revealed that BCR in all cases is greater than one (1.61-2.19) indicating profitable nature of the system. International rate of returns (IRR) is also higher than interest rate of 10 per cent charged by commercial banks in all cases suggesting competitive nature of returns. If soil is fertile, 5x5m spacing (400 trees/ha) can be maintained to get the highest yield and income. In poor fertility areas, 4x4m (625 trees/ha) is recommended (Yadukumar *et al.*, 2003).

In trials conducted at Regional Research Station, Vridhachalam under All India Coordinated Research Project on Cashew, planting cashew at a closer spacing of 5 x 4m (500 plants/ha) with a fertilizer dose of 225:75:75 kg/ha recorded high yield per unit area *i.e.*, 3000 kg nuts per hectare with a BCR of 4.5. The trees in the closer spacing yielded on par with the trees of wider spacing. Due to increased plant population per unit area, in high density planting system, the yield per hectare was higher. Partial budgeting analysis has revealed that a net return of Rs. 68,875.00 could be obtained every year from 6th year onwards. (Aneesa Rani and Jeeva, 2006). A close planting of 4x4 m or 5x5m or 6x4m maintained up to 9 years with proper pruning and training and thinning out the excess plants to provide a final spacing of 8m x 8m or 10m x 10m or 6m x 8m depending upon the canopy expansion rate enables higher returns during initial years (Pradeep Kumar, 2008). In another high density planting system at DCR, Puttur; the highest net profits of Rs. 49,367/ha/year was obtained from fertigation plots (RDF through fertigation) while it was only Rs.32,425/ha/year from absolute control (Yadukumar *et al.*, 2009).

References

Abdul Salam, M. (1999). Cashew varieties suitable for high density planting. *Cashew Bull.* 37(6): 3-5.

Abdul Salam, M., Mohanakumaran, N. and Balasubramanian, P. P. (1998). Cashew cultivation Science and Techniques.DCCD, Cochin, 9-12.

Aliyu, O.M., Genetic Diversity of Nigerian Cashew Germplasm, Genetic Diversity in Plants, Prof. Mahmut Caliskan (Ed.), ISBN: 978-953-51-0185-7.

Aneesa Rani, M.S. and Jeeva, S. (2006). High Density Planting in cashew plantation. In: Proceedings of the UGC sponsored National Seminar on Resource Management for Sustainable Agriculture. Department of Agronomy, Faculty of Agriculture, Annamalai University. pp. 236-239.

Asogwa, E. U., Hammed, L. A. and Ndubuaku, T. C. N. (2008). Integrated production and protection practices of cashew (*Anacardium occidentale*) in Nigeria. *African Journal of Biotechnology*. 7 (25): 4868-4873.

Balasimha, D. and Yadukumar, N. 1993. Effect of plant density on photosynthesis in cashew. *Indian J. Plant Physiology*, 36(1): 5-7.

Despande, S.B., Haldankar, P.M., Sapkal, B. B., Chavan, A.P., Joshi, G.D., Patil, B. P. and Rao, E. V. V. B. (2002). Evaluation of high density planting in cashew (*Anacardium occidentale* L.). *The Cashew*, 16(1): 23–28.

Ghosh, S.N., Chattopadhyay, N. and Hore, J.K. (2000). Effect of plan spacing on yield of cashew. Recent advances in Plantation Crops Research. Proc. of PLACROSYM – XIII, 1998, pp. 157 – 15.

Singh, G. (2010). Practical manual on Canopy management in Fruit crops. Department of Agri. and Coop. Ministry of Agri. GOI. New Delhi.

Kabitha Sethi, Tripathy, P. and Das, P.C. (2014). Evaluation of cashew germplasm under Odisha condition. *Agrotechnol.*, 2-4.

Kasuga, L. J. F. (2010). Status of Cashew industry at Tanzania. In. Proceedings of the second International Cashew Conference, Cashew, People and Environment. at Kampala, Uganda. 26-29 Apr. 2010.

Kurian, A. and Peter, K.V. (2007). Commercial crops technology. 2007. New India Publishing. pp. 83-84.

Martin P.J., C.P. Topper, R.A. Bashiru, F. Boma, D. De Waal, H.C. Harries, L.J. Kasuga, N. Katanila, L.P. Kikoka, R. Lamboll, A.C. Maddison, A.E. Majule, P.A. Masawe, K.J. Cashew nut production in Tanzania: Constraints and progress through integrated crop management. *Crop Protection*, 1997, 10(1): 5-14.

Millanzi, N.Q. Nathaniels, S.H. Shomari, M.E. Sijaona and T. Stathers.

Meena,R.K., J D Adiga, M G Nayak, P L Saroj and D Kalaivanan.2014. Effect of Paclobutrazol on growth and yield of Cashew (*Anacardium occidentale* L.) Vegetos. 27(1): 11-16.

Mini Poduval (2012). Research Highlights 2011-12, AICRP on Cashew, RRS, BCKV, Jhargram, West Bengal. p.4.

Oliviera, 2008. Cashew crop. Rev. Bras. Frutic. 30 (1).

Pradeep Kumar, T. (2008). Management of Horticulture crops. New India Publishing. p. 491.

Pushpalatha, P. B., Sobhana, A., Gavas Ragesh and Gregory Zachariah (2013). Scientific tools for increasing income from Cashew plantations. In National Symposium on Cashew. Cashew promotion in India- challenges and opportunities. 29-30 Oct. 2013, Shimoga, Karnataka.

Rupa, T. R., Saroj, P. L., Bhat, P. S and Naik, M. G. (2013). The role of research in reaching sustainable production of Cashew. In: National Symposium on Cashew. Cashew promotion in India: Challenges and opportunities. 29-30 Oct.2013, Shimoga, Karnataka.

Rupa, T.R., Rejani, R., Bhat, M. G. (2013). Impact of climate change on Cashew and Adaptation strategies. In. Climate Resilient Horticulture: Adaptation and Mitigation Strategies. H.P. Singh *et al.* (Eds). Springer India.

Salam, M. A, (1997). High density planting in cashew - principles and practices. *The Cashew,* 11(4): 12-20.

Saroj, P. L. and Nayak, M. G. (2013). Potential of high density plantation for enhancing quality production of cashew. In: Canopy Management and High Density Planting in Subtropical Fruit Crops. Pub. CISH, Lucknow, pp. 55-63.

Suma, B., Abdul Salam, M., Suma, A. and George, T. E. (1998). Effect of planting density on the nut yield of cashew (*Anacardium occidentale* L.). *The Cashew,* 12(2): 33-37.

Yadukumar, N., Rao, E.V.V.B. and Mohan, E. (2001). High density planting of cashew (*Anacardium occidentale* L.). *Tropical Agriculture,* 78(1): 19-28.

Yadukumar, N., Swamy, K.R.M and Bhaskara Rao, E.V.V. B. (2003). Projection on economics of establishment and maintenance of cashew plantations under different plant densities. *The Cashew,* 17(3): 6-16.

Yadukumar, N., Rejani, R., Prabhakar, B. (2009). Fertigation for efficient water and nutrient management in high density cashew plantation, *Journal of Plantation Crops,* 37 (2) pp. 102-110.

Yadukumar, N., Rejani, R., Nandan, S. L. And Prabhakar, B. (2013). Nutrient budgeting and nutrient balance under high density planting system in cashew (*Anacardium occidentale*). *Indian Journal of Agricultural Sciences,* 83 (1): 14–20.

Wait, A. J. and Jamieson, G.I. (1986). The Botany and Cultivation of Cashew. *Agricultural Journal,* September- October, 1986.

Cashew: Improvement, Production and Processing (2017) *Pages* **233–252**
Editor: **Prof. P.L. Saroj**
Published by: **ASTRAL INTERNATIONAL PVT. LTD., NEW DELHI**

13

Nutrient and Water Management

T.R. Rupa

1. Introduction

Cashew (*Anacardium occidentale* L.) is usually grown on a diverse range of soils, from the sandy seacoast to laterite hill slopes, pure sandy soils to sandy loam, laterite soil, deep loam and red latosols, but several of these soils have low effective cation exchange capacities and low exchangeable base status. Though, cashew is cultivated in numerous type of soils but it requires light textured and well drained soils rich in organic matter. Cashew tree is considered drought resistant to some extent and is generally grown as an unirrigated crop but the yield can be doubled, if irrigated. Low productivity is the main concern in cashew cultivation in India. Of several factors associated low productivity, poor soil fertility and low moisture availability during the fruiting season which normally coincides with the onset of dry season in the cashew growing areas are the major factors. There are 17 essential nutrients required for plant growth: carbon (C), hydrogen (H), oxygen (O), nitrogen (N), phosphorus (P), potassium (K), calcium (Ca), magnesium (Mg), sulfur (S), iron (Fe), manganese (Mn), zinc (Zn), copper (Cu), boron (B), molybdenum (Mo), chlorine (Cl) and nickel (Ni). Of these 17 nutrients, all except C, H and O are derived from the soil. When the soil cannot supply the level of nutrient required for adequate growth, supplemental fertilizer applications become essential. Owing to its extensive root system, cashew can draw nutrients from large volume of soil, and as a result it can perform reasonably well on poor soils where other crops fail to do so. In spite of cashew's ability to grow in poor soil, it does respond positively to improvements in soil fertility. Manures and fertilizers are the important inputs, account for 20-30 per cent of the total cost of production. Nitrogen is the mineral nutrient that cashew requires in the greatest amount. It has more influence on tree growth, production and quality of cashew than any other nutrient. Phosphorus is the second most limiting nutrient after N in the nutrition of cashew. It plays an indispensable role for many life processes such as photosynthesis, synthesis and breakdown of carbohydrates, and the transfer of energy within the plant. Potassium is the second largest nutrient next to N required by cashew. Potassium is necessary for several basic physiological functions like the formation of sugars and starch, synthesis of proteins, normal cell division and growth, and neutralization of organic acids.

In India, majority of the cashew growing regions are located on soils with poor fertility status. On other hand, nutrient mining has occurred in many cashew growing soils due to lack of affordable fertilizer/organic sources and where fewer or no biomass residues/leaf litter are recycled to the soils. Integrated nutrient management/site specific nutrient management practices involving combined application of chemical fertilizers, organic manures/green manuring and biofertilizers which constitute an efficient nutrient management strategy in cashew are crucial to improve soil quality for sustainable production. In cashew; flushing, flowering and fruit development stages are critical phases that decide the nut yield. Any form of stress, (biotic or abiotic) during these phases can decrease the yield substantially. Besides, canopy development; water availability strongly influences flowering and fruit set and can affect fruit drop, fruit size, nut yield and quality. Studies in India have shown that nut yield can be enhanced by providing protective irrigation with 200 L of water/tree once in 15 days from January to March during the summer season. Research results in Brazil and other countries have shown that irrigation could increase productivity by up to 300 per cent depending on the region. Drip irrigation is the most efficient irrigation system with savings of water between 40 to 60 per cent over other irrigation systems.

2. Nutrient Management in Cashew

Nutrient Removal

Theory of nutrient removal is one of the bases for integrated nutrient management. It has been estimated that annual removal of nutrients by a cashew tree of 30 years old was 2.847 kg N, 0.752 kg P and 1.265 kg K per tree (Mohapatra *et al.*, 1973). The nutrient requirement to produce one kg of cashew nut was 64.1 g N, 2.05 g P, 24.7 g K, 4.19 g Ca, 1.57 g S, 525.7 mg Fe, 63.6 mg Mn, 87.8 mg Zn and 26.5 mg Cu per tree (Beena *et al.*, 1995). The annual nutrient uptake required by cashew trees of 70 months of age in Australia were 2.1 kg N, 0.45 kg P, 1.32 kg K, 0.54 kg Ca and 0.57 kg Mg in order to maintain the tree structure (Richards, 1993). While, an eight-year old cashew tree removes 610 g N, 58 g P, 394 g K, 52 g Ca, 39 g Mg, 34 g S, 2.12 g Fe, 343 mg Mn, 390 mg Zn and 130 mg Cu in Australia (Grundon, 2001). Nutrient budgeting and balance studies in six-year old cashew plants under high density planting system (625 trees/ha) indicated a negative N, P and K balance of 113, 38 and 92 kg/ha in control plot where no fertilizer was applied. A strong positive N balance ranged from 137 to 251, P balance from 34 to 75 and K balance from 89 to 164 kg/ha/year was obtained in trees with two-thirds and full dose (750 g N and 150 g each of P_2O_5 and K_2O per tree/year) of fertilizer applications, respectively. While one third dose of fertilizer application resulted in a positive N, P and K balance of 40, 19 and 40 kg/ha/year (Yadukumar *et al.*, 2009). The constant removal of produce without or with inadequate replenishment of plant nutrients cause a steady decline of soil fertility. This mining of plant nutrients leads to severe deterioration of soil health. However, application of imbalanced and/or excessive nutrients led to declining nutrient use efficiency making fertilizer consumption uneconomical and producing adverse effects on soil and atmosphere.

Nutritional Management in Nursery

In cashew, initial growth of seedlings depends upon the nutrient reserve of cotyledon. Ximenes, (1995) reported that the macronutrients contained in the cotyledons can provide around 54 per cent of N, 45 per cent of P, 17 per cent of K, 1 per cent of Ca, 16 per cent of Mg and 36 per cent of S necessary for seedling development for up to 75 days after planting. The large size cashew nut (8 to 12 g) gave initial seedling growth advantage over that of small sized nuts (6 g) only at the juvenile stage but as the plants age advances the influence fizzles out (Ibiremo *et al.*, 2012). It has been reported that biofertilizers (*Azospirillum, Azotobacter* and VAM) increased the germination percentage of nuts, plant growth, number of graftable seedlings and reduced the incidence of fungal diseases in the nursery (Kumar *et al.*, 1998; Ramesh *et al.*, 1999; Sinish *et al.*, 2005). Inoculation of *Azotobacter* resulted in higher root growth (Oblisami *et al.*,1985) and yield (Singh, 1997) of cashew. Cashew has been described as a host plant for arbuscular mycorrhizal (AM) fungi (Sivaprasad *et al.*, 1992). Mycorrhizae have also been shown to help increase the resistance of plant root systems to soil-borne diseases (Perrin, 1990). Additionally, plants colonized by mycorrhizal fungi have been shown to survive better than uncolonized plants under suboptimal growing conditions and in marginal soils (Lioi and Giovannetti, 1987). VAM (25 g/bag) is helpful for better graft uptake at the time of grafting (Sridhar *et al.*, 1990). It has been demonstrated that inoculation of cashew with mycorrhizal inoculum had higher concentrations of K in both leaves and roots (Haugen and Smith, 1993). Among VAM, *Acaulospora laevis* and *Gigaspora mosseae* are better symbionts for inoculating cashew (Lakshmipathy, 2000). While Ananthakrishnan *et al.* (2004) reported, among VAM, *Glomus fasciculatum* is superior in terms of increased shoot length, internode number, number of leaves, stem diameter, root length and root number under nursery conditions.

A fertilizer dose of 150:20:100 (ppm) N:P:K to rootstocks and grafts of cashew at a rate of 100 ml/plant/week resulted in higher plant height, stem girth and number of leaves (Manjunatha, 2001). Addition of cocoa pod husk at 3 per cent by volume as substrate to soil is effective for raising cashew seedling in the nursery as it enhanced growth of cashew seedlings (Agele and Agbona, 2008).They also observed that cocoa pod husk improves the soil pH, organic carbon, N, K, Na and Ca contents of the soil besides increasing the contents of ash, N, K and Na in the cashew leaves. Diva Correia *et al.* (2003) evaluated the use of mature and immature coir dust for preparation of substrates for grafted dwarf cashew seedlings and reported that coir dust either from mature or immature fruit was suitable for seedlings growth, being able to replace the low humid glei soil at 20 per cent.

Nutritional Management in Cashew Plantations

Nitrogen and P were most important nutrients during the pre-bearing stage, but at the bearing stage, K together with N is also important. The response of cashew to applied N is tremendous and the same is observed universally. Increase in cashew yield due to N application was reported by several workers (Lefebvre, 1973; Reddy *et al.*, 1982; Rao *et al.*, 1984; Veeraraghavan *et al.*, 1985 and Ghosh, 1988). Ghosh (1990) reported that number of nuts/plant and nut weight was the highest at 600 g N/tree/year but Latha *et al.* (1994) obtained response to N up to 1000 g/tree. Urea is the most commonly used nitrogenous fertilizer in India. In Nigeria, urea and sulphate of ammonia are generally used. Falade (1984) reported that sulphate of ammonia was superior to urea particularly when medium or high doses of N were applied to cashew. Whereas, contradictory reports were observed regarding the response of cashew to P fertilizer. Rao *et al.* (1984) observed no response to P application in sandy loam soils. Also, Veeraraghavan *et al.* (1985) reported no effect of P on cashew in laterite soils of Madakkathara. It was observed that the main effect of P to increase the yield was limited to a dose of 25 kg/ha, but when applied with N fertilizer, P application increased yield up to a dose of 75 kg/ha (Sawke *et al.*, 1985). However, Kumar (1985) reported positive influence of P on nut yield. Richards (1993) reported that soil P is a major limiting nutrient in P deficient soils of Australia and P application increased the nut number and nut yield. Of phosphatic fertilizers for use in acid soils in India, the slow release and more efficient ground Mussoorie (rock) phosphate is popular. Application of K increased the production of cashew particularly in the presence of N (Lefebvre, 1973). Significant positive effects of K on growth and yield of cashew were reported by Ghosh (1988) and Ghosh (1990). While Veeraraghavan *et al.* (1985) reported no positive effect of K application in cashew. Kumar (1985) obtained linear response for K up to 150 g K_2O/tree. Phosphorus and K application at higher level improved the nut yield (Sawke, 1980). Increased nut weight and nut yield due to application of higher levels of N, P and K was reported by Ghosh and Bose (1986), Harishu Kumar and Sreedharan (1986), Ghosh (1990) and Kumar *et al.* (1995).

It is essential that level of micronutrients in soil and plant should be optimum for growth and development, since the micronutrients need is site specific. Foliar feeding is often the most effective and economical way to correct micronutrient deficiencies in horticultural crops. Foliar application of nutrients normally reduces the loss through adsorption, leaching and other processes associated with soil application. Deficiencies of Fe, Mn, Zn, Cu, B and Mo can be corrected by foliar sprays of 0.5-1.0 per cent ferrous sulphate, 0.5-1.0 per cent manganese sulphate, 0.5 per cent zinc sulphate, 0.1 per cent copper sulphate, 0.1 per cent solubor and 0.1 per cent Mo salts, respectively to cashew at the emergence of the flush, panicle initiation and fruit set stages.

Application of fertilizers, dosage and the time and its schedule under different agro-climatic zones has been standardized (Veeraraghavan *et al.*, 1985; Harishu

Kumar and Sreedharan, 1986; Grundon, 2001; Salam *et al.*, 2008; Yadukumar *et al.*, 2009). A summary of the recommended dose for the macronutrients for major cashew producing states is presented in Table 13.1. Application of farmyard manure (FYM) at a rate of 10-15 kg/plant in combination with N, P and K fertilizer is effective in enhancing the nutrient use efficiency. The requirement of nutrient dose may vary from location to location based on the initial fertility status of soil. The magnitude of yield response to fertilizer dose may vary with respect to tree canopy, initial soil fertility status, nature of soil, other management practices etc. Site specific nutrient management offers the most appropriate option to deal with the spatial variation in soil fertility using variable fertilizer rates as per soil test values.

Table 13.1: Recommended Dose of Fertilizers to Cashew

State	*Nutrient Dose for Mature Cashew Plantations (5th year of planting onwards) (g/tree/year)*		
	N	*P_2O_5*	*K_2O*
Kerala	500–750	125–325	125–750
Karnataka	500–500	250–125	250–125
Tamil Nadu	500	200	300
Andhra Pradesh	500–1000	125–125	125–125
Maharashtra	1000	250	250
Odisha	500	250	250
West Bengal	1000	250	250

Source: ICAR-DCR, Puttur and AICRP-Cashew Centres.

Integrated use of organic manures, inorganic fertilizers and biofertilizers assumes great importance for sustainable cashew production and maintaining soil health. The organic manures and biofertilizers not only supply essential plant nutrients, but also improve the soil physical, chemical and biological health. Inoculants of *Azotobacter* and *Azospirillum* either sole or in combination have been shown to improve N nutrition of plants through biological N_2 fixation and also secretion of some growth promoting substances which affect the growth, nutrition and microbial activity in the rhizosphere (Zayed, 1999). The phosphate solubilizing microorganisms (*Pseudomonas*) play an important role in conversion of unavailable forms of P into available P forms through secretion of organic acids and enzymes. Arbuscular mycorrhizal fungi (AMF), on the other hand are ubiquitous in soils throughout the world and play an important role in affecting the plant growth through mobilization of nutrients. Often, FYM is limited in supply, suggesting that green manure may be a more feasible substitute for fertilizer N. Green manuring maintains and improves soil structure by addition of organic matter, minimize P, K fixation in soils, produces humus, which enhances the utilization of fertilizer nutrients by plants and helps in reducing leaching losses by enhancing water retention ability of soil. Growing green manuring crops like *glyricidia*, sesbania, sunhemp and cover crops between two rows of cashew resulted in the nutrient addition of 186 kg N, 23.6 kg P_2O_5 and 126.2 kg K_2O through *glyricidia* and 141 kg N, 17.9 kg P_2O_5 and 162.3 kg K_2O/ha through sesbania (Yadukumar *et al.*, 2008).

Method of Fertilizer Application

Cashew trees are surface feeders with about 50 per cent of the root activity being confined to the top 15 cm of the soil and about 72 per cent of root activity was found within a 2 m radius from the tree trunk (Wahid *et al.*, 1993). It has been suggested to apply fertilizers within a radius of 2 m from the main stem for efficient utilization of the applied nutrients. During the 1st, 2nd, 3rd, 4th and 5th year of planting 1/5th, 2/5th, 3/5th, 4/5th and 5th year onwards full quantity is to be applied. George *et al.* (1984) reported application of N, P and K fertilizers in two circular trenches (1.5 and 3 m from the trunk) for sandy soils, a single trench method (25 cm wide and 15 cm deep circular trench at 3 m from the trunk) for sloping ground, and the band method (in a circular band 1.5-3 m from the trunk + soil incorporation) for flat ground are best suited. The root activity of cashew in relation to phenological phases studied by Beena *et al.* (1995) employing ^{32}P soil injection technique reported that flushing and early flowering phase (September to December) is the most appropriate time for fertilizer application in cashew orchard. The annual dose of fertilizers to cashew are to be applied in two split doses, the first split dose at the onset of the pre-monsoon period and the second split dose during the post-monsoon period when the soil moisture condition is at its optimum; if only one application is given, it should be in the post-monsoon period when enough moisture is available.

3. Water Management in Cashew

Water Requirement, Time and Method of Irrigation

Irrigation scheduling, knowing how much water to apply and when, has a direct influence on tree health as well as nut yield and quality. Right irrigation scheduling requires an understanding of how much water can be held in the plant root zone. In order to maintain the optimum level of water in crop and to schedule irrigation, an estimation of water requirement is prerequisite. Water requirement of a crop depends on crop, soil type and atmospheric demand for the water. Generally, crop water requirement is defined as the depth of water needed to meet the water loss through evapotranspiration (ET_{crop}) as a healthy crop under non-restricting soil conditions and attaining full production potential under the given growing environment. Fresh grafts when planted require sufficient soil moisture for initial establishment, hence cashew is planted during monsoon season. Whenever there is drought situation after planting they need protective irrigation. The irrigation through pitcher (hold pots) is recommended in dry land situations. Under drought situation, irrigation is one of the most important factors in establishing the newly planted grafts well. Care must be taken to keep the soil moist but not waterlogged. The root ball of a newly planted graft must be kept moist to supply the plant with water until its roots grow into the soil. Newly planted grafts need to be watered every three to seven days, depending on the soil type and weather conditions. Because of their deep taproot system, established cashew trees can survive the dry season without irrigation, but premature nut drop is a common problem. While not essential, irrigation could prove to be a major benefit to production, largely by preventing nut drop. Irrigation during critical phases of established cashew trees improves nut yield.

Field experiments conducted at ICAR-Directorate of Cashew Research (ICAR-DCR), Puttur on fertigation in cashew indicated that under normal density planting system (7 m x 7 m), the effective canopy coverage per tree is 12.56 m^2. The quantity of water requirement calculated based on the effective canopy area was 12.56 L/tree/day from December to January (Daily open pan water evaporation is 5 mm) and 20 L/tree/day from February to March (Daily open pan water evaporation is 6.5 mm) to meet 20 per cent of the cumulative pan evaporation (CPE). Similarly, for 40 per cent CPE and 60 per cent CPE, the water requirement was 24 L/tree/day and 38 L/tree/day from December to January and 36 L/tree/day and 58 L/tree/day from February to March, respectively. In order to meet 20 per cent CPE, four drippers of 2 L/hour discharge rate can be fitted at two equidistant points 1 m away from the base of the tree. Similarly, to meet 40 per cent and 60 per cent CPE, four drippers and six drippers of 4 L/h discharge rate can be fixed. Drip irrigation can be given for 1 h and 30 minutes during December-January and 2 h during February-March (Yadukumar and Rejani, 2008).

Estimating Water Requirement for Cashew

To meet 20 per cent CPE

Age of the tree: 5 years

Canopy spread: 4 m, Canopy spread = canopy diameter = mean of EW and NS length

Canopy area = πr^2 where r = radius of the canopy.

If the radius is 2 m, the total area covered by individual tree canopy is 3.14(π) x 2^2 = 12.56 m^2 (Ground coverage by canopy)

Daily CPE = 5 mm, 20 per cent of CPE = 1 mm, Therefore, quantity of water to be given to meet 1 mm of water in 12.56 m^2 area = 12.56 x 1/1000=0.01256 cubic M, 1cubic M =1000 L. Therefore, 0.01256 cubic M =12.56 L/tree/day. The water requirement of cashew can be calculated depending upon the canopy coverage and daily water evaporation.

Mishra *et al.* (2008) estimated the water requirement of cashew on daily basis for all months of a particular year and the net average seasonal water requirement of crop as per their estimate was 997 mm. Water requirement can be calculated based on evaporation rate in the given area and crop factor (K_c). For cashew, K_c varies from about 0.8 at peak flowering to 1.1 at peak nut set. Cashew needs irrigation from the onset of flowering to late nut set. Supplementary irrigation for younger plants may be required during drier parts of the year and to assist with fertilizer applications. Trees irrigated with sprinklers will need about 500 L/tree/week and 50 per cent of the water requirement can be saved with drip irrigation. Irrigation should be stopped before starting harvest, to avoid nut germination on the ground (Website: www.deedi.qld.gov.au). The optimum temperature for CO_2 assimilation is in the range 25–35°C. Progressive closure of the stomata occurs at saturation deficits of the air >1.5 kPa. In the field, differences in rates of gas exchange between irrigated and unirrigated cashew trees only become apparent three or four months after the end of the rains. The stomata playing an important role in maintaining a favourable leaf water status in dry conditions. Sap flow measurements indicate transpiration rates of 20–28 L d^{-1} $tree^{-1}$. Irrigation can be beneficial during the period from flowering to the start of harvest, but reliable estimates of water productivity have yet to be established. The best/only estimate is 0.26 kg (nut in shell) m^{-3} (irrigation water) (Carr, 2014).

The ability of unirrigated cashew trees to draw moisture below 1.8 m depth was apparent from the experimental results as such trees gave virtually zero water use over recorded depth at peak nut set, yet they were able to give yield ranging from 20 to 70 per cent of the highest irrigated yield (Richards, 1993). He further stated that this poses the question of the long term irrigation requirements of well-established cashew trees. Richards (1993) raised questions whether such mature trees with presumably dry root system require wet season fertilizer application only as in the case of no irrigation and or require irrigation reduced to a critical period only in order to produce economic yields. Schaper (1991) reported that leaf gas exchange was lower and water potential higher in leaves of unirrigated cashew trees near peak flowering compared to irrigated trees, but that both declined after flowering commenced compared with pre-flowering, wet season levels. Further, he reported that stomatal regulation in cashew appears to prevent leaves from losing water faster than their roots can replace it by absorption. Non production of new shoots in March, April and May in unirrigated tree was the mechanism to withstand stress situations whereas in irrigated trees new shoots were produced in March, April and May but then these shoots never flowered during that season. Advantage of these new shoots is mostly for the better production of flushing shoots after the rainy season (October onwards) thereby total canopy area increased and total flowering laterals increased/tree considerably.

In a study conducted in Australia by Schaper *et al.* (1996) indicated that the plant could be irrigated only between flowering and harvest without decreasing yield compared to irrigating during the entire drought period. This saves much water. The water needs of the plant vary with climate, the plant's foliar area, the growth phase of the plantation and with the irrigation method used. During periods of high evapotranspiration, 5 L of water/day are recommended for each square meter of soil surface shaded by the plant crown or area wet by the emitters (Table 13.2). The frequency of irrigation depends on the water retention capacity of the soil and should vary between two and four days, for sandy and clayey soils, respectively. With drip irrigation, the volumes of water recommended in Table 13.2 may be reduced by about 15 per cent. The number of drippers per plant should increase gradually, according to the age and stature of the plant, from one dripper during the first year to up to four, six or eight to mature plant in clayey, medium textured and sandy soils, respectively. Richards (1993) reported that water requirement by five-year-old cashew tree growing in sandy soil and under high evaporative demand is about 400 to 500 L/tree/week which equated about 30 L/m^2 of canopy area.

Table 13.2: Average Values of Crown Projection Areas, Percentage of Soil Covered by the Plant and Volume of Water to be Applied in Irrigation as a Function of Plant Age

Year of Crop	*Crown Projection Area* m^2	*Soil Covering per cent* [(1)]	*Volume of Water L/plant/d* [(2)]
1st	1	2	5
2nd	5	10	25
3rd	15	30	70
4th	25	50	120
5th +	30	60	145

1. Assuming the spacing between plants is 7 m x 7 m

2. If the area wetted by the nozzle is greater than the crown projection, the volume of water to be applied should be chosen as a function of the wetted area.

Studies on effect of black polyethylene mulch and supplemental irrigation (60 L of irrigation water/tree given manually at 15 days interval from the emergence of panicles and 6 to 8 total irrigations during each season) on fruit retention of cashew nut revealed that polyethylene mulch + irrigation treatment had maximum fruit retention of 66.15 per cent followed by irrigation alone (58.04 per cent) and polyetheylene mulch alone (52.83 per cent). The correspondence fruit retention in control (no polyethylene mulch and no irrigation) was 44.98 per cent (Nawale *et al.*, 1985). Studies on the effect of supplementary irrigation on fruit retention and nut yield at NRC-Cashew (Presently ICAR-DCR) on cashew trees of 13 years old with four different treatments, *viz.*, fortnightly irrigation @ 200 L/tree from (1) November to January; (2) January to March; (3) November to March; and (4) Control (No irrigation) revealed that there has been increase in fruit retention in irrigated plots as compared to control (Table 13.3). Irrigating cashew during November-March (10 irrigations) was significantly superior to other treatments of irrigations: November - January and January - March (5 irrigations each). Analysis of two years pooled data also showed similar results (Table 13.4). Irrigations has increased fruit retention (nut retention) which ultimately increased the yield (Yadukumar and Mandal, 1994).

Drip irrigation or more broadly known as micro-irrigation is mainly suited for orchard and plantation crops where it saves 30-70 per cent irrigation water and increase yield by 25-80 per cent. It is recommended to use micro-irrigation (spray or drip), because of certain advantages like decreased incidence of leaf sickness and weeds, water saving by decreasing losses by evaporation and greater efficiency of water use. Micro-irrigation can also be adapted to different soil and topographies; there is a saving in labour costs and efficient application of fertilizers via irrigation water (fertigation). The initial cost of a system of micro-irrigation for cashew varies from R$ 3,000 to R$ 4,500 (US$ 1,000 to US$ 1,500) per hectare. Where spraying is used it is recommended to have one jet per plant, with a nominal flow of 30 to 70 L/h and wetting diameter of 3.5 to 5.0 m. In dripping, a minimum of four drippers per plant ought to be used per mature plant in clayey soils, and up to eight drippers per plant in sandy soils. To choose between spray and dripping as a system of irrigation, the water availability (quantity and quality) should be considered. In dripping there is a greater savings in water and energy, because the loss of water by evaporation from the soil surface is less and the system operates at a lower pressure. On the other hand, the risk of emitter blocking is greater than with spray irrigation, thus better filtering, especially when surface water with a lot of organic matter is used. Dripping also offers the advantage of not wetting the fruits that fall onto the ground, allowing less frequent collecting where the primary product required is the nut (www.ipipotash.org).

Table 13.3: Fruit Retention as Affected by Irrigation

Treatments	*Fruit Set (Mean 5 panicles)*	*No. of Fruits Harvested*	*Fruit Retention (Per cent)*
(a) Irrigation once in 15 days @ 200 L/tree from			
November-January	27	9	33.0
January-March	16	6	37.5
November- March	25	11	44.0
(b) Control	13	4	30.4

Table 13.4: Cashew Nut Yield as Affected by Irrigation

Treatments	*Nut Yield (kg/tree)*		
	1987	*1988*	*Mean of Two Years*
(a) Irrigation once in 15 days @ 200 L/tree from			
November-January	5.33	4.53	4.93
January-March	5.82	4.93	5.37
November- March	5.33	7.32	6.25
(b) Control	2.91	3.54	3.23
CD (p=0.05)	NS	1.47	1.64

NS: Non significant.

The most serious problem in trickle irrigation is clogging of emitters or applicators. Recommendations and guidelines need to be followed for preventive maintenance which include water filtration, chemical treatment, pipeline flushing, and field inspection. A suitable type, size, and capacity of a filtration unit is required. Chemical treatment should be considered in terms of theory and field research on the reclamation and prevention of emitter clogging. Proper procedures for the flushing and field inspection of trickle irrigation systems are also essential (Bucks *et al.*, 1979). Salts in soil or water reduce water availability to the crop to such an extent that yield is affected. No specific quality standards apply to cashew irrigation, however, the general water quality standards can be used. Electrical conductivity of irrigation water should not exceed 0.8 dS/m and total dissolved ions should be less than 600 ppm.

In young cashew, application of 30 L of water/tree at 15 days interval increased the nut yield by 393 per cent when compared with unirrigated plants in West Bengal

(Ghosh, 1995). Drip irrigation @ 43 mm/week from April to October increased nut yield by 20 per cent in Australia (Schaper *et al.*, 1996). The response to irrigation varied among cashew genotypes. Irrigation increased nut yield of dwarf cashew clones CCP-76 and CCP-09. The highest nut yield was obtained with cashew clone CCP-09 irrigated when accumulated pan evaporation reached 10 mm. Cashew alternated years of high nut yield with years of low nut yield, even when irrigation was applied. Irrigation did not influence nut weight of clones CCP-76 and CCP-09 (Oliveira *et al.*, 2006). Mishra *et al.* (2008) evaluated the economic feasibility of drip irrigation with black Linear Density Poly Ethylene (LDPE) mulch on the loamy soil in coastal Odisha and reported that the drip irrigation is economical and cost effective as compared with conventional basin irrigation. It was shown that an increase of 108 per cent in nut yield and 122 per cent in net seasonal income obtained with 80 per cent irrigation requirement through drip irrigation + black LDPE mulch with the highest benefit cost ratio of 3:1 over the conventional ring basin irrigation. Mulch alone could increase the yield by 16 per cent even in the absence of drip.

4. Response of Cashew to Supplementary Irrigation during Flushing and Flowering Phases

Experiments conducted at All India coordinated Research Project on Cashew (AICRP-Cashew), Vengurla (Maharashtra) indicated that the growth and yield attributing characters of cashew variety Vengurla-7 did not vary significantly among irrigation levels using drip. The fruit set/m^2 was the highest in irrigation @ 60 per cent CPE. Mean No. of nut/panicle was the highest in irrigation @ 80 per cent CPE. While cumulative nut yield for nine harvests was maximum in irrigation @ 40 per cent CPE (Tables 13.5 and 13.6).

Experiments conducted at AICRP-Cashew, Vridhachalam (Tamil Nadu) revealed that irrigating the cashew plants variety VRI-3 at 80 per cent CPE enhanced plant height, trunk girth, canopy spread and canopy surface area. The flowering was early in trees receiving irrigation at 80 per cent CPE (Table 13.7).

Experimental results of AICRP-Cashew, Chintamani (Karnataka) revealed that, among different levels of irrigation, irrigating the crop at 80 per cent CPE resulted in significantly the highest plant height and stem girth while there was no significant difference in canopy spread. Maximum E-W and N-S canopy spread was obtained in irrigation at 80 per cent CPE. Nut yield varied significantly among the irrigation levels. The highest nut yield, nut weight, shelling per cent and cumulative yield of 5 harvests were obtained in irrigation at 80 per cent CPE (Table 13.8).

5. Irrigation and Fertilizer Effects on Cashew

The irrigation and fertilizer experiments in cashew conducted in Northern Territory of Australia indicated substantial reduction in yield, tree size and yield efficiency associated with absence of fertilizers and irrigation inputs (Richards, 1993). Application of fertilizer increased yield significantly compared to that of unirrigated and or unfertilized trees. He has reported that rate of irrigation had limited impact on yield, size and efficiency in similar size trees producing at high, medium and low rates of irrigations (600, 490, 290 L/tree once in a week) within irrigation and

Table 13.5: Effect of Drip Irrigation on Growth and Yield Attributes of Cashew at Vengurla (Maharashtra)

Irrigation Levels	*Mean Plant Height (m)*	*Mean Stem Girth (cm)*	*Mean Canopy Spread (m)*	*Mean Canopy Area (m^2)*	*Mean No. of Laterals/m^2*	*Mean No. of Flower Panicle/m^2*
No irrigation	7.30	86.24	7.64	46.84	31.90	15.97
Irrigation at 20 per cent CPE	7.11	91.08	8.27	54.44	31.60	17.23
Irrigation at 40 per cent CPE	7.47	90.83	8.34	55.15	30.30	16.62
Irrigation at 60 per cent CPE	7.41	85.41	7.83	48.72	32.0	17.95
Irrigation at 80 per cent CPE	7.14	91.91	8.03	51.49	28.93	16.80
SEm±	1.19	2.96	0.34	4.17	1.11	0.42
CD (p = 0.05)	NS	NS	NS	NS	NS	NS

NS = Non significant.

Table 13.6: Effect of Drip Irrigation on Yield Attributes and Yield of Cashew at Vengurla (Maharashtra)

Irrigation Levels	*Mean Flower Duration (days)*	*Mean Fruit Set/m^2*	*Mean No. of Nut Panicles*	*Mean Apple Weight (g)*	*Mean Nut Weight (g)*	*Cumulative Yield for 9th Harvest (kg/tree)*
No irrigation	102	69.07	14.77	66.00	9.75	26.81
Irrigation at 20 per cent CPE	105	79.37	15.20	67.75	9.17	28.30
Irrigation at 40 per cent CPE	108	76.55	14.67	70.00	9.22	29.84
Irrigation at 60 per cent CPE	110	89.87	15.55	67.75	9.10	26.46
Irrigation at 80 per cent CPE	101	84.65	16.65	69.00	9.00	28.74
SEm±	5.45	4.50	1.08	2.09	0.20	–
CD (p = 0.05)	NS	NS	NS	NS	NS	–

NS = Non significant.

Table 13.7: Effect of Drip Irrigation on Growth of Cashew at Vridhachalam (Tamil Nadu)

Irrigation Levels	*Plant Height (m)*	*Trunk Girth (cm)*	*Mean Canopy Diameter (m)*	*Canopy Height (m)*	*Canopy Surface Area (m^2)*
No irrigation	3.42	40.5	2.46	2.22	9.65
Irrigation at 20 per cent CPE	3.98	42.2	2.82	2.76	14.50
Irrigation at 40 per cent CPE	4.12	44.0	3.44	3.02	18.68
Irrigation at 60 per cent CPE	4.56	45.8	3.92	3.34	23.86
Irrigation at 80 per cent CPE	4.91	52.2	4.60	3.70	32.49
CD (p = 0.05)	0.18	0.26	0.64	0.32	5.12

Table 13.8: Effect of Drip Irrigation on Growth and Yield of Cashew at Chintamani (Karnataka)

Irrigation Levels	*Plant Height (m)*	*Stem Girth (cm)*	*Nut Yield (kg/tree)*	*Cumulative Yield (kg/tree) of 5 Harvests*	*Nut Weight (g)*	*Shelling (Per cent)*
No irrigation	4.42	78.54	8.25	31.81	6.9	29.50
Irrigation at 20 per cent CPE	4.61	79.46	10.10	39.00	7.1	30.10
Irrigation at 40 per cent CPE	4.67	85.67	12.85	44.91	7.2	31.40
Irrigation at 60 per cent CPE	5.03	86.83	14.20	51.63	7.2	31.30
Irrigation at 80 per cent CPE	5.11	88.08	14.75	54.55	7.4	32.10
SEm±	0.10	0.94	0.94	–	–	–
CD (p = 0.05)	0.30	2.90	2.89	–	–	–

fertilizer frequency treatments. At medium and high level of irrigation the yield and kernel recovery increased significantly compared to low irrigation levels for trees applied after commencement of flowering. Further, it was reported that the combination of irrigation treatments and fertilizer frequency interacted to effect tree size with smallest tree canopy in the absence of fertilizer with or without irrigation. Kernel recovery was influenced by the combined effects of irrigation and fertilizer as post-flowering irrigation at low rates, unirrigated and unfertilized treatments, gave significantly lower recovery rates.

It has been documented that irrigation in the absence of adequate nutrition is wasteful (Richards, 1993). Irrigation and nutrient application together gave larger tree canopy, greater yield and improved efficiency although the phenology cycle was not affected. Irrigation at low rates was the most efficient in terms of water use efficiency but carries the risk of reduced yield and kernel recovery rate. Richards (1993) reported that duration and intensity of flowering for both irrigated and unirrigated tree from 1988 to 1990 in cashew showed similar patterns in each year. In 1990 the unirrigated trees gave significantly lower peak flower level than irrigated trees. The pattern of 1990 flowering was different from that of previous years and showed a delay in flowering at the beginning before recovery in the seventh week and a quicker decline in flowering after the peak was reached. He reported that in cashew significant variations in water use and Kc (Crop coefficient) values range from 0.45 to 1.0 depending upon the ground cover and size.

The highest nut yield (3.8 kg/tree) was obtained from trees provided with 80 per cent recommended dose of fertilizers as water soluble fertilizers through drip irrigation, compared to trees supplied with recommended dose of NPK through soil without drip irrigation (Kumar *et al.,* 1998). Latha and Salam (2001) found that in rainfed trees, application of N 500 g/tree/year produced 0.77 kg nuts/tree while trees applied with no N resulted in zero yield. In irrigated trees (40 L/tree/day), N application of 1.5 kg/tree/year resulted in an increase in yield by 54 per cent compared to rainfed trees. When the irrigation level was increased to 80 L/tree, the yield increased was 124 per cent. In studies on drip irrigation and graded levels of N, P and K on cashew productivity, irrigation through drip @ 80 L/tree once in 4 days and a fertilizer dose of 750 g N, 187.5 g P_2O_5 and 187.5 g K_2O/tree/year resulted in 10 fold higher total root production (119.30 kg/tree) and 34.34 per cent higher nut retention (70.7 per cent) over untreated trees (no irrigation and no fertilizer) (Yadukumar, 2001).

6. Fertigation

The application of fertilizers through the irrigation water (fertigation) has the advantages of increasing the efficiency of the fertilizers and reducing the costs of labour and machinery for its application. Fertigation allows the application of nutrients with greater frequency, without increasing the cost of the application, minimizing losses by volatilization and leaching and optimizing nutrient absorption by the roots. The nutrients most frequently applied in fertigation are those with greater mobility in the soil, like N and K (Oliveira *et al.,* 2003). To apply nutrients by fertigation, tanks of the solution, where the fertilizers are pre-diluted in water, and an injecting device are necessary. The types of injectors most utilized in fertigation are: injector pumps, venturi and differential pressure tanks. There are many advantages to fertigation: a) uniform application of nutrients; b) application of nutrients according to the needs of the plant and the rate of uptake; c) greater efficiency of nutrient use due to its mobility in the wetted zone of the soil where the root system is concentrated; d) savings on labour and agricultural equipment; e) reduction in soil compaction from the use of heavy equipment; f) ability to apply nutrients more frequently thus reducing nutrient losses (Santos,1997). Fertigation needs to be carefully managed to avoid soil acidification and salinization in the root zone. To avoid blocking the emitters the fertilizers used should be fully soluble in water and should not form precipitates, especially calcium and iron phosphates.

With fertigation, the quantity of nutrients (through fertilizers and organic manure) to be applied can be reduced to half the quantity of recommended nutrients. It has been reported that an increase of 100 per cent and 226 per cent in yield with the application of half of the recommended dose of N, P and K in inorganic form (Recommended dose: 500 g N, 125 g each of P_2O_5 and K_2O/tree/year) of nutrients through fertigation and balance half applied in organic form through castor cake as compared to the above dose applied through soil and irrigated separately and absolute control (without fertilizer, manure and irrigation), respectively indicating better nutrient use efficiency (Yadukumar and Rejani, 2008). Fertigation saved 50 per cent in the fertilizer requirement and doubled the cashew yield (Richards, 1993; Yadukumar and Mandal, 1994; Mishra *et al.*, 2008).

7. Drainage

In poorly drained soils, cashew plants exhibit poor growth with the resultant low productivity. In order to get optimum yield, a proper drainage in cashew orchards is essential. In fact, removal of excess water from the active root zone during rainy season facilitates favourable soil moisture conditions for the growth of plants. High water table and soil moisture content during critical periods of reproduction phase may also affect flowering of trees and subsequent productivity. The water table in the orchard should be below the active root zone *i.e.*, one to one and a half meters below the ground level. With shallow water tables, a salinity problem may also exist due to upward movement of water and salts from the ground water as the water evaporates from the soil or is used by the crop. Such a salinity problem is related to high water tables and the lack of drainage; it is only indirectly related to salts in the irrigation water.

Therefore, adequate drainage both surface and subsoil drainage is essential. Surface drainage removes excess surface water quickly during and after heavy rains. This is achieved by making beds unless the natural topography has a slope of at least 0.5 per cent. Beds may contain from 1 to 4 rows of trees. The height and slope of beds need only be enough to move surface water (0.5 per cent or more), assuming sufficient profile drainage. Surface water can be diverted into collection ditches by means of drop pipes. Water that soaks into the soil and thereby resulting in raising the water table should be removed by subsurface drainage. This may be done through the use of subsurface drains or open ditches. Both surface and subsoil drainage are required for water management in poorly drained soils in high rainfall areas. Proper drainage not only improves soil aeration but also prevents salt accumulation in soils. However, lands with poor drainage and high water table can also be utilized for cashew cultivation by planting on mounds.

8. Future Line of Work

- Need to develop site-specific nutrient management strategy which enables cashew growers to apply the right quantity of required nutrients at right time.
- Need to promote balanced, efficient and environmentally sound nutrient management system in cashew based cropping systems.
- Need to develop optimum concentrations and/or optimum ranges of nutrients useful for correct diagnosis and improvements of nutrient status of cashew plants.
- Development of models for predicting nutrient response in cashew.
- Need to standardize scheduling of irrigation to cashew based on plant water balance in consonance with soil and climate.
- Need to develop the water requirements of cashew under different planting systems and agro-climatic zones.

References

Agele, S.O. and Agbona, A. I. (2008). Effects of Cocoa Pod Husk Amendment on Soil and Leaf Chemical Composition and Growth of Cashew (*Anacardium occidentale* L.) Seedlings in the Nursery. *American Eurasian Journal of Sustainable Ag riculture,* 2(3): 219-224.

Ananthakrishnan, G., Ravikumar, R. and Girija, S. (2004). Selection of efficient arbuscular mycorrhizal fungi in the rhizosphere of cashew and their application in the cashew nursery. *Scientia Hort.,* 100(1/4): 369-375.

Beena, B., Abdul Salam, M., and Wahid, P.A. (1995). Root activity of cashew (*Anacardium occidentale* L.) varieties in relation to phenological phases. *J. Plantation Crops,* 23(1): 35-39.

Bucks, D.A., Nakayama, F.S. and Gilbert, R.G. (1979). Trickle irrigation water quality and preventive maintenance. *Agricultural Water Management,* 2(2): 149–162.

Carr, M.K.V. (2014). The water relations and irrigation requirements of cashew (*Anacardium occidentale* L.): A review. *Experimental Agriculture,* 50 (1): 24-39.

Diva Correia, Morsyleide de Freitas Rosa, Elis Regina de Vasconcelos Norões, Fátima Beatriz de Araujo (2003). The use of coir dust for preparation of substrates for grafted dwarf cashew seedlings. *Rev. Bras. Frutic., Jaboticabal,* 25(3): 557-558.

Falade, J.A. (1984). Effect of lime on the efficiency of two nitrogen fertilizers applied to cashew in acid sandy soil. *J. Plantation Crops,* 12(2): 140.

George, T.E., Veeraraghavan, P.G. and Rao, D.S. (1984). Studies on the leaf nutrient content of cashew (*Anacardium occidentale* L.) in relation to methods of fertilizer application. *Indian Cashew J.*, 16(2): 11-13.

Ghosh, S.N. (1988). Effect of nitrogen, phosphorus and potassium on flowering duration, yield and shelling percentage of cashew (*Anacardium occidentale* L.). *Indian Cashew J.*, 19(1): 19-23.

Ghosh, S.N. (1990). Studies on the NPK requirement of cashew (*Anacardium occidentale* L.) in lateritic tract of West Bengal. *The Cashew*, 4(2): 6-9.

Ghosh, S.N. (1995). Studies on effect of watering during flowering and fruiting on yield of cashew. *The Cashew*, 9(3): 5-8.

Ghosh, S.N. and Bose, T.K. (1986). Nutritional requirement of cashew in laterite tract of West Bengal. *Indian Cashew J.*, 18(1): 11-16.

Grundon, N.J. (2001). A desktop study to predict the fertilizer requirements of cashew trees in Northern Australia. CSIRO Land and Water, Atherton, Technical Report 32/01.

Harishu Kumar, P. and Sreedharan, C. (1986). Nut characters as influenced by different levels of NPK in cashew (*Anacardium occidentale* L.). *Indian Cashew J.*, 18(2): 15-17.

Haugen, L.M. and Smith, S.E. (1993). The effect of inoculation of cashew with Nutrilink on vesicular arbuscular mycorrhizal infection and plant growth. *Australian Journal of Agricultural Research*, 44(6): 1211-1220.

Ibiremo, O.S., Olubamiwa, O., Agbeniyi, S.O and Akanbi, O.S.O. (2012). Response of cashew seedlings from different nut sizes to phosphate fertilizers and arbuscular mycorrhizal inoculation in two soils in Nigeria. *International Journal of Plant, Animal and Environmental Sciences*, 1(1): 147-158.

Kumar, D.P., Khan, M.M., and Venkataramu, M.N. (1995). Effect of NPK and growth regulators on harvesting, nut yield, shelling per cent and kernel grade of cashew. *J. Plantation Crops*, 23(2): 96-104.

Kumar, D.P., Hegde, M. and Guruprasad, T.R. (1998). Fertigation for higher nut production in cashew (*Anacardium occidentale* L.). *Cashew Bulletin*, 35(4): 2-4.

Kumar, P.H. (1985). Effect of NPK on seed progeny and air layers in cashew. M.Sc. (Ag.) thesis submitted to Kerala Agricultural University, Vellanikkara.

Lakshmipathy, R. (2000). Symbiotic response of cashew root stocks to different VAM fungi. *The Cashew*, 14(3): 20-24.

Latha, A., John, P.S. and George, M. (1994). Productivity of cashew as influenced by chlorophyll and leaf nitrogen content. *J. Tropical Agriculture*, 32(1): 86-88.

Latha, A. and Salam, M.A. (2001). Effect of N at different levels of drip irrigation in cashew: *The Cashew*, 15(3): 18-26.

Lefebvre, A. (1973). Mineral fertilization of Cashew. *Fruits* (France), 28(2): 19-20.

Lioi, L. and Giovannetti, M. (1987). Infection by the VA-mycorrhizal fungus Glomus caledonium in Hedysarum coronarium as influenced by host plant and P content of soil. *Plant and Soil,* 103: 213–219.

Manjunatha, D. (2001). Effect of nutrients (NPK) supplied through irrigation water on growth of rootstocks and grafts of cashew. *The Cashew,* 15(1): 13-18.

Mishra, J.N., Paul, J.C. and Pradhan, P.C. (2008). Response of Cashew to Drip Irrigation and Mulching in Coastal Odisha. *Journal of Soil and Water Conservation,* 7(3): 36-40.

Mohapatra, A.R., Vijayakumar, K. and Bhat, N.T. (1973). A study on nutrient removal by the cashew tree. *Indian Cashew J.,* 9(2): 19-20.

Nawale, R.N., Sawke, D.P., Deshmukh, M.T. and Salvi, M.T. (1985). Effect of black polyethylene mulch and supplemental irrigation on fruit retention in cashewnut. *Cashew Causerie,* 7(3): 8-9.

Oblisami, G., Santhanakrishnan, P., Pappaiah, C.M. and Shanmugavelu, K.G. (1985). Effect of Azotobacter inoculant and growth regulators on the growth of cashew. *Acta Horticulturae,* 108: 40-45.

Oliveira, V.H., Miranda, F.R., Lima, R.N. and Cavalcante, R.R.R. (2006). Effect of irrigation frequency on cashew nut yield in Northeast Brazil. *Scientia Horticulturae,* 108: 403–407.

Oliveira, V.H., Barros, L.M., Lima, R.N. (2003). Inuencia da irrigacao e do genotipo na producao de castanha em cajueiro-ana o-precoce. *Pesquisa Agropecuaria Brasileira,* 38: 61–66.

Perrin, R. (1990). Interaction between mycorrhizae and diseases caused by soil-borne fungi. *Soil Use and Management,* 6: 189–194.

Ramesh, N., Lingaiah, H.B. and Khan, N.M. (1999). Economics of bio-fertilizer application in cashew graft production under eastern dry zone of Karnataka. *Crop Research* (Hisar), 17(3): 336-341.

Rao, M.R., Reddy, S.K. and Rao, K.R. (1984). Effect of NPK on nut yield of cashew. *Cashew Causerie,* 6(1): 3.

Reddy, A.V., Narasimha Rao, P.V., Ankaiah, S. and Subba Rao, I.V. (1982). Cashew NPK nutrition in relation to growth under graded doses of nitrogen fertilization. *Indian Cashew J.,* 14(4): 15-21.

Richards, N.K. (1993). Cashew response to water and nutrients in a sandy red earth soil of the Northern Territory. Technical Bulletin Northern Territory, Department of Primary Industry and Fisheries. 202: 17-38.

Salam, M.A., John, P.J., Joseph, M., Poduwal, M., Kumar, P., Yadukumar, N. and Bhat, M.G. (2008). Quantitative estimation of soil fertility and fertilizer recommendations (QUEFC) for Cashew (*Anacardium occidentale* L.). *J. Plantation Crops,* 36(2): 86-94.

Santos, C.A.S. (1997). Distribuição espacial e absorção de água pelo sistema radicular da cultura da manga (*Mangifera indica* L.) irrigada por microaspersão. Piracicaba: ESALQ, USP, 51p. Dissertação Mestrado.

Sawke, D.P. (1980). Cashewnut Research Station, Vengurla. *Cashew Causerie*, 2: 15-16.

Sawke, D.P., Salvi, M.J. and Patil, M.M. (1985). Softwood grafting-A sure technique of clonal propagation in cashew. *Indian Cashew Journal*, 17(1): 17-18.

Schaper, H. (1991). Leaf gas exchange as influenced by environmental and plant internal factors in cashew (*Anacardium occidentale* L.). Inaugural Dissertation submitted to Rheinischen Friedrich-Willhelms Universitat zu Bonn.

Schaper, H., Chacko, E.K. and Blaikie, S.J. (1996). Effect of irrigation on leaf gas exchange and yield of cashew in northern Australia. *Australian Journal of Experimental Agriculture*, 36(7): 861-868.

Singh, S.N. (1997). Effect of bio-fertilizer (*Azotobacter chroococcum*) in cashew plantation. *Environment and Ecology*, 15(2): 482-484.

Sinish, M.S., Mercy, G. and John, P.S. (2005). Organic methods for cashew root stock production. *The Cashew*, 19(1): 8-15.

Sivaprasad, P., Sulochana, K.K., George, B., Salam. M.A. (1992). Growth and phosphorus uptake of cashew as influenced by inoculation with mycorrhizae. *The Cashew*, 6(1): 16-18.

Sridhar, R., Shah, H.A. and Manivannan, K. (1990). The effect of VA mycorrhizae inoculation on certain growth characters of cashew. *Cashew Bulletin*, 27(4): 15-16.

Veeraraghavan, P.G., Celine, V.A., Balakrishnan, S. (1985). Study on the fertilizer requirements of cashew (*Anacardium occidentale* L.). *Cashew Causerie*, 7(2): 6-8.

Wahid, P.A., Kader, K.V.A. and Salam, M.A. (1993). Rooting pattern of cashew. Rooting pattern of tropical crops (Eds. Abdul Salam, M.A. and Wahid, P.A.), Tata Mc Graw Hill Publishing Co. Ltd. pp. 223-234.

Ximenes, C.H.M. (1995). Adubação mineral de mudas de cajueiro anão precoce cultivadas em diferentes substratos. Fortaleza, Universidade Federal do Ceará. M.Sc. dissertation.

Yadukumar, N. and Nandan, S.L. (2005). Recycling organic wastes of cashew plantations by aerobic composting. *J. Plantation Crops*, 33(2): 99-102.

Yadukumar, N. (2001). Efficacy of drip irrigation and graded nitrogen, phosphorus and potassium on the productivity of cashew (*Anacardium occidentale* L.). Ph.D. Thesis, Department of Biosciences, Mangalore University, Mangalore – 574 199, India.

Yadukumar, N. and Mandal, R.C. (1994). Effect of supplementary irrigation on cashewnut yield. Water Management for Plantation Crops - Problems and Prospects. Centre for Water Resource Development and Management, Calicut, Kerala. pp. 79-84.

Yadukumar, N. and Rejani, R. (2008). Soil and Water Management in Cashew Plantations, NRCC Tech. Bulletin No. 10, National Research Centre for Cashew, Puttur, Karnataka.

Yadukumar, N., Rejani, R. and Nandan, S.L. (2008). Studies on green manuring in high density cashew orchards. *J. Plantation Crops*, 36: 265-269.

Yadukumar, N., Rejani, R. and Prabhakar, B. (2009). Fertigation for efficient water and nutrient management in high density cashew plantation. *J. Plantation Crops*, 37(2): 102-110.

Zayed, G. (1999). Evaluation of N_2 fixation efficiency of Azotobacter in alginate-encapsulated and free cell systems. *Egyptian Journal of Microbiology*, 34: 45-55.

Cashew: Improvement, Production and Processing (2017) *Pages* **253–264**
Editor: **Prof. P.L. Saroj**
Published by: **ASTRAL INTERNATIONAL PVT. LTD., NEW DELHI**

14

Plant Growth Regulators

R.K. Meena, J.D. Adiga and P.L. Saroj

1. Introduction

Cashew (*Anacardium occidentale* L) is an important commercial plantation crop of the country, grown in an area of 9.82 lakh ha with a production of 7.28 lakh tones of taw cashew nuts (2012-13). Globally, India's share in the cashew area and production is 20 per cent and 16 per cent, respectively (Cashew News, 2013). Moreover, productivity of Indian cashew is very low (772 kg/ha). The present low productivity is not only due to establishment of plantation with seedling of non-descript origin but also due to poor and irregular flowering because of adverse environmental conditions (Parameswaran *et al.*, 1984). There were reports of poor fruit set and excessive premature fruit drop in the crop (Patnaik *et al.*, 1985), low hermaphrodite flowers (Parameswaran *et al.*, 1984 and Aliyu 2006), nutritional deficiency (Subbaiah 1983 and Ghosh 1989), inefficient pollination (Heard *et al.*, 1990) and irregular and prolonged flowering (Masawe *et al.*, 1996; Aliyu and Awopetu, 2003 and Aliyu, 2006) where growth regulators can play an important role. In fact, Plant growth regulators (PGRs) are the most powerful tools for manipulating tree growth and yield in an existing orchard. Therefore, growth regulators are gaining importance in cashew cultivation for overcoming problems associated with plant growth, fruit set, development and, final retention.

Cashew is a polygamous tree and produces male and hermaphrodite flowers in different phases during flowering. The yield of cashew is very low owing to the production of low percentage of hermaphrodite flowers, poor fruit set, immature fruit drop and low fruit retention (Haribabu, 1982). Flowering in cashew is seasonal, producing flower buds in varying phases starting from September to March depending upon the cultivars and climatic conditions. Though, cashew produces innumerable flowers, only 1-2 per cent of the flowers set fruit and reach maturity. Production of more number of pistillate flowers and reduction in nut drop can be accomplished by the use of plant hormones. Fruit set and their retention are the major limiting factors for low yield in cashew which need due attention. The nuts those develop after pollination start drying followed by dropping, leading to very low percentage of matured nuts. Use of growth regulators like auxins, gibberellins and ethylene has resulted in improving the vegetative and reproductive parameters which are associated with high yield in many fruit crops (Lafer, 2008).

Plant growth regulators are the organic compounds other than nutrients which modify or regulate physiological processes in an appreciable measure in the plants when used in small concentrations. They are readily absorbed and these chemicals move rapidly through the tissues when applied to different parts of the plant. Plant hormones or phytohormones are also regulators but produced by the plants in low concentrations and these hormones move from the site of production to the site of action. Therefore, the difference between the plant regulator and plant hormone is that the former one is synthetic and the latter one is natural from the plant source. A plant growth regulator is also an organic compound, either natural or synthetic, that modifies or controls one or more specific physiological processes within a plant through translocation. The major growth regulators include, auxins, gibberellins, cytokinins, ethylene, abscisic acid, florigen, anthesins, vernalin, morphactins. Some of the works related to application of plant growth regulators in cashew are reviewed and presented in this chapter.

2. Seed Germination and Seedling Growth

Cashew is commercially propagated through soft wood grafting, wherein early attainability of seedling height can reduce the cycle of propagation. Young seedlings of cashew usually do not produce lengthy tap root and secondary roots and probably for want of adequate root system, seedlings suffer heavy causalities especially during summer. Better root system is particularly beneficial for hybrid seedlings when breeder raises seedlings for field planting. Many researchers have been used a variety of growth regulators to increase the germination rates of different crops (Ayfer and Serr, 1961; Frutos and Barone, 1988; AK *et al.*, 1995).

Effect of plant growth regulators on seed germination revealed that GA_3 induced better seed germination of cashew than other chemicals. Shanmugavelu and Rangasawamy (1970) have suggested that the seed of tree species contained natural occurrence of Auxin which might regulate the seed germination. Gibberellin was found to induce 100 per cent germination in cashew seeds. GA_3 induced the highest shoot length at 500 ppm, the increase being 77 per cent over the control, whereas, GA_3 failed in inducing the root growth which was less than the control. However, NAA @100 ppm was associated with highest root length which was 92 percent more over the control. In respect of production of secondary roots, NAA was found superior at 500 ppm. GA_3 did not influence the production of secondary roots of cashew. The IBA at 100 ppm and at 0.2 per cent was found to induce maximum root growth in cashew. It was reported by Roa and Hassan (1957) that no lateral roots could be seen in the case of one or two month old seedling but in the present study by treating the seed or seedling with growth regulators, the lateral root production was induced. The most striking response of GA_3 on cashew is the stem elongation. The number and length of cells in the cortex and pith region were influenced by GA_3 treatment. Therefore, it appears that the stem elongation is predominantly due to cell elongation supplemented with cell division in the cortex and pith region, which support the views of Stowe and Tamaki (1957). In cashew, increase in the fresh as well as dry weight of shoots and roots can be attributed to the increase in the overall assimilation and redistribution of materials within the plants.

An experiment was conducted at ICAR-Directorate of Cashew Research (DCR), (Lakshmipathi *et al.*, 2013) to study the influence of GA_3 on germination, shoot length, shoot fresh weight, shoot dry weight, seedling length, seedling girth, number of leaves per seedling, root length, root fresh weight, root dry weight and vigour index of cashew. Seeds treated with GA3 @ 200 ppm recorded highest values for germination percentage (100 per cent) shoot length, shoot fresh weight, shoot dry weight, seedling length, seedling girth, number of leaves per seedling, root length, root fresh weight, root dry weight and vigour index of cashew. This increase in shoot length with GA3 treatment might be related to the fact that gibberellic acid promotes stem and shoot elongation through the increase of both cell division from internodes and elongation in higher plant (Hartmann *et al.*, 2002; Hopkins and Huner, 2004; Harris *et al.*, 2004). The number and length of cells in the cortex and pith regions were influenced by GA_3 treatment. The increased weight of shoot was mainly attributed to enhanced germination, early seedling emergence and better seedling growth. This can also be attributed to the increase in the overall assimilation and redistribution of materials within the plant (Shanmugavelu, 1985). The increased dry weight of shoots and the data on dry matter content of different organs strongly support the view that effects of GA on photosynthesis are linked with a more effective translocation to shoots which will use them for their increased growth in length. From this, it can be concluded that GA_3 @ 200 ppm can be effectively employed in seed priming treatment for increasing the seedling growth and vigour index of cashew which would help in producing healthy seedlings at a short period of time for grafting operations and in turn ultimately reduce the duration of propagation.

3. Propagation

Semi-hardwood cashew cuttings, 20 cm long, prepared from non-juvenile plants, invigorated plants and 4-month-old seedlings, were basally dipped, for 15 seconds, in different concentrations of auxins (IBA, IAA or NAA) or phenolic compounds and then inserted into a sand and sphagnum moss medium in a bottom heated chamber (basal temperature 30±2°C), or into a sand and coconut coir dust medium in a mist chamber. In the mist chamber, 49.72 per cent of cuttings from the 4-month-old seedlings rooted; their survival rate was 100 per cent, and the number of roots per cutting ranged from 4.5 with no growth regulator treatment to 15.3 with NAA at 1000 ppm and the length of the longest root ranged from 8.5 cm with IAA at 500 ppm to 17.7 cm with IBA at 2000 ppm. In all the treatments in the bottom-heated chamber or with cuttings taken from older or invigorated plants, either rooting percentage, or the number of roots formed, was very low. The phenolic compounds had no effect on rooting in any of the treatments (Melanta *et al.*, 1990).

4. *In-vitro* Propagation

To study the embryogenic response of cashew cotyledon in BAP under *in vitro* conditions, the sterilized cotyledon segments (10 mm long, 5 mm width) were cultured on full strength MS basal medium containing various concentrations (0-10 mg/l) of BAP (benzyl amino purine). The result revealed that medium without BAP exhibited profuse root organogenesis which decreased with the increase in the concentration of BAP. Greenish or yellowish white nodular structures appeared on the edge of the surface concentrated towards the convex side of the explants. Consequently, these structures gave rise to somatic embryoids and the degree of embryoid formation increased with the increasing concentration of BAP. MS medium with 8 mg/l BAP was the most optimal concentration for the induction of somatic embryos among the tested. Cytological study confirmed the development of somatic embryoids directly from cultured cotyledon explants (Bavatharine *et al.*, 2010).

Nodule induction in cotyledon explants of cashew was higher (80 per cent) when MS medium was supplemented with 2 mg/l BAP (benzyl amino purine). The root formation was higher (47 per cent) in MS medium which contained 2 mg/l kinetin and 2 mg/l NAA (Naphthalene acetic acid) where longest roots (120 mm) were recorded. Further, it was noted that the medium which contained 2 mg/l BAP showed higher per cent of somatic embryoid formation directly from cotyledon explant. Subsequently somatic embryos were noted 2-3 weeks after culture in MS basal medium without growth regulators (Kamshananthi *et al.*, 2012). At Directorate of Cashew Research, Puttur, a repeatable axillary shoot-bud induction in cashew was obtained on the MS medium with half-strength major nutrients, 2.74 m*M* L-glutamine, 87.6 m*M* sucrose containing thidiazuron (TDZ) alone and in combination with BA. TDZ at 0.45 µ*M* was best for axillary shoot-bud proliferation (4.5 buds per shoot) with maximum response (100 per cent). Bud elongation could be stimulated in multiple shoots on medium containing 116.8 m*M* sucrose. *In vitro* rooting on auxin media and pulsing microshoots in 10 m*M* naphthalene acetic acid (NAA) was ineffective (Thimmappaiah *et al.*, 2002).

5. Control of Plant Growth

The control of cashew grafts growth in nursery was achieved using PP. 333 Pacobutrazol ((2 RS, 3 RS)-1-(4-chlorophenyl)-4, 4-dimethyl-2-(1, 2 4-triavol-1-yl) pentan-3-ol). The chemical was used as soil drench, foliar spray and stem swabbing to the grafts of cashew (variety Anakkayam-1). Application of 50 ppm as soil drench could restrict vegetative growth up to six months. Tap root growth was reduced and lateral root growth increased to a great extent. Treated grafts, when planted in field showed normal growth after three months (Misra *et al.*, 1991).

Studies on the effect of paclobutrazol on growth and yield of cashew were conducted at ICAR-DCR, Puttur. Paclobutrazol (PBZ) was applied @ 1, 2 and 3 g a. i. per plant as soil drench in the month of October before flushing. The frequency of application was every year, once in two years and once three years. The observations of morphometric parameters such as flowering, fruiting and nut yield were recorded after the imposition of PBZ. The application of PBZ reduced the plant height, canopy spread and internodal length, while the number of flushes increased with the application of paclobutrazol. The plant height among PBZ treated plant ranged from 229.2 to 261.1 cm in different doses and frequency of application, while in the untreated plants, plant height varied from 262 to 275.7 cm. The canopy spread in East-West direction among PBZ treated plants ranged from 223.7 to 252.5 cm while the untreated plants recorded 269.4 to 274.5 cm in same direction. The canopy spread in N-S direction among PBZ treated plants ranged from 234.5 to 268.2 cm, while the untreated plants varied from 234.5 273.6 to 277.7 cm in the same direction. The reduction in plant height and canopy spread on account of PBZ application could be due to suppression of growth by paclobutrazol as the compound blocks three separate steps in the terpenoid pathway for the production of gibberellins. One of the main roles of gibberellins in trees is the stimulation of cell elongation. When gibberellin production is inhibited, cell division still occurs, but the new cells do not elongate. This results in production of shoots with the same numbers of leaves and internodes compressed into a shorter length leading to reduced plant height. Similarly, reduction in tree height with the use of paclobutrazol might be due to the inhibitory effect of paclobutrazol on gibberellin biosynthesis pathway at the sub-apical meristem, which ultimately reduced cell elongation and rate of cell division and decreased the shoot growth which led to the trees of shorter height. The girth of collar region among PBZ treated plants ranged from 28.1 to 30.0 cm, while the untreated plants recorded an average girth of 29.8 to 30.2cm. The process of cell division (secondary growth) is associated with increment in stem girth. Since PBZ has little or no role to play in cell division, stem girth was not influenced by PBZ application. The internodal length in PBZ treated plants ranged from 0.6 to 3.1 cm, while the untreated plants recorded 3.4 to 3.5 cm indicating the role of PBZ in reducing the plant vigour (Anon., 2012-13).

6. Sex Ratio

Preliminary studies on improvement of sex ratio, fruit set, fruit retention and yield by using plant growth regulators have indicated beneficial effects in cashew. Increase in fruit set by 55 per cent in cashew with foliar application of NAA @ 10 ppm was reported by Murthy *et al.* (1975). Konhar and Mech (1988) reported highest fruit retention in cashew with nutron 500 ppm, followed by etherl 50 ppm and planofix 45 ppm. Similarly, Kumar *et al.* (1994) found ethrel @50 ppm, NAA 25 ppm and 2, 4-D 4 ppm to be most effective in improving sex ratio and yield.

The effects of ethephon (ethrel) and NAA (planofix) on sex expression and fruit set of cashew were studied in India. The best production of hermaphrodite flowers and the highest nut count and nut weight were found after application of NAA at 200 ppm, followed by ethephon at 100 ppm (Mariappan *et al.*,1995). In respect of flowering behaviour, Ethrel treatments @ 100, 200 and 400 ppm showed a positive impact and produced significantly higher number of flowering panicles, *i.e.*, 12, 10 and 11per m^2, than in control or water spray, *i.e.*, 7.3 per m^2. Per cent increase in perfect flowers over control was highest under 100 ppm ethrel, followed by ethrel 400 ppm. This is in agreement with previous investigations in cashew by Gajbhiye *et al.* (2007). Further, with water spray, there was a decline in number of perfect flowers as compared to control. Dorajeerao *et al.* (2001) reported that clones having broader sex-ratio were high yielders. Ethrel at 200 ppm recorded less number of perfect flowers (20.0) and higher percentage of staminate flowers (94.6), resulting in lower sex ratio. Ethrel at 100 ppm might have exerted its effect on sex expression by manipulating endogenous auxin corresponding to the reduction in staminate flowers as reported by Mariappan *et al.* (1995). Further, it was observed that the percentage of staminate and perfect flowers was negatively correlated and, therefore, higher percentage of perfect flowers (18 per cent) was observed with ethrel @100 ppm. Similarly, in the case of mean fruit set per m^2, it was significantly higher (28.8) with ethrel @100 ppm than in the remaining treatments. Finally, the mean yield per tree was also significantly higher with ethrel @100 ppm than in the remaining treatments. This can be attributed to the availability of significantly higher sinks with ethrel at 100 ppm in respect of perfect flowers, fruit set per m^2 and number of nuts per panicle. Gajbhiye *et al.* (2007) and Mohan and Rao (1995) also reported highest nut yield with ethrel spray. However, the mean nut weight was not significantly influenced by treatments.

The study of plant growth regulators on flowering, fruiting and yield on cashew cv. Bhaskara was also conducted at ICAR-Directorate of Cashew Research, Puttur. The trees sprayed with ethrel @ 50 ppm produced highest number of flowering laterals (19.50) and least number of non-flowering laterals (5.50) per square metre. The trees sprayed with NAA @ 25 ppm and the combination of NAA @ 25 ppm with GA_3 @ 50 ppm also influenced flowering and non-flowering laterals (Laxmipati *et al.*, 2014). A significant increase in number of flowering laterals per square meter with the spray of ethrel @ 50 ppm over control might be due to increased activity of peroxidase and α-amylase which ultimately release more sugar for induction of flowering (Yamdagni and Khangia, 1989). Application of ethrel @ 50 ppm recorded highest number of total laterals (23.3) per square metre whereas, lowest number of total laterals (14.8) per square metre was recorded in control. Duration of male and hermaphrodite flowers was also significantly increased with the spray of ethrel @ 50 ppm followed by NAA @ 25 ppm whereas, the duration of male and hermaphrodite flowers got decreased in control. Application of ethrel @ 50 ppm had positive effect in increasing number of hermaphrodite flowers (306) and male flowers (434) followed by NAA @ 25 ppm (278 and 397 number of hermaphrodite and male flowers, respectively). The least number of hermaphrodite flowers and male flowers were recorded with control and 2, 4-D @ 10 ppm. The total number of hermaphrodite and male flowers was significantly higher under ethrel @ 50 ppm than rest of the treatments which ultimately improved the sex ratio also. Improvement in sex ratio with application of growth regulators was mainly due to increased number of bisexual flowers. It is clearly evident from the above results that both auxin and ethrel had stimulating effect and caused the physiological changes in the tissues influencing the flowering characters (Salisbury and Ross, 1986).

7. Flowering and Fruiting

The PBZ application was found effective in increasing the sex ratio, number of panicles per plant while, the length and width of panicles and flowering duration got reduced. The number of rachis in flower panicles was unaffected by PBZ application. The sex ratio of flowers ranged from 0.17 to 0.27 in treated plants while untreated plants it was 0.15 to 0.21per cent. The number of flower panicles per plant ranged from 53.5 to 68.4 in treated plants while in untreated plants it was 53.5 to 62.7. The length and width of flower panicles ranged from 11.6 to 17.3 cm and 16.5 to 23.3 cm respectively, in treated plants while, the untreated plants recorded 17.1 to 18.1 cm and 22.8 to 23.6 cm respectively.

Exogenous Application of hormones (GA_3, IAA, IBA, NAA, 2, 4-D) at various concentrations (0, 10,25,50,100, 250, 500 mg/l) in two cashew genotypes CSO06 and CSO06 has shown to improve fruiting/yield significantly. The significant reduction in the period of flowering will considerably enhance flowering synchronization in cashew. These two events coupled with the increased number of hermaphrodite flowers were likely responsible for higher fruit set through improved pollination. In general, most of the cashew farms attained peak blooming at the peak of dry season (February to March), when average monthly temperature usually ranged between 36 and 40°C and water stress is at the peak. Such high temperature and excessive drought during flowering and fruiting will not only increase fruit drop (abortion), but result in fruits with wide variation in size with shrunken nuts due to incomplete kernel development. In addition to increased flowering and fruit set due to application of GA_3, significant improvement in fruit retention was recorded. Imbalance in the endogenous hormone is known to aid embryo/seed abortion and low fruit set (Nunez-Elisea and Davenport 1983; Ram *et al.*, 1976). Application of exogenous hormone might in a way enhance balancing of endogenous hormones. Furthermore, GA_3 is known to enhance photosynthetic production through an efficient use of nitrogen "N" (Khan *et al.*, 2004). The remarkable response recorded in cashew twigs treated with GA_3 over other exogenous hormones suggests hormone-specific nature of cashew tree (Davenport, 2003). Meanwhile, genotypic response, for example, in fruit and nut development, was concentration sensitive. Such genotypic variation has been attributed to differential rate of photo assimilation, breeding history of the cultivars (Chacko *et al.*, 1995; Searle *et al.*, 1995) and level of abscisic acid (Roitsch and Ehneb, 2000). Twigs treated with 500 mg/l consistently performed lower than untreated (0 mg/l) in flowering and fruiting traits, suggesting negative effect of higher concentration of the exogenous hormone on the floral initiation and development in cashew. It was observed that 100 mg/l was the ideal dosage for promoting flowering and 50 mg/l would increase fruit set. It seems that efficient pollination in cashew could only be achieved with treatment at lower concentrations (< 50 mg/l). It is probable that gibberellins may be playing important role in the germination of pollens during pollination and fertilization, while auxins are essential for seed development. The performances of auxins were comparatively low and less suitable for use in cashew. Perhaps, it would be better to combine GA_3 with any of the auxins, as the synergetic effect between auxins and other plant growth regulators has been reported (Fujioka *et al.*, 1983). Hence, application of Gibberellic acid (GA_3) applied at the rate of 50 mg L^{-1} improved flower precocity by 500 per cent, increased the numbers of hermaphrodite flowers by 500 per cent, fruit set by 500 per cent, fruit retention by 69 per cent and nut size by 25 per cent, which would significantly improve output/tree in cashew (Olawale Mashood Aliyu *et al.*, 2011).

The increase in length and number of secondary branches per panicle are important attributes for the production of high number of flowers which increases the yield. With respect to different growth regulators, spraying of ethrel @ 50 ppm increased the number of fruits set, number of fruits retention per panicle, nut weight (g), nut yield (kg) per tree and reduced the fruit drop per panicle over other treatments. Increased fruit set and fruit retention due to application of ethrel and other growth regulators could be attributed to the increased number of bisexual flowers and reduced pre mature fruit drop. Reduced fruit drop in cashew due to application of growth regulators was also reported by Konhar and Arun Mech (1988). Increased nut yield with application of growth regulators could be attributed to increased number of bisexual flowers, fruit set, fruit retention and total number of nuts per tree (Veeraraghavathatham and Palaniswamy, 1983).

A significantly higher fruit set and fruit retention in the paclobutrazol treated plants had a favourable impact on culminating higher final fruit number and yield per plant. Paclobutrazol has been reported to exert influence on partitioning the photosynthates to the sites of flowering and fruit production consequent to the reduction of vegetative growth. The application of hydrogen cyanamide, paclobutrazol, and their combination inhibited catalase enzyme action compared to control. The application of cytofex (10 ppm) had significant effect on the fruit number, the percentage of fruit set, and the percentage of fruit drops as compared to control (Notodimedjo *et al.*, 1999)

8. Nut Yield

Nut yield per tree (kg) was found to be most significantly and positively correlated with number of flowering laterals per square metre (0.84), total number of laterals per square metre (0.83), duration of male flowers (0.89), duration of hermaphrodite flowers (0.76), number of male flowers per panicle (0.85), number of hermaphrodite flowers per panicle (0.89), total number of flowers (0.90), number of fruits set per panicle (0.94), number of fruits retention per panicle (0.92) and nut weight (0.87). However, the correlation of number of non-flowering laterals per square metre (-0.64) with nut yield was significant and negative. Kumar *et al.* (1996) observed that number of perfect flowers per panicle was positively correlated with yield in cashew. Similar correlation was also reported by Lenka *et al.* (2001).

The PBZ application had increased the nut yield per plant but the weight, length, width and volume of nut decreased. The PBZ treated plants recorded nut length of 3.0 to 3.3 mm, nut width of 2.4-2.5 mm, nut weight of 6.63 to 7.56 g and nut volume of 6.84-7.74 cc, while untreated plants recorded nut length of 3.0 to 3.2 mm, nut width of 2.4-2.5 mm, nut weight of 7.02 to 7.36 g and nut volume of 7.25-7.64 cc respectively. The average yield of treated plants ranged from 756 to 1127 g/plant, while, untreated plants recorded a nut yield of 668 to 728 g per plant. This might be due to the concentration of paclobutrazol and the combined effect of dosage and frequency of applicat ionf PB2 which were noticed to be significant, while time of application had significant effect only on number of fruits, shelf life and yield (Anon., 2013)

9. Nutritional Value

The effect of GA_3 on free amino acid and protein contents vary according to the concentrations of the hormone used, but does not follow a uniform pattern.

Tryptophane, leucine (s), aminobutyric acid, valine-methionine, glutamic acid, alanine, aspartic acid, glycine and cystine occur in assayable amounts in the early kernel. There is a marked reduction in the total amount of free amino acid as the fruit matures, and this decline is steeper in the treated category. GA_3 treatment leads to qualitative and quantitative changes in the free amino acid: cystine, which is absent in the control kernels and hydroxyproline though present only in traces in control, can no longer be detected in treated ones. The application of growth regulators and nutrients does not influence the weight and juice content of the apple of the cashew nut, although the apple yield and the soluble solids of the apple juice increase. The effects of levels of NPK nutrients and growth regulators (such as 2,4-D, NAA and ethephon) on apple yield and quality were studied in 10-year-old plants of Ullal 1 cashew variety in Karnataka, India, during 1989-1991. Application of ethephon (ethrel) spray at 50 ppm and 500:250:250 g NPK/plant/year was found to be the most effective treatment. The decrease in weight and juice content of the apple, after the application of growth regulators and nutrients, was probably due to the increase in the number of fruits per panicle. The foliar application of growth regulators increased the total soluble solids and the apple yield when the nutrient levels were high. The increased yield could be attributed to the positive interaction between the nutrients and the growth regulators (Kumar *et al.*, 1996).

References

Ak, B.E., Ozguven., A. I. and Nikpeyma, Y. (1995). The effect of GA_3 applications on pistachio nut seed germination and seedling growth. *Acta Horticulturae,* 419: 115-120.

Aliyu, O. M and Awopetu, J.A. (2005). *In vitro* regeneration of hybrid plantlets of cashew (*Anacardium occidentale* l) through embryo culture. *African Journal of Biotechnology,* (6): 548-553.

Anonymous (2012-13). Effect of paclobutrazol on growth and yield of cashew, Directorate of Cashew Research, Puttur, *Annual Report*, pp. 31-35.

Ayfer, M. and Serr, E. F. (1961). Effects of gibberellins and other factors and seed germination and early growth in *Pistacia* species. *Journal of the American Society for Horticultural Science,* 77: 308-315.

Bavatharine, G. and Thayamini, H. Seran. (2010). *In Vitro* Response of Cotyledon Explants of Cashew (*Anacardium occidentale* L.) Cultured in Different Concentrations of BAP, Department of Crop Science, Faculty of Agriculture, Eastern University, Chenkalady, Sri Lanka.

Chacko, E. K., Kohli, R.R. and Randhawa, G.S. (1974). Investigation on the use of 2- chloroethyl phosphoric acid (ethephon, CEPA) for control of biennial bearing in mango. *Scientia Horticulturae,* 2: 389-398.

Chacko, E.K., Lu, P., Kohli, R.R. (1995). Photoassimilate production and distribution in relation to productivity in mango (*Mangifera indica* L.). In: Mango 2000 - Marketing Seminar and Production Workshop Proceedings. Department of Primary Industries, Brisbane, pp. 109-124.

Davenport, T.L. (2003). Management of flowering in three tropical and subtropical fruit tree species. *Hort Science,* 38: 1331-1335.

Dorajeerao, A. V. D., Ravishankar, C. and Reddy, M. L. N. (2001). Influence of flowering phases on nut yield of different cashew clones. *Journal of Plantation Crops,* 28: 55-60.

Frutos, D. and Barone, E. (1988). Germinacion de *Pistacia vera* L. primer crecimiento de las plantas de semilia tratadas can acido giberelico (GA_3). *Rapport EUR 11557. CEE. Colloque AGRIMED-GREMFA Reus;* 289-298.

Fujioka, S., Yamaguchi, I., Murofushi, N., Takahashi, N., Kaihara, S. and Takimoto, A. (1983). The role of plant hormones and benzoic acid in flowering of *Lemna aucicostata* 151 and 381. *Plant Cell Physiol.,* 24: 241-246.

Gajbhiye, R.C., Jalgaonkar, V.N., Bendale, V.W. and Gawankar, M. S. (2007). Effect of Growth Regulator to Improve Fruit Set and Yield in Cashew. National Seminar on Research, Development and Marketing of Cashew. In: Souvenir and Extended summaries, ICAR Research Complex for Goa, 20-21st November, pp. 136-137.

Ghosh, S. N. (1989). Effect of nitrogen, phosphorus and potassium on flowering duration, yield and shelling percentage of cashew (*Anacardium occidentale* L.). *Indian Cashew J.,* 19: 19-23.

Haribabu Sri, R. (1982). Increasing fruit set in cashew by growth substances. *Cashew causerie,* 4 (4): 14-15.

Harris, R. W., Clark, J. R. and Matheny, N. P. (2004). *Arboriculture integrated management of landscape trees, shrubs and vines* (4th edition). Prentice Hall, Upper Saddle, New Jersey.

Hartmann, H. T., Kester, D. E., Davies, F. T. and Geneve, Jr R. L. (2002). *Plant Propagation Principles and Practices* (7th edition). Prentice Hall, Upper Saddle River, New Jersey.

Heard, T. A., Vithanage, V. and Chacko, E. K. (1990). Pollination biology of cashew in the Northern Territory of Australia. *Aust. J. Agric. Res.,* 41: 1101-1114.

Hopkins, W. G. and Huner, N. P. A. (2004). *Introduction to Plant Physiology* (3rd Edition). John Wiely and Sons Inc, USA.

Kamshananthi, T. and Thayamini H. Seran. (2012). Induction of somatic embryogenesis from cotyledon explants of cashew (*Anancardium occidentale* L.). *Journal of Agricultural Technology,* 8(6): 2089-2099.

Khan, N.A., Mobin, M., Samiullah, M. 2004. The influence of gibberellic acid and sulfur fertilization rate on growth and S-use efficiency of mustard (*Brassica juncea*). *Plant Soil,* 270: 269-274.

Konhar, T., and Arun Mech. (1988). Effects of growth regulators on flowering, fruit set and fruit retention in cashew (*Anacardium occidentale* L.). *Indian Cashew Journal,* 18: 17-19.

Kumar D. P., Khan, M. M. and Melanta K. R. (1996). Effect of nutrition and growth regulators on apple characters and yield in cashew (*Anacardium occidentale* L.). *The Cashew,* 10 (2): 17-24.

Kumar, D. P., Khan, M. M. and Melanta, K. R. (1994). Effect of growth regulators on sex expression, fruit set, fruit retention and yield of cashew. In: Proceedings of the PLACROSYM XI held during 30 November to 3 December 1994 at Calicut, Kerala, India. 610-627.

Kumar, D.P., Khan, M. M. Melanta, K.R. (1996). Effect of nutrition and growth regulators on apple characters and yield in cashew (*Anacardium occidentale* L.), *The Cashew,* 10(2): 17-24.

Lakshmipathi, Dinakara Adiga, J., Kalaivanan, D. and Saroj, P.L. (2013). Response of cashew (*Anacardium occidentale* L.) to osmopriming with gibberllic acid (GA_3). *Journal of Plantation Crops,* 41(3): 455-459.

Lakshmipathi, Dinakara Adiga, J. and Kalaivanan, D. (2014). Influence of growth regulators on certain reproductive parameters of cashew (*Anacardium occidentale* L.) variety Bhaskara. *Journal of Plantation Crops,* 42(1): 113-116.

Lenka, P. C., Mohapatra, K. C., Dash, S. and Mishra, N. K. (2001). Genetic variability and character association in cashew (*Anacardium occidentale* L.). *Horticultural Journal,* 14(2): 105-110.

Marriappan, S., Prabhakaran, J. and Sambandamoorthy, S. (1995). Effect of growth regulators on sex expression and fruit set in cashew (*Anacardium occidentale* L.). *The Cashew,* 11: 11-13.

Masawe, P. A. L, Cundall, E. P, Caligari, P. D. S. (1996). Distribution of cashew flower sex-types between clones and sides of tree canopies in Tanzania. *Ann. Bot.,* 78: 553-55.

Melanta, K. R. Sulladmatho, U. V. (1990). Studies on propagation of cashew (*Anacardium occidentale* L.) by cuttings. *Mysore Journal of Agricultural Sciences,* 24: 79-82.

Misra, L. P., Singh, R. (1991). Effect of paclobutrazol on cashew (*Anacardium occidentale* L.) grafts in nursery. *Indian Journal of Plant Physiology,* 34(1): 102-105.

Mohan, E. and Rao, M. M. (1995). Effect of growth regulators and pruning on the growth and yield of cashew. *Environment and Ecology,* 13: 675-679.

Notodimedjo, S., Tampubolon, M. and Sunu, P. (1999). The effect of growth regulator application (Hydrogen cyanamide, paclobutrazol, cytofex) on the production and growth of cashew. *Agrivita,* 21: 32-36.

Nunez-Elisea, R., and Davenport, T. L. (1983). Abscission and ethylene production of mango (*Mangifera indica* L.) fruit cv. Tommy Atkin. Proc. *Florida State Hort. Soc.,* 96: 185-188.

Olawale Mashood Aliyu, Oluwayemisi Oluwatosin Adeigbe Joshua, Adedokun Awopetu. (2011). Foliar Application of the exogenous plant hormones at Pre-Blooming Stage Improves Flowering and Fruiting in Cashew (*Anacardium occidentale* L.) *J. Crop Sci. Biotech.* (2): 143–150.

Parameswaran, N. K., Damodaran, V.K. and Prabhakaran, P.V. (1984). Relationship between yield and duration of different phases in flower opening in cashew (*Anacardium oecidentale* L.). *Indian Cashew J.*, 16(4): 15-19.

Patnaik, H. P., Das, M.S. and Panda, J. M. (1985). Studies on the fruit set and fruit drop in cashew (*Anaeardium oeeidentale* L.) under Odisha Conditions. *Indian Cashew Causerie J.*, 7 (4): 7-8.

Ram, S. L., Bist, D., Lakhanpal, S. C., Jamwal, I. S. (1976). Search of suitable pollinizers for mango cultivars. *Acta Hort.*, 57: 253- 263.

Rao, V. N. M. and Hassan M. V. (1957). Preliminary studies on the floral biology of cashew (*Anacardium occidentale L.*). *Indian J. Agric. Sci.*, 27(3): 277-288.

Roitsch, T., and Ehneb, R. (2000). Regulation of source/sink relations by cytokinins. *Plant Growth Regulators*, 32: 359-367.

Salisbury, F. B. and Ross, C. W. (1978). *Plant physiology*. Wadsworth Publishing Company, Belmont, California, USA.

Salisbury, F. B., Ross, C. W. (1986). *Plant physiology* (Third edition). Wadsworth Publishing Company, Belmont, California, USA; 309-349.

Searle, C., Whiley, A.W., Simpson, D.R., Saranah, J. B. (1995). A preliminary phenol-physiological model for 'Kensington' mango in subtropical Australia. In Mango 2000-Marketing Seminar and Production Workshop Proceedings. Department of Primary Industries, Brisbane, pp. 109-124.

Shanmugavelu, K.G. (1985). Studies on the effect of plant growth regulators on cashew. *Acta Horticulturae*, 108: 35-43.

Singh, S. K., Syamal, M.M. and Maurya, A. N. (1992). Effect of NAA and Ethrel on vegetative growth, flowering, fruiting and yield of cashew nut (*Anacardium occidentale* Linn). *The Cashew*, 6(2): 11-12.

Subbaiah, C.C. (1983). Fruiting and abscission patterns in cashew. *J. Agric. Sci.* (Cambridge) 100: 423-427.

Thimmappaiah., Shirly, R. A., and Sadhana, P. H. (2002). *In-vitro* propagation of cashew from young trees. *In-vitro Cell. Dev. Biol. Plant*, 38: 152–156.

Veeraraghavathatham, D. and Palaniswamy, K. P. (1983). Effect of certain growth regulators on the yield of cashew (*Anacardium occidentale* L.) *Cashew Casuerie*, 3: 3-4.

Yamdagni, R. and Khangia, B. (1989). Effect of growth regulators on enzyme activities in shoot tips at various stages of bud differentiation in mango (*Mangifera indica* L.) cv. S. B. Chausa. *South Indian Horticulture*, 37(4): 194-198.

Cashew: Improvement, Production and Processing (2017) *Pages* **265–276**
Editor: **Prof. P.L. Saroj**
Published by: **ASTRAL INTERNATIONAL PVT. LTD., NEW DELHI**

15

Weed Management

D. Kalaivanan and P.L. Saroj

1. Introduction

The weed is a plant out of place not intentionally sown; whose undesirable qualities outweigh its good points. Some crop plants even can become weeds when they grow where they are not wanted. In contrast, a number of plants usually thought of as weeds may actually be helpful in controlling erosion or serving as food for wild animals and birds. Weeds are troublesome in many ways. They reduce crop yield by robbing water, light, space, and soil nutrients. Weeds can produce allelopathic substances that are toxic to crop plants. Weeds often serve as hosts for crop diseases and also provide shelter for insects and diseases.

The productivity of cashew plantations in India is quite low due to many factors including the problem of weeds. Therefore, the main task after establishment of cashew plantation is to control excessive weed growth which might otherwise severely deplete soil moisture and compete for nutrients along with cashew. Their effect becomes more severe in young seedling stage where the roots of the weeds and the cashew are in the same soil layer. It has been reported that the failure of most cashew programmes in West Africa was due to competition from local grasses (Ohler, 1988). In the mature plantations, weed growth could also impede nut picking in situations where nuts are collected from fallen apples. Normally the cashew plantations are severely infested with a wide variety of annual and perennial weeds, which may reduce nut yield considerably. Based on the severity of weed infestation the yield reduction is observed as high as 60-70 per cent in some cases. Considerable expenditure is required to keep the plantations weed free. Depending upon the type of weeds and intensity of weed growth, weeding is to be done manually, mechanically or chemically. The weed growth is seen maximum after the monsoon and harvesting becomes a problem if the weeds are not removed. At present, farmers manage weeds in cashew orchards through manual slashing. However, Ohler (1988) mentioned that slashing alone is a very unsatisfactory method of weed control since its effect is short-lived. Other methods for managing weeds in cashew plantation that have been investigated elsewhere include the use of herbicides (Komolafe, 1974 and Grundon, 1999) and intercropping. Some of the findings of these investigations have been positive and include higher plant stand with herbicide application (Komolafe, 1974) and reduced weed incidence, weed biomass, frequency of weed control and increased financial returns as a result of intercropping with food crops (Famaye and Adeyemi 2011). As the spacing between cashew trees are normally large, it might be necessary to develop cost effective weed control strategies which promote plant growth during establishment and also better yield in mature plantations.

2. Weed Flora in Cashew Orchards

The dominating weed flora in cashew plantations is mostly dicots. Monocots are also present in certain areas. The common dicot weed species identified in cashew plantations in India are; *Eupatorium odoratum, Tridex procumbens, Mimosa pudica, Lantana camara, Brachiaria* spp. Griseb., *Ischaemum indicum* (Houtt.) Merr., *Chromolaena odorata* L., *Synedrella nodiflora* Gaertn., *Elephantopus scaber* L., *Desmodium triflorum* DC., *Glycosmis arborea* DC., *Hemidesmus indicus* R. Br., *Ichnocarpus frutescens* R. Br., *Passiflora foetida* L., *Cyclea peltata* Cooke., *Tragia involucrata* L., *Triumfetta rhomboidea* Jacq., *Ziziphus* sp. Juss., *Narengamia alata* W and A., *Macaranga peltata.* Among the dicot weeds *Chromolaena odorata* (Communist weed) and West Indian daisy are most noxious weeds encroaching cashew plantations in the West coast.

Chromolaena odorata (Figures 15.1 and 15.2) is considered an invasive weed of field crops and natural environments in its introduced range. It has been reported to be the most problematic invasive species within protected rainforests in Africa.

Figure 15.1: *Chromolaena odorata* (Young plant).

Figure 15.2: *Chromolaena odorata* (with flower).

In Western Africa, it prevents regeneration of tree species in areas of shifting cultivation. It affects species diversity in southern Africa. In Sri Lanka it is a major weed in degraded areas and coconut plantations. In 19th century *Chromolaena odorata* escaped from the botanical gardens at Dhaka (Bangladesh), Java (Indonesia) and Peradeniya (Sri Lanka). In Western Africa the plant was accidentally introduced with forestry seeds. It was introduced as an ornamental in Southern Africa and was introduced to Ivory Coast in 1952 to control *Imperata* grasses. It was first found in Queensland, Australia in 1994 and was perhaps introduced with foreign pasture seeds. It was introduced to India from South Africa while importing grains and raw cashew nuts; Seeds stuck to this produce transported unknowingly. It was introduced to India when Kerala state was ruled by communist government. It was also called as communist weed in Kerala due to the above fact. From Kerala it started spreading to coastal Karnataka and coastal Maharashtra. As it is an invasive weed, it occupied all vacant spaces all along the road side and vacant fallow lands. This weed is rampant in arecanut, coconut, cashew and rubber plantations.

Figure 15.3: *Pennisetum polystachyon.*

Another noxious weed present in cashew plantations is *Pennisetum polystachyon* (Figure 15.3). This grass weed is robust growing and dissemination of seeds is very fast. This competes not only with other weeds but also competes with cashew. Another introduced weed is *Lantana camara* (Figure 15.4). This weed was introduced as ornamental plant. Initially this was grown as ornamental plant in big cities in India and later this spread to rural area where different crops were grown. This weed has also invaded cashew garden in coastal area and remained as noxious weed. Control of this weed can be done initially by uprooting during the onset of monsoon and later spot application of systemic weedicide such as glyphosate. Of late one more weed West Indian daisy belonging to compositae family invaded cashew field and this has occupied initially in coconut and Areca nut gardens in west coast and later this started spreading to cashew and rubber plantations.

Figure 15.4: ***Lantana camara.***

3. Weed Management

Manual Weeding

Manual method like hand pulling, sickling and mulching are extensively used wherever labour is available. Hand pulling is the most efficient in annual and biennial weeds which are propagated by seeds. Sickling is done to remove the tall growth of perennial weeds and to cut at the ground level for a follow up application of foliage applied herbicides on the regrowth. Sickle weeding is the most common method of weed control adopted in cashew plantations. The cost and availability of labour and the longer period of time to complete the task necessitated examining the possibility of chemical weed control. Noxious weeds like *Pennisetun polystachyon* and *Eupatorium odoratum*, West Indian daisy and *Lantana camara* weeds can be controlled by initially uprooting and later with spot application of weedicides.

Weeding around the young trees within 1.5 m radius by hoeing is necessary during 2-3 years after planting. In grownup orchards, two rounds of weeding are recommended for management of weeds in cashew orchard. The first round of weeding may be done before heavy rains start by uprooting the weeds within the radius of 2 m from the main stem. The remaining weeds in the interspaces are slashed to ground level. The second round of weeding may be done during November-December to facilitate spraying, harvesting and picking of nuts.

Mechanical Weeding

Weed control using manual labour is costly; therefore, farmers resort to mechanical and chemical means of weed control. A tractor-mounted slasher is a handy mechanical weeder which is efficient and economical for cashew plantation. *Mikania* is a broad leaved climbing weed, which smothers the tree completely causing heavy yield reduction. Mechanical hoeing may be carried over the whole-planted strip to a width of about 4 m, leaving 2 m in both sides, to avoid root injury. Tillage practices like hoeing, ploughing and mowing are done to clear off weeds, especially in newly established orchards where sufficient spaces may be available for inter-cultivation. Mulches like straw, hay and farm wastes are used to cover the soil. They produce heat which smothers or kills the weeds. However in undulating topography where cashew is grown mechanical weeding is difficult. This situation is prevalent in the West coast and partially East coast region. Motorized weed cutter can be used to some extent but not the hoeing machines or power tiller or tractor driven weed cutter

Chemical Weed Management

The weeding cycles are generally confined to the pre-monsoon and post-monsoon periods to coincide with the fertilizer application. Alternatively, weedicides may also be applied after slashing, well in advance of the rainy season if the under growth is too dense. The recommendation is to apply agrodar-96 (2-4 D) at the rate of 4 ml/litre of water followed by gramoxone at the rate of 5 ml/litre of water. Approximately 400 litres of spray is required to cover one ha. The spraying is repeated in the post monsoon season if the weed load is heavy (Bhaskara Rao, 1998). More effective weedicide spray is application of glyphosate @ 8ml/litre twice a year during the above mentioned period. Even the noxious weeds can be killed by spraying glyphosate @ 8 ml/litre twice a year. Chemical weed control is costly and thus reserved for exceptional cases, especially in areas where the labour wages are high. Effective chemical weed control methods are available for cashew plantations as mentioned below:

- Application of paraquat @ 0.4 kg/ha twice at bi-monthly intervals starting from July will effectively control all types of weeds.
- Application of glyphosate @ 0.8 kg/ha, in June–July can also control the weeds effectively.
- Abdul Salam *et al.* (1993) reported that application of glyphosate at 0.8 kg/ha (single spray) or glyphosate at 0.4 kg per ha + 2, 4-D @ 1.0 kg per ha (single spray) or paraquat @ 0.4 kg per ha + 2, 4-D @ 1.0 kg per ha (single spray) during August can effectively control the weed growth in cashew plantations. It was also found that chemical weed control is economical compared to the common method of sickle weeding.
- Abraham and Abraham (1999a) reported that either diuron @ 1 kg per ha (alone or with 2,4-D @ 0.25 kg per ha) was effective in selectively killing *Mikania*. In another study Abraham and Abraham (1999b), found that use of glyphosate @ 1.2 kg/ha or 2, 4-D @ 1 kg per ha controlled *Mikania* infestation effectively.

A field experiment was conducted at Cashew Research Station, Madakkathara, Kerala during 1990 to study the efficacy of certain herbicides in comparison with sickle weeding for controlling the weed population in adult cashew plantation (18-year-old). There were 9 treatments for the study. Total weed dry matter production five months after the treatments and economics of herbicide application as well as sickle weeding were also worked out.

There was considerable difference in weed control due to the treatments tried (Table 15.1). The data also indicated that herbicide application has resulted in considerable reduction in weed growth. The weed recurrence measured in terms of weed dry matter was considerably low in the plots applied with glyphosate 0.8 kg/ha (single spray), glyphosate 0.4 kg/ha + 2, 4- D 1.0 kg/ha (single spray), paraquat 0.4 kg/ha + 2, 4-D 1.0 kg/ha and sickle weeding once + one spray of paraquat 0.8 kg/ha 30 days after sickle weeding. Among these herbicide treatments, glyphosate 0.8 kg/ha (single spray) was found to be the best. The results indicate that the four herbicide treatments mentioned above can be effectively used for weed control in cashew plantations. The combined action of the contact herbicide (paraquat) and a systemic herbicide (2, 4-D) was also effective. It may be noted that the herbicide application was done during August, a period when rainfall was comparatively low. It is further encouraging to note that the treatment effects were observable for a period of 5 months (up to January) by which the rains were over and environment was dry. As a result the possibility of sudden regrowth of weeds was checked up to the commencement of the subsequent Southwest monsoon. Thus, it is possible to control the weeds of cashew plantations for a period of 10 months (August through May end) if herbicides can be applied during August.

Table 15.1: Effect of Herbicides on Weed Dry Matter Production and Cost of Weed Control

Treatment Details	*Weed Dry Weight (kg ha^{-1})*	*Cost of Weed Control Treatments (Rs/ha)*
Control (unweeded)	2266.68	
Sickle weeding twice	1100.00	2500.00
Paraquat 0.4 kg/ha – 3 sprays at monthly intervals	2000.00	1830.00
Glyphosate 0.8 kg/ha – single spray	800.00	1513.00
Glyphosate 0.4 kg/ha + 2,4-D 1.0 kg/ha – single spray	933.32	1195.60
Paraquat 0.4 kg/ha + 2, 4-D 1.0 kg/ha –single spray	933.32	874.00
Paraquat 0.8 kg/ha + Diuron 1.0 kg/ha – single spray	1166.68	1263.00
Sickle weeding once + one spray of glyphosate 0.4 kg/ha 30 days after	1400.00	2181.60
Sickle weeding once + one spray of paraquat 0.8 kg/ha	1066.68	2120.00
CD (0.05)	300.28	

Source: Abdul Salam *et al*., 1993.

A comparative study on the cost of the weed control treatments revealed that the treatment sickle weeding twice is the costliest. The cost of the weed control treatment paraquat 0.4 kg/ha + 2,4-D 1.0 kg/ha was the lowest (Table 15.1). The plots applied with glyphosate produced the lowest amount of weed dry matter which was only 35 per cent of that in the unweeded control. The systemic action of this herbicide might have helped to considerably reduce the recurrence of weeds in the treated plots due to its broad spectrum activity. Studies conducted in the AICRP during 1988 on weed control in plantation crops also indicated that glyphosate 0.8 kg/ha was effective to check the weed growth in cashew plantations.

Weed Control through Growing Cover and Green Manuring Crops

Weeds can also be controlled by growing cover crops in between two rows of cashew. The most common and successful cover crops are *Pueraria javanica, Calopogonium mucunoides, Mucuna bracteata and Mimosa invisa.* Growing of cover crops helps in controlling weeds as these cover crops grow faster than weeds. Growing of cover crops not only suppresses weeds but also reduces water evaporation thereby withholding more moisture in soil profile and soil temperature during summer season which in turn benefits crop growth. Cover crops also reduce soil temperature during typical summer season which is also beneficial for crop growth. Experiments conducted at Directorate of Cashew Research showed that *Mucuna bracteata* (Yadukumar, 2007) as cover crop in cashew increased soil moisture content as well as reduced soil temperature during typical summer season. This cover crop has also caused increase in soil organic matter content. Indirectly cover crops effectively controlled the weed growth. Growing green manure crops like *Gliricidia sepium* and dhaincha (Yadukumar *at al.*, 2008) in between two rows of crop not only controlled weeds but also increased soil organic matter content when incorporated in to the soil. This had increased soil fertility status particularly in degraded soil where cashew was grown.

Weed Control by Intercropping

In case of pure cashew plantation weed control incurs huge amount and contribution of weed control to the cost of cultivation goes up by 33 per cent. Instead of this the vacant space available in between two rows of crop can be effectively utilized by growing profitable intercrops like pineapple, turmeric, ginger and ground nut (Yadukumar *et al.*, 2006 and 2007).

4. Weed Management in different Countries

Sri Lanka

The *Pennisetum polystachyon* and *Imperata cylindrica* are the common weeds in cashew plantations which often become a fire hazard during the dry months. Slash weeding is usually practiced to control grass weeds. Herbicides such as gramoxone or roundup are also used to a limited extent, especially at nursery sites. Sri Lanka State Plantations also use rotary movers in between tree rows to control weed flora. Mulching with weed residue or tree loppings and sometimes with coconut husk is carried out by many growers to keep under growth down and also conserve moisture during the dry months (Surendra, 1998).

China

Up to three years of age, young cashew trees need weed-free conditions around the base and 3-4 weeding cycles have to be carried out per year. Mulching is also practiced to keep the basal area weed-free and to maintain moisture and temperature balance in the upper layers of soil. Mulching is however, not essential for mature trees as the root system becomes extensive. Weeding is usually carried out by hand and occasionally when obnoxious weeds such as twitch-grass are present, they are controlled with the use of glyphosate (Liu Kangde, Liang Shibang and Deng Suisheng, 1998).

Australia

Mowing between the rows of cashew encourages the development of low grass growth to protect the soil from erosion. Paraquat has been used successfully to control weeds, but the cashew leaves and green stems should be fully protected from spray drift. New herbicides under test may allow easier weed management in the future. During harvesting, the nuts are collected from the ground under the trees. Therefore, weed growth should be mown and raked away from under and around the trees before nut fall begins.

Myanmar

Weeding operations are essential in cashew orchards to minimize competition for water and light. Weeding practices vary from region to region. When orchards are inter-cropped, weeding is assured, in other instances slash weeding is carried out once or twice in a year. In other areas, there is total neglect of cashew orchards, rendering them almost inaccessible and difficult to manage (Maung Maung Lay, 1998).

Nigeria

The production of cashew in Nigeria is faced with serious problem of weed management because of the large spacing between the cashew stands which do not allow the crop to close canopy quickly despite the fast rate of growth (Adeyemi, 1989). Hoe weeding, a traditional method of weed control is very expensive and a times not readily available. Intercropping has reduced weeding period of cashew by about 50 per cent when compared to sole cashew (Adeyemi, 1989). Herbicides that should have been found to be more effective in weeds control are very expensive for peasant farmers who are the major cashew growers in Nigeria. There is need for farmers to maximize profit by reducing production cost. Hence, there is need for appropriate cropping system that will enhance good growth and reduce labour cost on weeding in cashew production at early stage of establishment. Field trials on the effects of intercropping food crops (maize, cassava, cowpeas and plantain bananas) with cashews for the control of weeds in cashew plantations within the early years of cashew establishment were conducted at Ibeku, Umuahia, Nigeria, during 1986-1988. The results showed that weed biomass was reduced in plots carrying the crop mixtures (499.5-949.1 kg/ha) compared with those of sole cashews (1165.5-1332 kg). The number of weedings carried out per annum in intercropped plots (2.5-5.0) was lower than that of cashew in pure stands (5.5-7.5). The degree of weed succession was markedly affected in some crop mixtures leading to alteration in the weed sequence and their population dynamics. The main problematic weeds were *Chromolaena odorata, Axonopus compressus, Euphorbia heterophylla, Tridax procumbens, Commelina nudiflora, Acanthus montanus* and *Fleurya aestuans*. The effect of intercropping on cashew plant height, girth and leaf area was variable (Adeyemi, 1989). Furthermore, when cashew was intercropped with maize, cassava, cowpea and melon, its morphological growth parameters were not affected except the girth that was slightly reduced under cashew/cassava mixture.

Another study to determine the effects of weed incidence and biomass on cashew intercropped with rice and plantain was conducted at Cocoa Research Institute of Nigeria (CRIN) Uhonmora Substation, Edo State, between 2007 and 2008. There were six treatments comprising of cashew sole, rice sole, plantain sole, cashew/rice, cashew/plantain and cashew/rice/plantain (Famaye and Adeyemi 2011). Cashew and plantain were planted 4.5m and 3.1m apart, respectively while rice 30cm apart using small teaspoonful as a standard measure adopted per hole in this trial. Result showed that cashew sole gave the highest weed biomass of 25 and 17 t/ha followed by rice sole 20 and 15 t/ha and plantain sole 15 and 10 t/ha in for the first and second year, respectively. The intercropped gave 13 and 10 for cashew/rice, 13 and 8 for cashew/plantain and 12 and 5 t/ha for cashew/rice/plantain in 2007 and 2008, respectively. This also indicated that intercropping reduces weed as the crops in the intercrops cover spaces that could be occupied by weeds. The least weed biomass was recorded in cashew/rice/plantain. Cashew sole gave the highest weed biomass that was significantly higher ($P<0.05$) than other treatments. Cashew/plantain was lower in weed incidence than cashew/rice intercropped. This reduction in weed biomass in the intercropped suggests a reduction in labour requirement than sole cashew. This is in consonant with the report of Olukotun (1983) that appropriate intercropping of cashew with food crop(s) would increase labour efficiency and reduce cost of production. Considering the morphological parameters (plant height, leaf, and plant girth and leaf area) no significance difference between the sole cashew, cashew/rice and cashew/plantain intercropped were observed. However, cashew/rice/plantain with lowest weed biomass was lower than other treatment in terms of these growth parameters. Therefore, cashew/rice and cashew/plantain intercrops with lower weed biomass and comparable vegetative growth could be recommended to cashew farmers instead of sole cropping.

Ghana

Three field trials, one for establishment and two for mature phases respectively, were conducted to develop weed management strategies for cashew cultivation in Ghana (Kwabena Opoku-Ameyaw *et al.*, 2012). The treatments for the establishment phase consisted of hoe weeding of whole plot, hoe weeding of cashew lines and slashing of inter-rows, herbicide application with glyphosate of whole plot or cashew lines with slashing of inter-rows, relay cropping of the cashew plot with sorghum and a legume and intercropping with a legume. In one trial of the mature

Table 15.2: Residual Effects of Weed Control Methods on Cashew Yield

Treatment	*Yield (kg/ha)*				
	1st Year	*2nd Year*	*3rd Year*	*4th Year*	*Cumulative*
Hoeing of whole plot	13.0	36.7	112.4	96.7	258.8ab
Clean line weeding + high slashing of inter-rows	7.2	22.6	59.4	80.6	169.8c
Glyphosate (1.5/ha) applied to whole plot	5.9	12.6	69.5	56.8	144.8c
Glyphosate (1.0 L/ha) applied to whole plot	10.1	19.0	57.8	90.6	177.5c
Ring application of glyphosate (1.0 L/ha) + high slashing of inter-rows	7.4	21.3	62.6	92.6	183.9bc
Rotational cropping with sorghum and legume + manual weeding	11.6	44.3	123.8	135.9	315.6a
Cultivation of legume* + manual weeding	15.3	47.9	88.2	123.6	275.0a
LSD ($P \leq 0.05$)	6.31	ns	ns	ns	88.0

* Cowpea in 2007/2008; soybean in 2008/2009; ns: not significant at 5 per cent probability.

Table 15.3: Nut Yield of Mature Cashew Under Different Weed Control Treatments

Treatment	*Yield (kg/ha)*			
	2006	*2007*	*2008*	*Cumulative*
T1 – Hoeing whole plot 2 times/year	186.0	109.3	189.1	484.4
T2 – Hoeing whole plot 3 times/year	223.6	95.2	184.3	503.1
T3 – High slashing whole plot 3 times/year	225.3	113.4	217.4	556.1
T4 – Whole plot application of glyphosate (1.5 l/ha) 2 times/year	281.3	141.4	235.8	658.5
T5 – Ring application of Glyphosate and high slashing,2 times/year	222.6	109.2	166.0	497.8
LSD ($p \leq 0.05$)	ns	ns	ns	ns

ns: Not significant at 5 per cent probability.

phase, the treatments were hoe weeding of whole plot two or three times a year, high slashing of whole plot three times a year, herbicide application (glyphosate) of whole plot and ring application of herbicide (glyphosate) plus high slashing of inter-rows twice a year. In the other trial, this was carried out on farmers' fields, glyphosate (roundup) at 2.0 l/ha was tested against manual slashing. In the study of the establishment phase, neither the girth nor height of the cashew seedlings was significantly affected by the treatments. The first year and four year cumulative cashew yields were significantly affected by treatments with the intercrops giving higher yields than the glyphosate-treated plots (Table 15.2). Intercropping as a weed control measure was generally more costly than glyphosate application or manual weeding. In the mature trials, chemical weed control slightly improved cashew nut yield and was cheaper to apply (Table 15.3).

References

Abdul Salam, M. Pushpalatha, P.B., Suma, A. and Abraham, C. T. (1993). Efficacy of chemical weed control in cashew plantations. *J. Plantation Crops,* 21 (1): 54-56.

Abraham, M. and Abraham, C.T. (1999a). Control of *Mikania micrantha* with herbicide. *Proceedings of the 11th Kerala Science Congress,* Kasaragod, India, pp. 268-270.

Abraham, M. and Abraham, C.T. (1999b). Control of *Mikania micrantha* with herbicide. In: Proceeding of 11th Kerala Science Congress, Kasaragod, India pp. 68-70.

Adeyemi A. A. (1989a). Cultural weed control in cashew plantation. Use of intercrops to reduce weed incidence in cashew plots. Proc. Int. DLG-symposium on integrated pest management in Tropical and sub-tropical cropping systems, Bad Darkheem Fed. Rep. Germany.

Adeyemi A. A. (1989b). Farming system in some CRIN scheduled crops. Growth and performance of cacao and cashew, intercropped with food crops. CRIN Annual Seminar May 17.

Adeyemi, A. A. (1989). Cultural weed control in cashew plantations: use of intercrops to reduce weed incidence in cashew plots. In: Proceedings: Integrated pest management in tropical and subtropical cropping systems '89, vol. 3, February 8-15 1989, Bad Dürkheim, Germany. pp. 827-842.

Bhaskara Rao E.V.V. (1998). Integrated Production Practices of Cashew in India. *Integrated Production Practices of Cashew in Asia* (Eds. Minas K. Papademetriou and Edward M. Herath). Food and Agriculture Organization of the United Nations, Regional Office for Asia and the Pacific, Bangkok, Thailand.

Famaye A. O and Adeyemi E. A. (2011). Effect of cashew/rice/plantain intercropped on weed incidence in Edo state, Nigeria. *ARPN Journal of Agricultural and Biological Science,* 6 (6): 62-65.

Grundon, N. (1999) Overview of Australian Cashew Literature, CSIRO Land and Water Technical Report 25/99, p. 47.

Komolafe, D.A. (1974) Herbicide trial in cashew, Report of the Cocoa Research Institute of Ghana, Ibadan, 1972-73, pp. 172-173.

Kwabena Opoku-Ameyaw, F.K. Oppong, F.M. Amoah, S. Osei-Akoto, E. Swatson (2012). Growth and early yield of cashew intercropped with food crops in northern Ghana, *Journal of Tropical Agriculture*, 49 (1-2): 53-57.

Liu Kangde, Liang Shibang and Deng Suisheng. (1998). Integrated Production Practices of Cashew in China. *Integrated Production Practices of Cashew in Asia* (Eds. Minas K. Papademetriou and Edward M. Herath). Food and Agriculture Organization of the United Nations, Regional Office for Asia and the Pacific, Bangkok, Thailand.

Maung Maung Lay. (1998). Integrated Production Practices of Cashew in Myanmar. *Integrated Production Practices of Cashew in Asia* (Eds. Minas K. Papademetriou and Edward M. Herath). Food and Agriculture Organization of the United Nations, Regional Office for Asia and the Pacific, Bangkok, Thailand.

Ohler J.G. 1988. Cashew, Koninklijk Instituut voor de Tropen, Amsterdam, ISBN 90-6832-074-2, 1988 (originally printed in 1979, p. 260.

Olukotun O.A. (1983). Intercropping of cashew with food crops. CRIN Annual Report. p. 42.

Surendra G. B. B. (1998). Integrated Production Practices of Cashew in Myanmar. *Integrated Production Practices of Cashew in Asia* (Eds. Minas K. Papademetriou and Edward M. Herath). Food and Agriculture Organization of the United Nations, Regional Office for Asia and the Pacific, Bangkok, Thailand.

Yadukumar, N, Rejani, R and Nandan, S.L. (2008). Studies on green manuring in high density cashew orchards. *J. Plantation Crops*, 36(3): 265-269.

Yadukumar, N. (2006). Resource use efficiency in cashew. In: National seminar on resource use efficiency in cashew held at CPCRI, Kasaragod, Kerala on 25th March, 2005.

Yadukumar, N. (2007). Development of suitable cashew based cropping systems–inter and mixed cropping with cashew. Proceedings of the National seminar on Research, Development and Marketing of Cashew, 20-21st November, ICAR, Research Complex, Goa, p. 62.

Cashew: Improvement, Production and Processing (2017) *Pages* **277–293**
Editor: **Prof. P.L. Saroj**
Published by: **ASTRAL INTERNATIONAL PVT. LTD., NEW DELHI**

Cashew Based Cropping Systems

J.D. Adiga and D. Kalaivanan

1. Introduction

Cropping system refers to the crops and crop sequences and the management techniques used on a particular field over a period of years. Cropping systems of a region are decided by a number of soil and climatic parameters which determine overall agro-ecological setting for nourishment and appropriateness of a crop or set of crops for cultivation. Nevertheless, at farmers' level, potential productivity and monetary benefits act as guiding principles while opting for a particular crop/ cropping system. These decisions with respect to choice of crops and cropping systems are further narrowed down under influence of several other forces related to infrastructure facilities, socio-economic factors and technological developments, all operating interactively at micro-level. The system must be compatible, practically feasible, socially acceptable, ecologically sound and economically viable. Moreover, appropriate management strategy with respect to canopy architecture, light interception, moisture, nutrition, pest-diseases, weed and their interaction is essentially required (Saroj and Arora, 1994).

In India, cashew is cultivated both in the West and East coast regions. In the West coast, lands are level at the seashore and undulated in the interior areas. The climate is warm and humid with temperature ranging from 22 to 38°C, relative humidity 75 to 85 per cent and mean annual rainfall above 3500 mm. The soils are mainly laterite or gravelly and porous with mostly acidic particularly in hilly tracts. Because of undulating topography, soil erosion is a major problem in exposed areas. In the East coast, land is level and soils are sandy, red sandy loam, red loamy and laterite. This area receives rainfall ranging from 600 to 1500 mm annually. Climate is warm and subhumid or dry with temperature ranging from 25 to 42°C and relative humidity from 40 to 60 per cent. In India, cashew is mainly grown by small holders in marginal lands despite the fact that it fetches valuable foreign exchange. Only in recent years, its economic importance has been realized and attempts were made to increase its productivity through improved management and better varieties. Intercropping was not viable earlier as cashew itself was grown on soils with low water availability and fertility. With establishment of large plantations and adoption of systematic package of practices, intercropping is practiced to obtain returns during the initial years of cultivation. Once cashew canopy becomes dense, intercropping (legumes and millets) is discontinued. Mixed cropping with other tree species is not widely practiced though there are some reports of *Casuarina* being planted along with cashew in Andhra Pradesh and Odisha.

2. Global Scenario

Cashew is grown in several tropical countries including Latin America, Africa and South-East Asia. Various intercrops are being grown in cashew orchards based on agroclimatic suitability and farmers' choice in different countries as given below.

China

Natural grass and legume cover is usually maintained at time of land clearing to conserve soil and water since the trees are planted far apart. In young orchards, cashew is also intercropped with green manure cover crops and/or short-term annual crops. Main cover-crops include *Calopogonium mucunoides, Macrophyllum atropurpureius* cv. Siratro, pasture grasses and inter-crops such as peanut, sweet potato and beans. In recent years, cashew is being inter-cropped in some areas with melons (watermelon and sweet melon) and vegetables such as hot pepper (Liu Kangde *et al.*, 1998).

Myanmar

Although recommendations have been made to establish cover crops, especially on the degraded slopes in many parts of the country, this practice is not yet popular among cashew farmers in Myanmar, with the result that soil erosion is severe in cashew plantations. Several inter-crops are taken by some farmers, predominantly, annual crops such as sweet potato, sesame, peanut, maize, cassava, pigeon pea *etc.* In some areas other tree crops such as jack fruit, mango and fuel wood trees are grown in association with cashew (Maung Maung Lay, 1998).

Sri Lanka

Creeping cover crops such as *Peuraria phaseoloides, Centrosema pubescens* and bush cover crops such as *Gliricidia maculata, Leucaena leucocephala* and nitrogen fixing trees such as *Acacia mangium* are the principal cover crops used in cashew growing areas. Due to frequent fire hazards, most large-scale plantations have fire gaps laid down when cover crops are grown. Banana is a popular inter-crop in many cashew plantations. Pineapple, papaya, pomegranate and coconut are also used as semi-perennial and perennial inter-crops in some areas. The common annuals grown in cashew plantations are Legumes (cowpea, black gram, and green gram), oil crops (sesame, ground nut) and condiments such as hot pepper and onion (Surendra, 1998).

Vietnam

Most farmers, who own a larger plot of cashew, often intercrop with different food crops such as beans, maize, and cassava for own use and in some instances, on better soil types for sales as well. Besides food supply, intercropping with leguminous species such as beans is also under practice to reduce the erosion risk on steep slopes and improve the soil structure.

Africa

Under wider spacing of 10 m apart, the crops such as millet, maize, groundnut, vegetables, sesame, and upland rice are among other crops that will not damage the sensitive cashew root system. Intercropping up to 4-5 years is best done when planting at least 3 m away from each cashew tree.

Nigeria

The crops like cocoa, oil palm, rubber or kola are planted in the Southern states, while cereals and pulses are planted beneath the cashew orchards in the Northern states (Aliyu and Hammed, 2008). The intercropping of cashew in association with maize, cassava and plantain has proved useful in reducing weed incidence in cashew plots (Adeyemi, 1986). Furthermore, when cashew was intercropped with maize, cassava, cowpea and melon, its morphological growth parameters were not affected except the girth that was slightly reduced under cashew/cassava mixture.

Ghana

Intercropping tree crops during the establishment phase with food crops is an age-long practice in the tropics. The benefits of such a practice may include food security for the household, income generation to partially offset the cost of establishment, weed control, and better use of growth resources (Rodrigo *et al.*, 2001; Opoku-Ameyaw *et al.*, 2000 and 2003). However, the success of cashew (*Anacardium occidentale* L.) + food crop intercropping system will depend on whether the food crops compete with the young cashew seedlings for growth resources. Ohler (1979) recommended the use of early bearing low growing food intercrops and suggested that tall intercrops like certain varieties of sorghum (*Sorghum bicolor* (L.) Moench), and millet should not be grown with cashew as they cast too much shade and affect the growth of the latter. However, in Nigeria crops that vary in height like maize, cassava (*Manihot esculenta* Crantz), bambara groundnuts (*Vigna subterranea* (L.) Verdc.), cowpea (*Vigna unguiculata* (L.) Walp)., melon (*Citrullus lanatus* (Thunb.) Mansf.), and plantain (*Musa* spp.) have been successfully grown along with cashew. Intercropping cashew with food crops during the establishment phase in the savannah and transitional zones of the country is also recommended.

The investigation carried out at Bole in the Guinea savannah agro-ecological zone which included sole cashew, cashew + groundnut (*Arachis hypogaea* L.), cashew + maize (*Zea mays* L.), cashew + sorghum (*Sorghum bicolor* (L.) Moench), cashew + yam (*Dioscorea alata* L.), cashew + sorghum in rotation with groundnut, and cashew + groundnut in rotation with maize (Opoku-Ameyaw *et al.*, 2011) revealed that intercropping significantly improved the girth of cashew seedlings in the cashew+ maize combination. Height of cashew seedlings was significantly higher in the cashew + maize and cashew + sorghum/groundnut system. Intercropping did not substantially alter early yield of cashew nuts. Economic analysis of the packages, however, indicated that maize and yam intercropping during the establishment phase was profitable. It is therefore advisable to intercrop cashew with yams and maize in northern Ghana during the early period of establishment to generate additional returns to the growers. Cropping systems did not significantly affect the nutrient content of the soil at the end of the intercropping period. Intercropping significantly ($p<0.05$) affected the girth and height of the cashew seedlings two years after the treatment application (Table 16.1). Plant girth was significantly higher in the cashew + maize and cashew + sorghum in rotation with groundnut than sole cashew. Plant height, on the other hand, was significantly improved when cashew was intercropped with maize and sorghum. This beneficial effect of intercropping on the growth of cashew may be due to the tillage and other crop husbandry practices such as weed control and fertilizer application (Warui, 1985; Abeysinghe, 2009).

Table 16.1: Effect of Cropping Systems on the Growth and Development of Three Year-Old Cashew Plants Two Years After Intercropping and the Proportion of Plants Flowering at Four Years of Age at Bole (Ghana)

Cropping System	*Girth (mm)*	*Height (cm)*	*Crown Spread (cm)*	*Proportion of Plants Flowered (per cent)*
Sole cashew	37.5[c]	125.0[c]	195.6	78.8 (64.1)
Cashew + groundnut	38.4[bc]	132.9[bc]	190.5	81.6 (65.9)
Cashew + maize	41.9[a]	159.1[a]	236.4	77.3 (66.0)
Cashew + sorghum	38.4[bc]	146.9[ab]	199.1	62.1 (53.3)
Cashew + yam	40.1[ab]	122.5[c]	194.2	87.9 (71.3)
Cashew + sorghum/groundnut	41.4[ab]	139.0[bc]	205.7	90.9 (75.1)
Cashew + groundnut/maize	37.4[c]	130.6[bc]	205.9	83.4 (67.3)

Values followed by the same letters in a column are not significantly different at p = 0.05; values in parenthesis are transformed data.

Cashew starts bearing four years after transplanting (*i.e.*, one year after the intercropping period). The first three years of cashew yields were not significantly affected by the cropping systems (Table 16.2). This indicates that food crop intercropping during the establishment phase may not have any adverse effect on cashew yield as it was the case with soil nutrient contents, which is similar to the findings of Abeysinghe (2009).

Brazil

In cashew plantations, on very large holdings which are mechanized, industrial catch crops such as grain sorghum, peanuts, sesame (*Sesamum indicum*) or cotton are planted. Once catch cropping is phased out as the trees increase in size, spontaneous grasses replace them or pasture grasses are planted as permanent ground cover.

Native ginger grass (*Paspalum maritimum*) and African guinea grass (*Panicum maximum*) have proven to be successful on cashew plantations. Thus in contrast to what occurs with native stands, when these perennial crops are formally cultivated, grazing replaces annual cropping after a few years, rather than coexisting with it. Occasional intercropping of cashew, coconut or carnauba with other perennial crops is also practiced in the region. Reportedly in Ceara, cashew trees are utilized to provide support for pepper vines. On very small farms, one can see coconut and cashew, cashew and citrus, and cashew, banana and coconut inter-planted.

Table 16.2: Effect of Cropping System on the First Three Years' Yield of Cashew Nuts at Bole (Ghana)

Cropping System	*Cashew Yield (kg ha^{-1})*		
	2008	*2009*	*2010*
Sole cashew	70.8	149.5 (21.5)	96.6 (29.4)
Cashew + groundnut	71.0	161.6 (10.6)	123.5 (32.4)
Cashew + maize	51.8	104.3 (20.8)	98.1 (33.7)
Cashew + sorghum	55.5	145.5 (36.2)	120.7 (16.1)
Cashew + yam	80.3	168.5 (10.5)	165.0 (68.3)
Cashew + sorghum/groundnut	74.9	161.3 (28.7)	88.7 (16.9)
Cashew + groundnut/maize	80.8	154.3 (59.7)	138.0 (51.5)
F test	*ns*	*ns*	*ns*

Values in parentheses are standard errors of the means; ns= not significant at p=0.05; values in parenthesis are transformed data.

3. Indian Scenario

Considering the pre-bearing period and low income in the early period of bearing and fluctuations in the yield and price from year to year, it is recommended to take up inter-cropping in cashew plantations. Even, inter-cropping has become popular with the systematic establishment of large-scale cashew orchards also. It is practiced in the first few years when there is sufficient space between crop rows with the main objective of deriving some income until the cashew starts giving economic returns. Inter crops which can replace weeds and as such would not compete for light, moisture and nutrients with cashew and can be harvested very early in the dry season or at the end of rainy season may be taken up. Tall growing intercrops like certain varieties of sorghum and millet should not be encouraged between young cashew, as they cast too much shade. Leguminous crops such as groundnuts and beans are suitable for inter cropping. Besides the annual crops, fruit like *Anona*, phalsa, etc., especially with less canopy spread can be thought of depending on the suitability. Popular cover crops for cashew plantations are *Peuraria javanica, Calapagonium mucunoides* and *Centrosema pubescens* which improve the fertility and moisture balance and help in conserving orchard soils. Leguminous cover enriches soil with the plant nutrients and adds organic matter, prevents soil erosion and conserves moisture. Cover crop seeds are generally sown with the advent of the monsoons at a seed rate of about 7 kg/ha. On degraded steep lands, cover crops are usually established on seed beds between tree rows. Prepare seed beds of 30 cm x 30 cm size in inter space in sloppy degraded soils by loosening soil and mixing a little quantity of compost. The seeds of these crops are sown in beds and covered with a thin layer of soil. The seeds should be soaked in water for 6 hr before sowing.

A systematic intercropping system involving compatible crops with varying morphological and rooting habits will enable better resource use and more returns. Biennial horticultural crops like pineapple, banana and papaya and annual food crops like cowpea and tapioca are suitable inter-crops in cashew (Mandal, 1992). In addition, pepper and pomegranate also come up well along with cashew. Cashew grafts are normally planted at a spacing of 7.0 m x 7.0 m to 10.0 m x 10.0 m. In the first year, when planted at a spacing of 7.5 m x 7.5 m (177 plants/ha) very little land area is utilized by the crop. The canopy development pattern of cashew in relation to age is illustrated in Figure 16.1 (spacing of 7.5 m x 7.5 m). During the first year, cashew uses hardly 1.2 per cent of the land area provided to the crop. During the second and third year, the canopy coverage is only 5 per cent and 13.4 per cent respectively. This pattern of canopy development offers tremendous scope for intercropping in cashew during early years.

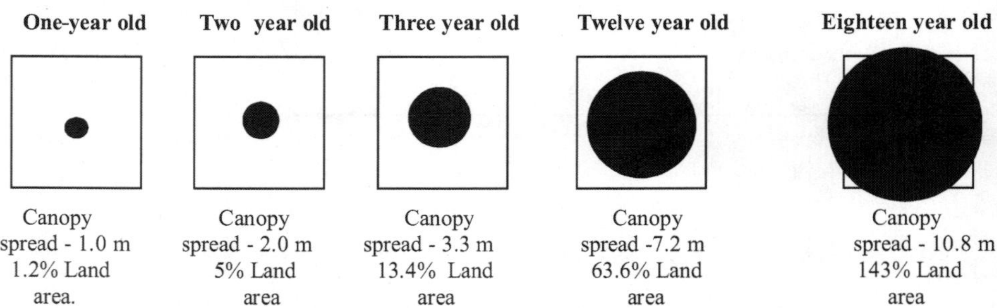

Figure 16.1: Canopy Development Pattern in Relation to Age.

West Coast

a. Karnataka

Pineapple as Intercrop

Pineapple can be grown as an intercrop in cashew garden profitably for the first 7 years. Both main and intercrop can be planted simultaneously. Normally pineapple can be grown in the inter spaces available between two rows of cashew plants planted at 8m x 8m (156 trees/ha) or 7.5 m x 7.5 m (175 trees/ha) or 10 m x 5 m (200 trees/ha). Pineapple is planted with the onset of pre-monsoon showers in three trenches dug out at 90 cm distance across the slope between two rows of cashew. In leveled land straight trenches can be opened between two rows of cashew. Each trench is of 1 m width, 0.5 m depth and convenient length. Pineapple suckers with 8-15 leaves are planted in each trench in two rows at 60 cm apart. The distance between two plants within a row is 40 cm. Before planting pineapple suckers, the trench should be half filled with mixture of top soil, farm yard manure and rock phosphate. Nearly 2.5 kg of farm yard manure and 160 g of rock phosphate/metre length of trench should be added and thoroughly mixed.

Fertilizers should be applied annually in two split doses (May-June and September-October). Fertilizers are applied at the rate of 25 g N, 7 g P_2O_5 and 25 g K_2O/pineapple sucker/year. Each time whenever fertilizer is applied in September-October period, earthing up of soil at the base of suckers in each row is done after weeding. This operation is most important for pineapple for better anchorage in addition to better rooting. Nearly 35 per cent of the plant population yields in the second year itself. Remaining 65 per cent yield can be realized during the subsequent years. In the fourth year, replanting of pineapple should be done in the freshly opened trenches at the adjacent areas between the two existing trenches. Normally in the fourth year, replanting is done and the same can be maintained for another 3 years. Beyond 7 years it is not advisable to take up pine apple cultivation as intercrop in cashew garden as by then, the shade cast over the pineapple is too much, resulting low recovery of yield which is not economical. Added to this, it will be very difficult to pick cashew nuts falling in pineapple trenches as cashew attains peak yielding stage during that period. Advantages of growing pineapple in trenches dugout across the slope between two rows of cashew are that there will be better soil and water conservation which will have indirect benefit to main crop (cashew). Studies conducted at ICAR-DCR, Puttur has shown that growing pineapple as intercrop across the slope between two rows of cashew in west coast region has indirect beneficial effect on main crop (cashew) resulting in 1.5 times increase in yield as compared to plot where cashew alone is grown. This was mainly due to increased soil moisture conservation and due to control of weed as interspace is used for pineapple cultivation.

Pepper as Inter/Mixed Crop

Trailing pepper on the stem and branches of grown up cashew trees (more than 6 years) is also adopted. Separate application of manure is necessary for pepper vine. With the onset of monsoon, pit size of 45 cm x 45 cm x 45 cm should be opened about 45 cm away from the cashew tree at the base and pits should be filled with top soil, 200 g rock phosphate, 200 g lime and 0.5 kg neem cake. Rooted pepper cuttings should be planted during June with the onset of monsoon (South-West Monsoon season). Once the vine establishes and starts growing, the vine should be tied to the stem of the tree at the base with jute thread. This practice should continue till vines grow to almost 1.5 m height all around the stem surface of the cashew tree. Cashew stem girth will be around 30 cm when it is 6 years old and because of the rough stem surface of the cashew tree, pepper vine easily clings to the plant and spreads to all the remaining thick branches. Pepper virus when exposed to filtered sunlight through cashew canopy, flower profusely during rainy season. From second year onwards, regular application of recommended doses of fertilizers in split doses to pepper (100 g N, 40 g P_2O_5 and 140 g K_2O g/tree/year) at 45 to 60 cm away from the base results in better growth and yield. One split dose (1/3rd of full dose) may be applied in May soon after the receipt of pre-monsoon showers and other split dose (2/3rd of full dose) during August when moisture in soil still remains towards the end of rainy season. Pepper as intercrop trained on to stem of grown up cashew tree can be grown in west coast region without irrigation. But with irrigation, both main and intercrop will be benefited. High price for pepper makes farmers in west coast region, particularly Kerala and coastal districts of Karnataka to go for pepper cultivation on any live or dead standards. Growing pepper on coconut and arecanut palms is a common practice in these regions. Even on cashew, pepper can be grown and good profit can be obtained. Nearly 5 to 6 years after planting with fairly good management a farmer can get 0.5 kg dry pepper per vine trained on to stem of cashew tree. Pepper vine can remain normally for 18 years and it can fit in very well as an intercrop in cashew gardens. Field experiment carried out at Ullal established that black pepper can be raised as mixed crop in well spaced cashew gardens under rainfed condtion (Kumar *et al.*, 1986).

Ginger as Intercrop

Ginger can be grown as an intercrop in the initial 3 to 4 years of cashew plantation. Particularly this is more suitable in the interior areas of west coast on hillocks with forest surroundings. Whenever forest surroundings are noticed, growing ginger in the initial years before taking up regular plantation is a common practice in west coast region. With the onset of pre-monsoon season (April-May), raised beds of 2.5 m length, 1.5 m width and 0.25 m height should be prepared across the slope between two rows of cashew. With this operation, all the jungle growth including small bushes will be removed and soil is loosened. Loose and made up soil present in the raised bed, acts as good soil and moisture conservation structure which also facilitates better penetration of cashew roots. Even the weeds will be controlled while preparing beds for ginger. Approximately 10 quintal disease free ginger rhizomes are required as seed material for planting one ha. Rhizomes of ginger should be planted in raised bed at 10 cm distance from one planting spot to another spot.

Farm yard manure or compost at the rate of 15 t/ha has to be applied initially to the bed after planting ginger rhizomes. Soon after planting, thick mulch has to be applied at the rate of 10 t/ha. Application of heavy mulch is necessary to conserve moisture during pre-monsoon and later to avoid soil erosion during rainy season. Heavy mulch also is necessary to conserve moisture soon after the cessation of rain so that this will continue for another 3 months without necessitating irrigation. Ginger crop is of 7 to 8 month duration and harvesting can be done in the month of January. Total cost of cultivation includes items like clearing of jungle, raised seed beds, cost of ginger seed rhizomes, application of FYM and weeding and mulching. One can expect a yield of around 60 quintals ginger per hectare. In the subsequent 2 to 3 years, part of rhizomes removed from the previous crop can be utilized as seed material each year and to that extent the cost of cultivation is reduced. Ginger is prone for deadly disease Rhizome rot, which needs proper care.

Fodder and Legume Forages

Experiments conducted at Agricultural Research Station, Ullal revealed that among the fodder crops as intercrops, NB-21 grass gave the highest green fodder yield (41.9 t/ha) over Guinea grass (27.6 t/ha) and para grass (18.0 t/ha). Among legume fodders, *Stylosanthus hamata* ranked first in green fodder yield (12.55 t/ha) followed by *Mimosa invisa* (10.25 t/ha) and *Lupinus* spp. (7.28 t/ha). The unutilized interspace of cashew in coastal Karnataka was successfully exploited for 1st three years of plantations by growing fodder and grasses like NB-21, Guinea grass and para grass with NB-21 yielding highest green fodder of 21 t per ha (Chalapathi, 1989).

Vegetables as Intercrops

Growing of bitter gourd as an intercrop during the *Kharif* season has been found most profitable with expected yield of 100 kg/ha from bitter guard.

b. Kerala

Tuber crops are found to be the most suitable intercrops with cashew. In terms of tuber yield, tapioca recorded the maximum yield (18.706 t/ha) followed by *Amorphophallus* (18.083 t/ha). Highest total returns and net returns were recorded by *Amorphophallus*, followed by tapioca. The highest cost of cultivation was in *Amorphophallus*, lowest cost of cultivation was recorded by tapioca but C: B ratio (2.38) was highest with tapioca followed by coleus (1.89). The lowest net return (Rs. 70080) and C: B ratio (1.67) was recorded by sweet potato. The experiment revealed that tapioca, coleus, *Amorphophallus*, colocasia and sweet potato could be cultivated profitably as intercrop in young cashew plantations. *Amorphophallus* ranked first in net income while tapioca recorded the highest C: B ratio (Table 16.3). Banana is another crop that can be successfully intercropped with cashew during the first four years. Locally adapted banana varieties may be planted in the inter-space at a distance of 2.5 to 3.0 m.

Table 16.3: Economics of Intercropping of Tuber Crops in Cashew at Madakkathara (Kerala)

Name of Intercrop	*Tuber Mean Yield*		*Cost of Cultivation (Rs/ha)*	*Total Return from Intercrop (Rs./ha)*	*Net Profit (Rs./ha)*	*C : B Ratio*
	(kg/plot of 22.68 m²)	*t/ha*				
Coleus	31.00	12.88	102120	193305	91185	1.89
Colocasia	32.00	13.30	115250	199530	84280	1.73
Tapioca	45.00	18.70	94480	224472	129992	2.38
Sweet potato	27.89	11.59	103830	173910	70080	1.67
Amorphophallus	43.50	18.08	185855	325494	139639	1.75

Anon, 2011.

Studies on cashew based intercropping suggested that the association between cashew and pepper look like a harmonious matrimony between the two dollar earners by achieving the objective of higher resource use efficiency, yielding greater agronomic and economic advantages to the farmer (Abdul Salam, 1999). Crop association involving cashew and pine apple in farming was proved as an excellent combination to the agro-ecological condition than prevailed in the area.

The association resulted in advantages in terms of sharing of nutrients, smothering of weeds, conservation of soil and water and enhanced growth of base crop *i.e.,* cashew. Due to intercropping, cashew also received regular attention, better care and management. As a result, 3 year old cashew tree showed growth equivalent to that of 3 year old tree. During the second year itself, 300 kg of raw cashew nuts per acre was obtained from cashew (Abdul Salam, 2001).

c. Gujarat

The highest yield of intercrops (3577 kg/ha) was recorded when cashew was intercropped with okra which was followed by cashew + cowpea (1849 kg/ha) and cashew + pigeon pea (1826 kg/ha). The highest net profit of Rs. 46,145/ha was found in cashew + okra and it was followed by cashew + pigeon pea. However, highest B: C ratio (3.17) was under cashew + pigeon pea which was followed by cashew + okra (1.82) (Table 16.4).

Table 16.4: Yield and Economics of Intercropping in Cashew at Paria

Treatment	Yield		Total Cost of Inter-crops (Rs./ha)	Total Returns from Inter-crops (Rs./ha)	Net Profit (Rs./ha)	B : C Ratio
	kg/plot	kg/ha				
Cashew + Pigeon pea (Vaishali)	3.99	1826	13,125	54,780	41,655	3.17
Cashew + Okra (GO2)	7.82	3577	25,395	71,540	46,145	1.82
Cashew + Indian bean (GW-2)	2.88	1316	9,160	19,740	10,580	1.16
Cashew + Indian bean (NPS-9)	3.33	1522	9,160	22,830	13,670	1.49
Cashew + Cowpea (GC-4}	4.04	1849	8,160	18,490	10,330	1.27
Cashew alone	0.00	0.00	0.00	0.00	0.00	–

Anon, 2012.

d. Konkan Region

For Konkan region, cucurbitaceous vegetables and tuber crops were found to be suitable as intercrops. The net returns from intercropping of cucumber was the highest (Rs. 33,659/ha) which resulted in C: B ratio of 1: 3.17. The other crops *viz.,* ridge gourd, snake gourd, bitter gourd, bottle gourd and cucumber also led to a C: B ratio of more than 1:2.0 (Table 16.5). Out of five different tuber crops, elephant foot yam recorded significantly higher yield (55.25 kg/plot and 7.29 t/ha) which was followed by lesser yam (31.50 kg/plot and 4.15 t/ha) and greater yam (20.75 kg/plot and 1.15 t/ha). The main crop, cashew recorded an average yield of 11.9 kg/tree and 1.86 t/ha (Table 16.6). In the Thane and Nasik districts, the cashew has been promoted successfully in the underutilized lands where possibilities of growing other cash crops are limited. The plantation of cashew has also prompted the farmers to grow floriculture crop like *'Mogra'*, vegetables and thus utilize the land intensively and earn additional income (Raj Ganguly, 2011).

Table 16.5: Performance of Intercrops in Cashew Plantation at Vengurle

Intercrop	*Yield kg/ha*	*Total Returns Rs./ha*	*Net Profit Rs./ha*	*C : B Ratio*
Snake gourd	47.88	33514.50	19939.90	1:2.46
Ridge gourd	48.64	34047.90	21793.00	1:2.77
Bitter gourd	32.31	25848.50	13666.10	1:2.12
Bottle gourd	56.71	28355.00	16245.50	1:2.34
Cucumber	70.23	49160.50	33658.70	1:3.17
SEm(±)	0.82			
C.D. at 5 per cent	2.53			

Anon, 2006.

Table 16.6: Economics of Intercropping Tuber Crops in Cashew at Vengurle

Intercrop	*Yield kg/plot*	*Yield (t/ha)*	*Local Market Rate (Rs./kg)*	*Income (Rs./ha)*
Lesser Yam (Kangar)	31.50	4.15	40/-	1,66,320/-
Greater Yam (Ghorkand)	20.75	1.15	40/-	46,200/-
Aerial Yam (Karanda)	8.75	2.73	40/-	1,09,560/-
Elephant foot Yam(Suran)	55.25	7.29	25/-	1,82,325/-
Tapioca	9.25	1.22	4/-	4,884/-
SEm±	2.372	0.311		
CD at 5 per cent	7.310	0.959		
Yield of Cashew (V1)	11.900 kg/tree	1.86	85/-	1,57,794/-

Anon, 2012.

East Coast

In east coast region possibility of growing different vegetables and oil seed crops as intercrops is more, as the inter spaces between rows of cashew can be easily ploughed by farm machines or bullock drawn inter cultivation implements due to the availability of level lands and, it is easy to take up such intercrops during Kharif season. In east coast region, growing pepper under unirrigated condition may not be feasible due to low rainfall ranging from 600 to 1500 mm annually. However, this can be grown under irrigated condition in east coast regions also using cashew plant as standard for pepper vine.

a. Andhra Pradesh

At Bapatla, groundnut, green gram and black gram can be grown as inter crops in rabi season. Groundnut recorded a maximum yield of 1400 kg/ha and had higher cost benefit ratio (1:1.96). Horsegram as an intercrop resulted in net profit of Rs.19900/- with a cost benefit ratio of 1:1.80 (Table 16.7). Whereas, inter cropping with flower crop like marigold with cashew recorded the highest net profit of Rs.65,967/- per hectare (Table 16.8). During the year 2011-12, cluster bean, marigold, amaranthus and mesta were grown as intercrops. Cluster bean recorded maximum yield of 9097 kg/ha and gave higher cost benefit ratio 3.7 and led to maximum net returns of Rs. 94,002/-. In the West Godavari district, cashew is grown in combination with *Casuarina* and coconut.

Table 16.7: Yield and Net Returns of Main Crop (Cashew) in Intercrop Trial at Bapatla

Name of Treatment	Mean Yield of Main Crop		Total Cost of Input (Rs./ha)	Total Returns (Rs./ha)	Net Profits (Rs./ha)	C : B Ratio
	Per Tree (kg)	Per ha (q)				
Cashew+Groundnut	5.30	8.40	18000	53440	35440	1.96
Cashew+Sesamun	4.92	7.16	13500	30155	16655	1.23
Cashew+Horsegram	4.25	6.80	11000	30900	19900	1.80
Cashew (Sole crop)	3.40	5.44	7500	16320	8820	1.17

* The Rate of raw cashew nuts taken at Rs.3000/- per quintal) (*Anon*, 2008).

Table 16.8: Yield and Net Returns of Intercrops in Cashew Inter Crop Trial at Bapatla

Treatment Details	Yield of Intercrop		Yield of Cashew		Cost of Cultivation (Rs./ha)	Returns (Rs./ha)	
	kg/plot	q/ha	kg/tree	q/ha	Cashew+ Intercrop	Total	Net
T1 - Cashew + Cluster bean	46.75	73.05	2.75	4.29	71,934	94,500	22,566
T2 - Cashew+ Marigold	38.50	60.16	2.63	4.10	74,853	1,40,820	65,967
T3 - Cashew alone	—	—	2.72	4.24	—	21,200	3,172

Sale price (Rs./kg) Raw cashew nuts = 50.00, Cluster bean = 10.00, Marigold= 20.00 (*Anon*, 2010).

b. Tamil Nadu

Among field crops, pulses and oil seeds were attempted as intercrops for cashew. Groundnut performed better as an intercrop in cashew in terms of total returns (Rs. 16,188/- per hectare) whereas, the highest cost benefit ratio for black gram (1:2.1) revealed black gram as a suitable intercrop for cashew (Table 16.9). Black gram led to the highest C: B ratio of 1:2.1 followed by groundnut (1:1.19). However, groundnut resulted in maximum net profit of Rs.16, 187/ha (Table 10). Further, medicinal plants are also found to be suitable intercrops for cashew. Intercropping of *Aloe vera* with cashew recorded higher BCR value of 4.1. *Ocimum sanctum* intercropped in cashew showed sustained performance for four years and *Aloe vera* + cashew for three consecutive years. Hence, *Ocimum* and *Aloe vera* could be promoted as profitable intercrops in cashew (Table 16.11).

Table 16.9: Performance of different Intercrops in Cashew Orchard at Vridhachalam

Treatment	*Yield from Intercrops*		*Total Cost of Intercrops (Rs./ha)*	*Total Returns from Intercrops/ ha (Rs.)*	*Net Profit (Rs./ha)*	*C : B ratio*
	Plot Yield (16 M²) (kg)	*Hectare yield (kg)*				
Cashew + Black gram	1.60	1000.50	5850	18000.00	12150	1 : 2.1
Cashew + Gingerly	0.45	281.25	4200	5343.75	1144	1 : 0.27
Cashew + Green gram	0.80	500.00	5450	8000.00	2550	1 : 0.46
Cashew + Groundnut	2.50	1562.50	13500	29687.50	16188	1 : 1.19

Anon, 2006.

Table 16.10: Performance of different Intercrops in Cashew Orchard at Vridhachalam

Treatments	*Yield (Intercrops) (kg/ha)*	*Net Profit (Rs./ha)*	*C:B ratio*
Cashew + Black gram	1000.50	12150.00	1:2.1
Cashew + Gingelly	281.25	1143.75	1:0.27
Cashew + Green gram	500.00	2550.00	1:0.46
Cashew + Groundnut	1562.50	16187.50	1:1.19

Anon, 2007.

Table 16.11: Economics of Intercropping Medical and Aromatic Crops in Cashew Orchard at Vridhachalam

Treatment	*Yield from Intercrops*		*Total Cost of Cultivation Intercrops + Cashew (Rs./ha)*	*Total Returns Intercrops + Cashew (Rs./ha)*	*Net Profit (Rs/ha)*	*BCR*
	(kg/plot 15 m²)	*kg/ha*				
Cashew + *Ocimum sanctum* (leaves and stem)	6.5	650	18700	63900	45200	3.4
Cashew + *Catharanthus roseus* (leaves and stem)	4.0	400	18040	59000	40900	3.2
Cashew+*Phyllanthus niruri* (leaves and stem)	2.5	250	14750	38400	23650	2.6
Cashew+*Aloe vera* (leaves)	14.25	1425	20000	82500	62500	4.1
Cashew alone	4.5	800	10000	54000	44000	4.4
SEm±CD(0.05)	0.012 0.056**					

Anon, 2012.

In some areas of Tamil Nadu, farmers intercrop pomegranate, a fruit plant, with cashew during early years of cashew. The spacing between pomegranate in the inter space ranges from 2 to 3 m. Under irrigated conditions, pomegranate start yielding from the third year of planting. A net return of Rs. 37,150/ha has been reported for the graft varieties and investments were found financially viable. With blackgram as an intercrop, an additional average income of Rs.6550/- per ha was also reported. When cashew was grown under agro forestry system with tropical nitrogen fixing trees, height of cashew was enhanced by 25 per cent when interplanted with *Casuarina,* while girth was maximum (10.8 cm) when interplanted with *Leucaena,* although the differences between 2 nitrogen fixing trees were non-signinficant (Swaminathan, 1999).

c. Odisha

Vegetable crops such as brinjal, cowpea, chilli, bhindi, pumpkin, colocasia were evaluated for their economic suitability as intercrops with different doses of fertilizer. The yield and total net returns per hectare from inter-crops as well as main crop after 4 years revealed that maximum return was received from colocasia (Rs. 66,216/-) followed by bhindi (Rs. 58,155/-), brinjal (Rs. 58,035/-), cowpea (Rs. 57,635/-), chilli (Rs. 56,815/-), pumpkin (Rs. 52,493/-) and control (Rs. 40,075/-) (Table 16.12).

Table 16.12: Yield and Returns from Main Crop and Intercrops at Bhubaneswar

Treatment	*Yield q/ha (20 per cent area)*	*Expen-diture (Rs/ha)*	*Total Return (Rs/ha)*	*Net Return (Rs/ha)*	*Net Return (Rs) (Main crop)*	*Net Return (Rs) (Main Crop + Intercrop)*
Cashew + brinjal	36.50	6000	7300	1300	19075	20375
Cashew + chilli	10.60	1200	3710	2510	16100	18610
Cashew + cowpea	16.30	1900	3260	1360	21105	22465
Cashew + bhindi	15.10	2000	3020	1020	20475	21495
Cashew + pumpkin	27.30	1600	3278	1676	16205	17881
Cashew + colocasia	42.70	2100	5978	3878	17430	21308
Cashew alone	Nil	Nil	Nil	Nil	18725	18725

Anon, 2009.

d. West Bengal

In the eastern humid region, intercrops such as bottle gourd, *Amaranthus*, pumpkin, cucumber and bitter gourd were evaluated at Jhargram. Maximum yield was obtained in cucumber (10.607 q/ha) followed by bottle gourd (9.615 q/ha) and amaranths (5.160 q/ha). The yield of cashew was only 6.00 q/ha without an intercrop while it was 9.87 q/ha with amaranthus. The benefit cost ratio (2.44) confirmed that cashew + bottle gourd was the most profitable practice followed by cashew + amaranths (benefit cost ratio of 1: 1.93) and cashew + cucumber (benefit cost ratio of 1: 1.81). Intercrops such as coriander, dill and fenugreek were grown in open canopy area under 5m x 4m spacing. Maximum yield was obtained from fenugreek (14.77 q/ha) followed by coriander (6.74 q/ha). Significant difference was noticed in the yield of cashew between cashew grown alone and cashew grown with intercrops. The yield of cashew was 1.28 q/ha without an intercrop, while it was more than 2 q/ha with intercrops. The cost benefit ratio (0.41) confirmed that cashew + fenugreek was the most profitable practice followed by cashew + dill (benefit cost ratio: 1: 0.14). When eucalyptus (*E. teretecornia*) was grown as intercrop in cashew, cashew yield was reduced from 200 g per tree for cashew grown alone to 50 g per tree for cashew grown with eucalyptus. The percentage of TMB infestation on cashew was greater (78.8 per cent) when *Eucalyptus* was intercropped with cashew than sole crop of cashew (20 per cent) (Ghose, 1993).

Madhya Pradesh

Under Bastar plateau zone of Madhya Pradesh, intercrops like cowpeas, bush type French bean, cluster beans, ricebeans, urd beans, moong beans, soya beans and ground nuts grown in 3 year old cashew orchard, recorded higher net returns than sole crop of cashew (Gupta, 1999).

References

Abdul Salam, M. (1999). Cashew and pepper live in harmony. *The Cashew*, 13(3): 54-55.

Abdul Salam, M. (2001). Cashew and pineapple live in harmony. *The Cashew*, 15(4): 12-13.

Abeysinghe, D. C. (2009). Effect of intercropping of young cashew (*Anacardium occidentale* L.) on land productivity. *Proceedings of the Cashew Research Workshop.* Wayamba University, 20 November 2009 Sri Lanka, p. 23.

Adeyemi, A. A. (1986). Cashew/arable crops intercropping experiment. CRIN Annual Report. pp. 60.

Aliyu, O. M. and Hammed, L. A. (2008). Nigerian cashew economy: A review of nut production sector. *Proceedings of the International Academy of African Buisiness and Development (IAABD) Conference.* May 20–24, 2008. University of Florida, Gainesville, USA.

Anonymous, (2006). Intercropping in Cashew. Annual Report 2004-05. All India Coordinated Research Project on Cashew. Directorate of Cashew Research, Puttur. pp. 66-69.

Anonymous, (2007). Intercropping in Cashew. Annual Report 2005-06. All India Coordinated Research Project on Cashew. Directorate of Cashew Research, Puttur. pp. 65-68.

Anonymous, (2008). Intercropping in Cashew. Annual Report 2006-07. All India Coordinated Research Project on Cashew. Directorate of Cashew Research, Puttur. pp. 66-70.

Anonymous. (2009). Intercropping in Cashew. Annual Report 2007-08. All India Coordinated Research Project on Cashew. Directorate of Cashew Research, Puttur. pp. 59-62.

Anonymous. (2010). Intercropping in Cashew. Annual Report 2009-10. All India Coordinated Research Project on Cashew. Directorate of Cashew Research, Puttur. pp. 61-63.

Anonymous. (2011). Intercropping in Cashew. Annual Report 2010-11. All India Coordinated Research Project on Cashew. Directorate of Cashew Research, Puttur. pp. 72-74.

Anonymous. (2012). Intercropping in Cashew. Annual Report 2011-12. All India Coordinated Research Project on Cashew. Directorate of Cashew Research, Puttur. pp. 73-77.

Chalapathi, K. (1989). Eucalyptus of fodder grasses as intercrop in cashew in Dakshina Kannada. *The Cashew,* 3(4): 17-18.

Ghose, S. N. (1993). Effect of eucalyptus plants as intercrop in the cashew plantation-A case study in West Bengal. *The Cashew,* 7(2): 17-19.

Gupta, C, R. (1999). Intercropping in cashew orchard under rainfed condition-a model for Bastar Plateau zone of Madhya Pradesh. *The Cashew,* 13(2): 18-22.

Kumar, D. P., Khan, M. M. and Hiremath, I. G. (1986). Black pepper as mixed crop in cashew. *Cashew Causerie,* 8(4): 17.

Liu Kangde, Liang Shibang and Deng Suisheng. (1998). Integrated Production Practices of Cashew in China. *Integrated Production Practices of Cashew in Asia* (edited by Minas K. Papademetriou and Edward M. Herath). Food and Agriculture Organization of The United Nations, Regional Office for Asia and the Pacific, Bangkok, Thailand.

Mandal, R.C. (1992). *Cashew Production and Processing Technology.* p.195. Agro Botanical Publishers (India), Bikaner.

Maung Maung Lay. (1998). Integrated Production Practices of Cashew in Myanmar. *Integrated Production Practices of Cashew in Asia* (edited by Minas K. Papademetriou and Edward M. Herath). Food and Agriculture Organization of the United Nations, Regional Office for Asia and the Pacific, Bangkok, Thailand.

Ohler, J. G. (1979). *Cashew.* pp. 260. Koninklijk Instituut voor de Tropen, Amsterdam.

Opoku-Ameyaw, K., Oppong, F. K., Amoah, F. M., Osei-Akoto, S. and Swatson, E. (2011). Growth and early yield of cashew intercropped with food crops in northern Ghana. *Journal of Tropical Agriculture,* 49 (1-2): 53-57.

Opoku-Ameyaw, K. and Appiah, M. R. (2000). Improving the growth of cashew (*Anacardium occidentale*) seedlings inter planted into mature sheanut stands in northern Ghana. *Ghana Journal of Agricultural Science,* 33: 159-164.

Opoku-Ameyaw, K., Oppong, F. K., Ofori-Frimpong, K., Amoah, F. M. and Osei-Bonsu, K. (2003). Intercropping robusta coffee with some edible crops in Ghana: agronomic performance and economic returns. *Ghana Journal of Agricultural Science,* 36: 13–21.

Raj Ganguly. (2011). Analysis of pro-poor agriculture value chains in Maharashtra: preliminary findings. *Proceedings of workshop on policy options and investment priorities for accelerating agricultural productivity and development in India.* November 10-11, 2011. India International Centre, New Delhi.

Rodrigo, V. H. L., Stirling, C. M., Teklehaimanot, Z. and Nugawela, A. (2001). Intercropping with banana to improve fractional interception and radiation use efficiency of immature rubber plantations. *Field Crops Research,* 69 (3): 237–249.

Saroj, P. L. and Arora, Y. K. (1994). Horticulture based agroforestry systems of Doon Valley. *Indian Journal of Soil Conservation,* 3: 60-65.

Surendra, G. B. B. (1998). Integrated Production Practices of Cashew in Myanmar. *Integrated Production Practices of Cashew in Asia* (edited by Minas K. Papademetriou and Edward M. Herath). Food and Agriculture Organization of The United Nations, Regional Office for Asia and the Pacific, Bangkok, Thailand.

Swaminathan, C. (1999). Early growth performance of mango and cashew coplanted with four nitrogen fixers in a tropical alfisol. *Journal of Sustainable Forestry,* 8(2): 15-26.

Warui, C. M. (1985). Development in cashew cultivation and research in Kenya. *Acta Horticulturae,* 108: 233-237 Available at http: / /www.actahort.org/books/108.

Cashew: Improvement, Production and Processing (2017) *Pages* **295–322**
Editor: **Prof. P.L. Saroj**
Published by: **ASTRAL INTERNATIONAL PVT. LTD., NEW DELHI**

Organic Cultivation of Cashew

D. Kalaivanan and T.R. Rupa

1. Introduction

The 'Green Revolution' technologies involving greater use of synthetic agrochemicals such as fertilizers and pesticides with adoption of nutrient responsive, high-yielding varieties of crops have boosted the productivity in most cases. However, this increase in production has slowed down and in some cases there are indications of decline in production and productivity on long run. Moreover, the success of industrial agriculture and the green revolution in recent decades has often masked significant externalities, affecting natural resources and human health as well as agriculture itself. Environmental and health problems associated with agriculture have been increasingly well documented, but it is only recently that the scale of the costs has attracted the attention of planners and scientists. Immense commercialization of agriculture has had a very negative effect on the environment. The use of pesticides has led to enormous levels of chemical buildup in the environment, in soil, water, air, in animals and even human bodies. Fertilizers have a short-term effect on productivity but a long term negative effect on the environment where it remains for years after leaching and run off, contaminating ground water and water bodies.

Use of chemical fertilizers is the easiest way of increasing crop yield. However, the increasing cost of fertilizers, removal of subsidies and environmental problems; the farmers are inclined to prefer organic manures over inorganic manures. The farmers of the developed and developing countries are realizing the harmful effects caused by excessive use of chemical fertilizers and pesticides on the food crops produced. Though agricultural production has continued to increase, but productivity has started to decline. Modern chemical based agriculture practices have led to several new challenges, *viz.* decline in productivity, degradation of soil and water resources, diminishing biodiversity and increase in environmental pollution. The increasing awareness of adverse effect of excess use of fertilizers and pesticides in modern system of agriculture has led to the considerable interest in the use of organics and microbes as source of plant nutrients and also botanicals or biocontrol agents for the management of pests and diseases.

The growing of crops by following organic practices in conformity with certain standards is a process beginning from land preparation to finally reaching the produce in the hands of consumers. In fact, organic and bio-dynamic farming signify the production of a kind of chemical residue-free organic labelled products to sensitive consumers. In other words, organic is a 'process claim' rather than a 'product claim'. This is because the agricultural products such as cereals, pulses, grams, fruits, spices or cotton cannot be distinguished into organic type or inorganic type. India is bestowed with lot of potential to produce all varieties of organic products due to its agro-climatic regions. In several parts of the country, the inherited tradition of organic farming is an added advantage. Currently, India ranks 33rd in terms of total land under organic cultivation and 88th position for agriculture land under organic crops to total farming area in the World. The cultivated land under certification is around 2.8 million ha in India. This includes one million ha under cultivation and the rest is under forest area (wild collection). India exported 86 items during 2007-08 with the total volume of 37,533 MT. The export realization was around 100.4 million US $ registering a 30 per cent growth over the previous year (APEDA, 2010).

2. Adverse Effects of Agrochemicals on Soil, Water, Food and Environment

(i) Fertilizer Pollution

Nitrate Pollution

Application of nitrogen fertilizers such as urea and ammonium sulphate to soils produces acid by two processes. Firstly, the natural process of oxidation of ammonia ions to nitrate ions release acid. Part of acid produced is neutralized by alkaline ions released by plants during the subsequent uptake of the nitrate ions. Secondly, since nitrate ions are not strongly absorbed by the soil they are liable to leach or move down through the soil. The negatively charged nitrate ions carry positively charged basic cations such as Ca, K, Mg and Na in order to maintain the electric charge on the soil particles. A high nitrate concentration indicates likely presence of harmful bacteria as well. In condition, to high enrichment, nitrate may produce as state known as methemoglobinemia (blue babies) which generally affects the infants under six months of age. Repeated heavy dose of nitrate on ingestion may likely to cause carcinogenic diseases. Consumption of high dose of nitrate may develop symptoms of dizziness, abdominal cramps, vomiting, weakness, convulsion, mental impairment and even nitrosamine which cause stomach cancer. Apart from this, over use of nitrogen fertilizers leads to swindling of earthworms from the particular area, earthworms have always been considered a farmers friends and their absence mean loss to the soil fertility.

Accumulation of Heavy Metals

Contamination of soil by heavy metal through fertilizers such as cadmium from phosphatic fertilizers is also getting increasing attention of environmentalists (Kostial, 1986). Health hazards associated with heavy metals entering the food chain through soil is demanding attention. Fertilizers contain heavy metals as impurities. The application of rock phosphate or its produce to soil always implies the addition of significant amount of lead and cadmium into the soil. Analysis of several commercial fertilizers commonly used revealed that a combination of low analysis and straight fertilizers can add more lead and cadmium to soil than high analysis and mixed fertilizers (Arora *et al.*, 1975; Bijay Singh and Sekhon, 1977).

Eutrophication of Water

It is the process of enrichment of surface water bodies like lakes, reservoirs and streams with nutrients. Nutrient enrichment of water bodies results in intense proliferation and accumulation of algae and higher aquatic plants in excessive quantities which can result in detrimental changes in water quality and can significantly interfere with the use of water resources. The excessive accumulation of dissolve nutrients such as phosphorus, nitrogen, silicon and other element in water leads to an excess production of algal biomass. This requires a corresponding increased supply of oxygen for decomposition of organic materials when the algae and their remnants sink to bottom, reduced O_2 content and eventually anaerobic condition may prevail. This is symptomatic a severe eutrophication and may leads to a serious loss of marine life, blockage and aquatic passages and a major reduction in real estate value of the affected areas. The normal N: P ratio in water is around 20:1. Algal growth increase when this ratio drops to around 7:1 (Asmed, 1993). In the developed countries like USA due to heavy use of fertilizers, more than 50 per cent of the community and rural wells are contaminated with nitrate nitrogen (NO_3N). The contamination has exceeded the standard (10 mg NO_3/lit.) of drinking water by 2 per cent (Dahama, 1997).

Stratospheric Changes

Many research studies involved in the atmosphere may increase the rate of reaction in the stratosphere that to be the distraction of ozone layer. The stratosphere ozone layer shields the biosphere from the harmful ultraviolet radiation and also influences the earth's temperature (Dahama, 1997). Nitrogen loss to the atmosphere through denitrification may contribute to "green house gases" in the atmosphere thereby exacerbating the problems of the breaking down of ozone layer. Nitrogen losses can be particularly high from intensively cultivated and fertilized land whether the fertlilzer is from organic or inorganic.

(ii) Pesticide Pollution

By the use of different types of poisonous substances as pesticide, it contributes towards imbalances in the ecosystem and polluting the environment. Pesticides are now extremely used to control various pests which are harmful to the crops raised by man for food, feed and fibre production. However, most of the chlorinated pesticides are non-biodegradable and leave residue which are detrimental to human beings, animals and the environment. Even small quantities of the residues ingested daily along with food can build up to high levels in body fat (Dhaliwal and Singh, 1993). It is, therefore, a constant fear that society is being slowly poisoned by the intake of food contaminated with pesticide residues. During the 1900's, organochlorine insecticides such as chlorine, dieldrin, DDT and heptachlor were widely used in agriculture to control insect/pests in different countries. Since 1960's, the organochlorine use was progressively restricted and finally banned now. Nevertheless, their residues still found in soil and continue to cause problems of food and food contamination. The extensive use of pesticides possess a more direct problem on farm workers. Epistemologies data shows that workers who handle pesticides more than 20 days a year have an increased risk of developing certain type of cancer (Dahama, 1997).

(iii) Herbicide Pollution

The use of chemicals for controlling weeds started with the introduction of 2,4-D in 1940's. The usage of herbicide is higher than any agro-chemicals. Farming has now realized the importance of herbicide usage for harvesting higher crop yields. Herbicide plays an important role for the disturbance of soil ecosystem when soil micro-flora and fauna lies in the breaking down of organic matter, incorporating it into the soil and releasing nutrients for plant growth. The herbicide can have direct effect upon decomposing micro-organisms, rhizosphere microorganisms, root pathogens and disease antagonists such as parasites and predators as well as organisms pathogenic to invertebrates. The increased use of herbicides in recent years causes more concern about their effect on farmer's health. While herbicide technology has made remarkable progress in terms of developing sage herbicides that are less toxic to human beings. The herbicides applied over many years not only affect many species of plants and animals in and around farmland but also cause pollution of underground as well as surface water. The new ecotypes of weed which are resistant to herbicides have been developed. Some species even have multiple resistant to all possible options.

3. Concerns of Intensive Agriculture

- ☆ Chemical fertilizers and herbicides are easily washed from the soil and pollute rivers, lakes and water courses.
- ☆ The prolonged use of chemical fertilizers results in soils with a low organic matter content which is easily eroded by wind and rain.
- ☆ Dependency on fertilizers. Greater amounts are needed every year to produce the same yields of crops.
- ☆ Artificial pesticides can stay in the soil for a long time and enter the food chain where they build up in the bodies of animals and humans, causing health problems.

- ☆ Artificial chemicals destroy soil micro-organisms resulting in poor soil structure and aeration and decreasing nutrient availability.
- ☆ Pests and diseases become more difficult to control as they become resistant to artificial pesticides. The numbers of natural enemies decrease because of pesticide use and habitat loss.

4. Concept, Definition and Objectives of Organic Farming

Organic farming does not mean going 'back' to traditional methods. Many of the farming methods used in the past are still useful today. Organic farming takes the best of these and combines them with modern scientific knowledge. Organic farmers do not leave their farms to be taken over by nature; they use all the knowledge, techniques and materials available to work with nature. In this way the farmer creates a healthy balance between nature and farming, where crops and animals can grow and thrive. To be a successful organic farmer, the farmer must not see every insect as a pest, every plant out of place as a weed and the solution to every problem in an artificial chemical spray. The aim is not to eradicate all pests and weeds, but to keep them down to an acceptable level and make the most of the benefits that they may provide. For example, the use of green manures and careful cultivation, together provide better control of weeds than if the techniques were used on their own. The primary goal of organic agriculture is to optimize the health and productivity of interdependent communities of soil life, plants, animals and people (Scialabba and Hattam, 2002).

A large number of terms are used as an alternative to organic farming. These are: biological agriculture, ecological agriculture, bio-dynamic, organic-biological agriculture and natural agriculture. According to the National Organic Standards Board of the US Department of Agriculture (USDA) the word 'Organic' has the following official definition (Lieberhardt, 2003): "An ecological production management system that promotes and enhances biodiversity, biological cycles and soil biological activity. It is based on the minimal use of off-farm inputs and on management practices that restore, maintain and enhance ecological harmony."

Organic farming is defined as a 'production system' that avoids or largely excludes the use of synthetic fertilizers, pesticides, and growth regulators. Instead, it relies on crop rotations, crop residues, animal manure, green manure, off farm organic wastes, mechanical cultivation, mineral bearing rocks, and aspects of biological pest control to maintain soil productivity and tilth, to supply plant nutrients and to control insects, weeds and other pests (USDA, 1980). Organic farming works in harmony with nature rather than against it. This involves using techniques to achieve good crop yields without harming the natural environment or the people who live and work in it.

The objectives of organic farming are to 1) increase long-term soil fertility, 2) control pests and diseases without harming the environment, 3) ensure that water stays clean and safe, 4) use resources which the farmer already has, so the farmer needs less money to buy farm inputs, and 5) produce nutritious food, feed for animals and high quality crops to sell at a good price.

5. Principles of Organic Farming

The principles of organic farming are to 1) organize the production of crops and livestock and the management of farm resources so that they harmonize rather than conflict with natural system, 2) use and develop appropriate technologies based upon an understanding of biological systems, 3) achieve and maintain soil fertility for optimum production by relying primarily on renewable resources, 4) use diversification to pursue optimum production, 5) aim for optimum nutritional value of staple food, 6) use decentralized structures for processing, distributing and marketing of products, 7) strive for equitable relationship between those who work and live on the land and 8) maintain and preserve wildlife and their habitats.

6. Components of Organic Farming

Crop and Soil Management

Organic farming system encourages the use of rotations and measures to maintain soil fertility. Carefully managed soil with a high production of humus offers essential advantages with respect to water retention, ion exchange, soil erosion and animal life in the soil. Green manuring and inter-cropping of legumes is another important aspect for biological farming systems not only in regard to weed control but also in reducing the leaching of nutrients and in reducing soil erosion. A green cover throughout most of the year is one of the main goals of such farming methods. Depending on the green manure mixture or the legumes used for under sowing, there may be an increased soil organic matter and soil N_2 as well as in other nutrients.

Waste Recycling

Increase price of chemical fertilizers have enables organic wastes to regain an important role in the fertilizer practices on the farm. Good manure management means improved fertilizers value of manure and slurry and less nutrient losses. Composting of all organic wastes in general and of Farm Yard Manure (FYM) or feedlot manure in particular and its recycling are important in organic farming. Sewage and sludge use for crop production after proper treatment can form an important component of organic farming.

Non-chemical Weed Management

Weed management is one of the main concerns in organic agriculture. The elements to consider in preventing weed problems are crop rotation, green manuring, manure management and tillage. Mulching on a large scale by using manure spreaders may also be useful in weed control.

Energy Use

In the energy requirement for production measured per rupees of produce for organic farms is only one third of what it is for their conventional counterparts because N-fertilizer and pesticides are not used in this approach.

Food Quality

Food quality is one of the main issues, which concerns both scientists and consumers. Nitrates in water and farm produce, desirable components, pesticides residues, keeping quality and physiological imbalances are some of the important aspects of food quality.

Ecological Agriculture

The growing concern about environmental degradation, dwindling natural resources and urgency to meet the food needs of the increasing population are compelling farm scientists and policy makers to seriously examine alternative to chemical agriculture. As reported by Venkataramani (1995) case studies shows that when chemical farm incurred about Rs. 11,250 towards the cost of cultivation of rice. Organic farm spend Rs. 10,590 to produce 5625 kg paddy and 8 tonnes of straw/ha. The net returns from the ecological farming system at the current cost of Rs. 3.34/kg paddy is Rs. 8,197.50. In chemical farming, the net profit is Rs. 7500. If one gets a premium price for the organically grown rice, the economic returns from the ecological farming system will highly be encouraging. Integrated intensive farming system (IIFS) involves intensive use of farm resources. To be ecologically sustainable, such intensification should be based on techniques which are knowledge intensive and which replace to the extent possible, market purchased chemical inputs with farm grown biological inputs (Venkataramani, 1995).

7. Value of Organic Farming

The value of organic materials as fertilizers and soil conditioners is often misunderstood and has been the source of some controversy. The simplest and the most common means of estimating the value of organic amendments is by assigning the market value of the potentially available plant nutrients, they contain nitrogen-phosphorus-potash (NPK) and other micro-organisms.

8. Benefits of Soil Organic Matter

- It serves as a slow release of nitrogen, phosphorus and sulphur for plant nutrition and microbial growth.
- It possess considerable water holding capacity and thereby helps to maintain the water regime of the soil.
- It acts as a buffer against changes in pH of the soil.
- Its dark color contributes to absorption of energy from the sun and heating of the soil.
- It acts as "cement" for holding clay and silt particles together, thus contributing to the crumb structure of the soil and resistance against soil erosion.
- It binds micro-nutrient metal ions in the soil that otherwise might be leached out of the surface soil.
- Organic constituents in the humic substances may act as plant growth stimulants.

9. Organic Certification

In recent years, the idea of a healthy lifestyle and natural food has grown both in North America and Europe due to genuine health concerns, and corporations' use of these concepts as marketing tools. There is no certification process or other requirement for use of terms such as 'natural', 'healthy' or 'whole food', so as consumers become more aware of their food consumption, they look for substantiation of producers' claims. Organic certification is one of the few claims which can be verified independently by one of the many agencies that certify organic production. There are a wide range of organic certification bodies, and governments have legislation in place covering organic food production and trade. The necessity for strict checks and certification measures has increased in recent years, as the global trend toward organic food has created unprecedented demand. Concerns have been voiced too about the future of organic productions. If the market becomes a mass market, then sustainable methods and environmentally responsible policies must be adapted to support large volume production. For example, questions such as how to monitor organic and non-organic production from the same farm need to be addressed. It is essential that any venture into organic exports is fully and properly certified, in conformity with legislation of the destination country, as well as the country of origin. Producers and processors need be able to demonstrate 'one step back' traceability through the sourcing and production line back to the grower. Standards have been developed for certification of organic farms to meet the requirements for international trade. Twelve certification agencies have been identified by the Accreditation Agencies which are now operational under the Ministry of Commerce for certification of organic farms. Government of India has designated the following six commodity boards/agencies to accredit certifying agencies to undertake inspection and certification of organic products which are meant for domestic and international trade:

1. Agricultural and Processed Food Products Export Development Authority (APEDA)
2. Coffee Board
3. Spices Board
4. Tea Board
5. Coconut Development Board
6. Directorate of Cashewnut and Cocoa Development (DCCD)

Within the nuts business, organic production is becoming increasingly more important, however, edible nuts are a small part of the overall food business. The entry into organic production and marketing has been late and disjointed. Organic almonds from California are available and have become an important niche market. Italy is another leader in organic almond production. Giuseppe Calcagni of Besana, Italy, reports that Italian organic hazelnut production will rise to 20,000 MT in the next five years – 15 per cent of national production. Imports of organic nuts from outside the EU are small, with the exception of Californian almonds. European producers report that prices for organic nuts are about 20-30 per cent higher than conventional nuts. The requirements in terms of quality, logistics, packing, and reliability are the same for organic as for conventional product. Therefore, the organic producer will always have the safety net of the conventional market regardless of developments in the organic sector. Unlike fresh fruit, vegetables, meat, and coffee which have shown rapid growth rates in the organic sector, organic cashew has not yet been taken up by supermarkets across Europe as an organic line. However, there are signs of substantial change with four of the big five UK supermarkets now offering organic lines and the major German discounters offering organic nuts (although not cashew as yet).

Previously, major buyers were reluctant to promote organic cashew because of the following reasons:

- ✰ Supply is often seen as erratic and unreliable.
- ✰ Small quantities are often offered, making logistics and management difficult.
- ✰ Quality is perceived as unreliable.
- ✰ In some markets (*e.g.*, the Netherlands) the consumer is not willing to pay the premium price.

All of these problems have been overcome in the fresh fruit and vegetable markets where European organic consumption continues to grow at a fast rate (last quoted as 23 per cent per year). The organic cashew industry must reach a critical mass so that buyers can be confident that if they successfully promote a retail item, it will be available in increasing volumes as their requirement grows. Overall, the outlook for organic foods is good. Consumers are prepared to pay the premiums, large retailers are supporting the development, and the traditional 'health and whole food' sector is still playing an important role. The European market for organic cashew is dominated by Germany, the Netherlands, France and the UK. Between them, these four account for 75 per cent of sales of organic cashew, just as they do for conventional cashew nuts. The organic market remains a small but growing sector. There is no authoritative figure on the size of the market as custom authorities do not distinguish organic from conventional imports. However, a compilation of trade estimates put the market in the range of 2,000-2,500 MT per year. This suggests that organic cashew is in the range of 3-3.5 per cent of the total import. It has been difficult to obtain an estimate of the US market size. The most informed opinions estimate the market in the range of 3,000 MT per year. The world market is estimated to be approximately 5,250 MT per year or about 2 per cent of the world import trade.

10. Organic Cashew

India is the largest processor, consumer and exporter of cashew in the world but low productivity is a serious concern. For increasing production of cashew nuts in the country, it is not only necessary to increase the area under the crop but also to increase per unit yield with the application of manures and fertilizers besides other cultural operations. However with increasing pressure on land establishment of new cashew plantations would greatly be limited and an increase in productivity is the only answer to overcome the crisis in our cashew industry. The cashew is generally grown on least fertile soil. Fertilizer management in cashew is very much crucial but totally neglected crop as far as nutrition is concerned. Thus, considering average annual removal of major plant nutrients at the present level of production by six years old and above cashew trees and total productive area under cultivation in India, the total nutrient removal of soil is up to the tune of 185 thousand tonnes of N, 40.24 thousand tonnes of P_2O_5 and 120 thousand tones of K_2O annually. Actual estimated addition of the nutrients through fertilizer application is meager for the reason that only 20 per cent of the total area under cultivation receives fertilizers. If we take this into account 11.17 thousand tonnes of N, 2.99 thousand tonnes of P_2O_5 and 2.7 thousand tonnes of K_2O are applied through fertilizers. This much quantity of fertilizers applied is too low compared to any other horticultural crop.

Though cashew tree sheds 20-40 kg leaves and apple annually and 50 per cent of this is used for composting and application to other important crops like, arecanut, coconut and rice in west coast and east coast regions where cashew is generally grown in addition to the above crops. There is a local practice of collecting cashew leaf litter at the time of harvest of the crop which coincides with summer season and apply the same to cattle shed where it undergoes half decomposition due to presence of cow dung and cow urine and trampling of the biomass. Occasionally cashew leaf litter is also collected for fuel purpose. These practices lead to general depletion of nutrients in cashew orchard every year thereby reduction in productivity. If this biomass is properly utilized for application to cashew orchard considerable quantity of nutrients can be retained for replenishment of nutrients. Appreciable quantity of nutrients present in the crop residue often not recycled into the system in a systematic manner. The potential nutrients in recycling of biomass generated in cashew has been quantified as 33.8 kg N, 17.6 kg P_2O_5 and 20.6 kg K_2O per hectare per year and considerable quantity of other essential elements including micronutrients (Anonymous, 2003). These organic forms of nutrients improve soil health through their effect on soil physico-chemical properties and microbial actions.

Renewed and intensified efforts are in progress during the past 10-15 years to grow cashew organically ever since the depleting soil fertility attained a serious concern with the practice of high density planting coupled with heavy use of chemical fertilizers in some cases. Organic cashew cultivation is often considered amongst one of the sustainable agricultural practices; if used appropriately, it promises to offer rich dividends on a long-term basis. Opinions vary greatly about organic farming being perceived as a part of sustainable agriculture. Concerns about improving nitrogen use efficiency, reducing nitrate pollution, contamination due to byproducts of various chemical pesticides in use, and continued gradual loss of soil organic matter have always been the major core issues in organic farming (Ferguson 1994). But the organic cultivation of cashew has yet not received the priority it deserves; as a result, soil physical, chemical, and microbiological health have deteriorated irrecoverably. In addition to changes in land use pattern, unfavorable climatic conditions have further enhanced the rate of decomposition of soil organic matter and its further depletion (Velayutham *et al.*, 1999). A major constraint to intensive production of cashew in west and east coast of India is the low level of inherent fertility of the acid sandy soils. Due to the porous nature of the soils, their low organic matter, pH and clay contents; native soil N, K and Ca are equally very low. The implication is that with the usually high rainfall experienced in this zone, the application of inorganic fertilizers is accompanied by a high rate of leaching losses. Organic farming has the capability to take care of each of these problems. Besides the obvious immediate and positive effects of organic or natural farming on the environment and quality of cashew kernels, it also greatly helps a farmer to become self-sufficient in his requirements for agro-inputs and reduces input costs. These problems warrant revision of ongoing agricultural practices, and adaptation of some alternative strategies whose origin is presumed to be age old, popularly known as organic farming or natural farming.

Potential of Organic Cultivation

Cashew is grown in most of the places as a naturally grown crop with less attention with little or no application of inorganic or organic inputs. Hence, the expected returns are not achieved. The major reason being the organic waste recycling is not properly dealt with. Leaf litter, dried twigs, branches, fruits etc. should be utilized as organic inputs for cashew. There is growing demand for organic food grains, fruits, vegetables and other products which would fetch premium price in the international market. Since dairying and rearing of animals are a part and parcel of most of the farming systems, the availability of organic materials in plenty determines the scope for organic cultivation of most of the crops, apart from converting a large number of smaller cashew holdings into organic. In organic farming, most important aspect is maintaining the soil fertility status. When organic materials are added to the soil, the soil microbes work on them and convert them into readily available nutrients for the plants besides improving the soil structure. Earthworms, millipedes, centipedes and many more soil macro-fauna also play major role in improving the soil properties.

Organic Cashew Markets

'Organic' is not a characteristic of the product, but a method of production. The market started to move into the mainstream in the 1990s and today trade is close to $30 billion worldwide, and still growing. The Organic Trade Association of the US anticipates growth rates of 18 per cent per year until 2010. Over the past decade, the land under organic production has risen from 5 to 31 million ha, in 120 countries. Whereas in the past organic products could only be found in health food or specialist stores, today they have moved into the mainstream and can be found in supermarkets throughout the world. In the US, almost 40 per cent of organic sales are through mainstream supermarkets. Organic food production is big business, especially for fresh food items such as fruits and vegetables, but also for a range of dried foods and food ingredients, as well as dairy and meat. The main markets for organic foods are Europe, the US and Japan. Rising demand has come from the rising desire for healthy eating, and to a lesser extent, rising concern for the environment. Organic food consumers are willing to pay a premium for certified organic foods. Growers receive a premium for their product, and it is likely that throughout the supply chain, each player makes a better margin on an organic product than on conventional ones.

World Organic Cashew Scenario

Brazil is the largest producer of organic cashew nuts in the world. It produces nearly 30 per cent of its total production as organic cashew. Consumer demand for healthier food has been considered one of the major factors influencing increased demand for organic food. Growth is natural and significant. The growth of organic sector is indicating towards what can be called at least a boom. In the last decade, Europe has recorded the highest growth and sales touched US $ 12 billion, followed by North America (US $ 10 billion), Japan the third largest, accounts for bulk of the revenues from Asian markets. Over the next 10 years the market for organic products is projected to multiply to $ 100 billion and major share of the growth promises to come from US and Japan. Considering the trend of organic growth in the world, definitely organic cashew will find better place in the organic food market. It is the high time to evolve feasible strategies to augment the production of organic cashew.

Organic Cashew Farming Areas in India

In major cashew growing areas of the country, natural farming is followed wherein no inputs (organic or inorganic) are being applied. Hence, the yield from such orchards is generally low. This is due to the improper utilization of cashew waste consisting of dry cashew leaves, cashew apple and other farm wastes. Thus organic farming could be a boon to a large number of farmers as a natural alternative for those who lack the means for applying the recommended technologies. Those with adequate resources could also adopt organic farming because there is a business opportunity due to growing awareness and increase in market for organic food. Since India is the major player in the global cashew scenario, organic cashew could be one of India's best options for competing in the global market for boosting export earnings and it could fetch a premium price in the world market.

Cashew Varieties for Organic Farming

Only mid and late season varieties are best suited for organic farming. The early season varieties flushes early and flowering early are more affected of Tea Mosquito Bug (TMB). Because it is during this period (November–January) the climate is cool and dew fall during early morning followed by cloudy day results in maximum spoilage of the flowers due to highest infestation by the Tea mosquito bug. Mid and late season (February-April) varieties escape this menace. Due to increased temperature during flowering and fruiting in mid and late varieties, the TMB population comes down thereby the crop damage by the pest is minimum or nil. As such, control of TMB does not arise under such situations. However, under extreme changes in weather conditions favourable to sudden outbreak of TMB, its control even with chemical means is difficult. Mostly, mid season (Bhaskara, Dhana, Amrutha, BPP-8 V-4,V-7, VTH 174, Dharashree and Priyanka) and late season varieties (Ullal-1, Chinthamani-1 and Madakkatara-2) are suggested for organic farming.

Potting Mixture for Propagation

The pot mixture used for raising seedlings meant for grafting should contain organic manure (cow dung), rock phosphate and biofertilizers. The potting mixture should have one third soil, one third cow dung and the remaining one third sand and 5 g each of *Azospirillum, pseudomonas* and *Aspergillus* biofertilizers and 5 g rock phosphate. Bags filled with potting mixture are kept for raising seedlings. These seedling rootstocks are used for grafting purpose of scion cultivars.

Planting

Planting grafts in $1m^3$ pits after filling two third depths with topsoil, farmyard manure of 5 kg or 2 kg poultry manure or 1 kg castor cake per pit is recommended. Planting of cashew grafts should be at $2/3^{rd}$ depth and after planting it should be properly mulched. Normal soil and water conservation measures like terracing with crescent bund should be taken up within three years after planting. The plant spacing should be maintained at 7.5m x 7.5m, 8m x 8m, 10m x 5m under normal density planting system and 5m x 5m, 6m x 4m under high density planting system.

11. Nutrient management

Manuring

In organic farming system, nutrient should be given in an organic way. A grown up cashew tree produces about 20 kg of biomass waste per year (biomass waste includes cashew leaf litter, prunings and waste cashew apples etc.). This should be returned to the soil. In addition, 5-8 kg castor cake or 10 kg FYM or 5 kg poultry manure with 50 g bio-fertilizers should be applied when there is optimum soil moisture. Manure application should be taken up in the beginning of monsoon (June) at low rainfall areas and in mid monsoon (August) in high rainfall areas. Addition of biofertilizers namely N-fixers and P-solubilisers also gives benefits. For young plants, the dosage starts from one fifth of recommendation and increases year after year till fifth year. After the application of organic manure around the trees, it should be covered by a thin layer of soil and properly mulched. Periodic spray of cow's urine (1:10 dilution) or compost tea (1:40 dilution) facilitates better growth. They also prevent pest and disease attack. When there are plants of more than 20 years age in an orchard, pits of 1.0 m length, 1.0 m breadth and 0.5 m depth should be dug at the centre of the four trees and cashew biomass with fruits are to be incorporated into the pits along with organic manure and bio-fertilizers. Cow dung slurry at 20 per cent of the total biomass should be added to it and covered by a thin layer of soil. May is the ideal month for carrying out this operation. Some of the roots of old trees may get exposed and become vulnerable to CSRB attack hence these roots should be covered by thin layer of soil. In an experiment conducted at NRCC, Puttur (2000-2003) it indicated that application of 20 kg poultry manure alone with soil conservation technique (with crescent bund) increased yield by 80 per cent compared to control plot. Subsequently to understand the performance of the crop for different combinations of organic and inorganic manures a trial was laid out in 2002 and continued up to 2003 on six and seven years old cashew variety NRCC Sel. 2. The results indicated that application of biofertilizers-*Azospirilum* with the compost of organically recyclable biomass (ORB) available in cashew orchard to meet 100 per cent N requirement gave the highest yield of 1.053 t/ha on the seventh year whereas in the previous year it gave minimum yield. This indicated that in the long run this treatment is good.

In view of rising demand for organically grown cashew, strategies need to be developed for organic farming for different agro-ecological zones because organic manure availability may vary according to locations. Some of the commonly available organic manures like poultry manure, farm yard manure (FYM), Vermicompost, oil cakes, coir pith, organic manure mixtures etc. hold promise in organic farming in cashew as per the results obtained at Directorate of Cashew Research. There is high potentiality to utilize cow dung slurry in major cattle rearing areas. In addition to the above manures, poultry litter and oil cakes are also available in huge quantities in most of the states. Green leaves which are in abundance in Kerala, Coastal Karnataka and Maharashtra, where cashew is mostly grown also serve as good organic manure to cashew. Green manure crops like *Glyricidia* and *Pongamia* can be grown very well under rain fed conditions on the border or inside cashew plantations to produce green manure.

By considering the quantity of nutrients available in soil, the approximate quantity of organic manure required to meet the recommended dose of Nitrogen of 500 g per tree is presented in Table 17.1. The recommended dose of Nitrogen has been arrived at based on the trails conducted in various research stations. In organic manures, availability of nutrients normally stretches for a longer period and in addition the loss of nutrients through leaching is minimum. If the nutrients are applied through chemical fertilizers hardly 50 per cent of nitrogen and 75 per cent of the applied potash (K_2O) will be made available to the plants due to nutrient loss through seepage, runoff and denitrification. Such nutrient loss is negligible in the case of organic manures. Hence we need not apply the quantity of organic manure to the extent of actual nutrients requirement, which we otherwise calculate, when we require applying in the form of inorganic manure. Detailed studies need to be taken up for suggesting options to farmer like, the exact quantity of different organic manures to be applied per tree depending upon the availability of these manures in the respective locality.

The quantity of organic manures mentioned in Table 17.1 will meet the full requirement of phosphorus (P_2O_5) and potash (K_2O) in all cases except in green manure and groundnut cake. When these are used, bone meal and wood ash may be suggested to supplement P_2O_5 and a K_2O requirement for producing cashew organically, which fetches almost 30 per cent increase in price

Use of Traditional Additives

Most countries have traditionally utilized various kinds of organic materials to maintain or improve the tilth, fertility and productivity of their agricultural soils. However, several decades ago, organic recycling practices in some countries were largely replaced with chemical fertilizer which were applied to high yielding cereal crops that responded to a high level of fertility and adequate moisture, including irrigation. Consequently, the importance of organic matter to crop production

Table 17.1: Quantity of Organic Manure Requirement by Adult Cashew Tree (kg/tree)

Manure	*Quantity (kg/tree)*	*Nutrient Supplied (g/tree)*		
		N	*P_2O_5*	*K_2O*
Poultry manure	17.4	500	180	226
Farm yard manure (FYM)	62.5	500	256	462
Vermicompost	31.3	500	689	210
Green manure gliricidia (dry weight)	12.5	500	89	183
Biogas slurry	27.8	500	256	222
Water hyacinth compost	25.0	500	250	575
Castor cake	8.6	500	155	138
Ground cake	6.4	500	95	90
Pongamia cake	14.7	500	147	191
Cashew waste compost	33.0	500	250	200
Neem cake	9.61	500	96	134

Note: Calculated based on nutrient concentration in each type of manure as worked out at DCR.

received less emphasis and its proper use in soil management sometimes neglected or even forgotten. With these changes and the failure to implement effective soil conservation practices, agricultural soils in a number of developed and developing countries have undergone serious degradation and decline in productivity because of excessive soil erosion and nutrient run-off and decreased soil organic matter levels. Traditional additives are classified into two groups which are given below:

a) Bulky organic manure: FYM, farm compost, night soil, sludge and green manure are bulky in nature and supply large quantities of organic matter but small quantities of plant nutrients in comparison to the inorganic fertilizers. Their average nutrient content of bulky manure is given in Table 17.2.

b) Concentrated organic manure: Concentrated organic manures are those materials that are organic in nature and contain higher percentage of essential plant nutrients such as nitrogen, phosphorus and potash as compared to bulky organic manure. The concentrated manures are made from raw materials of animal or plant origin. The concentrated organic manure commonly used is oil-cakes, blood meal, fish meal, meat meal and horn and hoof meal. Their average nutrient content is given in the Table 17.3.

Bio-gas Slurry as Manure

The dung and the farm wastages are increasingly being burned instead of being returned to the soil as manure. Technology is available for the conversion of the dung to fuel and at the same time retain fertilizer value of the material. The gas produced from cow dung and water as a result of anaerobic fermentation is called bio-gas. Bio-gas contains methane gas (50-65 per cent) as most useful component

Table 17.2: Average Nutrient Content of Bulky Manure

	Manure	*Per cent Content*		
		N	*P_2O_5*	*K_2O*
1.	Animal Refuse:			
	Cattle dung, fresh	0.3-0.4	0.1-0.2	0.1-0.3
	Horse dung, fresh	0.4-0.5	0.3-0.4	0.3-0.4
	Sheep dung, fresh	0.5-0.7	0.4-0.6	0.3-1.0
	Night soil, fresh	1.0-1.6	0.8-1.2	0.2-0.6
	Poultry manure, fresh	1.0-1.8	1.4-1.8	0.8-0.9
2.	Wood Ashes:			
	Ash, household	0.5-1.9	1.6-4.2	2.3-12.0
	Ash, wood	0.1-0.2	0.8-5.9	1.5-36.0
3.	Farm Factory and Habitation:			
	Rural compost, dry	0.5-1.0	0.4-0.8	0.8-1.2
	Urban compost, dry	0.7-2.0	0.9-3.0	1.0-2.0
	FYM, dry	0.4-1.5	0.3-0.9	0.3-1.9
4.	Plant Residues:			
	Rice hulls	0.3-3.5	0.2-0.5	0.3-0.5
	Groundnut husks	1.6-1.8	0.3-0.5	1.1-1.7
	Straw and stalks: Banana, dry	0.61	0.12	1.0
	Maize	0.42	1.57	1.65
	Paddy	0.36	0.08	0.71
	Wheat	0.53	0.10	1.10
5.	Green Manure, fresh			
	Cowpea	0.71	0.15	0.58
	Green gram	0.72	0.18	0.53
	Sunhemp	0.75	0.12	0.51

and the remaining part mostly being CO_2 with small amount of other gas (Khatri and Chhetri, 1991).

Use of Non-traditional Additives

A number of products are now available that are generally referred to as soil and plant additives of non-traditional nature. These products include i) Microbial fertilizers and soil inoculates which are reported to contain unique and beneficial strains of soil microorganisms, ii) Microbial activators that supposedly contain special chemical formulations for increasing the numbers and activity of beneficial micro-organisms in soil, iii) Soil conditioners that claim to created favorable soil physical and chemical conditions that result in increased growth and yield of crops and iv) Vermicompost which help in improving soil health and fertility.

Table 17.3: Average Nutrient Content of Major Oil Cakes and Slaughter House Wastes

	Manure	*Per cent Content*		
		N	*P_2O_5*	*K_2O*
1.	Oil Cake:			
a.	Non-edible type:			
	Castor Cake	4.3	1.8	1.3
	Cotton Seed Cake (Undecorticated)	3.9	1.8	1.6
	Mahua Cake	2.5	0.8	1.8
b.	Edible Oil Cakes:			
	Coconut Cake	3.0	1.9	1.8
	Groundnut Cake	7.3	1.5	1.3
	Cotton Seed Cake (Decorticated)	6.4	2.9	2.2
	Rape Seed Cake	5.2	1.8	1.2
	Sesame Cake	6.2	2.0	1.2
2.	Slaughter House Wastes:			
	Meat Meal	10.5	2.5	–
	Blood Meal	1.0-12.0	1.0-2.0	–
	Fish Meal	4.0-10.0	3.0-9.0	0.3-1.5

Biofertilizers are the biological active product called microbial inoculates containing active strain of selective microorganisms like bacteria, fungi, algae or in combination. The biofertilizers containing biological nitrogen fixing organisms are of upmost importance in agriculture in the view of the following advantages:

- ✰ Enhance bio-mass production and grain yield by 10-20 per cent
- ✰ Cheap and can help to reduce the consumption of chemical fertilizer
- ✰ Make available nitrogen directly to the plant
- ✰ Solubilize phosphorus and increase phosphorus uptake to the plants
- ✰ Enhance plant growth due to release of hormones, vitamins, auxins etc.
- ✰ Improve the soil properties and sustain the soil fertility
- ✰ Control and suppress soil borne diseases
- ✰ Suitable for organic farming

The following types of bio-fertilizers are available:

Azotobacter and *Azospirillum*

It has been shown that biofertilizers (*Azospirillum*, *Azotobacter* and VAM) increased the germination percentage of nuts, plant growth, number of graftable seedlings of cashew and reduced the incidence of fungal diseases in the nursery (Ramesh *et al.*, 1999; Sinish *et al.*, 2005).

Phosphorus Solubilizing Bacteria (PSB)

Phosphorus solubilizing bacteria are a group of beneficial bacteria capable of hydrolysing organic and inorganic phosphorus from insoluble compounds. It is generally accepted that the mechanism of mineral phosphate solubilization by PSB strains is associated with the release of low molecular weight organic acids, through which their hydroxyl and carboxyl groups chelate the cations bound to phosphate, thereby converting it into soluble forms. In addition, some PSB produce phosphatase like phytase that hydrolyse organic forms of phosphate compounds efficiently. One or both types of PSB have been introduced to agricultural community as phosphate biofertilizer. Phosphorus is one of the major essential macronutrients for plants and is applied to soil in the form of phosphate fertilizers. However, a large portion of soluble inorganic phosphate which is applied to the soil as chemical fertilizer is immobilized rapidly and becomes unavailable to plants. Currently, the main purpose in managing soil phosphorus is to optimize crop production and minimize P loss from soils. PSB have attracted the attention of agriculturists as soil inoculums to improve the plant growth and yield. When PSB used with rock phosphate, it can save about 50 per cent of the crop requirement of phosphatic fertilizer. The use of PSB as inoculants increases the P uptake by plants. Simple inoculation of seeds with PSB gives crop yield responses equivalent to 30 kg P_2O_5/ha or 50 per cent of the need for phosphatic fertilizers.

Mycorrhizae

Mycorrhiza is the symbiotic association of roots with roots of vascular plants. The main advantages of mycorrhiza to the host plants lies in the extension of the penetration zone of the root fungus system in the soil, facilitating an increased phosphorus uptake. *Mycorrhizal* fungi assist the uptake of phosphorus (Tinker, 1980) and trace metals and possibly influence water and nutrients via hormonal influences is not in doubt.

Green Manuring

Green manuring is a practice of ploughing or turning undecomposed green plant materials into the soil either *in-situ* or *ex-situ*. The main advantage of green manuring is to add organic matter to the soil which improves the structure of soil and other physical properties, facilitates penetration of rain water, thus decreasing run-off and erosion, hold plant nutrients that would otherwise be lost by leaching, leguminous plants add nitrogen to the soil and increases the availability of certain plant nutrients like phosphorus, calcium, potassium, magnesium and iron *etc.* Leguminous crops such as *Pueraria javanica, Calopogonium mucunoides, Sesbania, Sunhemp* and *Centrosema pubescens* and *Gliricidia maculata* are common green manuring crops. The seeds of these cover crops can be sown in the beginning of rainy season at the rate of 12 kg per ha in the interspaces of cashew orchard. Beds of 30 cm x 30 cm are prepared in sloppy degraded soils by loosening the soil and mixing compost or cow dung. Then the seeds are sown in the beds should be covered by a thin layer of soil. Presoaking of the seeds in water for six hours ensures better germination. Fencing is necessary to avoid cattle grazing. In addition to utilization of recyclable cashew biomass waste, *Gliricidia* can be grown in the inter spaces between two rows of cashew or all along the border. Three rows of *Gliricidia* can be grown in the interspace of two rows of cashew at 1 m spacing. *Gliricidia* may be grown by sowing seeds or planting stem cuttings of one-meter length during rainy season. Nearly 60 kg leaf and tender branches can be collected and applied to each cashew plant as green leaf manure (Yadukumar *et al.*, 2007 and 2008). In one-hectare area of cashew plantation nearly 7 tonnes of dried green biomass can be produced and applied to cashew that is equal to application of nearly 186 kg N, 40 kg P_2O_5 and 67 kg K_2O per hectare per year. This will meet entire nutrient requirement of cashew and it was found that the same yield could be achieved compared to the plot receiving recommended package of practices.

Intercropping

Leguminous crops like horse gram, cowpea, and other crops such as ground nut, tapioca, vegetable, elephant-foot yam and fodder crops, are grown as intercrops in cashew plantations. Among the various annuals, biennials, fruit crops and tree species; pineapple has been found to be the best when grown as inter or mixed crop in cashew orchard for the first seven years (Yadukumar, 2007). Between two rows of cashew three trenches have to be opened across the slope between them and size of trench should be 1 m x 0.5 m (width x depth) and of any convenient length. In each trench two rows of pineapple suckers should be planted at 60 cm between rows and 40 cm between two suckers within the row. For one meter length, half basket of compost should be added and mixed with soil before planting pineapple suckers. One hectare cashew orchard can accommodate 15,000 suckers. Pineapple starts yielding from second year and after fourth year, it should be replanted in a new trench dug out by the side of existing trench or the same could be retained till seventh year of cashew plantation. After seven years, because of heavy shade of cashew tree over the pineapple and due to difficulty in picking raw cashew nuts fallen over pineapple plants, it may not be feasible to grow this intercrop economically. Nearly 40 tonnes of pineapple fruits can be expected over a period of seven years with a gross return of about Rs 2.0 lakhs. Expenditure would be Rs. 1.0 lakh per ha of cashew orchard. Apart from giving an additional income, 30 per cent increase in cashew nut production was also observed mainly due to better soil and water conservation in the trenches, where pineapple is planted across the slope between two rows of cashew. Other crops like turmeric, ginger and elephant foot yam can be grown as intercrops organically for the first 5 years of planting cashew. The profit realized from these crops may range from Rs. 20,000 to 50, 000 per year.

Mixed Cropping

It is not possible to achieve sustainability by growing only one particular type of crop. Whatever is the size of the holding, growing a variety of crops within the available land contributes to the ecological balance. While selecting mixed crops its compatibility with the main crop, spacing, and availability of light, manure and water requirement should be properly studied. Mango, sapota, kokum, amla, jack fruit, vanilla and pepper etc. could be grown organically as mixed crops in cashew orchards. In countries like Malaysia, Indonesia and Cambodia, rambutan, longan and dwarf bamboo are grown with cashew by providing proper spacing for each crop according to canopy coverage. Considering various mixed crops grown in different cashew growing areas and from the experiments conducted at DCR, Puttur, some of the promising combinations of crops are listed below. Nevertheless, farmers are advised to study carefully various aspects of cultivating mixed crops in their locations namely suitability, irrigation requirements, marketing avenues etc. before adopting them in their orchards. Some of the crop combinations along with spacing recommended for each crop in high density multispecies situations are given below:

1) Between two rows of cashew, the spacing maintained is 30 m. In the middle of two rows planting mango, tamarind, silk cotton, jack, sapota, amla etc. at a uniform distance of 10 m would give additional income. These crops do not require irrigation. But timely pruning of these trees would be highly beneficial after seven years of planting. Catch pits help in the conservation of soil and water.
2) Between two rows of cashew the spacing maintained is 21 m. In the middle, two rows of planting with guava or longan and rambutan or bread fruit at 7 m distance between rows and 6 m between plants would be economical. These crops require light irrigation and are not suitable for dry areas.
3) Between two rows of cashew the spacing maintained is 10 m. In the middle, planting *Acacia* or *Casuarina* along with seedling mango on the borders at 12 m distance would give ecological sustainability. *Acacia* and *Casuarina* should be regularly pruned after 2-3 years and leaves may be incorporated into cashew basins. After 5-6 years these trees are cut and sold. For planting along on the borders, softwood grafted mango of modern varieties will give better return.
4) The other combination would be cashew, dwarf bamboo and kokum. Bamboo is increasingly gaining importance due to its suitability for fabrication of inexpensive and beautiful furniture and handicrafts. It also serves as a basic raw material to industry. Due to flowering, there is decline in bamboo crop over the years and abrupt shortfall of supply may increase the price in the future. Kokum is a fruit of various medicinal uses. The rind of the fruit is used in making juices and beverages. The juice prepared from the rind has a natural cooling effect on the body and is widely used in Maharashtra and Karnataka. It contains the chemical hydroxyl citric acid, which is known to have an anti-obesity effect. The fat extracted from the Kokum seeds has a special property of melting at higher temperature. Hence, it can be blended with cocoa, milk and sugar so that the chocolate can be kept stable at room temperature, unlike the chocolates prepared purely from cocoa butter. At DCR, Kokum is grown as mixed crop in cashew orchard accommodating 200 plants in one hectare area. About 20-30 kg yield from single tree (400-600 kg per ha) can be expected. The return from a kokum tree will be about Rs 100 to 150 per tree and Rs 20,000 to 30,000 per ha. The cost of cultivation is Rs. 30 per plant (Rs. 6,000 per ha.) and net profit will be Rs. 70 to 120 per plant or Rs. 14,000 to 24,000 per ha.

Nutrient Management in High Density Cashew Plantation

In the case of high density planting system, it is proved beyond doubt that the recommended doses of nutrients are not required at 6 or 7 years after planting since the crop canopy covers the ground completely at this age. It is at this stage crop does not perform in terms of yield and every year if we add lot of external nutrients as per the recommended doses the result will be unwanted increased vegetative growth and overlapping of branches causing heavy shade and consequent decreased flowering and fruiting resulting decreased yield. In high density planting system the leaf deposit per unit area is more than two times compared to normal density planting system. This huge deposit of leaf litter and apples contribute natural build up of nutrients in soil, which serves as rich source of nutrients. Every year external nutrient application has to be reduced. Studies on nutrient management indicated that 1/3rd of recommended doses will be enough for maintaining cashew plantation when grown in high density planting.

Organically Recyclable Biomass Waste (Crop Residue)

Organic production of cashew offers immense potential. Cashew plantations have vast potential of organic biomass available for recycling. The availability of cashew leaf litter from different age group plantations (10 to 40 years) ranged from 1.38 to 5.20 t/ha. Vermicomposting of cashew leaf litter and apple by using local earthworm *Eudrilus* spp. has been standardized at DCR, Puttur. About 5.5 tonnes of available cashew biomass waste per ha can be converted into 3.5 tonnes of compost or vermicompost and this helps in meeting nutrient requirement of cashew by 50 per cent.

In general, every year considerable quantity of leaf litter and cashew apple are available at the base of the cop if the same is not taken away from the location. The cashew leaf litter, apple waste and weed biomass collected at the base of the plant is to the tune of 40 kg per tree per year. When it is decomposed 65 per cent recovery can be expected and this amounts to 26 kg of biomass. This half decomposed biomass contains around 0.65 per cent N, 0.41 per cent P, 0.45 per cent K, 0.22 per cent Ca, 0.19 per cent Mg and 369 ppm Fe, 263 ppm Mn 14.6 ppm Cu and 16.5 ppm Zn (Yadukumar and Nandan, 2005). The total macronutrients added to the soil when above biomass is incorporated will be approximately 169 g N, 107 g of P_2O_5 and 117 g K_2O per tree. Soil mineralization also takes place in the system thereby releasing the nutrients in available form. In addition to the utilization of native soil nutrients cashew requires additional quantities of nutrients for economical yields.

Compost from Cashew Leaf Litter and Cashew Apple Waste

Cashew leaf litter, apple and other jungle growth available in cashew garden are collected before the pre-monsoon season during harvest of the crop. It is heaped in a chamber constructed specially for the preparation of compost. Chamber size should be 5 m length, 1.5 m width and 1.5 m height. This can be constructed with a shift able angle iron and rods or with local materials available in the garden. Only the walls can be erected by using dry coconut leaves and wooden poles and split stem of arecanut palm or bamboo whichever is available locally. One layer of cashew leaf litter and cashew apple waste is spread over the ground in the chamber and to that cow dung slurry equal to 20 per cent of total weight of biomass is sprinkled after sufficiently wetting approximately to 50 per cent moisture. Like this each layer of one above the other layer has to be treated with cow dung slurry. Top of the materials should be completely covered with either coconut leaves or arecanut leaves whichever available locally. Even local thatching materials can also be used. Once in 45 days turning of the biomass should be done for uniform decomposition. Since the chamber is open excess water in the rainy season may create problem. Excess water entering the chamber can be very well prevented by covering the heap with coconut leaves or any thatching material. If the rainy season lasts for 4 months, it is enough to decompose the material easily. If there is gap without rain for more than 15 days during rainy season, it is better to pour water to sufficiently wet the biomass. The compost can be removed after six months. When it is removed the nutrient concentrations will be 1.16 per cent N, 0.41-0.55 per cent P, 0.45 per cent K, 0.121 per cent Mg, 0.223 per cent Ca, 27 ppm Mn, 74.9 ppm Fe, 1.5 ppm Zn and C: N ratio is 10. Population of useful bacteria, fungi and actinomycetes are $21x10^6$, $23x10^4$, and $12x10^5$ respectively.

It was estimated that when recyclable biomass of cashew (cashew leaf litter and cashew apple) is left in the field for 6 months it partially decomposes and the weight reduces to 65 per cent. Towards the end of six months the recyclable biomass contained 0.65 per cent N, 0.34 per cent P, 0.398 per cent K, 0.208 per cent Ca, 0.11 per cent Mg, 22 ppm Mn, 0.66 ppm Cu, 68.68 ppm Fe and 1 ppm Zn. At rate of 40 kg biomass available from cashew tree after six months it reduces to 26 kg with 65 per cent recovery. This means totally 5.2 tonnes of partially decomposed biomass is available/ha/year which contributes 33.8 kg N, 17.6 kg P, 20.6 kg K, 10.4 kg Ca, 5.72 kg Mg, 0.114 kg Mn, 0.08 kg Cu, 0.352 kg Fe, 0.005 kg Zn/ha/year.

Vermicompost from Cashew Leaf Litter and Cashew Apple

Cashew leaf litter and apples are allowed to decompose for 20 days after mixing with 15 per cent cow dung slurry. Then the material is fed to earthworms that were developed previously in plastic bucket by feeding cow dung material. Vermicompost chambers can be prepared as per the following details. Simple structure of chamber with 1.5 m height, 5 m long with 1.5 m width measurement is erected. At the bottom of the chamber there should be a cement channel to fill water to prevent entry of ants. The chamber should be shaded during summer and protected from rain during rainy season by covering with shaded nets and silpaulin sheet or any local thatching materials like coconut or arecanut leaves spread over the roof constructed specially for the purpose. For drainage of vermi wash a hole should be provided which is connected to a tank with PVC pipes so that the Vermiwash can be utilized as liquid manure. Per cent recovery of compost will be 75 per cent. The compost should be sprinkled with sufficient quantity of water as and when moisture level reduces below 40 per cent. For this, once in 4 days the water should be sprinkled so that the biomass always in wet condition. Nearly 750 kg compost can be obtained from 1000 kg cashew recyclable biomass. The vermicompost contained 1.21 per cent N, 0.898 per cent P, 0.595 per cent K, 2.755 per cent Ca, 0.825 per cent Mg, 29.66 ppm Zn, 24.26 ppm Mn, 12.23 ppm Cu and 162 ppm Fe. Population of useful bacteria, fungi, actinomycetes, azospirillum are 39x10^6, 39X10^5, 28x 10^5, and 0.52x 10^4 colony-forming unit (CFU) respectively.

Mulching

Mulching cashew plantations with organic matter prevents weed growth, reduces evaporation during summer regulates the soil temperature, improves the soil fertility and also prevents soil erosion. Therefore, green matters obtained during weeding may be utilized for mulching the plantation at the base of the respective trees. The mulching of coir-pith (7.5 cm thickness) has reduced weed growth to the extent of 73.2 per cent in propagation plot followed by 46.00 and 45.16 per cent in top worked trees and nursery plot respectively (Kumar *et al.*, 1989). In low rainfall areas, mulching around the base of trees with grasses or slashed weeds help in the control of weeds, retention of water and modulation of soil temperature, especially in dry season.

Weed Management

Generally weeding in cashew can be done twice a year. One weeding in the month of August during manure application and another weeding can be done just before the start of flushing and flowering (October or November). The weed biomass can be effectively recycled as that of cashew biomass waste explained earlier.

Management of Tea Mosquito Bug (TMB)

Tea mosquito bug can cause yield reduction to the tune of 30-40 per cent by damaging tender shoots, inflorescences and nuts. Both adults and nymphs of TMB suck the sap from tender shoots, panicles, immature nuts and apples resulting in the formation of black lesions. These lesions on shoots and panicles coalesce causing shoot blight or blossom blight. The plants can escape TMB attack if the new flush delayed. Planting mid-season or late season flowering varieties would be the right strategy to escape TMB. Even early flowering varieties also flower 10-15 days late if the plants are grown organically compared to chemical fertilizer applied ones. The delay in flowering naturally minimizes the incidence by escaping multiplication of TMB population. The pest could be repelled by smoking the orchard by burning organic residues three times during flushing, flowering and fruiting. Care must be taken to see that small heaps of organic wastes in several places on the ground below the canopy of the tree are burnt slowly. This can be achieved by putting a thin layer of soil on the heaps and setting fire. At any chance the burning should not produce too much of heat lest flowers and shoots gets affected. Clean cultivation is another method to check multiplication of TMB considerably. All weeds should be removed and heaped as mulch. Encouraging multiplication of weaver ants (*Oecophylla smaragdina*) checks the spread of TMB by feeding on its eggs and nymphs. In the initial stages the colonies of weaver ants may be brought from nearby forest or by placing small pieces of jaggery at the cracks and crevices of the bark can attract the red ants.

At DCR, Puttur, a plot of 2.5 hectares is being maintained for the last four years and though initially the TMB damage was noticed in small pockets, no economic loss on yield was observed. Studies on the effect of various organic pesticides like phalada, phytozeal and phalini in late season variety NDR2-1 indicated that combination of phytozeal and phalada resulted in maximum yield with minimum TMB damage. Phalada IIIC is an herbal based liquid insecticide certified as 100 per cent organic insecticides by Skal International (Certifying agency based at Netherland). The contents of this herbal preparation are combination of water and extracts of *Annona reticulata* seeds (15 per cent), *Adhatoda vasica* leaf (12 per cent), *Sapindus trifoliatus* (30 per cent), *Pongamia pinnata* (20 per cent). *Phytozeal* is a herbal based liquid insecticide combined with vermiwash (90 per cent), plant extracts of *Lantana camara* flower (0.5 per cent), *Cynadon dactylon* grass (0.5 per cent), seaweed (5 per cent), *Calotropis procera* (2 per cent) and extracts of *Pseudomonas* fluorescens (0.1 per cent) and *Trichoderma viride* (0.1 per cent). Phalini is diluted form of cashew nut shell oil mixed with water. In another observational trial it was found that tea mosquito damage score is more in both Bhaskara (0.51) and Ullal-3 (2.10) varieties in conventional farming compared to organic farming *i.e.,* 0.21 and 0.30 respectively.

Seventeen types of plant products *viz.,* Neem oil, cotton seed oil, *Pongamia* oil, *Annona* seed extract, palmarosa oil, *Calophyllum* oil, tobacco decoction, *Butea* leaf extract (LE), *Adhatoda* LE, *Vitex negundo* LE, *Strychnous nuxvomica* LE, neem based commercial formulations (Limanool, Nimbicidine, Achook and Repellin), Dillapiole and garlic extract were evaluated against TMB at DCR and none was found to be effective. Even smoking of the canopy in the early morning with dried leaves of cashew or *Vitex negundo* neither caused any mortality of TMB nor reduced its damage on cashew. This pest was found to be regulated under natural conditions mainly through its egg parasitoids (*Telenomus* sp., *Ufens* sp. and *Chaetostricha* sp.). These parasitoids are specialist type and not amenable for mass multiplication under conventional laboratory rearing methods. Therefore, they could not be exploited under inundative release methods. However, their activities are naturally enhanced under favourable weather condition (increased minimum temperature and relative humidity) during vulnerable period (November-February). Among predatory fauna, different species of ants and spiders except weaver ant (*Oecophylla smaragdina*) play negligible role on suppression of this pest. Under field conditions, virgin TMB females exerted high level of male attraction and thus provide substantial scope in future for exploiting this sex pheromone behaviour for mass trapping and monitoring the pest population. Considering the changing consumer demand for organic products, organic cultivation of cashew is other important area of research.

Management of Cashew Stem and Root Borer (CSRB)

Older cashew trees are more prone to infestation by CSRB. Adult female beetles lay eggs in the bark crevices of the stem (near the collar region) or on the exposed roots. After hatching, the young grubs tunnel immediately into the bark of the trunk or root regions. Due to their extensive feeding by irregular tunneling in the bark, the translocation of the nutrients is hampered leading to drying of leaves and twigs. In the infested portions the gum and frass (chewed fibers and excreta pellets) start oozing out from the tree which is the initial symptom of pest attack. Curative measures given below should be taken up immediately. The infested portion, having frass below the bark on either the main stem or root portion (up to a depth of 60 cm) should be carefully chiseled to inflict minimum damage to the bark and tunnels should be tracked towards the fresher frass and to locate the CSRB grubs, which should then be removed and killed. In case white powdery fibres are seen it indicates that the grubs have entered the heartwood for pupation. In such cases a pliable wire (for example automobile gear wire) should be inserted deep into the pupation hole and pushed in till a slushy sound is heard or milky white fluid flows out indicating damage to the grub or pupa. Later chiseled portion of the bark should be swabbed with 10 per cent Bordeaux paste. When a paste of cow dung and ash is applied to the injured portion of the bark, it helps in curing of wounds. Swabbing neem oil is also quite effective.

Trees having more than 50 per cent of the bark circumferences damaged by the insect or with yellowing of the canopy hardly ever recover. However, the existence of different pest stages in such trees will act as inoculums for the next season. Hence phytosanitation should be adopted by uprooting such trees which are beyond recovery and disposed off immediately, along with removal of pest stages in other infested trees. Such operations prevent build up of pest inoculums. As the pest incidence occurs during the period of nut collection (Feb-May), the infested trees can be marked and curative measures could be adopted immediately, so that the grubs do not inflict more damage. As a preventive measure brushing waste motor oil or clay soil once or twice a year to the lower trunk region discourages egg laying. Minor pests like cashew apple and nut borer, leaf miner, leaf and blossom webber can be mechanically controlled at younger stages of plant growth or with organic sprays suggested for TMB. As the plant grows older the loss is insignificant.

Organic Sprays for Cashew

Before the arrival of chemical pesticides, there were several traditional methods of insect and disease control, which slowly disappeared. Off late, they are gaining importance once again. Such organic control measures have been collected from various sources and listed here for the benefit of organic farmers. Some of them have been tried in cashew cultivation.

i) Sprays Providing Nutrition and Imparting Protection from Insects/Diseases

- ☆ *Cattle urine:* for every litre of cattle urine, 8-10 litres of water is added and sprayed.
- ☆ *Compost tea*: one kg of well degraded compost powder is mixed in 40 litres of water, filtered and sprayed.
- ☆ *Jeevamrutha:* 10 kg fresh cow dung, 5-10 litres of cattle urine, 2 kg of black jaggary (or palm sugar), 1 kg legume seed powder(black gram/Bengal gram/green gram), one handful soil from the bunds of the field and 200 litres water. All these constituents are mixed in a barrel, kept in shade for 2 days stirring three times a day. The mouth of the barrel should be kept closed with a wet gunny bag. The solution is to be used for spraying within 7 days of its preparation and filter before use.
- ☆ *Beejamrutha:* 5 kg fresh cow dung, 5 litres cattle urine, 50 g CaO (lime), one handful of soil from the bunds of field and 20 litre water. All these constituents have to be thoroughly mixed and the mixture could be used

for the seed treatment. Treated seeds have to be shade dried and sown. The treatment enhances germination. Cashew nuts can be soaked for a day in this solution before sowing.

ii) Organic Sprays for Control of Pests

- ☆ *Strychnos nux-vomica* L.: seeds, bark, leaves and roots of this plant are used. Two kg of the plant part is crushed and boiled in 30 litres of water. The solution should be filtered and sprayed in 1:2 proportions with water. It is useful in avoiding TMB to certain extent. Spraying once in 15 days during the flushing period is suggested.
- ☆ *Neem seed concoction;* one kg of pounded neem seeds is soaked in 10 litres of water for 20 litres of water with 1 ml of liquid soap are added, filtered and sprayed. It is useful against some insect pests.
- ☆ *Tobacco concoction:* 200 g tobacco should be tied in a piece of cloth and boiled in 2 litres of water and allowed to soak for 24h and filtered. 100 ml of neem oil with 20 ml of liquid soap are added and diluted in 5 litres of water and sprayed.
- ☆ *Neem oil/castor oil:* 200 ml of neem/castor oil is mixed with 40 ml of liquid soap and diluted in 10 litres water and sprayed. This is effective against TMB if sprayed frequently.
- ☆ *Lantana:* One kg lantana (*Lantana camara* L.) leaves are finely crushed and diluted in 10 litres of water and sprayed. Effective against leaf miner.
- ☆ *Fish:* one kg of fresh fish is crushed and mixed with one kg of black jaggary or palm sugar and allowed to ferment for 15 days. This preparation may be used at 15 ml per litre of water for spraying. This is effective against bacterial diseases.
- ☆ *Annona* (*custard apple-Annona squamosa L.*) + *Chilli* + *Neem Seed:* Two kg of *Annona* leaves should be shredded and made into paste by adding some water. 500 g of dry chillies has to be soaked for 8-10 hour in water. All the ingredients are mixed and water is added to prepare a solution. Then diluted to 50 l with water and sprayed, controls thrips and caterpillars.
- ☆ *Chilli* + *Garlic:* One kg of green chilli along with 3 garlic bulbs is crushed to make a paste and soaked in five litres of water for 8-10 h. this solution is dissolved in 18 l water and sprayed. Effective against leaf hoppers and grass hoppers.

  Note: all the sprays are to be carried out during either morning or evening hours.

iii) Preparation of Bordeaux Mixture

One kg lime (CaO) should be dissolved in 5 litres of hot water and a solution is prepared. One kg copper sulphate ($CuSO_4$) crystals are tied in a cloth and suspended in 5 litres of water in another vessel. When both the solutions are ready, lime solution should be added slowly to the copper sulphate solution by stirring to get 10 litres of thick solution. This is 10 per cent Bordeaux paste. If diluted further by adding additional 90 liters of water to make the volume to 100 litres results in 1 per cent Bordeaux solution, which is useful for spraying cashew nurseries in rainy season. The Bordeaux paste is used to smear the cut surfaces of cashew trees.

References

Anonymous. (2003). Developing integrated production package for enhancing productivity of cashew. NATP final report, NRCC, Puttur, DK., Karnataka, p: 55.

APEDA. 2010. www.apeda.com

Arora, C. L., Nayaar, V. K. and Randhuwa, S. S. (1975). Note on secondary and micro nutrient content of fertilizers and manures. *Indian Journal of Agricultural Sciences*, 45: 80-85.

Asmed, S. (1993). Agriculture-fertilizer interference in Asia. Issue of growth and sustainability, Oxford and IBH Publishers, New Delhi, India.

Bijay Singh and Sekhon, G. S. (1977). Some measures of reducing leaching loss of nitrates beyond potential rooting zone. III. Proper crop rotation. *Plant and Soil*, 47: 585-591.

Dahama, A. K. (1997). *Organic farming for sustainable agriculture.* Ashila Offset Printers, Daruagung, New Delhi, India.

Dhaliwal, G. S. and Singh, B. (1993). *In Pesticides: their ecological impact in developing countries.* Commonwealth Publishers, New Delhi, India.

Ferguson, J J. (1994). Organic citrus production: certification and production guidelines. *Citrus Industry, 75*(3): 40-44.

Khatri and Chhetri, K. C. (1991). *Introduction to soil and soil fertility.* Institute of Agriculture and Animal Science, TU, Rampur, Nepal.

Kostial, K. (1986). Cadmium. In: *Trace elements in human and animal nutrition*, pp. 319-325. W. Mertz (Ed.). Academic Press, London.

Kumar, D. P., Hegde, M. and Guruprasad, T. R. (1998). Fertigation for higher nut production in cashew (*Anacardium occidentale* L.). *Cashew Bulletin*, 35: 2-4.

Lieberhardt, B. (2003). What is organic agriculture? What I learned from my transition. In. Organic Agriculture, Sustainability, Markets and Policies, Organization for Economic Cooperation and Development (OECD) and CAB 1, Wallingford, UK. pp. 31-44.

Ramesh, N., Lingaiah, H. B. and Khan, N. M. (1999). Economics of bio-fertilizer application in cashew graft production under eastern dry zone of Karnataka. *Crop Research* (Hisar), 17(3): 336-341.

Scialabba, E. N. and Hattam, C. (2002). Organic Agriculture Environment and Food Security. FAO, Rome pp. 252.

Sinish, M. S., Mercy, G. and John, P.S. (2005). Organic methods for cashew root stock production. Cashew 19(1): 8-15.

Tinker, T. (1980). Agriculture and World Development, Overseas Development Council, India, pp. 228-240.

USDA. (1980). *Report and Recommendations on Organic Farming*. United States Department of Agriculture, Washington DC, USA.

Venkataramani. (1995). *Role of bio fertilizer in organic farming: Organic farming in rainfed agriculture*. Central institute for dry land agriculture, Hyderabad, India, pp. 85-95.

Velayutham, M., Pal, D. K. and Bhattacharya, T. (1999). Organic carbon stock in soils of India. In: *Global Climatic Change and Tropical Ecosystems* (R Lal, J M Kimble, H Eswaran and B A Stewart, Eds.), Lewis Publishers, Boca Raton, Florida, USA, pp. 71-96.

Yadukumar, N. (2007). Organic farming in cashew. (in) *Proceedings of the National Seminar on Research, Development and Marketing of Cashew* held during 20-21st November 2007 at Goa, pp. 36-46.

Yadukumar, N. and Nandan, S. L. (2005). Recycling organic wastes of cashew plantations by aerobic composting. *Journal of Plantation Crops*, 33: 99-102.

Yadukumar, N., Rejani, R. and Nandan, S. L. (2007). Green manuring for high density cashew orchards. (in) Proceedings of the *National seminar on research, development and marketing of cashew* held during 20-21 November 2007 at Goa, pp. 59-61.

Yadukumar, N., Rejani, R. and Nandan, S. L. (2008). Studies on green manuring in high density cashew orchards. *Journal of Plantation Crops*, 36: 265-269.

Section IV: Plant Health Management

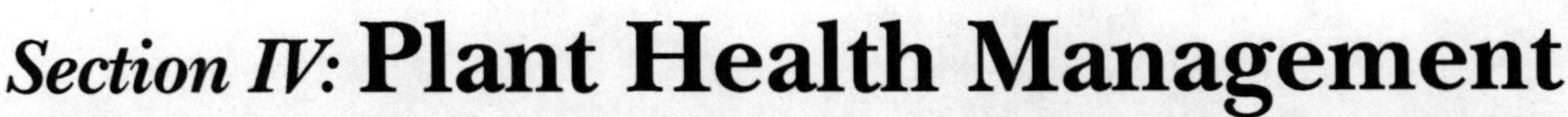

Cashew: Improvement, Production and Processing (2017) *Pages* **325–330**
Editor: **Prof. P.L. Saroj**
Published by: **ASTRAL INTERNATIONAL PVT. LTD., NEW DELHI**

Rejuvenation of Old Plantations

V. Pandey

1. Introduction

In India, most of the cashew plantations have been established from seedling progenies which are usually genetically poor in quality and highly erratic in bearing, owing to poor productivity. Cashew being a highly cross-pollinated crop, variation in seedling progenies is a natural phenomenon. During early period of cashew introduction, non-availability of adequate clonal material of high yielding varieties for massive planting has been largely responsible for such a situation. Large area under cashew plantations exists in the coastal region, which are not only seedling in origin but also senile. Besides, most of the plantations are on degraded lands and with poor management practices, which are also responsible for poor productivity. Adoption of better management practices like proper nutrition, irrigation and timely plant protection measures no doubt, would increase the yields but boosting cashew production from the existing old plantations would perhaps be possible only if the existing seedling plantations can be converted into high yielding varieties by *in-situ* top working. However, sudden decline of beheaded cashew stumps due to infection by *Collectotrichum gleoeosporiodes* Penzig and infestation by stem borer are the other limiting factors during rejuvenation. Post rejuvenation management is also crucial. Therefore, rejuvenation of old cashew plantations requires good skill and proper management for desired success.

2. Purpose of Rejuvenation

- ☆ To enhance the yield performance of senile poor yielding trees.
- ☆ To change the variety of existing plantations.

- ✰ To include several cashew varieties in a single tree.
- ✰ To conserve large number of genetic stock in smaller piece of land.
- ✰ To redevelop the exhausted canopy.
- ✰ To improve the aesthetic performance of plantation.

3. Conditions for Rejuvenation

- ✰ Tree should be healthy and free from cashew stem and root borer (CSRB) infestation.
- ✰ Trees should not be very old and unrecoverable.
- ✰ Only senile and very low yielders need to be rejuvenated.
- ✰ Beheaded tree should have sufficient exposure to sun light.
- ✰ Area should be free from bushes and trees competing for nutrition and space.
- ✰ Ensure appropriate after care of rejuvenated trees.

4. Technique of Rejuvenation

For rejuvenation of cashew plantations, the initial attempt made by Krishnamurthy *et al.* (1985) on side/crown grafting on stumps of old cashew trees had not made any encouraging response. The first systematic study on the rejuvenation of unthrifty cashew plantations by top working were initiated at Agricultural Research Station, Ullal, Dakshina Kannada distt., Karnataka during 1983, (Khan *et al.*, 1986). Poor bearing trees (yielding less than 3 kg/tree/year) which normally had very sparse and lanky growth were selected for rejuvenation. In order to find out the most desirable period for beheading of trees and determine the most suitable type of grafting and optimum season for getting maximum success, as many as 240 trees (20 during each month) were beheaded throughout the year from September 1983 to August 1984. Four methods of grafting, *viz.*, soft wood grafting; veneer (cleft) grafting; air layering, approach grafting and patch budding (each method on 20 shoots/tree for 5 trees) were adopted to determine the most optimum method and month of grafting. Out of the four types of grafting tried all throughout the year, December to February months were found to be ideal for beheading, criteria for selection of trees, height and technique of beheading, selection and pre-curing of scion material, grafting and aftercare of successful grafts, and management practices like nutrition, mulching and plant protection to be adopted for the rejuvenated trees were refined and standardized during 1985-88 and recommended this viable and feasible technology for rejuvenatioin of old cashew plantations by top working.

Among various methods tried for *in-situ* grafting; softwood and veneer grafting were found to be the most suitable techniques for *in-situ* grafting with 70 to 80 per cent of success. February to May was found to be the best period for *in-situ* grafting. For soft wood grafting, scion sticks of 0.8 to 1.0 cm girth, 10-12 cm length, bearing mature leaves and un-sprouted apical buds are selected from the past season's growth. They are first pre-cured on the plant, by clipping the leaves just above the petiole one week before grafting operation. On shoots of 30 to 45 cm length and 0.8 to 1.0 cm thick aroused on beheaded stump, scions are grafted and covered with inverted polythene caps to provide protection from severe sun and to minimize desiccation. Once the sprouting of scion is seen usually after a week, the polythene cap is removed for further growth. About 10 shoot to be grafted so as to ensure at least 4 to 5 successful grafts per stump to provide better framework and canopy for the tree.

There are several factors responsible for success of top working. Studies on the suitability of plants of different age groups for rejuvenation of cashew indicated that trees in age group of 5 to 20 years could be successfully rejuvenated while trees beyond 25 years of age and above showed reduced sprouting and stunted growth probably because of ageing. Though, superiority of beheading at 0.5 m height was observed but beheading can be done at 1 to 1.5 m height to maintain better frame work. However, more height may pose problem of damage of new growing branches by wind and grafting operation is also cumbersome.

5. Management

5.1 Care of Beheaded Stump

After beheading, there is every possibility of severe infestation by stem borer and incidence of *Collectrotrichum gleoeospoiodes* Penzig disease. Therefore, trees already affected with stem borer should never be selected for top working. On the other hand, mortality was more when beheading was done from May to September (being the period of high precipitation) mainly due to the infection by *Colletotrichum gloeosporiodes* Penzig, which is responsible for the die-back disease in cashew. Gradual death of the sprouts beginning from the cut ends of the stumps are the usual symptoms of the disease. Therefore, selection of trees free from stem borer infestation, beheading with a power saw to get clean cuts followed by pasting with Bordeaux paste or any copper fungicide on cut ends is very essential for cashew rejuvenation programme. If possible, covering the top cut end portion of stumps with poly bags for about one month are important to protect the stumps from the infection of *Colletorichum gleoeosporiodes* Penzing, and infestation by stem borer and bark splitting. Besides, judicious application of manures and timely plant protection measures are some of the important management practices to be adopted to overcome mortality of beheaded trees.

5.2 Thinning of Shoots

In general, sprouting is seen between 30 to 45 days after beheading. Sprouting was found to be relatively early and profuse in young trees compared with old trees. After 15 to 20 days, thinning of the excess sprouts is necessary to reduce overcrowding and encourage better growth of the shoots to be grafted. Initial retention of 15 to 20 shoots in young trees (<15 years age) and 20 to 25 shoots in older trees (15-25 years age) is desirable. Excessive thinning is also however found to be undesirable as this causes exposure of the stumps to severe sun resulting in bark splitting and sometimes poor development of canopy frame due low success of grafting.

5.3 Pre-curing of Scion

For better graft intake precuring (defoliating leaves one week before grafting operation) of scion shoots is essential. Pre-cured scion sticks of 8-10 cm length and 0.8 to 1.0 cm girth, brownish in colour and plump without any sprouts at the tip from high yielding desired quality varieties should be used for higher success. It was also observed that scion sticks separated even 3 to 4 days earlier can also be used for grafting, the success being however inversely proportional to the number of days the scions were stored (Khan *et al.*, 1988).

5.4 Success of Top Working

The success of top working largely depends on the quality of scion material used for grafting, conditions of the sprouts and time of grafting and skill of grafter. About 90 per cent success in top working by grafting on juvenile shoots of the beheaded stumps was observed. After grafting is done on 8 to 10 shoots and even with 50 percent success, there would be 4 to 5 successful grafts per tree (Guruprasad *et al.*, 1988). In younger trees, even 3 to 4 successful grafts per tree were enough to provide sufficient frame work and canopy to the rejuvenated tree.

5.5 Care of Top-worked Trees

Thinning of unsuccessful grafts and extra shoots from the stumps after grafting, and subsequently any auxiliary sprouts coming from the stump should also be removed for better growth of the successful grafts. The successful grafts may be provided support to avoid lodging and damage by heavy winds. Usually, grafts are vigorous in nature due to the well established root system of the beheaded trees. The growth would however, be very rapid in the initial years and slowing down subsequently, commensurate to the age of the trees. As the top worked trees would have the potentiality to give high yields, judicious application of fertilizers (500: 250: 250 g NPK per tree) should be done. Proper management of moisture also helps healthy growth and better yield.

5.6 Pest Management

Top working operation requires serious attention towards pest management. Cashew stem and root borer is considered as major pest after beheading of the old trees. Besides, top worked trees behave like juvenile trees during the first year of their growth, putting forth vegetative flush almost throughout the year, thus effective management of tea mosquito bug and leaf minor infestation is also required.

5.6.1 Cashew Stem and Root Borer

Cashew stem and root border (CSRB) is the most serious pest attacking rejuvenated cashew trees. Due to extensive feeding by irregular tunneling in the bark by the grub, the translocation of the nutrients is hampered leading to drying of leaves and twigs. In the infested portion, the gum and frass start oozing out from the tree; which is the initial symptom of pest attack for taking up curative measures. The infested portion having frass below the bark on either the main stem or the root portion should be carefully chiseled causing minimum damage to the bark and tunnels should be tracked towards the fresh frass to locate the grubs, which should then be removed and killed. The chiseled portion of the bark should be swabbed or sprayed with 0.2 per cent chlorpyriphos solution (10 ml/l water). Trees having more than 50 per cent of the bark circumference damaged or with yellowing of the canopy do not recover, however the pest stages in such trees will act as inoculums for the next season. Hence, phytosanitation should be adopted by uprooting such trees which are beyond recovery and they should be disposed off immediately. As a preventive measure, beheaded stumps are to be sprayed with 0.2 per cent chlorphyriphos twice or thrice in the initial 1 to 2 years.

5.6.2 Tea Mosquito Bug and other Foliage Pest

After new sprouts, both adult and nymph of tea mosquito bug (TMB) suck the sap from tender leaves and shoots and when rejuvenated tree starts bearing, besides leaves and shoots; panicles, immature nuts and apples are also affected which results in formation of black lesions. These lesions on shoots and panicles coalesce causing shoot blight or blossom blight. The insecticides recommended for spraying are λ-cyhalothrin (0.003 per cent *i.e.* 6 ml/10 litre of water) or carbaryl (0.1 per cent *i.e.* 2 g/litre of water).

During early monsoon, the tender shoots are sometimes infested by red coloured beetles, which skeletonize the leaves. In severe case of damage, spraying of above mentioned pesticide can minimize the problem.

6. Development of Scion Bank

Transforming a few of the existing unthrifty trees into high yielding types for serving as scion banks, has been found to be the quickest and most viable method of mass production of scion material. Production of scion material was found with this system to be almost five times more than that of from conventional methods (Khan *et al.*, 1988).

7. Alternate Method of Grafting

Soft wood grafting is suggested method of cashew rejuvenation which require new succulent region and grafting has to be done only at a specific point below the apical portion when shoots are of 30 to 45 cm in length. When grafting is delayed for any reason, it is possible only at a higher level from the base of the shoots. As a consequence, shoots above the grafted joint become more prone to wind damage if soft wood grafting technique is adopted. Since new shoots are very fast in growth, soft wood grafting would not be feasible if they are overgrown and for such overgrown shoots, veneer grafting technique is ideal as it could be done on overgrown shoots with good success. Therefore, based on situation specific technique can be adopted.

8. Yield Response

Fourteen-year old cashew trees yielding less than 3kg/tree/year, top worked during April, 1984 have not only put forth a canopy of 5 to 6 metres in height and 3 to 4 metres in width in a period of six years but have also virtually given an average yield of 4.21 kg/tree (3.57 kg/tree during 1986; 4.65 kg during 1987, 5.92

kg during 1988, 9.58 kg during 1989; 12.32 kg during 1990) indicating the viability and profitability of the technology.

9. Economics of Rejuvenation

The technology of cashew rejuvenation is practically feasible and can be adopted by small and marginal farmers with proper training. The approximate cost of rejuvenation and management of top worked trees during the first two years (including the loss of yield of 2 to 3 kg nuts/tree) has been worked out to Rs. 50/tree, part of which (Rs. 25 to 30) could be realized by the sale of pruned wood during the first year itself. A minimum increased yield of even 5 kg/tree every year, for the remaining life span of the plantation would not only compensate for the initial expenditure and provide additional and sustained income to the grower for a considerable period of time but will also help in boosting cashew production as a whole in the country. Even if one-third of the area under cashew in each state wherever possible is rejuvenated with high yielding varieties, the increased yield and production will not only provide additional and sustained income to the poor farmer who is growing this crop but will also revolutionize cashew industry in the country and will help in attaining self- sufficiency in raw cashewnut production (Khan *et al.*, 1994)

References

Guruprasad, T. R., Khan, M. M., Iquals Ahmed, Hanumanthappa, H. and Jagadesh, G. B. 1988. Success of cahsew top working technology at farmer's doors. *The Cashew,* 1 (1): 10-11.

Khan, M.M., Hegde, M., Mallik, B., Hiremath, I.G., Hanumasheti, S.I., Madhava Rao, V.N. and Krishnamurthy, K. 1986. Rejuvenation of old cashew trees by top working. *Indian Cashew Journal,* 18(3): 9-25.

Khan, M.M., Hiremath, I.G. and Krishnamurthy, K. 1991. Cashew Research and Development in Karnataka. *The Lalbaugh,* 32 (3 and 4).

Khan, M. M., Kumar, D.P. and Hegade, M. (1994). Rejuvenation of old cashew plantations by top working. In: Advances in Horticulture (Eds. K.L. Chadha and P. Rethinum), Mahotra Publishing House, New Delhi, pp. 689-696.

Khan, M.M., Kumar, D.P., Hiremath, I.G. and Subrayappa, A. 1988. Top working in cashew-a rapid method of producing scion material. *Cashew Bull.,* 25 (9).

Krishnamurthy, K., Khan, M.M, Hiremath, I.G., Nagaraju, A.P, Mallik, Band Hegde, M. 1985. Three Decades of Cashew Research at Agricultural Research Station, Ullal (1954-1984). *Tech. Bull.,* No.3.

Cashew: Improvement, Production and Processing (2017) *Pages* **331–346**
Editor: **Prof. P.L. Saroj**
Published by: **ASTRAL INTERNATIONAL PVT. LTD., NEW DELHI**

19

Biotechnological Approaches for Management of Cashew Pests

R. Asokan, N.K. Krishna Kumar,
K.B. Rebijith and V.V. Ramamurthy

1. DNA Barcoding

Insects are the numerous life forms that captured the attention and attention of human beings since ancient times. In the same context proper classification and identification of life forms has been a challenge and a plausible method of classification was established by Carlous Linnaeus, a Swedish botanist who published *Systema Naturae* in 1758. But the Linnaeus system of classification was not based on evolutionary relationships among the target groups. Later Darwin's publication "The origin of species" in 1859 changed the way of classification of life forms, where identification, description and explanation of diversity of organisms had come to be known as Systematics. According to Mayr and Ashlock (1991) systematics is the scientific study of the kinds and diversity of organisms and any and all relationships among them and Taxonomy as theory and practice of classification of organisms. It took 200 years for taxonomists to describe 1.7 million species which is only 10 per cent of the total number of species estimated. In this context identification of insects has been a monumental task where it calls for availability of more number of specialists and funding. But with the dwindling interest in taxonomy and fund availability, classification and identification of various life forms particularly insects has been major challenge to the scientific community. With the advent of molecular biology and molecular tools identification of life forms including insects has become quick, precise and easy. Development of species-specific markers enables even a non-specialist to identify insects to species level.

As mentioned above to classify the remaining 90 per cent of the unidentified organisms we need about 15000 taxonomists working for centuries to complete the task. Economic development and increased international commerce are leading to higher extinction rates and introduction of invasive pest species. Therefore there is a need for faster species identification and information about their biodiversity for conserving them before they vanish from the face of Earth. No doubt! Contributions of morphological taxonomy is enormous but has also some draw backs like

- ☆ incorrect identification due to phenotypic plasticity and genetic variability
- ☆ there are many morphologically cryptic taxa which are common in many groups
- ☆ morphological keys are often effective only for a particular life stage or gender and many individuals cannot be identified
- ☆ although modern interactive versions represent a major advance the use of keys require high level of expertise that often lead to misidentification (Hebert *et al.*, 2003)
- ☆ taxonomists have always looked for discontinuous character variations that could signal divergence between species. The debate on threshold values employing molecular identification for interspecific divergence is also true with morphology based identification

Hence there is a need for an adjunct tool that facilitates rapid identification of species where molecular identification popularly called "DNA barcoding" becomes handy. A first International conference in this direction was organized at Munich in the year 2002 by German Science Association (DFG). In this regard, Paul Hebert of University of Guelph, Canada developed the use of one mitochondrial gene, mitochondrial cytochrome oxidase I (mtCOI) as Universal identification marker for identification of animal species which include insects.

The following are the set of Universal primers for the identification of species by Folmer *et al.* (1994).

LCO 1490 5′-GGT CAA ATC ATA AAG ATA TTG G-3′

LCO 2198 5′-TAA ACT TCA GGG TGA CCA AAA AAT CA-3′

For identifying species a sequence length of 617 nucleotides is recommended and 669 nucleotides for phylum analysis.

1.1. Uses of DNA Barcoding

- ☆ Uniform, practical method of species identification
- ☆ Utility in conservation biology and biodiversity surveys
- ☆ Identification in all developmental stages
- ☆ Determination of food webs

- ✰ Facilitate basic biodiversity inventories like molecular phylogenetics to assembling tree of life and identifying clades and evolutionary relationships
- ✰ Apart from identifying species the number of distinct DNA sequences in environmental sampling and reconstruction of phylogenetic trees which help in inventories of cryptic biodiversity
- ✰ Useful in Biosecurity

The core idea of barcoding is based on the fact that short pieces of DNA that vary only to very minor degree within the species and the variation is much less than between species. Therefore a threshold value of variation could be characterized for each taxonomic group (2-12 per cent), above which groups of individuals do not belong to the same species but form supra-species taxon. Therefore unknown individuals could be assigned to species level.

1.2. Targets for Molecular Identification

a) Mitochondria

Mitochondria are the cells' respiratory power plant for the generation of ATP. Animal mitochondria are small (6 to 20 kb), circular and lack introns with the genes arranged on both DNA strands. Animal mitochondrion has 37 genes coding for small and large subunit rRNA, 13 proteins and 22 *t*RNAs. All mitochondria have at least one non-coding region which contains regulatory elements for replication and transcription, but intergenic regions are small or absent. There are thousands of mitochondria in each cell. Therefore the small size, relatively rapid rate of evolutionary change and maternal inheritance make it suitable for examining population history and evolution among closely related taxa as well as deeper evolutionary relationships. In addition to the above gene order in the mitochondria could also be used as phylogenetic tool.

The rates of molecular evolution vary between different segments of the genome, across taxa. Therefore analysis of rapidly evolving gene regions or taxa will aid in the diagnosis of lineages with brief histories of reproductive isolation, while the reverse will be true for rate decelerated genes or species.

1.3. Advantages of Using Mitochondrial Genome

- ✰ it does not have introns.
- ✰ haploid mode of inheritance.
- ✰ supports less recombination.
- ✰ universal primers are robust which can amplify 5′ end in most of the animals including insects.
- ✰ the 13 protein coding genes in the animal mitochondrial genome are better targets because of rare insertion and deletions (*indels*).
- ✰ the third position nucleotide in the codon of mtCOI show a high incidence of base substitutions, leading to a rate of molecular evolution that is greater than three times as compared to 12S, 16S rDNA sequences.

- ☆ rapid evolution allow discrimination of not only closely related species but also phylographic groups within a single species.
- ☆ identifying amino acid substitution patterns of mtCOI it is possible to assign any unidentified organisms to a higher taxonomic group (phylum, order) before examining nucleotide substitutions to determine its species identity.

There are about 52 million sequence records available currently and it is expected that barcoding initiative of animal kingdom will produce about 100 million sequences and will be available through GenBank. An International data base called BOLD (barcode of life data system) organizes the sequence data on species identification, which was initially developed as on informatics work bench for a single, high volume DNA barcode facility. Later the same has been selected by the Canadian Barcode of Life Network (www.bolnet.ca) for barcode all eukaryotic life of Canada and subsequently it has been adopted by major barcode community like birds, fishes, lepidoptera etc. BOLD provides an integrated bioinformatics platform that supports all phases of analytical pathway from specimen collection to tightly validated barcode library. It also provides a vehicle for collaboration across research communities by coupling flexible security and data entry features with web based delivery. A copy of all sequences in BOLD also sent to NCBI, DDBJ, EMBL as soon as the results are ready for public release.

1.4. Molecular Identification of Mirids

Mirids of the genus *Helopeltis* are serious pests of various cultivated plants in the Old World tropics. Over a century ago, Peal (1873) and Wood-Mason (1884) documented the damaging effect of these insects on tea plants in India. The common name 'tea mosquito bug' was established, because of the damage that resulted from feeding, including 'teablight', 'mosquito blight' and 'spot blight' (Stonedahl, 1991). Of the 41 recognized species of *Helopeltis*, 26 are restricted to Africa, 15 are distributed in Austrasian region (Sundara Raju and Sundara Babu, 1999). However, In India, 3 species of *Helopeltis* are reported; *viz. Helopeltis antonii* Signoret, *H. bradyi* Waterhouse and *H. theivora* Waterhouse that infest various crop plants (Sundara Raju and Sundara Babu, 1999). Over 100 species of plants have been reported as hosts for *Helopeltis* spp. including cashew (*Anacardium occidentale*) black pepper (*Piper nigrum*), tea (*Camellia sinensis*), cocoa (*Theobroma cocoa*), cinchona (*Cinchona spp.*) (Stonedahl, 1991), neem (*Azadiracta indica*) (Omkarappa and Kumar, 1997) and guava (*Psidium guajava*) (Karmavati, 2007). In general, symptoms observed by *Helopeltis* spp. infestation are the discolorations and presence of necrotic areas or lesions around the entry point of labial stylet of the insect into plant tissue. In India, prior to the use of chemical pesticides crop loss on tea plantations sometimes reached 100 per cent and in commercially grown cashew, crop losses sometimes reaching 30-40 per cent (Stonedahl, 1991).

Many species of oriental *Helopeltis* are very similar in coloration and general structures and accurate identification is possible only by examining the male and female genital structures. In this regard, external morphological identification can lead to widespread misidentifications, as in case of *H. antonii* in South India and

Srilanka, since some of these almost certainly pertain to the externally similar species *H. bradyi* (Stonedahl, 1991). While traditional systematic has its strength, molecular identification is not limited by polymorphism, sex and stages of development of the target species and development of suitable species- specific markers will help even a non-specialist too to identify a given species quickly, with a high degree of accuracy. The molecular studies could also elucidate the prevalence of biotypes, cryptic or sub species etc. In mirids very less molecular work has been done. Molecular markers that produce information on genetic diversity are valued for planning pest eradication programmes, for the management of germplasm, etc (Latip *et al.*, 2010). For species identification and molecular diversity studies, researchers have employed various molecular markers and methods *viz.* cytochrome b (Raboudi *et al.*, 2005), 16SrRNA (von Dohlen and Moran, 2000), 18SrRNA, 28SrRNA, 5.8SrRNA, internal transcriber spacer, EF1α (Ji *et al.*, 2003), CO-I, etc. Due to maternally inherited characteristics and reliable inter-specific variation as compared to other markers (Salvolainen *et al.*, 2005) mitochondrial genes have been widely employed in the molecular systematics of insects (Simen *et al.*, 1994).

The concept of DNA barcoding for species using mitochondrial Cytochrome oxidase I (CO-I) gene was first reported by Hebert *et al.*, 2003 a,b. Later a number of workers used the CO-I primers to study the molecular diversity, to prove host associated genetic differences, biotypes, sub (cryptic) species etc. Recently Jung *et al.* (2011) and Park *et al.* (2011) used the CO-I gene for species identification of the true bugs (Insecta: Heteroptera). Molecular identification using CO-I is not limited by life stages, polymorphism, and sex of the target species (Asokan *et al.*, 2011). DNA sequence-based analyses have provided the opportunity for evolutionary biologists and ecologists to address questions they were unable to answer using other morphology data. In this regard, large scale DNA barcoding of species using the CO-I gene was introduced by Hebert *et al.*, 2003 a,b, where the 5′- 658 base pair segment of mitochondrial cytochrome c oxidase subunit 1 (CO-I) was proposed as a 'DNA barcode' for discriminating most animal species. Later, Hebert *et al.* (2004) differentiated ten cryptic species in a single morphologically identified skipper butterfly species which differed in their larval appearance, host and habitat preferences by employing the CO-I gene fragment. DNA barcoding requires only PCR and sequencing both of which are readily available, rapid and inexpensive (Glover *et al.*, 2010). Molecular studies using the CO-I region of mtDNA revealed cryptic species in many insects' *viz. Baetis vernus* (Stahls and Savolainen 2008), *Mordellistena convicta* (Blair *et al.*, 2005) and *Thrips tabaci* (Brunner *et al.*, 2004).

The primary utility of DNA Barcoding is to identify unknown specimens at the species-level by comparing the query sequence to a DNA barcode reference library built based on known species, for the quarantine purpose, phytosanitory application, food quality and ecological analysis. In addition, patterns of sequence variations can be used for the recognition of cryptic (sub) species, haplotypes, biotypes, *etc.*, which can be effectively used for the insect pest management programmes of mirids. Various studies unambiguously demonstrated the development of species-specific markers based on CO-I, a significant step in the identification of *H. antonii, H. thievora, H. bradyi and Pachypeltis maesarum* from many other species of mirids that occur on cash crops in India (Asokan *et al.*, 2012, Rebijith *et al.*, 2012, Figure 19.1). Colour morphs and size variations in *Helopeltis* species are influenced by many factors, which could be an impediment for species identification (Stonedahl, 1991). Species-specific markers require only conventional PCR, followed by agarose gel electrophoresis by which even a non specialist can identify the species in any of the life stages based on the amplicon size. Hence the molecular identification using species-specific marker is an advantage where there is polymorphism in the target species. Jung and Lee (2012) and many other researchers showed that CO-I specific primers could successfully differentiate mirid species, but the additional cost was sequencing and time requirement, both of which can easily trounced by using the species-specific markers.

Crop losses due to *Helopeltis* species is being increasingly felt in Old World countries, especially in India on a range of plantation crops. *Helopeltis* is typically a low intensity pest with high damage and the plants get affected by the toxin injected during the feeding by the nymphs and adults. While on crops such as cashew, cocoa, tea, neem and guava, the chemical nature, potency of the toxin and extend of damage may vary depending upon the species of *Helopeltis*. Endosulfan, the present potent chemical control against mirids, has initiated an unabated debate on issues *viz.* environmental-impact assessment, Resistance, Residues, and Resurgence and frequently the health hazards. At present lack of precise species determination has resulted in designating most of the *Helopeltis* species occurring on cashew, tea, neem, guava as *H. antonii*, in many part of the country. Thus valuable data on insecticide resistance, host-natural enemy interactions, distribution, host-range etc. is all attributed to *H. antonii* which may prove costly in its management. Besides it has quarantine and other issues both within and outside the country. It will also strengthen the hands of traditional taxonomists in adding validity to their description. In this regard, recent work on molecular identification of mirids has succesfully identified *Helopeltis antonii, H. bradyi, H. thievora* and *Pachypeltis maesarum* employing CO-I marker (Rebijith *et al.*, 2012, Figure 19.2). Such studies will stand as a connecting link for developing such a decisive tool in imperative for identification of *H. antonii, H. thievora, H. bradyi* and *Pachypeltis maesarum* in any of the developmental stages *viz.* eggs, nymphs and adult and also for formulating an IPM strategy in cashew. Future management of these pests can now be more accurately implemented with the accurate identification between these species, and other mirids.

2. RNAi in Insect Pest Management

The noble cause of feeding 110-crore hungry stomachs is the Herculean task faced by the Indian agriculture today and the population is slated to increase by leaps and bounds in the ensuing years. This calls for more land to be brought under cultivation and also increasing productivity of crops. While there is no scope to further increase the area under cultivation, increasing the productivity is only the plausible, viable option to meet the demand. It has been estimated India's population will have roughly 200-crore by 2040 and we have to go a long way from the current production level of 800 million tons of major crops. Among

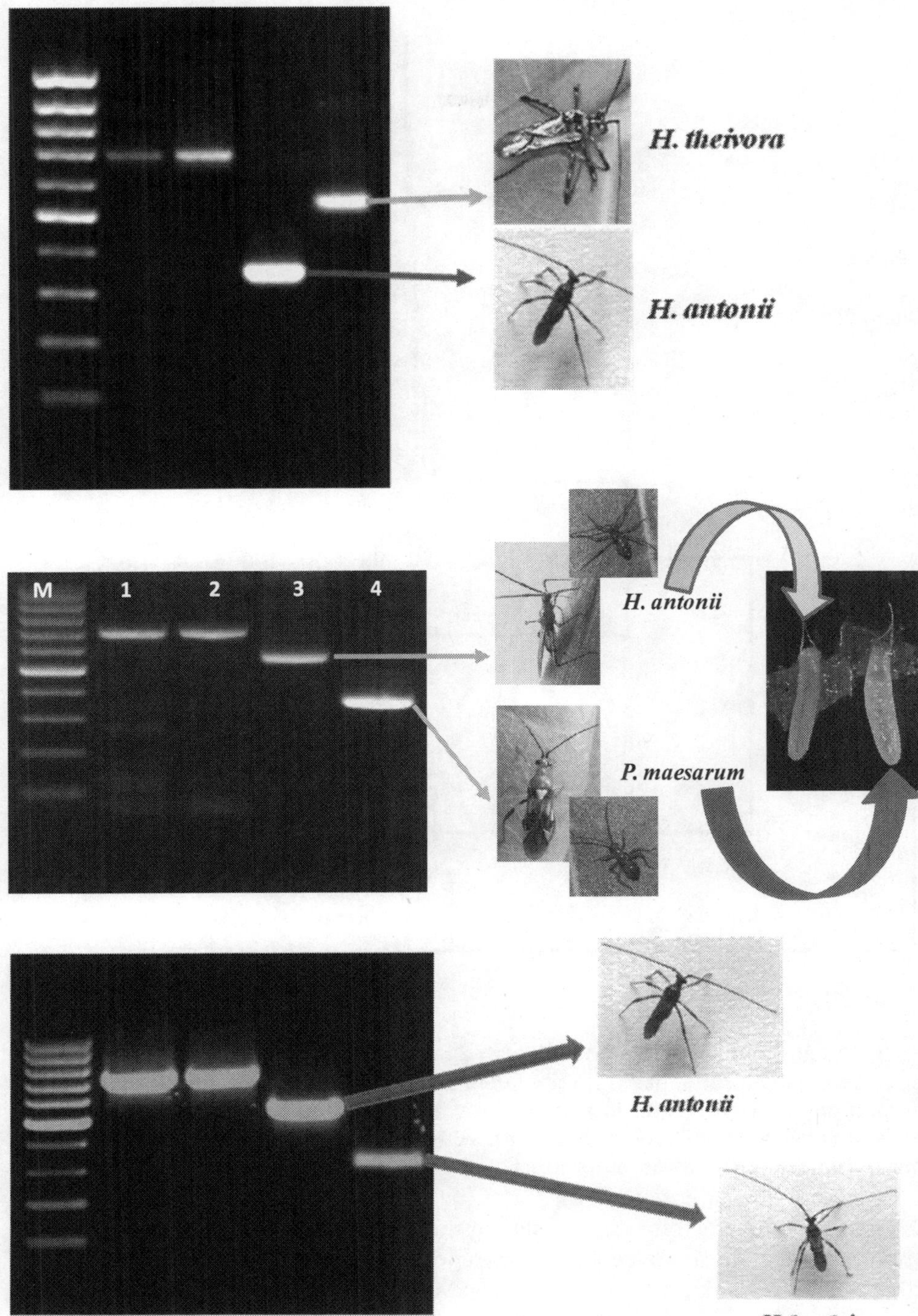

Figure 19.1: Species-Specific Markers Developed for Various Mirids Species in India. (Asokan *et al.,* 2012, Rebijith *et al.,* 2012).

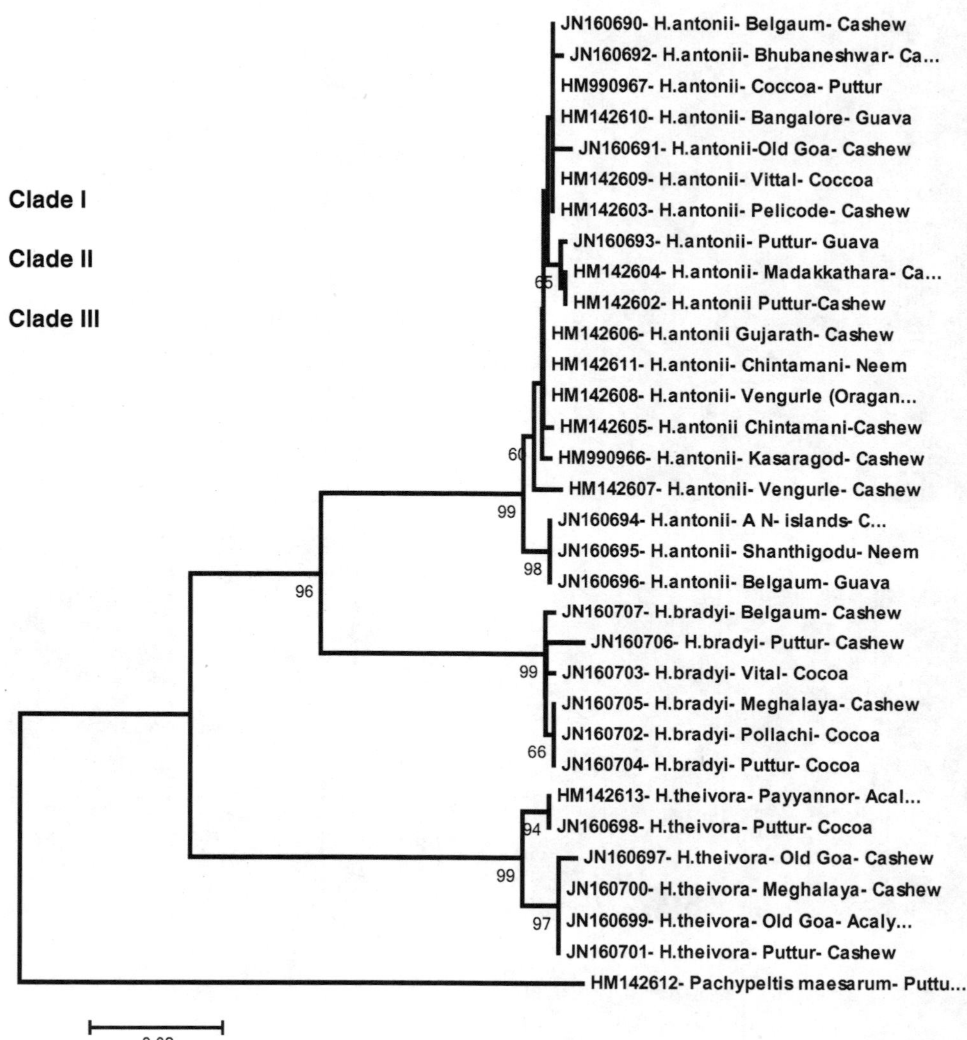

Figure 19.2: Neighbour-Joining Tree with Bootstrap Support Showing Clustering of Species for CO-I Sequences. Three clades can be seen, (Clade I, associated with *H. antonii* populations, Clade II, associated with *H. bradyi* populations and Clade III, consists of *H. theivora* populations), with > 90 per cent bootstrap support. *Pachypeltis maesarum* was used as an outgroup.

the various biotic factors insect pests limit the crop productivity to a larger extent and the agrochemical policy group has reported that the crop loss during the year 2007 was Rs.1.40 lakh crores due to pests. Management of insect pests Insecticides account for and. excessive reliance on chemical insecticides (61 per cent of the total pesticides usage) has already resulted in many control failures due to accelerated development of resistance. Further it has also resulted in contamination of soil, water and affected non-target pests including human beings.

Therefore, there is an urgent need to look for an effective, ecofriendly alternative for insect pest management, where a new and novel approach called RNA interference (RNAi), popularly called Gene Silencing is poised to play a vital role. Employing this RNAi approach, it is now possible to control insect pests by silencing some of the vital genes that play an important role in insect-host plant interaction, growth and development, flight, reproduction etc by delivering cognate double stranded RNA (*ds*RNA) either as spray or through transgenic plant. But RNAi approach has to be tailor made and cannot be universal, as the mechanism of RNAi is highly species specific. Hence the basic mechanisms underlying RNAi for selected species of insect pests that belong to three important orders of insects has to be studied thoroughly before venturing into field level application.

RNA interference (RNAi)/gene silencing is a rapidly evolving tool in molecular biology, cellular process by which an mRNA is targeted for degradation by a small interfering RNA that contains a strand complementary to a fragment of the target mRNA, resulting in sequence specific inhibition of gene expression. The discovery of RNAi enabled the use of loss-of-function analyses in many non-model insects other than *Drosophila* to elucidate the roles of specific genes. The RNAi approach has been widely used on insects in several fields, including embryogenesis, pattern formation, reproduction, biosynthesis and behavior. The increasing availability of insect genomes has made the RNAi technique an indispensable technique for characterizing gene functions in insects.

The specificity is sequence-based and depends on the sequence of one strand of the dsRNA corresponding to part or all of a specific gene transcript (for recent RNAi reviews see). RNAi is a post-transcriptional control mechanism involving degradation of a target mRNA. This degradation is mediated through the production of small interfering RNAs (siRNAs) from the dsRNA, which is cleaved by dsRNA-specific endonucleases referred to as dicers (from the dicer gene identified in Drosophila melanogaster, reviewed in). The siRNAs are 21 bp dsRNA fragments carrying two base extensions at the 30 end of each strand; one strand of the siRNA is assembled into an RNA-induced silencing complex (RISC) in conjunction with the argonaute multi-domain protein, which contains an RNaseH-like domain responsible for target degradation. The process is closely related to post-transcriptional gene regulation by microRNAs (miRNAs), where the end-result is inhibition of translation initiation, and shares many of the same components. In plants and nematodes, RNAi can have systemic effects on gene expression, so that gene knockout spreads throughout the organism and persists over development. The basis of this effect is thought to lie in the presence of an RNA-dependent RNA polymerase (RdRP) that is able to interact with the RISC complex and generate new dsRNA based on the partially degraded target template by using the hybridised *si*RNA strands as primers. The synthesized *ds*RNA is then acted on by the dicer enzymes to generate new *si*RNAs (secondary *si*RNAs), thus acting as an amplification step. In this way, once a *ds*RNA is introduced into a cell, its effect can persist over development; in addition, the *ds*RNAs can be exported to neighbouring cells and thus spread the gene knockout effect through the organism.

RNA interference was first observed in petunias, when Napoli *et al.* (1990) discovered that introduction of a pigment-producing gene under control of a powerful promoter suppressed expression of both the introduced gene and the homologous endogenous gene, a phenomenon they called cosuppression. Cosuppression was subsequently found to occur in many species of plants and fungi (where it was called quelling) and to occur at the post-transcriptional level.

Such post-transcriptional gene silencing (PTGS) was shown to be mediated by a diffusible, trans-acting molecule in both *Neurospora* and plants. The crucial 1998 discovery by Fire *et al.,* that injection of dsRNA - a mixture of both sense and antisense strands of the target mRNA, rather than either strand alone - into gonads of the nematode *Caenorhabditis elegans* resulted in extremely potent silencing unequivocably identified dsRNA as the inducer of RNA interference. Short (~25 nt), antisense RNAs were first implicated in PTGS in plants. Further work in *Drosophila*-derived in vitro systems showed that long dsRNA was processed to short interfering RNAs (*si*RNAs), 21–23 nt long, that could mediate target mRNA cleavage in the region corresponding to the introduced siRNA. Current thinking about the mechanism of RNAi and components of the RNA-induced silencing complex is described more fully in the text. In *C. elegans,* the demonstration that silencing could be induced by simply feeding nematodes with bacteria that had been engineered to express *ds*RNA homologous to the target gene, or by soaking worms in such dsRNA, drove rapid adoption of RNAi as a technology for reverse genetics at the organismal level. Likewise, technology was quickly developed for inducing RNAi in *Drosophila* cell culture by bathing or transfection with dsRNA. Efforts to use RNAi in mammalian cells were hampered at first because of a nonspecific, interferon-mediated response to dsRNA longer than 30 bp that is seen in most mammalian cell lines. With the demonstration that 21–22 nt siRNAs, either chemically synthesized or expressed from a plasmid vector, could efficiently induce RNAi in mammalian cells without inducing the interferon response, the door was opened for development of RNAi tools in mammalian systems.

While some researchers were studying the phenomenon of RNAi and working to exploit its use as a tool for reverse genetics, others were trying to understand a distinct but related set of small RNAs—microRNAs (miRNAs). The first miRNAs to be discovered, lin-4 and let-7, were identified through loss-of-function mutations affecting control of postembryonic development in *C. elegans*. Other miRNAs were soon identified in C. *elegans, Drosophila,* and mouse by combinations of forward genetics, cDNA cloning, bioinformatics, and reverse genetics. By 2002, miRNAs were firmly established as a large class of conserved regulatory molecules in animals and plants. miRNAs have been shown to play a role in developmental timing, cell death, cell proliferation, and oncogenesis. This class of small RNA may represent 2–3 per cent of the total number of genes in humans, and estimates of miRNA target-binding sites indicate that *mi*RNAs may play a role in regulating as many as 30 per cent of mammalian genes.

A formal connection between the phenomenon of RNA interference and miRNAs was made when the same enzyme, Dicer, was shown to process both long dsRNA into siRNAs and cytoplasmic miRNA precursors (pre-miRNAs) into mature miRNAs. Other components of the RNAi and miRNA silencing pathways have been shown to be closely related, most prominently Argonaute, which is a key component of the RNA-induced Silencing Complex (RISC). It now appears that miRNAs function in a silencing complex that is similar, if not identical, to RISC to regulate expression of target genes either through cleavage of mRNA or translational repression: if the miRNA exhibits perfect complementarity to its target mRNA, the mRNA is cleaved (typically in plants); if there is only partial complementary, translational repression occurs (typically in animals). The further discovery of endogenous small RNAs, distinct from miRNAs, that function in transcriptional silencing and genome stability has driven the adoption of the more general terms ¯RNA silencing and ¯small RNAs to describe the collection of related silencing pathways and their RNA guides. The term ¯RNA interference (RNAi) continues to describe mRNA cleavage that is induced by a dsRNA trigger.

Mechanism of RNAi

RNAi is an innate immune response that retard the gene expression by degrading specific mRNA molecules. RNAi mechanism is generally employed for maintaining genome integrity against the invasion by transposons and viruses. The RNAi pathway is initiated by the cellular enzyme called Dicer (RNase III), which cleaves the long double stranded RNA (dsRNA) into short fragments of 20-25 base pairs. One of the two strands of each fragment known as guide strand is incorporated into the RNA induced silencing complex (RISC) and base pair with complementary mRNA sequences and cleaves the cognate mRNA thus arresting the target gene expression (Figure 19.3). This mechanism is also known as post transcriptional gene silencing (PTGS) in plants. In some organisms like *Caenorhabditis elegans* and ticks the silencing signal generated in a cell is communicated to other cells which are exposed to the dsRNA which is known as *systemic silencing*. Two components which are needed to elicit systemic silencing are RdRP (RNA dependent RNA polymerase) and *sid* 1 and 2 genes (systemic interference defective). Presence of at least one of them has been documented in various insects except for *Drosophila melanogaster* (Figure 19.4).

The use of long dsRNAs (>400bp) has been successful in generating RNA interference effects in many organisms including *Drosophila* (Misquitta and Paterson, 1999), zebrafish (Wargelius *et al.*, 1999), *Planaria* (Sanchez-Alvarado *et al.*, 1999) and numerous plants (Jorgensen, 1990, Fukusaki *et al.*, 2004, Jensen *et al.*, 2004). In mammalian systems, siRNA molecules of 21–22 nucleotides or short hairpin RNAs (shRNAs) are used to avoid endogenous nonspecific antiviral responses that target longer dsRNAs (Caplen *et al.*, 2001, Elbashir *et al.*, 2001a). Yu *et al.* (2002) and others (Brummelkamp *et al.*, 2002b; McManus *et al.*, 2002; Sui *et al.*, 2002; Xia *et al.*, 2002; Barton and Medzhitov, 2002) demonstrated that shRNAs bearing a fold-back, stem-loop structure of approximately 19 perfectly matched nucleotides connected by various spacer regions and ending in a 2-nucleotide 32-overhang can be as efficient as siRNAs at inducing RNA interference. si/shRNAs can induce specific gene silencing in a wide range of mammalian cell lines without leading to global inhibition of mRNA translation (Caplen *et al.*, 2001; Elbashir *et al.*, 2001a; Paddison. *et al.*, 2002).

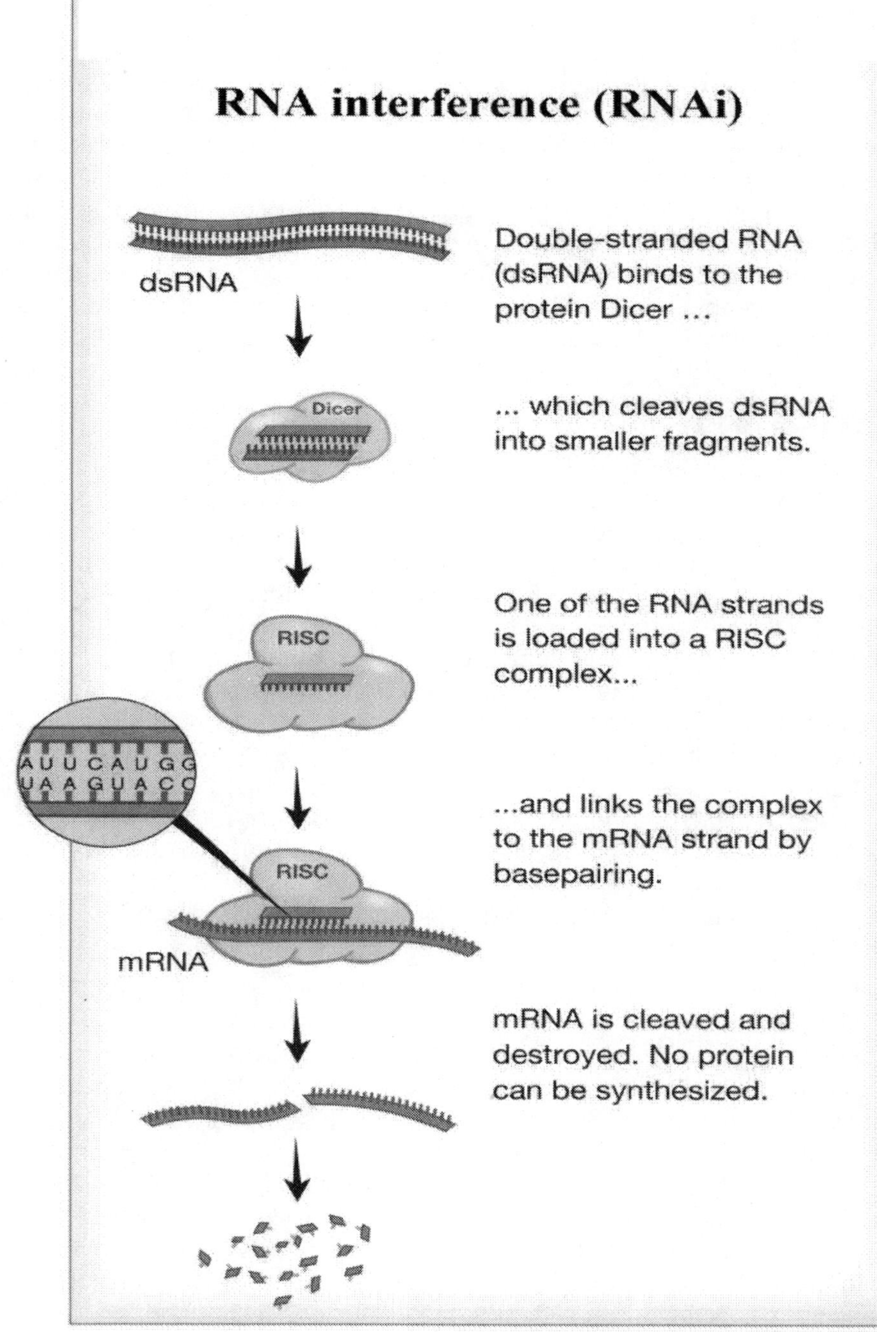

Figure 19.3: General Mechanism of RNAi.

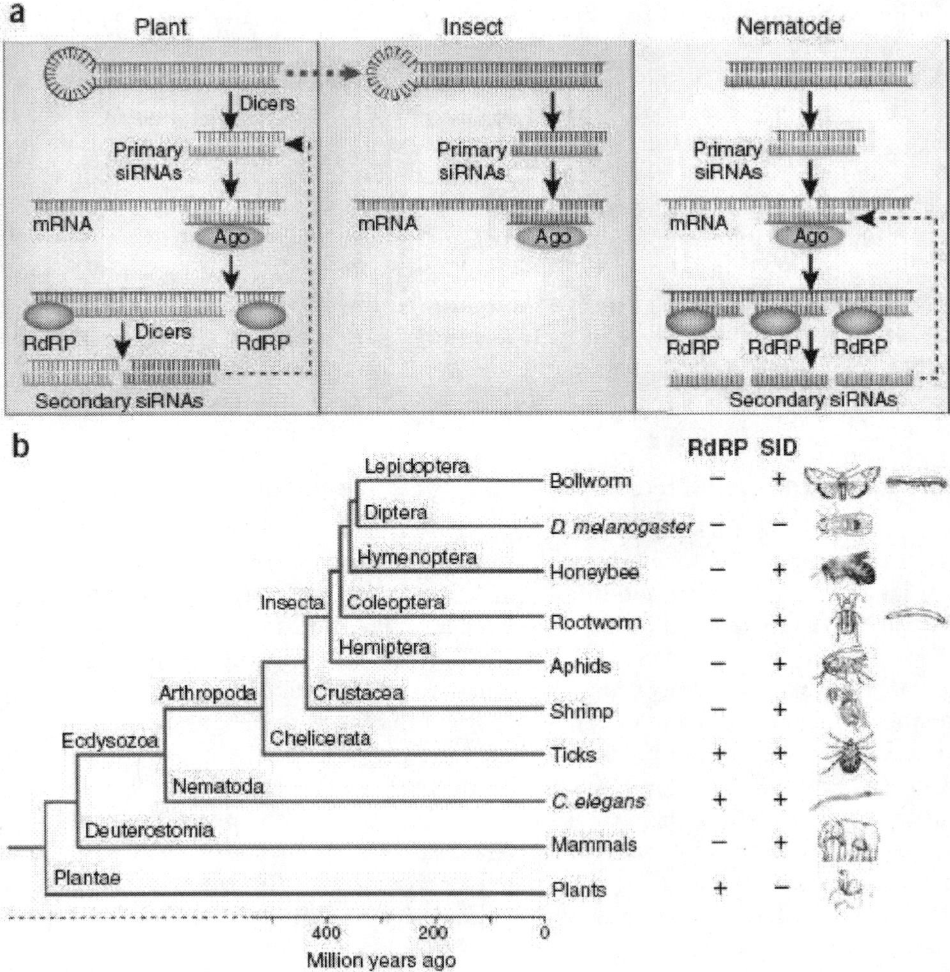

Figure 19.4: Variation in Theme of RNAi.

***Courtesy*: Gordon and Water house, 2007**

RNAi involves a multistep process (Figure 19.3). dsRNA is recognized by an RNase III family member (*e.g.*, Dicer in *Drosophila*) and cleaved into siRNAs of 21–23 nucleotides (Agrawal *et al.*, 2003; Elbashir *et al.*, 2001b; Bernstein *et al.*, 2001; Hammond *et al.*, 2000). These siRNAs are incorporated into an RNAi targeting complex known Gordon and Water house, 2007 as RISC (RNA-induced silencing complex), which destroys mRNAs homologous to the integral siRNA (Hammond *et al.*, 2000; Bernstein *et al.*, 2001). The target mRNA is cleaved in the center of the region complementary to the siRNA (Elbashir *et al.*, 2001c), with the net result being rapid degradation of the target mRNA and decreased protein expression.

RNAi has revolutionized the study of gene function, and is being explored as a therapeutic tool (for reviews, see Dorsett and Tuschl, 2004; Hannon and Rossi, 2004). For example, RNAi has been used to identify gene products essential for cell growth (Harborth *et al.*, 2001), to cause subtype and species-specific knockdown of various protein kinase C (PKC) isoforms in both human and rat cells (Irie *et al.*, 2002), and to specifically target degradation of an oncogene product (Wilda *et al.*, 2002). RNAi has also been used to specifically target and prevent viral infections by HIV-1 and HCV in cell culture (Park *et al.*, 2002) and intact animals (McCaffrey *et al.*, 2002). These observations open the field for further studies toward novel gene therapy approaches for anti-cancer or anti-viral treatments using siRNAs or shRNAs. dsRNA processing proteins (RNase III-like enzymes) bind to and cleave dsRNA into siRNA. The siRNA forms a multicomponent nuclease complex, the RNA-induced silencing complex (RISC). The target mRNA recognized by RISC is cleaved in the center of the region complementary to the siRNA and quickly degraded.

Applications of RNAi in Insect Pest Management

Most of the studies on RNAi in entomology were on establishing the function of genes that are involved in various metabolic processes (Kramer and Bentley, 2003., Fabrick *et al.*, 2004., Dong and Friedrich, 2005., Amico and Nijhout, 2006., Arakane *et al.*, 2006., Collinge *et al.*, 2006., Lagos *et al.*, 2006., Mutti *et al.*, 2006). But many researchers envisaged the potential of RNAi for field level insect pest management, but success was not forthcoming due to many reasons like poor understanding on the mechanism of RNAi in different orders of insects, amplification and systemic spread of the silencing signal in the treated insects, amenability of different orders of insects for the RNAi approach etc. In the year 1998 an experimental proof that introduction of dsRNA into the cell evoked RNAi mechanism in the free living nematode, *C. elegans* (Fire *et al.*, 1998) opened a new exciting area of RNAi in various spheres of research including entomology. In the year 2007 two research groups from China and USA experimentally validated the utility of RNAi in insect pest management. The choice of the gene for RNAi mediated silencing could be many and generally falls into two broad categories *viz.* silencing of genes that results in quick control *e.g.* genes involved in insect-plant interactions, digestion, moulting etc and silencing of genes that results in long term management *e.g.* genes involved in pheromone biosynthesis, pheromone reception, migration, flight, diapause etc. Mao *et al.* (2007) demonstrated that it is possible for a no chemical management of the cotton boll worm, *Helicovera armigera* by transgenic cotton mediated silencing of cytochrome P 450. Similarly Baum *et al.* (2007) proved that RNAi is approach is feasible in the management of the coleopteran pest the corn root worm (*Diabrotica virgifera virgifera*). Likewise different sets of genes could be targeted depending upon the requirement.

The recent demonstration of the potential of RNAi in insect pest management has opened a new avenue which will fuel a futuristic approach where application of chemical insecticides will no longer be needed. Identification of suitable target genes and delivery method will usher a new ecofriendly approach that is safer to non-target organisms including humans. In addition to its role in field level insect pest management RNAi is also promising in other entomological investigations such as validation of gene functions, control of insect vector transmitted plant viruses, management of insecticide resistance, etc. (Asokan, 2007).

References

Hebert, P. D.W., Cywinska, A., Ball, S. L and deWaard, J. R. (2003). Biological identification through DNA barcodes. *Proceedings of the Royal Society of London,* 270: 313-321.

Folmer, O., Black, M., Hoch, W., Lutz and Vrijenhock, R. (1994). DNA primers for amplification of mitochondrial cytochrome C oxidase subunit I from diverse metazoan invertebrates. *Molecular Marker Biology and Biotechnology*, 3: 294-299.

Rubinoff, D., Cameron, S and Will, K. (2006). A genomic perspective on the shortcomings of mitochondrial DNA for "Barcoding" identification. *Journal of Heredity*, 97: 581-594.

Amico, L. J. and Nijhout, H. F. (2006). Isolation and characterization of a bombyxin-like cDNA from *Manduca sexta*. *Journal of Insect Science*, **6**: 14.

Arakane, Y., Lomakin, J., Muthukishnan, S., Beeman, R. W., Gehrke, S. H., Kanost, M. R. and Kramer, K. J. (2006). An RNAi study of the metabolic pathways for cuticle sclerotization and pigmentation in *Tribolium castaneum*. *Journal of Insect Science*, 6: 4.

Asokan, R. (2007). Genetic Engineering of Insects. *Resonance*, 12: 47- 56.

Baum, J. A., Bogaert, T., Clinton, W., Heck, G. R., Feldman, P., Ilagan, O., Johnson, S., Plaetinck, G., Munyikwa, T., Pleau, M., Vaughn, T. and Roberts, J. (2007). Control of coleopteran insect pests through RNA interference. *Nature Biotechnology*, 25: 1-5.

Collinge, D., Gordon, K., Behm, C. and Whyard, S. (2006). Stable transformation and RNAi in *Helicoverpa armigera*. *Journal of Insect Science*, 6: 10.

Dong, Y. and Fredrich, M. (2005). Nymphal RNAi: systemic RNAi mediated gene knockdown in juvenile grasshopper. *BMC Biotechnology*, 5: 5-13.

Fabrick, J. A., Kanost, M. R. and Baker, J. E. (2004). RNAi-induced silencing of embryonic tryptophan oxygenase in the pyralid moth, *Plodia interpunctella*. *Journal of Insect Science*, 4: 1-9.

Fire, A., Xu, S., Montgomery, M. K., Kostas, S. A., Driver, S. E. and Mello, C. C. (1998). Potent and specific genetic interference by double-stranded RNA in *Caenorhabditis elegans*. Nature, 391: 806-811.

Gordon, K. H. J. and Waterhouse, P. M. (2007). RNAi for insect proof plants. *Nature Biotechnology*, 25: 1231-1232.

Kramer, S. K. and Bentley, W. E. (2003). RNA interference as metabolic engineering tool: potential for in vivo control of protein expression in an insect larval model. *Metabolic Engineering*, 5: 183-190.

Mao, Y. B., Cai, W. J., Wang, J. W., Hong, G. J., Tao, X. Y., Wang, L. J., Huang, Y. P. and Chen, X. Y. (2007). Silencing of a cotton boll worm P450 monooxygenase gene by plant mediated RNAi impairs larval tolerance of gossypol. *Nature Biotechnology*, 25: 1307-1313.

Mutti, S., Park, Y., Reese, J. C. and Reeck, G. R. (2006). RNAi knockdown of a salivary transcript leading to lehatlity in pea aphid, *Acyrthosiphon pisum*. *Journal of Insect Science*, 6: 1-7.

Cashew: Improvement, Production and Processing (2017) *Pages* **347–365**
Editor: **Prof. P.L. Saroj**
Published by: **ASTRAL INTERNATIONAL PVT. LTD., NEW DELHI**

20

Tea Mosquito Bug and its Management

P.S. Bhat, T.N. Raviprasad, K. Vanitha, and K.K. Srikumar

1. Introduction

Tea mosquito bugs (*Helopeltis* spp). belong to family Miridae, sub family Bryocorinae and tribe Monaloniini. Signoret (1858) raised genus *Helopeltis* mainly to differentiate it from the closely resembling genus *Pachypeltis*. *H. antonii* Sign is the first described species based on the specimen provided by Antonii Dohrn from Sri Lanka (Mann, 1907). In old world tropics, *Helopeltis* spp. are serious pests of many economically important crops. The damaging effects of these insects on tea plants in India were documented over a century ago by Peal (1873) and Wood-Mason (1884). These early records established the common names 'tea bug' and 'tea mosquito'. The damage was reported as 'tea blight', 'mosquito blight' and 'spot blight'. Waterhouse, (1886) subsequently, identified this species as *H. theivora*. In Southern India, the outbreak of *H. theivora* Wat. was reported around 1920 on tea (Rao, 1970), whereas, the damage of *H. antonii* was first reported on neem and guava (Rao, 1915) and its damage on other crops *viz.*, cashew (Ayyar, 1932), cocoa (Fletcher, 1914; Abraham and Padmanaban, 1967), cinchona (Fletcher, 1914), drumstick (Pillai, *et al.*, 1979) and ber (Sundararaju, 1996). The record of *H. bradyi* Wat. occurring on cocoa, guava and cashew was brought out by Stonedahl (1991) and Sundararaju (1996). Occurrence of *H. cinchonae* Mann as a pest of cinchona was also reported in Bhutan (Mann, 1907). However, in India, at present only three species of *Helopeltis viz.*, *H. antonii, H. bradyi and H. theivora*, infest various crop plants (Sundararaju, 1996)

(Figures 20.1–20.3). As *Helopeltis* spp. are key pests of certain plantation and horticultural crops, the vivid account on information available on *H. antonii, H. theivora* and *H. bradyi* with particular reference to cashew has been presented through subsequent compilations by Devasahayam and Nair (1986), Stonedahl (1991) and Sundararaju and Sundarababu (1999).

Figure 20.1: ***H. antonii.***

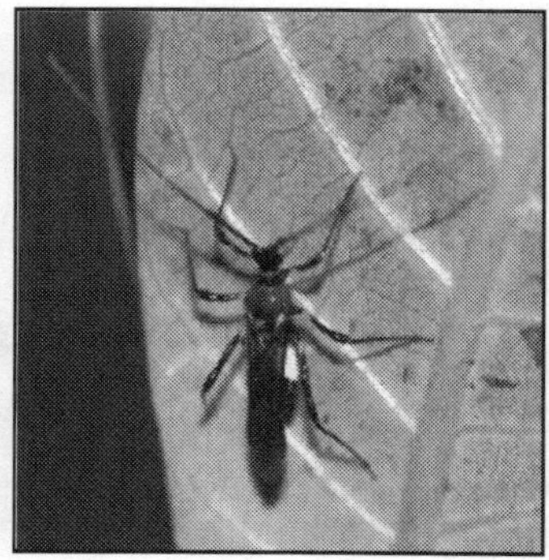

Figure 20.2: ***H. bradyi.***

Figure 20.3: ***H. theivora.***

2. Distribution

Of the 41 recognized species of *Helopeltis*, 26 are restricted to Africa and 15 are distributed in Australasian region (Stonedahl, 1991and Stonedahl, *et al.*, 1995). Only three species (*H. antonii*, Sign., *H. theivora* Wat. and *H. bradyi* Wat.) are recorded from India. *H. antonii* occurs in Southern and Eastern India, whereas, *H. theivora* exists in West coast, Western Ghat and Northeast India. *H. bradyi* is found in West coast and Western Ghat regions (Sundaraju and Sundarababu, 1999). Cashew and cocoa are damaged by all the three species. Tea is damaged only by *H. theivora*. Since, the late 1800s, over 100 species of plants have been reported as host plants for Helopeltis. Stonedahl (1991) clearly demarcated the distribution of *H. antonii, H. theivora* and *H. bradyi. H. antonii* is restricted only to Indian subcontinent. The earlier records of *H. antonii* occurring in Indonesia were the incorrect identity (Roepke, 1916; Leefmans, 1916; Sudarmadji and Gunawan, 1994). Its correct identity was *H. bradyi.* In spite of the earlier information, *H. antonii* was being reported as a pest of tea in South India (Ayyar, 1940). As such in the oriental region, *H. antonii* is confined only in South and East India, Andaman Islands and Sri Lanka and *H. theivora* in South India, (Western Ghats and West coast), Northeast India, Sri Lanka and South East Asia, whereas, *H. bradyi* is confided to South India (Western Ghats and West coast), Sri Lanka, Indonesia and Malaysia (Stonedahl, 1991; Sundararaju, 1996). Variations in colour pattern of head and pronotum of *Helopeltis* spp. occurring in South India, Assam, Vietnam were described by Stonedahl (1991), Sathiamma (1981) and Sundararaju (1996).

3. Host Range

Helopeltis antonii

Cocoa, cinchona, Persian neem and annatto (Fletcher, 1914), Mahogany

(Fletcher, 1914; Rao, 1915); guava (Cherian, 1937; Puttarudriah, 1952; avocado (Puttarudriah, 1952), apple and grapevine (Puttarudriah and Appanna, 1955); cotton (Puttarudriah, 1958) cocoa (Abraham and Remamony, 1979; camphire (*Lawsonia alba* Lam.) (Sundararaju, 1984); drumstick (*Mringa oleifera* Lam.) (Pillai *et al.*, 1979); rose apple, mango, all spice and black pepper (Devasahayam and Nair, 1986); poria tree (*Thespesia populnea* L.) (Sundararaju and Baktavatsalam, 1994); *Ailanthus excels* Roxb. (Satapathy, 1993; Sundararaju, 1996); ber (*Zizyphus mauritiana* Lam.), Indian gooseberry (*Emblica officinalis* L.), cotton (*Gossypium barbadense* L. and *G.hirsutum* L.), cowpea and a Compositae weed plant (*Lactuca runcinata* DC.) (Sundararaju, 1996); neem (Sundararaju and Sundarababu, 1996; Onkarappa and Kumar, 1997) Singapore cherry (*Muntingia calabura* L.) (Srikumar and Bhat, 2013); *Annona* spp. (Venkata, 2009).

Helopletis theivora

Tea (Mann, 1907), *Solanum* spp., *Mussaenda frondosa* L., *Ochlandra travancorica* Benth., guava and cinchona (Anstead and Ballard, 1922); cinchona, cocoa, *Cinnamomum camphora* L., *Erythrina* sp., *Melia azadirachta* L., *Tephrosis* sp., guava, *Maesa* sp., *Polygonum* sp., jack, *Jasminum* sp., ornamental *Camellia* sp., red gum, coffee, *Bidens pilosa* L., morning glory, sweet potato, mango, cashew and *Passiflora* sp. (Rao, 1970); wild rhododendron (*Melastoma malabathricum* L.), *Maesa ramentacea* Roxburgh., *Eurya acuminate* DC., *Jasminum scandens* Vahl. and *Mikania micrantha* (Das, 1984); *Similax* sp. and *Phlogacanthus* sp. (Somchoudhury *et al.*, 1993); *Chromolaena odorata* L. (NRCC, 1993; Srikumar and Bhat, 2013), golden dew drops (*Duranta repens* L.), tita phool (*Phlogacanthus thrysiflora* Nees), jungli pan (*Piper* sp.,) pepal tree (*Ficus* sp.), bor sonborial (*Sida cordifolia* L.), sorn tree (*Persea bombycina* King ex Hook.f.), bhang leaf (*Cannabis sativa*), ixora, missi (*Melastoma malabathricum* L.), Mikania (*M. micrantha*), *Acalypha* sp., Malati (*Jasminum scandens* Vahl.), Bortengeshi (*Oxalis acetosella* L.), *Eurya acuminata* DC., *Morus* sp., kadam (*Neolamarckia cadamba* Roxb.), jamun (*Syzygium cumini* L.), boal (*Cordia dichtoma* G.Forst.) and China rose (*Hibiscus* sp.) (Tocklai, 2010).

Helopeltis bradyi

Tea, cinchona, *Dioscorea* sp., *Oxalis* sp. and *Cephaelis angustifolia* Ridl. (Miller, 1941); cocoa, cashew, cinchona, tea, *Capsicum* sp. and *Citrus* sp. in Java, gutta percha (*Palaquium gutta* Hook.) in West Malaysia, coffee (*Coffea arabica* L.) in Sri Lanka and *Eucalyptus saligna* Sm. in Sumatra (Lever, 1949), cocoa (De Silva, 1957); coffee in Sri Lanka, *E.saligna* and *Acacia mangium* Willd. in Sarawak (Hamid, 1987); cashew (Sundararaju, 1996).

4. Damage Symptoms

Helopeltis antonii is the dominant species on cashew and reviews by Devasahayam and Nair (1986), Sundararaju and Baktha-vatsalam (1994), Sundararaju (1996) and Sundararaju and Sundarababu (1999) cover information on distribution, nature and extent of damage, biology, natural enemies and host plants. Neem is the primary host especially in Tamil Nadu and Southern parts of Karnataka and Andhra Pradesh for *H. antonii* (Pillai and Gopi, 1990). It spreads to cashew from neem as a regular

Figure 20.4: Damaged Shoot.

Figure 20.5: Damaged Panicle.

Figure 20.6: Damaged Immature Nut and Apple.

pest. In the West Coast region, neem is not an endemic plant species and *H. antonii* confines mainly to cashew as an adaptive change (Sundararaju, 1996).

Helopeltis antonii is distributed in most of cashew growing regions of Kerala, Karnataka, Goa, Maharashtra and Tamil Nadu (Pillai *et al.*, 1976); Goa (Sundararaju, 1984a); Andhra Pradesh (Rao and Srinivasan, 1984); Odisha (Jena *et al.*, 1985). Loss in nut yield of 25 to 50 per cent has been reported from Karnataka, Goa, Kerala and West Bengal (Abraham and Nair, 1981; Chatterjee, 1989). The closely related mirid species *Pachypeltis maesarum* Kirk, also cause similar damage as a minor pest (Pillai *et al.*, 1976; Sundararaju, 1996).

On cashew, the nature of damage caused by *H. antonii* has been described by various workers (Pillai *et al.*, 1976; Ambika and Abraham, 1979; Devasahayam and Nair, 1986; Satapathy, 1993; Sundararaju, 1996; Sundararaju and Sundarababu, 1999). Nymph and adult stages of the pests suck sap from tender shoots, panicles and immature nuts and apples.

A large number of crops are attacked by *Helopeltis* spp. as mentioned above. Typical feeding damage by *Helopeltis* spp. appears as a discoloured necrotic lesion around the point of entry of the labial stylets into the plant tissue. The lesion can be elongate or spherical, and becomes darker with age as the tissue around the stylet puncture dies, in response to the enzymatic action of the insects salivary secretions (Stonedahl, 1991). Nymphs and adults of *H. antonii* initially tap the plant surface with labial tip followed by immediate insertion of stylet intracellularly with a minimum depth of 0.3mm. Water soaked lesion appear within a minute after insertion of stylet indicating rapid diffusion of salivary secretion. Feeding on a single site lasted with a maximum of 17 min and evacuated the contents of plant cells leaving initially the cell walls intact and uncollapsed. Subsequently, the melanization and necrosis of feeding lesions appear (Sundararaju, 1996). In the salivary gland of *H. antonii*, hydrolytic enzymes (protease and lipase), oxido-reductase enzymes (catechol oxidase, catalase and peroxidase) and free amino acids were detected. The salivary enzymes cause phyto-toxaemia on various host plants as well as detoxification of defensive chemicals especially in the neem. The role of free amino acids existing in salivary secretion was to protect the salivary enzymes from denaturing by defensive chemicals of host plants (Sundararaju and Sundarababu, 1996).

The lesions on shoots and panicles coalesce and ultimately result in shoot blight or blossom blight. During the emergence of new flushes, the severely infested trees bear a scorched appearance. Each nymph/adult during its life time has got potential to damage atleast three tender shoots/panicles. A single nymph can cause blighting of emerging tender shoot/panicle within 3-4 days of feeding. In general, 1-3 nymphs were seen in each emerging tender shoot/panicle, whereas, in each flowering panicle with immature fruits, maximum of upto 20 nymphs were found (Sundararaju *et al.*, 1993; Sundararaju and Bakthavatsalam, 1994). In 24 h feeding, the mean number of feeding lesions produced by a single nymph/adult on tender shoot and leaves of cashew seedlings varied from 87.4±41.4 to 139.4±60.5. However, the hardened (matured) shoots of cashew were least preferred both for feeding and oviposition, whereas, the hardened shoots of neem were preferred both for feeding and oviposition similar to tender shoots of cashew (Sundararaju, 1996).

The fungi *Gloeosporium mangiferae* and *Phomopsis anacardii* have been reported to cause blossom blight in association with *H. antonii* (CCRS, 1966; Nambiar *et al.*, 1973). The feeding injury by the bug was attributed as one of the causes of infection and manifestation of die-back disease caused by *Colletotrichum gloeosporoides* and *Botryodiplodia theobromae*. The die-back disease progressed beyond the feeding injury region on the lower part of shoots and panicles leading to wilting of whole leader shoot or branches. The die-back disease was much rampant especially in west coast on accessions having thin size shoots (*e.g.*, VRI-1 and 2) resulting in 100 per cent loss in yield in the plantation where outbreak of *H. antonii* coupled with North East monsoon rain.

5. Seasonality

Helopeltis spp. exhibit a continuous cycle of generations throughout the year like other tropical mirids, (Stonedahl, 1991; Sundararaju, 1996). On cashew, the buildup of the *H. antonii* commences during October/November synchronising with the emergence of new flushes, after the cessation of the monsoon. The population reaches a peak during January, when the trees are in full bloom. The pest prevails in the field till May and is absent during the monsoon (June-September) especially in older plantations (Pillai *et al.*, 1984; Rai, 1984; Sundararaju, 1984a; Satapathy, 1993). However, the population exists in a negligible number during monsoon period (NRCC, 1993). But in young plantations, the pest is noticed continuously with a higher intensity during February and March (Sathiamma, 1977). Negative relationship between rainfall and incidence of *H. antonii* was reported (Senguttuvan and Bhaskaran, 1993), whereas, in guava no such relationship was observed between incidence of *H. antonii* and weather factors (Gopalan and Perumal, 1973). However, in Tamil Nadu, the outbreak of *H. antonii* depends on its severity on neem. Once, severe outbreak occurs on neem, the population rapidly spreads to cashew, guava, drumstick and ber, wherein damage on cashew will be severe on panicles and tender apple and nuts. It spreads like storm. Similar situation also prevails in west coast condition in a different form. The population build up is regular in young plantations in which all the plants have infestation. Subsequently, it will be patchy as the age of plantation progress and in the plantations of the age group of 10 years and above, low level of population prevails. But, whenever, old plantations are situated near to young plantations, under severe outbreak condition, the population from young plantation rapidly spreads to older plantations, resulting in severe outbreak. The life table studies carried out with *H. antonii* on cashew, neem and guava revealed that rapid buildup of population is more likely on neem and cashew than on guava (Sundararaju, 1996; Sundararaju and Sundarababu, 1999). The demographic studies of *H. antonii* on cocoa, neem and henna suggested that the pest can multiply more rapidly on cocoa and neem than on henna and also sustained under unfavourable conditions on both the host plants. *H. antonii* had a fine λ value (1.1295) on henna; suggesting the importance of this host plant for the pest (Srikumar and Bhat, 2013).

Fluctuations in the weather parameters were found to influence the buildup of tea mosquito bug under field conditions. Minimum temperature and sunshine hours were highly correlated ($p < 0.01$) with increased *H. antonii* populations. The maximum temperature and afternoon humidity also had significant effect ($p < 0.05$) on *H. antonii* population (Sundararaju, 2005). The largest populations were observed in December and January. The *H. bradyi* population showed significant correlation ($p < 0.01$) with minimum temperature. The observations of minimum temperature and sunshine hours can be an indicative to predict the behavior of the field trend of *Helopeltis* spp. populations and planning control measures.

6. Biology

Morphometrics of eggs of *Helopeltis* spp. revealed that they are white, ovoelongate (slightly narrower apically) and laterally compressed apically, 1.0 to 1.31 mm long (Ambika and Abraham, 1979; Satapathy, 1993; Sundararaju, 1996). Extra-chorionic processes (a long ventral one and shorter dorsal one) arise from the anterior end of the eggs; those usually are 0.29-0.67 mm in length. The eggs are embedded in plant tissue singly or in small groups usually with the operculum and extra-chorionic processes exposed. Initially, after deposit of egg in the plant tissue, the shape of egg is almost cylindrical. After two days, expansion of eggs to a tune of 0.04 to 0.08 mm both longitudinally and laterally, particularly at the distal region is seen. The colour of the operculum is hyaline for the first two days and it later turns light brown in colour. Similarly, the eggs remain whitish for first two days and subsequently turn to orange colour with a distinct dark orange band at the distal region (indicating the legs of the embryo). The size of the egg also varies with species of *Helopeltis*. Longitudinally it is in the range of 1.09 to 1.21, 1.27 to 1.35 and 1.37 to 1.40 mm for *H. theivora, H. antonii,* and *H. bradyi,* respectively and laterally it is almost uniform in size (0.35 to 0.39 mm). The shape and length of extra-chorionic process varied in all three species. In *H. theivora,* they are comparatively shorter and erect/angular whereas, in both *H. antonii* and *H. bradyi,* they are wavy and longest in *H. bradyi.* Aeropyles or air tubes that open on the apical region of the horn provide a direct route for the exchange of gases between the ambient air and intra-chorionic air. The intra-chorionic air forms continuous layer that is held between vertical struts about O.7μ long that separates a thin inner sheet of chorion from the main body of the shell (Hinton, 1962).

Helopeltis spp. prefers oviposition sites depending on the host plant. On cocoa, *H. theivora* prefers the pods, but will occasionally oviposits on young shoots (Stonedahl, 1991) and on tea, it prefers new shoots and rarely the petioles and midribs of leaves (Das, 1984). In guava *H. antonii* lays eggs primarily on ventral midribs of leaves, flower bud and pea size guava fruit either singly or in groups It rarely lays eggs on the fruits of marble size and above.

Whereas in cashew its eggs are laid on the young shoots, inflorescence stalks, developing nuts, petioles and ventral midribs of leaves of cashew (Figures 20.7 and 20.8). The fecundity of *H. antonii* on cashew and guava varies from 10 to 220 and 48 to 185 respectively, whereas, the fecundity of *H. theivora* on tea and cocoa is 172 and 50-63 respectively and of *H. bradyi* is 235 (Sundararaju, 1996).

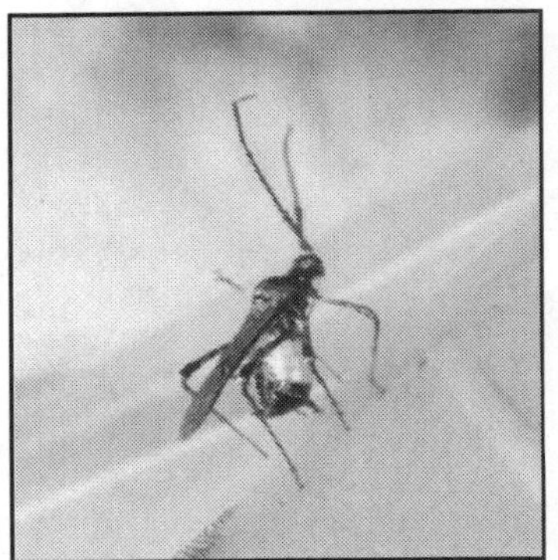

Figure 20.7: Ovipositon Behaviour of *H. antonii.*

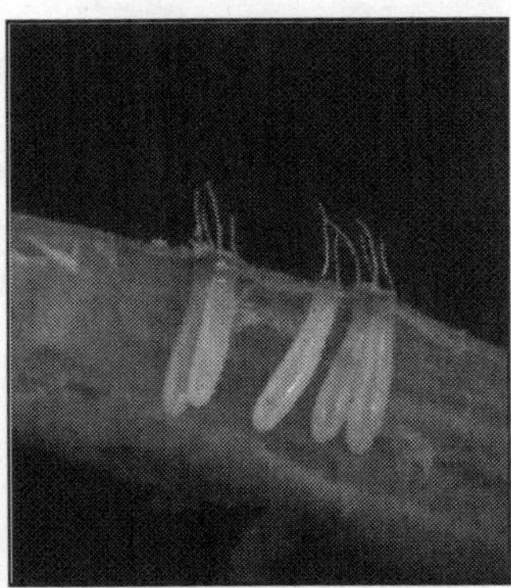

Figure 20.8: Eggs Laid in Groups by *H. antonii* (Extra-chorionic process exposed).

The egg production and ovarial development depend considerably on environmental condition. The abrupt interruption and resumption oviposition depending upon weather conditions and host plants explains the erratic nature of the pest problem (Roepke, 1916). Lowest incubation period of 05 days in June-August and longest incubation period of 27 days in January were observed for *H. theivora* (Das, 1984), whereas, the incubation period of *H. antonii* and *H. bradyi* varies from 6-15 days. However, in all species for proper hatching of eggs, the plant parts containing eggs should be always alive and remain turgid (Devasahayam and Nair, 1986; Sundararaju, 1996).

The nymphal stage of *Helopeltis* spp. consisted of five instars that vary in size, colour and the development of certain body parts *viz.*, antennae and wings. Description and illustration of the nymphal instars are provided by Ambika and Abraham (1979) and Satapathy (1993) for *H. antonii*. The rate of development is affected by weather factors especially temperature (Sundararaju and John, 1992; Satapathy, 1993). Least development period of (egg to adult) around 14-22 days has been observed in all three species of *Helopeltis*. The nymphs of *H. antonii*, *H.*

bradyi and *H. theivora* are light brownish, greenish yellow and greenish in colour, respectively. The size of nymphs of *H. antonii* and *H. bradyi* are almost similar in size except colour variation, whereas, the nymphs of *H. theivora* are comparatively smaller and slender in size (Sundararaju, 1996; Srikumar and Bhat, 2012).

The general colouration of adults of *H. antonii* is brown to dark reddish brown with white and black patches in the abdomen and about 6-8 mm in length, whereas, the adults of *H. bradyi* are slightly bigger than *H. antonii* showing same colour pattern and close resemblance with later except with minute colour variation in the hind femur and abdominal region. A pale coloured band in the proximal end of hind femur and dark patch as the lateral surface of abdominal sterna of I-III can be distinctly seen in *H. bradyi*. Further, in *H. bradyi*, the first antennal segment is slightly longer than *H. antonii*. The adult of *H. theivora* is smaller and 7.6 mm in length. It has longer antennae and yellowish pronotal ring and greenish white patch in the abdomen. It is much more active than *H. antonii* and *H. bradyi*. The longevity of adults varies from 7-46 days (Stonedahl, 1991; Sundararaju, 1996). The mass culturing techniques for *H. antonii* have also been standardized (Sundararaju and John, 1992; Sundararaju, 1996).

Devasahayam (1985) described the mating process and oviposition behaviour of *H. antonii*. Presence of sex pheromone in virgin female of *H. antonii* has been demonstrated by observing attraction of males towards three/four day old virgin female. The attraction was continued upto its death and confined during day time with maximum attraction during forenoon period. However, no such attraction of males towards mated females was seen under field condition (Sundararaju, 1996; Bhat and Srikumar, 2013).

In plantations where all three species of *Helopeltis* coexisted, interspecific attraction of males of *H. theivora* and *H. bradyi* towards virgin females of *H. antonii* was not observed and attraction of males of the *H. theivora* and *H. bradyi* confined to intraspecific virgin females and thus indicating highest intraspecies specificity. However, very rare interspecific mating between female of *H. antonii* and male of *H. theivora* and also between female of *H. bradyi* and male of *H. antonii* were observed under field condition. Interspecific attraction in *H. theivora* in the laboratory was observed by (Somachoudhury *et al.*, 1993). But such interspecific mated females were found to lay nonviable eggs and exhibited sex pheromonal attraction towards intra-specific males till intraspecific mating was restored. The preliminary GC-MS analysis of virgin female extract of *H. antonii* revealed the presence of 17 significant compounds (Sundararaju, 1996). Studies were taken up at Directorate of Cashew Research (DCR), Puttur to identify the compounds having pheromonal activity for *H. antonii*. In the electroantennaogram studies, the antenna dissected out from male insect showed highest response of -1.058mv for the volatiles from virgin females collected in dichloromethane. Four day old male TMB had higher response for the extracts/volatiles. GC-MS analysis of abdomen extract and volatiles from virgin female revealed the existence of large number of compounds. Methyl butyrate, a compound that has been shown to exhibit pheromone activity in other insects of Miridae family was one of the compounds detected in the analyses (Bhat and Raviprasad, 2008).

Gas chromatography-mass spectrometry (GC-MS) analysis of female thoracic extracts and dynamic head space samples of virgin females showed the presence of five compounds: (Z)-3 hexenyl acetate, (Z)-3 hexenyl butanoate, (E)-2 hexenyl pentanoate, 2,4 dimethyl pentanal, and (E)-2-hexenol. Male tea mosquitoes were attracted to blends of (Z)-3 hexenyl acetate and (E)-2-hexenol in the wind tunnel with a 1:5 ratio eliciting the greatest response. Electroantennogram (EAG) recordings of male antenna confirmed the ability of this blend to evoke antennal responses in male insects. This female sex pheromone blend may be useful for TMB control and management in future (Sachin *et al.*, 2008).

7. Management Strategies

Susceptibility of Cashew

Cashew accessions *viz.*, VTH 153, Kunthur 24, Goa 11/6, VTH 153/1, 9/78 and 51 different cashew types in Karnataka, accession No. 665 in Kerala and BLA - 39-4 in West Bengal were reported as least susceptible to *H. antonii* (Sathiamma, 1979; Ghosh and Chatterjee, 1987; NRCC, 1988; Sundararaju and John, 1993). But, none of the workers confirmed the intensity of typical resistance. However, the least susceptible accessions *viz.* Goa 11/6 and VTH 153/1 reported by Sundararaju and John (1993), when raised separately as plantation were damaged severely in the initial years. But in the later years, incidence was noticed in patchy distribution pattern which is peculiar to *Helopeltis* spp. infesting cashew as reported by Stonedahl *et al.* (1995). Least susceptible types to *H. antonii* contain higher phenols (Annapoorna and Nagaraja, 1988) which cannot be implicated towards resistance, since *H. antonii* has potential salivary detoxification mechanism (Sundararaju and Sundarababu, 1996). Besides, the existence of other antibiosis mechanism is also remote, since Kunthur 24 and Goa 11/6 accessions had not shown any inhibitory effect on the growth of *H. antonii* (NRCC, 1994). Further, Goa 11/6 was released in as variety 'Bhaskara' for cultivation in Karnataka State (Sundararaju *et al.*, 2006). The studies conducted by Beevi and Mahapatro (2007) indicted that among released varieties, Amrutha, Damodar and Raghav were least susceptible while, Priyanka and Anagha were highly susceptible. Damage analysis indicated that, none of the accessions exhibited damage below 10 per cent.

Biological Control Agents

Biological control programmes against *Helopeltis* spp. have not been attempted, but high levels of parasitism provided by some egg parasitoids certainly warrant further investigations (Stonedahl, 1991). In most instances, it appears that natural enemies, while playing an important role in control, cannot maintain populations of *Helopeltis* spp. below economic thresholds. However, integrated control programmes with reduced pesticide use and the monitoring of natural enemies have been suggested as reasonable alternatives to blanket spraying (CIBC, 1983). This would allow population of natural enemies to increase and provide more suitable environments for investigations of improved biological control (Stonedahl, 1991).

Egg Parasitoids

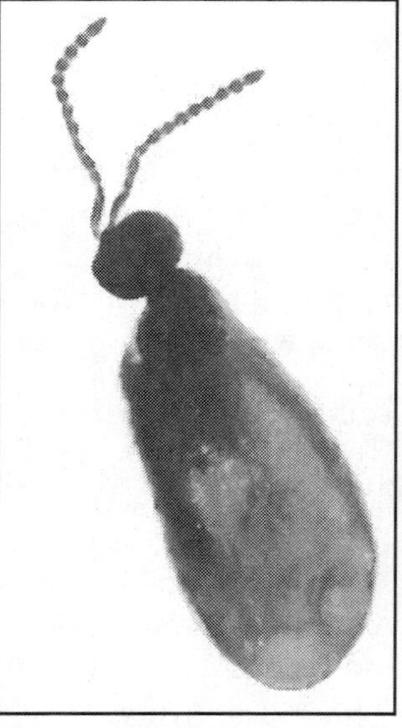

Figure 20.9: *Telenomus* sp. Emerging from *Helopeltis* Egg.

Though the eggs of *Helopeltis* spp. are laid deep and concealed, they are often attacked by a range of Hymenopteran parasitoids of *Telenomus* sp., *Ufens* sp. and *Chaetostricha* sp. (Trichogrammatidae) and *Erythmelus helopeltidis* Gahan and *Gonatocerus* sp. (Mymaridae) (Sundararaju, 1996; Sundararaju and Sundarababu, 2000). All these species are solitary parasitoids. Egg parasitoids of the genus *Telenomus* (Hymenoptera: Platygastridae) are particularly promising. *Telenomus* paraitization of mirid eggs has been reported by Silvestri (1932) and Chang (1982). In Ghana 10 to 73 per cent of *H. bergrothi* eggs were parasitized by *Telenomus* sp. (CIBC, 1983). *Telenomus* sp. has been reported in India on *H. cinchonae* Mann (Simmonds, 1970), on *H. theobromae* Miller from Malaysia (Ibrahim, 1989) and on *H. antonii* Signoret from India (Sundararaju, 1993). On cashew, *Telenomus* sp. (Figure 20.9) is predominant and acted as a constant mortality factor in the population of *H. antonii* and the extent of parasitization was observed almost throughout the year with a range of 1.8 to 45.5 per cent, whereas *Chaetostricha* sp. showed low level of parasitization of 0.7 to 4.3 per cent (Sundararaju, 1993). The highest parasitism up to 70.8 per cent by *Telenomus* sp. in certain months was recorded by Sundararaju (1996).

Moreover, parasitism by *Telenomus* sp. is negligible in the eggs of *H. antonii* laid on neem in east coast (Tamil Nadu), whereas, *Ufens* sp., thelytokous type is the dominant species in the neem ecosystem of Tamil Nadu. The study on relative intensity of parasitism indicated that *Telenomus* sp. might have evolved in the eggs of *H. bradyi* (Sundararaju, 1996). *Erythmelus helopeltidis* has been reported as egg parasitoid of *H. antonii* (Devasahayam, 1989); *H. theivora* on tea (Sudhakaran and Muraleedharan, 2006). Dissection of host plant parts containing eggs of *Helopeltis* and examining the host eggs are the most reliable method for estimation of extent of parasitism. On examination of *Helopeltis* spp. eggs showing the external whitish chorion enclosing dark brownish, chitinous indicates parasitization by *Telenomus* sp., eggs of creamy or orange white with whitish cup shaped deposit shows trichogrammatid group parasitism and creamy or orange-white without any cup shaped white deposit indicates mymarid group parasitism (Sundararaju and Sundarababu, 1999; 2000a). Among abiotic factors, temperature has shown positive influence on extent of parasitism under field condition (Sundararaju, 1996). Seasonality of egg parasitoids of *H. antonii* in relation to weather parameters was reported by Sundararaju (2004a).

Nymphal Parasitoids

Nymphal parasitoids of genus *Leiophron* (Hymenoptera: Braconidae) were reported as promising biological control agents of *Helopeltis*. Parasitism levels of

6 to 66 per cent have been reported for L. sahlbergellae Wilkinson on *Sahlbergella singularis* Haglund (Heteroptera: Miridae) in West Africa (CIBC, 1983). In India, Leiophron like parasitoid has been reported from *H. theivora* from tea (Rao, 1948). For the first time grub of was encountered from adult females of *H. antonii* at coastal Karnataka region from cashew ecosystem (Sundararaju, 2002).

Predators

Predators also appear to play an important role in the natural control of *Helopeltis* spp. Predators of *Helopeltis* include spiders, reduviids, ants and mantids. Most of the predators are not specific to *Helopeltis* or even the Miridae and their potential value as biological control agents are questionable (CIBC, 1983; Stonedahl, 1991). *Crematogaster wroughtonii* Forel (Formicidae) has been recorded as a predator on nymphs of *H. antonii* on cashew (Ambika and Abraham, 1979). A list of arthropod predatory fauna recorded from cashew panicles was given by Sundararaju (2003) comprising 6 spider species. A total of 35 spiders were recorded from the cashew fields of Madakkathara belonging to 9 families and 29 genera. Feeding trials with 19 spider species revealed that, nine spiders were found to be feeding on TMB adults in captivity. Most of the studied spiders preferred adults than nymphs, while *O. sunandae* Tikader, *Telamonia elagans* Thorell and *Hyllus diacanthes* preferred nymphs (Beevi and Mahapatro, 2008). Consecutive survey for three years at DCR, Puttur, revealed occurrence of 117 species of spiders belonging to 18 families in cashew. Field observation revealed *Telamonia dimidiata* and *Oxyopes shweta* were the major predators of *Helopeltis* spp. The spiders *viz., Argiope pulchella, Cyclosa fissicauda, Eriovixia laglazei, Neoscona mukerjeri, Nephila pilipes, Oxyopes sunandae, Bavia kairali, Carrhotus viduus, Epocilla aurantiaca, Hyllus semicupreus, Achaearanea mundula, Camariacus formosus* and *Thomisus lobosus* were also recorded as superior predators of *Helopeltis* spp. (Bhat *et al.*, 2013) (Figure 20.10).

Reduviids (Hemiptera: Reduviidae) also constitute an important group of predatory insects that could be most successfully harnessed as effective biological control agents. Five species of reduviid bugs [*Sycanus collaris* (Fab.), *Sphedanolestes signatus* Dist., *Endochus inornatus* Stal., *Irantha armipes* Stal., and *Oecamus typicus* Dist.] have also been recorded as predators of *H. antonii* (Sundararaju, 1984a; Sundararaju, 1996) (Figure 20.11). Use of *Chrysoperla* sp. as predator of *H. theivora* has been. reported by Somchoudhury *et al.* (1993).

Nematodes and Pathogens

The occurrence of *Agamermis paracaudatu* Steiner (Mermithidae) on *H. theivora* infesting tea was reported by Rao (1948). Two fungal pathogens *viz. Aspergillus flavus* and *A. tamarii* can cause infection in *H. antonii* (Karthikeyan, 1992; Satpathy, 1993).

Chemical Control

Use of chemicals is one of the short term strategies suggested for increasing cashew yield for the effective control of *H. antonii* (Nambiar *et al.*, 1973). Spraying of Cuman (0.1 per cent) in combination with phosphomidan (0.03 per cent) was earlier suggested for control of the pest (CCRS, 1966). The insecticides like DDT belonging to organo- chlorine group recommended against TMB are banned. The

Figure 20.10: *Telamonia dimidiata* Predating on *H. antonii.*

Figure 20.11: Reduviid Predating on *H. antonii.*

recommendation of endosulfan has been withdrawn in all the cashew growing States in the country.

Insecticides *viz.* monocrotophos (0.05 per cent), carbaryl (0.1-0.15 per cent), quinalphos (0.05 per cent), dimethoate (0.05 per cent), fenthion (0.05 per cent), phosalone (0.07 per cent), phosphomidan (0.03 per cent) were also reported to be effective against *H. antonii* (Sundararaju, 1984b; Chatterjee, 1989; Godse *et al.*, 1993). However, considering the residual action of various insecticides, carbaryl (0.1 per cent) and monocrotophos (0.05 per cent) had maximum residual action for seven days and found to be superior than other insecticides (Sundararaju *et al.*, 1993). In Tamil Nadu, in the major cashew tract (Panruti area) insecticidal sprayings with various formulations were being undertaken indiscriminately to avoid the potential damage of *H. antonii* leading to resurgence of other sucking pests like mealy bug (*Ferrisisa virgata* Cock.) (Sundararaju, 1996).

Effectiveness of synthetic pyrethroids *viz.*, decamethrin (0.002 per cent), permethrin (0.01 per cent), cypermethrin (0.0075 per cent) against *H. antonii* was also reported (NRCC, 1988; Godse *et al.*, 1993). Although, cashew is an insect pollinated crop, spraying of these insecticides during flowering season did not influence the fruit set (Pillai *et al.*, 1984; Rai, 1984, Sundararaju *et al.*, 1993). Superiority of carbaryl dust formulations in bringing down TMB infestation was reported by Bakthavatsalam *et al.* (1993). Under severe outbreaks of *H. antonii* on cashew, carbaryl (0.1 per cent) or monocrotophos (0.05 per cent) can be used (Sundararaju, 1996).

All the insecticides tested shown no ovicidal action, but λ-cyhalothrin followed by carbaryl and monocrotophos exhibited highest residual action for seven days against late instar nymphs and adults of TMB (Sundararaju, 2004b; Raviprasad *et al.*, 2005; Bhat and Raviprasad, 2007). The sequential sprays of monocrotophos, λ-cyhalothrin and carbaryl registered the least per cent TMB damage and higher nut yield (Naik and Chakravarthy, 2013).

References

Abraham, C. C. and Nair, G. M. (1981). Effective management of the tea mosquito bugs for breaking the yield barriers in cashew. *Cashew Causerie,* 3: 6-7.

Abraham, C.C. and Remamony, K.S. (1979). Pests that damage cocoa plants in Kerala. *Indian Arecanut, Spices and Cocoa Journal,* 2: 77-81.

Abraham, E.V. and Padmanaban, M. D. (1967). Pests that damage cacao in Madras. *Indian Horticulture,* 11: 11-12.

Annapoorna, K.R. and Nagaraj, K.V. (1988). Tea mosquito bug (*Helopeltis antonii*) induced chemical changes in cashew (*Anacardium occidentale*). *Cashew Bulletin,* 25: 1- 4.

Anstead, R.D. and Ballard, E. (1922). Mosquito blight of tea. *Planters Chronicle,* 17: 443-447.

Ayyar, T.V.R. 1932. Annotated list of the insects affecting the important cultivated plants in South India. Agricultural Department of Madras, Bulletin No. 27. p. 95.

Ayyar, T.V.R. (1940). Hand Book of Economic Entomology for South India. Govt. Press, Madras. p. 518.

Beevi, P.S. and Mahapatro, G. K. (2008). Species-spectrum and inter-relationship between ant and spider fauna in cashew agro-ecosystem. *Journal of Plantation Crops,* 36: 375-381.

Beevi, S. P. and Mahapatro, G. K. (2007). A new field screening methodology for cashew genotypes against tea mosquito bug (*Helopeltis antonii* Sign.), *Journal of Plantation Crops,* 35: 139-145.

Bhat, P.S., Srikumar, K. K and Raviprasad, T. N. (2013). Seasonal diversity and status of spiders (Arachnida: Araneae) in cashew ecosystem. *World Applied Sciences Journal,* 22 (6): 763-770.

Bhat, P. S. and Srikumar. K. K. (2012). Record of *Erythmelus helopeltidis* Gahan (Hymenoptera: Mymaridae), an egg parasitoid from *Pachypeltis maesarum* (Heteroptera: Miridae) infesting cashew. *Pest Management in Horticultural Ecosystem,* 18 (1): 103-104.

Bhat, P. S. and Srikumar. K. K. (2013). Genitalia structure examination for interspecific mating of *Helopeltis antonii* and *H. theivora. Current Science,* 105(1): 23-25.

Bhat, P.S. and Raviprasad, T. N. (2007). Management of tea mosquito bug *Helopeltis antonii* Sign. with newer insecticide/products. National Seminar on Research Development and Marketing of cashew, 20-21 November. p. 56.

Bhat, P.S. and Raviprasad, T. N. (2008). Sex pheromone of tea mosquito bug *Helopeltis antonii* Signoret (Miridae: Heteroptera). *Journal of Plantation Crops,* 36 (3): 451-453.

Bin, F. (1994). Biological control with egg parasitoids other than *Trichogramma*. In: Biological control with egg parasitoids (Eds: E. Wajnberg and S.A.Hassan). CAB International, Borkshire, U.K. pp.146-153.

Cadou, J. (1994). Les Miridae du cotonnier en Afrique et a Madagascar, CIRAD, Montpellier, France, p. 74.

CCRS (Central Cashew Research Station). (1966). Annual Progress Report, 1965-66. Ullal, Karnataka.

Chang, Y.C. (1982). The life history, population density and control of the bamboo mirid *Mecistocelis scirtetoides* Reuter (Miridae: Hemiptera). Natl. Sci. Counc. Monthly, ROC. 10(4). pp. 335-347.

Chatterjee, M. L. (1989). Insect pests of West Bengal and status of some important pests. *The Cashew*, 3: 19-20.

Cherian, M.C. (1937). Administrative Report of the Govt. Entomologist. Report of Department of Agriculture, Madras. pp. 126-133.

CIBC (Commonwealth Institute of Biological Control). (1983). Possibility for the use of natural enemies in the control of *Helopeltis* spp. (Miridae). *Biocontrol News and Information*, 4: 7-11.

Das, S.C. 1984. Resurgence of tea mosquito bug (*Helopeltis theivora* Waterh.), a serious pest of tea. *Two and a Bud*, 31: 36- 39.

De Silva, M. D. (1957). A new species of *Helopeltis* (Hemiptera- Heteroptera, Miridae) found in Ceylon. *Bulletin of Entomological Research*, 48: 459-461.

Devasahayam, S. (1985). Seasonal biology of tea mosquito bug *Helopeltis antonii* Signoret (Heteroptera: Miridae) a pest of cashew. *Journal of Plantation Crops*, 13: 145-147.

Devasahayam, S. and Nair, C. P. R. (1986). The mosquito bug, *Helopeltis antonii* Sign. on cashew in India. *Journal of Plantation Crops*, 14: 1-10.

Devasahayam, S. (1989). *Erythmelus helopeltidis* Gahan (Hymenoptera: Mymaridae)-a new egg parasite of *Helopeltis antonii* Signoret on cashew. *Journal of the Bombay Natural History Society*, 86: 113.

Fletcher, T. B. (1914). *Some South Indian Insects*. Government' Press, Madras. pp. 488-489.

Ghosh, S. N. and Chatterjee, M. L. (1987). Study on performance of some cashew types and their susceptibility levels against Tea Mosquito at Cashew Research Station, Jhargram, West Bengal. *The Cashew*, 1: 21-23.

Godse, S.K., Dumbre, R.B. and Kharat, S.B. (1993). Evaluation of some insecticides for control of tea mosquito bug on cashew. *The Cashew*, 7: 13-14.

Gopalan, M. and Perumal. R.S. (1973). Studies on the incidence of tea mosquito bug (*Helopeltis antonii* S.) on some varieties of guava. *Madras Agricultural Journal*, 60: 81-85.

Hamid, A.A. (1987). Insect pests of *Acacia mangium* Willd. In Sarawak. Sarawak Forest Department, Research Report FE 1. pp. 1-10.

Hinton, H. E. (1962). Respiratory systems of insect egg shells. *Annual Review of Entomology,* 14: 343-368.

Ibrahim, W.A. (1989). *Erythmelus helopeltidis* - an egg parasitoid of *Helopeltis theobromae* in cocoa. *Planter,* 65: 211-215.

Jena, B.C., Patnaik, N.C. and Satapathy, C.R. (1985). Insect pests of cashew. *Cashew Causerie,* 7: 10-11.

Karthikeyan, K. (1992). Pests of Neem (*Azadirachta indica* A. Juss) and their management. M.Sc. (Ag.) Thesis, AC and RI, TNAU, Killukulam. pp. 96.

Leefmans, S. (1916). Bijdrage tot net *Helopeltis* - Vraagstuk Voor de thee (contribution to the *Helopeltis* problem of tea). Mededeelingen van het laboratorium voor plantenziekten No. 26). (Communication of Institute for Plant Diseases No.26). Summary of investigations. pp. 201-208.

Lever, R. J. A.W. (1949). The Tea Mosquito Bugs (*Helopeltis* spp.) in the Cameron Highlands. *Malayan Agriculture Journal,* 32: 91-108.

Mann, H.H. (1907). Individual and seasonal variations in *Helopeltis theivora,* Waterhouse, with description of a newspecies of *Helopeltis. Memoirs of the Department of Agriculture in India* (Entomological Series) 1: 275-337.

Miller, N.C.E. (1941). Insects associated with cocoa (*Theobroma cocoa*) in Malaya. *Bulletin of Entomological Research,* 32: 1-15.

Naik, C. M. and Chakravarthy. A. K. (2013). Sustainable management practices for tea mosquito bug *Helopeltis antonii* Signoret (Miridae: Hemiptera) on cashew. *Karnataka Journal Agricultural Sciences,* 26: 54-57.

Nambiar, K. K.N., Sarma, Y. R. and Pillai, G. B. (1973). Inflorescence blight of cashew (*Anacardium occidentale* L.). *Journal of Plantation Crops,* 1: 44-46.

Nambiar, K.K.N., Sarma, Y. R. and Pillai, G. B. (1973). Inflorescence blight of cashew (*Anacardium occidentale* L.). *Journal of Plantation Crops,* 1: 44-46.

NRCC (National Research Centre for Cashew) (1988). Annual Report for 1987, Puttur, Karnataka, India. p. 41.

NRCC (National Research Centre for Cashew). (1993). Annual Report for 1992-93, Puttur, Karnataka, India. p. 87.

NRCC (National Research Centre for Cashew). (1994). Annual Report for 1993-94, Puttur, Karnataka, India. p. 108.

Peal, S.E. (1873). The tea-bug of Assam. *Journal of the Agricultural and Horticultural Society of India,* 4: 126-132.

Pillai, G. B., Dubey, O. P. and Singh, V. (1976). Pests of cashew and their control in India. A review of current status. *Journal of Plantation Crops,* 4: 37-50.

Pillai, G.B., Singh. V., Dubey, O.P. and Abraham, V.A. (1984). Seasonal abundance of tea mosquito bug, *Helopeltis antonii* on cashew in relation to meteorological factors. In: *Cashew Research and Development* (Eds): Rao, E.V.V.B. and Khan, H.H. Indian Society for Plantation Crops, CPCRI, Kasaragod, India. pp. 103-110.

Pillai, K.S., Saradamma, K. and Nair, M.R.G.K. (1979). *Helopeltis antonii* Sign. as a pest of *Moringa oleifera*. *Current Science*, 49: 288-289.

Pillai, S. R. M. and Gopi, K.C. (1990). Seasonal drying of the distal shoots of neem (*Azadiracta indica* A. Juss) and important insect pests associated with it. *MYFOREST*, 26: 33-50.

Puttarudriah, M. (1952). Blister disease "Kajji" of guava fruits (*Psidium guava*). *Mysore Agricultural Journal*, 28: 8-13.

Puttarudriah, M. (1958). A dangerous potential pest of cotton. *Indian Cotton Growing Review*, 12: 406-407.

Puttarudriah, M. and Appanna, M. (1955). Two new hosts of *Helopeltis antonii* in Mysore. *Indian Journal of Entomology*, 17: 391-392.

Rai, P.S. (1984). *Handbook on Cashew Pests.* Researchco Publications, Delhi. pp. 124.

Rao, D.V.S. and Srinivasan, S. (1984). Tea Mosquito (*Helopeltis antonii* Sign.)-a new pest of cashew in Andra Pradesh. *Cashew Bulletin*, 21: 10-12.

Rao, S.A. (1948). Report of the Entomologist. Annual Report of UPASI Tea Science 1947-48. p.9.

Rao, S.A. (1970). *Helopeltis*: a breakthrough in its control. UPASI Tea Science Department Bulletin, 28. pp. 21-28.

Rao, Y.R. (1915). *Helopeltis antonii* as a pest of neem trees. *Agricultural Journal of India*, 10: 412-416.

Raviprasad, T.N., Sundararaju, D. and Bhat P.S. (2005). Efficacy of botanicals against *Helopeltis antonii* Sig. infesting cashew. *The Cashew*, 19: 9-14.

Roepke, W. (1916). Zwei neue Gambir-schadliche Capsiden aus Sumatra. *Tijdschrift voor Entomologie*, 59: 180-183.

Roepke, W. (1916a). Het *Helopeltis*- vraagstuk, in het bijzonder met betrekking tot cocoa. (The Helopeltis – problem with particular reference to coca) Mededeelingen Proefstation Midden- Java, 21. p. 1-40.

Sachin, J. P., Selvasundaram, R., Babu, A. and Muraleedharan, N. (2008). Behavioral and electroantennographic responses of the Tea Mosquito, *Helopeltis theivora*, to female sex pheromones. *Environmental Entomology*, 37: 1416-1421.

Satapathy, C. R. (1993). Bioecology of major insect pests of cashew (*Anacardium occidentale* Linn.) and evaluation of certain pest management practices. Ph.D. thesis, U.A.S., Bangalore, p. 224.

Sathiamma, B. (1977). Nature and extent of damage by *Helopeltis antonii* S., the tea mosquito on cashew. *Journal of Plantation Crops*, 5: 58-62.

Sathiamma, B. (1979). Variety reaction of cashew to tea mosquito bug, *Helopeltis antonii* Sign.- a major pest of cashew. In: *Proceedings of Placrosym II*, 26-29 June. Ootacamund, India. pp. 530-534.

Sathiamma, B. (1981). Occurrence of colour variants in *Helopeltis antonii* S. (Hemiptera: Miridae) on cashew. *Cashew Bulletin,* 18: 5-6.

Senguttuvan, T. and Baskaran, M. R. K. (1993). Influence of weather factors on the incidence of cashew pests. *Cashew Bulletin,* 30: 22-26.

Signoret, V. (1858). Descriptions de nouvelles especes d'Hemipteres. II. Note sur les Hemipteres Heteropteres de la famille des unicellules. *Annales de la Societe Entomologiquede France* (3)6: 499-502.

Simmonds, F.J. (1970). A memorandum on the possibilities of biological control of cocoa - infesting mirids. Commonwealth Institute of Biological Control Report (Mimeographed). p. 9.

Sivestri, F. (1932). Contribuzione alla consceza del Lopus lineolatus (Brulle) e di un suo parassita. Societe Entomogique de France, Livre du Centenaire. pp. 561-565

Somchoudhury, A.K., Samanta, A. and Dhar, P. (1993). Proc. International Symposium on tea science and human health (Teatech, 1993). Tea Research Association, India. pp. 330-338.

Srikumar, K. K. and Bhat, P. S. (2012). Field survey and comparative biology of tea mosquito bug (*Helopeltis* spp.) on cashew (*Anacardium occidentale* Linn.). *Journal of Cell and Animal Biology,* 6(14): 200-207.

Srikumar, K. K. and Bhat, P.S. (2013). Demographic parameters of *Helopeltis antonii* Signoret (Heteroptera: Miridae) on neem, cocoa and henna. *African Journal of Agricultural Research,* 8(35): 4466-4473.

Srikumar, K. K. and Bhat, P.S. (2013). Biology and feeding behaviour of *Helopeltis antonii* (Hemiptera: Miridae) on Singapore cherry (*Muntingia calabura*)-a refuge host. *Journal of Entomological Research,* 37(1): 11-16.

Srikumar, K. K. and Bhat, P.S. (2013). Biology of the tea mosquito bug, *Helopeltis theivora* Waterhouse on *Chromolaena odorata* (L.). *Chilean Journal of Agricultural Research,* 73 (3): 309-314.

Stonedahl, G.M. (1991). The Oriental species of *Helopeltis* (Heteroptera: Miridae): a review of economic literature and guide to identification. *Bulletin of Entomological Research,* 81: 465-490.

Stonedahl, G.M., Malipatil, M.B. and Houston, W. (1995). A new mirid (Heteroptera) pest of cashew in northern Australia. *Bulletin of Entomological Research,* 85: 275-278.

Sudarmadji, D. and Gunawan, S. (1994). Patogenisitas fungi entomopathogen *Beaveria bassiana* terhadap *Helopeltis antonii* (Pathogeniticty of entomopathogenic fungus *Beaveria bassiana* to *Helopeltis antonii*) *Menara Perkebunan,* 62: 1-5.

Sudhakaran, R. and Muraleedharan, N. (2006). Biology of *Helopeltis theivora* (Hemiptera: Miridae) infesting tea. *Entomon,* 31: 165-180.

Sundararaju, D. (1984a). Cashew pests and their natural enemies in Goa. *Journal of Plantation Crops,* 12: 38-46.

Sundararaju, D. (1984b). Chemical control of tea mosquito bug (*Helopeltis antonii* Sign.) *Cashew Causerie* 6: 10-11.

Sundararaju, D. (1993). Studies on the parasitoids of the mosquito bug, *Helopeltis antonii* Sign. (Heteroptera: Miridae) on cashew with special reference to *Telenomus* sp. (Hymenoptera: Scelionidae). *Journal of Biological Control,* 7: 6-8.

Sundararaju, D. (1996). Studies on *Helopeltis* spp. with special reference to *H.antonii* Sign. in Tamil Nadu. Ph.D. thesis. T.N. A.U., Coimbatore, India. p. 210.

Sundararaju, D. (2002). Description of endoparasitism in nymph and adults of *Helopeltis* spp. infesting cashew. *Journal of Plantation Crops,* 30: 66-68.

Sundararaju, D. (2003). Record and cumulative effect of recommended insecticidal spray schedule on arthropod predatory fauna occurring on cashew. *The cashew,* 17: 30-33.

Sundararaju, D. (2004a). Influence of weather parameters on oviposition and egg parasitism of tea mosquito bug in the tender flush stage of cashew. *Journal of Plantation Crops,* 32: 61-63.

Sundararaju, D. (2004b). Evaluation of promising newer insecticides in large plots for management of tea mosquito bug on cashew. *Journal of Plantation Crops,* 32: 285-288.

Sundararaju, D. (2005). Seasonal abundance and extent of damage of tea mosquito bug on cashew. *Journal of Plantation Crops,* 33: 53-58.

Sundararaju, D. and Bakthavatsalam, N. (1994). Pests of cashew. In: *Advances in Horticulture* (Eds: Chadha K.L. and Rethinam. P) Malhotra Publishing House, New Delhi- 10. pp. 759-785.

Sundararaju, D. and John, N.J. (1992). Mass rearing technique for *Helopeltis antonii* Sign. (Heteroptera: Miridae) - An Important pest of cashew. *Journal of Plantation Crops,* 20: 46-53.

Sundararaju, D. and John, N.J. (1993). Susceptibility of cashew accessions to *Helopeltis antonii* (Heteroptera: Miridae), in the preflowering phase. *Journal of plantation crops,* 21: 50-53.

Sundararaju, D. and Sundarababu, P. C. (1999). *Helopeltis* spp. (Heteroptera: Miridae) and their management in plantation and horticultural crops of India. *Journal of Plantation Crops,* 27: 155-174.

Sundararaju, D. and Sundarababu, P.C. (1996). Neem pest not a mystery. *Nature,* 381: 108.

Sundararaju, D. and Sundarababu, P.C. (2000). Egg parasitoids of neem mosquito bug from cashew. *Cashew,* 14: 32-34.

Sundararaju, D. Yadukumar, N. Bhat, P.S. Raviprasd, T.N., Venkattakumar, R. and Sreenath Dixit. (2006). Yield performance of "Bhaskara" cashew variety in coastal Karnataka. *Journal of Plantation Crops,* 34: 216-219.

Sundararaju, D., Bakthavatsalam, N. and Bhat, P.S. (1993). Ovicidal activity and residual toxicity of certain insecticides of tea mosquito bug *Helopeltis antonii* Sign. (Heteroptera: Miridae) infesting cashew. *Pestology*, 17: 14-17.

Tocklai (Tea Research Association). (2010). Special bulletin on Tea Mosquito Bug, *Helopeltis theivora* Waterhouse. p. 10.

Venkata, R. (2009). Record of *Helopeltis antonii* (Homoptera: Miridae) on the fruits of *Annona* spp. *Pest Management in Horticultural Ecosystems*, 15: 74-76.

Waterhouse, C.O. (1886). Some observations on the tea-bugs (*Helopeltis*) of India and Java. Transactions of the Entomological Society of London. pp. 457-459.

Wood-Mason, J. (1884). Report on the tea-mite and tea- bug of Assam. London, Taylor and Francis. p. 20.

Cashew: Improvement, Production and Processing (2017) *Pages* **367–376**
Editor: **Prof. P.L. Saroj**
Published by: **ASTRAL INTERNATIONAL PVT. LTD., NEW DELHI**

21

Management of Cashew Stem and Root Borer

K. Subaharan and A. Verghese

1. Introduction

The high value commercial crop cashew (*Anacardium occidentale* L.) provides employment and valuable foreign exchange to countries in Asia, Africa, South and Central America (Andrighetti *et al.*, 1994; Martin *et al.*, 1997). In India, cashew is grown in Maharashtra, Goa, Karnataka and Kerala along West coast and Tamil Nadu, Andhra Pradesh, Odisha and West Bengal along East coast. It is also grown to a limited extent in non-traditional areas such as Bastar region of Chhattisgarh, Gujarat, Jharkhand and in NEH region (Saroj and Rupa, 2014). It is grown in an area of 9.82 lakh ha with a production of 7.28 lakh tonnes and a productivity of 772 kg/ha (DCCD, 2013). India is the largest processor of raw cashewnuts, but the domestic production is insufficient to meet raw nut demand of the processing units in the country. This *necessitates* the import of cashew nut with shell to a tune of 7.98 lakh tonnes form Africa and South East Asian countries (FAO, 2012) causing a dent on the country's foreign exchange reserves. Hence, there is a need to increase production of raw nuts in the country.

Among the factors that limit cashew production, infestation by insect pests is a vital component (Sahu and Sharma, 2008). Being a perennial crop, cashew is subjected to attack by an array of pests round the year. All parts of tree *viz.*, roots, stems, twigs, inflorescence and the pseudo-apple are subjected to attack by insect pests (Eguagie, 1972; Pillai *et al.*, 1976; Devasahayam and Nair, 1986; Xianli and Van Der Geest, 1990). Among 50 species of insects that infest cashew in India, cashew stem and root borer (CSRB), *Plocaederus ferrugineus* L., is of serious concern. Wide spread occurrence of this pest has led to decline in the tree density in cashew growing tracts of the country (Prasad *et al.*, 2009). Two other species of borer's *viz.*, *P. obesus* and *Batocera rufomaculata* were reported in association with *P. ferrugineus* (Abraham, 1958; Uthaiah *et al.*, 1989 and Rai,1984).

2. Distribution

In India *P. ferrugineus* occurs in cashew growing states of Andhra Pradesh, Goa, Karnataka, Kerala, Maharashtra, Chattishgarh, Odisha and Tamil Nadu (Bhaskara Rao, 1998; Mohapatra, 2004; Sahu and Sharma, 2008). *P. ferrugineus* infests cashew in Myanmar (Maung Maung Lay, 1998), Phillipines (Concepcion Magboo, 1998), Sri Lanka (Surendra, 1998), Thailand (Suwit Chaikiattiyos, 1998), Vietnam (Nguyen Minh Chau, 1998), China (Liu Kangde *et al.*, 1998),Tanzania and Nigeria (Omole, 1972;Gerini, 1976; Asogwa *et al.*, 2009).

3. Host Range

The principal host of *P. ferrugineus* is cashew. Under induced feeding condition *P. ferrugineus* completed its life cycle when fed on drumstick (*Moringa oleifera*) and silk cotton (*Ceiba pentandra*) (Mohapatra and Jena, 2007). In Nigeria *hog plum / junglee aam* (*Spondias mombin*) plants growing in the viscinity of the cashew plantation were a preffered host by CSRB (Asogwa *et al.*, 2009).

4. Nature and Extent of Damage

The long dark brown longicorn beetles of *P. ferrugineus* lay eggs singly in the crevices of loose bark or on the trunk and occassionally in the parts of the roots near the soil surface (Gerini, 1976). Grubs bore into the bark of the stem and root forming irregular tunnels and feed on the phloem and xylem tissues resulting in exudation of gum (gummosis) and extrusion of fibrous frass from damaged portion (Asogwa *et al.*, 2009). Feeding by grubs on vascular tissues arrests the sap flow. Obstruction of the nutrient flow to the leaves results in yellowing and premature leaf shedding, drying of twigs and gradual death of the tree (Ayyanna and Remadevi, 1986). At advanced stage, the leaves falloff and branches start drying leading to death of the tree. In neglected plantations, on an average 5 to 10 per cent trees die every year due to stem borer damage (Abdul Salam and Peter, 2010). Grub makes a calcareous cocoon. Intensity of damage depends on the pest load, circumfrence of the trunk and age of tree (Misra and Basu Choudhary, 1985).

Infestation of CSRB is high on trees with branches originating at bottom-most position of the trunk, *i.e.*, 0-50 cm above collar region. Trees below five years of age were less prone to attack (Mohapatra *et al.*, 2008). CSRB damage is positively correlated with the age of the cashew trees, the infestation level increased with age. Trees over ten years of age were more susceptible than young trees (Senguttuvan and

Figure 21.1: Trunk Damage by Cashew Stem and Root Borer.

Biostages of Cashew Stem and Root Borer.

Figure 21.2: *Plocaederous* Adult.

Figure 21.3: *Plocaederous* Grub.

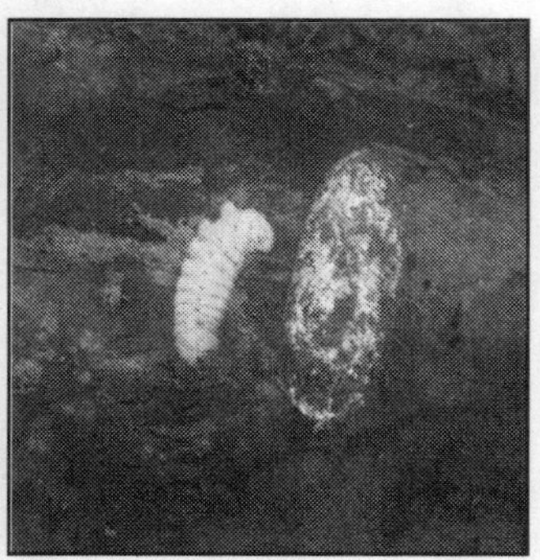

Figure 21.4: Cashew Stem Borer Grub and Pupae.

Mahadevan, 1999). Infested cashew trees have higher egg load as they are preffered by adults for oviposition (Senguttuvan and Mahadevan, 1997).

Mortality of cashew trees due to damage by CSRB ranged from 2-5 per cent (Rai, 1984). Damage by CSRB in cashew plantations ranged from 7–20 per cent in Kerala and 30–35 per cent in Tamil Nadu (Misra and Basu Choudhary, 1985). Forty per cent infestation was recorded from Guntur and Prakasam districts of Andhra Pradesh (Arjuna Rao, 1978; Ayyanna and Rama Devi, 1986). In Nigeria infestation by CSRB ranged from 13.3 to 18.13 per cent (Asogwa *et al.*, 2009).

5. Seasonal Incidence

P. ferrugineus L. is active during warmer part of the year (March-May) with two generations per year. Maximum temperature showed positive influence on activity of the pest. Positive correlation was observed between maximum temperature and fresh infestation. In Tamil Nadu, pest had two peak oviposition periods (December to February and April to May). Higher larval populations were observed during all months (except February) in trees with early, intermediate and advanced stages of infestation. Early instar larvae were at a higher side from January to July, whilst intermediate and late-instar larvae were high during March to November and August to January, respectively. Severely infested trees had overlapping of generations (Senguttuvan and Mahadevan, 1997).

6. Bionomics

P. ferrugineus adults mate a day after eclosion. Mating occurs 3 to 6 times during night. The mean pre-oviposition and oviposition periods are 3.6 and 7.2 days, respectively. On an average a female lays 62 eggs with an incubation period of 6.9 days. The larval period is 163 days; pupal period is 60 days. The total life cycle spreads fom 240 – 250 days (Senguttuvan and Mahadevan, 1998). Under laboratory conditions, the mean grub period of *P. ferrugenius* was 163.10±20.55 days, mean pupal period was 144.33±29.57 days for males and 145.16±29.10 days for females.

Figure 21.5: Chipping off the Damaged Portion of the Trunk to Remove the Grubs.

Figure 21.6: Treating the Damaged Portion of Trunk with Sealer-cum-Healer.

The total development period ranged from 261 to 363 days for male and 242 to 356 days for female.

7. Strategies for Management of CSRB

Various strategies are involved in managing the pest. No single method is effective in isolation. The cryptic behaviour of the CSRB grubs is a hurdle in managing the pest. Integration of the measures like phytosanitation, mechanical extraction, prophylactic treatment, chemical control method, botanical insecticides and bio agnets are beneficial to check the damage.

Phytosanitation

Incidence of *P. ferrugineus* is severe in poorly managed plantations. Infested trees act as source of inoculum for re-infetstation and/or fresh infestation (Misra and Basu Chowdhury, 1985: Jena, 1990). Phytosanitory measures involving cleaning base of trees, extraction of grubs from infested trunk, cutting of dead branches, and removal and destruction of dead trees is effective to reduce pest population (Mohapatra and Jena, 2008). Plantation hygiene is to be ensured to prevent harbouring of the insect (Abdul Salam and Peter, 2010). Removing infested-dead cashew trees and alternate host plant *Spondias mombin* is recommended in cashew growing region of Nigeria (Anikwe *et al.*, 2007). Removal of infested trees with more than 50 per cent of bark circumference damage reduced the level of fresh incidence of the pest during the subsequent years. Regular adoption of phytosanitation caused reduction in pest load (Ravi Prasad *et al.*, 2009).

Mechanical Control

Mechanical extraction of the grubs and swabbing the tree trunk with 2 per cent carbaryl solution minimizes the damage (Abdul Salam and Peter, 2010). Post extraction prophylaxis by treating trees with 0.2 per cent chlorpyriphos after extracting pest stages from infested trees helped to prevent reinfestation (Ravi Prasad *et al.*, 2009). The cryptic nature of pest makes it difficult for control using insecticides, as they fail to reach target site in sufficient quantity. Scrubbing the trunk at ten days interval with hard nylon brush dislodges eggs and 1st instar larvae (Chakraborti, 2011). In Nigeria a flexible wire or a bicycle spoke is pushed into the tunnels to kill larvae or adults in the trunk. Cotton wool attached on a flexible wire or a bicycle spoke end and soaked in insecticide could also be pushed into the holes to kill the insects (Asogwa *et al.*, 2009).

Cultural Control

The dead trees should also be cut down with their root extension exhumed and burnt to prevent further spread of this pest of cashew.

Botanical Insecticides

Application of neem oil 5 per cent (50 ml of Neem oil + 1,000 ml of water + 0.5 ml of teepol or 5 g of bar soap) on collar portion upto 1 m height and on exposed roots keep the insects away for a period of 3 months (Abdul Salam and Peter, 2010).

Application of neem oil 5 per cent three times as prophylactic measure is effective in management of cashew stem and root borer (Chakraborti, 2011). Neem oil at 5 per cent caused 66.7 per cent recovery of infested trees (Senguttuvan and Mahadevan, 1997) and prevented trees from being reinfested by the root and shoot borer (Mohapatra *et al.*, 2004). Smearing of cashew nut shell liquid (CNSL) on basal trunks of 2 to 4 unproductive trees during the period of *P. ferrugineus* adult emergence in December to February and April to May, help to control *P. ferrugineus* on cashew (Senguttuvan and Mahadevan, 1999).

Chemical Control

Pouring 2 per cent carbaryl solution in to soil at tree basin minimises incidence of stem borer. Soil incorporation of 75 g of sevidol (4 G) in tree basin also minimises attack of stem and root-borer (Abdul Salam and Peter, 2010). Injecting quinalphos 2 per cent into galleries made by the borer, caused mortality to larvae.

Swabbing the pest infested portion after removal of different stages of pest occurring internally and drenching the root zone with chlorpyriphos (0.2 per cent) as postextraction prophylaxis (PEP) reduces reinfestation upto 70.7 per cent. Swabbing the affected portion of trunk with chlorpyripos at 0.2 per cent did not have harvest time residues in the kernals (Bhat and Raviprasad, 2008).

Swabbing the affected portions and drenching soil around basal trunk after mechanical removal of larvae with monocrotophos at 0.2 per cent @ 2 litres per tree were effective. Swabbing mud slurry with carbaryl four times and basal application of lindane decreased incidence of CSRB (Mohapatra *et al.*, 2000). Applying phorate and spraying monocrotophos and chlorpyriphos was effective against CSRB grubs and prevented the egg laying by adults (Ayyana and Ramadevi, 1986; Pillai,1976; Senguttuvan and Mahadevan,1997 and Mohapatra, 2004. Holes in the infested trees probed with a thin wire killed the larvae. Cotton swab soaked in a mixture of carbon disulfide, naphthalene and creosote when plugged and then sealed with clay or cattle dung was effective in managing the CSRB (Misra and Basu Choudari, 1985).

A special powder formulation called sealer-cum-healer (S/H) was tried at Vridhachalam in Tamil Nadu. The base of an infested trunk and the subterranean regions of the main root system were examined. Infested portions were scrapped of the frass and doped with a mixture of copper oxychloride and dichlorvos. On this slurry of S/H was dabbed and allowed to dry. There was a recovery of more than 60 per cent of the trees. Highly infested trees required a second application, while trees which showed bad damage were beyond recovery.

Biological Control

Use of entomofungal pathogens is an ecofriendly measure particularly for concealed pests like stem borers owing to prevalence of congenial environment within the larval tunnels for mycosis of the grubs by *Metarhizium anisopliae* and *Beauveria bassiana*. Biostages of *P. ferrugineus* in dead cashew trees infected by *M. anisopliae* (Metsch.) Sorok., and *B. bassiana* (Bals.) at enzootic levels have been observed (Sundararaju,1984; Bhat and Raviprasad, 1995; Ambethgar *et al.*, 1999 and Ambethgar 2001).

Entomofungal pathogen when mixed with organic carriers like farm yard manure, vermicompost and cashew apple chunk increases spore load and improved survival of *M. anisopliae* for three months under field condition. Delivery of bioagents in target area is important to achieve maximum benefit. Pouring/injecting conidial suspension into the borer's entry hole is effective over swabbing conidial slurry or soil application (Saminathan *et al.*, 2004 and Ambethkar, 2010).

Placing the viral suspension of *Rhabdionvirus oryctes* (Huger) mixed with sawdust at the base of the cashew tree infested by root and stem borer caused establishment and infection of the virus in the beetle population in Sri Lanka (Rajapakse and Jeevaratnam, 1982). The efficacy of *Oryctes baculovirus* in *P. ferrugineus* was confirmed by Bakthavatsalam and Sundararaju (1990). Bioagents apart from being safe to non-target organisms are also effective in the long term pest control programme as they are self-perpetuating.

Ethological Approach

Adult borers prefer the infested tree for laying the eggs. Observations revealed that 70 – 74 per cent of the trees having the egg laid were in the frass and this was prefered over the basal portion of the trunk. The kairomones that are essential for attracting the adults to deposit the eggs may be exploited for management of this pest. Presence of bitter trees (*Holarrhena antidysenterica* Wall.) in cashew plantations wards of the cashew stem and root borer during its blooming period. There was no fresh/reinfestation during this period. Volatile organic compounds from bitter tree flowers when mixed up with volatiles emitted from cashew stem and root borer infested cashew trees caused disorientation of cashew stem and root borer adults (Chakraborti, 2010).

Therefore, it can be concluded that understanding the diversity of *P. ferrugienus* using the morphometrics and molecular tools would aid to develop location specific management techniques. Keeping in view the importance of the pest and crop, the cashew stem and root borer can be managed or suppressed in the plantations using the combinations phytosanitory measures, or mechanical collection of the biostages of the pest, use of botanochemicals and chemicals with little harvest time residues and bioagents. Barks are the oviposition sites for the beetles. Identifying the physical toxicants like inert clays if swabbed over the trunk can have impact on the eggs of CSRB. Previous experience on the ethology of the pest established the orientation of the adult beetle to oviposit in the damaged portion of the stem or in the frass. Identifying the volatile organic compounds emating from the frass can be exploited to develop oviposition attractants or repellents. Herbivory induced volatiles from the damaged plant can be charecterized and the compounds responsible for the orientaion of the beetle can be established by electrophysiological and behavioural approach and can be exploited for pest management. Botanochemicals that are used in the management of CSRB are neem based formulations. Biorationals like essential oils contain terpenoids that possess vapour toxicity, ovicidal and repellency used in tandem with efficient delivery devices could be exploited. The local area management of cashew stem and root borer aims mainly at suppression, rather than eradication. Area wide management by using the IPM tools will aid to protect against reinvasion.

References

Abdul Salam, M. and Peter, K. V. (2010). Cashew -A Monograph, pp. 1-257

Abraham, E. V. (1958). Pests of Cashew (*Anacardium occidentale* L.) in South India. *Indian Journal of Agricultural Sciences,* 28: 531-544.

Ambethgar, V. (2001). Indigenous entomofungi for biological control of stem and root borer infesting cashew. *South Indian Horticulture,* 49: 281-284.

Andrighetti, L., Bassi, G. F., Capella, P., De Logu, A. M., Deolalikar, D., Haeusler, G., Malorgio, G. A., Franca, F. M. C., Rivoria, G., Vannini, L. and Deserti, R. (1994). The world cashew economy. Edizioni L'inchiostroblu, Bologna, Italy.

Anikwe, J. C., Okelana, F. A., Otuonye, H. A., Hammed, L. A., Aliyu, O.M. (2007). The integrated management of an emerging insect pest of cashew: A case study of the cashew root and stem borer, *Plocaederus ferrugineus* in Ibadan, Nigeria. *Journal of Agriculture Forestry and Social Sciences*, 5(1): 55-64.

Arjuna Rao, P. (1978). Pest complex of cashew, *Ancardium occidentale* in Guntur and Prakasm district of Andhra Pradesh. *Cashew Bulletin*, 15 (11): 7-11.

Asogwa, E. U., Anikwe, J. C., Ndubuaku, T. C. N., Okelana, F. A. and Hammed, L. A. (2009). Host Plant Range and Morphometrics Descriptions of an Emerging Insect Pest of Cashew, *Plocaederus ferrugineus* L. (Coleoptera: Cerambycidae) in Nigeria: A Preliminary report. *International Journal of Sustainable Crop Production*, 4(3): 27-32.

Ayyanna, T. and Rama Devi. (1986). A study of the distribution and status of stem and root borer, *Plocaederus ferrugineus* dreadful pest on cashew in the coastal districts of Andhra Pradesh and its control. *Cashew Causerie*, 8 (1): 6-8.

Bakthavatsalam, N. and Sundararaju, D. (1990). Pathogenicity of *Oryctes* baculovirus to cashew stem and root borers. *Journal of Biological Control*, 4(2): 127-129.

Bhaskara Rao E.V.V. (1998). Integrated production practices of cashew in India. RAP FAO Corporate Document Repository. www.fao.org/docrep/005/ac451e/ac451e04.htm. p. 12.

Chakraborti, S. (2011). Management of the early stages of cashew stem and root borer, *Plocaederus ferrugineus* L. (Cerambycidae, Coleoptera). *Journal of Entomological Research*, 35 (3): 203-207.

Concepcion Magboo. (1998). Integrated Production Practices of Cashew in The Philippines – RAP FAO Corporate Document Repository. www.fao.org/docrep/005/ac451e/ac451e04.htm. p. 37.

DCCD (2013). Directorate of Cashewnut and Cocoa Development, Ministry of Agriculture, Kera Bhavan, Kochi, Kerala.

Devasahayam, S. and Nair, C. P. R. (1986). The mosquito bug, *Helopeltis antonii* Sign. on cashew in India. *J. Plantation Crops*, 14: 1-10.

Eguagie, W. E. (1972). Insects associated with cashew *Anacardium occidentale* in Nigeria. *Ann, Rep. CRIN,Ibadan, Nigeria*. pp. 134-137.

FAO. (2012). Production Yearbook, FAO, Rome.

Gerini, V. (1976). I coleotteri presenti sull'anacardio con particolare riferim ento a *P. ferrugineaus* nello Sri-Lanka. *Rivista di Agricoltura Subtropicale Tropicale* (Italy), 70(7-9/10-12): 185-190.

Jena, B. C. (1990). Cashew stem, root and wood borer and their management. *The Cashew*, (1): 5-7. *Journal of Plantation Crops* 5: 58-62.

Liu Kangde, Liang Shibang and Deng Suisheng. (1998). Integrated Production Practices of Cashew in China. RAP FAO Corporate Document Repository. www.fao.org/docrep/005/ac451e/ac451e04.htm. p. 4.

Martin, P. J., Topper, C. P., Bashiru, R. A., Boma, F., De Wall D., Harries, H. C., Kasuga, L. J., Katanila, N., Kikoka, L. P., Lamboll, R., Maddison, A. C., Majule, A. E., Masawe, P. A., Millanzi, K. J., Nathaniels, N. Q., Shomari, S. H., Sijaona, M. E., and Stathers, T. (1997). Cashew nut production in Tanzania: constraints and progress through integrated crop management. *Crop Protection,* 16: 5–14.

Maung Maung Lay. (1998). Integrated Production Practices of Cashew in Myanmar - RAP FAO Corporate Document Repository. www.fao.org/docrep/005/ac451e/ac451e04.htm. p. 26.

Misra, M. P. and Basu Choudhury, J. C. (1985). Control of *Plocaederus ferrugineus* Linnaeus (Coleoptera: Cerambycidae) through field hygiene. *Indian Journal Agricultural Sciences,* 55: 290-293.

Mohapatra, L. N. (2004). Management of cashew stem and root borer *Plocaederus ferrugineus* L. *Indian Journal of Plant Protection,* 32 (1): 149-150.

Mohapatra R, Jena B C and Panda D. 2004. Effect of insecticides and phytosanitation in management of cashew stem and root borer. *Journal of Plantation Crops,* 32(Suppl.): 349-350.

Mohapatra, R. N., Jena, B. C. (2007). Biology of cashew stem and root borer, *Plocaederus ferrugineus* L. on different hosts. *Journal of Entomological Research,* 31(2): 149-154

Mohapatra, R. N., Jena, B. C. and Mohapatra, L. N. (2008). Population dynamics, survey and behaviour of cashew stem and root borer, *Plocaederus ferrugineus* L. in Odisha. *Journal of Plantation Crops,* 36(2): 145-148.

Nguyen Minh Chau. (1998). Integrated Production Practices of Cashew in Vietnam. RAP FAO Corporate Document Repository. Available online at www.fao.org/docrep/005/ac451e/ac451e04.htm. p. 55.

Omole, M. M. (1972). Insects associated with cashew *A. occidentale* in Nigeria. *Annual Report, CRIN.* pp. 134-137.

Pillai, G. B., Dubey, O. P. and Singh, V. (1976). Pests of cashew and their control in India: A review of current status. *Journal of Plantation Crops,* 4: 37-50.

Rajapakse, H. S. R. and Jeevaratnam, K. (1982). Use of a virus against the root and stem borer *Plocaederus ferrugineus* L. (Coleoptera: Cerambydiae) of the cashew. *International Journal of Tropical Insect Science,* 3 (1) : 49 -51

Rai, P. S. (1984). Handbook of Cashew Pests. Researchco Publications, Delhi.

Raviprasad, T.N., Bhat, P. S. and Sundararaju, D. (2009). Integrated pest management approaches to minimize incidence of cashew stem and root borers (*Plocaederus* spp.). *Journal of Plantation Crops,* 37: 185-189.

Sahu, K. R. and Sharma, D. (2008). Management of cashew stem and root borer, *Plocaederus ferrugineus* L. by microbial and plant products. *Journal of Biopesticides,* 1(2): 121 – 123.

Saminathan, V. R., Mahalingam, C. A., Ambethgar, V., Vanitha, J., Ananthanayaki, D. and Prabaharan, G. (2004). Efeect of fungal pathogens on cashew stem and root borer, *Plocaederus ferrugineus* L. (Coleoptera: Cerambycidae) under field condition. *Pest Management in Horticultural Ecosystems,* 10 (1): 61-65.

Saroj, P. L. and Rupa, T. R. (2014). Cashew Research in India: Achievements and Strategies. *Progressive Horticulture,* 46 (1): 1- 17.

Senguttuvan, T. and Mahadevan, N. R. (1999). Influence of tree characteristics on the incidence of cashew stem and root borer, *Plocaederus ferrugineus* L. *Journal of Plantation Crops,* 27(1): 59-62.

Senguttuvan, T. and Mahadevan, N. R. (1997). Studies on the population fluctuation and management of cashew stem and root borer, *Plocaederus ferrugineus* L. *Pest Management in Horticultural Ecosystems,* 3(2): 85-94.

Sundararaju, D. (1984). Cashew pests and their natural enemies in Goa. *Journal of Plantation Crops,* 12: 38–46.

Surendra. (1998). Integrated Production Practices of Cashew in Sri Lanka. RAP FAO Corporate Document Repository. www.fao.org/docrep/005/ac451e/ac451e04.htm. p. 41.

Suwit Chaikiattiyos. (1998). Integrated Production Practices of Cashew in Thailand. RAP FAO Corporate Document Repository. www.fao.org/docrep/005/ac451e/ac451e04.htm. p. 49.

Uthaiah, B. C., Rai, P. S., Khan, M. M., Hiremath, I. G., Kumar, D. P. and Rao, K. B. (1989). Prebearing performance of some cashew type in coastal Karnataka. *The Cashew,* 3: 9-11.

Xianli, P. and van der Geest, L. P. S. (1990), Insect pests of cashew in Hainan, China, and their control. *Journal of Applied Entomology,* 110: 370–377.

Cashew: Improvement, Production and Processing (2017) *Pages* **377–393**
Editor: **Prof. P.L. Saroj**
Published by: **ASTRAL INTERNATIONAL PVT. LTD., NEW DELHI**

Minor Pests and their Management

K. Vanitha, P.S. Bhat, T.N. Raviprasad and D. Sundararaju

1. Introduction

Cashew is an important commercial edible plant grown in India. Globally, more than 300 species of arthropods are associated as pests of cashew (Vanitha and Saroj, 2015). Among them, certain pests affect the cashew plant only temporarily, while others live for several generations on the plant. In some instances, the attack may result even death of the plant, but in most cases damage results in yield reduction. All parts of the plant *viz.*, leaf, stem, bark, root, flower, apple and nut are infested by one or other pest species, resulting in yield loss. A yield loss of 50 per cent was reported annually due to pests and diseases in cashew (Hari Babu *et al.*, 1983). Apart from two important major pests *viz.*, tea mosquito bug and cashew stem and root borer, there are few insect pests which cause economic damage in certain occasions.

Depending on the climate, location, age and management measures of the plantation, each geographic region has its own distinctive pest complex. Documentation of pest diversity across different cashew growing regions is necessary to formulate effective management strategies. Though, each geographical region has its own distinctive pest fauna composed mainly of indigenous species, a small number of key pests are found common throughout the world *viz.*, tea mosquito bug (*Helopeltis* spp.), cashew stem and root borer (*Plocaederus* spp.), leaf miner (*Acrocercops* spp.) and thrips (*Selenothrips rubrocinctus* Giard). Damage by different insect species, at different intensities, is generally widespread throughout all cashew growing countries. Many insect pests have the ability to move long distances and could become a pest at a new place in future. In addition, as the area of cashew expands, pest dynamics also change. In India alone, more than 190 pests are reported on cashew in different cashew growing states including Andaman and Nicobar. This chapter deals with the advances made in understanding of minor pests of cashew, their biology, ecology and management.

2. Biology, Ecology and Management

Among different cashew pests, some may not be problematic in certain regions but could be a problem in other regions. In general, management actions taken for TMB could manage these pests also, but separate spraying may be required during certain periods to prevent economic loss. The details of important minor insect pests of cashew are given below. Management measures suggested are based on several reports and results of trials conducted at AICRP on Cashew Centres.

a. Shoot Tip Caterpillars

Anarsia epotias M. and *Hypotima haligramma* M. (Lepidoptera: Gelechiidae)

Caterpillars of *A. epotias* and *H. haligramma* damage the shoot tips of cashew during post monsoon flushing period. Larvae of *A. epotias* are pale yellowish green with black head and turn pinkish-brown later. While, larvae of *H. haligramma* are tiny, yellowish to greenish brown in colour. The pest is presently distributed in Goa, Karnataka, Kerala, Tamil Nadu, Andhra Pradesh and Odisha. This pest is regularly reported in the east coast tracts of India (Mohapatra *et al.*, 1998; Jena *et al.*, 1985). The egg, larval and pupal period of *H. haligramma* last for 3-4 days, 12-16 days and 7-10 days respectively, and the life cycle is completed within 25-29 days. Pupation takes place inside the bore holes on the shoots or inflorescences. While moths of *A. epotias* lay upto 60 eggs/female either singly or in groups of 10-20. The egg period lasts for 3-4 days and the larva feeds inside the tunnel. Pupation takes place in larval tunnels of the attacked shoot, in crevices of the branches, twigs or at the cut end of branches. Pupal period is 7-10 days and the total life cycle is completed within 27-29 days (Remamony, 1965).

During active growth period, caterpillars of *A. epotias* at the early stage web together the tender leaves and feed within. Later on, they bore in to the terminal shoots and tunnel inside up to 2-3 cm. A gummy substance oozes out from the infected tips and finally the attacked shoots dry up. Larvae of *H. haligramma* also damage shoot tips by folding the fresh leaves and feed within and can tunnel upto 2 cm. The larvae may also damage inflorescences subsequently. Exudation of gummy web like substances mixed with faecal pellets from the injury site can be noticed. Later, the terminal shoots turn black and perish, which results in production of auxiliary shoots.

In west coast, up to 13.0 per cent damage in newly emerging post-monsoon flushes (in October) and 10.5 per cent in pre--monsoon flushes (in May) have been reported (Pillai *et al.*, 1976) indicating the significance of this pest. Abiotic factors do not exert a significant influence on infestation and population of this particular pest (Mohapatra *et al.*, 1998).

Management Measures

Spraying may not be required, since natural enemies including parasitoids take care of this pest. Four larval parasitoids, *viz., Pristomerus* sp. (Ichneumonidae), *Apanteles* sp. (Braconidae), *Elasmus* sp. (Elasmidae) and *Sympiesis* sp. (Eulophidae) have been recorded on shoot tip caterpillar larvae causing parasitism up to 25 per cent. But, under severe incidence, spraying of either quinolphos (2 ml/lit) or phosalone (2 ml/lit) or phosphamidon (2 ml/lit) or monocrotophos (1.5 ml/lit) or lambda-cyhalothrin (0.6 ml/lit) or profenophos (1.5 ml/lit) can manage this pest effectively.

b. Leaf Miner

Acrocercops syngramma M. (Lepidoptera: Gracillaridae)

Leaf miner is one of the serious pests of cashew during post monsoon period throughout the country. Its occurrence was recorded in Goa, Karnataka (Sundararaju, 1984, Sundararaju and Bakthavatsalam, 1994), Odisha (Jena and Satapathy, 1988) and also Andaman Islands (Jacob and Belvadi, 1990). Eggs are tiny, transparent and glue like, laid on the upper side of tender leaves. Eggs hatch in 5-7 days. During the developmental period, larvae are dull white in colour. Full grown caterpillars measure about 5-7 mm in length, are reddish brown in colour. After full development, the larvae fall off to the soil for pupation emerge as adults after 7-9 days (Pillai *et al.*, 1976). The adult is a tiny silvery grey moth with red eyes.

The incidence is most common in post monsoon flushes during August-December reaching the peak during October- November, but incidence can be seen upto April at low level. The mining injury by caterpillars occurs both on tender leaves as well as shoots. Young plants are observed to be prone to attack by this pest. The caterpillars mine and feed below the epidermal layer of the tender leaves and shoots causing extensive blisters, which later dry up resulting in distortion, browning and curling of the leaves. As the attacked leaf, holes develop due to drying out of the damaged matures portion. Generally, 1-5 larvae are seen in a single leaf (Rai, 1984; Jena, 1985), but up to 45 larvae have been observed during peak infestation at Puttur, Karnataka (Vanitha *et al.*, 2015). Abraham (1959) estimated around 26 per cent leaf miner damage in severely infested cashe leaves. Varied leaf damage levels were recorded in different states *viz.*, Kerala (70-80 per cent), Karnataka (60 per cent), Andhra Pradesh (6-20 per cent), Odisha (8.4 per cent) and West Bengal (18-20 per cent) respectively (Basu Choudhuri, 1962; Rai, 1984; Ayyana *et al.*, 1985; Jena *et al.*, 1985; Chatterjee, 1989). The leaf miner pest status was assessed on medium aged common cashew varieties at Puttur during 2011-2014 and found low to medium (Vanitha *et al.*, 2015).

Management Measures

Two larval parasitoids *viz., Chelonus* sp. (Braconidae) and *Sympiesis* sp. (Eulophidae) have been recorded on leaf miners in Kerala and Goa. Recently, three larval parasitoids namely, *Chyrsocharis* sp., *Aprostocetus* sp. and *Closterocerus* sp. (Eulophidae) were recorded in Puttur region of Karnataka. Spraying may not be required since these larval parasitoids manage this pest even up to 50 per cent. But under severe incidence in nursery and young plants, spraying is required. Spraying of quinolphos (1.5 ml/lit) or monocrotophos (1.5 ml/lit) or profenophos (1.5 ml/ lit) or lambda-cyhalothrin (0.6 ml/lit) can effectively manage leaf miners.

c. Leaf Folder and Leaf Rollers

***Hypatima haligramma* M. (Gelechiidae), *Caloptilia tiselaea* M. (Gracillaridae), *Dudua aprobola* M. (Tortricidae), *Sylepta derogatta* F. and *S. auranticollis* (Pyralidae)**

Larva of *H. haligramma* is yellowish green with black head and prothoracic shield, pupates inside leaf roll, pupa is small and brown. Adult is a small, narrow shiny moth, with forewings having two whitish transverse bands. While, larva of *C. tiselaea* is sluggish, light yellowish with creamy head, grows up to 7 mm and pupates inside a silken cocoon by making a small folding on lamina. Adult is a tiny, narrow plume moth, dirty white in colour. But, larva of *D. aprobola* looks dark green, very active with black head and prothoracic shield, grows upto 1.5 cm long, pupates inside leaf roll while, adult is a tiny narrow grey moth. Wings are smoky green with ornamentation having black spot towards apical margin. Whereas, larva of *S. derogatta* looks glistening green with dark brown head and prothorax, grows up to 2.5 cm and pupates inside the leaf roll. Adult is a small moth with light yellowish wings having brown wavy margins. In Kerala, pink leaf roller, *Anigraea albomaculata* (Noctuidae: Lepidoptera) occurs on cashew (Pathummal beevi *et al.*, 1993). Incidence of leaf folder and leaf roller has been reported on cashew in Kerala, Karnataka, Andhra Pradesh and Odisha.

These pests cause considerable damage to post-monsoon tender foliage of cashew between August and January. The incidence of *C. tiselaea* commences during the first fortnight of November and prevails up to the end of January and peak population is seen during the second fortnight of December. Higher leaf infestation was recorded in early cashew types compared to mid and late cashew types (Mohapatra, 2007). At Andhra Pradesh, *D. aprobola* and *C. tisilea* occur from second fortnight of August to first fortnight of March and maximum population is seen during November. Higher population and damage were observed in the plantation of 8-11 years' age (Ayyana and Ramadevi, 1987).

In general, early cashew types record higher infestation by leaf folders compared to late cashew types. Larva of *H. haligramma* rolls up the leaf margin towards ventral side. Roll is narrow, tight and the larva feeds the leaf margins, as a result portions of the leaf dry up. Similar kind of roll but big sized is made also by *D. aprobola*. Larva of *C. tiselaea* folds loosely the leaf either from the top or from lateral margin towards the midrib ventrally. The larva remains inside the roll and feeds from inside by scraping the green portion of the leaf lamina which later dries up. Larva of *A. albomaculata* damages tender leaves by making spindle shaped folds. Two to four terminal leaves are folded longitudinally one above the other and fastened with silken threads to form a tight tubular roll at the growing point thus delays emergence of inflorescence. Larvae of *S. auranticollis* during their early stages roll the tender leaves and scrape the green matter, but later defoliate entire leaves. Whereas, a one more leaf webber namely, *Macalla albifusa* (Pyralidae) join the leaves one above the other by silken threads and feed on them and the damaged portion gradually dries up. The larva is very active, remains inside a tunnel formed of excretory matter and silk, wriggles out when disturbed.

Management Measures

Since damage resulted by these pests is very less, spraying may not be required. Besides, natural enemies including parasitoids take care of these pests. *Cotesia* sp. (Braconidae) and *Chrysocharis* sp. (Eulophidae) parasitize larvae of *C. tiselaea* to a great extent. But under severe incidence, spraying of quinolphos (2 ml/lit) or profenophos (1.5 ml/lit) or monocrotophos (1.5 ml/lit) or lambda-cyhalothrin (0.6 ml/lit) on affected plants manage this pest.

d. Thrips

(i) Leaf Thrips (*Selenothrips rubrocinctus* Giard (Thysanoptera: Thripidae), *Rhipiphorothrips cruentatus* Hood and *Retithrips syriacus* (Mayet))

Thrips are minute worm like insects damage the crop by sucking the plant sap and present throughout cashew growing regions. Attack on tender foliage of cashew by thrips results in silvery leaves (Ananthakrishnan, 1984, Ayyanna *et al.*, 1985; Jena *et al.*, 1987). Adults of *S. rubrocinctus* are dark brown and about 1-2 mm long. Nymphs are pale yellowish with a red band around the middle of their body. Tiny eggs are laid singly into the lower epidermis and covered with black excrement of adults. Egg period lasts between 3-10 days depending on the temperature. Nymphs are wingless, vermiform, carry a drop of excrement at the anal end, while adults are winged. The nymphal, pre-pupal and pupal period lasts for 10, 1 and 2-3 days, respectively (Jena, 1990). Immature stages of *S. rubrocinctus* are abundant on the older leaves than on young leaves while adults are present both on young and old leaves (Cavalcante *et al.*, 1975). In Ivory Coast also, *S. rubrocinctus* is serious and egg, nymphal and adult period lasts for 10, 12 and 25 days, respectively. The nymphs of *R. cruentatus* are white when they hatch, but pale red markings develop soon. The female thrips are 1.2 to 1.5 mm long, blackish-brown in colour, with yellow legs while antennal segments and the forewings are pale with yellowish veins. Whereas, nymphs of *S. dorsalis* are pale or cream coloured and adults are about 1.2 mm long with dark wings.

Among the thrips, *S. rubrocinctus* is very serious in nursery and young cashew plantations. In seedlings, initially it attacks lower leaves and cause premature leaf fall, stunting and finally drying of seedlings (Pillai *et al.*, 1976). In grown up plants, it damages young leaves, shoots, inflorescence and flowers and is more active during summer months (Mutter and Bigger, 1962). The infested leaves become pale brown and slightly crinkled with roughening of the upper surface. Honeydew excretory products from the thrips give rise to black sooty mould. In severe cases, shedding of leaves occurs. Leaves infested by *R. syriacus* become silvery white initially, later turn into pale brown, crinkle with roughening of upper surface and fall prematurely.

(ii) Flower thrips: (*Scirtothrips dorsalis* H. (Thripidae), *Rhynchothrips raoensis* G. (Phlaeothripidae), *Haplothrips ganglbaueri* (Schmutz) (Phlaeothripidae), *H. ceylonicus* Schmutz (Phlaeothripidae), *Thrips hawaiensis* (Morgan) (Thripidae), and *Frankliniella schultzei* (Trybom) (Thripidae))

Flower thrips attack flowers, apples and nuts. Infestation causes shedding of flowers, immature fruit drop, formation of scabby as well as, malformed apples and nuts (Gowda *et al.*, 1979 and Rai, 1984). Occurrence of different species of thrips, seasonal incidence and extent of damage on cashew are studied *viz.*, *R. roaensis* (Abraham, 1958, Ayyanna *et al.*, 1985, Patnaik *et al.*, 1987, Thirumalaraju *et al.*, 1991), *S. dorsalis, H. ganglbaueri, T. hawaiiensis* (Ayyanna *et al.*, 1985), *H. ceylonicus* and *F. schultzei* (Patnaik *et al.*, 1987). Among the flower thrips, *S. dorsalis* and *R. raoensis* are prevalent in the East coast regions of India, whereas in West coast regions, *H. ceylonicus* and *F. schultzei* are prevalent (Patnaik *et al.*, 1987). Up to 15-25 per cent fruit drop is noticed (Gowda *et al.*, 1979). Apart from cashew, it breeds on a number of annual crops, including *Calycopteris floribunda* Lamk. (Combretaceae) a common shrub in many cashew plantations (Sundararaju, 1984). In Karnataka, *S. dorsalis* (75.4 per cent) is dominant than *R. raoensis* (24.6 per cent). The initiation of population buildup of both the species occurs at the time of flower bud initiation in cashew *i.e.*, first fortnight of November reaches its peak during February and completely disappears after May. Cashew varieties that bear off-season flowers have continuous infestation of thrips. In Odisha, there is an increase in population of flower thrips from October and reaches its peak during first fortnight of December (Jena *et al.*, 1987).

Management Measures

There are several predators like various syrphids (*Paragus* sp.), coccinellids (*Pseudospidemerus circumflexa* Mots., *Menochilus sexmaculata, Coccinella transversalis, Scymnus* sp., *Illeis cincta*), lace wing bugs etc that take care of this pest. Spraying of monocrotophos (1.5 ml/lit) or lambda-cyhalothrin (0.6 ml/lit) or dimethoate (2 ml/lit) or carbaryl (1 g/lit) is effective for managing thrips.

e. Leaf and Blossom Webber

Lamida (Macalla) moncusalis Wlk. (Pyralidae: Lepidoptera)

Male moths are dark, fuscous and the females are green. Eggs are deposited ventrally on tender leaves and occasionally on tender shoots either singly or in groups of 5-6. This pest is distributed in all cashew growing areas especially East coast tracts of India. Infestation occurs during post monsoon flushing period (September - October) and flowering period (February- March). The egg, larval, pre-pupal and adult stages last 4-7, 16-22, 9-15 and 3-6 days, respectively (Murthy *et al.*, 1974, Rao *et al.*, 2002, Panda, 2013). The caterpillars feed on the terminal leaves of new shoots and blossoms after webbing them. Presence of webbing on terminal portions, with clumped appearance, and drying of webbed shoot/inflorescences are noticed. Galleries of silken webs reinforced with plant scraps and castings, indicate the presence of caterpillars (Ayyanna *et al.*, 1985; Satapathy and Panigrahi,

1995; Sundararaju, 2007). It occurs sporadically and can cause damage between 25-60 per cent (Dharmaraju *et al.*, 1975). In Andhra Pradesh, maximum infestation occurs during second week of May (Subba Rao *et al.*, 2006), whereas in Karnataka, it is noticed during January (Thirumalaraju *et al.*, 1991). Increase in temperature and fall in relative humidity is congenial for the pest outbreak (Ramadevi and Radhakrishnan, 1991, Subba Rao *et al.*, 2006).

Management Measures

In East coast regions, leaf and blossom webber is parasitized by braconids (*Apanteles* spp.), elasmid (*Elasmus* sp., *Elasmus johnstonii* F.) and a tachinid fly (*Blepharella lateralis*) in Andhra Pradesh and Odisha and a maximum of 50 per cent parasitism has been reported. While in Kerala, *Apanteles* sp. and *Avga choaspis* (Nixon) (Braconidae) occur as parasitoids on leaf and blossom webber. The green lace wing, *Chrysoperla* sp. also predate on this pest. Spraying of monocrotophos (1.5 ml/lit) or lambda-cyhalothrin (0.6 ml/lit) can manage this pest.

f. Hairy Caterpillars

Metanastria hyrtaca Cram (Lasiocampidae: Lepidoptera) and *Lymantria ampla* Wlk. (Lymantridae: Lepidoptera)

Plenty of hairy caterpillars damage cashew, in which, *M.hyrtaca* and *L. ampla* cause severe sporadic defoliation in cashew. The caterpillars are capable of defoliating the cashew trees completely leaving only bare branches. *M. hyrtaca* occur sporadically and attack isolated trees. Egg, larval, pupal and adult period are 9, 33-35, 12 and 1-6 days, respectively (Nair *et al.*, 1974; Arjuna Rao *et al.*, 1977). The early instars of *M. hyrtaca* are gregarious feeders on tender foliage and the full grown caterpillars feed voraciously on mature leaves as well. During day time they congregate in large numbers on the ground under dry leaves near the base of the tree, in crevices of bark or lower parts of well shaded branches and are capable of complete defoliation of young trees (Nair *et al.*, 1974).

The other hairy caterpillars like *Euproctis* spp. (Lepidoptera: Lymantridae) *viz.*, *E. fraterna* Moore, *E. scintillans* Walker and *E. subnotata* scrape the green tissues when young and start defoliating the leaves and inflorescence branches and also feed on the shell of the nut in the tender green stage and tender apples (Rai, 1984; Sundararaju, 1984). Full grown larvae of *E. fraterna* are stout, dark reddish-brown thickly covered with whitish hairs with a pair of dark tufts on either side of the head and one on the anal segment. While full grown larvae of *E. scintillans* are stout, dark-brown with tuft of fine hairs. A pale yellow strip runs down the back and on the first abdominal segment a thick tuft of blackish hairs is seen. The light brown moths of *Diacrisia obliqua* Walker (Lepidoptera: Arctiidae) lay 400- 1200 spherical, pale yellow eggs in small clusters that hatch in about 8-10 days. Larva is black and yellow with long, black and white hairs, and several yellow bands are seen on the body. While, *Estigmene lactinea* Cramer (Lepidoptera: Arctiidae) batches on the leaves or in soil. The caterpillars are active, dirty black and the body is suffused with black and yellow hairs arising from warts set in a ring around the middle of each segment.

Management Measures

Braconids such as *Aleiodes* spp., *Apanteles oblique*, the chalcid, *Brachymeria poithetrialis* and the tachinid flies such as *Blepharipa* sp., *Carcolia* sp., *Exorista* sp. and *Palexorista* sp. are the parasitoids recorded on *L. ampla*. While, *Perilampus microgastri* Ferr (Perilampidae) is a parasitoid recorded on *M. hyrtaca* Cram. Larvae of *E. scintillans* are parasitized by *Apenteles euproctisiphagus* (Braconidae). Spraying is required only when severe defoliation and marked infestation on panicles is noticed. Insecticides including Monocrotophos (1.5 ml/lit) or lambda-cyhalothrin (0.6 ml/lit) effectively manage this pest.

g. Loopers and Semilooers

Oenospila flavifusata Walker, *Thallasodes quadraria*, *Hyposidra talaca* (Walker) and *Pingasa ruginaria* Guenee (Geometridae)

The defoliating loopers are very common during the new flush period from August to January causing severe defoliation (Rai, 1984). Tiny reddish eggs of *O. flavifuscata* are laid on margin of tender leaves and they hatch in about 5 days. Young caterpillars have reddish tinged body, green in colour. Larval period lasts for about 15 days and they pupate within leaf folds. Pupal stage lasts for 10 days. The eggs of *T. quadraria* are laid on the leaves that hatch in 3-5 days. The larvae are pinkish, slender and assume a characteristic pose oblique to the stem on the twigs and are mistaken for part of a twig or leaf petiole. Pupal period lasts 6-8 days. Larva of *P. ruginaria* is green having white cross markings throughout the body. In adults, the patterns on its wings would enable it to blend against the bark of a tree, especially one with lichen growth. *H. talaca* is recorded as an emerging pest in cashew (Chutia *et al.*, 2011) having wide host range including various forest trees, crops and weeds. Egg period of *H. talaca* is 6-7 days while, larval and pupal period lasts for 15-18 days and 7-10 days, respectively. The geometrids are sporadic in occurrence but they do cause considerable damage. They defoliate the tender leaves from the margins. The tender shoots damaged by *O. flavifusata* are left only with portions of midribs. Besides leaves, the larvae of *H. talaca* damage inflorescences also.

Management Measures

Since these pests cause sporadic defoliation, spraying is generally not required against this pest. But if necessary, insecticides like monocrotophos (1.5 ml/lit) or profenophos (1.5 ml/lit) can be sprayed to manage the pest.

h. Leaf Beetles and Weevils

Monolepta longitarsus Jac., *Neculla pollinaria* Baly, *Lypesthes* sp. (Chrysomelidae: Coleoptera), *Deporaus marginatus* (Coleoptera: Attelabidae)

Several chrysomelid beetles attack cashew especially during monsoon as well as post monsoon flushing period. Among them, *M. longitarsus* and *N. pollinaria* are important defoliators. While, *D. marginatus* occur occasionaly on tender shoots and nursery plants. During rainy season (June-August), the chrysomelid leaf beetles and weevils defoliate cashew voraciously. *M. longitarsus* is a regular pest in the west coast regions during the south west monsoon. They appear abundantly especially on young trees and skeletonise the leaves which gradually dry up. Tender shoots are also attacked that finally dry off. When nursery seedlings are attacked, the entire seedlings dry up. A group of 60-75 beetles are capable of causing complete drying of shoots in 2-3 days. It is also severe on current season limb pruned trees. These beetles also damage tapioca (Rai, 1984), *Terminalia arjuna* and *T. paniculata* Roth (Sundararaju, 1984). While, adults of *N. pollinaria* and *Lypesthes* sp. also attack the postharvest flushes causing defoliation and drying up of the shoots. Scrapping of the bark of tender shoots by the beetles appears as linear depressions. Adults of *D. marginatus* weevils remain on the lower side of the leaves and scrape the leaf surface making minute feeding holes which appear as 'windowpanes' on young leaves.

In Odisha, another beetle, *Microserica quadrinotata* Moscr (Melolonthidae: Coleoptera) occurs on cashew from June-October causing up to 30 per cent leaf damage during peak infestation on September (Jena *et al.*, 1985; 1986a). The adults, cause skeletonization of the leaf by scraping chlorophyll that turn red and finally dries up. The beetles also infest mango, neem, acacia, basal, Crotalaria and cucurbits. Similarly, in Odisha an ashy weevil, *Peltotrachelus pubes* Faust. (Curculionidae: Coleoptera) feeds on the tender leaves causing up to 15 per cent leaf damage (Jena *et al.*, 1986b). Besides, leaf rolling weevil, *Arodepus marginatus* (Attelabidae), Curculionid weevils *viz.*, *Amblyrrhinus poricollis* Schoenherr, *Apion amplum* Faust, *Apoderus tranquebaricus* Fab., *Myllocerus discolour*, cetoniid beetle, *Oxycetonia versicolor* Fabricius and other chrysomelid beetles *viz.*, *Basilepta flavicorne* Jac., *Hyperaxis albostriata* Mots and *Pagria costatipennis* Jac also defoliate cashew occassionally.

Management Measures

Spraying of any systemic or contact insecticide manages these pests. Spraying of chlorpyriphos (1.5 ml/lit) or monocrotophos (1.5 ml/lit) or quinolphos (2 ml/lit) or lambda- cyhalothrin (0.6 ml/lit) could cause mortality of the beetles in a short period.

i. Other Defoliators

Bombotelia jacosatrix Guenee (Noctuidae: Lepidoptera) is a leaf eating caterpillar that feed tender leaves from the margins. Early instars are gregarious and at the later stages they feed on the entire leaf, leaving only the midribs. Eggs are laid in rows on leaf margins of tender leaves; they hatch in 3-5 days. Larvae are greenish, striped with reddish-brown spots and larval period lasts for 11-18 days. Pupation takes place in a silken cocoon inside the leaf folds and adult emerges in 9-15 days.

Caterpillars of *Orthaga exvinacea* Hampson (Pyralidae: Lepidoptera) web together tender shoots and leaves, live within the webs and feed on the leaves. Several caterpillars are found in a single webbed-up cluster of leaves. Presence of silken webs reinforced with pieces of plant parts on terminal portions and blossoms as well as dried up appearance are the symptoms of its infestation. Eggs are yellowish green that hatch in 4-5 days. Caterpillars are slender, pale-green with dark bands, pupates in 28-33 days. Pupation takes place in silken cocoon. The pupa is reddish-brown and lasts for 11-14 days. The larvae of *Spodoptera litura*

F., *Helicoverpa armigera* Hubner (Noctuidae: Lepidoptera) and bag worm, *Dappula tertia* Templeton (Psychidae: Lepidoptera) also defoliate cashew in certain pockets of India. Besides, the caterpillars of Tassar silk moth, *Antherea paphia* L., *A. mylitta* Drury and *Circula trifenestrata* Helfer (Saturniidae: Lepidoptera) are also sometimes, noticed on cashew that defoliates voraciously (Rai, 1984, Vanitha *et al.*, 2015). The wild silk moth *C. trifenestrata* occurs in swarms during September-October causing severe defoliation. They feed voraciously for more than a month and pupate in golden yellow hairy silken cocoons which are found in masses inside group of leaves during November (Sundararaju, 1992).

Management Measures

Spraying of monocrotophos (1.5 ml/lit) or lambda-cyhalothrin (0.6 ml/lit) can effectively manage these defoliators.

j. Stem and Bark Feeders

Apart from CSRB, there are also other pests damaging cashew stem and bark. The caterpillars of *Inderbela tetraonis* Moore (Arbelidae: Lepidoptera) make a small residential hole on the wood normally where the branches fork and make superficial galleries inside which they feed on the tissues. The presence of winding galleries on the bark made of powdered bark, faecal pellets and silk webbed together indicates this pest attack. Feeding damage in small branches results in drying up of branches. The eggs are laid under loose bark that hatch in 8-10 days. Larvae are pale brown with dark head move along the branches concealed under the gallery. Larval period lasts even up to 10-11 months while, pupal period lasts for 15-25 days (Rai, 1984). Adults are stout, pale brown moths with wavy grey markings on the wings.

In addition, adult beetles of *Paranaleptes reticulate* Thomson and *Stenias grisator* Fabricius (Cerambycidae: Coloeptera) girdle the cashew branches with their strong mandibles. Hence, xylem and phloem tissues are damaged; the branches above the ringed portion dry up. Adult *P. reticulata* beetle lays eggs in series of irregular incisions made in the dead wood during the dry season. Hatching grubs are yellowish, tunnel into the dead wood of the branch and cause damage. Whereas, adults of *S. grisator* are stout greyish- brown beetles having elliptical greyish spots and an eye-shaped patch on the elytra. While, coleopteran grubs of *Xystocera globosa* Oliver (Cerambycidae), *Xylothrips flavipes* (Bostrichidae), *Belinota prasina* Thunberg *Lampetis fastuosa* F. (Buprestidae), *Analeptes trifasciata* Fabricius (Cerambycidae), *Coptos aedificator* Fabricius (Cerambycidae), *Mecocorynus loripes* Chevrolet (Curculionidae) also cause bark and sap wood damage in cashew.

Management Measures

To manage bark eating caterpillar, removal of galleries plastered on tree trunk or pouring of kerosene during early stage of infestation is suggested. In the chemical method, application of quinolphos (2 ml/lit) or dichlorvos (2 ml/lit) either by injection or by inserting a cotton swab soaked in the insecticide is the most widely used method.

k. Apple and Nut Borer

Thylacoptila paurosema Meyrick, *Hyalospila leuconeurella* Ragonet, *Nephopteryx* sp. and *Anarsia epotias* Meyrick (Pyralidae)

T. paurosema lays eggs on the immature fruits or nuts, and the incubation period is 3-5 days. The caterpillars are very active, dark pink in colour, measures 2-2.5 cm and the larval period lasts for 15-20 days. Larvae initially damage flowers by webbing the panicles and feed the unopened flower buds. Then they tunnel near the junction of apple and nut and bore inside the tender nuts and developing apples resulting in shrivelling and premature fall. Damaged fruits can be easily located as they have frass hanging externally at fruit and nut joint. Variable degrees of damage by this species have been reported from different cashew-growing tracts of India and damage of 10 per cent was reported by Nair *et al.* (1979). This pest also damages stored cashew nuts (Rai, 1984).

The caterpillars of *H. leuconeurella* bore through the apple from one end to the other but remain inside the apple till the fruit drops. Attacked apples generally fall down from the trees. Infested nuts become severely deformed. The egg, larval and pupal period lasts for 4-5, 12-17 and 9-12 days, respectively (Jena, 1990). The adult is dark-brown moth with a wing expanse of 16 mm. In a single apple, up to 6 caterpillars of different sizes are seen (Basu Choudhuri and Misra, 1973). Similarly, *Nephopteryx* sp. (Pyralidae: Lepidoptera) is common in Tamil Nadu and Andhra Pradesh (Ayyanna *et al.*, 1985, Dharmaraju *et al.*, 1975) attacking fruits at all stages of development causing up to 60 per cent of nut damage. Initially, the larvae scrape the epidermis of tender nuts and apples. Then, the young larvae move to the point of attachment of nut and apple and bore into apples and nuts. The entry hole is minute and plugged with the excreta. The infested fruits shrivel and drop prematurely, while, the nuts do not develop further and dry up. The larval period is 15-33 days. In one fruit, around 3-5 larvae were found (Rai, 1984). Larva pupates in earthen cocoon adult emerges in 8-10 days (Jena, 1990).

In south India, *Anarsia epotias* Meyr. (Gelechiidae) and *Helicoverpa armigera* (Noctuidae) were found as apple and nut feeders (Basu Choudhuri and Misra, 1973; Ramadevi and Ayyanna, 1988). The larva of *A. epotias* binds dry inflorescences to the side of the apples and nuts hanging adjacently and nibbles them continuously. At times, the caterpillars make galleries even inside the nut. The female lays 50-60 eggs singly or in groups of 10-20 that hatch in 3-4 days. Pupation takes place in larval tunnels of the attacked shoot, in crevices of the branches, twigs, at the cut end of branches or within the galleries in the apples. Pupal period is 7-10 days and the total life cycle is completed within 27-29 days (Basu Choudhury and Misra, 1973). Besides, larvae of *Orthaga exvinacea, L. moncusalis* and *Euproctis* spp. also damage tender nuts and apples, however, they are considered to be external feeders. If proper management is taken up against flower and fruit pests, more than 60 per cent yield loss can be avoided (Sundararaju, 2007).

Management Measures

Removal and destruction of infested portions of inflorescences as well as infested apples and nuts having larvae can be followed to prevent spread of the pest. Three larval parasitoids *viz., Panerotoma* sp. (Braconidae), *Trathala flavorbitalis* (Ichneumonidae) and one unidentified tiny dipteran fly occur on apple and nut borer larvae and a maximum of 46.2 to 50 per cent parasitism has been recorded under field conditions. Spraying of carbaryl (1 ml/lit) or lambda cyhalothirn (0.6 ml/lit) or quinolphos (2 ml/lit) or dichlorvos (1 ml/lit) is found effective for apple and nut borers.

l. Mealybugs

Planococcus citrii Risso, *Planococcus lilacinus* Cockrell and *Ferrisia virgata* Cockrell, *Planococcoides robustus* Ezzat and Meconnel

Mealybugs are potential pests in case of cashew. *F. virgata* are small plumpy, soft insects, body is covered with number of white waxy filaments all over, with two pronounced long waxy processes at the posterior end. A pair of dark submedian stripes is also present. Nymphs of *P. citri* are pale yellow without waxy coating. Adult females are slightly elongate, ovate, 5-7 mm long but covered with white mealy wax including the appendages. It is presently distributed in Karnataka, Kerala, Tamil Nadu, but severe in Konkan region and Goa. Egg masses of *F. virgata* remain under the females till the young ones hatch out. A female lays 100-300 eggs in 3-4 weeks. Eggs hatch in 3-4 hours and the nymphs develop as adults in about 26-47 days. Adults live for 14-21 days. The pest is prevalent during December- April. Mealybug colonies develop on young vegetative shoots, leaves, inflorescence, and tender nuts and fruits. Damaged flowers wither and dry, while the fruits shrivel, under develop or sometimes dry up. Due to honey dew secretion by mealy bugs, sooty mould develops on the affected portions.

Management Measures

Removal and destruction of mealy bug infested plant parts help to minimize their infestation and spread. *Apanteles* sp. has been reported as a parasitoid on *F. virgata*, besides *Blepyrus insularis* Cameron. In Kerala, up to 35 per cent parasitism has been reported in *F. virgata* to be caused by *Aenasius advena* Campere (Encyrtidae). In Karnataka also, *A. advena* could cause upto 50 per cent parasitism in *F. virgata*. If essential, spraying of profenophos (2 ml/lit) or chlorpyriphos (2ml/lit) or dimethoate (2 ml/lit) or thiamethoxam (0.6 g/lit) or imidacloprid (0.6 ml/lit) may be followed to manage mealy bugs in combination with fish oil Rosin soap at 20 g/lit.

j. Aphids

Toxoptera odinae van der Goot and *Aphis gossypii* (Aphididae: Hemiptera)

Aphids are tiny (1-3 mm), soft, brownish in colour with prominent cornicles and found in groups during December – April and June. They reproduce mostly by parthenogenetic vivipary and also by sexual reproduction. Nymphs are tiny, complete four instars in a period of 9-16 days and become adults. A single female can produce around 35-50 young ones in a short period. Though aphids are seen in huge numbers on leaves, shoots, inflorescence sand immature nuts, the feeding seldom results in damage.

Management Measures

This is not a serious pest at present, but intense damage could cause drying of infested parts and sooty mould growth. Removal and destruction of aphid infested plant parts help to minimize their infestation and spread. There are several predators like syrphids (*Paragus* sp.,), coccinellids (*Pseudospidemerus circumflexa* Mots., *Cryptolaemus montrouzieri, Menochilus sexmaculata, Coccinella transversalis, Scymnus* sp., *Illeis cincta*), lace wing bugs, mantispid flies etc. control this pest. If necessary, spraying of dimethoate (2 ml/lit), phosphomidon (2 ml/lit), or methyl demeton (2 ml/lit) can be taken up.

3. Storage Pests

In general, regular practice of proper sun drying of raw cashewnuts for 2-3 days after harvest kills most of the pest stages. Besides, raw cashewnuts are seldom stored more than two to three months and are rarely damaged by insects. But, processed cashew kernels are damaged by many insect pests during storage. Infestation by insects causes losses in terms of quality and quantity of cashew kernels besides changing the chemical composition affecting its nutritive value. Around twenty species of beetles, five species of caterpillars and some psocids and mites are reported to infest cashew kernels during storage. Among these, *Ephestia cautella* (Wlk.), *Corcyra cephalonica* (St.) (Pyralidae), *Tribolium castaneum* (Herb.) (Tenebrioniidae) and *Necrobia rufipes* (De C.) (Cleridae) damage directly the kernels. While other insects contaminate the cashew kernels by their excreta (Nair *et al.*, 1985, Vijay Singh, 1988a, 1988b; Vishnu Priya, 2011).

The biology of *T. castaneum, C. cephalonica* and *Oryzaephilus surinamensis* on cashew kernels have been studied. Egg, larval, pupal and adult period lasts for 4, 29.36, 10.04 and 28.6 days, respectively for *T. castaneum*. Egg, larval, pupal and adult period of *C. cephalonica* are 5, 29.59, 12.24 and 10.67 days respectively, while 4, 19.7, 7.09 and 100 days, respectively for *O. surinamensis*. Grubs and adults of *Trogoderma granarium* infest the kernels damaging the entire soft portion. Infestation of *T. castaneum* reduces the levels of proteins, fat, carbohydrates and ash contents of cashew kernels. Upon pest damage, energy values of kernels drop down significantly and the moisture content increases. Hence, strict sanitation of the processing sheds and premises, keeping processed nuts in closed containers, undertaking peeling, grading and packing in quick succession are suggested as measures to control these pest infestations.

4. Cashew Pest Management: Some Considerations

- ☆ In general, plant protection measures taken up against tea mosquito bug take care of the infestation of most of the foliage pests. Spraying is required, only under severe infestation.

- ✰ In young cashew plants, removal of different stages of pests like egg masses/egg laden leaves or shoots, caterpillars or grubs, pupa or cocoons from the infested plants gradually reduces the pest population.
- ✰ Under unsprayed conditions, an array of predators *viz.*, spiders, ants, reduviids, coccinellids, neuropterans, hemipteran bugs and praying mantids take care of many of the cashew pests. But so far, very little progress could be made in biological control options.
- ✰ Red ants (*Oecophylla smaragdina*) are the potential biocontrol agents in cashew plantations that feed on bugs, caterpillars, hoppers, moths etc. Apart from predators, there are many parasitoids on several cashew pests. Hence, indiscriminate spraying may be avoided.
- ✰ Tree to tree spraying is to be advocated, instead of whole plot spraying to avoid environment pollution. Besides trees harbouring ant nests especially red ants should be spared of spraying to allow ants to take care of pests naturally. Avoiding spraying on the non-target areas such as trunk, tree bases etc can help to protect some natural enemies.
- ✰ Neem oil (3-5 per cent), Karanj (*Pongamia*) oil @ 2 per cent and Fish Oil Rosin Soap and Neen seed kernel extract (NSKE) (5 per cent) are some of the botanical preparations can be effective against many of the cashew foliage pests like leaf miners and leaf feeding caterpillars. While using botanicals, emulsifiers (soap water/bar soap 0.5 per cent @ 5 g/lit or teepol (0.1 per cent) should be used in the spray fluid.
- ✰ Spraying should be done before 9 am or after 4 pm in order to save non-target pollinators. Rotation of insecticides between sprays is advised to prevent development of resistance to any particular pesticide.
- ✰ Monitoring pests of weeds in cashew plantations and their mangemnet is required, since plants like *Terminalia paniculata, Chromolaena odorata, Melastoma malabathricum* and *Getonia floribunda* are not only competitors of cashew but also serve as host plants for few cashew pests including TMB.

References

Abraham, E. V. (1958). Pests of cashew in South India. *Indian Journal of Agricultural Sciences,* 28: 531-534.

Abraham, E. V. (1959). Killer insects of cashew. *Indian Farming,* 9 (3): 14-15.

Ananthakrishnan, T. N. (1984). Host relationships and damage potential of thrips infesting cashew. *In*: Cashew research and development (Eds. E.V. S. B. Rao and Khan) Indian Society for Plantation Crops, CPCRI. Kasargod, India, pp. 132-134.

Ayyanna, T. Tejkumar, T. and Ramadevi, M. (1985). Distribution and status of pests on cashew in coastal districts of Andhra Pradesh. *Cashew Causerie,* 7 (2): 4-5.

Basu Choudhury, J. C. (1962). Preliminary investigations on the insect pests of cashew plantations in Kerala. *Indian Forester,* 88: 516-522.

Basu Choudhuri, J. C. and Misra, M. P. (1973). *Hyalospila leuconeurella* Ragnot (Lepidoptera: Pyralidae) and *Anarsia epotias* Meyrick (Lepidoptera: Gelechiidae), two new pests of cashew apples in South India with brief notes on their control. *Journal of Plantation Crops* (Suppl). 1: 168-170.

Cavalcante, R. D., Santos, O. M. de L. and Castro, Z. B. De. (1975). Population study of cashew-tree thrips *Selenothrips rubrocinctus* (Giard). *Biologico*, 41(12): 355-356.

Chatterjee, M. L. (1989). Insect pests on cashew in West Bengal and status of some important pests. *The cashew*, (3): 19-20.

Chutia, B. C., Rahaman, A., Sarmah, M., Borthakur M., Barthakur, B. K. (2011). Biology of *Hyposidra Talaca* (Walker)-A Major Defoliating Pest of Tea. In: North East India. *Indian Journal of Entomology*, 72 (2): 106-109.

Dharmaraju, E., Ayyanna, T. and Sreeremulu, G. (1975). *Nephopteryx* xp. (Lepidoptera: pyralidae) a serious pest of cashew. *Indian Cashew Journal*, 10 (3): 9-11.

Gowda, G., Ramaiah, E. and Reddy, C. V. K. (1979). *Scirtothrips dorsalis* (Hood) (Thysanoptera: Terebrantia: Thripidae) a new pest on cashew (*Anacardium occidentale* L.) *Current Research*, 8 (7): 116-117.

Hari Babu R., Rath, S. S. and Rajput, C. B. S. (1983). Insect pests of cashew and their control. *Pesticides*, 4: 8-16.

Jacob T K and Belavadi V V. (1990). The Cashew leaf miner, *Acrocercops syngramma* M. - its pest status and larval size relationship with leaf area damage in the Andamans (India). *The Cashew*, (4): 7-9.

Jena, B., Patnaik, N. C. and Satapathy, C. R. (1985). Insect pests of cashew. *Cashew Causerie*, 7 (3): 10-11.

Jena, B. C., Satapathy, C. R. and Satpathy, J. M. (1986a). Seasonal varietal response of cashew to foliage feeding beetle, *Microserica quadrinotata* M. *Cashew Causerie*, 8 (4): 9.

Jena, B. C., Satapathy, C. R., Patnaik, N. C. and Mishra, S. K. (1986b). The weevils *Pelto-trachelus pubes* F. infesting cashew varieties. *Cashew Causerie*, 8 (3): 10-12.

Jena, B. C., Satapathy, C. R., Kanungo, A. P. and Nanda, S. K. (1987). Geographical distribution of cashew pests. *The Cashew*, (1): 18-20.

Jena, B. C. (1990). Pests of cashew apple and nuts and their control. *The Cashew*, (2): 19-20.

Mohapatra, L. N., Behara, A. K. and Satapthy, C. R. (1998). Influence of the environmental factors on the cashew nut shoot tip caterpillar, *Hypotima haligramma* Meyr. *Cashew Bulletin*, 35: 17-18.

Mohapatra, L. N. (2007). Seasonal activity of leaf folder *Caloptilia tiselaea* M. on some cashew varieties at Bhubaneswar. *Journal of Plant Protection and Environment*, 4 (2): 151-153.

Murthy, M. M. K., Sayi, I. V., Rao, S. V. and Rao, B. H. K. (1974). Note on the biology of shoot and blossom webber, *Macalla moncusalis* Walker (Lepidoptera: Pyralidae) of cashew. *Indian Journal of Entomology*, 36 (1): 76-77.

Mutter, E. S. and Bigger, M. (1962). Cashew Bulletin, Ministry of Agriculture, Tanzania. 11: 1-5.

Nair, R. C. P., Abraham, V. A. and Pillai, G. B. (1974). Biology of *Metanastria hyrtaca* Cram., A defoliator of cashew. *Journal of Plantation Crops,* 32-33.

Nair, M. R. G. K., Oommen, C. N. and Mohandas, N. (1985). Pests of processed cashew and their control. *Acta Horticulturae,* (ISHS) 108: 125-130

Panda, H. (2013). The complete book on cashew (Cultivation, processing and by-products). Asia-Pacific Business Press Inc. pp. 1-600.

Pathummal Beevi, S., Jagadeesh Kumar, T. N., Suma, A., Usha, K. E. and George Mathew. (1993). New record of *Anigraea albomaculata* Hamp. (Noctuidae: Lepidoptera) as a pest of cashew in Kerala. *The Cashew,* 7 (3): 14-16.

Patnaik, H. P., Satapathy, C. R., Sontake, B. K. and Senapathy. (1987). Flower thrips of cashew (*Anacardium occidentale* L.), their seasonal incidence and assessment of damage in coastal Odisha. *The Cashew,* 1 (4): 11-13.

Pillai, G. B., Dubey, O. P. and Vijaya Singh. (1976). Pests of cashew and their control in India- a review of current status. *Journal of Plantation Crops,* 4 (2): 37-50.

Rai, P. S. (1984). Hand book on cashew pests. N. K. Sharma of Research Co Publications. 124p.

Ramadevi, M.and Radhakrishnan, Y. (1991). Influence of climatic factors on major pests of cashew. *Cashew Bulletin,* 28 (4): 15-20.

Ramadevi, M. and Ayyanna, T. (1988). *Heliothis armigera* Hubner as a serious pest on cashew. *The Cashew,* 2(2): 11.

Rao, A. R., Naidu, V. G., Prasad, P. R. (2002). Studies on the biology of cashew shoot and blossom webber (*Lamida moncusalis* Walker). *Indian Journal of Plant Protection,* 30: 167–171.

Remamony, K. S. (1965). Biology of *Anarsia epotias* Meyr. a shoot borer on cashew in Kerala. *Agricultural Research Journal of Kerala,* 3(2): 46-47.

Satapathy, C. R. and Panigrahi, P. K. (1995). Effect of plant height on incidence of *Lamida moncusalis* Wlaker infesting cashew. *Odisha Journal of Horticulture,* 23: 30-32.

Subba Rao, Y., Rajasekhar, P., Rama Subba Rao, V. and Srinivasa Rao, V. (2006). Seasonal incidence and control of cashew shoot and blossom webber, *Lamida moncusalis* Walker in Andhra Pradesh due to insect infestation along with the fungal incidence, *Archives of Phytopathology and Plant Protection,* 44 (4): 346-351.

Sundararaju, D. (1984). Studies on cashew pests and their natural enemies in Goa. *Journal of Plantation Crops,* 12: 38-46.

Sundararaju, D. (1992). Pest problems of cashew plantations and control measures. In: Proceedings of National Workshop on Cashew. Kannur. Kerala. Organized by Directorate of Cashew and Cocoa Development. Cochin, India. pp. 86-89.

Sundararaju, D. and Bakthavatsalam, N. (1994). Pests of cashew. In: *Advances in Horticulture*, Vol.10 (Eds. K L Chadha and P Rethinam), Malhotra Publishing House, New Delhi, India, pp. 759-785.

Sundararaju, D. 2007. Pests and diseases management in cashew. In the souvenir of 6th Nation Seminar on 'Indian cashew in the next decade- challenges and opportunities'. 18-19 May, 2007. Raipur. Organized by Directorate of cashewnut and cocoa development, (DCCD), Cochin. pp. 53-63.

Thriurmalaraju, G. T., Krishnappa, K. S., Venkateshgowda, T. N., Chandregowda, M. and Shankaranarayana, V. (1991). Seasonal occurrence of major pests of cashew in maiden parts of Karnataka. *The Cashew*, 5 (2): 12-13.

Vanitha, K., Bhat, P. S. and. Raviprasad, T. N. (2015). Pest status of leaf miner, *Acrocercops syngramma* M. on common varieties of cashew in puttur region of Karnataka. *Pest Management in Horticultural Ecosystems*, 21, (1): 55-59.

Vanitha, K., Saroj, P.L. and Bhat, P.S. (2015). Indian tropical tasar silkworm on cashew plantations. *ICAR newsletter*, 22 (1): 8-9.

Vanitha, K. and Saroj, P.L. (2015). Insect pests of cashew and their management. DCR Technical Bulletin No. 27. pp. 1-68.

Vijay Singh. (1988a). Biology of *Corcyra cephalonica* on cashew kernel. *The Cashew*, 2 (3): 9-10.

Vijay Singh. (1988b). Biology of *Tribolium castaneum* on cashew kernel. *The Cashew*, 2 (2): 7-8.

Vishnupriya Sharma. (2011). Deterioration of cashew nuts in Kerala state of India. *Archives of Phytopathology and Plant Protection*, 44 (4): 346-351.

Cashew: Improvement, Production and Processing (2017) *Pages* **395–403**
Editor: **Prof. P.L. Saroj**
Published by: **ASTRAL INTERNATIONAL PVT. LTD., NEW DELHI**

23

Pollination for Higher Yield

D. Sundararaju

1. Introduction

Cashew (*Anacardium occidentale* L., Anacardiaceae), an ever green tree is native of tropical Central and South America; but now distributed all over tropics and part of warm tropics. It is widely cultivated for its kernel and apple (pseudo-fruit formed by swollen peduncle). The cashew trees are polygamo-andromonocious, bearing both, staminate (male) and hermaphrodite (bisexual) flowers on the same inflorescence. Cashew inflorescence is an open panicle which is usually borne at the end of current season shoot (lateral shoot). For this reason, the majority of flowers are found on periphery of the tree on current season lateral shoots. The panicles are conical, pyramidal or irregular in shape. The peak period of flowering varies in different parts of world based on latitude and temperature and again among cashew types; three different types exist as early season, mid-season and late season flowering types in the same region. Generally, flowering season is from October to March in India, from June to November in Tanzania, from December to March in Central America and from December to March in West Africa, whereas, in Kenya, flowering occurs twice, the first one from September to November and second one from December to January. Cashew requires pollination for proper fruit set, but inadequate and under-pollination appeared to be main reason for low yield along with excessive fruit drop in most of the countries. Therefore, the status of pollination biology of cashew has been reviewed and presented in this chapter in order to device appropriate pollination technologies for enhancing productivity of cashew.

2. Floral Biology of Cashew

As the cashew tree is andromonoecious, presenting male and hermaphrodite flowers in the same panicle, it is amenable for xenogamous (fertilization of a flower by pollen from a flower on a genetically different plant) or geitonogamous (fertilization of a flower by pollen from another flower on the same or a genetically identical plant) pollination. Both types of flowers produce pollen and nectar and both have five white petals with two pink stripes on each petal at anthesis that gradually become red over a period of five days before withering. Each flower has 6-10 pollen-bearing structures, one (large or long stamen) being much longer than the others (staminoids or short or small stamen). In the hermaphrodite flower, the style assumes the same position as the large stamen of male flowers, and its own large stamen is shorter than its style. The ovary bears one single ovule and is rudimentary in male flowers (Northwood 1966, Free 1993, Freitas 1995, Wunnachit *et al.*, 1992a). Panicles last around 100 days in a flowering season that varies from 5 to 7 months, with each tree producing hundreds of panicles across the flowering season. Panicles consisted of approximately 10 per cent hermaphrodite flowers to 90 per cent male flowers (N = 500 panicles) (Free 1993, Freitas 1994, Freitas *et al.*, 2002). The exact proportion of hermaphrodite flowers per panicle ranges from <0.5 per cent to circa 25 per cent (Rao and Hassan 1957, Damodaran *et al.*, 1979).

The hermaphrodite flowers have extremely strong floral scent and are self-fertile but not self-pollinating. Peak anthesis for male flowers is between 6.00 h and 10.00 h and for hermaphrodite flowers, it is between 11.00 h and 13.00 h. High proportion of anther dehiscence in the stamen of male flowers and hermaphrodite flowers occurs between 10.30 h and 12.00 h and 12.00 h and 14.00 h respectively. Thus high pollen build up in male flowers can be found timing with the peak anthesis of hermaphrodite flower. Cashew pollen grains are sticky in nature and most of them will be held as clumped. Because of this, test for pollen presence in the air is negative (Mohan *et al.*, 1981). The stigma is receptive immediately after anthesis and receptive for 24 h and as the dehiscence of anther occurs later in hermaphrodite flower, occurrence of xenogamous pollination is quite possible. Pollen vigour and composition in relation to andromonoecy in cashew had been studied in detail. The cashew trees produce four types of pollen from the large and small stamens of hermaphrodite and male flowers (H-hermaphrodite large, HS-hermaphrodite small, ML-male large, MS-male small). Anther and pollen grain numbers (range of pollen grains in HL, HS, ML and MS was 885-1047, 131-201, 752-1044 and 116-188 respectively) and dimension of the pollen types were similar, as were pollen structure and staining characteristics. The fluorescein diacetate test showed that HL pollen had the highest percentage of fluorescence and viability of all pollens had declined by 48 h after dehiscing of anthers. Following controlled hand pollination, ML pollen had the highest capacity to germinate on stigma and penetrate ovule, followed by MS, HL and HS pollens. Glucose and fructose and 19 free amino acids were present in all pollen types with higher levels in hermaphrodite than male flower pollens. Proline was the major amino acid detected in all four pollen types and the proline in HL pollen represented half (150 nmol/mg fresh wt.) of the total amino acid content. Thus, the pollen of male flower is specialized for pollination and fruit set, whereas that of hermaphrodite flower may be specialized for insect attraction (Wunnachit *et al.*, 1992a).

The nectar volume of one-day old cashew flower was significantly higher than two days old flowers. Genetic differences, initial volume of nectar at anthesis and patterns of evaporation and reabsorption of nectar determine the nectar available for collection. These factors would affect the insect visitors' attractiveness to the cashew crop. The nectar volume and sugar concentration were recorded from Australia (0.04-0.07 ml; 71.3-71.9 per cent), Brazil (0.31-1.10 ml; 14.01-19.85 per cent) and India (0.04-1.16 ml; 0.4-5.1 per cent) respectively (Wunnachit *et al.*, 1992b; Freitas 1996a; Bhattacharya, 2004). The main sugar identified in the nectar was glucose with small amounts of sucrose and fructose. The male flowers of high yielding selections contained higher level (2.0 per cent) of glucose than low yielders (1.2 per cent), whereas, hermaphrodite flowers contained relatively lower level of glucose (0.1-0.55 per cent) (Moncur and Wait 1986).

3. Pollinating Agents of Cashew

Earlier studies in cashew have suggested that wind and many insect species, such as ants, thrips, butterflies, flies, wasps and honey bees, are the pollinating agents of cashew (Bigger 1960, Damodaran *et al.*, 1966, Free and Williams 1976, Reddi 1991, Freitas and Paxton 1996, Ambethgar 2002, Bhattacharya 2004). However, their importance has not been critically determined. Indeed, only recently attention has been paid to the natural pollinators of cashew in its native range (Freitas and Paxton 1998). Wind and night flying insects played no role in pollination (Heard *et al.*, 1990) and the experiments incorporating the bagging of flowers confirmed that wind was not important as a pollinator (Freitas *et al.*, 2002). But, the floral ontogeny or morphogenesis of cashew flowers provides equal or greater scope to geitonogamy than, that of xenogamy (Moncur and Wait, 1986). Under natural condition, movement of panicles by wind results in geitonogamous pollination by the way of gravitational fall of pollen grains from upper flowers (both from male and hermaphrodite flowers) to lower hermaphrodite flowers in addition to regular insect pollination. Pollen clump from hermaphrodite flowers is more vulnerable for gravitational fall than from male flowers (Sundararaju, 2011a).

The honey bee, Apis *mellifera* L. has been described by many pollination biologists as the most effective pollinator of cashew (Free and Williams 1976, Heard *et al.*, 1990). The study in Ghana showed that honey bee workers infrequently foraged on cashew flowers, but when they did, they spent relatively short time collecting only nectar. On any hermaphrodite flower there was a 50: 50 chance of the honey bee worker effecting pollination. This suggests that *A. mellifera* is not the most effective pollinator of cashew orchards in Ghana as found in other cashew growing areas of the world. The study identified two stingless bees: *Dactylurina staudingeri* (Gribodo) and *Liotrigona parvula* Darchen and seven solitary bees *Brausapis* sp., *Ceratina* sp., *Compsomelissa* sp., *Halictus* sp., *Lasioglossum* sp., *Lipotriches* sp. and *Thyreaus* sp., whose foraging activities indicated their effectiveness in the pollination of cashew. These two groups of bees are known to be heavy feeders of pollen, and were found to be persistent foragers, and exhibited behavioural characteristics which demonstrated their strong affinity to cashew flowers. The cashew plant's ability to produce large amounts of pollen is probably to attract pollen foragers rather than nectar foragers. These bees therefore are the effective pollinators of cashew flowers in Ghana. The abundance and diversity of the right pollinators in the cashew agro-ecosystem gives Ghana a comparative advantage over other growing countries. The presence of various forms of natural landscapes provided good sanctuaries for pollinating bees and other beneficial insects. As such, seven bee species effectively pollinate cashew trees in Ghana. Their pollinating activities were supplemented by *A. mellifera* and occasional flower visitors such as ants, butterflies, sun birds and wasps (Aidoo, 2009).

Studies in Brazil recorded only two species of bees, *A. mellifera* and *Centris tarsata* Smith as effective pollinators of cashew flowers and very small insects play little or no role in cashew pollination (Freitas *et al.*, 2002). Heard *et al.* (1990) found *A. mellifera* as the only bee pollinating cashew efficiently in Northern Australia. In the West coast of India, Devasahayam (1986) and Sundararaju (2000, 2003) have reported various species of bees (*Pseudapis oxybeloides* (Smith), *Ceratina smaragdula* (F.), *Apis* spp. (*A. cerana indica* F., *A.florea* F. and *A. dorsata* F.), *Lasioglossum* sp., *Trigona iridipennis* Smith and one unidentified halictid) visiting cashew. In India, all above *Apis* spp. are exclusive nectar feeders like *A. mellifera* in Ghana (Sundararaju 2003). The diversity of solitary bees which maintained consistency on cashew flowers by foraging both nectar and pollen were identified and it consisted of six species from coastal Karnataka (*Pseudapis oxybeloides* (Smith), *Lasioglossum* sp. and *Halictus* sp., Halictidae; *Braunsapis* sp., *Ceratina smaragdula* (F.) and *Ceratina* sp., Apidae) and four species from coastal Tamil Nadu (*Ceratina* (*Pithitis*) *binghami* Cockerell, *C. smaragdula, Braunsapis* sp. and *P. oxybeloides*) (Sundararaju, 2011b).

In Brazil, individuals of the two species (*A. mellifera* and *C. tarsata*) were observed for their visits to flowers, and the average number of bees visiting each cashew tree at 10 h over several days, the average number of flowers visited by each bee species within a minute and the average time spent per flower visit for each bee species were compared. The efficiency of each bee species in terms of fruit set after single visit to flowers was also studied. *C. tarsata* individuals were almost twice as abundant as *A. mellifera* foragers at 10 h and visited almost twice as many flowers per minute (4.22±0.24 *versus* 2.18±0.19 and 11.94±0.12 *versus* 6.20±0.17 respectively). Furthermore, visits to flowers by *C. tarsata* were half as long as visits made by *A. mellifera* (5.07±0.04 seconds *versus* 10.59±0.34 seconds respectively). There was no significant difference between the numbers of flowers pollinated with single visit by each bee species, but *C. tarsata* pollinated more flowers per time unit. It is concluded that *C. tarsata* shows potential to be used for cashew pollination (Freitas, 1996b). In India, the total time spent/flower of above bees including *A. cerana indica* ranged from 0.8 to 25.1 seconds per flower. Whenever, pollen alone is collected, the time spent was found to be least (range of 0.8 to 5.0 seconds per flower). At the time of combined collection of pollen and nectar or collection of nectar alone, the time range was highest (5.0 to 25.1 seconds) (Sundararaju, 2011b).

4. Pollinator Problems of Cashew

Cashew is an entomophilous crop. Its yield is often pollinator dependent. Owing to the paucity of flower visiting insects, pollinator limitation was identified as one of the major constraints of cashew nut production. Scanty and unrewarding forage offered by cashew was blamed for the neglect of this crop by honeybees. As a result, wide gap was noticed between the realized and the potential yields of cashew nuts. Knowing that there is ample scope to boost cashew yields, the idea of managing fly pollinator species was recommended to cashew farmer. Several management techniques have been developed and tested in order to increase fly populations in cashew orchards. But, the use of *Euphorbia antiquorum* as hedge to the cashew crop was proved economic and environmentally acceptable. Insect pollinators are fast declining and the day is not far away to the cashew farmer to treat pollination as a major issue over the routine things (Reddi, 2003).

For maximum fruit set, hermaphrodite flowers need to be pollinated within 3-6 h of anthesis (Wunnachit *et al.*, 1992c). Since maximum flowers open during forenoon hours, visit of pollinators during late forenoon and early afternoon was reported in Australia (Heard *et al.*, 1990). In India, under- pollination is the major production constraint (Reddi, 1987), but, the extent of pollination is in the range of 16.9 to 72.0 per cent (Reddi, 1993; Sundararaju, 2011b) and more than 25 per cent pollination is considered to be optimum (Reddi 1993). However, the problem of under-pollination/insufficient pollination persists in the Southeastern coast of India (Sundararaju, 2011a). Similarly, lowest levels of natural pollination of 18, 26 and 28 per cent in Malaysia, Jamaica and Kenya, respectively are also reported (Free, 1993). In Brazil, cashew commercially grown in orchards also suffers from low crop yields, in large part due to inadequate pollination. There are a few or no visits to orchard-grown cashew flowers by the appropriate bee species that are competent pollinators. One seemingly suitable pollinator, *A. mellifera,* does not readily visit cashew flowers even when brought into orchards in large numbers because it is attracted to competing weeds in bloom (Freitas *et al.*2002). Among the bee species managed for pollination of commercial cashew orchards, both *A. mellifera* and *C. tarsata* would need to visit cashew flowers more than once to assure a high level of fruit set. Because of short period of receptivity of cashew stigmas, visit of appropriate pollinators are required within 4 h of peak stigma receptivity (Freitas and Paxton, 1998).

In India, native bees especially solitary bees are main pollinators of cashew. During every cropping season of cashew, native bees could get both pollen and nectar forage for 5- 7 months period from both cashew flowers and other floral sources which consist of weeds growing under the canopy of cashew plants and in the remaining period of the year, these bees survive only on other floral sources having bloom. It is interesting to note that in West coast region of India, these bees maintain two separate strong constancies on cashew and other floral resources simultaneously during the cropping season of cashew. But in the Southern East coast region of India, strong constancy was observed only on other floral resources within the cashew plantation, whereas on cashew, it was very weak (similar to attraction of *A. mellifera* to competing weeds in Brazil). The population abundance on the early

(NRCC Selection-2), mid (Bhaskara) and late season (Chinthamani -1) flowering varieties had recorded visit of bees as 0.35, 0.16 and 0.09 bee/panicle, respectively, at a constant period of 10 minutes duration within continuous observational period of 30 days (Sundararaju 2011b). In general, the visit of bees is very low (Ohler 1979, Free 1993, Moncur and Wait 1986) and bees rarely visit cashew flowers (Roubik 1995). Low yielding trees are less attractive to insects than high yielding trees (Free and Williams 1976). Further, the volume of nectar secreted by cashew flower is relatively small, sufficient only for fairly small and metabolically undemanding pollinator (Kevan and Baker, 1983).

It has for long been recognised that insecticide spraying can be harmful for pollinating insects in general (Kevan 1975, Heard *et al.*, 1990), and ploughing is particularly damaging because it may destroy nests of fossorial (ground-nesting) bee species such as those of *C. tarsata*. *Centris* bees nest in sandy ground beneath the bushes of *Byrsonimia verbascifolia*. Unfortunately, there is a lack of direct information on the impact upon *C. tarsata* of insecticide spraying and ploughing on *C. tarsata*, though the absence of the bee from commercial cashew orchards may in part be related to both forms of agricultural practices (Freitas *et al.*2002). But, the use of recommended insecticides for management of cashew pests has not affected the required level of pollination in the West coast of India (Sundararaju 2000, 2003, 2004).

Thus, cashew requires pollination for proper fruit set and pollination occurs through xenogamy or geitonogamy. Adequate pollination is provided by constant visits of solitary bees (native bees). The managed pollinator, *A. mellifera* is not an ideal pollinator for cashew orchards, but it can supplement for solitary bees. As solitary bee species survive on weeds existing under the canopy of cashew plant parts, their diversity needs to be encouraged as a pollinator management practice. In spite of existence of diversity of solitary bees in cashew plantation, under-pollination or insufficient pollination is common in some of the countries. For this purpose, better cashew genotypes/varieties having higher pollinator attraction have to be identified and recommended to enhance the productivity of cashew.

References

Aidoo, K. S. (2009). Boosting cashew production in Ghana. *Bees for Development Journal*, 91: 8-9.

Ambethga, V. (2002). Insect visitors of cashew inflorescence in north-eastern zone of Tamil Nadu, India. *Progressive Horticulture*, 34: 223-229.

Bhattacharya, A. (2004). Flower visitors and fruit set of *Anacardium occidentale*. *Annales Botanici Fennici* 41: 385-392.

Bigger, M. (1960). *Selenothrips rubrocinctus* Giard and the floral biology of cashew in Tanganyika. *The East African Agricultural Journal*, 25: 229-34.

Damodaran, V. K., Abraham, J. and Alexander, K. M. (1966). The morphology and biology of the cashew flower. II: anthesis, dehiscence, receptivity of stigma, pollination, fruit-set and fruit-development. *Agricultural Research Journal of Kerala*, 4(2): 78-84.

Damodaran, V. K., Vilasachan, Y. and Valsalakumari, P. K. (1979). Research on cashew in India. *Technical Bulletin:* 10-35. KAU,Vellanikara, Kerala.

Devasahayam, S. (1986). Some observation on insects visiting cashew inflorescence. *Cashew Bulletin,* 23 (5): 1-5.

Free, B. J. (1993). *Insect Pollination of Crops,* pp. 122-124. Academic Press, London.

Free, J. B. and Williams, I. H. (1976). Insect pollination of *Anacardium occidentale* L., *Mangifera indica* L., *Blighia sapida* Koenig and *Persea americana* Mill. *Tropical Agriculture* (*Trinidad*), 53(2): 125-39.

Freitas, B. M. (1994). Beekeeping and cashew in North-eastern Brazil: the balance of honey and nut production. *Bee World,* 75(4): 160-8.

Freitas, B. M. (1995). The pollination efficiency of foraging bees on apple (*Malus domestica* Borkh) and cashew (*Anacardium occidentale L.*) [*PhD thesis*]. Wales: University of Wales.

Freitas, B. M. (1996a). Bee plants: volume, concentration and total sugar of nectar secreted by flowers of distinct genetic material of cashew (*Anacardium occidentale* L.). *Anais da XXXIII Reuniáo Anual da Sociedade Brasileira de Zootecnia, Fortaleza, 21 a 26 de julho, 1996 b. Volume 2 - Forragicultura, Sociedade Brasileira de Zootecnia* 395-397.

Freitas, B. M. (1996b). Foraging behaviour and pollination efficiency of the bee species *Apis mellifera* and *Centris* (*Hemisiella*) *tarsata* visiting cashew flowers. *Anais da XXXIII Reuniáo Anual da Sociedade Brasileira de Zootecnia, Fortaleza, 21 a 26 de julho, Volume 2 - Forragicultura, Sociedade Brasileira de Zootecnia* 666-668.

Freitas, B. M. and Paxton, R. J. (1996). The role of wind and insects in cashew (*Anacardium occidentale*) pollination in NE Brazil. *Journal of Agricultural Science,* 126: 319-26.

Freitas, B. M. and Paxton, R. J. (1998). A comparison of two pollinators: the introduced honey bee *Apis mellifera* and an indigenous bee *Centris tarsata* on cashew *Anacardium occidentale* in its native range of NE Brazil. *Journal of Applied Ecology,* 35: 109-21.

Freitas, B. M., Paxton, R. J. and Holanda-Neto, J. P. (2002). Identifying pollinators among an array of flower visitors, and the case of inadequate cashew pollination in NE Brazil. *Pollinating Bees - The Conservation Link Between Agriculture and Nature,* pp. 229-244. Kevan P and Imperatriz Fonseca V L (Eds). Ministry of Environment, Brasília.

Heard, T. A., Vithanage, V. and Chacko, E. K. (1990). Pollination biology of cashew in the northern territory of Australia. *Australian Journal of Agricultural Research,* 41: 1101-14.

Kevan, P. G. (1975). Pollination and environmental conservation. *Environmental Conservation,* 2: 293-8.

Kevan, P. G. and Baker, H. G. (1983). Insects as flower visitors and pollinators. *Annual Review of Entomology*, 28: 407-53.

Mohan, K. V. J., Kumar, K. M., Murthy, K. N. and Nayar, N. M. (1981). Anthesis and anther dehiscence of cashew flower. *Genetics, plant breeding and horticulture*, pp. 360-374. Vishveshwara S (Ed). Narosa Publ., New Delhi.

Moncur, M. W. and Wait, A. J. (1986). Floral ontogeny of cashew, *Anacardium occidentale* L. (Anacardiaceae). *Scientia Horticulture*, 30: 203-211.

Northwood, P. J. (1966). Some observations on flowering and fruit setting in the cashew *Anacardium occidentale* L. *Tropical Agriculture (Trinidad)*, 31 (1): 35-42.

Ohler, J. G. (1979). *Cashew*, 260p. Koniklijk Institute Voor de Tropen, Amsterdam.

Rao, V. N. M. and Hassan, V. M. (1957). Preliminary studies on the floral biology of cashew (*Anacardium occidentale* L.). *Indian Journal of Agricultural Sciences*, 27: 277-88.

Reddi, E. U. B. (1987). Under-pollination: a major constraint of cashewnut production. *Proceedings of the Indian National Science Academy*, 53(3): 249-52.

Reddi, E. U. B. (1991). Pollinating agent of cashew-wind or insects? *Indian Cashew Journal*, 20(4): 13-8.

Reddi, E. U. B. (1993). Pollination studies of cashew in India: an overview. *Pollination in tropics*, pp. 321-324. Veeresh G K, Umashaankar R and Ganeshaiah K N (Eds). University of Agricultural Sciences, Bangalore.

Reddi, E.U.B. (2003). Poilinator Problems of Cashew Farmer in India - A Review. Ecology, Environment and Conservation, 9 (3): 295-296.

Roubik, D. W. (Ed) (1995). *Pollination of cultivated plants in the tropics*. pp. 67. FAO, Food and Agriculture Service Bulletin Number 118, Rome, Italy.

Sundararaju, D. (2000). Foraging behaviour of pollinators on cashew. *The Cashew*, 14 (4): 17-20.

Sundararaju, D. (2003). Occurrence of bee fauna and extent of pollination in insecticide sprayed ecosystem of cashew. *Journal of Palynology*, 39 : 121-125.

Sundararaju, D. (2004). Evaluation of promising new insecticides in large plots for management of tea mosquito bug on cashew. *Journal of Plantation Crops*, 32(Suppl.): 285- 288.

Sundararaju, D. (2011). Studies on extent of pollination and fruit set in cashew. *Journal of Plantation Crops*, 39 (1): 157-160.

Sundararaju, D. (2011). Diversity of pollinators and flora in cashew. *Journal of Horticultural Science*, 6 (1): 52-55.

Wunnachit, W., Jenner, C. F. and Sedgley, M. (1992a). Pollen vigour and composition in relation to andromonoecy in cashew (*Anacardium occidentale* L.: Anacardiaceae). *Sexual Plant Reproduction*, 5: 264-269.

Wunnachit, W., Jenner, C. F. and Sedgley, M. (1992b). Floral and extrafloral nectar in *Anacardium occidentale* L. (Anacardiaceae): an andromonoecious specious. *International Journal of Plant Science,* 153 (3): 413-420.

Wunnachit, W., Pattison, S. J., Giles, L., Millington, A. J. and Sedgley, M. (1992c). Pollen tube growth and genotype compatibility in cashew in relation to yield. *Journal of Horticultural Science and Biotechnology,* 67 (1): 67–76.

Cashew: Improvement, Production and Processing (2017) *Pages* **405–419**
Editor: **Prof. P.L. Saroj**
Published by: **ASTRAL INTERNATIONAL PVT. LTD., NEW DELHI**

24

Major Diseases and their Management

M. Loganathan and P.L. Saroj

1. Introduction

Cashew (*Anacardium occidentale* L.) is considered to be an important foreign revenue earning horticultural crop in India. Low productivity of the crop is the major concern in the country (Bhat *et al.*, 2009). The productivity of the crop is challenged by many biotic and abiotic factors. Among the biotic factors, apart from insect pest, diseases especially caused by fungal pathogens, have played a significant role in causing yield loss in the crop worldwide (Table 24.1), which could be probably one of the reasons for the low productivity of cashew in India. However, much attention was not given to identify the causal agent of the diseases of cashew and to develop management practices. Probably due to the facts that earlier, cashew was considered as forest crop and it was grown to protect erosion. Now the crop is under intensive cultivation and the problem due to diseases is also aggravating. Diseases such as anthracnose (*Colletotrichum gloeosporioides*), gummosis or inflorescence blight/ floral shoot and twig die back (*Lasiodiplodia theobromae*), powdery mildew (*Oidium anacardii*), black mold (*Pilgeriella anacardii*), angular leaf spot (*Septoria anacardii*), pink disease (*Corticium salmonicolor=Pellicularia salmonicolor*), shoot rot or leaf fall (*Phytophthora nicotianae*), foot and root rot (*Pythium splendens*) and damping off (*Fusarium* sp.) have been documented in different parts of the world (Thankamm, 1974; Freire *et al.*, 2002; Hammed *et al.*, 2011; Adeigbe *et al.*, 2015). Anthracnose, gummosis or shoot die back and inflorescence blight are the major diseases in India (Singh *et al.*, 1967; sundararaju 2002). Hence considering the importance of diseases

Table 24.1: Yield Loss Due to Important Diseases of Cashew

Sl.No.	*Disease*	*Causal Organism*	*Yield Loss*	*Country*	*Reference*
1.	Inflorescence blight	*Lasiodiplodiatheobromae*	40-45 per cent	Nigeria	Olunloyo, 1979
2.	Floral shoot and twig die back	*L. theobromae*	Up to 70 per cent nut yield. 50 per cent vegetative shoot death.	Nigeria	Hammed *et al.*, 2008
3.	Anthracnose	*Colletotrichum gloeosporioides*	Up to 50 per cent apart from affecting apple and kernel quality.	Brazil	Cardoso *et al.*, 1994.
4.	Powdery mildew	*Oidium anacardii*	50 per cent - 70 per cent	Tanzania	Sijaona and Shomari,1987.
5.	Black mold	*Pilgeriella anacardii*	33 per cent nut yield	Brazil	Cardoso *et al.*, 2000a

in cashew, in this chapter, updates on status of different diseases of cashew and their management, and future strategies have been described.

2. Diseases Caused by Fungal Pathogens

2.1. Anthracnose

Symptom

Anthracnose caused by *Colletotrichum gloeosporioides* is a serious disease and its occurrence has been reported from India (Singh *et al.*, 1967; Venkateswaralu *et al.*, 2003), Brazil (Freire *et al.*, 2002) and other cashew growing areas of the world (Cardoso *et al.*, 1994). The disease is noticed on both young and adult plants. Anthracnose infected plants show oval, reddish brown shiny water soaked lesions spread over the leaves. Normally spots begin from tip or margin and proceeds to the centre. Plant parts such as leaves, twigs, inflorescences, young apples and fruits are attacked by the pathogen. Severe cases, tender leaves may crumple and form necrotic spots. Other symptoms include necrotic spots on nuts and black mummification on apple. In advanced conditions, both leaves and fruits may completely blight and drop.

Favourble Conditions and Mode of Spread

The disease is favoured by high relative humidity and rainfall, and optimum temperature (22 to 28°C). The pathogen enters the fruit via floral stigma. The infected tissues and plant parts serve as source of inoculums. It produces conidia when rain going side with high humidity. Wind and rain splashes normally spread the conidia in plantations hence intensity of the disease tends to increase during the rainy season.

Management

- ✰ **Cultural method**: Growing *Casuarina* and *Eucalyptus* around plantation may act as wind break and prevent the spread of disease (Mandal, 2012).
- ✰ **Chemical method:** Application of copper fungicides (Teixeira *et al.*, 1991) or Bordeaux mixture (1.0 per cent) or Mancozeb (0.2 per cent) at 3 weeks interval has been found effective to control the disease. Certain chemicals such as 2.6mM salicylic acid, 12.6 µM a.i. acibenzolar-S-methyl and 12.6 µM a.i. 2,6-dichloro-isonicotinic acid have induced the plant defense system to act against the disease (Lopez and Lucas, 2002).
- ✰ **Host plant resistance:** Brazil has identified few dwarf clones *viz.*, CAP2, CAP5, CP06, CAP6, CAP7, CAP9, CAP12, CAP13, CAP14, CAP15 CAP16, CAP17, CAP18, CAP19, CAP20, CAP24, CAP22, CP76 and CP1001 resistant to the disease (Cardosa *et al.*, 1999).

2.2. Inflorescence or Blossom Blight/Floral, Shoot and Twig Die Back/Gummosis

Symptom

Different causal agents have been reported for inflorescence or blossom blight/ floral, shoot and twig die back/gummosis disease. However, global status on the disease is unique while India it has not been clearly identified. Probably, it indicates that in India no clear cut experiments have been conducted to find etiology of the disease.

Gumosis or exudative disease was in severe form in Kerala and Tamil Nadu. Initially, it was reported to cause by three fungi, *Pellicularia salmonicolor*, *Ceratocystis* sp. and *Diplodia natalensis* and later, *Phytophthora nicotianae* was identified as a causal agent (Mishra *et al.*, 1993). The symptoms observed were brownish liquid exudation from longitudinal cracks of the affected trunks and branches. Inflorescence blight disease has been associated with infection by both fungi (*Gloeosporium mangiferae* and *Phomopsis anacardii*) and Tea mosquito bug (TMB) *Helopeltis antonii* (Nambiar *et al.*, 1973; Nair *et al.*, 1979). Typical symptoms noted are initial water soaked lesions on rachis and secondary rachis and later the lesions become pink to brown colour. Also it produces gummy exudations from lesion and drying of floral branches. In many instances, the inflorescence blight is reported to cause primarily by TMB (Pillai and Abraham 1975; Pillai, 1980) and secondary saprophytic infection due to fungi (Nambiar and Brahma 1979; Sundararaju 2002). Inflorescence die back has been associated with infection by *L. theobromae* and it was reported as serious disease in India (Lima *et al.*, 1950; Rao, 1969; Naik *et al.*, 1988). However, die back of shoots and braches believed to be caused by TMB predisposed *L. theobromae* infection (Sundararaju, 2008).

In Brazil, it is known in the form of gummosis or exudative disease and *Lasiodiplodia theobromae* (*Botryodiplodia theobromae*) was identified as causal agent of the disease (Freire, 1991). Typical identification symptoms are gummy resin like exudation from swollen and cracked trunk or woody part of the plant (Freire *et al.*, 2002). In advanced stage, the infected plant may die as its water transport and photosynthetic activities are affected. In some cases, in Brazil, the pathogen is also associated with floral or shoot die back disease (Cardoso *et al.*, 2000). Colonization of the pathogen in internal cells and disruption of cell walls have been observed (Muniz *et al.*, 2011). Sometimes even after chemical spray for controlling the disease, when branches are affected reinfection occurs on trunk after 2 to 3 months (Cardosa *et al.*, 1995a). In Nigeria, *L. theobromae* is widely associated with inflorescence blight and die-back (Olunloyo and Esuruoso, 1975). Symptoms on inflorescence show progressive die back and withering of the floral parts. Die back also noticed on shoot and twigs. On leaves, yellowing and dropping are the major symptoms. On stems, the infected portions show dark colour cracking. The disease is predisposed by both biotic and abiotic factors as wounds created by insects especially *Marshallus bondari*, the root beetle (Olunloyo1979; Adejumo 2005) and water stress (Freire, 1991; Freire *et al.*, 2002; Cardoso *et al.*, 2006) respectively favoured the disease development. The pathogen also survives endophytically and spreads through vegetative progative materials (Cardoso *et al.*, 2009).

Management

- ✰ Spraying of Maneb (0.2 per cent) or Mancozeb (0.2 per cent) or Bordeaux mixture (1.0 per cent) was effective to control the disease (Naik *et al.*, 1988). Unseasonal rain at fruiting stage causes more damage to inflorescence hence timely spraying during the period was most effective (Sundararaju, 2002).

- ☆ Pruning the infected shoots followed by spraying of copper sulphate or benomyl or chlorothalonil+methyl thiophanate have been suggested to control shoot/inflorescence die back (Cardoso *et al.*, 2000b).
- ☆ Removal of infected twigs and alternate hosts (*Adansonia digitata* and *Ficus mucosa*), and insecticidal spray to control cashew stem girdler (*Analeptes trifasciata*) have also been suggested (Adeigbe *et al.*, 2015).
- ☆ Care should be taken while collecting scions and they should be collected from disease free plants as the disease also spreads through vegetative propagative materials (Cardoso *et al.*, 2009). Surgical removal of gummosis and canker infected portion and plugging of the wounds with copper fungicides were effective (Freire, 1991).
- ☆ It has been suggested to do periodical sprays of insecticides (lambda cyhalothrin @ 0.003 per cent or monocrotophos @ 0.05 per cent or carboryl @ 0.1 per cent) for controlling TMB followed by removal of infected shoots/branches and spraying of Bordeaux mixture (1.0 per cent) for controlling die back of shoots and branches (Sundararaju 2002, 2008).
- ☆ Certain genotypes/clones *viz.*, 1 Eruwa 188/276, 2Sc Eruwa, 1sc Iwo 79/89 and Iwo 245/262 in Nigria (Olunloyo, 1994), CAPC 42 and BRS 226 in Brazil (Cardoso *et al.*, 2006; Cysne *et al.*, 2010; Cipriano *et al.*, 2015), and Vengurla 36/3, Vengurla 37/3, Sawantwadi, K19-1, UI-28-1, M-6/1, Hybrid 2/11, NLR 2/1 and Hybrid 2/12 in India (Patnaik *et al.*, 1987) were tolerant to the disease.

2.3. Powdery Mildew

Symptom

Powdery mildew caused by *Oidium anacardii* (Castellani and Casulli, 1981) was first reported from Sao Paulo, Brazil (Noack, 1898). Symptoms include initial whitish or grayish–whitish mildew colonies on upper surface of leaves and later purplish to brown patches on main leaf veins. It is a serious threat to cashew cultivation in African countries particularly in Tanzania (Sijaona and Shomari, 1987; Agboton *et al.*, 2013) as the countries face epidemic nature of infection on immature shoots and flowers (Shomari and Kennedy, 1997). However, the disease is not a serious one in Brazil as it affects only matured leaves while infloresces were rarely infected. So far the disease is not been reported from India.

Management

The disease is effectively managed by dusting sulfur products (dusting of sulphur @ 2.0 per cent) or spraying of Dinocap or benomyl (Cardoso *et al.*, 1995b; Menezes, 1997).

2.4. Black Mold

Symptom

Black mold disease caused by obligate parasite *Pilgeriella anacardii* is a major foliar disease in cashew. The name black mould was proposed by Aquino and Melo (1974). Symptoms of the disease include black to dark brown colonies with chlorotic spots on leaves. Normally matured leaves are affected by this disease while young leaves remain unaffected. It also manifests yellowing and defoliation of leaves, and shortening of internode.

Management

Different management practices such as application of plant defence activator (acibenzolar-S-methyl 50 WG @ 0.03g L-1,) at 30 days interval twice (Viana *et al.*, 2012), use of resistance dwarf clones *viz.*, CAP8, CAP20, CAP17, CAP11, CAP12 and CP06 (Cardosa *et al.*, 1999) and exploring the potentiality of biocontrol agent *Acremonium* (Freire, 1999) have been suggested.

2.5. Angular Leaf Spot

Symptom

Angular leaf spot is caused by *Septoria anacardii* (Freire, 1997). Disease attacks both young and adult plants however; the young ones are severely damaged by this disease. The pathogen produces light brown or grey or cream colour spots with dark brow margin on the leaves mostly confined to angular region of the veins.

Management

No specific chemicals have been recommended for management of this disease. But common fungicides like Mancozeb 75 per cent WP (0.2 per cent) may be used to manage the disease. In Brazil, dwarf clones such as CAP2, CAP5, CAP4, CAP14, CAP3, CAP20, CAP1, CAP23, CAP24, CAP26 and CAP21 were identified as resistant clones to angular leaf spot disease (Cardosa *et al.*, 1999).

2.6. Pink Disease

Symptom

Pink disease is caused by *Corticium salmonicolor* (Syn. *Pellicularia salmonicolor*). It produces white or pinkish growth on bark and a film of silky thread of fungus patches on branches. It also causes bark splitting and peeling off, die-back of shoots, yellowing of leaves and defoliation. Rainy season favours fast spread of the disease. In advanced stage, the fungus may enter into deep layer and causes severe damage to the branches.

Management

Common management practices such as pruning of affected portion and plugging the cut end with Bordeaux paste, spraying of Bordeaux mixture (1.0 per cent) during pre-monsoon (May-June) and post monsoon (October) (Mandal, 2012), and application of Vitavax (1.0 per cent) were reported.

2.7. Shoot Rot or Leaf Fall

Symptom

Phytophthora nicotianae is a causal agent of the disease and first it was observed in India from Kerala state (Thankamma, 1974). Symptoms include, immature leaves show black lesions on midrib initially and later the lesions spread to veins and lamina. Blighting symptom is produced when all lesions coalesce and defoliation occurs at final stage. The pathogen produces typical white mycelial growth on the under surface of affected leaves especially between blighted and healthy area. The affected shoot produces black linear lesions and gum exudations. It also causes death of shoots and shriveling of tender leaves.

Management

General management practices such as removal of affected portions and spraying with Bordeaux mixture (1.0 per cent) may be useful.

3. Nursery Diseases

Young seedlings are prone to attack by the pathogens. Several diseases have been documented in different stages of establishment of seedlings. Nursery diseases, their identification symptoms and management practices have been summarized in Table 24.2.

4. Disease Caused by Bacterial Pathogen

4.1. *Xanthomonas* Spot

Identification symptoms for *Xanthomonas* spot are darkening of veins and surrounding tissues with occasional angular leaf spots on leaf blade. Other symptoms include, spread of dark colour infection from the main vein to secondary veins and petioles. In advanced stage, it causes dry necrosis of twigs and their tips. On fruits, initially it produces oily spots and later they become necrotic spots whereas pseudofruits never get infection. *Xanthomonas citri* pv. *anacardii* in Brazil (Gama *et al.*, 2011) and *Xanthomonas citri* pv. *mangiferaeindicae* in Burkina Faso, Western Africa (Zombre *et al.*, 2015) were reported as causal agents of the disease. Management practices for the disease have not been reported.

5. Minor Diseases

Certain diseases in cashew such as sooty mold (*Capnodium brasiliensis*, *Capnodium citri*, and *Capnodium* sp.), leaf spot (*Pestalotia dichaeta*, *P. paeoniae*, *Phyllosticta brasiliensis* etc.) and red rust (*Cephaleuros virescens*) were observed but they do not cause severe damage to the crop hence considered them as minor diseases.

6. Nutritional Deficiency

6.1 Yellow Leaf Spot

During monsoon season, minute yellow spots are noticed on leaves. Later they enlarge and coalesce to form yellow blotches. In severe conditions, the affected plant

Table 24.2: Nursery Diseases of Cashew and their Management

Disease	*Causal Agent*	*Identification Symptoms*	*Management Practices*
Foot and root rot	*Pythium splendens* in Brazil (Freire *et al.*, 2002). *P. ultimum* in Nigeria (Hammed *et al.*, 2011)	Yellowing of seedlings. Wilting and death of seedling within 2-5 days.	General management practices such as providing proper ventilation, appropriate shading, judicious watering, and use of well decomposed organic manures may reduce the incidence. Application of dexon in soil at113.6 kg/ha is also suggested (Mandal, 2012).
	Sporadic incidence of root and foot rot also caused by *Sclerotium rolfsii* in Brazil.	In addition to the above symptoms it produces white mustard like sclerotial bodies in the infected regions.	General standard nursery cares should be followed.
Damping off	*Phytophthora palmivora* (Kumararaj and Bhide, 1962)	Initially, seedlings with water soaked girdles of darkened tissues in the collar region of the stems. Later, the seedlings turn pale yellow and topple down. Eventually they rot and die.	Regular nursery management care with application of Bordeaux mixture (1.0 per cent) may control the disease.
Anthracnose	*C. gloeosporioides*	Same as described earlier.	Spraying of copper oxychloride (3 g/l) or benomyl (1 g/l) at weekly interval (Freire *et al.*, 2002).
Blight	*Phytophthora heveae* and *P. nicotiana* reported in Brazil	Initial water-soaked spots on leaves later they become necrotic dark brown spots and finally the spots spread to whole seedlings and the infected seedlings die quickly.	Prophylactic spray of metalaxyl (1 g/l) (Freire *et al.*, 2002).
	Cylindrocladium scoparium reported in Kottarakkara, Kerala state, India (Philip, 1973)	Wilting and withering of leaves, and rotting of stem at and little below the ground. The disease was severe during rainy season. Continuous rain with high humidity may cause die back of apical shoot and stem rotting.	Prophylactic application of carbendazim (@500 ppm) before the onset of monsoon (Mandal, 2012). Keeping nursery site away from *Eucalyptus* trees as it is a susceptible host for the pathogen (Sharma and Mohanan, 1982).
Rotting of graft union	*L. theobromae*	Rotting of grafted seedlings.	Periodical cleaning of the nursery knives with sodium hypochlorite solution (1.0 per cent) and dipping scions in benomyl solution (1.0 per cent) for 15 min (Freire, 1996 and 2001).

Table 24.3: European Union MRL's for Commonly Used Fungicides in Cashew Nut (EU registration code 0120030) in Europe

Fungicide	*Registration No (Date of registration)*	*MRL (mg/kg)*
Azoxystrobin	Reg. (EU) 2015/1040 (01-07-2015)	0.01
Captan	Reg. (EU) No 2016/452 (30-03-2016)	0.02
Carbendazim and benomyl (sum of benomyl and carbendazim expressed as carbendazim)	Reg. (EU) No 559/2011 (11-06-2011)	0.1
Chlorothalonil	Reg. (EU) 2016/67 (22-01-2016)	0.01
Copper compounds (Copper)	Reg. (EC) No 149/2008 (01-03-2008)	30
Thiophanate-methyl	Reg. (EU) No 559/2011(11-06-2011)	0.2
Tridemorph	Reg. (EU) No 899/2012 (06-10-2012)	0.02
Trifloxystrobin	Reg. (EU) 2016/67 (22-01-2016)	0.02
Ziram	Reg. (EC) No 149/2008 (01-03-2008)	0.1
Tebuconazole	Reg. (EU) 2016/1 (05-01-2016)	0.05
Metalaxyl and metalaxyl-M (metalaxyl including other mixtures of constituent isomers including metalaxyl-M (sum of isomers))	Reg. (EU) 2016/486 (06-04-2016)	0.05
Oxycarboxin	Reg. (EU) No 899/2012 (06-10-2012)	0.02
Pencycuron	Reg. (EC) No 149/2008 (01-03-2008)	0.05
Sulphur	Reg. (EC) No 459/2010 (28-05-2010)	No MRL required
Dithiocarbamates (dithiocarbamates expressed as CS2, including maneb, mancozeb, metiram, propineb, thiram and ziram)	Reg. (EU) 2016/567(15-04-2016)	0.05
Iprodione	Reg. (EU) 2015/400 (14-03-2015)	0.01
Kresoxim-methyl	Reg. (EU) 2016/486 (06-04-2016)	0.01
Cymoxanil	Reg. (EU) No 978/2011 (04-10-2011)	0.05
Dinocap (sum of dinocap isomers and their corresponding phenols expressed as dinocap)	Reg. (EU) No 1127/2014 (24-10-2014)	0.05

Source: http://ec.europa.eu/food/plant/pesticides/max_residue_levels/index_en.htm.

show leaf fall symptoms. The disease was not caused by pathogens as no fungal or bacterial pathogen could be isolated from the yellow spot infected leaves (Menon *et al.*, 1978). It has been associated with low soil pH (4.5-5.0), excessive amount of manganese, and deficient amount of molybdenum and Kjeldahl nitrogen (Subbaiah *et al.*, 1986).

Management

Foliar spray with molybdenum salts and adjusting the soil pH with lime have been recommended (Subbaiah *et al.*, 1986).

7. Policy Issues

In India only chemical/biological agents registered under Central Insecticide Board and Registration Committee (CIBRC) and having label claims, are to be used to avoid residue problems and to ensure safe use of the products. Unfortunately, as on date (last latest update from CIBRC on 31.08.2015), no single fungicide has been registered under CIBRC for use in cashew crop. The crop product is being an export oriented one, monitoring residue level is so essential. European Union is having maximum residue limit (MRL) for most of commonly used fungicides in cashew (Table 24.3), which may be useful the growers/end-users for monitoring the residue levels.

8. Future Line of Action

- ☆ **Intensive research on diagnosis and management of diseases is required:** Though TMB has been proposed as main cause of inflorescence or blossom blight/die back, in Brazil and Odisha (India) the TMB attacks were not noticed but still the blight incidences were recorded in severe form (Patnaik *et al.*, 1987; Freire *et al.*, 2002), indicating that possibility of involvement of fungal pathogen as primary causal agent of the disease. Hence research focus in this line should be intensified
- ☆ **Strict quarantine for non-existing diseases:** For an instance powdery mildew disease in cashew has not been reported in India hence measures should be taken to avoid introduction of such disease from outside.
- ☆ **Exploring for host plant resistance for combating major diseases:** As chemicals are costlier and due to socio economic reasons, farmers are unable to adopt chemical method of disease control. Hence attempts should be made to identify resistance sources and to utilize them in resistance breeding.

References

Adeigbe, O. O., Olasupo, F. O., Adewale, B. D. and Muyiwa, A. A. (2015). A review on cashew research and production in Nigeria in the last four decades. *Scientific Research and Essays*, 10: 196-209. DOI: 10.5897/SRE2014.5953.

Adejumo, T. O. (2005). Crop protection strategies for major diseases of cocoa, coffee and cashew in Nigeria. *African Journal of Biotechnology*, 4 (2): 143-150.

Agboton, B.V., Salifu D., Seguni, Z., Sijaona, M. E., Shomari, S., Ekesi, S. and Maniania, N. K. (2013). Bioecology of some key cashew insect pests and diseases in diverse habitats and landscapes in Tanzania. *Journal of Applied Entomology*, 137: 782–789.

Aquino, M. L. N. and Melo, G.S. (1974). Uma nova enfermidade do cajueiro (*Anacardium occidentale* L.) (A new disease of cashew, Anacardium occidentale L.). IPA, Boletim T!ecnico 71, Recife-PE,Brazil, pp. 16.

Bhat, M.G., Yadukumar, N., Nagaraja, K.K. and Nayak, M.G. 2009. Current status of cashew research and technology in India. *The Cashew and Cocoa Journal*, (April-June): pp. 28-42.

Cardoso, J. E., Bezerra, M. A., Viana, F. M. P., Sousa, T. R. M., Cysne, A. Q. and Farias, F.C. (2009). Endophyte occurrence of *Lasiodiplodia theobromae* in cashew tissues and its transmission by vegetative propagules. *Summa Phytopathologica*, 35: 262-266.

Cardoso, J. E., Aragão, M. L., Bleicher, E. and Cavalcante, M. J. B. (1995a). Efeito de práticas agronômicas na ocorrência da resinose do cajueiro. Fitopatologia Brasileira, 20: 242.

Cardoso, J. E., Aragao, M. L. and Cavalcante, M. J. B. (1995b). Avaliacao de produtos quimicos no controle das principais doencas do cajueiro no semi arido (Evaluation of fungicides on the control of the main cashew diseases in the northeastern semi-arid). Proceedings of the 30th Meeting of the Brazilian Phytopathological Society. Ilheus-BA, Brazil (Abstract, 443).

Cardoso, J. E., Cavalcante, M. J. B., Cavalcanti, J. J. V. and Vieira, A. C. F. (1994). Identifica cao de fontes de resistencia em clones de cajueiro anaoprecoce as principais doen-cas foliares (Identification of sources of resistance in dwarf clones of cashew to important leaf diseases). Proceedings of the 13th Annual Meeting of the Brazilian Society of Fruit Crop. Salvador-BA, Brazil (Abstract, 123).

Cardoso, J. E., Cavalcanti, J. J. V., Cavalcante, M. de J. B., Aragao, M. do L. and Felipe, E. M. (1999). Genetic resistance of dwarf cashew (*Anacardium occidentale* L.) to anthracnose, black mold and angular leaf spot. *Crop Protection*, 18: 23-27.

Cardoso, J. E., Paiva, J. R., Cavalcanti, J. J. V., Apoliano dos Santos, A. and Vidal, J. C. (2006). Evaluation of resistance in dwarf cashew to gummosis in north-eastern Brazil. *Crop Protection*, 25: 855–859.

Cardoso, J. E., Vidal, J. C., Santos, A. A., Viana, F. M. P., Freire, F. C. O. and Souza, R. N. M. (2000). Ocorrencia da podridao-preta dos ramos do cajueiro no Ceara e Piaui (Occurrence of shoot black-rot of cashew in the States of Ceara and Piaui). Comunicado Tecnico No. 52, EMBRAPA/CNPAR, Brazil, 3p.

Castellani, E. and Casulli, F. (1981). Osservazioni preliminari su Oidium anacardii Noack agent del mal bianco dell anacardio (Preliminary observations on *Oidium anacardii* Noack, causal agent of cashew powdery mildew). *Riv. Agric. Subtrop. Trop.*, 75 (2/3): 211–222.

Cipriano, A. K. A. L., Gondim, D. M. F., Vasconcelos, I. M., Martins, J. A. M., Moura, A. A., Moreno, F. B., Monteiro-Moreira, A. C. O., Melo, J. G.M., Cardoso, J. E., Paiva, A. L. S., Oliveira J T. A., Cipriano, A. K. A. L., Gondim, D. M. F., Vasconcelos, I. M., Martins, J. A. M., Moura, A. A., Moreno, F. B., Monteiro-Moreira, A. C. O., Melo, J. G. M., Cardoso, J. E., Paiva, A. L. S. and Oliveira, J. T. A. (2015). Proteomic analysis of responsive stem proteins of resistant and susceptible cashew plants after *Lasiodiplodia theobromae* infection. *Journal of Proteomics*, 113: 90-109.

Cysne, A. Q., Cardoso, J. E., Maia, A. H. N. and Farias, F. C. (2010). Spatial-temporal analysis of gummosis in three cashew clones at Northeastern Brazil. *J. Phytopathol.*, 158: 676–82.

Freire, F. C. O., Cardoso, J. E., dos Santos, A. A. and Viana, F. M. P. (2002). Diseases of cashew nut plants (*Anacardium occidentale* L.). in Brazil. *Crop Protection*, 21: 489–494.

Freire, F. C. O. and Cardoso, J. E. (1995). Doencas do cajueiro (Disease of cashew tree). In: Araujo, J.P.P., Silva, V.V. (Eds.), Cajucultura- Modernas Tecnicas de Producao (Cashew crop- Modern Production Techniques). Embrapa/CNPAT, Fortaleza-CE, Brazil, pp. 249-267.

Freire, F. C. O. (1991). A resinose do cajueiro (The cashew gummosis). Caju Informativo, EMBRAPA-CNPCa 4(1): 1–2.

Freire, F. C. O. (1996). Ocorrencia de *Cylindrocladium scoparium*, *Pythium splendens* e *Phytophthora* sp. em mudas de cajueiro (*Anacardium occidentale* L.). no Brasil (Occurrence of *Cylindrocladium scoparium*, *Pythium splendens* and *Phytophthora* sp. on cashew seedlings in Brazil). *Agrotropica*, 8(3): 69–72.

Freire, F. C. O. (1997). Angular leaf spot of cashew (*Anacardium occidentale* L.). caused by *Septoria anacardii* sp. nov. *Agrotropica*, 9(1): 19–22.

Freire, F. C. O. (1999). Mycoparasitism of the cashew black mould agent by *Acremonium* sp. *Agrotropica*, 11(1): 25–30.

Freire, F. C. O. (2001). Queima de Mudas do Cajueiro (Blight of Cashew Seedlings). In: Luz, E.D.M.N., Santos, A.F., Matsuoka, K., Bezerra, J.L. (Eds.), Doen-cas causadas por Phytophthora no Brasil (Diseases Caused by *Phytophthora* in Brazil). Livraria Editora Rural, Campinas-SP, Brazil, pp.267–282.

Gama, M. A. S., Mariano, R. L. R., Viana, F. M. P., Ferreira, M. A. S. V and Souza, E. B. (2011). Polyphasic characterization of pigmented strains of *Xanthomonas* pathogenic to cashew trees. *Plant Dis.*, 95: 793-802.

Hammed, L. A., Lawal, B.A. and Kolapo, K. A. (2011). Growth and nutrient uptake of cashew (*Anacardium occidentale* L.). seedlings as affected by nut size in the nursery. *Afr. J. Res.*, 6(17): 3962-3971.

Hammed, L.A., Anikwe, J. C and Adedeji, A. R. (2008). Cashew nuts and production development in Nigeria. Am-Euras. *J. Sci. Res.*, 3(1): 54-61.

Kumararaj, K and Bhide, V.P. (1962). Damping off of cashewnut (*Anacardium occidentale* L.). seedlings caused by *Phytophthora palmivora* Butler in Maharashtra state. *Curr. Sci.*, 1(Jan.): 23.

Lima, J. E., Menezes, de A., Karam, M., Oilvera, de and Rodriques, M. G. R. (1950). List of common names of Indian Plant disease. *Indian Journal of Agricultural Sciences*, 20(1): 107-162.

Lopez, A. M. Q and Lucas, J. A. (2002). Effects of plant defence activators on anthracnose disease of cashew. *European Journal of Plant Pathology*, 108: 409–420.

Mandal, R. C. (2012). Diseases of cashew. In: Cashew production and processing technology. Published by AGROBIOS (INDIA), Jodhpur, India, pp108-119.

Menezes, M (1997). Doen-cas do Cajueiro (Diseases of Cashew). In: Kimati, H., Amorim, L., Bergamin Filho, A., Camargo, L.E.A., Rezende, J.A.A.M. (Eds.), Manual de Fitopatologia (Plant Pathology Manual). Editora Agron#omica Ceres Ltda, Sao Paulo, pp. 201–206.

Menon, M. A., Chandramohanan, R., Harishu Kumar, P., Rakhiyappan, P. and Nair, B. P. (1978). Yellow leaf spot of cashew. *Indian Cashew J.*, XII (3).

Mishra, S. K., Pattanayak, B. and Naik, R. P. (1993). Control of gummosis disease in cashew (*Anacardium occidentale* L.). under field condition. *The Cashew*, II (2): 10-12.

Muniz, C. R., Freire, F. C. O., Viana, F. M. P., Cardoso, J. E, Cooke, P., Wood, D. and Guedes, M.I. F. (2011). Colonization of cashew plants by *Lasiodiplodia theobromae*: Microscopical Features. *Micron*, 42: 419–428.

Naik, R. P., Mishra, S. K., Mishra, K. C. and Patnaik, H. P. (1988). Fungicidal effects of certain chemicals against *Botryodiplodia theobromae* causing die back of cashew. *The Cashew*, (January - March): 19-20.

Nair, M. K., Rao, E. V. V. B., Nambiar, K. K. N. and Nambiar, M. C. (1979). Monograph on plantation crops I. Cashew (*Anacardium occidentale* L.). Central Plantation Crop Research Institute, Kararagod, p131.

Nambiar, K. K. N and Brahma, R. N. (1979). Important diseases of cashew and their control. *Indian Farming*, 28(12): 19–20.

Nambiar, K. K. N., Sarma, Y. R. and Pillai, G.B (1973). Inflorescence blight of cashew (*Anacardium occidentale* L.). *J. Plant Crops*,1: 44-46.

Noack, F. (1898). Cogumelos parasitos das plantas de pomar, horta e jardim (Fungi parasitic on plants of orchard, garden and back-yards). *Bol. Inst. Agronom Sao Paulo Brazil*, 9: 75–88.

Olunloyo, A. O. (1979). The role of fruit flies in rot disease of cashew apples in the plantation. *Ann. Rept, Cocoa Res. Inst. Nigr.*, 1978/79: 101-103.

Olunloyo, O. A. and Esuruoso, O. F (1975). *Lasiodiplodia* floral shoot dieback disease of cashew in Nigeria. *Plant Dis. Rep.*, 59(2): 176–179.

Olunloyo, O. A. (1994a). Further studies on the relative stability of field tolerance of ten cashew genotypes to inflorescence blight disease. Ann. Rept. Cocoa Res. Inst. Nigeria, Ibadan, 33-34.

Patnaik, H. P., Senapathy, B. and Mishra, S. K. (1987). Inflorecence blight incidence in cashew varieties-A preliminary study under east coast conditions. *The Cashew*, 1(2): 10-11.

Phillip, S. (1973). Seedling blight of cashew (*Anacardium occidentale* L.). due to *Cylindrocladium scoparium*. *Curr. Sci.*, 42: 440.

Pillai, G. B. and Abraham, V. A. (1975). Tea mosquito a major menace to cashew. *Indian Cashew Journal*, 10(1): 5-7.

Pillai, G. B. (1980). Pest problems of cashew. *Cashew Causerie*, 2(2): 3-10.

Ponte, J. J. (1986). Mapeamento e importaöncia das doencas do cajueiro no Nordeste do Brasil. *Nematologia Brasileira*, 10(1): 59-68.

Rao, A. R. (1969). Pest and diseases on cahew. *Lalbagh*, 14(4): 9-10.

Sharma, K. K. and Mohanan, C. (1982). *Cylindrocaldium* spp. associated with various diseases of Eucalyptus in Kerala. *Eur. J. For. Path.*, 12(3): 129-136.

Shomari, S. H. and Kennedy, R. (1997). Field and laboratory investigations on the development of Oidium anacardii in relation to environmental factors. Proceedings of the International Cashew and Coconut Conference, Dar Es Salaam, Tanazania, pp.260–265.

Sijaona, M. E. R. and Shomari, S. H. (1987). The powdery mildew disease of cashew in Tanzania. *TARO Newslett.*, 11(3): 4–5.

Singh, S., Sehgal, H. S., Pandey, P. C and Bakshi, B. K. (1967). Anthracnose disease of cashew (*Anacardium occidentale* L.). its use, epidemiology and control. *Indian Forester*, 93: 374-376.

Subbaiah, C. C., Manikandan, P. and Joshi, Y. (1986). Yellow leaf spot of cashew: A case of molybdenum deficiency. *Plant and Soil*, 94: 35-42.

Sundararaju, D. (2002). Pest and disease management of cashew in India. *The Cashew*, XVI(4): 32-38.

Sundararaju, D. (2008). Pests and diseases management of cashew in India. *The Cashew*, XXII(1): 8-20.

Teixeira, L. M. S., Melo, Q. M. S., Mesquita, A. L. M. and Freire, F. C. O. (1991). Recomendaco es para o controle de pragas e doencas do cajueiro (Recommendation for controlling cashew pests). Embrapa/CNPCa Comunicado Tecnico 2, Fortaleza-CE, Brazil.

Thankamma, L. (1974). *Phytophthora nicotianae* var. *nicotianae* on *Anacardium occidentale* in South India. *Plant Dis Reptr.*, 58: 767-768.

Venkateshwarlu, D., Krishnamurthy, K.V.M., Reddy, M.L.N. and Mahalakshimi, M.S. (2003). Effect of anthracnose disease of the cashew seedlings on graft success in cashew nurseries. *The Cashew,* XVII (2): 22-24.

Viana, F.M.P., Lima, J.S., Lima, F.A. and Cardoso, J.E. (2012). Control of cashew black mould by acibenzolar-S-methyl. *Tropical Plant Pathology,* 37(5): 354-357.

Zombre, C., Sankara, P., Ouedraogo, L., Wonni, I., Boyer, K., Boyer, C., Terville, M., Javegny, S., Allibert, A., Verniere, C. and Pruvost, O. (2015). Natural infection of cashew (*Anacardium occidentale*). by *Xanthomonas citri* pv. *mangiferaeindicae* in Burkina Faso. *Plant Disease*. http: //dx.doi.org/10.1094/PDIS-10-15-1137-RE.

Section V: Postharvest Technology

Cashew: Improvement, Production and Processing (2017) *Pages* **423–432**
Editor: **Prof. P.L. Saroj**
Published by: **ASTRAL INTERNATIONAL PVT. LTD., NEW DELHI**

25

Maturity Indices

Mini Poduval

1. Introduction

Cashew nut is one of the most nutritious and delicious dry fruits of the world. More than 20 countries are engaged in cashew nut production, processing and export. India was the first country exporting cashew kernels and cashew nut shell liquid. India is the pioneer in cashew research and development. Intensive research carried out in the country, particularly during the past three decades leading to the development of more than 46 high yielding varieties and efficient technologies capable of yielding 2 to 3.5 tonnes of raw nuts per hectare. Normally, cashew trees can grow up to a height of more than 8m and the canopy spread varies from 8 to 12 m. Salam and Peter (2010) reported that flowering lasts for 2 to 3 months. The duration of nut maturity from flowering depends on the season and variety. The cashew fruit has two distinct parts, *viz.*, the swollen and pear shaped stalk (cashew apple) and kidney shaped nut attached to the lower end of the apple. The cashew nut is the true fruit, which contains the edible kernel. The apples (pseudo fruit) in general are yellow or red in colour. Under the tropical conditions of India, it takes 50 – 70 days from pollination to maturity of the fruit. In a week after pollination, it may attain the peanut stage. Within a month it will develop in to a tender nut and in another 30–40 days the nuts will mature. Cashew nut is typically reniform. At three weeks after pollination, the nut length will be approximately 3 cm and nut width around 2.5 cm. About 33 per cent of the available space inside the nut will be filled with kernel, at this stage of nut development. The developing tender nut is green. The shell of the mature nut contains a volatile liquid called cashew nut shell liquid in abundance. About six weeks after pollination the nut length will be 4.0 cm and nut width around 3.5 cm. The immature nuts at this stage will be filled with white kernel with lot of moisture. The nut size will be reduced thereafter due to desiccation and shrinking. Similar shrinkage occurs in the kernel also. The mature nut will be grey, strong and sturdy. The matured kernels will be white and hard with relatively less moisture compared to the earlier stages of development. The nut size also varies with the variety. Certain varieties are cluster bearing with large number of fruits in a single panicle. From fertilization to nut set it takes about seven days. Nuts attain maximum size in about 35 days after fertilization. Grey nut stage is attained in about 47 days after fertilization. Fruits attain full maturity within 60 – 70 days after fertilization. The kernel development is slow during initial stages of development. The green nuts will be full between 5 to 6 weeks after fertilization. Kernel development becomes faster thereafter. At grey nut stage, kernel becomes more firm. The nuts attain physiological maturity between 8 to 10 weeks, after fertilization. During advanced stage of maturity, the nut size decreases due to shrinkage and apple size increases due to enlargement. Apple becomes swollen, juicy and sweet. The stalk end of the apple becomes soft. Due to increase in weight of the apple, stalk end fail to hold the apple and nut. At this stage, even a mild wind will result detachment of fruit from the stalk. Thus the fruit fall off.

2. Fruit Growth, Development and Maturation

The cashew apple and nut grow independently (Pratt and Mendoza, 1980). Following anthesis the growth of nut is fast and uniform, reaching maximum size around 30 days (Nut weight around 14 g). After this fast growth phase, nut weight decreases around 15 – 40 per cent as the shell dehisces and hardens. Initially the cashew apple grows slowly and then growth rapidly increases after the nut has reached maximum size. Mature cashew apple colour can vary from yellow to red and reaches maximum size in about 50 days. The edible portion represents about 90 per cent of the entire cashew and weighs 70 – 90 g (Filgueiras *et al.*, 1999). Figueiredo *et al.* (2002) also found that after the development of the cashewnut was completed (Stage- I) the apple began to grow and develop with significant changes in physical characteristics. During maturity stages one to seven the cashew apple increased approximately five folds in weight (32. 2 to 169.8g) two fold in diameter (3.31 to 6.41cm) and 1.5 fold in length (6.6 to 8.24 cm). The most accentuated increase in cashew apple weight occurred between stages 6 to 7. For nut harvest and/or extraction of cashew nut shell liquid the apples are left on the tree to assure full maturity of the kernel, dropping to the ground, 7 to 8 weeks following apple maturity (Acland, 1971).

3. Respiration and Ripening

Cashew apple has a high respiration rate (62 to 72 ml/kg/hr at 20°C). However, the pattern is non- *climacteric* (Biale and Barcus, 1970). According to Pratt and Mendoza (1980) at 20°C respiration rates were approximately 50 ml/kg/hr and ethylene production was steady but very low (200 to 400 ml/kg/hr). To achieve optimal quality the cashew apple must remain on the tree until it reaches maturity. During ripening, cashew apple changes colour due to chlorophyll loss and pigment synthesis, acids and astringency decrease, while soluble solids, reducing sugars and ascorbic acids increase. The tender skin becomes exceedingly waxy as the cashew apple ripens.

4. Maturity Indices

Fresh market cashew apple must be the correct shape and have uniform external appearance with no signs of physical injury. Pear shapes are preferred and red fruit are increasingly preferred to yellow fruit. Other important quality characteristics for fresh market are low astringency, low acidity and sweet taste. Although physical appearance is important to consumers, no direct correlation has been found between externel characteristics (*e.g.*, colour, size, shape) of different cashew apple varieties/selection and variations in composition (Filgueiras *et al.*, 1999). For processing, misshapen cashew apple is tolerated provided there are no signs of disease or insect damage. The most important attributes for processing are juice yield, low astringency and sugar/acid ratio. Ripe cashew apple yields approximately 80 per cent. Figueiredo *et al.* (2002) found that between maturity stages 1 to 7 cashew apple pulp firmness decreased. Anthocyanins, total carotenoids, tannins, vitamin C and reducing sugars increased continuously during ripening. Sixty-three volatiles were detected in the head space of cashew apple juice (Garruti *et al.*, 2001). From the 63 compounds, 49 were identified. The volatile profile was characterized by a variety of esters (38 per cent), nine alcohols, 7 aldehydes, two ketones, three acids, one terpene and fourteen unidentified compounds. Volatile compounds were predominantly esters, mainly methyl and ethyl esters of saturated C2 and C6 carboxylic acids (Garruti *et al.*, 2001). Broinizi *et al.* (2001) determined that cashew apple contained up to 50 per cent β carotene and 95 per cent antioxidant activity. Yahaya *et al.* (2010) studied the quality attributes of the different cultivars of cashew. Parameters investigated include pH, refractive index, specific gravity, vitamin C content, juice astringency, chlorophyll and juice yield. Results showed that percentage juice yield increases linearly with maturity for both cultivars. A maximum of 79.5 per cent is obtained for the yellow apple cultivars, while 74.1 per cent juice yield is obtained in the red apple cultivar. The pH ranged from 3.73-4.5 and 3.64-4.55 for red and yellow apple cultivars, respectively, with a marked fall between seven and eight weeks after pollination. Specific gravity and refractive index are of the order 0.999-1.007 and 1.34-1.89 for both selected samples. Highest refractive index was obtained in red and yellow cultivars at seven and eight weeks after pollination. The changes in ascorbic acid followed a linear pattern and the maximum values of 249.5 mg/100 ml and 185.5 mg/100 ml were obtained for yellow and red apple cultivars, respectively. Meanwhile, the juice astringency level decreased steadily with maturity and at the ninth weeks after pollination, values obtained are 0.47 and 0.86 mg/100 ml tannin for yellow and red apple cultivars, respectively. Generally, pH, astringency, refractive index, juice content, ascorbic acid and chlorophyll contents showed marked variations between six and eight weeks after pollination, confirming physiological maturity and ripening in the crop species. Cashew farmers and processors will know the best time to harvest the crop when the quality is at its peak.

Table 25.1: Physical Characteristics of different Varieties of Cashew Apple

Variety	*Shape*	*Colour*	*Ridges on Cashew Apple*	*Apex of Cashew Apple*	*Grooves on Cashew Apple Base*	*Cavity at Cashew Apple Base*	*Skin of Cashew Apple*	*Shape of Cashew Apple Base*	*Attachment of Apple to Nut*
VRI - 2	Conical to Obovate	Pink	Entire	Oblique	Shallow	Shallow	Smooth and Glossy	Angular	Intermediate
BPP – 5	Conical to Obovate	Yellow red	Entire	Oblique	Absent	Shallow	Smooth and Glossy	Obliquely flattened	Intermediate
Kanaka	Conical to Obovate	Yellow	Entire	Oblique	Shallow	Shallow	Smooth and Glossy	Obliquely flattened	Tight
V- 2	Conical to Obovate	Red	Entire	Oblique	Shallow	Deep	Smooth and Glossy	Flattened	Intermediate
V – 6	Cylindrical	Yellow	Broken	Oblique	Absent	Deep	Smooth and Glossy	Flattened	Tight
NRCC - 1	Cylindrical	Yellow red	Broken	Oblique	Shallow	Shallow	Smooth and Glossy	Obliquely flattened	Intermediate
Dhana	Cylindrical	Yellow	Entire	Oblique	Deep	Shallow	Smooth and Glossy	Rounded	Tight
Madakkathara -1	Conical to Obovate	Yellow	Broken	Oblique	Absent	Deep	Smooth and Glossy	Flattened	Loose
V – 4	Cylindrical	Pink	Entire	Oblique	Deep	Deep	Rough and Dull	Obliquely flattened	Intermediate
BPP – 1	Conical to Obovate	Yellow	Broken	Oblique	Shallow	Shallow	Smooth and Glossy	Flattened	Intermediate
Anakkayam – 1	Conical to Obovate	Red	Broken	Oblique	Shallow	Deep	Smooth and Glossy	Obliquely flattened	Loose
Ullal – 3	Conical to Obovate	Pink	Broken	Oblique	Absent	Shallow	Smooth and Glossy	Obliquely flattened	Intermediate
BPP – 3	Cylindrical	Yellow	Broken	Oblique	Shallow	Shallow	Smooth and Glossy	Obliquely flattened	Intermediate
V – 3	Cylindrical	Yellow	Broken	Oblique	Shallow	Shallow	Smooth and Glossy	Obliquely flattened	Tight

Contd...

Table 25.1–*Contd...*

Variety	*Shape*	*Colour*	*Ridges on Cashew Apple*	*Apex of Cashew Apple*	*Grooves on Cashew Apple Base*	*Cavity at Cashew Apple Base*	*Skin of Cashew Apple*	*Shape of Cashew Apple Base*	*Attachment of Apple to Nut*
NRCC – 2	Cylindrical	Red	Broken	Oblique	Absent	Shallow	Smooth and Glossy	Obliquely flattened	Tight
Jhargram – 1	Cylindrical	Yellow	Broken	Oblique	Deep	Deep	Smooth and Glossy	Obliquely flattened	Intermediate
Priyanka	Pyriform	Pink	Broken	Oblique	Absent	Shallow	Smooth and Glossy	Pointed	Intermediate
BPP – 8	Conical to Obovate	Yellow	Broken	Oblique	Shallow	Shallow	Smooth and Glossy	Obliquely flattened	Loose
H – 1600	Conical	Red	Entire	Oblique	Shallow	Shallow	Smooth and Glossy	Obliquely flattened	Intermediate
H – 1610	Conical	Yellow red	Broken	Oblique	Shallow	Shallow	Smooth and Glossy	Flattened	Intermediate
H – 2/15	Pyriform	Yellow	Broken	Oblique	Absent	Shallow	Smooth and Glossy	Obliquely flattened	Loose
VTH – 59/2	Cylindrical	Red	Entire	Oblique	Shallow	Deep	Smooth and Glossy	Angular	Intermediate
VTH – 30/4	Conical to Obovate	Red	Entire	Oblique	Shallow	Deep	Smooth and Glossy	Obliquely flattened	Intermediate
T.N. 40	Pyriform	Yellow	Entire	Oblique	Deep	Shallow	Smooth and Glossy	Flattened	Intermediate
T.N. 129	Cylindrical	Yellow	Broken	Oblique	Shallow	Shallow	Smooth and Glossy	Obliquely flattened	Loose
M – 33/3	Pyriform	Red	Entire	Oblique	Deep	Deep	Smooth and Glossy	Rounded	Loose
VRI – 3	Cylindrical	Red	Entire	Oblique	Shallow	Shallow	Smooth and Glossy	Flattened	Loose
KGN - 1	Cylindrical	Yellow	Broken	Oblique	Shallow	Shallow	Smooth and Glossy	Rounded	Tight

Table 25.2: Physico-chemical Parameters of Cashew Apple

Variety	Length of Cashew Apple	Breadth of Cashew Apple	Weight of Cashew Apple	Specific Gravity	T.S.S. (º Brix)	Acidity (Per cent)	Total Sugar (Per cent)	Vitamin C (mg/ml of Juice)	Fibrous Residue (Per cent)	Sugar : Acid
VRI - 2	5.90	4.33	40.67	0.920	14.43	0.433	8.00	4.213	17.60	18.50
BPP – 5	5.50	4.40	34.33	0.953	14.13	0.450	9.06	2.990	14.30	20.13
Kanaka	6.97	5.53	78.00	0.960	16.33	0.617	9.51	3.383	17.23	15.41
V- 2	5.40	4.07	32.33	0.940	12.83	0.383	8.34	3.317	15.03	21.77
V – 6	6.47	5.50	67.00	0.920	14.53	0.400	9.54	5.473	15.03	23.85
NRCC - 1	7.03	4.57	49.33	0.927	16.03	0.517	12.53	3.090	19.80	24.24
Dhana	7.80	5.50	85.67	0.970	16.03	0.600	10.25	5.058	17.20	17.08
Madakkathara -1	6.03	4.47	38.00	0.975	15.10	0.517	11.76	5.147	20.43	22.75
V – 4	7.50	5.00	59.67	0.957	15.83	0.317	10.00	3.412	18.40	31.55
BPP – 1	6.17	4.97	51.33	0.910	14.73	0.733	9.76	2.833	21.30	13.32
Anakkayam – 1	6.63	4.15	41.66	0.743	16.10	1.180	9.76	4.360	23.70	8.27
Ullal – 3	6.87	5.70	62.33	1.007	15.70	0.483	8.87	5.377	21.77	18.36
BPP – 3	6.77	4.67	55.00	1.060	14.13	0.550	8.40	3.163	14.08	15.27
V – 3	5.87	4.93	50.00	0.960	19.07	0.500	10.27	4.273	24.53	20.54
NRCC – 2	5.40	5.00	42.00	0.913	15.80	0.537	10.54	5.107	21.83	19.63
Jhargram – 1	5.47	4.00	29.00	1.140	14.30	0.350	8.19	5.613	22.03	23.40
Priyanka	7.57	4.13	36.00	0.747	16.10	1.183	9.76	5.571	37.43	8.25
BPP – 8	5.93	4.40	40.33	0.943	15.37	0.467	9.09	4.140	15.30	19.46
H – 1600	6.77	4.57	45.00	1.127	17.37	0.650	11.08	3.347	28.40	17.05
H – 1610	6.23	3.77	39.00	0.933	22.47	1.167	12.12	5.177	23.60	10.39
H – 2/15	6.50	4.20	39.33	1.193	20.03	0.700	10.81	3.960	23.60	15.44

Contd...

Table 25.2–*Contd...*

Variety	Length of Cashew Apple	Breadth of Cashew Apple	Weight of Cashew Apple	Specific Gravity	T.S.S. (º Brix)	Acidity (Per cent)	Total Sugar (Per cent)	Vitamin C (mg/ml of Juice)	Fibrous Residue (Per cent)	Sugar : Acid
VTH – 59/2	5.97	4.73	50.67	0.957	16.40	0.450	7.28	3.133	22.40	16.18
VTH – 30/4	5.83	4.43	44.67	1.233	12.40	0.700	10.00	1.523	24.40	14.28
T.N. 40	6.53	4.17	39.00	0.980	18.40	0.783	10.49	2.747	33.27	13.40
T.N. 129	5.80	4.17	0.34	0.950	17.77	0.500	8.33	1.507	28.80	16.66
M – 33/3	7.00	4.97	52.33	0.963	15.83	0.533	12.12	4.337	24.47	22.74
VRI – 3	5.17	4.90	41.00	0.963	18.43	0.533	11.48	1.417	21.40	21.54
KGN - 1	8.50	5.83	93.33	0.927	18.20	0.900	9.51	3.927	29.60	10.57

Aliyu and Hammed (2000) reported an initial similar growth patterns and a later differential growth patterns between the nut and apple of cashew. The growth of the nuts showed a simple sigmoid growth curve in terms of length and circumferences of the lower and upper ends, against weeks from pollination, thus comprising four distinct stages of development (embryo-cotyledonary, nut-filling, physiological maturity and drying). The developmental stages of apple showed asigmoidal growth curve comprising four stages (receptacular, slow-growth, rapid growth and ripening). The per cent dry matter and the fresh weight of the apple are negatively correlated (r=-0.92).

5. Pre Harvest Factors Affecting Nut Yield

Cashew trees produce a significant number of horizontal roots which led to the erroneous, assumption that the tree requires a period of drought to achieve proper flowering and that rainfall is sufficient for maximum fruit yield. It was also thought that cashew could only withstand drought if tree spacing was wide enough to prevent overcrowding (Acland, 1971). The premise was based on the fact that cashew orchards worldwide were planted with common rootstocks with supposed drought resistance. However studies have demonstrated that the low productivity of cashew trees grown under drought stress (220 Kg nuts/ha) is commercially unprofitably. Dwarf cashew trees have attained yield of 5000 kg of nuts per ha with high density planting and appropriate cultural practices with proper fertilization and irrigation (Oliveira *et al.*, 1998).

6. Postharvest Handling Factors Affecting Apple Quality

Harvested fruits are highly perishable and become unmarketable after 24 hours under ambient conditions. Several factors affect postharvest quality of fresh cashew apple including use of improper harvest containers, delays between harvest and cooling, absence/or ineffective use of rapid cooling, poorly designed packing, temperature fluctuations during transport, high temperature during retail. As a non- climacteric fruit, the cashew apple must be harvested ripe for the fresh market, which imposes limitations on postharvest life and quality. Cashew apples are very bruise sensitive and should be hand harvested with the nut attached before abscission from the tree (Rosengarten, 1984). Several indices can be used for identifying harvest maturity such as colour, firmness, composition and specific gravity. However, the most common method of harvesting cashew apple is when it is still firm without sign of green colour and easily detachable from the tree by hand. At this stage the flavour, aroma and sugar content are at the maximum while acidity and astringency are the minimum. The cashew apple skin is very delicate and care must be taken to prevent damage.

7. Physico-Chemical Characteristics of Cashew Apple in Different Varieties

Cashew apple contains sugars, amino acids, tannins, ascorbic acid (vitamin C) and fiber. Cashew apples are very rich in ascorbic acid (240 mg/100 g) which is almost six times more than that of citrus fruits (40 mg/100 g). Besides vitamin C, cashew apple contains free soluble sugars most of which are reducing sugars. The apples vary in size, colour, taste and juice content depending upon the variety. Nair *et al.* (1979) reported the existence of yellow apples tended to be heavier, softer and less astringent than red apples. It means that the biochemical as well as physical characteristics of cashew apple varies in different cultivars. On maturity the physico-chemical changes occur within the cashew apple, though the biochemical changes are same for the different varieties but the quantity of accumulation of different constituents vary with the varieties. Physical and physico-chemical properties of different varieties of cashew apples are given in Tables 25.1 and 25.2.

References

Acland, J. (1971). *East African Crops*, London, Longman/FAO, 252.

Aliyu, O. M. and Hammed, L. A. (2000). A study on nut and apple development in cashew (*Anacardium occidentale*, Linn). *Nigerian Journal of Tree Crop Research*, 4(2): 1-10

Biale J, and Barcus, D. (1970). Respiratory patterns in tropical fruits of the Amazon basins, *Tropical Science*, 12, 93-104.

Broinizi, P., Andrade-Wartha E., Silva, A., Novoa, A. and Tores, R. (2007). Evaluation of the antioxidant activity of phenolic compounds naturally contained in by-products of the cashew apple (*Anacardium occidentale* L.) Cienc Technol Aliment, *Compinas*, 27(4): 902 – 908.

Figueiredo, R., Lajolo, F., Alves, R. and Filgueiras, H. (2002). Physical-chemical changes in early dwarf cashew pseudo fruits during development and maturation. *Food Chemistry*, 77: 343-347.

Figuiredo, R.W., Lajolo, F. M., Alves, R. E., Filgueiras, H. A. C., Maia, G. A. and Sousa, P. H. M. (2007). Qualidade de pedunculus de caju submetidos a aplicacao pos-colheita de calico e armazenados sob refrigeracao, *Pesq Agropec Bras*, 42(4): 475-482

Filgueiras, H., Mosca, J., Alves, R. and Menezes, J. (1999). Cashew apple for fresh consumption. Research on harvest and postharvest technology in Brazil, *Acta Horticulturae*, 485 (I): 55–160.

Garruti, D., Franco, M. Janzantin. and Awes, R. (2010). Effect of harvest season on the volatile profile of cashew apple (*A. Occidentale* L.) juice. *Proc. Interamer Soc. Trop. Hort.*, 45: 48–52.

Nair, M. K., Rao, E. V. V. Bhaskara, Nambiar, K. K. N. and Nambiar, M. C. (1979). *Cashew* (Monograph on Plantation Crop– I). Pub: CPCRI, Kasaragod, Kerala.

Oliveira, V., Crisostomo, L., Miranda, F. and Almeida, J. (1998). Produtividade de clonescomerciais de cajuciro anao precoce (*Anacardium occidentale* L.) irrigados no municipio de Mossor-RN, Fortaleza, EMBRAPA-CNPAT, *Comunicado Tecnico*, 14,6.

Pratt, H. and Mendoza, D. (1980). Influence of nut removal on growth and ripening of the cashew apple. *J. Am. Soc. Hort. Sci.*, 105: 540–542.

Rosengarten, F. (1984). *The Book of Edible Nuts*, Pub. New York, Walker.

Yahaya, L. E., Aliyu, O. M., Hammed, L. E. Yahaya L. A. and Aroyeun, L. A. Hammed S. O. (2010). Variation in the physicochemical characteristics of cashew (*Anacardium occidentale* L.) apples with maturity. S. O. Aroyeun Remove suggestion. *British Food Journal*, 112(1): 93-99.

Cashew: Improvement, Production and Processing (2017) *Pages* **433–450**
Editor: **Prof. P.L. Saroj**
Published by: **ASTRAL INTERNATIONAL PVT. LTD., NEW DELHI**

26

Mechanization in Indian Cashew Processing

D. Balasubramanian

1. Introduction

Cashew processing is deemed to be sunrise sector of the Indian economy owing to its exponential growth since its inception and socio economic impact, specifically on employment, income generation and gender equity. Industrialized agriculture and food production rely on efficient utilization of energy to carry out the desired operations and obtain high processing output through mechanization towards safe storage of agricultural products and conversion processes that create new forms of food. Cashew (*Anacardium occidentale* L.) is one of the horticultural crops and is often referred to as 'wonder nut' having potential to provide source of livelihood for the cashew growers, create employment opportunities and generate foreign exchange through exports. Cashew kernels are a high value luxury commodity with sales growing steadily at an annual rate of seven per cent, with every expectation that the market will remain strong. The cashew industry ranks third in the world production of edible nuts with a total production of 2.8 million tonnes of raw nuts during the last fiscal and its estimated value in excess of US$ 4.19 billion (FAO, 2012).

The engineering phase of cashewnut processing deals with the processes involved to change or alter the raw cashewnuts by the application of various unit operations and the machinery to the desired level. Traditionally, processing operations were performed manually by the skilled personnel in India, the world's largest producer of cashew kernels. Although various mechanized equipments were developed globally for different operations in the line of processing during 60's, it could not attract Indian processors primarily due to twin reasons *viz.*, availability of adequate work force, possessing required skill to extract whole kernels which fetched premium price at consumer level and also due to lower performance of the processing machines. Cashew business expanded tremendously over a period of time, providing more opportunity for the new entrepreneurs to plunge in to this business. Spurt in the cashew value chain and employment creation in other sectors led to non availability of skilled work force in this system and presently these industries are gradually progressing towards mechanization. This chapter encompasses the status of mechanization in Indian cashewnut processing system and analyses the challenges ahead.

2. Cashew Processing

Harvesting and Collection of Nuts

Cashewnut is a seasonal crop and is harvested from March to May in the east and west coast of India. Presently, harvesting of raw nuts is done in four different ways, they are: (i) Collection after natural drop (ii) Thrashing by stick and collection (iii) Climbing on the tree and thrashing and (iv) Shaking the tree. Mechanization of harvesting cashewnuts is yet to be practiced not only in India, but also in cashew growing countries in the world. An experiment conducted in the farms of Sindudurg district of Maharashtra (Dalvi *et al.*, 1992) clearly indicates that about 72 per cent of the farms follow harvesting the nuts by thrashing by stick and collection method. Certainly, this will increase the percentage of immature nuts supplied to the industries. Only 16 per cent of farms follow the collection of nuts after natural drop. Besides climbing on the tree and thrashing and shaking the tree to harvest the nuts amounts to 10 per cent and 2 per cent respectively. These two methods undoubtedly increase the immature nuts harvest and leads to inferior quality in final produce. Harvesting by thrashing or climbing also results in dropping of young flowers and ultimately reduces the production. Besides, the nuts have to be harvested at right stage. The colour of the nuts changes from brownish green at fruit set to light green at ¼th growth apple stage and thereafter turns grey irrespective of varieties. This is the right time to harvest the nuts. The different harvesting period of cashew in the world ensures the supply of nuts around the year. This helps to have control on imported nut price and the Indian processors are assisted by not carrying long term stock.

The season of peak out put for cashew growing countries are:

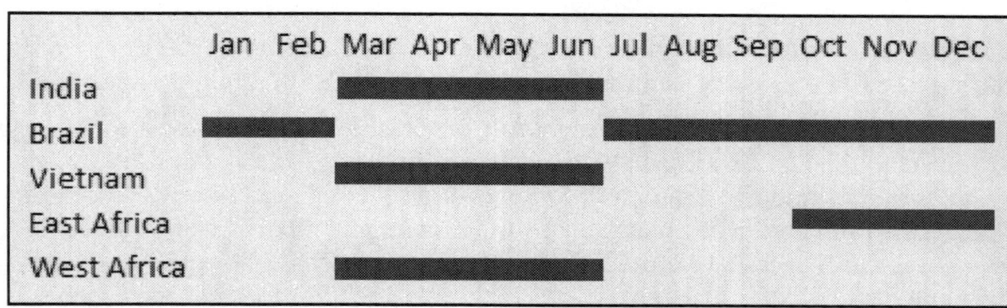

Grading of Cashew

Figure 26.1: Rotary Sieve Grader for Raw Cashewnuts.

Raw nut grading is the first step in mechanization of cashewnut processing. Size based distribution of nuts to be processed is a reasonable indicator towards product standardization. Grading is a process of separation of mixtures in to their more homogeneous parts (fractions). In fact, providing graded nuts can improve efficiency of subsequent operations. Grading nuts at preliminary level can ensure control over kernel drying cycle for better white whole kernel recovery and minimize workload at kernel grading stage. Compartmental sieve grader developed by Oltremare, Italy is the first of its kind (Oltermare, 2012). The kernel yield is primarily a function of raw nut, control over which is not much exerted prior to processing. Quality indicators of raw nut are moisture content, fractions of defective nuts, size and extraneous matter. Screens of various designs are widely used for size separation process. Considering the irregular shape of raw cashewnuts, a rotary sieve cylinder grading assembly is developed in view of minimizing clogging of nuts, compactness, self agitation and more opportunity to register with screen holes to grade raw cashewnuts based on its size. Concentric drum type rotary sieve grader developed at Directorate of Cashew Research, India has the advantage of compactness and higher grading efficiency (Balasubramanian, 2012). This unit consists of feed hopper with controller, concentric rotary sieve drums and drive mechanism. Blinding of nuts inside the sieve wholes can be prevented by rotary motion due to gravitational fall. Hopper bottom facilitate collecting the graded nuts just below each concentric rotary drum. The effective width of the nut is used as criteria to size grading. This grader is operated by a 2.0 hp electrically powered motor. Various machine parameters *viz.*, sieve and cylinder diameter, sieve cylinder length, slope of the cylinder and its rotational speed and material parameter *viz.*, bulk density and feed rate were optimized for better grading efficiency. Operational capacity and grading efficiency were found to be 292 kg h^{-1} and 97.58 per cent respectively.

Drying of Cashew

Certain cashew growing region like North Eastern Hilly regions, harvesting of cashewnuts coincides with monsoon rains, affecting its quality, if remain in the field for longer period. Freshly harvested cashewnut can be successfully preserved by reducing their moisture content to a level that will discourage the activities of

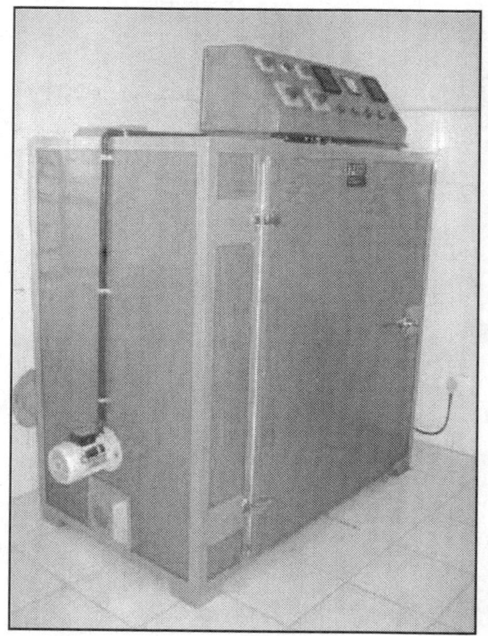

Figure 26.2: Dual Mode Dryer – Drying Chamber and Heat Exchanger.

micro-organisms and fungi from deteriorating them. Generally, microbes can be inactivated by reducing the moisture content of agricultural produce to less than 10 per cent. Ohler (1966) reported that moisture content above 8.0 per cent (d.b) for raw cashewnut will result to deterioration. Hence, harvested cashewnuts either affected by rain or not, need to be stored in dried condition holding moisture content less than 8 per cent to prevent attack and deterioration by external agents like micro organisms and fungi.

Raw cashewnuts after harvest has moisture content in the range of 14 to 17 per cent which should be dried under sun for 2 to 3 days to reduce its moisture to safer level (Ohler, 1966). Direct sun-drying and the use of solar driers depend on the intensity of the sun energy to heat up the air and effect drying. Dual mode dryer exclusively developed for raw cashewnuts is suitable for cashew growing regions wherein harvesting coincides with monsoon rains to extend its shelf life (Balasubramanian and Sandeep, 2011). Dual mode dryer consisted of three major components *viz.*, drying chamber, aspirator and heat exchanger. Its performance was evaluated by exposing raw cashewnuts at an average drying temperature of 65°C for safe drying. Rapid diffusion of moisture observed in the beginning and slowed towards end of drying. Time required to reduce its moisture level from initial moisture content of 20.44 per cent to safer level of 8.00 per cent d.b was in the range of 2.84 to 4.51 h irrespective of source of power, size of the nuts and position of the tray. Energy required for desired moisture reduction was found to be 32.93 MJ for electrical system and 201.71 MJ in the case of cashew shell cake as bio-fuel. Cost of drying raw cashewnuts using dual mode dryer is worked out to be Rs. 0.44 per kg.

3. Conditioning of Raw Cashewnut

Mechanized Drum Roasting Machine

Kernels obtained by drum roasting mode of processing are very much preferred in certain countries for its taste. Moreover, drum roasted nuts ease kernel peeling as evident from qualitative (whole kernel recovery) and quantitative efficiency. Drum roasting process for conditioning raw cashewnuts is followed in Kerala, Tamil Nadu, Andhra Pradesh, Odisha, West Bengal and NEH regions for several years. About 64 per cent of the total processing units available in India follows drum roasting process (Balasubramanian, 2000). Highly skilled personnel's are engaged in drum roasting operations *viz.*, feeding the nuts, rotate the drum, control fire beneath and to pull out burning nuts from heat source and the coordination of these operators plays vital role in proper roasting and extraction of white whole kernels. Processors are interested towards mechanization of drum roasting process for raw cashewnuts due to growing problem of labour scarcity especially skilled work force to operate drum roasting.

Raw cashewnuts exposed to various temperatures in muffle furnace revealed that better roasting *i.e.*, whiteness and yellowness index for cashew kernels could be achieved in the range of 350°C to 450°C. Based on the physical properties of the raw cashewnuts and roasting characteristics, a compact type drum roasting machine was developed and processing parameters were optimized for better kernel quality for various sized nuts. It consists of revolving drum, heating element, a bucket elevator, drive assembly and smoke outlet. This roasting machine could be operated by 1.0 hp 3-phase electric motor and its estimated output is around 100 to 120 kg h^{-1}.

Oil Bath Roasting

Oil bath units consists of a large rectangular bath filled with CNSL mounted over a grate furnace which uses spent cashew shell as fuel. Pre-conditioned nuts are lifted by bucket elevator and transferred in such a way that it forms a thin layer on an endless conveyor chain with cross ribs. These nuts are carried all along the bottom of the oil bath where CNSL is maintained at a temperature of 190 -195°C and discharged above the oil level at the other end in a short period of 10-15 sec. Residual oil on the surface of the roasted nuts is removed by spinning action in the centrifuge. This mode of processing is now obsolete in India due to huge investment on machinery and discoloration of cashew kernels.

Mechanized Steam Conditioner

Conditioning of raw nuts *i.e.* alternative wetting and drying process, facilitate extraction of whole kernel during mechanical shelling (Balasubramanian, 2006). Majority of the processing units utilizes 'twin bottle type steam boilers' for the purpose of conditioning raw nuts. Steam boiling of raw cashewnuts is a batch process. It consists of a cylindrical drum with a hopper on the top to feed raw cashew nuts. A boiler is used to generate steam and fitted with pressure gauge. Generated steam is passed into the cylinder through a pipe at the base and connected to a perforated central stem and laterals inside the drum. A valve provided facilitates to control the supply of steam to the nut chamber. At present, centralized steam generation system with multi chamber (3 nos.), increased the production rate and utilize the fuel economically. Manual transfer of raw cashewnuts in to the nut chamber for conditioning process is now replaced by bucket elevator, saving time and labour. Rotary type 'Steam kettle' ensures uniform conditioning process and preferred for handling bulk quantity (Fitzpatrick, 2011).

Figure 26.3: Single Cylinder Steam Boiler.

4. Shelling of Conditioned Cashew

Demand of highly skilled work force and non availability of adequate labour for shelling *i.e.* extraction of edible kernel from conditioned raw cashewnuts either by roasting or steaming led to mechanization of this operation. The mechanization introduced in cashew industry has been in the form of cutting apparatus for nuts in Mangalore (Subarao, 1981). Various decorticating methods, which have been tried in the past, are i) Explosive decompression of the nuts by which the kernel emerge through a natural point of weakness of the shells; ii) Freezing the nuts and break by standard centrifugal nutcracker and iii) Cutting the shell around its major axis to allow the kernel to be exposed.

During the development of single nut cutter, combined cutting and wedging operation in the plane of symmetry of the nut *i.e.* previously roasted under controlled conditions was employed for splitting along the natural plane of weakness in the shell. Later, the hand lever operation is converted to foot operation, leaving both hands free, one for the transfer and location of the nut and the other for the adjustment of the pillar height to suit the size of the nut. In oltremare system, Italy, the shells were cut longitudinally and separated by a pair of gripers freeing kernel with the outturn of 80 per cent of whole kernels. In another version, the foot operated lever pushed the anvil upwards against the cutting blade, which is held in an adjustable slide and on release returned upwards by a spring (Hall, 1965). The shelling machines used in Thailand and Myanmar, has two sets of blades, one concave and the other convex, with a divider in the middle. The concave blade with frame is fixed vertically to the table with the sharp edge on top. The other blade is fixed to a metal rod, which is capable of moving up and down through a holder above. When the roasted nut is introduced on bottom blade with the convex side of raw nut down and pressing the pedal, the blades cut the shell. Two halves of shells are plied apart from the right side to extract the kernel by means of pin fixed to the wooden handle (Mathew, 1995).

A power-operated cashew nut sheller is developed based on the principles of compression and shear. It consists of four sections: (i) feeding section, (ii) shelling section, (iii) discharging section and (iv) power transmission section. Feed hopper and horizontal screw conveyor provided positive feed of the roasted nuts to the shelling section. Roasted nuts are shelled between two wooden discs, of which, one is stationary *i.e.* fixed to the casing and the other one mounted on the shaft. The rotating disc is a spring loaded one to exert sufficient pressure to compress and shear the roasted cashewnut against the stationary disc. The rated capacity of the sheller was observed to be 18 kg h^{-1} of roasted nuts with shelling efficiency of 70 per cent. Yield in terms of whole and broken kernels are 22 per cent and 28 per cent respectively.

The Kerala Agricultural University (KAU) model is a manually operated cashew nut decorticator having top blade assembly, 2-blade bottom assembly and a linkage assembly fitted on a worktable. The 2-blade bottom assembly is made to slit the convex portion of the cashew nut and split the shells after cutting (Joby bastian and Jippu Jacob, 1994). The steamed cashew nut is placed in between the top and bottom blades with its notched portion facing the top blade by the left hand. Then the pedal is actuated by foot, which in turn operates the bell crank lever and pushes down the top blade assembly. Later, the release cam is operated by the right hand through the handle provided to move the bottom blades apart to tear out the shell and thus to release the kernel. Shelling efficiency is 88 per cent of whole kernels and the operational capacity is 900 nuts h^{-1}.

The single nut cutter developed by Mechanical Engineering Research and Development Organization (MERADO), Council of Scientific and Industrial Research, Kalamassery, Kerala to shell the roasted nuts (Anonymous, 1997), separates kernel by hand cum foot-operated mechanism. The decorticator can conveniently be mounted on a work table and worker can sit in front to operate it with minimum drudgery. The unit splits the nuts between two horizontally mount blades, specially shaped to suit the contour of the raw nut. Engaging a pair of horizontally mounted blades, of which one side is splitting type and other side in pair, can conveniently separate the outer shell. The upper blade is positively fixed in the same axis to that of sliding blade and the lower one is splitting type. A continuous centrifugal type cashewnut sheller developed at Indian Institute of technology (IIT), Kharagpur has a positive feeding mechanism. The shelling section consists of two wooden discs, of which one is stationery and the other is rubber padded. Shelling occurs due to differential speed of the discs and discharged by gravity force (Jain and Kumar, 1997). This has operational capacity of 18 kg h^{-1} and shelling efficiency is 70 per cent, but whole kernel recovery is 50 per cent only.

A power driven cashewnut detacher developed at Rukumani Adithya Memorial Polytechnic uses cam as the major component to transmit the mechanical energy to vertically reciprocating blade to cut open the shells. Cams are linked to Geneva crank and keyed to main shaft. The cams are keyed in such a way that eccentric holes are opposite to each other, thus pushing the bottom blade up while the top blade moves down for better penetration of nut. Two pairs of sprockets with a rubber flat belt in between are provided with cashew nut holders at equidistance.

The holders are designed for bigger size nuts. A 0.25 HP electrically operated motor operates pulley through reduction wheels. When the main shaft is driven,

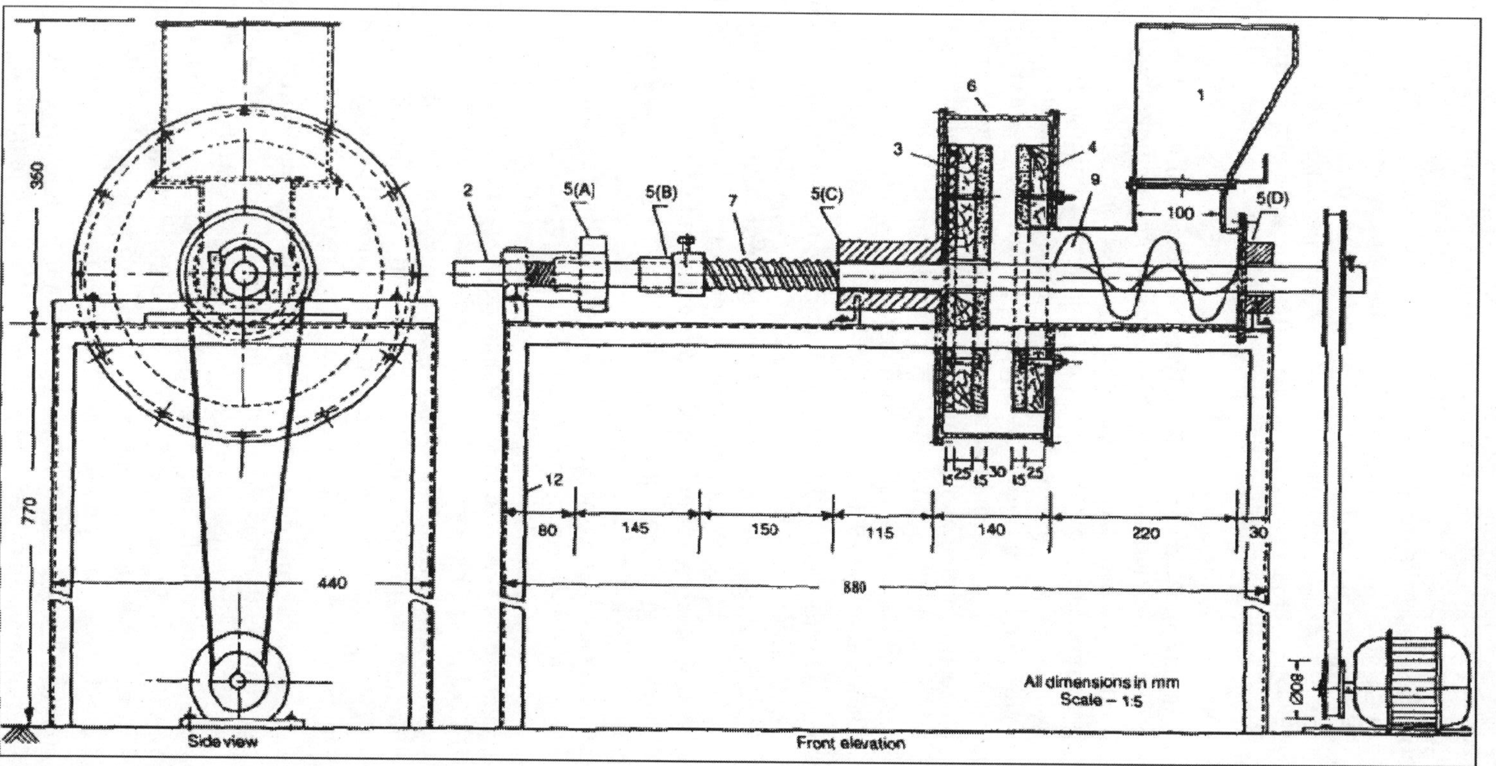

Figure 26.4: Continuous Type Cashewnut Desheller.

the Geneva crank engages the wheel slots and drives the sprocket for indexing the cashew nut holder. After indexing, the periphery of the crank wheel engages with the wheel radius on the sprocket and locks the indexing mechanism. Later, the cams are actuated to make the top blade to penetrate the shell. During this process, the bottom blade is positioned to give support at the bottom of the nut for better penetration. Upon completion of cutting, the cam moves back, releasing the nut and making the mechanism for next indexing. Due to the tension developed while moving on sprocket periphery the split nuts will be ejected. Since the cashew holders are designed for big size nuts, it is not suitable for medium and small nuts, which forms major portion in the cashew lot. The output of the unit is theoretically calculated as 32 nuts per minute.

In radial arm type cashew kernel extracting machine developed at Directorate of Cashew Research, Karnataka, single operation enables cashewnut holding and subsequent splitting. This consists of i) Radial type holding arm; ii) Radial type splitting arm; iii) Twin blades; iv) Split blade link; v) Drive assembly and vi) Nut size adjuster. Raw cashew nut is placed in between blades with its notched portion facing the holder blade and convex side resting on the edges of twin blades. Slipping of nuts lead to damage operator's hand during cutting as it is difficult to position the nut on the sharp edge of single blade. The lower part of the twin blade aids in proper positioning of the nut. When the pedal is pressed, sliding arm moves front and holds the nut. Further pressing make the cam to slip and allow the second cam to push the sliding rod to rotate the disc to a set angle. This helps the outward movement of twin blades resulting in splitting of shells. The compression spring provided ensures the whole assembly back in position after the release of force applied on pedal (Balasubramanian, 2010).

Figure 26.5: Radial Arm Type Cashew Kernel Extractor Tomatic Desheller.

Automatic cashew desheller developed by Aneesh enterprises, Kollam, Kerala consists of the following principal components i) Hopper with disc assembly; ii) Conveyor belt; iii) Gear box; iv) Cutting and splitting blade assembly; v) Nut holder and vi) Pusher and catcher mechanism. Deshelling of conditioned cashewnuts *i.e.* either Steam Boiled (SB) or Drum Roasted (DR) is basically done by holding between two blades assembly made out of high carbon

Figure 26.5: Automatic Desheller.

steel and split open by lifting one of the twin blades. Nuts are regulated from hopper through revolving off-set disc assembly Singulated nuts from hopper are conveyed forward towards deshelling point by a moving belt (various size belt need to be replaced to suit specific grades of nuts). Each nut is held between two blades and split open by lifting one of the off-set blades. Shelled nuts and kernels are cleared by a pusher to clear the space between two blades for the next nut in conveyor line to be shelled. Basically impact and shearing force is used to deshell the treated cashewnuts. Cutting and split blade assembly is adjusted according to the size of the nut *i.e.* width of the nut for better shelling efficiency.

Figure 26.6: Mechanized Desheller.

In the top driven mechanized sheller, conditioned nuts in the feed hopper are lifted up vertically by sliding component and transferred to the funnel. After positioning the nuts in the channel, nut is pushed down by a shoe against V- shaped blade. Due to the application of high force, nuts slits around its contour and split open by the blades provided on either side. Each machine is provided with two such shelling mechanisms deriving mechanical power from 0.37 kW single phase electric motor (Balasubramanian, 2012). Operational capacity is found to be in the range of 7.93 to 10.18 kg h^{-1} with whole recovery 98.63 per cent.

5. Unpeeled Kernel Drying

Drying of unpeeled kernels loosen 'Testa' a brown layer surrounding the edible kernel. Dryers employed in majority of the Indian cashewnut processing units have either natural or forced convective system (Anonymous, 2012). Accounting the problems of non uniform drying, high thermal energy loss, increased cost of drying, reduce environmental degradation due to the existing drying system, a novel dryer which works on the basic principle of heat exchange of super heated steam through radiator and fins assembly is developed. It controls air temperature more precisely ensuring uniform drying and minimize the energy during the

Figure 26.7: Steam Assisted Cashew Kernel Dryer.

process. Assessment on the performance of this steam assisted dryer indicated that it improves peelability and quality of cashew kernels in terms of surface colour. Significant difference was found in the peeling efficiency for cashew kernel dried up to 9 h. Total energy required to operate steam assisted cross flow dryer was found to be 16.49 per cent lesser than existing convective type dryer.

6. Peeling

Peeling is the process of removal adhering testa layer surrounding the edible kernel. Various types of automatic peeling machines *viz.*, brush, abrasion and impact type are the recent introduction in the Indian cashewnut processing industry. Brush type peeling machine is indigenously developed and has rotor with stiff bristles to peel off the kernel testa after steaming. Steam treated and dried unpeeled cashew kernels are mechanically peeled and its operation capacity is 300 kg h^{-1}. Peeling efficiency recorded as 84 per cent with whole kernel recovery of 69 per cent (Anonymous, 2012).

Figure 26.8: Brush Type Mechanized Peeler.

Abrasive type peeling machine is also developed in India and it consists of feed hopper, conveyor, peeling drums, compressor and products outlets. Dried unpeeled kernels are humidified to increase its moisture content to specified level. During peeling operation, unpeeled kernels are subjected to abrasion against inner surface of the drum due to centrifugal force. After steam treatment, moistened cashew kernels are dried in borma dryer for 2 h by maintaining at air temperature at 75°C. Peeling efficiency and whole kernel recovery are found out to be 65.28 and 56.34 per cent respectively.

Figure 26.9: Abrasive Type Mechanized Peeler.

Rotational speed of the abrasive drum and time of peeling are prefixed to control peeling operation inside abrasive drum. High pressure air generated through compressor aided in removal of loosened testa. Immediately after kernels are discharged to outlet, inlet opened automatically to allow preset quantity of unpeeled kernels to pass in to abrasion drum for peeling operation for continuous operation. Performance of the unit is worked out to be in terms of peeling efficiency (70.54 per cent) and whole kernel recovery (59.30 per cent).

An automatic peeling machine imported from Vietnam consists of feed hopper, bucket elevator, peeling drum, pneumatic separation, and a kernel grader. Peeling shaft is fixed eccentrically inside a perforated drum and revolves in opposite direction. Humidified unpeeled cashew kernels are subjected to impact and shear force by numerous spring loaded hook like structure mounted on central shaft. Afterwards, kernels are passed in to the pneumatic system to remove loosened testa and aspirated by a cyclone separator. Peeled and unpeeled kernels are conveyed to rotary kernel grader for segregation. Peeling capacity and whole kernel recovery found to be 273 kg h^{-1} and 80 per cent respectively.

Figure 26.10: Shear Type Mechanized Peeler.

7. Grading and Packaging

Cashew kernels are graded based on the wholesomeness, size and colour as per the specifications prescribed by the by Govt. of India under the export (Quality control and inspection) Act 1963, which recognizes 23 different export grades of kernels. Although manual grading is practiced in India, mechanized divergent fluted rollers are also used for grading the kernels based on size. Rotary or reciprocating sieve grader are mainly used for segregating low grade kernels *viz.*, Large White Pieces (LWP), Small White Pieces (SWP) and Baby bits (BB), Colour graders which works on the principles of spectral reflectance of light indicating the surface colour of kernel, is making breakthrough in cashew processing system in India.

Figure 26.12: Moulded Vacuum Packaging.

Figure 26.11: Reciprocating Sieve Grader.

Packaging

As far as, packaging is concerned, kernels are packed in 25 lbs, equivalent to 11.54 kg, tin containers and flushed with CO_2 called as 'Vita Pack' is followed by cashew processing industries for domestic market. In vitapack system, the tins are kept on the vibratory plot form and cashew kernels are filled by jerk movements. These tins are weighed and tightly packed. They are placed under a 'bell jar' and vacuumized initially and flushed with CO_2 as a protective measure against microbial attack. These tins are then sealed with tin lid and lead soldering. Afterwards, sealed tins are immersed in water to test leakage. Apart from 'Vita Pack' system, commercially available pouch packaging machines are also used for retail marketing.

Accounting the problem of strength to withstand the pressure of handling and transport, full proof against leakage of containers, shelf life of cashews in the containers, price of bulk packs, investments on machinery and demand of importing countries Moulded Vacuum Packaging (MVP) or Bag-in-Box is introduced in the line of processing. Flexi Packaging is gradually gaining momentum over 'Vita Packaging' and found to be cost effective mainly for export. Mechanised MVP system uses flexi pack made up of aluminium foils and high density polythene extruded in four layers as kernel containers. Cashew kernels are filled in flexi bag of 11.34 kg and sealed after infusion of N_2 and CO_2. Two such flexi packs are placed inside a carton box and strapped.

8. Cashewnut Shell Liquid Extraction

Cashewnut shell liquid (CNSL) is one of the by-products of cashewnut processing and extracted from steam treated shells to get additional revenue. Mechanized screw expellers of various capacities are utilized for the purpose of extracting CNSL. During the process, cashewnut shells are initially softened by exposing to steam and its pressure and period depend on the origin of raw cashewnuts. These steam treated shells are then transferred to screw expellers and high pressure exerted on the shell to release CNSL. Crude CNSL, thus extracted are filtered and allowed in large settling tanks for 48-72 h for sedimentation of associated solid matter. Later, clear CNSL is transferred to heating chamber and exposed to high temperature of 100-110°C to release the adhering moisture and cooled quickly. Heating and cooling process depends on the final product to be prepared from the CNSL as raw material. Research results indicated that the extraction of CNSL by means of tapered compression screw speed of @ 7-13 rpm, feeding rollers of transversal zigzag surface type and cylindrical casing @ 54-95 kg h^{-1} is in the range of 20.65-21.04 per cent. However, the residue from this method still contained significant proportions of CNSL, around 10 to15 per cent. Moreover, this method of extraction had higher levels of impurity, higher viscosity, lower thermo-oxidative stabilities and lower ebullition temperature.

9. Cashew Apple Juice Extractor

Preparation of alcoholic beverage from cashew apple is not encouraged in all cashew growing states other than Goa. Cashew farmers are unable to get additional revenue through this by-product due to its short shelf life and the presence of astringency. Juice should be extracted from cashew apple immediately after harvest and the hydraulic juice extractor developed serves the purpose. Non alcoholic products such as juice, squash, syrup and jam are simple to prepare and gradually gaining commercial importance at consumer level. A 2-Ton load hydraulically powered cashew apple juice extractor developed at Directorate of Cashew Research, India consists of a cylindrical vessel having capacity up to 10 kg of cashew apple, a piston and ram assembly with wooden flange to apply required load and a power pack system for fluid regulation operated by 0.25 hp single phase electrical motor. Preliminary trials revealed that juice extraction efficiency ranged from 81.6 to 89.2 per cent after 2nd pass. It is estimated that clear juice can be extracted from 200 kg h^{-1}. Cashew apple is highly fibrous in nature, blocking of perforations by pomace or fibrous material needs periodical cleaning for better juice extraction.

Figure 26.12: Cashew Apple Juice Extractor.

10. Impact of Mechanization in India

Selection of appropriate machines in the line of processing based on the operating capacity, electric power, skilled man power and market effects on profitability in a long run. As the cost incurred on raw material is more or less 70 per cent of the total processing cost, machinery purchased need to be utilized to its fullest capacity to make it economically viable. In the case of manually operated plants this is not as serious problem as the equipment is minimal and additional work force can be deployed for extra quantity as and when required.

Location of the plant is important in terms of access to raw material, factors influencing the operation of sophisticated equipment installed, storage of raw material, access to spare parts and access to utilities. A good knowledge on basic information about raw material *viz.*, expected yield at packaging, size of the nuts, the moisture and variation from time to time, the ease of peeling, the thickness of the shell and the oil content of the nuts in order to decide on the type of machines to be procured. Besides, careful planning is required to install sophisticated equipment in remote areas keeping in mind the communication facility and electric power supply. Technical personnel should be available or trained to operate the sophisticated machines without problem and attend to repair and maintenance work for continuous operation. It is important to assess the equipment options to consider workers ability and mind set to work with the equipment in the factory.

Purchasing of equipment involves capital investment, planning and management of resources and logistics. Further, it is very important to consider the economics of cashew equipment to assess long term market and not short term trends.

11. Issues and Strategies

In general, cashew processing developed as a low investment activity with minimal use of technology in India. In the past, it relied on low cost labour relative to the value of the product. The profitability of cashew processing depends largely on the proportion of kernels extracted without being broken or damaged. As for as the mechanization is concerned, Indian cashew processing system can be classified as i) Manual processing with semi automatic machine in certain stages; ii) Upgraded cashewnut processing and iii) Fully automatic processing with advanced techniques. There has been reluctance in the past to invest in capital intensive cashew processing in India. Technology was unable to provide a commercially viable solution to this problem for long time. Key issues which led to changes in the present system are i) Changes in the labor cost and demand for better conditions; ii) Decline in the operational efficiency of the labor with age; iii) Compensatory wages during non-working days; v) Non-preference of traditional industry by the younger generation; vi) Improvement in the country's economy providing opportunities in newer areas; vii) Wages insufficient to lead a urbanized life; viii) Monotonous job in the factory and these issues are central to developments in processing and new mechanized or automated technologies.

Mechanized processing expectedly increases the production and ultimately reduces the cost of operation. Technology will eventually make processing easier and will handle any kind of nuts, reducing premium price for nuts of some origin. But, presently the machinery introduced in the line of processing could not recover whole kernels equivalent to either manual drum roasting of steam treatment mode of processing which are quiet prevalent in Indian cashew processing sector. It requires high capital cost, high power requirement and technical knowledge to operate, repair and maintenance. It needs to be operated in shifts to make it economically viable and constant upgradation of technology is required for better performance. At this juncture, it is imperative that a systematic investigation on assessing the performance of these machinery and refinement would develop confidence on mechanization in lieu of manual processing in this country.

Although mechanization is gradually introduced at various levels of material separation, certain challenges are to be faced to remain as global leader in the cashewnut processing. In view of development of processing facilities in African countries, India would starve for raw cashewnuts. Research and development sectors should gear up their activities to address this long standing problem. Domestic production volumes must be sufficient to ensure a consistent supply of raw material. Significant difference in the investment requirements, knowledge on operation and maintenance of mechanized machines, qualitative and quantitative efficiency of machinery, energy requirement, depreciation costs, constant up gradation of technology and continuous monitoring, play vital role to make mechanization in cashew processing a feasible one. Management of calibrated nuts in shelling, control on humidification of unpeeled kernels and standardizing machine vision for grading cashew kernels are the major issues to be tackled. Sustainability of small scale processors against corporate sector needs policy changes for implementation.

Major constraints in developing mechanical devices for shelling cashewnuts are the tough and elastic nature of the outer shell wall and the peculiar shape of the cashewnut whose curvature of the sides varies considerably with individual cashewnuts. Avoiding the contamination of kernel by the corrosive CNSL and minimizing kernel breakage are the two major considerations while shelling cashew cashewnut. Primarily, the operational capacity needs to be as high as possible and the shelling efficiency is expected to be in the range of 85 to 90 per cent. As extraction of kernels in whole form is the aim in shelling process, due care has to be taken for increasing the whole kernel recovered and lesser output of broken kernel for the given quantity of conditioned nuts. Kernel intact obtained after shelling ought to be as low as possible. Design of shelling machine should accommodate a mechanism to position nuts precisely so that the shelling blades penetrate and split open exactly the line connecting two halves of the nut.

Blade penetration inside the shell while shelling has to be set very accurately, otherwise, it damages the edible kernel within the shell. Slit in the shelled kernel may not be visible at the end of shelling due to the presence of moisture, but it will break after drying and peeling of brown layer called 'testa'. In all respects, grading of nuts precisely based on the width is essential prior to shelling. Optical sensors may be used to adjudge the size of the nut and actuate the blade for calculated depth within the shell for effective shelling. Shelling operation need to be continuous operation for better production rate, therefore, kernel extracted is expected to mix up with shells having CNSL oozing out. In order to avoid kernel contamination with CNSL, conveying system should be incorporated at the outlet of the shelling machine for online separation of kernel and shell manually or mechanically. Besides, a pad should be provided to clean the adhering CNSL on the surface of the conveyor. Cost, power supply and installation space are other factors needs to be considered towards the economics of the shelling process.

Moisture conditioning prior to mechanical peeling is the major problem irrespective of various techniques adopted for peeling testa. Due to irregular shape of the kernel, mechanical peeling led to additional manual work. At present, processors are compromised with existing price range for the broken kernels as whole kernel recovery by machine is lower than manual system. Investigations on optimizing processing parameters related mechanical peeling need to be attempted to make mechanized peeling system economically viable in India. Single grading machine for cashew kernels accounting all the quality parameters *viz.*, size, surface colour and degree of smoothness can improve the qualitative and quantitative efficiency of cashew industry.

Cashew processing is beginning to develop into a modern industry in India. The increased competition for labour and advances in mechanization has changed the processing scenario in this country. Facilities provided to labourers in cashew factories may be better now, but are still not attractive for workers having alternatives in newer field due to the newer economy in this country. Investment and capital cost is rising and financial barriers for small and medium level processing is growing. As far as food safety, traceability and sustainability are concerned, strict enforcement of food safety standards by buyers is continuing to impact processors. Failure to meet the standards in the medium term may close markets. Compliance will involve investment, planning and good management. The trend toward the use of technology even in the limited form of cashew processing machines is making new demands of the owners and managers of small and medium size cashew processing plants. It is important that managers should enhance knowledge on working with machines and the factors which influence its success or failure. Environmental concerns will impact the cashew industry owing to energy and environmental audit on processing plants. Emissions into the air and water in particulate will be of more concern and development of highly efficient boilers and CNSL extraction plants will become imperative.

References

Anonymous (1997). Cashew Decorticator-Technology for Rural Development. Mechanical Engineering Research and Development Organisation, Cochin, Kerala.

Anonymous (2012). Directorate of Cashew Research, Annual Report, Puttur, Karnataka, India.104 pp.

Balasubramanian, D. (2000). Cashew processing industries in India- An overall analysis. *The Cashew*, 15(2): 14-20.

Balsubramanian, D. (2006). Improving whole kernel recovery in processing – Specific to Cashew nuts of Nigeria origin. Agricultural mechanization in Asia, Africa and Latin America, 37 (1): 58-64.

Balasubramanian, D. (2010). Design and development of radial arm type cashew kernel extracting machine" Agricultural Mechanization in Asia, Africa and Latin America, 42(2): 49-55.

Balasubramanian, D. and Sandeep, T. N. (2011). Performance evaluation of dual mode dryer for raw cashewnuts. *Journal of Plantation Crops*, 40(1): 28-39.

Balasubramanian, D. (2012). Performance Evaluation of Mechanized Shelling Machine. In. PLACROSYM XX. "Mechanization in Plantation Crops" conducted by United Planters Association of South India, Tamil Nadu from 12-15 December, 2012.

Bhuvana Chandran Pillai, (2009). Patent application No. 201/CHE/2009. India

Dalvi, V. D., Thakore, C. G. and Borude, S. G. (1992). Economics of production of cashew nut in Sindhudurg district in Maharashtra. *The Cashew*, 5(3): 15-17.

FAO, (2012). http: //www.fao.org/corp/statistics 2012. *Production statistics of in-shell cashewnuts in the world.*

Fitzpatrick, J. (2011). Cashewnut processing equipment study - Summary. African Cashew Initiative. Eds.Rita Weidinger, African Cashew Initiative, Accra, Ghana. 44 pp.

Hall, F.J. (1965). Cashew Nut Processing – Part II: Trials of equipment in Western Nigeria. *Tropical Science*, 8(4): 160-171.

Jain, R.K. and Kumar, S. (1997). Development of a cashew nut sheller. *Journal of Food Engineering*, 32: 339-345.

Joby Bastian and Jippu Jacob (1994). Cashew nut decarticator. *Cashew Bulletin*, 31(9).

Mathew, A.G. (1995). Improved shelling machine. *Cashew Bulletin*, 32 (4).

Ohler, J.G. (1966). Cashew processing. *Tropical Abstracts*, 21(9): 1792-2007.

Oltremare (2012). http: //www.oltremare.biz. Raw cashewnut cleaning and calibration system.

Subarao, M.V. (1981). Welcome address. In. National Seminar on Cashew Industry. *Cashew causerie*. Vol 3(4).

Subbarao, N.V., Krishna Prasad, K.M.M and Prasad, V.S.R.K. (2011). Review on Applications, Extraction, Isolation and Analysis of Cashew Nut Shell Liquid (CNSL). *The Pharma Research Journal*, 6(1): 21-41.

Cashew: Improvement, Production and Processing (2017) *Pages* **451–463**
Editor: **Prof. P.L. Saroj**
Published by: **ASTRAL INTERNATIONAL PVT. LTD., NEW DELHI**

27

Cashewnut Processing as Microenterprise

D. Balasubramanian

1. Introduction

Large-scale industries serve as backbone of national economy, while small-scale processing in rural area act as a survival backbone of majority of people. Rural India is now a mere producer of raw material instead of a producer cum primary processor. Rural industrialization contributes a lot to creation of employment and strengthening of regional economy. In the event of shrinking of farm size and migration of rural youth, it is the need of the hour to introduce micro level cashewnut processing either on or off farm as one of the strategies for the development of rural India. In view of balanced development of agriculture and industry to lift the economy from the vicious circle of poverty and unemployment, agro industries are the only viable alternative to promote rural development in general and cashew sector in particular. Of the world production, India produced approximately 7.57 lakh tones (28 per cent) and processed around 15.87 lakh tonnes (56.78 per cent) during the year 2013-14 (Anonymous, 2014). India exported 1.26 lakh tones of kernels earning foreign exchange to the tune of Rs.5096 Crores (FAO, 2002). Cashew industries are located either in urban area or well away from production catchments paving way for the intermediaries to involve in the cashew business chain. In fact, market oriented production often goes through several level of processing and carried out at or near the source of raw material so that more product value can be retained locally. Although the focus of cashewnut processing is taking the dimension of multiple product development an attempt to improve opportunities for processing at rural area *i.e.* production catchments is imperative.

2. Mode of Cashewnut Processing in India

India is the foremost in processing cashewnuts and majority of processing facilities are located in the east and west coast region of this country. These units are categorized under medium to large scale industries based on the capital investment and installed capacity. In early methods, the dried nuts are shelled manually after open pan roasting. The kernels extracted by this process had scorched tip. Oil bath method is a semi-mechanized one and requires huge investment for machinery and infrastructure development. In drum roasting process, the nuts are fed into an inclined rotary drum, wherein the exuding oil ignites and burns, thus charring the shell (Balasubramanian, 2000). The total loss of the shell liquid and exposure of workers to heat and acrid fumes are the major shortcomings of this method. Steam boiling is a cost effective process wherein raw nuts are subjected to regulated pressure and comparatively low temperature sufficient enough to bring out the shell liquid. The dexterity of women laborers in extracting whole kernels fetching premium price is one of the major reasons for the phenomenal growth of cashew industries in this country. A cross flow drier is commonly used to reduce kernel moisture to facilitate peeling operation. The highly mechanized industries in other countries suffer due to lower whole kernel recovery, which fetches premium price (Ohler, 1979).

3. Rawnuts and Farmers Role

In processing sector, irrespective of mode of extracting edible kernels, the raw nut share amounts to 70 per cent of the total cost excluding purchase tax (Balasubramanian and Rema, 1996). The current price of raw nuts has crashed to the value a decade ago. But the price trend of kernel in the international market reveals that the value is stable for several years and the decrease in price is compensated by the positive exchange value for US dollar. Unlike other crops, price structure is not followed for cashew based on quality standards. This has led the farmers to harvest nuts before full immaturity resulting quantitative and qualitative loss in the end product after processing. Accounting these problems, an urgent step has to be taken to improve the economic status of the cashew farmers exploiting strong domestic market.

4. Demand of Cashew Kernels

The consumption of kernels at domestic level showed an increasing trend over the last decade. Domestic consumption recorded an all time high of 2.59 lakh tonnes *i.e.* 72.62 per cent of total kernel production in India and it is a prime factor for promoting micro level cashew processing in this country. Besides, cashew market data reveals that kernel demand in India is unexpectedly growing at the rate of 8.75 per cent per annum. Share of domestic production in exported kernels also declined from 45.5 per cent in 1991-92 to 33.3 per cent in 2013-14, accounting a sharp fall of 27 per cent. It is evident that this changing scenario in domestic consumption favour micro level processor to continue in the value chain of cashew.

5. Processing at Microenterprise

Potential and Opportunities

Processing of cashew in micro level normally confined to use locally available raw materials and small machinery giving emphasis on integration of rural man power. It requires less power and infrastructure and can operate as everything under one roof. It is a simple and low cost technology for processing either at farm or in the proximity of production catchment (Table 27.1). Technological design maintains a balance between human energy and machine energy to strengthen work opportunities and asset-based creation in the event of modernization. The major emphasis is given on integration of the rural manpower at farm or household in to overall system.

Operation

Raw nuts are either sourced from own or neighbouring farm and/or from local traders. Procured nuts are dried under sun in single layer for a day or two to reduce

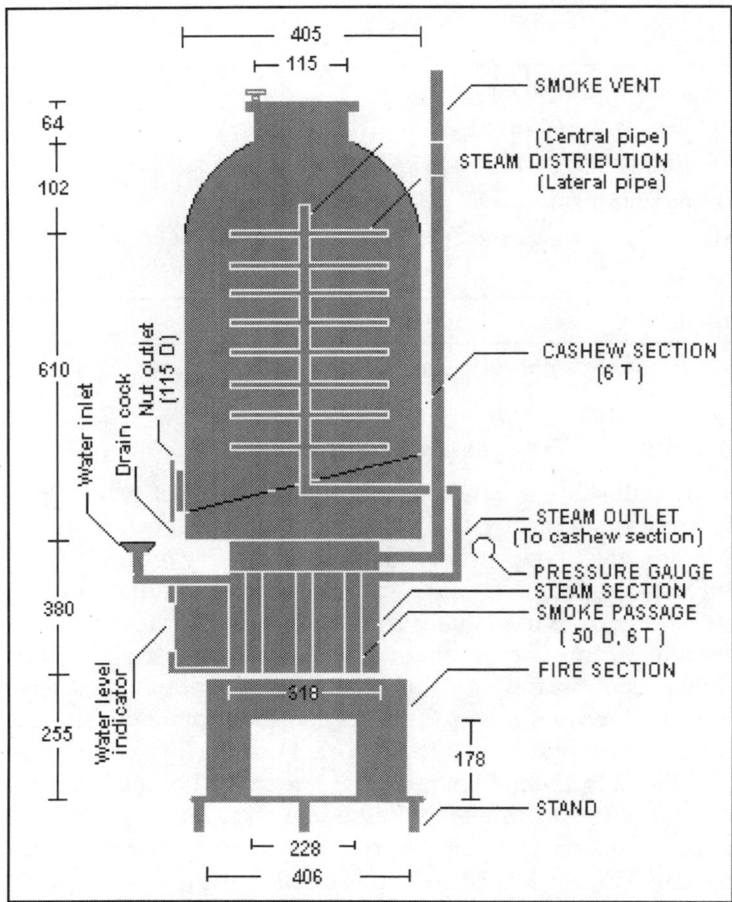

Figure 27.1: Sectioanal View of Steam Boiler - Vertical Type.

the moisture to safer level. Steam boiling technique is predominantly followed and four types of boilers *viz.*, direct steaming vertical type, Indirect steaming horizontal type, Indirect steaming vertical type and compact type are being used for nut conditioning. After nut conditioning, it is dried in ambient environment to make nut hard to suit for hand cum pedal operated type shelling machine. The shelling capacity is around 20 kg per head per day and the whole kernel recovery ranges between 80 to 90 per cent. Kernels after extraction need to be exposed in hot air to make it amenable for manual peeling. Trained personals are engaged with conventional type driers to avoid kernel scorching. An electrically powered tray drier with capacity up to 25 kg per batch is also used for the purpose. Kernels are subjected to hot air maintained at 70°C for 6 to 8 hours. Peeling is done manually and the operational capacity ranges between 4 to 6 kg/head/day depending on their skill and experience. After skin removal, the kernels are graded into wholes and broken and packed in polythene bags. List of machinery required for running a micro level cashewnut processing with cost is given in Table 27.1.

Table 27.1: Cost of Machinery for Micro Level Cashewnut processing

Sl.No.	*Machine/Tools*	*Cost/Unit (Rs.)*	*Required Qty.*	*Total Cost (Rs.)*
1.	Cashew steam boiler(40 kg/batch)	36500	1	36500
2.	Hand cum pedal operated Sheller	5000	2	10000
3.	Cashew kernel drier (500 watts) 25 Kg capacity	46000	1	46000
4.	Weighing scale (2 kg)	3500	1	3500
5.	Plastic crates and accessories	2500	–	2500
6.	Plastic sealing unit	1500	1	1500
	Total			**100000**

Value excludes packaging and forwarding cost and with reference to year 2012.

Optimum Values of Processing Parameters

In view of high share of raw cashew nut in the cost of processing, declining trend in the unit value of raw nut price, nut harvest at immature stage, socio economic position of cashew farmers, on farm processing technology is optimized for critical processing parameters to obtain better whole kernel recovery. Product value method is employed to evaluate the on farm cashewnut processing technique. In this method, quantity of kernel obtained is valued by multiplying the prevailing kernel price. Based on the index value obtained and optimization technique, the critical values at which better kernel grades could be obtained using steam boiler and kernel drier were worked out. Data revealed that the nuts could be conditioned at 70 to 75 PSI for 13 to 15 minutes in small scale steam boiling unit for maximum recovery of whole kernels in shelling. In the same way, extracted kernels could be exposed in hot air in electrically operated tray drier maintained at temperature in the range of 70 to 75°C for 6 to 7 hours to yield high value end product.

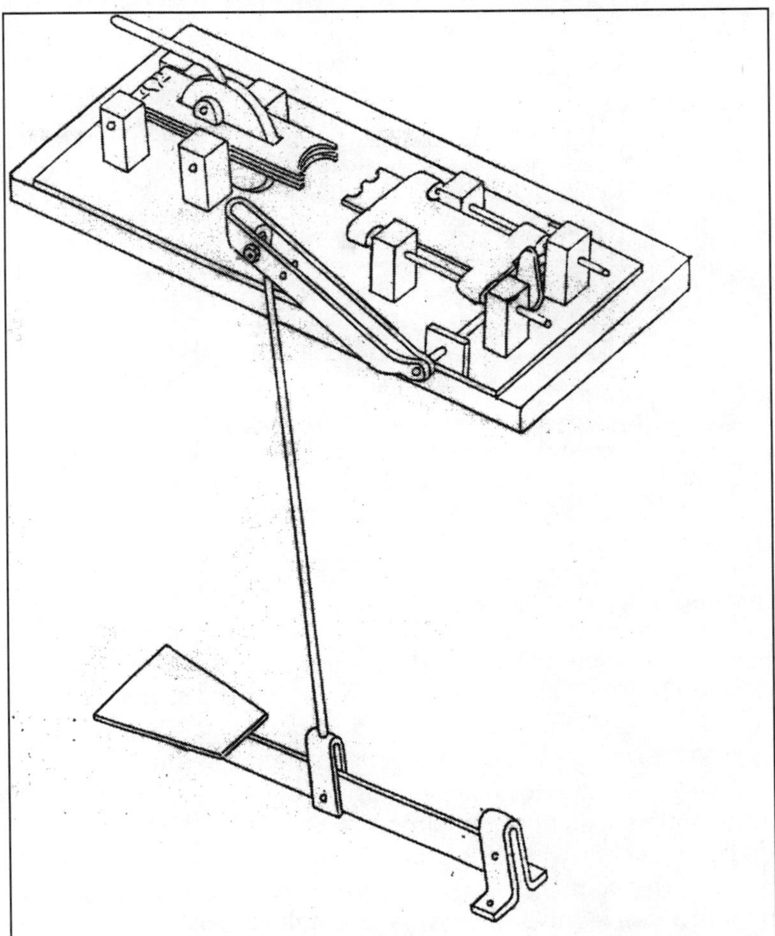

Figure 27.2

Energy Budget

Cost of production plays a vital role in cashewnut processing industry to make it economically sustainable level of production. Major components of manufacturing cost are utilization of fuel, electricity and labour. Data on energy consumption enable the management of cashewnut industry to develop strategies for better control of their production operations and serve as a tool for carrying out budgeting, forecasting energy requirements and planning plant expansion. Investigation carried out on computation of energy consumption of cashew processing system reveals that the total energy intensity in the cashewnut processing units varied between 0.21 and 1.161 MJ/kg depending on capacity utilization up to 1.0 T per day. Electrical energy intensity varied between 0.0052 and 0.029 MJ/kg, while thermal energy intensity varied from 0.085 to 1.064 MJ/kg. The two identified

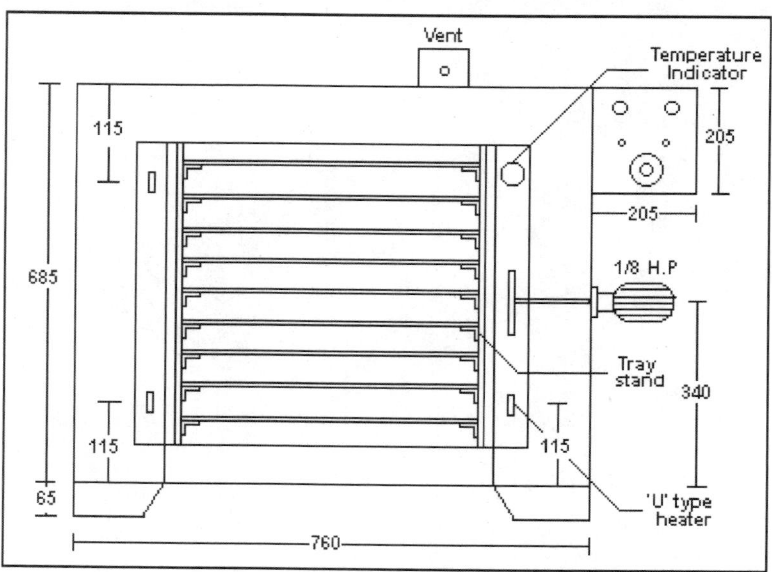

Figure 27.3: Sectional View of Kernel Dryer.

energy intensive operations in cashewnut processing are cashewnut conditioning and cashew kernel drying, altogether accounting for over 85 per cent of the total thermal energy consumption (Jekayinfa and Bamgboye, 2005). Similar investigation is carried out in Maharashtra, India which follows steaming as conditioning process following different methodology depending on the raw material, technological mechanization and availability of secured power supply (Mohad *et al.*, 2010) for micro level enterprise of cashewnut processing. It is revealed that on an average these units operate for about 227 days in a year depending on the rainy season and availability of raw cashewnuts. The average installed production capacity of the surveyed units is in the range of 60 to 120 kg per day and the women employees constituted the total work force. It is observed that, the energy intensity in the cashewnut processing units varied from 2.15 MJ/kg to 3.80 MJ/kg suggesting scope for better energy budgeting.

6. Kernel Market

Cashew kernels are generally consumed in its natural form but it has greater demand in confectioneries and bakeries. The cluster based cashew program (promoted by NABARD) also helps to trade kernels produced from small scale units. Kernel market is on upswing from March as processing starts during this period. The rates in the wholesale market speculate from Rs. 230/- to Rs. 300/- per kg depending on the grade. Other than kernels, the processing byproduct, cashew shell fetches Rs. 2.50 per kg. The kernel market is promoted through traditional fairs and helps to market the product directly to the consumers waiving commission agents and the distributors and their monopolized trade practice.

7. Merits of Running Micro Level Processing

Comparative advantages of establishing micro level cashewnut processing over large-scale industries are as follows.

- ✰ High capital investment for establishing large-scale industries (infrastructures and machineries) whereas processing is done in small-scale level requires less capital investment.
- ✰ Cashew nuts are available for 3 to 4 months due to seasonality of crop. Therefore, raw nuts have to be purchased during this period and hoarded for better utilization of processing facility throughout year. In case of small scale cashew processing, the installed capacity is around 12 TPA, requiring relatively lesser working capital.
- ✰ More technical manpower is requisite in large-scale industries, whereas in small-scale level, family members get employment in the non-farming season. Moreover, scattered manpower in a village can be utilized effectively.
- ✰ High level and effective management is a requisite in large-scale industries for better rate of return. In small scale processing, management and labour cost are less reducing cost of processing.
- ✰ Generate rural employment and improve economic growth of farming sector.

8. Economic Feasibility

Appropriate technology is not only the technology relevant to the needs for whom it is designed. It is socially appropriate, economically accessible and can be used and maintained in the local environment without causing damage. Down sized scale cashewnut processing is a household technology package tailored to escalate the economic status of farm sector. The feasibility of small-scale cashew processing units reveals that the net value addition per unit at farm level is worked out to be Rs. 23,700 and the value addition per unit weight of raw cashew nut is around Rs. 38. The economics for the unit having utilization capacity of 3 TPA is found out as Rs. 32,983 per month (Table 27.2). Besides, about 300 man-days per unit is generated during season.

9. Issues and Challenges

Procuring raw cashew is the largest component of the operating costs in cashew processing sector, a slight increase in cashew price adversely affects the entire economics of cashew processing. Traders and middlemen dominate the market for raw cashewnuts and kernels. The individual farmers are in a disadvantageous position as they are forced to sell the produce at a price determined by the traders/ leaseholders. Farmers are not interested to use the regulated markets due to taxes/ cess and as such quality based pricing system is not developed. Establishing supply chain for raw cashewnuts and developing quality standards could address this issue.

Table 27.2: Economic Feasibility and Cost Economics of Micro Level Cashewnut Processing

Feasibility		*Cost Economics*	
Number of cashew trees (Average)	100	Processing capacity/day (Kg)	40
Raw cashew nut production @ 6 kg/tree (Average)	600	Quantity of processing/year Kg)	3000
Receipt from RCN @ Rs 70/kg (Rs)	42000	Market price/kg (moisture loss – 3 per cent)	72
Shelling per cent (KPR)	25	Raw material cost (Rs)	216300
Receipt after processing (Wholes: Broken ratio is 4:1) (Rs)	63000	Labour cost (Rs)	12500
Receipt through Cashew shell @ Rs 7/kg	2700	Electricity, fuel and other expenses (Rs)	3000
Total income (Rs)	65700	Total processing cost (Rs)	231800
Net value addition (Rs)	23700	Sales (Rs)	330750
Unit weight benefit (Rs)	110	Profit (Rs)	98950
Value addition per unit weight (Rs)	38	Profit/month (Rs)	32983

Working capital is required for purchasing raw cashewnut inventory. Small processors have constraints in accessing working capital limits from the banks. Access to working capital would induce entrepreneurs to invest in cashew processing facilities. In this regard, new credit instrument need to be developed to tied over the collateral security of tiny enterprises. Awareness on quality and food safety standards need to be instlled in the mind of small scale processors. Machinery provider should be strengthened towards improving technical specifications. A low cost cashewnut shell liquid technique suiting to cooperative level to be developed and linked to the these processor. Cost effective package technique for better shelf life of kernels and setting up common service centre or facility provider for small-scale processors in respect of value addition and packaging has to be encouraged. Presently marketing is undertaken only through private channels where numbers of intermediaries are involved. In this context, state Government can set up Rural Marketing Development Corporation (RMDC) to handle the purchase and sale of kernels.

10. Status of Microenterprise: Indian Context

Cashewnut Processing under Women Self Help Group (SHG) in Maharashtra

The small-scale cashew nut processing technology followed in Sindhudurg district is developed in consideration of resources and need. The growth of micro level units (<3000 nos.) having processing capacity between 2 to 12 tonnes per season *i.e.* 3 months, indicates the technology adoption in this region. The financial support by the Government of Maharashtra and Khadi and Village Industry Commission for processing machineries is the key factors for the promotion of this sector in this region. Access to the supporting services brings success to the technology adoption.

Certain organizations *viz.*, Hegdewar prakalp and Ratnagiri Zilla Khadi Sangh are acting as an effective instrument to impart training skills. Training is conducted as vocational level through "learning by doing" to become proficient and confident in application of technology at work. Training program focuses on improving group activity, technical skill development, market support, value addition of low-grade kernels and cashew apple utilization (Balsubramanian, 2003b). The comparative advantages over large-scale industries, integration of rural man power and socio-economic improvement of farmer cum processor indicate that these micro enterprise are quiet successful and highly contribute to the rural development.

Micro-Enterprises in Cashew Processing in Tamil Nadu

'Desire to start own enterprise' is the main reason for investment on operating cashewnut processing as microenterprise by the SHG in Cuddalore district of Tamil Nadu. All the SHGs are women groups with 20 members in each group and all the members worked together to make a successful enterprise. These SHGs, managing the processing units are credit linked to a branch of Regional Rural Bank in the area. Marketing of the graded kernels is arranged through tie up with the local export houses and traders. The whole graded kernels are sold to the local export houses and the other low grade kernels like splits and butts are sold to the traders locally. The monthly net income worked out to Rs. 18107 and Rs. 30950 for the SHG units without and with boilers respectively (Yadav *et al.*, 2008). This variation is due to the difference in the number of bags of raw nuts processed per month. The profit is shared among the members of the SHGs over and above the wages paid to the members for the processing operations. Thus, the resource poor women could get access to financial services in terms of savings and institutional credit through SHGs. Convergence of efforts by stakeholders like financial institutions, State Government departments and credit cum savings groups led to creation of livelihood opportunities for the rural poor women. Potential for similar micro enterprises needs to be tapped by women SHGs in the cashew growing areas.

Women Cooperative Processing Unit in Kerala

Cashew, the multifaceted commercial crop has made significant contribution to the agriculture as well as socio economic scenario of Kerala state. Micro enterprises in cashew processing under Kudumbashree in Kasaragod district registered under independent society category. Twelve micro cashew processing units and a centralized grading and packaging unit are started as cluster under 'Safalam Cashew Processors' with financial support from Kudumbashree, banking sector and respective panchayats of the district. Cashew processing machinery and raw materials have been provided to 12-units functioning at different Panchayats. Comprehensive training program on primary processing offered by Directorate of Cashew Research, Puttur, Karnataka acted as an effective instrument to impart skills to all the employees of the 'Safalam cashew processors group'. Steam boiling is the pretreatment technique being followed owing to easy operation and less floor space requirement. Eight women personnel employed in the unit operates the processing unit. The financial support by the Government of Kerala for processing machineries and the systematic training offered by Directorate of Cashew Research,

Puttur facilitated towards the promotion of this sector in this region. The marketing channels established helps to sale the produce for better price. Value addition to unit weight of raw cashew nut is around Rs 11 and the economic return for the unit having utilization capacity of 5 tonnes per season is found out as Rs 9577 per month (Balasubramanian, D. 2010).

Average outturn of the kernels at packaging is worked out to be 24.66 per cent. Shelling capacity is in the range of 8 to 20 kg per head per day with shelling efficiency ranged from 70 to 92 per cent depending on the degree of skill. Peeling is done manually after hot air treatment. Peeling capacity found to be in the range of 1.5 to 5.0 kg per head per day and its efficiency worked varied from 70 to 85 per cent. Pre grading of kernels practiced while peeling in to White Whole kernel (WW), Dessert Whole kernel (DW), Speckled or dotted Whole kernel (KW), Shriveled Wholes (FW), Scorched Whole kernel (SW), Splits, Large White Pieces (LWP), Small White Pieces (SWP), Bitts and Scorched Pieces. Peeled kernels are transported to grading and packaging centre, a common facility available at Chatanchar, Kasaragod periodically. Women labourers are provided with apron to maintain hygienic condition and protection against the ill affect of Cashewnut Shell Liquid (CNSL). Repair and maintenance of shelling machine is taken up at regular interval. This cooperative sector having satellite system of micro enterprise in cashewnut processing become a successful model in India.

Small Scale Cashewnut Processing in Karnataka

Cashewnut processing units in Kolar district of Karnataka state follows down sized system of steam boiling the raw cashewnut and conventional type drier to make unpeeled kernel amenable for peeling. Kernel peeling is done at home level and finally sorted in to 15 different grades. The cost benefit ratio of processing in this region has been worked out as 1:1.28 (Balasubramanian, D. 2003a). Considering under utilization of established facility *i.e.* up to 12.5 per cent of total production in this region, inefficient technical and management aspects and the huge investment incurred on raw nuts and machines, existing large-scale processing units are not successful in this region. As small-scale processing units generates employment for 3177 persons and profitability of business *i.e.* Earnings before income tax (EBIT) is up to Rs. 88,390 per unit during season, value addition to raw nuts harvested is Rs.11.04 per kg which is 16 per cent more than existing system, it can be promoted to improve economic status and rural employment.

Potential users and manufacturers of technology have been convinced about the technical and economic feasibility of products. Local participation of processors indicated that they are mentally prepared and accept the change. The relative merits of micro level cashewnut processing over large-scale industries justifies that it ideally suits for farm level/homestead processing. Considering the role of women in the small-scale enterprises at farm level entrepreneurs can be encouraged to plunge actively in to this sector. These processing units have outnumbered large-scale industries in Sindhudurg district and the growth of these industries clearly indicates the technology adoption and the profitability of business.

References

Anonymous (2014). Cashew hand book – 2014. Global Prospective. Published by 3F Group, Chennai. India. p.140.

Balasubramanian, D. (2000). Performance of cashew nut processing in Andhra Pradesh-A case study. *The Cashew.* 14(4): 7-16.

Balasubramanian, D. (2003a). Cashew nut processing and its feasibility in Chintamani district of Karnataka-A Case Study. *The Cashew,* 17 (4): 6-11

Balsubramanian, D. (2003b). Small-scale Cashew nut processing – An Appropriate technology for Rural Development. *In. Proceedings of International seminar on "Downsizing of technologies for Rural Development,* 7-10 October, Bhubaneshwar, Odisha, India.

Balasubramanian, D. (2010). Success of small scale cashewnut processing in Kasragod of Kerala state. *The Cashew and Cocoa Journal,* 11(3): 9-16.

Balasubramanian, P. P. and Rema, M. (1996). Pricing and transaction trend of raw cashew nut in India. *The Cashew.* 10(4): 13-19.

FAO, (2002). Statistical database. http: / /www.fao.org on 21.02.2003.

Jekayinfa, S.O. and Bamgboye, A. I. (2005). Estimating energy requirements in cashewnut processing operations. *Energy,* 31: 1305-1320.

Mohod, A., Jain, S., Powar, A., Rathore, N. and Kurchania, A. (2010). Elucidation of unit operations and energy consumption pattern in small scale cashew nut processing mills. *Journal of Food Engineering.* 99: 184–189.

Ohler, J.G. (1979). Processing of the nuts. In. *Cashew.* Communication 71. Department of Agricultural Research. Kaninklijk. Institute voor de Tropen, Amsterdam pp. 201-213.

Yadav, Shalini and Kumbhare, S. L. (2008). Economics of Micro Enterprises: A Case of Cashew Processing Units in Tamil Nadu in Entrepreneurship, Growth and Economic Integration. Himalaya Publishing House, Mumbai. p. 97.

Annexure I: Specifications of Machinery Used in Farm Level Cashewnut Processing

Cashew Steam Boiler		*Cashewnut Sheller*		*Cashew Drier*	
General		*General*		*General*	
Capacity	40 Kg per batch	Type	Single cutter manually operated (Standing posture)	Type	Electrically operated
Mode of action	Direct steaming type (Vertical)	Dimension	1200 L x 600 W x 900 H	Capacity	25 kg per batch
Conventional or electrical	As per choice	Material of fabrication	MS (2.5 m gauge)	Dimensions (Internal)	550H x 650W x 450 D (Gauge: 1.5 mm)
		Footrest	7.5 D with grip	Dimensions (External)	725H x 750W x 550 D (Gauge: 1.25 mm)
		Weight	14 Kg	Material of construction	SS and MS
				Number of heating coil	Minimum 3 coils (500 watts each)
				Total power consumption	1500 Watts
Nut Chamber Vessel		*Cutting Blade*		*Kernel Holding Trays*	
Dimension	600H x 450D	Type	Profile cutting (Replaceable)	Type	Wire mesh -18 gauge perforated
Material of fabrication	MS – 6mm	Material of fabrication	EN –47 (2.5mm)	Capacity	2.5 kg per tray
Unit weight	70 Kg			Material of fabrication	SS and MS (1.25mm)
Accessories	Perforated steam pipe (GI -3 mm) and Drain cock			Space between trays	45 mm
				Insulation	Glasswool slab (75 mm)
				Accessories	Aircirculationfan, thermostat, indicator and 1/8 hp motor
				Unit weight	90 Kg

Steam Vessel		*Guide Rods*	
Dimension	425H x 375D	Material of fabrication	MS – polished
Material of fabrication	MS – 6mm	Dimension	180L x 10D
Unit weight	69 Kg	Accessories	Cutting blade (1 set), Compression spring (2 nos) and Tension spring (1 no)
Accessories	Pressure gauge, safety valve, water level indicator and exhaust pipe		

Cashew: Improvement, Production and Processing (2017) *Pages* **465–477**
Editor: **Prof. P.L. Saroj**
Published by: **ASTRAL INTERNATIONAL PVT. LTD., NEW DELHI**

28

Cashew Apple Feni

D. Balasubramanian and P.L. Saroj

1. Introduction

Cashew apple having tremendous potential for utilization in the food processing sector. Cashew apple is rich source of sugars, proteins, fats, carbohydrates, crude fibre, minerals (calcium, phosphorus, iron) and vitamins (thiamine, riboflavin, nicotinic acid). It is one of the good sources of ascorbic acid, B-complex and other vitamins (Vijaya Kumar, 1991). In general, cashew apple is lest utilized because of its astringent taste, due to the presence of high tannin content (0.06 to 0.22 g/100 g fruit pulp). Tannin content can be minimized by pre-treatments with steeping fruits in 3 to 5 per cent salt solution for three days or by using clarifying agents such as gelatin, poly vinyl pyrolidone (Carvalho *et. al*, 2003). After the removal of astringency, cashew fruits can be converted into several products like juice beverages, blended juice beverages, candy, jam, pickles etc. Cashew apple feni and wine are the alcoholic beverages having commercial significance.

'Feni' is distilled from the fermented juice of cashew apples. Feni is originated in Goa. Sattari which is sub-district of Goa is known as 'Feni capital' of the region, because of the more number of distilleries present in this area. The feni industry is an important source of revenue to the excise department and contributes to the revenue receipts of the state (Pranab *et al.*, 2008). Maria de Lourdes Brava da Costa Rodrigues, the librarian, *Central* Library, *Zulejos* (Portuguese ceramic tile work) inside Panjim indicated that Goa's signature drink, *kaju* (cashew) Feni begins with the Portuguese reaching India, with reference to arrival of Vasco da Gama at King Zamorin's during the year 1498.

Initially, Portuguese missionaries taught the Goan population, the art of distillation of cashew apple mostly for medicinal purposes rather than as an intoxicating drink. Composition and exotic flavour of the fruit gives a unique taste and medicinal properties which is not found in any other available liquor. Cashew feni is being used for centuries in Goa to cure various ailments of infants and aged as well. Cashew Feni can be taken with aerated water and ice. A half peg liquor with fresh lemon juice, sugar, ice and aerated water makes an excellent drink during summer (Bhakta, 1980).

2. Historical Perspective of Feni Preparation

Method of preparation of 'Caju feni' in Goa was described by Manekar (1975). It is to mention that Goa is the only state in India wherein cashew apples are utilized completely for the manufacture of alcoholic beverages. The secret of making cashew feni lies with Goan ingenuity. Owing to expertise and experience gained over several years, the Goan mastered the know-how to manufacture feni. In the traditional method the juice of cashew apples is collected in vats which when allowed to stay, is acted upon by the bacteria present in the apple causing fermentation. The fermented juice is distilled in pot stills to give "Arrack" which on further distillation produces "Feni", an alcoholic beverage. It is further matured in wooden barrels to give "Fineness" to the product. As an estimate, the total yield of cashew apples in Goa was estimated to be approximately 3.5×10^6 kg which could give about. 1.75×10^7 litres of juice and about 2.188×10^3 litres of cashew feni when distilled by modern techniques. Nair *et al.* (1979) reported that after extracting juice from Cashew apple by mechanical means, it needs to be filtered, detanned, pasteurized and cooled. Thereafter, it could be inoculated with a strain of pure yeast. After fermentation and racking, the distillation can be done by modern techniques.

Nanjundaswany *et al.* (1985) indicated that cashew apples are an important by-product of cashewnut industry and presently not utilised in the country to the fullest potential. Yield of cashew apple is 4 to 8 times that of cashewnut and the fruit weight varies from 50 g to 140 g and the yield of the juice is 60-70 per cent. Cashew apples contain astringent and acrid principles which produce an unpleasant biting sensation on the tongue and throat when eaten in natural condition. The fruits are highly susceptible to injury and rapid microbial deterioration. These factors together with the difficulties experienced in collection of fruits have come in the way of commercial utilization of this nutritious fruit. It has been found possible to remove astringent principles by simple methods like steaming, brine curing or by chemical treatment and then utilize the fruit for conversion into several edible products. These products include clarified and cloudy cashew apple juice, cashew apple juice blended with other fruit juices and pulps like lime, pineapple, orange, mango and papaya, cashew apple juice concentrate, preserve, candy, jam,mixed fruit jam, pickle and chutney *etc.*

It is estimated that about 3.5 kg of apples in the traditional process gives about a liter of juice. Modem pulp squeezing systems like mechanical crushers/squeezers are able to extract juice more efficiently almost twice that of the traditional method *i.e. kolmbi* having efficiency of 70 per cent. The extracted juice is kept for 2 to 3 days to get fermented which is verified ·on the formation of the film coating over the juice. Optimum values of the controlling factors like yeast, temperature, aeration of juice leads to 98 per cent of fermentation efficiency. The feni is obtained by distilling 'Uraq', mixed with fermented juice at 1:2 ratio, which contains 75 per cent alcohol. It is estimated that the cost of manufacturing 15 litres of uraq and feni are estimated to be around Rs.70 and Rs.140 respectively (Mandal *et al.*, 1985). It was also highlighted that during the year 1975, about 11 lakh litres of cashew liquor was produced from 70,000 tonnes of cashew apples in Goa, earning revenue to the tune of Rs. 17 lakh as excise duty.

Various steps involved in the preparation of cashew liquor and its feasibility was investigated at Kerala Agricultural University and it was standardised (Patent No. *195/MAS/82*). Augustin (1987), listed out the important steps in the production of cashew liquor are collection, washing extraction of cashew apple juice, de-tanning, fermentation, distillation and ageing process. Removal of astringent smell involves one more step adding to the cost of the final product. Fermentation, distillation and ageing were conducted in polyethylene barrels, glass distillation unit, glass vessels and wooden casks respectively. Hygienic condition of the processing unit was also important as the fermentation need to be becarried out under anaerobic condition.

Pranab *et al.* (2008) reported that the Feni industry is an important source of revenue to the excise department and contributes to the revenue receipts of the state. However, the production of beer, Indian-made foreign liquor (IMFL), appears to be upstaging the tax contribution of feni. This may partly have to do with consumer tastes and preferences, excise policies, and production techniques. Shuklajasha *et al.* (2006) prepared an undistilled alcoholic beverage (wine) by fermenting cashew apple juice with wine yeast, *Saccharomyces cerevisiae* var. *bayanus*. It was highlighted that the wine was slightly yellowish, acidic in taste (1.21±0.02 g tartaric acid 1100 ml), high in tannin (1.9±0.22 mg/100 ml) and low in alcohol (7 per cent) concentration.

3. Production and Utilization of Cashew Apple in India

Equivalent production of cashew apple is generally calculated as 8 to 10 times of the production of raw cashewnuts (Ohler, 1979). Accordingly, availability of cashew apple in different cashew growing state of India is worked out and presented in Table 28.1.

Although cashew apple production in Goa amounts to 3 x 10^5 MT, it is the only state in which preparation of feni is initiated and continuing due to favourable policies and support from Government of Goa. In all other cashew growing states, preparation of alcoholic beverage *i.e.* Feni is banned and cashew apple based non-alcoholic beverages are encouraged. As the cashew apple products such as Ready to serve juice (RTS), squash, syrup, pickle, jam, chutney etc. could not get consumer acceptance, cashew apple utilization is very much limited in these states. Accounting non-utilization of huge quantity of cashew apples, other than Goa state, various research organizations have initiated research and development activity on economically viable production of 'Biofuel' from cashew apple.

Table 28.1: State-wise Production of Cashew Apple in India (2011-12)

State		*Cashewnut*		*Cashew Apple*	
	Area ($x\ 10^6$) ha	*Production ($x\ 10^5$ MT)*	*Productivity (MT ha^{-1})*	*Production ($x\ 10^5$ MT)*	*Productivity (MT ha^{-1})*
Kerala	1.22	1.00	0.85	10.00	8.50
Karnataka	0.91	0.60	0.70	6.00	7.00
Goa	0.54	0.30	0.61	3.00	6.10
Maharashtra	1.21	1.25	1.47	12.50	14.70
Tamil Nadu	0.85	0.45	0.54	4.50	5.40
Andhra Pradesh	1.00	1.00	1.10	10.00	11.00
Odisha	0.84	0.40	0.62	4.00	6.20
West Bengal	0.09	0.08	0.90	0.80	9.00
Others *	0.17	0.12	0.80	1.20	8.00
Total	**6.83**	**5.20**	**0.84**	**52.00**	**8.43**

* Includes Assam, Meghalaya, Tripura, Manipur, Gujarat and Chhattisgarh.

Source: DCCD, Cochin, India (2013).

4. Physio-chemical Composition and Nutritive Value of Cashew Apple

Cashew apple (pseudo fruit) contains sugars, tannins, phenols, amino acids, ascorbic acid, minerals and fibre (Table 28.2). Compared to citrus fruit, ascorbic acid content is five to six times higher (Ohler, 1979) in cashew apple. The quantity of ascorbic acid, soluble solids, reducing sugars and total acid were found to vary between periphery and centre of the cashew apple (Arraes and Braga, 1970). Yellow coloured apples are less astringent, heavier and softer compared to red apple (Albuquerque *et al.*, 1960) (Table 28.3). High yielding varieties of cashew have been analysed for cashew apple composition and varietal variation has been observed (Table 28.4). Cashew apples of the released varieties contained total sugars 5.5-7.7 g/100g, reducing sugars 5.5-7.5 g/100g, ascorbic acid 144-269 mg/100g, phenols 8.2-23.0 mg/100g and tannins 27.4-153.4 mg/100g (Nagaraja and Nampoothiri, 1986). Free amino acids present in cashew apple have been identified as aspanic acid, asparagine, glutamic acid, serine, glycine, threonine, alanine, valine, proline, tryptophan and leucine (Ventura and Hollanda Lima 1959). Carotenoid and ascorbic acid contents have been shown to vary with the colour of the fruit and geographical regions in Brazil (Assuncao and Mercadante, 2003).

Volatile flavour components of cashew apple contained aldehydes and terpene hydrocarbons. The important constituents of volatile flavour are hexanol, carotene, limonene, and trans-hex-2-enal (Alexander Macleod and Nola Gonzalez, 1982). Hexane extract of cashew apple aroma distillate has been analysed by GC-MS and the compounds detected are cyclohexanol, dichlorobenzene, limonene, undecane, naphthalene, dodecane, benzotriazone, decanoic acid, hexadecane, heptacosane and heptadecane (Anonymous, 2001). Blending of cashew apple with orange, grape, pineapple, mango and lemon improved the acceptability of cashew apple juice in terms of taste and flavour. The vitamm C content of cashew apple juice was high (203.5 mg/100 ml), compared to other tropical fruits (Toyin, 2000). Major constituents of essential oil from fruits of red varieties are palmitic and oleic acids (19.6 per cent each) while, palmitic acid (11.4 per cent), furfural (10 per cent), hydroxydodecanoic acid lactone (8.2 per cent), E-hex-2-enal (7.2 per cent), Zhex-3-enal (6.2 per cent) and hexa decanol (6.2 per cent) were the principal components identified in the oil of yellow cashew (Jose *et al.*, 2000). Cashew apple juice serves as a good substrate for growing *Saccharomyces cerevisiae.* The yield of cells grown with and without nutrient supplementation was 0.5 and 0.39 g cells/g reducing sugar consumed (Layokun *et al.*, 1986).

Table 28.2: Nutritive Value of Cashew Apple

Constituents	*Cashew Apple*
Moisture (g)	87.8
Proteins (g)	0.20
Fat (g)	0.10
Carbohydrate (g)	11.6
Crude fiber (g)	0.90
Calcium (mg)	10.0
Phosphorus (mg)	10.0
Iron (mg)	0.20
Minerals (mg)	200
Vitamin C (mg)	261
Thiamine (mg)	0.02
Riboflavin (mg)	0.05
Nicotinic Acid (mg)	0.40
Vitamin A (IU) value	39.0

Source: Vijayakumar (1991).

Table 28.3: Physico-chemical Composition of Cashew Apples

Constituents	*Red*	*Yellow*
Fruit weight (g)	26.1	46.2
Juice yield (per cent)	69.8	72.2
pH	3.90	4.10
Total soluble solids (per cent)	12.3	12.2
Acidity (as Malic acid) per cent	0.35	0.32
True ascorbic acid (mg/100g)	200	153
Total sugars (per cent)	8.40	8.00
Total Tannins (per cent)	0.55	0.45

Source: Nanjundaswamy (1984).

Cashew apple dipped in mustard oil at 1 per cent concentration could be stored without any microbial decay at ambient temperature for 6 days (Chattopadhyay and Ghose, 1993). Benomyl treatment combined with storage in sealed polythenebags increased the shelf life of cashew apple at ambient condition from 4 to 12 days (Ogunwolu, 2000). Dipping of cashew apple after washing thoroughly with water in low concentration of antibiotics such as griseofulvin or cycloheximide has been shown to prevent microbial spoilage (Bopaiah, 1983).

5. Method of Preparation of Feni from Cashew Apple

Traditional Method

In the traditional method of making cashew fenni, the cashew apples are manually crushed in a rock on the hill which is carved or shaped like a basin with an outlet for the juice, called a Kolambi. In some places, cashew apples are crushed with legs to form a cake which were then tied by strings. A stone was placed on the top of the cake and there was an outlet to collect the juice called 'Neero'. The juice is collected in a huge earthen pot called a Kodem, which is buried in the ground for fermentation. The juice is then distilled in earthen or copper pots. Cashew apple juice is transferred to big pot called Bhann and serves as a closed boiler and connected to a smaller pot called 'Launni' by means of a conduit. The Launni serves as a receiver or collector. Juice in big pot is boiled with the help of firewood and process of vaporisation and distillation take place result in the concentrated liquid and collected in the smaller pot. The pressure in the receiver is kept in check by pouring cold water on it, typically with a wooden ladle. The first stage of processing (vaporisation) is done on big fire but the later stage of distillation has to be done on low fire to keep the pressure and heat under control. The process of distilling fenny with such apparatus takes about 8 hours and is locally called Bhatti. The liquor produced from cashew is of three grades *viz.*, Urrac, Cazulo and Feni. The Urrac (Arrack) is the product of first distillation. It is light and can be consumed neat. Its strength ranges between 14 and 16 grao. The Cazulo is the product of second distillation. It is moderately strong and no longer seen in the market now-a-days. Fenny is the product of the third distillation and its strength ranges between 20 and 24 grao having a long shelf life.

Commercial Method

Preliminary Treatment of Cashew Apple

Quality cashew apples are collected and washed in running tap water. Cleaned fruits are processed through mechanical juice extractor and mixed with potassium metabisulphate (KMS) to obtain 100-150 ppm of SO_2.

Fermentation

Cashew apple juice is sterilized at 1.05 kg cm^{-2} pressure for 15 min, After cooling, sterilised juice is inoculated with *saccharomyces cerevisiae* var *ellipsoides* and incubated at 30°C for 24 h. After 48 h, this inoculum was added to 1 litre of sterilized juice and incubated at 30°C for 24 h. Again the inoculum is added into sterilized juice in a ratio of 1: 10 and incubated at 30°C for 24 h. The above pure yeast culture of 10

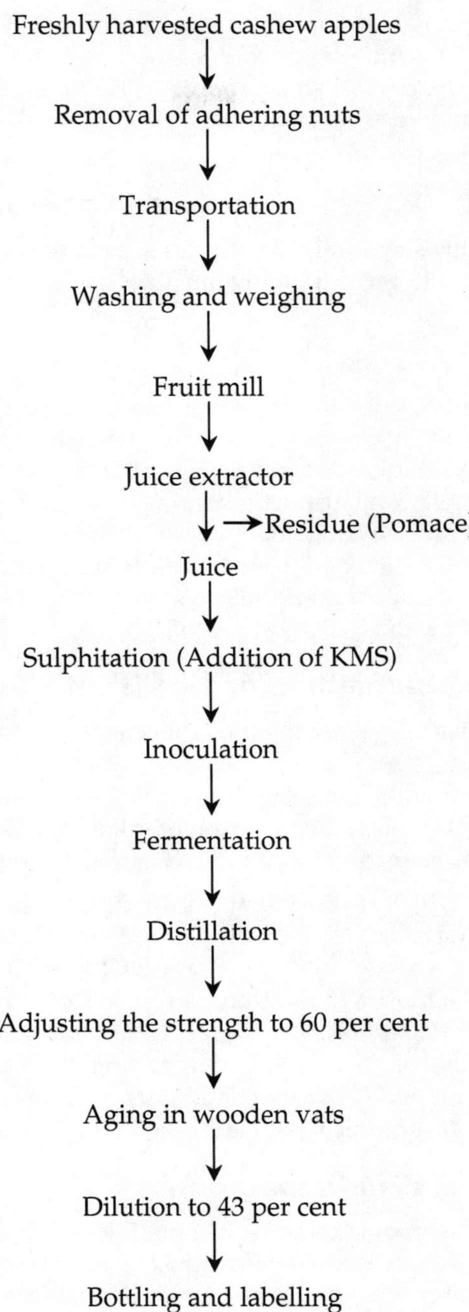

Flow Chart of Preparation of 'Feni' from Cashew Apple

litres thus propagated will be again inoculated into 100 liters of boiled and cooled juice. Next day, about 100 litres inoculum was added into 1000 liters of cashew juice and incubated for 72 h. During the process, sugar present in the cashew apple juice is converted into ethyl alcohol and carbon dioxide by the enzymic reaction of the yeast cells. This is known as fermentation.

$$C_6H_{12}O_6 \rightarrow 2C_2H_5OH + 2CO_2$$

Glucose → Ethyl alcohol + Carbon dioxide

The yeast consumes some quantity of glucose for its growth and hence the yield of alcohol will be around 48 per cent v/v. Fermentation will be practically accomplished within 72 h.

Distillation and Maturation

The fermented brew is fed into a pot and distilled to recover alcohol. The liquor obtained immediately after distillation is harsh, therefore it is adjusted to contain 60 per cent alcohol and kept for ageing in oak-wooden barrels in the cellar at 15°C for 3-5 years. During ageing, ethanolysis of lignin takes places giving rise to aromatic congeners like syringaldehyde, sinapaldehyde. vaniline and B-methyl, r-octalactone which are extracted from the oak-barrels. After sufficient maturation the strength of the distillate is adjusted to 43 per cent by the addition of water. It is filtered, bottled and labelled. (Patel *et al.*, 1984).

6. Geographical Indication for Goa 'Feni'

Cashew feni got its GI registration as a specialty alcoholic beverage from Goa, under the name "Feni", and described as a colourless clear liquid when matured in wooden barrels getting golden brown tint. This designation was achieved through the efforts of the Goa Cashew Feni Distillers and Bottlers Association and Department of Science, Technology and Environment, Government of Goa.

Some countries which are the member of WTO have G I for liquor as in Europe - Tequila (spirits), Jaffa (oranges) and Bordeaux (wines). Such as Tequila and Mezcal from Mexico, Scotch whiskey from the UK and Champagne from France, now in India also a liquor" GOA FENI " registered as GI. India is a member of WTO's rules on intellectual property. Trade-Related Aspects of Intellectual Property (TRIPS) and as of its obligations; the country has passed its own act. The Geographical Indications of Goods (Registration and Protection) Act, 1999 provides for the registration and better protection of geographical indications.

7. Economics of Feni Preparation

Economics for the production of 'feni' in small scale is worked out and presented in Table 28.4. The cost calculated for the product *i.e.* Rs. 80/- per 650 ml bottle is quite low for low alcoholic beverage available in the market. Further reduction in the cost could be made while scaling up the production. Project feasibility is prepared to run 'Manufacture of 'Cashew apple based liquor' in the cashew growing regions and given in Table 28.5. Quantity of cashew apple 'Feni' that could be extracted from certain varieties grown in India is given in Table 28.6. Among the various varieties

Table 28.4: Composition of Cashew Apple Juice of Released Varieties (mg/ml juice)

Sl.No.	*Variety (Unclarified)*	*Tannin (Unclarified)*	*Flavarols (Clarified)*	*Sugars (Unclarified)*	*Ascorbic Acid*
1.	NRC Sel-1	2.47	0.19	11.01	3.09
2.	NRC Sel-2	7.11	0.35	14.11	4.15
3.	NDR2-1	1.51	0.03	25.25	4.10
4.	BLA 139-1	2.60	0.21	6.56	4.75
5.	Kanaka	6.23	0.26	15.35	2.89
6.	K22-1	1.74	0.09	23.30	2.68
7.	BLA 39-4	2.81	0.52	81.81	1.25
8.	Dhana	2.81	0.31	86.34	0.87
9.	Priyanka	4.08	0.26	88.36	1.81
10.	Ullal-I	2.25	0.23	134.13	2.37
11.	Ullal-2	4.04	0.17	16.45	4.18
12.	Ullal-3	4.89	0.36	20.85	4.28
13.	Ullal-4	4.80	0.38	17.00	4.18
14.	VN-50	1.65	0.19	72.67	2.51
15.	VRI-1	4.68	0.30	30.22	4.08
16.	VRI-2	0.62	0.06	55.30	1.76
17.	VRI-3	1.40	0.20	84.08	2.02
18.	BPP-1	0.81	0.02	36.67	2.52
19.	BPP-2	6.04	0.28	74.08	4.42
20.	BPP-3	7.50	0.52	11.80	2.93
21.	BPP-4	2.68	0.33	73.70	0.55
22.	BPP-5	7.96	0.65	38.I3	2.76
23.	BPP-6	4.11	0.10	44.76	3.I0
24.	BPP-8	1.31	0.26	75.73	1.53
25.	V-I	3.36	0.I0	25.44	5.11
26.	V-2	3.84	0.24	34.05	1.60
27.	V-3	1.41	0.08	14.49	3.09
28.	V-4	6.08	0.I4	40.11	3.63
29.	V-6	4.30	0.33	11.80	3.60
30.	V-7	11.55	0.81	121.62	3.01
31.	Goa-1	1.18	0.07	13.64	2.06
32.	Jhargram-1	1.74	0.24	96.I9.	3.01
33.	Bhubaneswar-1	0.93	0.50	70.84	2.06
	CD 5 per cent	0.31	0.02	8.19	0.10

Source: Annual Report: Directorate of Cashew Research, Puttur (2010).

released, production potential of cashew apple is found to be higher than 2.5 MT for the varieties *viz.*, Ullal 1, VRI 3, Bhaskara, Ullal 3, Madakkathara and NRCC Selection 2 under normal density planting technique.

Table 28.5: Cost of Production of Cashew Apple 'Feni'

Sl.No.	*Material Required*	*Quantity*	*Rate (Rs.)*	*Amount (Rs.)*
1.	Cashew apple	100 kg	0.50 per kg	50.00
2.	Juice extraction	One man day	250 per day	250.00
3.	Sugar	10 kg	30 per kg	300.00
4.	Di-ammonium hydrogen phosphate	70 g	LS	50.00
5.	Potassium metabisulphite (KMS)	15 g	LS	50.00
6.	Bottles	100	10 per bottle	1000.00
7.	Crown corks	100	1 per cork	100.00
8.	Total cost			1800.00
9.	Overhead charges	@30 per cent		540.00
10.	Additional charges for fermentation, distillation and ageing	@25 per cent		450.00
11.	Excise duty	19 per bottle		1900.00
12.	Grand total			4690.00
13.	Sales tax	@30 per cent		1407.00
14.	Profit	@40 per cent		1876.00
15.	Sale price per bottle			79.73

* Juice yield is about 70 per cent in cashew apple, **Cost of bottle is on non-returnable basis.

Table 28.6: Project Feasibility of Preparing Cashew Apple Based 'Feni' in Goa

Process	*Manufacture of 'Feni' frorn Cashew Apple*
Product consumption pattern	Cashew apple is culitivated in India especially in coastal areas. Cashew apple refers to the fleshy portion to which the nut is attached. Due to problems of strong flavour of the fruit, larger levels of tannin which leaves an astrigent taste in the product and small amount of anacardic acid which is irritating to the throat, cashew apple cannot be utilized for various processed fruit products like jams, jellys, candy, fruit juice and squash. Accounting above mentioned reasons, cashew apple based brandy (called as feni) can be prepared.
Market information	Cashew apple is about 8-10 times weight of cashewnut, possessing a reddish or yellowish colour. On fermentation it yields alcoholic beverage from which brandy could be distilled. In Goa state, around 7-8 lakh gallons of liquor is prepared from cashew apple every yar and this finds a ready market.
Raw material	It is estimated that the area under cashew crop in India is around 6.95 lakh ha. At present about 52.55 lakh MT of cashew apple are produced every year during the season (April to May) in India.

Contd...

Table 28.6–*Contd...*

Process	*Manufacture of 'Feni' frorn Cashew Apple*	
Production process	Cashew apples are crushed and the juice is expressed by either screw press or basket press. Juice is then suphited to destroy the undesirable microorganism. The clear juice is inoculated with desired strain of yeast and the fermentation is allowed to proceed. After 7 days period, the fermentation is almost complete and the yeast settles down. The clear portion can be decanted and used for the preparation of brandy. Decanted juice is distilled in a batch or continuous distillation column and the distillate is allowed to age in wooden vats. The aged material is blended, filtered, alcohol level adjusted, bottled and labelled.	
Equipments	Juice extractor, boiler, distillation unit, bottle washing, filling and sealing units, filtration unit, filtration unit, storage vessels, holding tanks *etc.*, are the major equipments which are indigenously available.	
Economics	*Capital Investment*	*Cost (Rs.)*
	Land and building	20,00,000
	Plant and machinery	6,50,000
	Working capital (100 days)	
	Total capital investment	28,30,000
	II. Cost of production	
	Raw material and packaging material	3,72,593
	Utilities	30,000
	Labour and supervision	1,24,000
	Overheads	1,90,000
	Cost of production of 12000 bottles of 650 ml capacity	7,16,593
	Cost of production per bottle	59.72

Table 28.7: Projected Extraction of Cashew Liquor Production from Certain Indian Cashew Varieties

Varieties	*Number of Trees ha^{-1}*	*Nut Yield (kg ha^{-1})*	*Cashew Apple Produced (kg)*	*Juice Extracted (per cent)*	*Cashew Liquor Extraction (lit)*
VRI-3	200	610	4880	3416	427
NRCC Sel2	200	438	3504	2453	307
V-7	200	362	2896	2027	253
Ullal-1	200	744	5952	4166	521
Dhana	200	562	4496	3147	393
Madakkathara-2	200	438	3504	2453	307
Ullal-3	200	592	4736	3315	414
V-4	200	240	1920	1344	168
Bhaskara	200	618	4944	3461	433

Apple: Nut ratio is assumed as 8; Percentage of juice extracted is about 70 and 12.5 per cent of liqour extraction from juice

References

Albuquerque, S. D. S., Hassan, M. V. and Shetty, K. R. (1960). Studies on the apple character of cashew. *Mysore Agric J.*, 35: 2-8.

Alexander, J. MacLeod and Nola Gonzalez (1982). Volatile flavor components of cashew apple (*Anacardium occidentale* L.). *Phytochemistry*, 21(10): 2527-2530.

Anonymous (2001). Annual Report. National Research Centre for Cashew, Puttur, India. p. 119.

Anonymous (2010). Annual Report. Directorate of Cashew Reseach, Puttur, India. p. 57.

Arraes, M. G. and Braga, S. J. (1970). Determination of acid, sugars and ascorbic acid in cashew apple. *Bioi Cear Agron*, 2: 25-29.

Assuncao, R. B. and Mercadante, A. Z. (2003). Carotenoids and ascorbic acid from cashew apple (*Anacardium occidentale* L): variety and geographic effects. *Food Chem.*, 81 : 495-502.

Augustin, A. (1987). Fermented products of cashew apple. *Cashew Bulletin*, 24(7): 12-16.

Bhakta, K. P. D. (1980). Cashew feni-A unique liquor from cashew apple: its prospects and problems. *Indian Cashew Journal*, 12(2): 19-21.

Bopaiah, B. M. (1983). Microbial spoilage of cashew apple and its prevention. *Indian Cashew J.*,16: 15-17.

Carvalho, M. G. R., Raju, V. K., and Pushpalatha. P, B. (2003). Sensory quality of cashew apple wine as influenced by pre-treatments and clarifying agents. *The Cashew*, 17(1): 10-14.

Chattopadhyay, N. and Ghose, S. N. (1993). Extension of storage life of cashew apple- Note on manufacture of tannin extract. *The Cashew*, 7: 12-14.

Jose, G.S., Eloisa, H.A.A. and Maria, GBZ. (2000). Volatile constituents of the leaves, fruits, flowers of cashew (*Anacardium occident*ale L.). *J. Food Composition and Analysis*, 13: 227-232.

Layokun, S. K., Obawole, A. B., Fatile, L. A. and Solomon, B. O. (1986). Investigation of cashew apple juice as a substrate for single cell protein production. *J. Food Sci.*, 51: 237-238.

Mandal, R. C., Dhandar, D. G., and Dhume, C. V. (1985). Cashew apple liquor industry in Goa. International Cashew Symposium. *Acta Horticulturae* (ISHS), 108: 173-176.

Manerkar, K. M. (1975). Caju feni, *Indian Cashew Journal*, 9(2): 9-11.

Nagaraja, K. V., and Nampoothiri, V. M. K. (1986). Chemical characterization of high yielding varieties of cashew. *Qual Plant Food Hum Nutr.*, 36: 201-206

Nair, M. K., Bhaskara Rao, E. V. V., Nambiar, K. K. N. and Nambiar, M. C. (1979). Cashew products, *The Cashew*. pp. 96-97.

Nanjundaswamy, A. M., Radhakrishniah settee, G., and Patwardhan, M. V. (1985). Utilization of cashew apples for the development of processed products, International Cashew Symposium, *Acta Horticulturae* (ISHS), 108: 151-159.

Nanjundaswamy, A. M. (1984). Economic utilization of cashew apple, *Cashew Causerie.* 6(2): 2-7.

Ogunwolu, S. O. (2000). Effects of hot water dip-chemical treatment and polyethylene film packaging on storage life of cashew apple. *Nigerian J. Tree Crop Res.*, 4: 85-93

Ohler, J. G. (1979). Cashew. Comm 71 Amsterdam, The Netherlands: Dept. Agric Res: Koninkijik. Institute voor de Tropen, pp. 72-78.

Patel, J. D., Venkataraman, K., Prasad, M. S. and Venkatanarayana. (1984). Production of feni. *Cashew Causerie,* 3(1): 8-9.

Mukhopadhyay, Pranab, Santosh Mourya and Suryabhan Mourya (2008). Baseline Survey of the feni industry - A Report, for the ESRC Project on GIS and Feni, University of Warwick, Coventry, UK.

Shuklajasha, M., Pratima Ray., Swain, M. R., and Ray, R. C. (2006). Cashew apple into wine, *Journal of Food Processing and Preservation,* 30: 314-322.

Toyin Oluyemisi Akinwale (2000). Cashew apple juice: its use in fortifying the nutritional quality of some tropical fruits. *Eur Food Res Technol.,* 211: 205-207.

Ventura,M. M., and Hollanda Lima, I. (1959). Free amino acids of cashew apple (*Anacardium occidentale* L.). *Phyton Rev Inter Botany Exp.,* 12: 31-34.

Vijaya Kumar P. (1991). Cashew apple utilization a novel method to enhance the profit, *The Cashew,* 5(4): 17-21.

Section VI: Technology Dissemination and Economics

Cashew: Improvement, Production and Processing (2017) *Pages* **481–493**
Editor: **Prof. P.L. Saroj**
Published by: **ASTRAL INTERNATIONAL PVT. LTD., NEW DELHI**

Dissemination of Production Technologies: Approaches and Methods

M.V. Sajeev and M.G. Nayak

1. Introduction

Extension has traditionally been defined as the delivery of information and technologies to farmers. This leads to the technology transfer model of extension, identified by many organizations as the main purpose of agricultural extension (Moris, 1991). This notion is based on the idea that 'modern' knowledge and information is transferred through extension agents to recipient farmers. Swanson *et al.* (1997) defined agricultural 'extension', as extending relevant agricultural information to people. The World Bank defines extension as 'the process of helping farmers to become aware of and adopt improved technology from any source to enhance their production efficiency, income and welfare' (Purcell and Anderson, 1997).

Agricultural extension in India is largely deployed by government, implemented mainly through government institutions and to some extent through non-government agencies. In India, the extension efforts, particularly transfer of technology efforts, have largely been taken up by the state departments of agriculture and other disciplines as a state subject. The Indian Council of Agricultural Research (ICAR) as the apex body to provide new technologies in agriculture and allied aspects has its own transfer of technology activities. The extension efforts of ICAR, particularly through its institutes are largely of frontline extension programmes.

Transfer of technology efforts in cashew by ICAR are also mostly frontline in nature while large scale field level extension work being taken up by line departments of various states. This chapter looks at the various approaches and models used in transfer of technology in general and summarises experiences of ICAR-DCR, Puttur in utilizing various methods in cashew extension.

2. Approaches for Cashew Extension

There are various approaches in agriculture extension. In general we have identified eight different approaches to extension work (Axinn, 1988 and Rivera *et al.*, 2001). These approaches are also applicable for cashew extension and are briefly summarized below.

2.1 The General Agricultural Extension Approach

This approach assumes that technology and knowledge that are appropriate for farmers exist but are not being used by them. The purpose is to help farmers increase their production. The approach is usually fairly centralized and government-controlled. Planning is done on a national basis by the central government 'which knows better than farmers'. This is a typical case of top–down planning. Field personnel tend to be large in number and high in cost, with the central government bearing most of the cost. The rate of adoption of important recommendations and increases in national production are the measures of success. A survey of agricultural extension programs indicated that agricultural extension generally was part of the Ministry of Agriculture, with field extension officers at the bottom of the hierarchy and a minister at the top. This approach lacks two-way flow of information. It fails to adjust messages for each different locality. Only farmers who seek advice benefit and these tend to be large-scale wealthier farmers. This approach does provide farmers with information on a number of production alternatives from one single source.

2.2 The Commodity Specialized Approach

The key characteristic of this approach groups all the functions for increased production—extension, research, input supply, marketing and prices—under one administration. Extension is fairly centralized and is oriented towards one commodity or crop and the agent has many functions. Planning is controlled by a commodity organization for the purpose of increasing production of a particular commodity. Highly trained scientific personnel equipped with expensive vehicles and field scientific apparatus are employed. Techniques recommended must produce financial benefits for farmers, and be demonstrable on farmers' own fields. New inputs must be accessible, a credit scheme established, and the ratio between farm-gate inputs and commodity prices considered. Technology tends to be appropriate and distributed in a timely manner because it focuses on a narrow range of technical concerns. Interests of farmers, however, may have less priority than those of commodity production organizations. Successful examples in India vary from Rubber board, Spices board and that of Coffee board but cashew extension is yet to be taken up in this approach.

2.3 The Training and Visit Approach

Training and Visit (T and V) is one of the best known of the recent approaches, which was adapted by many countries to support the development of state extension services. It provides continuous feedback from farmers to extension agents and to research staff; it allows for continuous adjustment to the farmers' needs. The purpose of the training and visit approach (often called T and V) is to induce farmers to increase production of specified crops. This fairly centralized approach is based on a rigorously planned schedule of visits to farmers and training of agents and subject matter specialists. Close links are maintained between research and extension. Agents are only involved in technology transfer. Planning is controlled centrally and field personnel tend to be numerous and dependent on central resources. Success is measured in terms of production increases of the particular crops covered by the program. The T and V approach is again a top–down approach. The emphasis is on disseminating unsophisticated, low-cost improved practices, and teaching farmers to make best use of available resources. This builds on a combination of the individual and group approaches. In this system, the extension staff is trained every fortnight on relevant extension issues for that time of the year and the staff then extends these messages to contact farmers who receive special attention. Field days and other visits are arranged on the farms of contact farmers so that their neighbours can also benefit from the knowledge they have gained. Under T and V, the extension system changed its way of reaching out to farmers by using agents who focused mainly on technology diffusion (Picciotto and Anderson, 1997). T and V extension agents would meet with a small group of 'contact' farmers who were expected to disseminate information to the members of their respective communities and convey farmer's opinions back to the agents, thus creating a feedback mechanism absent in the prior system (Birkhaeuser *et al.*, 1991). T and V did, however, have its critics. With continued budgetary crises of less developed countries, some argued that it was too expensive and impossible to implement over extensive regions. Highly dispersed farmers could never establish frequent contact with extension agents. And their needs varied widely and could not be addressed with a single, inflexible technology package (Picciotto and Anderson, 1997; Feder *et al.*, 2001).

With T and V, the frontline worker becomes the vital link in a chain, which ensures two-way communication between research institutions and farmers. The T and V system focused on regular visits and making the extension agent responsible for diffusion in target farmers and region. The system also upgraded the technical capacity of the extension service through the creation of the regional specialist positions to give inputs to the field agents in direct contact with the farmers. Unfortunately, the T and V system took the emphasis off the adequacy of the technology and put it on organizational questions and increased the costs of operation by focusing on increasing field visits and upgrading the technical capacity.

2.4 The Farming Systems Development Approach

Farming systems research and extension evolved during 1965–80 (Hildebrand, 2001). This system was marked by greater emphasis on smallholder farmers and their livelihood systems, as researchers realized that such people were not being reached effectively with the traditional extension approaches. They thus began to use what is known as the farming systems approach to research and extension. This approach assumes that technology which fits the needs of farmers, particularly small-scale farmers, is not available and needs to be generated locally. A key characteristic of this type of extension is its systems or holistic approach at the local level. This approach is implemented through a partnership of research and extension personnel using a systems approach. Close ties with research are required and technology for local needs is developed locally through an iterative process involving local people. Analyses and field trials are carried out on farmers' fields and in homes. The measure of success is the extent to which farm people adopt technologies developed by the program and continue to use them over time. Control of the program is shared jointly by local farm families, extension officers, and researchers. Advantages of this system include strong linkages between extension and research personnel, and the commitment of farmers to using technologies they helped to develop. All on-farm trials (OFTs) conducted by Krishi Vigyan Kendras can be categorized under this approach. Cashew being a plantation crop, technology development process using this approach is fairly long. However the results from this approach are used widely for technology assessment and refinement and for providing micro-location specific recommendations.

Farming systems is a holistic approach that looks at the entire farm as a system with various subsystems. It provides for greater dialogue with and input from farmers, and for enhanced linkages between research, extension and farmers. This model was marked by participation at the farm level (through farmer input on research and on-farm trials) and by interdisciplinary linkages and a systems approach to extension. A model farming systems approach in cashew will be characterized by viewing the farm as a whole, involvement of cashew farmers and their priorities, research reflecting the various subsystems' interactions and linkages and reliance on participatory data collection and analysis.

2.5 The Participatory Agricultural Extension Approach

This approach assumes that cashew farmers are skilled in production from their crop, but their levels of living could be improved by additional knowledge. Active participation by cashew farmers themselves is necessary and produces a reinforcing effect in group learning and group action. Much of the work is through group meetings, demonstrations, individual and group travel, and local sharing of appropriate technologies. This approach often focuses on the expressed needs of farmers' groups and its goal is increased production and improved quality of rural life. Implementation is often decentralized and flexible. Success is measured through numbers of farmers actively participating, and the continuity of the program.

The participatory agricultural extension approach costs less, fits needs well, and is more efficient. However, it is more work for extension agents to organize and motivate farmers. It requires agents to live and to socialize with farmers. Where a government job is seen as a reward, the 'hardship' implied by this approach dooms it to failure. The agent is present only 'part time' and has no personal stake in the outcome. This explains why the approach is highly successful with cashew extension work done by various NGOs like BAIF, SKDRDP etc and not successful under government extension system.

3. Characteristics of Participatory Cashew Extension Approach

Community-based extension and joint learning is central to this approach. The process emphasizes the participatory facilitation role of extension staff. Main characteristics of participatory cashew extension approach are that it integrates mobilization for planning and action with cashew development, agricultural extension and research; is based on an equal partnership between cashew farmers, researchers and extension; aims to strengthen cashew farmer's problem-solving, planning and management abilities; promotes cashew farmers' capacity to adopt and develop new and appropriate cashew production technologies; encourages cashew farmers to learn through experimentation, building on their own knowledge and practices; and recognizes that farming communities are not homogeneous but consist of various social groups with conflicts and differences in interests, power and capabilities.

The role of extension is to facilitate this process. Under the emerging model, good extension work means talking with farmers, working with farmers, learning from farmers and suggesting new approaches to farmers. A number of lessons have been developed based on past experiences. However, if one considers the FSA to technology development and transfer and participatory extension methods, the differences are not that significant.

(i) The Project Approach

This approach concentrates efforts on a particular location, for a specific time period, often with outside resources. Part of its purpose is often to demonstrate techniques and methods that could be extended and sustained after the project period. It uses large infusions of outside resources for a few years to demonstrate the potential of new technologies. Control is at the central government level and there are often considerable financial and technical inputs from an international development agency. Short-term change is the measure of success. In the NAIP program of India, for example, financial support from the World Bank was used by the Ministry of Agriculture to support extension work in many different projects throughout the country. They were able to introduce innovations through an effort which combined the project approach with the specialized commodity approach. However this approach is rarely found utilized for cashew extension in India. One problem with this approach, however, is that a flow of ideas outside the project rarely occurs (Axinn, 1988).

Integrated approaches aim at influencing the entire rural development process. Extension is only one though often crucial element in this strategy which targets the entire population in a given area but emphasizes work with disadvantaged groups. Tribal sub plan projects on cashew are examples under this approach. Integrated approaches are generally implemented in the form of large-scale and foreign-funded projects aiming at alleviating mass poverty in rural areas on the basis of 'a simultaneous improvement in the utilization of natural resources and of human potential'(Rauch, 1993). Measures to promote production are coupled with a strong emphasis on self-help. The underlying concept is typically multi-sectoral.

Evaluations of more than a decade of such projects have revealed serious shortcomings in reaching the goal of mass poverty alleviation (IBRD 1987; BMZ 1990). Sizeable numbers of the farmers were not reached by project activities, nor were positive effects consolidated on a sustainable basis. Project deficiencies were in part management related and very often due to a serious underestimation of the great complexity of multi-sectoral programs with ambitious goals. The disregard of the target group principle and of due consideration for framework conditions (economic and institutional) played an even more important role, as did the lack of compatible technical solutions. Recent efforts to improve regional rural development (RRD) projects and enhance chances for a broad and sustainable impact (Rauch, 1993) are relevant for all general extension approaches. The key concept is the availability of locally adapted solutions established on a common basis.

(ii) The Cost Sharing Approach

This approach is based on cashew farmer sharing part of the cost of the extension program. Its purpose is to provide advice and information to facilitate farmers' self-improvement. It assumes that cost-sharing with farmers (who do not have the means to pay the full cost) will promote a program that is more likely to meet local situations and where extension agents are more accountable to local interests. Control and planning is shared by various entities and is responsive to local interests. Success is measured by farmers' willingness and ability to provide some share of the cost, be it individually or through local government units. Problems may arise if local farmers are pressured into investing in unproven enterprises. The approach is found utilized in potato and some spice and medicinal crops but yet to come to cashew sector.

(iii) The Educational Institution Approach

This approach is widely used in cashew extension in India and uses educational institutions which have technical knowledge and some research ability to provide extension services for farmers. Planning is controlled by those in the educational institution. Implementation is through non-formal instruction in groups or individuals through a college or university. Attendance and the extent of participation by cashew farmers in extension activities are the measures of success. While the Cooperative Extension Service (CES) of the United States is still the only system in which the main extension function remains within the university, some developing countries, notably India, have integrated educational institutions into practical extension work. Within the United States of America, state universities have traditionally cooperated with local counties and the US Department of Agriculture in doing extension besides education and research. Within the last 130 years, extension goals of the land-grant colleges have shifted from practical education to technology transfer and, more recently, to a much broader concept of human resource development.

While in most countries, the main contribution of educational institutions to extension will be the training of qualified, dedicated, and responsible personnel, some Indian agricultural universities have come close to the US model with All India Coordinated Research Project (AICRP)s (Ex: AICRP on Cashew) without taking over the full load of extension work. In the field, they have taken over few functions thus supporting general extension work. Remarkable features are direct assessment of clients' needs, user-oriented research, quality training for state personnel, and a strong linkage between academic education and field practice. This model varies from state to state. For example, the Punjab Agricultural University (PAU) has its own multidisciplinary extension team in each district, engaged in adaptive research, training and consultancy. Backed up by extension specialists on campus, they are transmitters and receivers of experiences from researchers, farmers and state extension workers. At PAU, a unique system of processing these experiences is practised. Regular workshops are held which unite university and department staff from research and extension together with outstanding farmers. New findings and feedback are presented, evaluated and published as a 'Package of Practices' to be used by all extension staff for the next season (Nagel, 1980).

4. Transfer of Technology Methods Adopted by DCR, Puttur

The extension efforts of ICAR, particularly through its institutes are largely of frontline extension programmes. DCR Puttur, over its past 25 years of existence has also made such efforts which are mostly frontline in nature. This article takes a look at the technology backstopping efforts made by the institute since its existence.

4.1 Participatory Technology Demonstration

To provide technology backstopping to farmers on frontline cashew production technologies developed by Directorate of Cashew Research and other cashew research stations, cashew demonstration plots based on Farmer Participatory Technology Development (PTD) model were laid out at selected farmers' plots. These PTD cashew gardens serve as demonstration plots to prove the effectiveness of the recommended cashew cultivation practices to farmers while for scientists, these demonstration plots are the tools for assessment and refinement of the recommended practices based on the performance of technologies at micro location level. Laying out of farmer participatory demonstration plots started from the year 1988 as part of the research project; "Transfer of Technology Programmes in Cashew" with the following objectives:

i) To provide technology backstopping on frontline cashew production technologies.

ii) To assess, refine and demonstrate the frontline cashew production technologies.

This research project is externally funded by Directorate of Cashew nut and Cocoa Development, Cochin. Under this programme, every year interested farmers are being selected for demonstrating the frontline cashew production technologies. Interested farmers are requested to apply for the demonstration plots. The application should be addressed to 'The Director, Directorate of Cashew Research',

Puttur. In their application, the farmers should mention the root map to their plots. The farmers should have at least two acres of land for demonstration. The plot should be located by the road side, so that the performance of the crop can be observed by the fellow cashew farmers. A team of scientists from DCR assess the plots for their suitability as demonstration plots, after checking the application. Based on report of the team, demonstration farmers will be selected for the particular year. The newly selected demonstration farmers are provided with subsidy and are being trained every year at DCR, Puttur on Cashew Production Technology.

The farmer participatory technology demonstration gardens are regularly visited by the project team and other scientists. The conditions of these plots are assessed based on the criteria *viz.*, removal of forest trees, gap filling, initial training, pruning, soil and water conservation practices, application of manures and fertilizers and plant protection measures. Technical advice is being given to the demonstration farmers based on conditions of the plots.

So far in Dakshina Kannada district a total of 162 model cashew demonstration gardens were laid out. Among the five taluks of the districts, in Puttur taluk more number of plots were laid out and 78 plots were under normal density system whereas remaining 84 plots were under high density planting system. The varieties/ types which were distributed as planting material are Bhaskara, VRI-1, VRI-2; Ullal-1, Ullal-3; H-3-13; VTH 174; BLA 139-1; H-3-17; BPP-3; VTH 59/2; VTH 30/4; T.No.40; Selection-1, 2; Goa 11/6; H 32/4; V-4; NDR-2-1 and Ullal-4. Presently 75 new FLD plots are taken up in tribal farmer plots under Tribal Sub Plan scheme. The technologies demonstrated include recommended varieties and planting densities.

4.2 Thematic Campaigns

In order to provide intensive training to the cashew demonstration farmers and cashew growers on important aspects of cashew cultivation *viz.*, soil and water conservation measures (terracing and opening of catchpits), plant protection measures (control of TMB and CSRB) and initial training and pruning, thematic campaigns are being organized. These thematic campaigns were organized in collaboration with NGOs like SKDRDP, Dharmasthala; KJP Research Foundation; Sri Durga Charitable Society and Development Departments.

Prior to the campaign wide publicity was given to attract large number of cashew growers. During campaign, the cashew growers were being exposed technologies through different extension methods *viz.*, exhibition, specially prepared literature; lecture-cum-discussion by experts and field demonstration. Mostly these campaigns were organized at the cashew demonstration plots.

All together, 19 campaigns on SWC measures, 17 campaigns on plant protection measures and 4 campaigns on initial training and pruning were organized and the impact of these campaigns were very high. Attempts were also made by the Directorate to assess the effectiveness of the campaigns by assessing the knowledge level of the participants on SWC and PP measures before and after the training. It was found that the campaigns were effective in imparting knowledge to the respondents on SWC/PP measures were the number of respondents having higher level of knowledge was more after the training compared to before the training.

Lessons Learnt

- ☆ Collaboration with NGO's and village institutions like schools and panchayats encourages participation of the farmers.
- ☆ Intensive training using various modes of training increases effectiveness of the programmes.
- ☆ These types of programmes serve as a platform in which problems about utilization of cashew and other plantation crops can be discussed and solution is suggested.

4.3 Cashew Field Day/Cashew Days

In order to inform the cashew growers about the latest research developments in cashew and to get the feedback from the farmers about the usefulness of the cashew production technologies disseminated, Cashew Field Days/Cashew Days are being organized either at DCR, Puttur or at the Cashew demonstration plots of progressive cashew growers. The farmers/cashew growers on the programme day are explained about the research results at respective plots during the field visit. An exhibition depicting the latest developments in cashew research and useful information about cashew also will be arranged using photos, panels, charts, specimens etc for the benefit of the farmers.

After the field visit and exhibition, seminar-cum-discussion are arranged during which experts in cashew research development and progressive cashew growers share their experience with the other farmers. At the end, question answer session is arranged to clarify the doubts of the farmers on various aspects of cashew cultivation and to express their feedback about the usefulness of the technologies. DCR, Puttur has so far organized 20 such programmes for the benefit of cashew growers, of which a total of 3000 farmers were benefited.

5. Capacity Building Programmes

5.1 Cashew Production Technology

In order to educate the extension personnel who are working in development departments in cashew growing states training programmes on "Cashew Production Technology" is being organized regularly. This programme is of two types' *viz.*, Refresher course and Special course. Refresher course is being organized for three days whereas special courses are being organized for five days. These programmes are open to extension personnel from all cashew growing states.

During the training, the participants will be trained on the various aspects of cashew cultivation through lecture-cum-discussion mode, visit to various experimental plots, visit to farmers' participatory demonstration plots, visit to private cashew nursery and processing industry. So far 50 such training programmes were conducted at this centre, out of which 1300 participants from all over the cashew growing states participated and benefited.

5.2 Vegetative Propagation

To meet out the demand for quality cashew planting material (softwood grafts), which is increasing in all cashew growing states, training programme on vegetative propagation of cashew is being organized regularly. During these training programmes theory and practical sessions on softwood grafting technique and nursery management are being imparted to the participants through intensive training for two days. Participants (nursery assistants/supervisors) from development departments and government owned plantations are regularly being deputed for this programme. So far at DCR, Puttur, 51 such training programmes were organized out of which around 650 participants were trained.

5.3 Cashew Apple Utilization

Cashew apples are rich in Vitamin C and minerals and have many advantageous effects on human beings. Interestingly, only 10-15 percent of the cashew fruits are utilized by processing, that too to produce country made liquor called 'Fenny' only in Goa. The fruit is wasted mainly due to its disliking qualities such as astringency and arid principles since the fruits are rich in tannin which causes throat and tongue irritation after eating. But the fruits can be made suitable for consumption by removing the undesirable tannin and preparing value added products such as clarified juice, syrup, squash, canned fruits, pickles, jam, jelly and chutney. With this aim DCR, Puttur initiated training programmes on 'Cashew apple utilization' with farm women as the target audience.

During the training the trainees will be imparted training through participatory learning (*i.e.*) 'learning by doing'. In addition to this literature on cashew apple utilization will be provided to the trainees. The method demonstrations so far organized were on cashew apple preparation *viz.*, juice/squash, jam and chutney. So far four such programmes were organized out of which a total of 293 farm/rural women were trained.

When the impact of these programmes were assessed by using post-exposure questionnaire (31 respondents) and through mailed questionnaire (20 respondents) it was found that the home level adoption of the demonstrated products was at a average of 50.00 per cent whereas the large scale adoption was poor. The reasons quoted by the respondents for non adoption of the products at a large scale were no awareness about the information on establishment of small scale units, no adequate facilities, non availability of adequate quantity of cashew apples, no adequate time, poor economical background and marketing problems.

5.4 Other Training Programmes

Apart from these, training on IPM in Cashew, pruning in cashew and composting of biomass, cashew apple utilization, soil and water conservation in cashew and high density planting and pruning in cashew are also provided. So far, fifteen such trainings were conducted covering more than 500 farmers and officials. The Directorate also conducts training programmes on request, apart from the trainings provided under the routine schedule.

6. Production and Supply of Planting Material

In order to produce and supply quality planting material for cashew (grafts) to the farmers and development departments, two revolving fund schemes were started at Directorate of Cashew Research, Puttur. These are ICAR Revolving Fund Scheme (at Kemminje campus) and DCCD Regional Nursery scheme (at Shantigodu campus). The ICAR revolving fund scheme is functioning from 1990-91 onwards whereas DCCD revolving fund scheme is in place from 1996-97. So far around 17,00,000 grafts were produced and distributed to farmers and development departments from the two revolving fund schemes.

7. Other Extension Efforts

7.1 Participation in Exhibitions

Apart from the above routine approaches DCR also participates in exhibitions/ expos at important venues across the country by putting up exhibition stall with posters/charts/live specimens depicting the news, information, technologies developed/available on cashew in the country thus aiding in technology dissemination to visiting farmers and other stakeholders. Scientists and technical staff of this Directorate also participated in the exhibitions for explaining technical information to the visitors.

7.2 Field Visits/Consultancy/Evaluation Programmes

Team of scientists has visited the cashew plots of farmers in all cashew growing states of India upon request. Scientists also participated in various workshops and seminars on cashew organized at various parts of India and presented paper and delivered lectures on various aspects of cashew cultivation. Periodically team of scientists from the Directorate participated in the phone-in programme of Doordarshan (Kannada) on Plant Protection and other aspects of cashew. Various Radio talks were delivered on cashew varieties and other cultivation practices and production technologies. Scientists of Directorate delivered talk on various aspects of cashew cultivation and processing aspects. Television interviews are also given on cashew varieties in other states which were telecast in Hindi also by D.D. News. The Scientists also participated in various State Level Workshops cum Seminars on Cashew and delivered talk on topics like cashew production technology, plant protection and training and pruning in cashew. DCR Scientists also visit the Frontline demonstration plots set up under NHM in other districts and states of the country for evaluation on the same.

In recent years, two trends of institutional context of research and extension can be observed. One trend is that central government agencies change their roles and are no longer the sole providers of research and extension services: this is due to public sector reforms, decentralization policies, privatization of formerly public agencies and, often enough, also due to financial crises which limit the range of activities that governments are still able to fund. The other trend is private sector enterprises become more important, *i.e.* private investment in cash crop technology is on the rise. The growing self-organization of farmers in producers' associations and community-based organizations as well as the increasing role of non-governmental organizations (the so-called 'Third Sector') in rural areas is frequently encouraged by development policies and international agencies.

In cashew sector, it is hard to find situations where farm innovations are brought about by private industry and market mechanisms alone. Normally, public entities (*e.g.* research institutes, universities), as well as private enterprises and third sector organizations (agricultural chambers, promotion councils, manufacturers associations, farmers organizations *etc.*) all can play a role in the innovation process and frequently have to work together to identify and introduce a new farming technique. A collaborative arrangement bringing together several (private and public) organizations working for technical change in cashew farming can be called a 'cashew innovation system'. Innovation systems are networks which organize around a common issue. Within the network, innovations can be promoted by different participants, not only by research institutes or extension services. What we see is in fact an increasing pluralism of (private and public) organizations with a potential role in cashew development. There are multiple sources of innovation and often several (public, semi-public or private) providers of research, information and advisory services and therefore a wealth of possibilities of 'cashew innovations' that can be initiated.

8. Future Line of Action

- ☆ Utilising advances in knowledge management using ICTs towards better technology realization and transfer of technology. Platforms like mobile telephony are to be utilised by implementing mobile advisory services to cashew farmers.
- ☆ Retaining interest of new generation cashew farmers and attracting more of them towards cashew farming is a big challenge. Utilising social media marketing by reaching them through innovative social media platforms like Facebook and Twitter is to be explored in this regard. Transfer of technology efforts of DCR has to be made available on social media platforms for increased reach and effective transfer of technology.
- ☆ Better utilization of advances in social capital research has to be made towards social mobilization for formation of cashew farmer groups aiding in participatory transfer of latest cashew production technologies.
- ☆ Identification of sustainable cashew based farming systems through extension research will prove to be beneficial towards planning meaningful outreach activities for overall improvement of cashew farmer livelihoods.
- ☆ Knowledge about technology fatigue existing in the Cashew sector is presently not much available. More emphasis has to be given for extension research and the knowledge about technology fatigue has to be made available in the coming years. This, along with advances in participatory research has to be used towards successful technology backstopping in the future.

References

Axinn, G. H. (1988). Guide on alternative extension approaches. FAO (Food and Agriculture Organization of the United Nations), Rome, Italy.

Birkhaeuser, D., Evenson, R. E. and Feder, G. (1991). The economic impact of agricultural extension: A review. *Economic Development and Cultural Change,* 39(3): 607–650.

BMZ (Bundesministerium für Wirtschaftliche Zusammenarbeit).1990. Querschnittsanalyse von Projekten der Landlichen Regionalentwicklung. BMZ, Bonn, Germany.

Feder, G., Willett, A. and Zijp, W. (2001). Agricultural extension: Generic challenges and the ingredients for solution. In: Wolf S and Zilberman D (Eds), Knowledge generation and technical change: Institutional innovation in agriculture. Kluwer, Boston, Massachusetts, USA.

Hildebrand, P.E. (2001). A personal history in FSR. In: Collinson M (Ed), A history of farming systems research. CABI (Commonwealth Agricultural Bureau International) Publishing, London, UK.

IBRD (International Bank for Reconstruction and Development). (1987). World Bank experience with rural development, 1965–1986. The World Bank, Washington, DC, USA.

Moris, J. (1991). Extension alternatives in tropical Africa. ODI (Overseas Development Institute), London, UK.

Purcell, D. L. and Anderson, J.R. (1997). Agricultural research and extension: Achievements and problems in national systems. World Bank Operations Evaluation Study, World Bank, Washington, DC, USA.

Rauch, T. (1993). Regional rural development RRD update: Elements of a strategy for implementing the RRD concept in a changed operational context. Deutsche Gesellschaft für Technische Zusammenarbeit (GTZ), Eschborn, Germany.

Rivera, W.M., Qamar, M.K. and Van Crowder, L. (2001). Agricultural and rural extension worldwide: Options for institutional reform in the developing countries. FAO (Food and Agriculture Organization of the United Nations), Rome, Italy.

Swanson, B., Bentz, R. and Sofranko, A. (1997). Improving agricultural extension: A reference manual. FAO (Food and Agriculture Organization of the United Nations), Rome, Italy.

Cashew: Improvement, Production and Processing (2017) *Pages* **495–521**
Editor: **Prof. P.L. Saroj**
Published by: **ASTRAL INTERNATIONAL PVT. LTD., NEW DELHI**

Economics of Cashew Cultivation

P.B. Pushpalatha, P. Indira Devi and A. Sobhana

1. Introduction

Cashew is valued as one of the most important commercial horticultural crop in India, which can facilitate rural development. More than six lakh farming households derive benefit from cashew through providing employment to 1.5 million people directly and indirectly (Hubballi, 2013). The economic importance of cashew cultivation is mainly based on the major produce raw nut and associated production of cashew apple, cashew nut shell liquid and to a small extent the timber and firewood value. It has a unique role in industrial and economic performance, livelihood support to farming communities, export earnings, agricultural and rural development and sustainable agriculture. Apart from protection of environment and income generation that enhances food security, cashew also contributes to the nutritional balance of people (Masawe and Kapinga, 2010).

The development efforts of the past have helped in increasing the area and production of cashew in India. The area and production of cashew in 1965-66 was 2.41 lakh ha and 1.04 lakh MT respectively with a productivity of 400 kg/ha. Due to the concerted efforts during 2nd and 3rd five year plan periods and the support during subsequent plan periods, the area and production of cashew increased to 3.75 lakh ha and 1.62 lakh MT respectively in 1975-76. Later on, the increase was steady and it was 5.18 lakh ha with 2.40 MT of production. From 1985-86 to 2011-12 the area has been registering a growth rate of 3.50 per cent per annum, where as the increase in production was to the tune of 7.50 per cent per annum. Currently, India is the largest producer of raw cashew nut in the world with 9.82 lakh ha under cultivation and 7.28 lakh MT production with a productivity of 772kg/ha (Hubballi, 2013).

The increase in nut production can be attributed to the adoption of scientific technologies evolved due to the research bestowed in cashew cultivation (Balasubramanian *et al.*, 2013). The release of high yielding varieties suitable for different areas coupled with the innovation of softwood grafting paved the path for production of planting materials with assured yield potential. Moreover, the High Density Planting (HDP) method which accommodates 625 plants/ha (4mx4m) as against 178 plants/ha (7.5mx7.5m) under the conventional practice, has also facilitated the increase in output.

However, some of the inherent issues limit the wider adoption of the crop and modern technologies. Most of these technologies demand better management which necessitates higher investment. Sivanatham *et al.* (1990) reported that the labour cost for maintenance of improved variety accounted 40.24 per cent of total cost and material cost was 51.88 per cent. Moreover, there is no income from cashew plantations for the initial 3-4 years. Simultaneously, there are signals from the consumer side that reflects increasing demand for cashew nut as well as quality products. The health conscious consumers are demanding organic produce.

There are research outcomes that address these issues. For instance, the research undertaken in different parts of the country have proved the potential of intercropping in cashew garden during the initial period to compensate for the lean period. Organic cashew production technology is gaining increased acceptance owing to the premium price and higher returns. Shetye *et al.* (2012) worked out the benefit cost ratio in inorganic and organic cashew orchards as 1.59 and 1.66 respectively. The technology for HDP and utilization of cashew apple for harnessing additional income are some of the techniques that make the enterprise more farmers friendly. The information on the economic aspects of these technologies and practices can facilitate rational decision making, at micro and macro levels. This paper details these scientific aspects and assesses the relative economics of some of these technologies. The analysis is done based on the data compiled from the basic records of Cashew Research Station, Madakkathara, Thrissur, under Kerala Agricultural University. The prices and costs correspond to respective period of time.

2. Cashew Cultivation: Management and Economics

2.1 Selection of Site for Planting

Performance of cashew is mainly dependent upon the soil and climatic factors of the area. A significant yield variation in cashewnut production among different agro climatic zones of the country is reported by Prasadarao and Gopakumar (2009). Suitable sites ensure higher productivity through the selection of appropriate variety. The weather condition also influence on the pest and disease attack which is also a factor deciding the yield level.

2.2 Establishment of Plantation

The recommended planting method for cashew is square system at spacing of 7.5m x 7.5m, in poor to marginal soils. This accommodates 178 plants/ha. According to the nutrient status of soil and terrain of land, the system of planting can vary and hence the planting density also varies. In fertile soils, the spacing is 10m x 10m and planting density is 100 per hectare under the square system. With the same spacing the planting density can go up to 116 plants in hexagonal system (KAU, 1996). The initial investments for the planting and related activities (clearing the area, making pits, planting and after care) when the planting density is 178 is estimated as Rs.47, 300 per hectare, as these are labour intensive activities where major cost component is the wages. Gap filling is necessary if the grafts fail to establish. Mulching, manuring and pest management are the other management aspects.

2.3. Management

2.3.1. Vegetative Phase

The grafts take three years to attain yielding stage. During this period, weeding, pruning, training and plant protection operations are to be done. Weed control in cashew plantation is very vital in ensuring better returns. Salam *et al.* (1993) reported 60- 70 per cent yield loss due to weed competition alone. Manual weeding is the common practice in cashew plantations. Recently, the labour scarcity and wage factors have prompted the farmers to opt for chemical weed control practices. The relative cost of manual methods as compared to chemical methods proved the latter to be cost effective. The cost of chemical weed control was Rs. 874 per hectare while that of manual method was nearly three times (Rs. 2500/-). However, the ecosystem damage due to chemicals is a major cause of concern and hence manual/mechanical methods are generally preferred in a social perspective.

Training and pruning in cashew plantations are done during the initial years (1-3 years). This ensures proper shaping of canopy for adequate sunlight which ensures better flowering and nut set. Lack of sunlight restrict flowering and production of bisexual flowers resulting in reduction of yield (Pushpalatha, 2000). Proper pruning and training also facilitate intercultural operations such as terrace making, weeding, fertilizer application, nut collection and plant protection. The management costs do not vary much during the first three years, except in the case of fertilizer dose. The dose of fertilizer nutrients varies according to growth phase.

2.3.2 Productive Phase

During the productive phase, the major management practices include nutrient, weed and pest management activities. Cashew responds very well to application of nutrients. Trials conducted in different parts of the country have shown a substantial increase in productivity of cashew with timely application of fertilizers. As cashew is mainly cultivated in laterite soils, the deficiency of major nutrients are often noticed. Veeraraghavan *et al.* (1985) observed significant yield increase in cashew with the application of nitrogenous fertilizers. He also observed significant increase in shelling percentage from 29.5 per cent to 35.6 per cent by application of N at the highest level of 500g/tree/year. According to Kumar (1985) the vegetative characters, yield as well as apple and nut characters were better at 450 g N/tree/year. Mathew (1990) reported a nut yield of 8.34 kg t^{-1} at 1000 g N tree/year against 6.91 kg t^{-1} at 250 g N tree/year. Latha (1992) reported a high yield of 8.7 kg t^{-1} with early application of 1000 g N t^{-1}. It was seven fold higher than the yield where no fertilizers were applied since planting.

Phosphorus also plays significant role in the early growth and nut development stage. Application of 250 g Phosphorous per tree could yield highest number of nuts per panicle (Kumar, 1982).A yield increase of 73 percent was achieved when P was raised from 125 to 500 g per tree (Mathew, 1990). He also confirmed the effect of potassium on improving yield of cashew.

Cashew is largely seen as a component in home gardens or as orchards on marginal lands. Hence, scientific management is not usually followed. Most of the cashew plantations are rainfed, though irrigation during fruit development stage helps in improving production. Therefore, economic analysis is done considering rainfed condition only.

2.3.3 Pest Management

Pest infestation is a serious menace in cashew plantation. Tea mosquito bug (TMB) (*Helopeltis antonii*), stem and root borer (*Plocaederus ferrugineus*), flower thrips (*Selenothripis rubrocinctus*), apple and nut borer (*Thylocoptila paurosema*) and leaf and blossom webber (*Macalla moncousalis*) are important pests of cashew. Of these, tea mosquito bug (TMB) and cashew stem and root borer (CSRB) are the serious pests. About 30-40 per cent crop loss is reported due to tea mosquito attack alone. For the management of TMB; Lambda Cyhaothrin (0.6ml/l) at flushing stage, Quinolphos (2ml/l) at flowering stage and Carbryl (2g/l) at nut initiation and fruit development stage is recommended. Whereas for CSRB,application of Chlorpyriphos (10ml/l) during April-May and December-January was found effective.

Continuous pest attack reduce the longevity of trees and hence result is a shortened yielding period. Adoption of plant protection measure ensure a yield increase of 532kg/ha amounting to Rs 39,900/ha. Deducting the cost of chemical and application charges of Rs 11030/- including Rs. 7690/ha for TMB management and Rs. 3340/ha for CSRB management; an amount of Rs 28870/- more can be expected from a hectare of plantation. Thus, scientific pest management result in an increase of 61 per cent in net returns, than the control (Table 30.1).

Table 30.1: Economics of Pest Management in Cashew Plantations in Kerala

Sl.No.	Particulars	Control	Chemical Pest Control
1.	No. of trees/ha	178	178
2.	Cost of plant protection	–	11030*
3.	Yield (kg/ha)	625	1157
4.	Income (Rs.)	46875	86775
5.	Net income (Rs./ha)	46875	75745
6.	Additional income	28870	

* Rs. 7690/ha for TMB and Rs. 3340/ha for CSRB management.

2.4. Yield and Returns

The productive phase of cashew trees starts at the age of 3 years and the yield gets stabilized at 8/9th year. From 9th to 20th year the yield is steady and then it starts declining. Usually, trees are felled at 25th year. The cashew timber have good demand as the base portion upto1½ - 2m height is mostly straight and good for making packing cases and carving into sculpture. Rest of the wood is mainly used as fuel wood. On an average the returns from the timber and firewood portion amounts to Rs. 4000-5000/- per tree.

Table 30.2: Economics of Cashew Cultivation Under Normal Density Planting (7.5 m x 7.5 m; 178 plants/ha)

Age (Year)	*Expenditure (Rs)*	*Yield/Tree (kg)*	*Total Yield (kg)*	*Income (kg)*	*Net Return (kg)*
Establishment	47300				
1	12650	–	–	–	–
2	12800	–	–	–	–
3	17350	0.5	89	6675	–10675
4	23850	1.0	178	13350	–10500
5	27150	3	534	40050	12900
6	30300	5	890	66750	36450
7	30300	7	1246	93450	63150
8	33600	9	1602	120150	86550
9	33600	10	1780	133500	99900
10	37400	10	1780	133500	96100
11	37400	10	1780	133500	96100
12	37400	10	1780	133500	96100
13	37400	10	1780	133500	96100
14	37400	10	1780	133500	96100
15	37400	10	1780	133500	96100
16	37400	10	1780	133500	96100
17	37400	10	1780	133500	96100
18	37400	10	1780	133500	96100
19	37400	10	1780	133500	96100
20	37400	10	1780	133500	96100
21	37400	8	1424	106800	69400
22	37400	8	1424	106800	69400
23	37400	6.5	1157	86775	49375
24	37400	4.0	712	53400	16000
Salvage value				8010000	

(Cost of cashew nut Rs. 75/kg), Yield of cashew nut calculated as follows; *Yielding stage start at 3rd year after planting.

*Stabilized yield from 10th year after planting; *Stabilized yield expected up to 20th year; Salvage Value: Rs 4500 x 178 trees=Rs.8,01,000; Net Present Worth (14 per cent) =Rs.2,00,486; Benefit Cost ratio (BCR) =1.88; Internal Rate of Return=29 per cent.

Appendix 1 details the year wise expenses for cultivation of the crop and Table 30.2 summarizes the cash inflows and outflows from the plantation and the financial viability of the investment. Assuming a discount rate of 14 per cent (being the average lending rate of commercial Banks for agricultural term loans), the Net Present Value of the enterprise was estimated at Rs.2,00,486 with a Benefit Cost Ratio of 1.88. The Internal rate of Return (IRR) was 29 per cent, which is fairly higher than the opportunity cost of capital. The analysis justifies the investment made in cashew and the crop ensures economically attractive returns to the farmer.

3. Economics of Emerging Technologies

3.1. High Density Planting (HDP)

HDP is a technique recommended for enhancing the productivity of cashew plantations (Salam, 1997). This technique involves planting more number of grafts per unit area. Later, monitoring the canopy pressure of the plantation, selective felling or pruning is done. In the conventional method, planting density vary from 100 to 177 plants per hectare (spacing ranging from 7.5 to 10m) in the square system of planting. It may be 74 to 204 plants (spacing ranging from 7.5 to 10m) in the triangular system of planting. In high density planting, 312 to 625 grafts are planted. As more number of plants are planted and managed per unit area, the cost of management is more. The establishment cost under this system is Rs.1, 40,750 which is nearly thrice that of conventional system. The maintenance expenditure is more than two times during the stabilized periods. The nutrient management in high density system is same as that of the conventional system on a per tree basis.

During the initial stage (3-6 year after planting) the yield (per tree) from normal and high density system is almost same. Yield gets stabilized from 8/10th year onwards. But selective pruning is necessary during this period. The stabilized yield continues till 20th year. Based on an experiment conducted at National Research Centre Puttur Yadukumar *et al.* (2001) reported that yield levels were maintained by pruning to 80 per cent light interception at early stages and later stages by diagonal thinning. However, Suma *et al.* (1998), based on her research at Cashew Research Station, Madakkatahara, Thrissur, Kerala, reported differing results that there is no necessity on thinning to promote higher yield.

Experiments conducted at Ratnagiri, Maharashtra to evaluate the optimum crop density for maximizing profit from cashew during initial years of production revealed that the gross income, gross expenditure and net profit were considerably higher in the high density orchard planted at 5mx5m, spacing (Despande *et al.*, 2002). The study on high density planting conducted at Cashew Research Station; Madakkathara during 1996-2008 revealed that cumulative yield per tree of eight years was marginally high under normal density planting over high density. The cumulative per hectare yield was 3.8 times high under high density system (Annual Research Report, 2008). Mathew and Mini (2008) conducted the economic analysis of the high density system and found significant increase in net income and benefit cost ratio.

Sustained yield could be obtained up to 20th year after planting in this system. Afterwards the yield start to decline and 23 years after planting the maintenance of plantation become uneconomical. The longevity of trees may get reduced due to the continuous pruning of plantation. The disposal of trees after economic yielding time can fetch more income as there is more number of trees per unit area in high density system. The income from disposal of trees ranges from Rs 4000 to 5000 per tree. Appendix 2 details the cost components in this system, on a year wise basis. Table 30.3 furnishes the results of the analysis to assess the economics of this system. The analysis justifies the technology with an NPW of Rs.2,16,527 and BC Ratio of 1.49. The IRR is higher than the opportunity cost of capital. The investment efficiency however is better in conventional system, as revealed by the better economic signals.

Table 30.3: Economics of Cashew Cultivation (High density planting - 625 plants/ha)

Age (Year)	*Expenditure*	*Yield*	*Total Yield*	*Income*	*Net Return*
Establishment	140750				
1	21650	–	–	–	–
2	22100	–	–	–	–
3	36350	0.5	313	23438	–12913
4	40200	1.0	625	46875	6675
5	66450	3.0	1875	140625	74175
6	75900	3.5	2188	164063	88163
7	75900	4.0	2500	187500	11600
8	81700	4.5	2813	210938	129238
9	81700	4.5	2813	234375	152675
10	81700	4.5	2813	210938	129238
11	81700	4.5	2813	210938	129238
12	81700	4.5	2813	210938	129238
13	81700	4.5	2813	210938	129238
14	81700	4.5	2813	210938	129238
15	81700	4.5	2813	210938	129238
16	81700	4.5	2813	210938	129238
17	81700	4.5	2813	210938	129238
18	81700	4.5	2813	210938	129238
19	81700	4.5	2813	210938	129238
20	81700	4.5	2813	210938	129238
21	71200	3.0	2813	140625	69425
22	69800	2.0	1250	93750	23950
Salvage Value				2812500	2812500

Salvage Value = Rs. 4500 x 625 trees =28,12,500; Net Present Worth (14 per cent)=Rs. 2,16,527; Benefit Cost ratio (BCR)= 1.49; Internal Rate of Return= 23 per cent.

3.2 Organic Farming

Being an export oriented crop, organic farming in cashew has more relevance since it can revitalize exports by exploiting the premium price for "organic cashew" in niche markets. A thorough understanding of the cashew agro-ecosystem is imperative in formulating an organic module. In view of the growing demand for organically grown cashew, strategies need to be developed for organic farming. Most of the cashew growing countries are in an advantageous position as nearly 90 per cent of the cashew plantations are organic by default where in chemical fertilizers are not usually applied. Most of the countries can effectively exploit this situation to their credit and can initiate organized production of organic cashew under approved certification, without any yield decline normally associated with crops where a transition is made from chemical farming to organic farming.

The major components of organic farming in crops, including cashew are crop rotation/cropping system involving suitable crops, use of crop residues, use of animal manures, inclusion of legumes in the cropping system, use of green manures, use of off-farm organic wastes, mechanical weed management, use of mineral bearing rocks and biological pest control. In this background, the standardized procedure to convert cashew leaf litter and apple waste available in cashew gardens into suitable compost material, assumes significance (Nampoothiri, 2001).Organic nutrient management in cashew involves the use of around 25kg of poultry manure or 60 kg of FYM or 30 kg of vermicompost per annum per tree. Apply $1/5^{th}$ dose of the organic manure during the first year, $2/5^{th}$ dose during the second year, progressively reaching full dose from fifth year onwards. Cashew apple residue and leaf litter could be effectively used for vermi composting. Combined application of *Azotobacter* and *Azospirillum* each @ 150g/plant is beneficial for increased yield.

3.3 Intercropping in Cashew Plantations

Being a crop with a gestation period of three years, cashew farmers often prefer to cultivate intercrops in their garden. The choices of the crop depend on the soil and climate of the region and the age of the plantation. The intensity of light penetration and root volume differs with age of the tree and accordingly the suitability of the intercrop also varies. Intercropping is highly significant as it reduces the weed competition and helps in maintaining fertility of orchards apart from generating income during the gestation period. It is advised to grow leguminous crops as it enhance the fertility of cashew orchards (Adeyemi, 1998).

During first year, about 90 per cent of the area is available for intercropping which reduces to 80-70 per cent during second and third years. Short duration crops like pineapple, papaya, cowpea, banana, ginger, turmeric vegetables and tapioca can be successfully grown in cashew plantation of 1-3 years. (KAU, 1996). Highest returns from intercrops were from tapioca cultivation (Mathew *et al.*, 2008) followed by Coleus and colocasia. Ginger is reported to be a suitable intercrop in cashew garden as reported by Vikram and Hegde (2014). Adewole and Adeoye (2014) reported maize as a suitable intercrop in early years of cashew growth. The relative economics of different intercrops in cashew plantations is given in Table 30.4. The detailed statement on the costs and returns is furnished as Appendix.III

While turmeric realizes a net gain of Rs.70000 within a short span of four months, it is nearly five times in six months if ginger is grown (Rs. 3,29,300). Pine apple farming in cashew plantations generates an income of Rs.3,0,2500 per ha over the life period of 18 months. Further, this facilitates to achieve food security and the crop residues help build up the soil structure, through crop residue inclusion.

Table 30.4: Relative Economics of Intercropping in Cashew Plantation (Rs/ha) in Kerala

Crop	*Duration*	*Total Cost (Paid out)*	*Total Returns*	*Net Returns*
Turmeric	4 months	192500	262500	70,000
Pineapple	18 months	373500	675000	3,01,500
Ginger	6 months	270700	600000	3,29,300

3.4 Top Working

Top working is a technique evolved to rejuvenate unproductive/senile cashew trees (KAU, 1990). This technology can successfully rejuvenate poor yielders in the age group of 15-20 years. The unproductive trees are to be beheaded at a height of 0.75 to 1.00 m from ground level. Sprouts emerge by 30-45 days after beheading. At one month age, these can be grafted with scions of desired varieties. Top working is a labour consuming process as the trees are to be beheaded and care is to be bestowed to protect the stumps. Large number of sprouts emerging from the stumps is to be removed retaining only the ideal ones. To get good canopy cover 6-7 grafted sprouts are to be retained, on four sides of the tree. For this, double the number of grafting is to be done. Accordingly, expenditure for preparation of scion is more and cost of scion sticks required for grafting is also on the higher scale. Obviously the labour for grafting is more as the grafting is done in the field on stumps of 0.75-1m height. However, this saves additional cost and efforts towards new planting.

The top worked trees start to yield in 2nd year after top working ie.one year ahead of normal ones, which is due to the well established root system. Yields of young top worked trees, grafted trees and seedling trees were compared in cashew orchards by Rossetti *et al.* (1998). At the end of fourth year cumulative yield was almost same in top worked plantations (968.90kg/ha) and grafted ones (1057.05kg/ha). Both were higher than the cumulative yield in seedling trees (257.76kg/ha). However, over years the yield is observed to be less because of low canopy volume of top worked trees.

The yield starts to increase steadily from 3rd year after top working and get stabilized at 8th year after grafting. When grafted with scions of high yielding varieties an yield of 6-8 kg/tree can be expected. The economic life of properly managed plantation is reported to be 18-20 years after top working. The present analysis is based on the yield data for 13 years from trees top worked at 20th year at Cashew Research Station, Madakkathara. The salvage value of the timber is not included, as the timber quality is observed to be of comparatively poor.

An account of the expenditure towards top worked orchad and the management costs and the benefits realized is given in Table 30.5. Appendix 4 provides the details of the costs and returns. The economic analysis does not indicate an attractive picture

Table 30.5. Economics of top worked cashew plantation/ha

Age (Year)	*Expenditure (Rs.)*	*Yield/Tree (kg)*	*Total Yield*	*Income (kg)*	*Net Return (kg)*
Establishment	47300				–47300
1	12650	–	–	–	–12650
2	12800	–	–	–	–12800
3	17350	0.5	89	6675	–10675
4	23850	1.0	178	13350	–10500
5	27150	3	534	40050	12900
6	30300	5	890	66750	36450
7	30300	7	1246	93450	63150
8	33600	9	1602	120150	86550
9	33600	10	1780	133500	99900
10	37400	10	1780	133500	96100
11	37400	10	1780	133500	96100
12	37400	10	1780	133500	96100
13	37400	10	1780	133500	96100
14	37400	10	1780	133500	96100
15	37400	10	1780	133500	96100
16	37400	10	1780	133500	96100
17	37400	10	1780	133500	96100
18	37400	10	1780	133500	96100
19	37400	10	1780	133500	96100
20	39250	–	–	–	–39250
21	8350	–	–	–	–8350
22	26100	3	534	40050	13950
23	28900	3	534	40050	11150
24	29600	5	890	66750	37150
25	30450	6	1068	80100	49650
26	30800	6	1068	80100	49300
27	33200	6	1068	80100	46900
28	33200	8	1424	106800	73600
29	33200	8	1424	106800	73600
30	33200	8	1424	106800	73600
31	33200	8	1424	106800	73600
32	33200	8	1424	106800	73600
33	33200	8	1424	106800	73600
34	33200	8	1424	106800	73600
35	33200	8	1424	106800	73600

Net Present Worth (14 per cent)=Rs.5071; Benefit Cost ratio (BCR) =1.02; Internal Rate of Return =29 per cent.

of the technology. However, in the absence of resource scarcity or management inabilities, the technology can be attempted when the existing plantations are at the risk of senility due to pest/disease incidence or due to any other weather factors. The technology ensures an extended period of returns in that case.

Conclusion

Cashew productivity can only be improved by adopting scientific technologies. It is therefore important that the technologies are to be demonstarted before the farmers with their economic feasibility. Those technologies which are economically not viable need refinement and only cost effective technologies are to be popularized. Apart from production and processing of raw nut, the related sectors like processing of CNSL, cashew apple utilization and nursery production should also get attention and their economics may be worked out. The present chapter highlighted economic analysis of cashew cultivation under rainfed condition along with high density planting system, intercropping in cashew orchard, organic cultivation and orchard rejuvenation through top working.

References

Adeyemi, A.A. (1998). Effects of intercropping on weed incidence in cashew (*Anacardium occidentale* L.) plantations. *Nigerian Journal of Tree Crop Research*, 2(1): 83-94.

Adewole, M.B. and Adeoye, G.O. (2014). Assessment of soil properties and crop yield under agroforestry in the traditional farming system. *African Journal of Argriculture Research*, 9 (27): 2119-2123.

Annual Research Report. (2008) Cashew Research Station, Madakkathara, Thrissur, Kerala.

Balasubramanian, D. Saroj, P.L., Meena, R.K. and Vanitha, K. (2013). Research status for technological development of cashew in India. Jnanadevan, R. (Ed.), *In*: *Souvenir of National Conf. on Cashew*. Ranchi, Jharkhand. November 2013: pp. 32-37.

Deshpande, S.B., Haldankar, P.M., Sapkal, B.B., Chavan, A.P., Joshi, G.D., Patil, B.P. and Raao, E.V.V.B. (2002). Evaluation of high density planting in cashew (*Anacardium occidentale* L.) *The Cashew*. 16 (1): 23-28.

Hubbali, V. N. (2013). Cashew Development in India; Strategies for non-traditional states. Jnanadevan, R. (Ed.), In: *Souvenir of National Conf. on Cashew*. Ranchi, Jharkhand. November 2013: pp. 22-29.

KAU (1990). Annual Report, All India Co-ordinated Research Project on Cashew. Madakkathara.

KAU (1996). Packages of Practices Recommendation-Crops. Kerala Agricultural University, Thrissur, p. 267.

Khan, M.M., Hedge, M., Allik, B., Hiremath, I.G., Hanamashetti, S.I., Rao, V.N.M. and Krishnamurthy, K. (1985). Rejuvenation of old cashew trees by top working. *Indian Cashew Journal*, 17 (3): 9-25.

Latha, A. (1992). Growth and yield of cashew in relation to foliar and soil nutrient levels. M. Sc (Ag) thesis submitted to Kerala Agricultural University, Vellanikkara, Thrissur.

Mathew, R. (1990).Yield prediction in cashew based on foliar nutrient levels. M. Sc. (Ag) thesis, Kerala Agricultural University, Vellanikkara, Thrissur. p. 210.

Mathew J, and Mini, C. (2008). Manipulation of tree densities for yield and income maximization in cashew. *Proc.3rd Indian Horticulture Congress,* Nov 6-9, 2008, Bhubaneswar, p.203.

Masawe, P.A.L. and Kapinga, F.A. (2010). Review of Research and Development Activities on Cashew in Eastern and Southern Africa.In: *Proceedings of the Second International Cashew Conference,* Kampala, Uganda-26-29, April, 2010 (Eds.Masawe, P.A.L., Esegu, J.F.O., Kasuga, L.J.F., Mneney, E.E. and Mujuni, D.) CAB International,Wallingford, U.K. pp. 133-139.

Mathew J, Gregory Zachariah and Mini, C. (2008). Economic potentials of tuber crops for intercropping in young cashew plantaions. *J. Plantation Crops,* 36 (3): 366-367.

Nampoothiri, K. U. K. (2001). Organic farming it's relevance to plantation crops. *Journal of Plantation Crops,* 29(1): 1-9.

Prasadarao, G.S.L.H.V and Gopakumar, C.S. (2009). Agroclimatology of cashew. In *Cashew Research and Development in Humid Tropics.* (Eds. Jose Mathew, C. Mini and Mareen Abraham). Directorate of Extension, KAU, Mannuthy, Thrissur, Kerala, pp. 26-36.

Pushpalatha, P.B. (2000). Morphophysiological analysis of growth and yield in cashew (*Anacardium occidentale L.*) Ph.d Thesis. Kerala Agricultural University, Thrissur.

Rossetti, A.G., Correa, M.P.F. and Pinheiro, D. M. (1998). Recovery of young dwarf cashew orchards propagated by seed. *Revista Brasileira de Fruticultura,* 20 (2): 201-205.

Salam, M. A, Pushpalatha P.B., Suma, A and Abraham, C.T. (1993). Efficacy of chemical weed control in cashew plantations. *J. Plantation Crops,* 21 (1): 54-56.

Salam, M. A. (1997). High density planting in cashew- principles and practices. The *Cashew,* 11 (4): 12-20.

Shetye, A.S., Talathi, J.M., Naik, V.G., Torane, S.R. and Wagale, S. A. (2012). Cost returns and profitability in organic and inorganic cashew production. *International Research Journal of Agriculture Economics and Statistics,* 3 (2): 333-336.

Sivanatham, M., Pugalendhi, L., Jeeva, S. and Somasundaran, D. (1990). An economic analysis of cashew cultivation in South Arcot districts of Tamil Nadu. *The Cashew,* 4 (3): 17-19.

Suma, B., Salam, M.A., Suma, A. And George, T. E. (19980. effect of planting densities on the nut yield of cashew (*Anacardium occidentale* L.). *The Cashew,* 12 (2): 33-3.7

Salam, M.A.1999. Cashew varieties suitable for high density planting. Cashew, 13 (4): 15-19.

Veeraraghavan, P.G., Celina, W.A and Balakrishnan, S. (1985). Studies on the fertilizer requirement of cashew (*Anacardium occidentale* L.). *Cashew Causere,* 7 (2): 6-8.

Vikram, H.C. and Hedge, N.K. (2014). Performance of ginger in cashew plantation (as intercrop) compared to sole cropping. *Asian Journal of Horticulture,* 9(1): 187-189.

Yadukumar, N., Rao, E.V.V B. and Mohan, E. (2001). High density planting of cashew (*Anacardium occidentale* L.). *Tropical Agriculture,* 78 (1): 19-28.

Appendix I: Cost of Cultivation of Cashew Plantation (Planting density-178/ha)

A. Establishment cost (Rs./ha)

B. Maintenance cost

B1 : Year 1

Sl.No.	Work	Labour		Cost (Rs.)	Amount (Rs.)
		Men	Women		
	Input cost				
1.	Cost of fertilizer(quantity 1/5th of recommended dose)				1500
	Labour cost				
1.	Multching @ Rs. 350/woman		3	1050	1050
2.	Weeding of plant basins		5	1750	1750
3.	Initial training and pruning @ Rs. 500/man	3		1500	1500
5.	Application of fertilizer	1	1	850	850
6.	Weeding (4 times/year)	12		6000	6000
	Total	**16**	**9**	**10300**	**12650**

B2 :Year 2

Sl.No.	Work	Labour		Cost (Rs.)	Amount (Rs.)
		Men	Women		
	Input cost				
1.	Cost of fertilizer (2/5th of recommended dose)				3000
2.	Weeding (4 times/year)	12			6000
	Labour cost				
1.	Weeding of plant basins		6	2100	2100
2.	Application of fertilizer	2	2	1700	1700
	Total	**2**	**8**	**3800**	**12800**

B3 : Year 3

Sl.No.	Work	Labour		Cost (Rs.)	Amount (Rs.)
		Men	Women		
	Input cost				
1.	Fertilizer cost (3/5th of recommended dose)				4500
	Labour cost				
1.	Weeding of plant basins		8	2800	2800
3.	Application of fertilizer	3	3	2550	2550
4.	Weeding (4 times/year)	15		7500	7500
	Total 18	**11**	**12850**	**17350**	

B4 : Year 4

Sl.No.	Work	Labour		Cost (Rs.)	Amount (Rs.)
		Men	Women		
	Input cost				
1.	Cost of plant protection chemicals				1250
2.	Cost of fertilizer (4/5th of recommended dose)				6000
	Labour cost				
1.	Weeding of plant basins		6	2100	2100
2.	Application of fertilizer 4/5th of recommended dose	4	4	3400	3400
3.	Application of plant protection chemicals	2		1000	1000
4.	Weeding	16		8000	8000
5.	Harvesting (2 woman 3 times)		6	2100	2100
	Total	**22**	**16**	**16600**	**23850**

B5 : Year 5

Sl.No.	Work	Labour		Cost (Rs.)	Amount (Rs.)
		Men	Women		
	Input cost				
1.	Application of fertilizer				3400
2.	Cost of Plant Protection				1500
3.	Cost of fertilizer (full dose)				7500
	Labour cost				
1.	Weeding of plant basins		6	2100	2100
2.	Application of PP chemicals	3			1500
3.	Weeding (4 times/year)	16			8000
4.	Harvest (3 women 3 times)		9	3150	3150
	Total	**19**	**15**	**5250**	**27150**

B6 : Year 6

Sl.No.	Work	Labour		Cost (Rs.)	Amount (Rs.)
		Men	Women		
	Input cost				
1.	Cost of fertilizer				7500
2.	Application of fertilizer				3400
3.	Cost of plant protection chemicals				2000

Contd...

Contd...

Sl.No.	Work	Labour		Cost (Rs.)	Amount (Rs.)
		Men	Women		
	Labour cost				
1.	Weeding of plant basins		8	2800	2800
2.	Application of plant protection chemicals	2		1000	1000
3.	Weeding (4 times/year)	16		8000	8000
4.	Harvest (4 women 4 times)		16	5600	5600
	Total	**18**	**24**	**17400**	**30300**

B7 : Year 7

Sl.No.	Work	Labour		Cost (Rs.)	Amount (Rs.)
		Men	Women		
	Input cost				
1.	Application of fertilizer				3400
2.	Cost of fertilizer				7500
3.	Cost of plant protection chemicals				2000
	Labour cost				
1.	Weeding of plant basins		8	2800	2800
2.	Application of plant protection chemicals	2		1000	1000
3.	Weeding (4 times/year)	16		8000	8000
4.	Harvest (4 women 4 times)		16	5600	5600
	Total	**18**	**24**	**17400**	**30300**

B8 : Year 8

Sl.No.	Work	Labour		Cost (Rs.)	Amount (Rs.)
		Men	Women		
	Input cost				
1.	Application of fertilizer				3400
2.	Cost of plant protection chemicals				2000
3.	Cost of fertilizer				7500
	Labour cost				
1.	Weeding of plant basins		8	2800	2800
2.	Application of plant protection chemicals	2		1000	1000
3.	Weeding (4 times/year)	16		8000	8000
4.	Harvest (4 women 4 times)		16	5600	5600
	Total	**18**	**24**	**17400**	**30300**

B9 : Year 9

Sl.No.	Work	Labour Men	Labour Women	Cost (Rs.)	Amount (Rs.)
	Input cost				
1.	Application of fertilizer				3400
2.	Cost of fertilizer				7500
3.	Cost of plant protection chemicals				2500
	Labour cost				
1.	Weeding of plant basins		8	2800	2800
2.	Application of Plant protection chemicals	2		1000	1000
3.	Weeding (4 times/year)	16		8000	8000
4.	Harvest (6 women 4 times)		24	8400	8400
	Total	**18**	**32**	**20200**	**33600**

B10 : Year 10

Sl.No.	Work	Labour Men	Labour Women	Cost (Rs.)	Amount (Rs.)
	Input cost				
1.	Application of fertilizer				3400
2.	Cost of plant protection chemicals				3500
3.	Cost of fertilizer				7500
	Labour cost				
1.	Weeding of plant basins		8	2800	2800
2.	Application of plant protection chemicals	2		1000	1000
3.	Weeding (4 times/year)	16		8000	8000
4.	Harvest (8 women 4 times)		32	11200	11200
	Total	**18**	**40**	**23000**	**37400**

Year 10 to year 24 the management costs are same.

Appendix II: Cost of Cultivation of Cashew Plantation (HDP-625 plants/ha)

A. Establishment Cost

Sl.No.	*Work*	*Labour*		*Cost (Rs.)*	*Amount (Rs.)*
		Men	*Women*		
	Input cost				
1.	Cost of planting material -625 Nos (No of plants at 4 x4 m spacing)				18750
2.	Cost of fertilizer and FYM				20000
	Labour cost				
1.	Cost of labour for weeding the area	20	50	27500	27500
2.	No of laboures for taking pit	50		25000	25000
3.	Labour cost for planting	30	30	25500	25500
4.	Cost of labour for fertilizer application	20	20	17000	17000
5.	Management of plot		20	7000	7000
	Total	**120**	**120**	**102000**	**1,40,750**

B. Maintenance Cost
B1 : Year 1

Sl.No.	*Work*	*Labour*		*Cost (Rs.)*	*Amount (Rs.)*
		Men	*Women*		
	Input cost				
1.	Cost of fertilizer(quantity 1/5th of recommended dose)				3000
	Labour cost				
1.	Multching @ 350/women		6	2100	2100
2.	Weeding of plant basins		10	3500	3500
3.	Initial training and pruning @500/men	6		3000	3000
5.	Application of fertilizer	3	3		2550
6.	Cleaning of plantation (4 times/year)	15		7500	7500
	Total 24	**19**	**16100**	**21650**	

B2: Year 2

Sl.No.	*Work*	*Labour*		*Cost (Rs.)*	*Amount (Rs.)*
		Men	*Women*		
	Input cost				
1.	Cost of fertilizer (2/5th of recommended dose)				6000
	Labour cost				
1.	Weeding of plant basins		12	4200	4200
2.	Application of fertilizer	4	4	3400	3400
3.	Weeding of plantation (4 times/year)	17		8500	8500
	Total	**19**	**16**	**15100**	**22100**

B3 : Year 3

Sl.No.	Work	Labour		Cost (Rs.)	Amount (Rs.)
		Men	Women		
	Input cost				
1.	Fertilizer cost (3/5th of recommended dose)				9000
	Labour cost				
1.	Weeding of plant basins		32	11200	11200
2.	Application of fertilizer	9	9	7650	7650
3.	Weeding of plantation (4 times/year)	17		8500	8500
	Total	**26**	**41**	**27350**	**36350**

B4: Year 4

Sl.No.	Work	Labour		Cost (Rs.)	Amount (Rs.)
		Men	Women		
	Input cost				
1.	Cost of plant protection chemicals				2500
2.	Cost of fertilizer (4/5th of recommended dose)				12000
	Labour cost				
1.	Weeding of plant basins		6	4200	4200
2.	Application of fertilizer 4/5th of recommended dose	8	8	6800	6800
3.	Application of plant protection chemicals	4		2000	2000
4.	Cleaning of plantation	17		8500	8500
5.	Harvest (4 women 3 times)		12	4200	4200
	Total	**29**	**26**	**25700**	**40200**

B5: Year 5

Sl.No.	Work	Labour		Cost (Rs.)	Amount (Rs.)
		Men	Women		
	Input cost				
1.	Application of fertilizer recommended dose				10200
2.	Cost of plant protectionchemicals				4500
3.	Cost of fertilizer (full dose)				22500
	Labour cost				
1.	Cleaning of plant basins		18	6300	6300
2.	Application of plant protection chemicals	9		4500	4500
3.	Weeding of plantation	18		9000	9000
4.	Harvest (9 women 3 times)		27	9450	9450
	Total	**27**	**45**	**29250**	**66450**

B6 : Year 6

Sl.No.	Work	Labour		Cost (Rs.)	Amount (Rs.)
		Men	Women		
	Input cost				
1.	Cost of fertilizer				22500
2.	Application of fertilizer				10200
3.	Cost of plant protection chemicals				6000
	Labour cost				
1.	Weeding of plant basins		24	8400	8400
2.	Application of plant protection chemicals	6		3000	3000
3.	Weeding of plantation	18		9000	9000
4.	Harvest (12 women 4 times)		48	16800	16800
	Total	**24**	**72**	**37200**	**75900**

B7 : Year 7

Sl.No.	Work	Labour		Cost (Rs.)	Amount (Rs.)
		Men	Women		
	Input cost				
1.	Application of fertilizer				10200
2.	Cost of fertilizer				22500
3.	Cost of plant protection chemicals				6000
	Labour cost				
1.	Weeding of plant basins		24	8400	8400
2.	Application of plant protection chemicals	6		3000	3000
3.	Weeding of plantation	18		9000	9000
4.	Harvest (12 women 4 times)		48	16800	16800
	Total	**24**	**72**	**37200**	**75900**

B8: Year 8

Sl.No.	Work	Labour		Cost (Rs.)	Amount (Rs.)
		Men	Women		
	Input cost				
1.	Application of fertilizer				10200
2.	Cost of plant protection chemicals				6000
3.	Cost of fertilizer				22500

Contd...

Contd...

Sl.No.	*Work*	*Labour*		*Cost (Rs.)*	*Amount (Rs.)*
		Men	*Women*		
	Labour cost				
1.	Weeding of plant basins		24	8400	8400
2.	Pruning 10		5000	5000	
3.	Application of plant protection chemicals	2		1000	1000
4.	Weeding of plantation	18		9000	9000
5.	Harvest (14 women 4 times)		56	19600	19600
	Total	**30**	**80**	**43000**	**81700**

B9: Year 9

Sl.No.	*Work*	*Labour*		*Cost (Rs.)*	*Amount (Rs.)*
		Men	*Women*		
	Input cost				
1.	Application of fertilizer				10200
2.	Cost of plant protection chemicals				6000
3.	Cost of fertilizer				22500
	Labour cost				
1.	Cleaning of plant basins		24	8400	8400
2.	Pruning 10		5000	5000	
3.	Application of plant protection chemicals	2		1000	1000
4.	Weeding of plantation	18		9000	9000
5.	Harvest (14 women 4 times)		56	19600	19600
	Total	**30**	**80**	**43000**	**81700**

B10 : Year 10

Sl.No.	*Work*	*Labour*		*Cost (Rs.)*	*Amount (Rs.)*
		Men	*Women*		
	Input cost				
1.	Application of fertilizer				10200
2.	Cost of plant protection chemicals				6000
3.	Cost of fertilizer				22500
	Labour cost				
1.	Weeding of plant basins		24	8400	8400
2.	Pruning 10		5000	5000	

Contd...

Contd...

Sl.No.	Work	Labour		Cost (Rs.)	Amount (Rs.)
		Men	Women		
3.	Application of plant protection chemicals	2		1000	1000
4.	Weeding of plantation	18		9000	9000
5.	Harvest (14 women 4 times)		56	19600	19600
	Total 30	**80**	**43000**	**81700**	

Year 10 to year 20: Maintenance cost is same

B21 : Year 21

Sl.No.	Work	Labour		Cost (Rs.)	Amount (Rs.)
		Men	Women		
	Input cost				
1.	Application of fertilizer				10200
2.	Cost of PP chemicals				10500
3.	Cost of fertilizer				22500
	Labour cost				
1.	Weeding of plant basins		24	8400	8400
2.	Application of plant protection chemicals	6		3000	3000
3.	Pruning 10		5000	5000	
4.	Cleaning of plantation	12		6000	6000
5.	Harvest (4 women 4 times)		16	5600	5600
	Total	**34**	**40**	28000	**71200**

B22 : Year 22

Sl.No.	Work	Labour		Cost (Rs.)	Amount (Rs.)
		Men	Women		
	Input cost				
1.	Application of fertilizer				10200
2.	Cost of plant protection chemicals				10500
3.	Cost of fertilizer				22500
	Labour cost				
1.	Weeding of plant basins		20	7000	7000
2.	Application of plant protection chemicals	6		3000	3000
3.	Pruning 10		5000	5000	
4.	Weeding of plantation	12		6000	6000
5.	Harvest (4 women 4 times)		16	5600	5600
	Total 28	**36**	26600	**69800**	

B23 : Year 23

Sl.No.	Work	Labour		Cost (Rs.)	Amount (Rs.)
		Men	Women		
	Input cost				
1.	Application of fertilizer				10200
2.	Cost of plant protection chemicals				10500
3.	Cost of fertilizer				22500
	Labour cost				
1.	Weeding of plant basins		20	7000	7000
2.	Application of plant protection chemicals	6		3000	3000
3.	Pruning 10		5000	5000	
4.	Weeding of plantation	12		6000	6000
5.	Harvest (4 women 2 times)		8	2800	2800
	Total	**28**	**28**	23800	**67000**

B24 : Year 24

Sl.No.	Work	Labour		Cost (Rs.)	Amount (Rs.)
		Men	Women		
	Input cost				
1.	Application of fertilizer				10200
2.	Cost of plant protection chemicals				10500
3.	Cost of fertilizer				22500
	Labour cost				
1.	Weeding of plant basins		20	7000	7000
2.	Application of plant protection chemicals	6		3000	3000
3.	Pruning 10		5000	5000	
4.	Weeding of plantation	12		6000	6000
5.	Harvest (4 women 2 times)		8	2800	2800
	Total	**28**	**28**	23800	**67000**

B25 :Year 25

Sl.No.	Work	Labour		Cost (Rs.)	Amount (Rs.)
		Men	Women		
	Input cost				
1.	Application of fertilizer				10200
2.	Cost of plant protection chemicals				10500
3.	Cost of fertilizer				22500

Contd...

Contd...

Sl.No.	*Work*	*Labour*		*Cost (Rs.)*	*Amount (Rs.)*
		Men	*Women*		
	Labour cost				
1.	Weeding of plant basins		20	7000	7000
2.	Application of plant protection chemicals	6		3000	3000
3.	Pruning 10		5000	5000	
4.	Weeding of plantation	12		6000	6000
5.	Harvest (4 women 2 times)		8	2800	2800
	Total	**28**	**28**	23800	**67000**

Appendix III: Cost Involvement in Intercropping in Cashew Plantation/ha

1. Turmeric

Sl.No.	Work	Labour		Cost (Rs.)	Amount (Rs.)
		Men	Women		
A.	**Cost of Cultivation**				
	Cost of Input				
1.	Turmeric – 2000 kg @ Rs.50/kg				100000
2.	Fertilizers/Manures				15000
	Cost of Labour				
1.	Weeding the area	8	20	11000	11000
2.	Taking beds for planting turmeric	60		30000	30000
3.	Planting turmeric		40	14000	14000
5.	Fertilizer application		20	7000	7000
6.	Mulching and other management practices		20	7000	7000
7.	Harvesting	10	10	8500	8500
	Total	**78**	**110**	**77500**	**192500**

B. Yield and Returns

Production: 17500 kg; Price: Rs. 15/kg; Returns: Rs. 2,62,500;
Net income: Rs. 2,62,500 – Rs.1,92,500=Rs. 70,000

2. Pineapple

Sl.No.	Work	Labour		Cost (Rs.)	Amount (Rs.)
		Men	Women		
A.	**Cost of cultivation**				
	Cost of input				
1.	No of pineapple suckers@ Rs. 5/- 30000 Nos				15000
2.	Fertilizers/Manures				20000
3.	Cost of hormone				2400
	Cost of Labour				
1.	Cleaning the area	8	20	11000	11000
2.	Taking trench for planting pineapple	120		60000	60000
3.	Planting pineapple	40	50	37500	37500
5.	Fertilizer application	16	20	15000	15000
6.	Cost of hormone application		24	8400	8400
7.	Weed management		150	52500	52500
8.	Harvesting	20	20	17000	17000
	Total	**204**	**284**	**201400**	**373500**

B. Yield and Returns

Production: 45000 kg; Price: Rs. 15/kg; Returns: Rs. 675000; Total cost: Rs. 373500; Net income: Rs. 675000-; Rs.373500 = Rs. 301500/-

3. Ginger

Sl.No.	*Work*	*Labour*		*Cost (Rs.)*	*Amount (Rs.)*
		Men	*Women*		
A.	**Cost of cultivation**				
	Cost of input				
1.	Ginger 1500 kg				112500
2.	Fertilizers/Manures				20000
3.	Cost of plant protection chemicals			2000	2000
	Cost of Labour				
1.	Weeding the area	8	20	11000	11000
2.	Taking bed for planting ginger	60		30000	30000
3.	Planting ginger	40		14000	14000
5.	Fertilizer application		20	7000	7000
6.	Cost of plant protection chemicals Application cost	6		3000	3000
7.	Weed management		150	52500	52500
8.	Mulching and other management practices		20	7000	7000
9.	Harvest 15	12	11700	11700	
	Total	**133**	**222**	**138200**	**270700**

B. Yield and returns

Production: 30000 kg; Price: Rs. 20/kg; Returns: Rs. 6,00,000; Total cost: Rs. 2,70,700; Net income: Rs. 6,00,000 – Rs. 2,70,700 = Rs. 3,29,300/-

Labour cost @ of Rs 500/- per men of Rs 350/- per women

Appendix IV: Cost Involvement in Cashew Irchard Rejuvenation (Top working/ ha)

Sl.No.	Work	Labour		Cost (Rs.)	Amount (Rs.)
		Men	Women		
1.	Beheading and removal of stumps	30		15000	15000
2.	Thinning of shoots	5		2500	2500
3.	Preparation of scion collection		5	1750	1750
4.	Grafting 35		17500	17500	
5.	After care (Thinning of extra growth on stock and scion, putting polythene cap etc.)	5		2500	2500
	Total		**39250**	**39250**	

Cashew: Improvement, Production and Processing (2017) *Pages* **523–534**
Editor: **Prof. P.L. Saroj**
Published by: **ASTRAL INTERNATIONAL PVT. LTD., NEW DELHI**

31

Yield Prediction Model

P.D. Sreekanth and P.L. Saroj

1. Introduction

Cashew is essentially a tropical crop and found Indian soil more homely than its home land. The cashew tree is native to Brazil and it was introduced to Mozambique and then India in the sixteenth century by the Portuguese. Today India is the largest processor and exporter of cashew kernels in the world. Over 65 per cent of the world export of cashew kernels is accounted for by India. Indian cashews are consumed in as many as 60 countries all over the world. Foreign exchange earned by the country through export of cashew kernels during 2011-12 touched Rs. 4,390 crore. The country exported 1.31 lakh tonnes of kernel during 2011-12 compared to 1.05 lakh tonnes during 2010-11. Currently, domestic production is hardly half of the processing requirement. The domestic cashew processing industry requires 12 lakh tonnes of cashew a year for fully utilizing the installed capacity. Remaining half of the requirement we are importing from other cashew growing countries. But these countries are establishing their own processing units and in future it is a great threat for Indian cashew processing units. Keeping in mind the scarcity of raw cashew nut for our cashew processing requirement, we need a very good yield forecasting models to estimate our yields well in advance and based on this we can procure raw cashewnuts from different countries for our processing units to fill up the processing requirement. Some studies were conducted earlier to predict cashew production well in advance using biometrical characters and yield attributes (during peak flowering phase), which contribute for the final yield of cashew (George *et al.*, 1979, 1989). However, recording biometrical characters *viz.*, canopy area, gradation of flowering, estimated nuts etc., in advance is cumbersome overlarge areas. Therefore, predicting cashew yield using secondary data *viz.*, area under cashew crop and cashew production will be highly useful. Attempts have been made earlier also by the authors to employ secondary data for estimation of cashew yield (Sreekanth *et al.*, 2003, 2008).

In this direction several yield forecasting models were tried using conceptual, mathematical and physical models that are not only laborious but also have practical limitations. In the recent past, soft computing tools like artificial neural networks (ANNs) have been used increasingly in various fields of science and technology for prediction purposes (Gail *et al.*, 2002). In particular, ANNs have been found useful in the area of forecasting modeling. The ANN is a general-purpose model with a limited set of variables, and is used as a universal functional approximator (Hornik *et al.*, 1989). It can forecast many nonlinear time series events over conventional simulation methods (Guan *et al.*, 2004, Hill *et al.*, 1996, Tang *et al.*, 1993, Zhang 2003, French *et al.*, 1992). Basically, ANNs are intelligent systems that are related in some way to a simplified biological model of the human brain. They are composed of many simple elements called neurons operating in parallel and connected to each other in the forward path by some multipliers called connection weights. Usually, ANNs are trained by adjusting the values of these connection weights between the network elements. These networks have self-learning capability and are fault-tolerant as well as noise-immune, and have applications in various fields like forecasting, system identification, pattern recognition, classification, speech recognition, image processing, etc. (Ramalingam *et al.*, 2006 and Sreekanth *et al.*, 2009). A reliable forecasting model for predicting the cashew yield using field variables through ANNs with added accuracy over the current methods is the need of the hour.

The major cashew growing states in India are Maharashtra, Kerala, Andhra Pradesh (AP), Odisha, Tamil Nadu (TN), Karnataka and Goa. The state wise data on area under cashew and production for past 45 years were used to develop the state-wise yield prediction models for above mentioned states. The secondary data were collected form Directorate of Cashew nut and cocoa Development (DCCD), Cochin and Cashew Export Promotion Council of India (CEPC), Cochin from 1965-66 to 2009-10. The identified independent variables for predicting cashew yield are given below;

- ☆ Area under cashew (X_1)
- ☆ Last five years area under cashew growth rate (X_2)
- ☆ Previous six years incremental area under cashew (X_3)
- ☆ Initial yield from 1st, 2nd and 3rd harvest (X_4)
- ☆ Previous five years moving average for production growth rate (X_5)

Yield of a new cashew plantation will stabilize only after six years of planting due to well establishment of plant and developed plant canopy as compared to canopy attained by the earlier age states (Nawale *et al.*, 1991). For this reason, variable X3 variable was incorporated to minimize the error.

$$X_3 = \sum_{i=1}^{6} X_{(n-i)}$$

where $X_{(n-i)}$ is $(n-i)^{th}$ year cashew cultivated area

After planting a cashew it will give minimal yield in 4th, 5th and 6th year i.e 1st, 2nd and 3rd harvests respectively, with this reason the variable X_4 was incorporated.

$$X_4 = \sum_{i=4}^{6} X_{(n-i)}$$

where $X_{(n-i)}$ is $(n-i)^{th}$ year cashew cultivated area

To nullify error, area under cashew and production moving average growth rate variables (X_2 and X_5) were incorporated. By using the above derived variables as explanatory variables, soft computing analysis was carried out to estimate the cashew yield.

The term Soft Computing indicates a collection of techniques which uses the human mind as a model and aims at formalizing our cognitive processes. These methods are meant to operate in an environment that is subject to uncertainty and imprecision. It exploits the tolerance for imprecision and uncertainty to achieve tractability, robustness, and low solution cost (Zadeh, 1994). In recent past, soft computing tools like Artificial Neural Networks (ANNs) have been used increasingly in various fields of science and technology for prediction purposes (Gail *et al.*, 2002). The ANN is a general-purpose model with a limited set of variables, and used as a universal functional approximator (Hornik *et al.*, 1989). It can forecast many nonlinear time series events over conventional simulation methods (French *et al.*, 1992; Tang *et al.*, 1993; Hill *et al.*, 1996; Zhang 2003 and Guan *et al.*, 2004).

2. ANN Architecture

In this study, Elman recurrent neural networks were used and trained with Levenberg-Marquardt backpropagation algorithm for all major cashew growing states except Odisha and Tamil Nadu where Resilient backpropagation algorithm were used to predict state-wise cashew yield. Its performance is further enhanced using the Log-sigmoid, Hyperbolic tangent sigmoid and Linear activation functions (Table 31.1). The identified ANN model structure was 5-12-6-1 *i.e.*, one input layer with five above identified nodes (variables), two hidden layers with twelve and six nodes respectively and one output layer. The mapping for the hidden layer/ respective nodes was carried out based on trial and error method, as there is no standard methodology for selecting the same (Abdusselam, 2007).

3. Elman Neural Network (ENN)

ENN consists of a two-layer back propagation network with an additional feedback connection from the output of the hidden layer to its input (Figure 31.1). The advantage of this feedback path is that it allows the ENN to recognize and generate temporal patterns and spatial patterns. This means that after training,

Table 31.1: ANN Architecture for State-wise

State Name	*Network*	*Algorithm*	*ANN Structure*	*Activation Functions*
Maharashtra	Elman recurrent	Levenberg-Marquardt	5-12-6-1	Hyperbolic tangent sigmoid, Hyperbolic tangent sigmoid, Linear
Kerala	Elman recurrent	Levenberg-Marquardt	5-12-6-1	Hyperbolic tangent sigmoid, Hyperbolic tangent sigmoid, Linear
Andhra Pradesh	Elman recurrent	Levenberg-Marquardt	5-12-6-1	Hyperbolic tangent sigmoid, Log-sigmoid, Linear
Karnataka	Elman recurrent	Levenberg-Marquardt	5-12-6-1	Log-sigmoid, Log-sigmoid, Linear
Goa	Elman recurrent	Levenberg-Marquardt	5-12-6-1	Log-sigmoid, Hyperbolic tangent sigmoid, Linear
Odisha	Elman recurrent	Resilient	5-12-6-1	Log-sigmoid, Log-sigmoid, Linear
Tamil Nadu	Elman recurrent	Resilient	5-12-6-1	Hyperbolic tangent sigmoid, Hyperbolic tangent sigmoid, Linear
Cumulative	Elman recurrent	Levenberg-Marquardt	5-12-6-1	Hyperbolic tangent sigmoid, Log-sigmoid, Linear

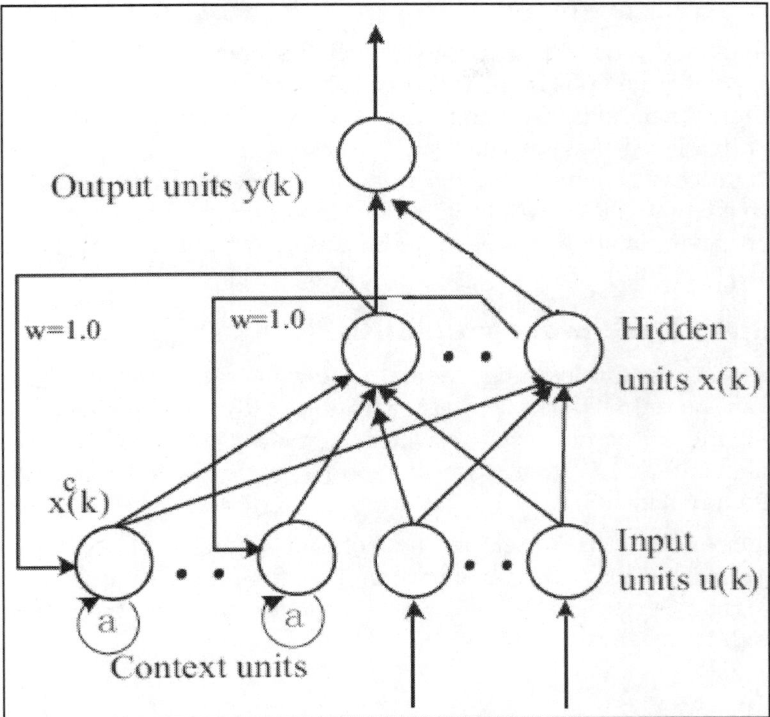

Figure 31.1: Structure of Elman Neural Network Model.

interrelations between the current input and internal states are processed to produce the output and to represent the relevant past information in the internal states. As a result, the ENN has been widely used in various fields, in which prediction is an important application direction (Fessant *et al.*, 1995; Koskela *et al.*, 1996; Stagge *et al.*, 1997 and Lim *et al.*, 2007).

4. Levenberg-Marquardt Back-propagation (LMB)

The LMB algorithm was designed to approach second-order training speed without having to compute the Hessian matrix. When the performance function has the form of a sum of squares (is typical in training feed-forward networks), then the Hessian matrix can be approximated as

$H = J^T J$ nd the gradient can be computed as

$g = J^T e$

Where **J** is the Jacobian matrix that contains first derivatives of the network errors with respect to the weights and biases, and **e** is a vector of network errors. The Jacobian matrix can be computed through a standard back-propagation technique that is much less complex than computing the Hessian matrix. Therefore, the LMB algorithm uses this approximation to the Hessian matrix in the following Newton-like update:

$x_{k+1} = x_k - [J^T J + \mu I]^{-1} J^T e$

When the scalar μ is zero, this is just Newton's method, using the approximate Hessian matrix. When μ is large, this becomes gradient descent with a small step size. Newton's method is faster and more accurate near an error minimum, so the aim is to shift toward Newton's method as quickly as possible. Thus, μ is decreased after each successful step (reduction in performance function) and is increased only when a tentative step would increase the performance function. In this way, the performance function is always reduced at each iteration of the algorithm (Sreekanth *et al.*, 2009).

5. Resilient Back-propagation (RBP)

It is a local adaptive learning scheme, performing supervised batch learning in feed-forward neural networks. The basic principle of RBP is to eliminate the harmful influence of the size of the partial derivative on the weight step. As a consequence, only the sign of the derivative is considered to indicate the direction of the weight update (Ried miller, 1993).

The performance of the selected network for accurate output was measured using statistical indices *viz.*, Error Variation (EV), Root Mean Square Error (RMSE) and Regression Coefficient (R^2) and were calculated based on the corresponding measured data as per Eq. (1), (2), and (3).

$$EV = ((y - w)/y)*100 \tag{1}$$

Where y is observed data and w is calculated data.

$$RMSE = \sqrt{\sum_{i=1}^{n} (y_i - \bar{y}_i)^2 / n} \tag{2}$$

$$R^2 = 1 - \frac{\sum \left(y_i - \bar{y}_i \right)^2}{\sum y_i^2 - \frac{\sum \bar{y}_i^2}{n}} \tag{3}$$

Where, y_i is the observed data $\bar{y}_i$ is the calculated data and n is the number of observations.

RMSE indicates the discrepancy between the observed and calculated values. The lower the RMSE, the more accurate is the prediction. The best fit between observed and calculated values, which is unlikely to occur, would have R^2 as 1 and RMSE as 0 (Ioannis *et al.*, 2005).

In the present study the historic data (1965-2010) were used to develop a state-wise cashew prediction models. For training the ANN model, data from 1965-66 to 2003-04 were used and for testing six years (2004-05 to 2009-10) data were used.

The training performance of the ANN model for state-wise and cumulative is given in Table 31.2 and Figures 31.2a and 31.2b.

Table 31.2: ANN Model performance

State	*RMSE*	R^2
India (cumulative)	0.47	0.99
Maharashtra	0.20	0.99
Kerala	0.13	0.96
Andhra Pradesh	0.13	0.98
Karnataka	0.07	0.98
Goa	0.04	0.98
Odisha	0.08	0.99
Tamil Nadu	0.07	0.99

Maharashtra

The projected yield for Maharashtra which is calculated using above model is found to be closer to actual yield for years *viz.*, 2004-05 (1.98); 2005-06 (-0.76); 2006-07 (2.22); 2007-08 (5.27); 2008-09 (19.87); 2009-10 (7.96). The values in parenthesis represent error variation in per cent (per cent).

Kerala

The estimated yield for Kerala is found to be closer to actual yields for years *viz.*, 2005-06 (-7.44); 2006-07 (19.94); 2007-08 (6.96); 2008-09 (14.79); 2009-10 (6.70) except year 2004-05 there is deviation from the trend.

Andhra Pradesh

The best fit ANN model revealed that, projected yield is very closer to actual yield during 2004-05 (5.23); 2005-06 (-5.79); 2006-07 (-1.55); 2009-10 (0.29) except during the years, 2007-08 (18.94) and 2008-09 (20.68).

Karnataka

The projected yield for Karnataka is closer to actual yield and error variation is almost negligible (Figure 31.2). It is ranging from -4.55 to -9.94 per cent except for the years, 2004-05 and 2009-10, where the error variation found to be 14.68 and -24.05 respectively.

Goa

The calculated projected yield for Goa using ANN model is found to be closer to actual yield for the years, 2005-06 (-6.70); 2006-07 (2.68); 2007-08 (-0.28); 2008-09 (-4.47); and higher side error variation was found during 2004-05 and 2009-10 *i.e.* -26.12 and -18.61 respectively.

Odisha

From the best fit ANN forecast model, it is observed that the estimated cashew

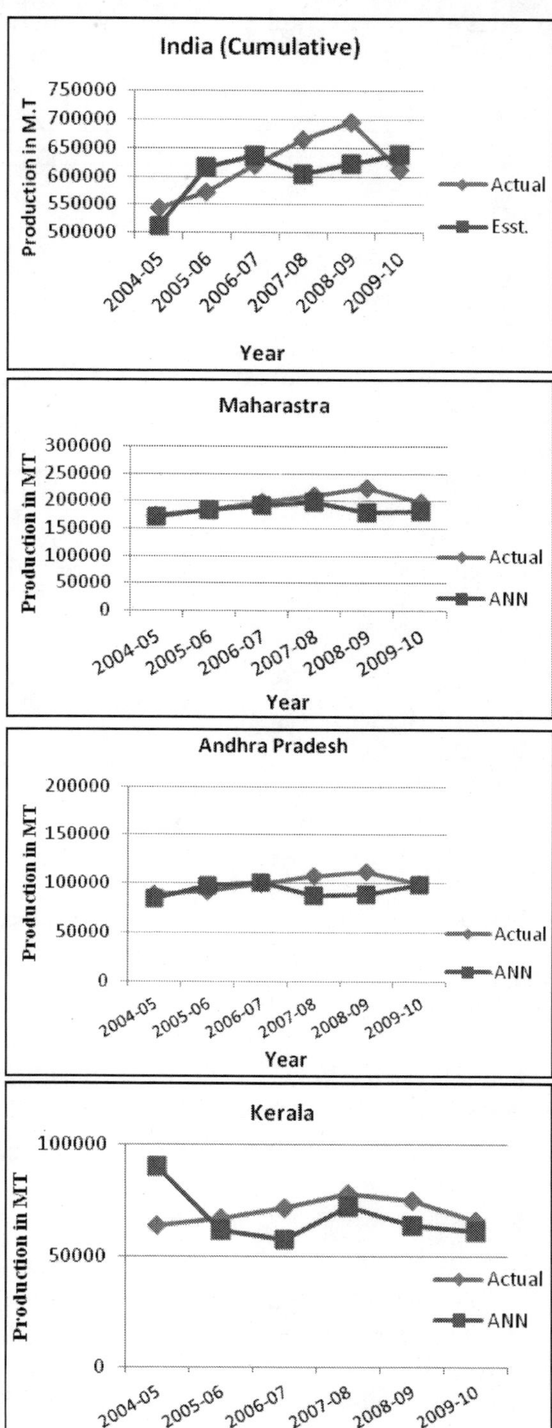

Figure 31.2a: Comparison of ANN Model Predicted Yield Values with Observed Data.

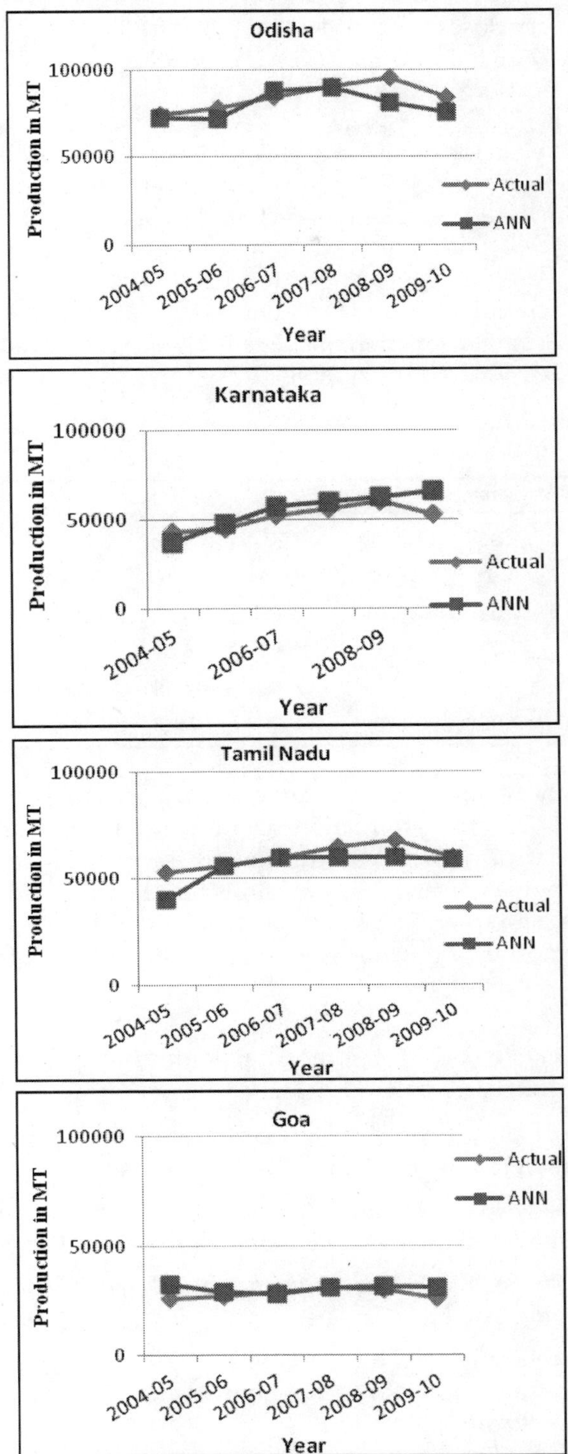

Figure 31.2b: Comparison of ANN Model Predicted Yield Values with Observed Data.

yield found to be very close to actual yield for the years 2004-05 (2.70); 2005-06 (8.16); 2006-07 (-4.45); 2007-08 (0.79); 2008-09 (15.21) and 2009-10 (10.80).

Tamil Nadu

The estimated yield for Tamil Nadu is closer to actual yield and error variation is almost negligible. It is ranging from 0.50 to 7.60 per cent except for the years 2004-05 and 2008-09, where the error variation was found to be 24.60 and 11.68 respectively.

India (Pooled)

The projected cumulative yield for India was found to be closer to actual yield. The cumulative yield error variation calculated between actual and estimated yields was ranged between -7.67 to 10.46 (Table 31.3).

Table 31.3: Per cent Error in Prediction of Cashew Production in different States and India as a Whole

Year	*India*	*Maharashtra*	*Kerala*	*AP*	*Karnataka*	*Goa*	*Odisha*	*T.N.*
2004-05	5.97	1.98	–40.91	5.23	14.68	–26.12	2.70	24.60
2005-06	–7.67	–0.76	7.44	–5.79	–5.76	–6.70	8.16	0.60
2006-07	–2.84	2.22	19.94	–1.55	–9.94	2.68	–4.45	0.50
2007-08	9.22	5.27	6.96	18.94	–6.84	–0.28	0.79	7.60
2008-09	10.46	19.87	14.79	20.68	–4.55	–4.47	15.21	11.68
2009-10	–4.32	7.96	6.70	0.29	–24.05	–18.61	10.80	2.23

By using the above developed formula it could be possible to work out reliable yield estimation of cashew production scientifically from available secondary data *viz.*, area under cashew and production. These models are highly useful in estimating the cashew yield and saves precious time by avoiding labour intensive field observations. However, future studies to incorporate climate data and field observations for improving the precision of yield estimates will be envisaged.

References

Abdusselam Altunkaynak (2007). Forecasting surface water level fluctuations of Lake Van by artificial neural networks, *Water Resource Management*, 21: 399-408.

Fessant, F., Bengio, S. and Collobert, D. (1995). On the prediction of solar activity using different neural network models, *Annales Geophysicae*, 14: 20-26,

French, M. N., Krajewski, W. F. and Cuykendall, R. R. (1992). Rainfall forecasting in space and time using a neural network. *Journal of Hydrology*, 137: 1-31.

French, M. N., Krajewski, W. F., and Cuykendall, R. R. (1992). Rainfall forecasting in space and time using a neural network, *J. Hydrol.*, 137: 1-31.

Gail, M., Brion, Neelakantan, T. R. and Lingireddy, S. (2002). A neural-network-based classification scheme for sorting sources and ages of fecal contamination in water. *Water Research*, 36: 3765-3774.

George, M.V., Amarnath, C.H., Bhagavan, S., and Vijayakumar, K. (1989). Field evaluation of forecasting model for cashew yield in large plantations. *The Cashew*, 3(1): 8-10.

George, M.V., Vijayakumar, K. and Amarnath, C.H. (1979). Forecasting of cashew yield from plantations. *Journal of Plantation Crops*, 16: 493-497.

Guan, P., Huang, D. and Zhou, B. (2004) Forecasting model for the incidence of hepatitis A based on artificial neural network. *World Journal of Gastroenterology*, 10: 3579-3582.

Guan, P., Huang, D., and Zhou, B. (2004). Forecasting model for the incidence of hepatitis A based on artificial neural network, *World Journal of Gastroenterology*, 10(24), 3579-3582.

Hill, T., O'Connor, M. and Remus, W. (1996). Neural network models for time series forecasts. *Management Science*, 42, 1082-1092.

Hill, T., O'Connor, M., and Remus, W. (1996). Neural network models for time series forecasts. Management Science, 42(7): 1082-1092.

Hornik, K., Stinchcombe, M. and White, H. (1989). Multilayer feed-forward networks are universal approximators. *Neural Networks*, 2: 359-366.

Hornik, K., Stinchcombe, M., and White, H. (1989). Multilayer feed-forward networks are universal approximators., *Neural Networks*, **2:** 359-366.

Ioannis, N., Daliakopoulos., Paulin Coulibaly., Ioannis K., and Tsanis. (2005) Groundwater level forecasting using artificial neural networks, *Journal of Hydrology*, 309: 229-240.

Koskela, T., Lehtokangas, M., Saarinen, J. and. Kaski, K. (1996). Time series prediction with multilayer perceptron, FIR and Elman neural networks, Proceedings of the World Congress on Neural Networks, 1996.

Lim, C. P. and W. Y. Goh. (2007). The application of an ensemble of boosted Elman networks to time series.

Nawale, R. N. and Salvi, M. J. (1991). The yield potential in relation with age of improved cashew varieties. *Cashew Bulletin*, 28: 13-17.

Prediction: A benchmark study, *International Journal of Computational Intelligence*, 3: 2.

Ramalingam, V., Palaniappan, B., Panchanatham, N. and Palanivel, S. (2006). Measuring advertisement effectiveness – a neural network approach. *Expert Systems with Applications*, 31: 159-163.

Ried miller (1993). A direct adaptive method for faster backpropagation learning: The RPROP algorithm, in H. Ruspini, editor, Proceedings of the IEEE International Conference on Neural Networks (ICNN), San Francisco, USA, 586-591.

Sreekanth, P. D., Geethanjali, N., Sreedevi, P. D., Shakeel Ahmed, N., Ravi Kumar and P. D. Kamala Jayanthi. (2009). Forecasting groundwater level using artificial neural networks, *Current Science*, 96(7): 933-939.

Sreekanth, P.D. (2008). Empirical prediction model for regional cashew production in India. *Journal of Plantation Crops*, 36(3): 508-511.

Sreekanth, P.D. and Bhaskara Rao E.V.V. (2003). Model for forecasting yield in cashew. *The cashew*, 17(4): 23-31.

Stagge, P. and Sendhoff, B. (1997). An extended Elman net for modeling time series, International conference on Artificial Neural Networks, LNCS, 1327: 427-432.

Tang, Z. and Fishwick, P. A. (1993). Feedforward neural nets as models for time series forecasting. *ORSA J. Computing*, 5: 374-385.

Tang, Z., and Fishwick, P. A. (1993). Feed forward neural nets as models for time series forecasting. *ORSA Journal on Computing*, 5: 374-385.

Zadeh, L. A (194). Soft computing and fuzzy logic. *Software IEEE*, 11(6): 48-56.

Zhang, G. (2003). Time series forecasting using a hybrid ARIMA and neural network model. *Neurocomputing*, 50: 159-175.

Zhang, G. (2003). Time Series forecasting using a hybrid ARIMA and neural network model. *Neurocomputing*, **50**, 159-175.

Cashew: Improvement, Production and Processing (2017) *Pages* **535–548**
Editor: **Prof. P.L. Saroj**
Published by: **ASTRAL INTERNATIONAL PVT. LTD., NEW DELHI**

32

Indigenous Technical Knowledge

M.V. Sajeev and P.L. Saroj

1. Introduction

The advent of the concept of sustainable agriculture in late eighties in Indian agricultural scenario has evoked interest on indigenous technical knowledge (ITK) that uses knowledge of natural products to solve problems pertaining to agriculture and allied activities. Indian farmers, over centuries, have learnt to grow food and to survive in difficult environments, where the rich tradition of ITK has been interwoven with the agricultural practices followed by them. India, a country spreading over 3287.26 thousand sq. km. and inhabited by about 120 crores people has been nurturing a tradition of very rich civilization over a period of five thousand years. The enhancement of the quality of life of the Indians who in great majority live in and depend on agricultural production systems would be impossible by keeping ITKs aside. The special features of indigenous knowledge are that:

1) they are local,
2) it is rooted in a particular community and situated within broader cultural traditions,
3) is tacit knowledge and, therefore, not easily modifiable,
4) transmitted orally, or through imitation and demonstration,
5) experiential rather than theoretical knowledge and also
6) learned through repetition and is constantly changing.

Indigenous Knowledge (IK) refers to the unique, traditional, local knowledge existing within and developed around the specific conditions of women and men indigenous to a particular geographic area. IK is the local knowledge - knowledge that is unique to a given culture or society. IK contrasts with the international knowledge system generated by universities, research institutions and private firms. It is the basis for local-level decision in agriculture, health care, food preparation, education, natural resource management, and a host of their activities in rural communities. Indigenous knowledge is the information base for a society, which facilitates communication and decision-making. Indigenous information systems are dynamic, and are continually influenced by internal creativity and experimentation as well as by contact with external systems.

Cashew (*Anacardium occidentale* L.), introduced to India by Portuguese travelers was growing in wild in coastal and western ghat regions of India till people started cultivation of cashew by their own knowledge and experience. The produce was known differently in different local languages. The cashew is known as 'Caju' in Hindi, 'Kasu andi' in Malayalam, 'Mundiri' in Tamil, 'Gerubeeja or Godambi' in Kannada, 'Jidipappu' in Telugu, 'Kaju' in Gujarati and 'Hijali Badama' in Bengali. The cashew, little known as crop plant in the past, had unique importance for its varied uses. Though it was growing wild in the region, people used to harvest the fruits with bamboo sticks and used to consume the apple and nut. The nuts were sold to the retail purchasers in the local market for a meager price.

Once the economic importance of cashew was realized, traditional cultivation practices were adopted particularly with respect to sowing, harvest and postharvest handling of the produce. The curry prepared from kernels of tender nuts and the locally made alcoholic drink 'Feni' are quite popular among villagers in cashew growing regions. The cashew nut shell liquid extracted by crude way from the shells of tender nuts find uses in medicines, wood industries, and also in varnishes for curing floors and walls of village houses. Similarly, cashew apple and kernels were processed indigenously for consumption.

The medicinal uses of leaf, bark, shell, cashew nut shell liquid and testa are well documented. Vast references can be seen in a book known as "Ajji Maddu" authored by well known Ayurvedic doctors "the Panaje Pandiths" near Puttur, Karnataka. The cashew wood, however, had negligible use in construction and timber industry. At the same time the wood had significant use in the preparation of agricultural implements besides its use as firewood. Similarly leaves both fresh and dry and twigs were used as raw material for organic manure by the farmers.

The information regarding ITK and the unscientific beliefs practiced in different parts of the country are compiled and summarized in this chapter. However the ITKs need validation for large scale adoption.

2. Indigenous Cashew Cultivation Practices

2.1 Orchard Management Practices

2.1.1 Raising and Planting of Seedlings

The fallen nuts from selected high yielding plants are sun dried for 2-3 days. During rainy season dried seeds are directly dibbled into soil. This is done either in bushes or in termetoria for protection from birds and animals during germination. Whole coconut husks are also used for raising seedlings. Coconut husks are tied along with the potting mixture and seed is sown inside. After the emergence and sufficient growth of the seedling the whole husk is planted along with the seedling in a pit.

2.1.2 Cashew as Trailing Support

Cashew trees are used for trailing black pepper, ridge gourd and snake gourd without affecting yield of each other. Seasonal vegetables provide additional income to cashew farmers without affecting cashew yield since they are trailed after cashew harvest. Black pepper is successfully grown in cashew but is found to affect cashew yield due to exhaustion of nutrients.

2.1.3 Preventing Flower Drying

Farmers believed that spraying water in the morning hours of winter months on the foggy days during early flowering period of cashew helps in preventing flower drying resulting in increased yield. Spraying is done using rocker sprayer. The practice is continued even today by small farmers.

2.2 Pest Management Practices

2.2.1 Control of Cashew Stem and Root Borer (CSRB)

It is claimed by the farmers that application of wood ash (15-20 kg per tree) to the base of cashew trees reduces the stem borer incidence. The ash mixed with tea waste was applied during July-August months around the base of the plants while manuring the plants. Farmers also claim that application of common salt to tree bases controlled cashew stem and root borer (CSRB).

2.2.2 Control of Tea Mosquito Bug (TMB)

Red ant (*Oecoyphylla smaragdina*) colonies which usually appear in old cashew plantations are used by farmers to improve the cashew yield by way of feeding on TMB and other harmful insect pests. Presence of red ants prevents loss by pilferage also as it is difficult to approach and climb such trees for stealing the nuts. Cashew farmers also collect the red ant colonies and allow it to spread in their cashew plantations. This practice is believed to help in pollination in cashew during flowering season.

2.3 Cashew Harvest

2.3.1 Harvesting, Storage and Sale of Nuts

Farmers usually collect fallen fruits from the ground. Also, mature fruits are harvested from smaller trees using bamboo sticks. Many a times, from high yielding trees the fully mature fruits are forced to fall by shaking the trees at regular intervals (once in 2-3 days). This practice is allows preventing pilferages and also reduce labour needs in collection of nuts. The picked fruits are separated from nuts by twisting the fruits.

The matured nuts are separated from apple, given a water wash and sun dried for 2 days. Storage of the nuts in large quantity is rarely practised as the kernel gets spoiled during rainy season. Only a small quantity (5-10 kg) required for family use used to be stored in earthern pots and placed in the upstairs of kitchen made of wooden reapers. This practice is followed to prevent the spoilage of kernels due to excess moisture during rainy season. Usually, small farmers dispose nuts immediately after drying in retail markets.

2.3.2 Seasoning Cashew Apple

Seasoning of cashew apple is done for using them as cattle feed and in alcohol preparation. For this, farmers harvest mature nuts leaving behind the peduncle for normal ripening as such in trees. Small or bushy plants are selected for this purpose for ease of collection. This practice also prevents the pilferage of the matured nuts.

2.3.3 Transport of Produce

Fallen apples are collected in bamboo baskets and brought home. Later nuts are separated from the apple. Sometimes when produce is of less quantity, fruits are carried one above the other using wild vines in the form of garlands.

3. Beliefs

3.1 Effect of Rainfall

There is a general belief that if it rains just before or during flushing cashew tree would not bear fruits. Vice versa, it is believed that a good rainfall after flowering takes care of pest and disease problems associated with fruiting and gives the best yield.

3.2 Scaring off Thieves

Ixora flowers, leaf bits of Palmyrah palm and muli grass were used to be tied together and offered to superstitial bodies by manthrikas and later were tied to a visible branch of the cashew tree. People who see this structure (called 'Kodi' in Kannada) would never touch the produce as it was believed that the 'manthra' or such powers to whom it was offered will harm the people who touch the produce. This belief was in practice till late seventies and then onwards people gradually did not care for such threats and this practice has almost disappeared. Even now in some under developed regions of Dakshina Kannada, Karnataka the practice is being followed.

3.3 Bamboo Thorns and Rotten Fish against Monkeys

Usually, monkeys' break and destroy the young cashew seedlings or grafts immediately after planting. At times, over large areas the menace was observed. Generally, grown up seedlings are planted and individually fenced with bamboo thorns to protect from monkeys' menace. Also, farmers keep rotten fish near the base of the planted graft which gives foul smell for quite a longer period thus preventing monkeys destroying the plant. However, there is risk of attracting wild boars if rotten fish is applied to scare away monkeys.

3.4 Human Hair against Porcupines

Porcupines and wild boars damage the cashew grafts at early stages of planting. While the wild boars uproot the plants during rainy season in search of earthworms, the porcupines cut the grafts at collar region and destroy the plants. Farmers' claim that spreading of small pieces of cut hair collected from barber's shop around the grafts scares the animals as these hairs get in while breathing causing irritation in nostrils thus forcing the animal to leave the plant.

3.5 Broom and Sickle against Ghosts

Many villagers believe that ghosts and other superstitious powers consume feni (alcohol from cashew apple) while in the stage of preparation itself. So they keep a broom and a sickle on the top of the vessel to save alcohol by scaring such superstitial powers. In reality, the feni is stolen and pilfered out during night by interested people who want to consume free liquor without undertaking the pains of processing.

4. Cashewnut Trade

4.1 Until early eighties, it was believed that the prices of raw cashewnut and rice would go hand in hand. These two commodities were used for barter exchange in those days. It was believed that the market price of both the commodities was on par and the trend continued upto early eighties. Later the trend did not last as the price of cashew went much higher during late eighties.

4.2 People who collect nuts through pilferage or those who are having very small holdings are found to sell the produce on the same day of collection without even drying for a day. At times, people pour water to dried nuts so that seed imbibes water and will contribute to more weight while selling in the retail market.

4.3 The immature/green cashewnuts/tender nuts fallen during harvest or obtained through pilferage are dried by mixing with wood ash which attains the normal colour on drying. This also prevents shrinking of nuts during drying. These immature dried nuts are then mixed with mature dried nuts and sold to retail purchasers. Also, the reject nuts (shriveled or shrunken) due to improper maturity are also mixed along with mature nuts to get more weight to the produce. This practice is fuelled by non availability of premium price for better quality nuts.

5. Uses of Cashew Produce

5.1 Uses of Cashew Apple

5.1.1 Alcohol from Cashew Apple

The alcohol from cashew apple (fenny) distilled by adopting local methods is believed to have several medicinal uses. It helps in overcoming shooting diarrhea and cold fever. It is recommended at a daily dose of 1-2 spoons to women after delivery for the speedy regain of normal body conditions. Fully ripened cashew apples are squeezed and fermented in earthern pots for about 7-9 days. Occassionally, apples are also mixed with jaggery for better yield of alcohol. After full fermentation the material is poured into a bigger vessel and a cloth is tied loosely inside the vessel. A collecting vessel is kept inside the bigger vessels. An earthen or aluminum pot with cold water is kept over the vessel and the distillation is done. As the water on the top vessel gets heated the hot water is replaced with cold water. The alcohol vapours escapes and gets cooled due to the cold water on the top and get collected in the vessel kept inside the bigger vessel. Likewise; 7-8 times the hot water has to be replaced and the complete alcohol extracted is condensed in small collecting vessel inside the bigger vessel. In recent days, vessels with some modifications (pipe fitted vessels) are used and the alcohol is directly collected outside in bottles through rubber tubing. Also, the distillation is done using pressure cookers. The strength of the alcohol content in extracted liquid is tested by holding it to the fire either through a stick or at times directly by dipping finger in the prepared liquid and holding it to flame. Cashew Feni so prepared is either consumed afresh or after preserving for 1 year (until next season of fruiting) or several years of ageing.

5.1.2 Cashew Apple 'Ada'

The apples are ground into a paste along with rice. Later the paste or wet flour is folded in banana leaves or teak (*Tectona grandis*) leaves and are steam cooked. The dish folded in teak leaves gives a typical reddish colour to the dish. In local tulu language (DK District, Karnataka) this dish is known as 'Madith Deedi Adye' and in Konkani language as 'Panege'. This dish used to be consumed as breakfast in villages.

5.1.2 Cashew Apple Sweet (for making 100 sweets)

Take 10 cashew apples and grind them thoroughly in mixer. Boil this semi-solid in a pan for 20-25 minutes on medium flame. Do not add water. When it is boiling, add half cup of milk. When the solution starts getting thick, add double the amount of sugar. After some time, add a quarter cup of ghee followed by cardamom powder. Put off the flame and when it is still hot, start making chocolates out of it.

5.1.3 Cashew Apple Halwa

Grind cashew apples and take the semi solid to a pan for boiling. No need of water or milk here. When it starts getting thick, add double the amount of sugar followed by cardamom powder, clove powder and cashew pieces. About half cup of ghee should be added. Compared to chocolate, the boiling time should be less here.

5.1.4 Cashew Apple Appa or Suttavvu

Grind cashew apples and add a quarter kg of wheat flour followed by Jaggery to taste. Fry this dough in a pan by making small balls.

5.1.5 Cashew Apple Payasam

Grind cashew apples and take the semi solid to a pan for boiling. When it starts getting thick, add double the amount of sugar, followed by milk, ghee and cashew pieces. This payasam resembles date Payasam.

5.1.6 As Fruit

Cashew apple is not a table fruit but some sweet types are used for fresh consumption. Usually fruits (whole apple) are consumed or otherwise fruits are cut into pieces, mixed with table salt and eaten.

5.1.7 As Maandala

Fully ripe cashew apples are partially cooked (boiled) and pulp is prepared by crushing the semi-boiled apples. The pulp is spread on carpets and dried. It is dried for several days, and preserved for off-season. It is locally known as 'Maandala' in Dakshina Kannada, Karnataka and is used for culinary uses particularly for the preparation of chutneys.

5.1.8 Medicinal Uses

The juice of cashew apple is extracted and bottled and kept for a week's time. The liquid is used as a drink after decanting and separation of floating pulp. This is usually practiced by financially weaker sections who cannot always manage to get alcohol. It is also believed that the warmed juice (in sun light for a fortnight) can be preserved for 2-3 years which would have medicinal uses in treating fever and diarrhea of both human beings and domestic animals. It is believed that cashew apple juice induces sleepiness, when given along with medicine to the patients affected by fever. The practice helps in fast recovery from fever.

5.1.6 Cattle Feed

The ripened apples after separation from nuts usually (if bulk) used as cattle feed. Apples are also dried and preserved as cattle feed for rainy season. Daily 3-4 kg of fresh apples can be fed to a cow which will be useful along with normal feed. Cattle will have stomach problem causing diarrhea, when fed excess quantity of cashew apples.

5.1.7 Biogas Material

Ripened fruits are also used as raw material for biogas plants. During fruiting season fresh cashew apples are used as one of the excellent materials for biogas production but its period of availability is limited.

5.2 Use of Cashew Nut

5.2.1 Dry nuts are burnt in open fire or at times burnt using a pan or tin containers having small holes. After burning, the kernel is extracted by gently hitting the outer shell and breaking it open. The kernels are directly consumed after the separation of testa. The dried nuts are usually used in this manner during rainy season. But, this is not a preferred practice as due to uneven heating it results in uneven colouring of the kernel. Recently, farmers started burning the nuts in holed containers tied with a handle for uniform burning.

5.2.2 The tender nuts or green nuts fallen during harvest of nuts are used for culinary purposes. The green nuts are cut and the immature kernel is removed and put in boiling water for easy separation of the kernel from the testa. After separation of the kernel it is cooked and fried with little grated coconut powder. This is most popular and one of the tasty dishes (curry) specially prepared in DK District of Karnataka and adjacent Kasaragode district of Kerala. The tender kernels are also used in preparation of curries (sambar) either with dhal or separately.

5.2.3 Tender nuts are usually eaten by school children as the cashew trees are grown all along the roads and school premises. Children, rub the ventral side of the tender nuts to the black topped roads or the tree trunk and later it is pressed by foot to remove the kernels. The kernel that comes out is wiped with dry leaves to remove CNSL and eaten as such by school children.

5.2.4 The matured and fallen nuts during late season (during May or June), may imbibe water due to rains and germinate in the field itself. Usually, germinating nuts with just emerging cotyledons are collected particularly for the preparation of curry. People believe that preparation (curry) made out of such germinating seeds is very rich in nutrients.

5.3 Uses of Cashew Nut Shell Liquid (CNSL)

5.3.1 Blackening and Hardening of Earthen Pots

Cashew nut shell liquid is used in blackening of fresh earthen vessels meant for cooking. This is mainly the process of hardening of the earthen vessels. Initially the pots are burnt in fire for 2-3 hours and the vessels are smeared with CNSL. Again the CSNL smeared pots are filled with fresh paddy husk and burnt for a day. Initially only waste material used to be cooked in such pots and later these pots are used for regular cooking. Only after blackening with CNSL, the pots are being used for cooking the food.

5.3.2 Smearing of Walls or Floors

Earthen floors of village houses are smeared with CNSL to give shining appearance and also to prevent the fungal growth during rainy season. Coating of walls with CNSL gives protection from fungal growth for several years and maintains the shiny appearance of the wall.

5.3.3 Preventing Termite Attack

The wood which are used for construction purposes, particularly the wall plates or rafters, wooden items which touch the soil or walls are smeared with CNSL. Also the reapers which help in holding the roof tiles are smeared with CNSL for seasoning as to prevent from termite damage and fungal decay. Cashew nut shell liquid is also used to smear bamboo baskets and winnows for strengthening and ensuring long life of the material.

5.4 Use of Cashew Gum

Cashew tree gives out gummy exudation during injury to its bark or stem. The exudation or oozing of gum is more with larger cuts or wounds. In olden days, this gum was collected and used as adhesive for pasting papers and books, labels *etc.*

5.5 Use of Cashew Leaves

5.5.1 Cashew Leaves as Tooth Brush

Green cashew leaves (leaf blades) were being used for brushing and cleaning the teeth by villagers. The leaf blades are folded and used as brush. The mid rib of the leaf after removal of leaf blades was used as tongue cleaner. It is believed that the leaf had medicinal property which reduces tooth decay and toothache and strengthens the teeth. It is also claimed that the powder made out of dry cashew leaves and arecanut used as tooth powder is helpful in overcoming tooth decay and pain.

5.5.2 Dry Cashew Leaves for Cucumber Cultivation

The dry cashew leaves are heaped in the furrows during rabi season and are burnt one day before sowing of cucumber. It is believed that the contents of the burnt cashew leaves gets accumulated in soil and helps in prevention of soil borne diseases.

5.5.3 Dry Cashew Leaves as Manure

Just before fruiting season or in early fruiting season all the fallen dry leaves are collected from the plantations or from individual plants and stored for rainy season to be put in cowshed as bedding material for cows. The leaves would be later transferred directly to areca gardens to be applied as manure. At times, the cashew dry leaves thus collected are heaped in paddy fields which would be directly incorporated into soil during ploughing and puddling.

The green cashew leaves are used for mulching (green manure) to coconut palms during rainy season. After the base clearing, about 50-60 kg of green leaves are put along with cattle manure and salt and covered with soil. It is believed that this practice helps in preventing button shedding in coconut. The cashew leaves are the second choice if leaves of *Nuxvomica stricta* are not available.

Dry leaves collected are at times heaped in large quantity in the field by placing coconut fronds vertically in the side and round pyramid like structures are made. On the top again coconut fronds are kept as roof to prevent the entry of water during monsoon season.

5.5.4 Green Leaves as Insect Repellent

The green cashew leaves applied as mulch to cowpea and bhendi, was found to help in overcoming the insect problem particularly aphids, and the sap sucking bugs of cow pea.

5.5.5 Cashew Leaves in Culinary Preparations

Cashew leaves are used for the dish prepared out of rice and jack fruit. The ground rice paste is to be folded in between two cashew leaves and to be cooked. The dish is locally known as "atte" in Dakshina Kannada, Karnataka. The dish gets a special characteristic odour with a specific taste due to use of cashew leaves.

5.5.6 Tender Cashew Leaves for Chutney

Tender cashew leaves (reddish in colour) are plucked along with tender leaves of mango, wild jamun, *Holigrana* sp., *Ixora* sp. *etc.* and ground into chutney and consumed once or twice a year as medicinal chutney. This chutney is medicinal and is useful in curing ulceration (Boils in the mouth).

5.6 Cashew Wood

5.6.1 As Agricultural Implements

Wherever the *Nuxvomica stricta* tree is not available the cashew tree is preferred for making of ploughs. The matured cashew wood is also used quite often in the preparation of handles for sickles as it is light in weight and also easy for carpenters to work. The planks made from old cashew trees are also used for puddling of paddy fields. It is believed by farmers that such fields have less weed growth.

5.6.2 As Timber

Generally, the cashew wood is used only as fire wood in villages. Villagers feel that the tree never grows straight, therefore the wood do not have any timber value. Incidentally, if a straight piece is available the wooden planks are prepared and seasoned by smoking which has reasonable timber value as it is also very light in weight. Wood can be used in construction of roof of low cost houses.

6. Medicinal Uses

6.1 As Pain Killer

The thorn pricks in foot sole can be treated with hot (warm) CNSL after the removal of the thorn and pus. Treatment helps in fast healing of the wound and pain will not be felt for 1-2 days after the treatment. Similar treatment can also be given for the injury due to pricking of glass pieces.

6.2 Treatment for Corns

The dry cashew nut is to be cut into two halves. After the removal of the kernel, the shell is to be heated on flame holding with the help of a sharp stick inserted into shell. The corn on the foot is to be cut and exposed. The hot (warm) liquid (CNSL) oozing out of shell when held over the flame should be applied on the corn and may be repeated several times (3 to 4 times in a week). This helps in complete healing of the corn.

6.3 Treatment for Diarrhea

Cashew apple juice is extracted and filtered through cotton cloth and the supernatant liquid is stored in glass bottles. In summer, once in a fortnight or so the bottles are kept in hot sun and seasoning is done. Every time it is filtered and the supernatant liquid is stored. The liquid could be stored for years as claimed by villagers. Liquid thus prepared is used as medicine for treating diarrhea, fever and cough in cattle during rainy seasons. This is also used in the treatment for fever and diarrhea of human beings.

6.4 Treatment for Fungal Infection

The animal and human feet (gaps between fingers) get infected with fungus during rainy seasons if they continuously walk bare footed on muddy, slushy soils. In villages this is usually tackled with smearing of a mixture of CNSL and coconut oil mixed (1:1 v/v). Smearing is done during night particularly before going to the bed.

6.5 Treatment against Skin Diseases

The live bark of matured cashew tree is removed and the outer scales are scraped off and bark is cut into pieces. The pieces of such scraped bark are cooked with boiled rice. The medicinal rice thus prepared is consumed as breakfast with common salt. It is believed that consumption of such medicinal rice for 3-4 days in the morning would help in curing the skin diseases of human beings. Also, at times the bark is boiled with water and a "Kashaya" (decoction) is prepared. This kashaya is used for bathing and is believed to heal skin diseases.

6.6 Treatment for Scabies or Skin Diseases of Young Calves of Cattle

The skin diseases of cattle particularly the young calf is being treated with shell liquid. Fresh cashew shell is boiled with ganji of rice (rice gruel) and fed on first day. On second day two shells are boiled in the same way and 3rd day again 3 shells are boiled in rice gruel and fed. This treatment helps in curing the skin disease of calf during initial 1-2 years.

6.7 Treatment for Reduction of Swelling

The skin diseases of cattle particularly, the young calf is being treated with shell. In certain locations, termataria of several species of *Microtermes* occur at base of cashew trees. These are formed by secretions of termites with chewed wood and fine clay, which makes them virtually water resistant. Along with several other herbs, the powder of this termatoria when smeared on the swollen part of human and animal organs, swelling can be reduced. These termetoria are preferred for usage in treatment of mumps, shoulder sprain, and ankle sprain by smear banding/smearing externally. Also the termetoria are thoroughly powdered and soaked overnight, mixed with red clay and applied for treating shoulder and neck sprain of draught animals as it has good shrinking property. These termetoria are also are also used to make firewood stoves by mixing them with clay and cowdung in plain areas.

7. Games Played using Cashew Nut

7.1 Guessing the Number of Nuts

This is a kind of gambling played by young children either as individual or as a team. Nuts are held between two hands by one kid. The other kid (opponent team) has to guess the number of nuts held by the person. At times, the guessing is restricted only to say odd or even numbers. If the opponent team answers correctly whatever the nuts held in the hands will go to the opponent. Likewise, the gambling continues. Usually, for these games the nuts used are collected through pilferage from the trees grown along the road side during fruiting season.

7.2 Moogoddi (Moogubeeja or Jubili)

Generally, young children play this game. The game starts by placing a nut on the ground at a known point by a kid. The other kid has to hit at it by throwing another nut from a specified distance. Usually for this game bigger nuts are preferred. If the aiming kid hits the nut with the nut he throws, he will get the nuts. The game continues all along the village roadside while children return home in the evenings from school. This is seen particularly during fruiting season.

7.3 Spreading Nuts and Striking

Handful of nuts is spread on the ground from a known height (1/2 foot) and a single finger is to be moved within the spread on the ground without touching the nuts. If happen to touch the nuts, game will be repeated by the opponent. After moving the finger completely in the space, any two nuts will be struck each other which should not touch the other nuts available nearby. Thus all the nuts are to be collected by the player. If he fails, the opponent will start a fresh game.

7.4 Striking Nuts with Fingers

At a known distance (2-3 m away) two children sit facing each other with a cashew nut placed before them. Later, one after another would aim with the nut placed before them and strike with finger to the opposite side nut. If the aiming kid strikes the opposite side nut while throwing it will go to him/her. Game continues by keeping a new nut.

8. Regional ITKs

8.1 Odisha

- ☆ The Cashew Nut Shell Liquid is used in treatment of leprosy, dormant skin, skin spots, pimples, wart and cures mumps in chicks.
- ☆ Cashew apple is used as an appetizer, stimulant for sweating and having medicinal values for cough, leprosy pain, anemia and different types of epilepsy.
- ☆ When cashew kernel is fried with ghee and eaten, it is believed to enhance semen count and also used in treatment of epilepsy.
- ☆ Application of common salt is found to reduce the incidence of Cashew Stem and Root Borer in cashew orchards.

8.2 Karnataka

- ☆ It is believed that cashew nut is nobody's crop after a local festival called 'Pathanaje' in the Tulu nadu (Tulu speaking region extends the entire Dakshina Kannada district. Part of Kodagu, part of Udupi and part of Kasaragod districts). Thus, the nuts either green or matured could be collected by anybody after the festival which almost coincides with the onset of monsoon. Reason may be that the cashew was grown by farmers mainly in unfenced lands including encroached government lands. So, people believed that it is God's/Nature's crop after the peak harvest (after the festival). Economically weaker section people used to collect the nuts after this festival from such plantations. Even today, once the cashew contractors vacate the place from plantations all the villagers (people of weaker section) go to such plantations of Karnataka Cashew Development Corporation (KCDC) or other commercial organizations and collect the remaining produce and no owner of the plantations would normally object to this practice.

- ✰ In Chintamani region, growing cashew all along the mango garden as wind break and additional source of income is being practiced.

8.3 West Bengal

- ✰ The CNSL is extracted by boiling the nuts in tin containers and used for lubricating the shaft of cow cart.
- ✰ For all fruit crops including cashew, sprinkling of cow dung slurry for protecting plants from grazing is a very common practice.

8.4 Kerala

- ✰ Cashew nut shell liquid is being used to apply in between and around fingers of the foot to prevent/cure ring worm infection.
- ✰ Cashew apple juice is extracted and filled in glass bottle and made airtight by corking and is kept under the soil for 5-6 months and used as a medicine for stomach problem.
- ✰ Cashew branches are considered equally good to *Artocerpus integriflia* as a mulch over the beds of ginger/turmeric immediately after planting which will hasten germination and prevent damage due to scavenging/foraging by poultry.
- ✰ Fresh cashew leaves and cashew nut shell are applied at the time of preparation of land for cultivation of vegetables and others crops in order to reduce pest incidence.
- ✰ The gum exudation from the cashew tree used as adhesive.
- ✰ Cashew shell liquid has been used to brush on the outer and inner surface of country boards together with or without fish oil to protect the wood from insect attack/decay.

8.5 Maharashtra

- ✰ Kernels obtained from immature raw nuts are coated with mud slurry and dried in sun light for about one week and stored in air tight containers. Such stored kernels are used during rainy season for mixing with fish/vegetable dishes.

8.6 Tamil Nadu

- ✰ There is a saying that if a person speaks out at once giving information without thinking he is referred as "Mundirikottai" (Cashewnut).

- During Pongal (harvest festival) after performing all poojas and rituals, neem leaves are thrown scatteredly in the cashew field and turmeric powder mixed with water is sprinkled all over the field. The festival usually coincides with the flushing phase of cashew. This is usally done with a belief to ward-off the evil sprits, but, in turn it also acts as disinfectant which would prevent diseases/pests.
- Powdered ash is dusted to cashew during flushing period to prevent sucking pests.
- Soaking of cashewnuts in cowdung slurry for germination purpose is an old age practice which softens seed coat and also acts as disinfectant.
- In the traditional cashew growing tracts during fertilizer application, common salt (Sodium chloride) is applied invariably to all the trees.

References

Chopra, R.N. (1956). Glossary of Indian Medicinal Plants, Council of Scientific and Industrial Research – p.17.

Nadkarni, A.K. (1954). Indian Materia Medica. Popular Book Depot, Bombay, p. 17.

Nayak M.G., Yadukumar N., Venkattakumar R and Raviprasad T.N. (2004). Indigenous technical knowledge in cashew, NRCC Technical Bulletin No.9, National Research Centre for Cashew, Puttur, Karnataka.

Rangaswamy, T.M. (1994). Vanavasudhi Guna dharma. Ananthasevashrama Viswatha Samithi, Malladi Halli.

Shankaranarayana Bhat (2001). Ajjimaddu. M/s Ayurveda Prakashana, Panaje, Puttur, Dakshina Kannada – 574 259 (Third Edition).

Section VII: Impact Analysis

Cashew: Improvement, Production and Processing (2017) *Pages* **551–563**
Editor: **Prof. P.L. Saroj**
Published by: **ASTRAL INTERNATIONAL PVT. LTD., NEW DELHI**

33

Climate Induced Effects: Adaptation and Mitigation Strategies

T.R. Rupa

1. Introduction

Agricultural productivity is sensitive to two broad classes of climate-induced effects: (i) direct effects from changes in temperature, precipitation, or carbon dioxide (CO_2) concentrations, and (ii) indirect effects through changes in soil moisture and the distribution and frequency of infestation by pests and diseases (IPCC 1996; 2001). Climate change impacts on agriculture are being witnessed all over the world, but countries like India are more vulnerable in view of the high population depending on agriculture and excessive pressure on natural resources. The warming trend in India over the past 100 years (1901 to 2007) was observed to be 0.51°C with accelerated warming of 0.21°C per every 10 years since 1970 (Krishna Kumar, 2009). Climate change projections made up to 2100 for India signify an overall increase in temperature by 2-4 °C with no substantial change in precipitation quantity. However, different regions are expected to experience differential change in the amount and distribution of rainfall that is likely to be received in the coming decades. It is projected that some parts of country will receive higher amount of rainfall. Another significant aspect of climate change is the increase in the frequency of occurrence of extreme events such as droughts, floods and cyclones. The projected impacts are likely to further aggravate yield fluctuations of many crops with impact on food security and prices. Climate change impacts are likely to vary in different parts of the country. Parts of Western Rajasthan, Southern Gujarat, Madhya Pradesh,

Maharashtra, Northern Karnataka, Northern Andhra Pradesh, and Southern Bihar are likely to be more vulnerable in terms of extreme events.

Agriculture is considered to be one of the major anthropogenic sources of atmospheric greenhouse gases (GHGs) (Lal, 2000). The atmospheric concentration of carbon dioxide (CO_2), methane (CH_4) and nitrous oxide (N_2O) and other GHGs has increased since the industrial revolution (1750 A.D.) due to natural and anthropogenic activities. The atmospheric concentration of CO_2 has increased from 280 parts per million by volume (ppmv) in 1750 to 394 ppmv in 2012 and is currently increasing at the rate of 1.9 ppmv year^{-1} (NOAA, 2012). Atmospheric CH_4 concentration has increased from about 715 to 1826 parts per billion by volume (ppbv) in 2012 over the same period and is increasing at the rate of 7 ppbv year^{-1}. Similarly, the atmospheric concentration of N_2O has increased from about 270 ppbv in 1750 to 325 ppbv and is increasing at the rate of 0.8 ppbv year^{-1} (lPCC, 2007; EPA, 2013). In order to compare emissions from different sources, the global warming potential (GWP) of each gas is equated to the GWP of CO_2, For example, the GWP of 1 tonne of CH_4 is 24.5 times more potent than 1tonne of CO_2 over a 100 year period (IPCC, 2007).

All these expected changes will have adverse impacts on climate sensitive sectors such as cashew. Cashew is mostly grown as a rainfed crop and the yield and quality of nut is highly dependent on climatic/weather conditions. Any variability in climate has direct impact on reproductive phase of cashew. With climate variability, there will be various impacts on cashew production such as reduction in yields, variation in flowering, fruit setting, nut development and kernel quality, incidence of pests and diseases and water stress. The increase in weather extremes like torrential rains, heat waves, mid waves and floods, besides year-to-year variability in rainfall affects cashew productivity considerably and leads to stagnation/decline in production across various agro-climatic zones. The challenge for cashew crop will be to improve yields in marginal lands under rainfed conditions where the harsh environment strongly limits crop growth, productivity and quality of the produce. It is therefore, important that adaptation and mitigation strategies need to be formulated to minimize the climate change related impact on cashew.

2. Climatic Suitability

Cashew, a tropical nut crop can be found growing between 28°N and 28°S latitude. In India, cashew is grown in the coastal belt between 8°N and 28°N. Flowering time depends on the latitude. In Brazil and Tanzania, peak flowering is between August and September. The highest flowering occurs in October in Mozambique, while in Philippines it is March. In the west coast of India, flowering is from October to March. Whereas in the east coast of India, flowering is delayed by about 2-3 months. However, the crop is ready for harvest in summer, both in the north and south of the equator. The phenology of cashew is very much influenced by the altitude of the region. Suitability studies for cashew using GIS at ICAR-Directorate of Cashew Research (ICAR-DCR), Puttur indicated that cashew grows at an elevation ranging from 0 to 1000 m above mean sea level (MSL). However, the productivity is the highest up to the altitude of 750 m above MSL. Low temperatures at higher altitudes have adverse effect on the crop. Flowering and fruiting are delayed irrespective of latitude at higher altitude.

Though cashew can tolerate wide range of temperature but the optimum monthly temperature is between 24 °C and 28 °C. Cashew grows in the semi-arid regions like northern Mozambique where daily maximum temperatures exceeds 40 °C and in Assam, cashew survives up to 7 °C. It has been reported that cashew cultivation is not economical in regions where annual temperature falls below 20 °C for prolonged periods. In major cashew growing regions, the mean daily maximum temperature varies between 25 °C and 35 °C and the mean daily minimum temperature vary between 15 °C and 25 °C. The productivity of cashew is higher in regions where the mean annual temperature ranged from 22.5 to 27.5 °C and the minimum temperature ranges from 10 °C to 22 °C. The productivity is lower in regions where the minimum temperature drops below 10 °C. For flowering, cashew requires mild winter, that is low minimum surface air temperature ranging between 16 °C and 20 °C coupled with more dew nights. Cashew is highly sensitive to light and produces more foliage, flowers and fruits on braches exposed to sunlight than on the shaded branches. Bright sunshine (greater than 9 h/day) with moderate dry weather is good for flowering.

Cashew can survive under very high and low rainfall conditions. Although cashew can tolerate drought conditions but for proper vegetative development and regular fruit-setting, the average annual rainfall of 1,300 to 2,500 mm is considered to be suitable. The average annual rainfall distribution in cashew areas ranged from low rainfall (300-600 mm in Gujarat) to high rainfall (2700-3000 mm in west coast and NEH region) but the productivity is highest in regions with a mean annual rainfall distribution of 600-1500 mm. Cashew needs a clearly defined dry season of at least 4 to 5 months. A dry spell from January to May with occasional light summer rains ensures better cashew production. A well distributed North-East monsoon rainfall of about 500 mm during September to December and about 100 mm during February to April are good for successful production of cashew. Light rains during flowering do not affect the production but heavy rains during flowering affects the yield. Year to year variation in time of flowering of a variety is common even under uniform cultural and management practices. It signifies the influence of weather factors on flowering behavior of cashew. Unusual heavy rainfall between January and March may encourage high incidence of pest like tea mosquito bug (TMB) and reduce yield and quality. Cashew is highly sensitive to high relative humidity (more than 80 per cent) during the harvesting season. High relative humidity adversely affects the nut quality.

Based on the information available on cashew production in relation to weather, soil and yield potential, an attempt has been made to demarcate agroclimatic zones for cashew across the west and east coasts of India by Prasada Rao (2002). Cashew productivity can be improved towards north across both the coasts of west and east while it may decline towards inland plateau of Peninsular India. The northeast region may not be favorable due to occurrence of frequent cold waves during fruiting season.

3. Climate Change Effects

The rainfed cashew crop is highly sensitive to changes in climate and weather, especially during reproductive phase. High temperature (>34.4 °C) and low RH (<20 per cent) during afternoon cause drying of flowers. The maximum temperature, humidity and rainfall are the major climatic factors that influence the productivity of cashew. According to Prasada Rao *et al.* (2010), the maximum temperature plays a crucial role on nut size and kernel weight of cashew during the nut development stage. Haldankar *et al.* (2003) reported that the relative humidity during pre-flowering phase is the main factor which explains yield variation in cashew plantations. The unusual rains between November and December inordinately delay reproductive phase of the late flowering varieties. Unseasonal rainfall and heavy dew during flowering and fruiting intensify the incidence of pests and diseases as well as deterioration of nut quality. For example, a rainfall of about 201 mm received at ICAR-DCR experimental farm during 15-25 March, 2008 adversely affected the yield and quality of nuts. Large quantity of nuts which could not be picked up in time germinated in the field itself. The nuts could not be dried in time due to non-availability of good sunshine hours resulting in poor quality of nuts. The excess moisture content in raw cashewnut damages the kernel inside by changing its colour from white to cream. The nut yield losses due to unseasonal rains were estimated to be between 50 and 65 per cent in March, 2008.

Cashew genotypes vary noticeably in their heat units (day °C) requirement. Early variety (Anakkayam-1) requires only 1953 heat units for reproductive phase, while late variety (Madakkathara-2) requires 2483 heat units. Cashew kernel weight is positively correlated with heat units especially for mid and late varieties. Continuous rains without critical dry spells and late winter rains delay the bud break in cashew. A dry spell of 7 days is usually necessary 30 days prior to the bud break. Late and extended winter rains reduce the number of bright sunshine hours invariably which results in delaying of bud break and better availability of soil moisture during flowering (December and January). For example, incessant rains received at Pilicode (Kerala) until November in 1998 delayed the bud break. The delay in bud break was prominent in early varieties. In late varieties, bud break was normal, because the required dry spell of 7 days was met 22-26 days before bud breaking (Prasada Rao *et al.*, 2001).

Cashew is infested by several insect pests, thereby limiting the production considerably. One of the main reasons for reduction of cashew nut yield is the occurrence of an important sucking insect pest, tea mosquito bug (TMB) (*Helopeltis antonii* Signoret) during the cropping season. The production loss from the TMB alone is estimated to be about 30 per cent. This pest has got potential to cause 100 per cent loss in yield in severe cases. The incidence and severity of the pest is highly dependent on climate and weather factors. The minimum temperature plays a vital role in the incidence of pest population and is negatively correlated with the TMB pest incidence (Godse *et al.*, 2005). The favourable minimum temperature for TMB incidence ranges between 13-18 °C. Low temperature (12 °C) is antagonistic for pest build up.

Apart from the damage caused by the infestation of TMB, infection of the panicles by the fungal pathogens, *viz., Colletotrichum gleosporoides* and *Gleosporium mangiferae* have been reported to cause drying up of young shoots, inflorescence and immature nuts in cashew. The characteristic symptom is the drying of floral branches. The symptoms appear as minute water soaked lesions on the main rachis and secondary rachis. The incidence is very severe when cloudy weather prevails. The incidence of this disease is being reported from different new locations in which it was not prevalent earlier. High relative humidity in forenoon during December - February both in 1997 and 1998 and the minimum temperature of 18-20 °C were found to be favourable for sporulation by fungi. A significant increase in dewfall was one of the most important factors which favoured the growth, sporulation and spread of fungi. Cloudiness leading to low bright sunshine hours (2 h/day) followed by dewfall triggered the growth, sporulation and spread of fungal pathogens causing inflorescence blight during 1998-99 in Kannur and Kasaragod districts (Prasada Rao, 2002).

4. Weather Extreme Effects

Changes in temperature and precipitation patterns together with occurrence of extreme events due to climate change are a major threat to future cashew sector. The Tsunami in coastal Tamil Nadu on 26 December, 2004 severely affected on cashew crop. Most of the standing crop was inundated with salty sea water ingressing in the mainland. Salinity is a major environmental stress and is a substantial constraint to crop production. Salt affected soils are the soils that contain considerable amounts of soluble salts and or sodium on the exchange complex. The high amount of soluble salts (in saline soils) and of sodium on the exchange complex (in sodic soils) hinder crop growth and have rendered them barren (Gupta and Abrol, 1990). In saline soil the available moisture range is low and crop has to spend extra energy to extract water from the soil because of high osmotic potential of the soil solution. These also adversely affect water and nutrient availability. Most of the cashew varieties are sensitive to salinity. Rise in sea water level due to climate change conditions may adversely affect the cashew plantation. Electrical conductivity of 1.48 dS m^{-1} in irrigation water is a threshold tolerance for precocious cashew during the initial growth (Carneiro *et al.*, 2002).

'Thane' cyclone with a very high wind speed created havoc in Cuddalore district of Tamil Nadu and Pondicherry on 30 December 2011. The entire cashew area of Panruti, Cuddalore and Kurinjipadi taluks of Cuddalore district were totally devastated. The extent of damage to cashew trees in Vridhachalam taluk was 60-70 per cent due to decrease in the wind speed away from the coast. A team of scientists of AICRP-Cashew Centre at Regional Research Station (Tamil Nadu Agricultural University), Vridhachalam visited and surveyed the 'Thane' cyclone affected area and the recommendations of the team to cope up with the situation include i) Removal of uprooted trees for new planting, ii) Mass replanting programme, iii) Government support for laying out a borewell for every 10 acres for establishment of the new plants, iv) Government support programmes for raising annual crops like pearl millet, blackgram, horsegram, groundnut and kodo millet upto five years *etc.*

5. Adaptation and Mitigation Strategies to Cope with Climate Change

There are both threats and opportunities for cashew sector with climate change, the key is in understanding the likely impacts and then managing the risks. Some of the adaptation measures in general that the cashew sector can undertake to cope with future climate change include: Changing planting dates, planting different varieties, developing new drought and heat resistant varieties, salt tolerant varieties, adoption of intercropping, using sustainable fertilizer, weed management, more use of water harvesting techniques, supplemental irrigation, drip irrigation, fertigation, better pest and disease management etc. Climate change mitigation refers to efforts to reduce/prevent emission of GHGs. Protecting natural carbon sinks like forests and oceans, or creating new sinks through silviculture or green agriculture are also elements of mitigation. In order to mitigate the ill effects of climate change on soil quality and to protect the soil and land resource, it is important to give more focus on adoption of soil and water conservation practices, mulching, cashew leaf litter retention and recycling, addition of animal based manures *etc.*

Soil and Water Conservation Practices

Major cashew growing area in India is under undulated topography (steep slopes). Cashew is grown as a rainfed crop, experiences severe moisture stress from January to May. It adversely affects flowering and fruit set. The main aim of soil and water conservation practices is to reduce or prevent either water erosion or wind erosion, while providing the desired moisture for crop growth/production. Erosion by water is the most serious soil degradation problem in the humid tropical and sub tropical India. In India, soil erosion has been taking place at an average rate of 16.35 t ha^{-1} $year^{-1}$ (Dhruvanarayana and Rambabu, 1983), totaling an annual loss of 5334 million tonnes, nearly 29 per cent of the total eroded soil is permanently lost into the sea, and nearly 10 per cent is deposited in reservoirs. Remaining 61 per cent of the eroded soil is transferrable from one place to another. Soil erosion by water and wind result in loss of top soil and depletion of soil organic matter. Depletion of organic matter in soil discourages the activity of soil microflora responsible for decomposition of organic matter to enrich soil fertility. It becomes imperative for the land use planners to adopt appropriate conservative measures.

In order to harvest the rainwater and to make it available to the cashew plant during critical period, an *in situ* soil and water conservation and rainwater harvesting are very important. Research work carried out at ICAR-DCR indicated that soil and water conservation measures such as modified crescent bund (at 2 m radius having a crescent shaped bund of 6 m length, 1m width and 0.5 m height on the upstream of the plant so that a trench of 6 m length and 50-75 cm deep will be formed while making the bund) and coconut husk burial (Trenches of size 5 m length, 1 m width and 0.5 m depth in the middle of 4 plants with coconut husk buried) were found effective in terms of higher yield, reduction in annual runoff and conserving moisture (Rejani and Yadukumar, 2010). It was shown that reduction in peak runoff and increase in recession time and groundwater recharge due to soil and water conservation practices (Deshmukh *et al.*, 1992). According to Badhe and Magar (2004), trapezoidal shaped staggered trenches (230/ha) having dimensions of 4.5 m length, 0.60 m top width, 0.30 m bottom width and 0.30 m depth were effective for reducing runoff and conservation of soil and nutrients. Mane *et al.* (2009) demonstrated that continuous contour trench (0.50 m x 0.60 m) is the best soil conservation practice for cashew on areas having 7 to 8 per cent slope.

Mulching

Cashew is often planted in areas which are totally dry and less fertile. Under such situations mulching is highly useful to conserve the soil moisture for a long period, protecting the soil from erosion and maintaining, restoring and improving soil organic carbon status. The basin area of cashew plants can be mulched either with green leaves or dry leaves and weeds soon after planting. Black polyethylene mulch was helpful to conserve soil moisture (Nawale *et al.*, 1985). Using coconut coir pith as soil mulch in cashew plantations resulted in 14.15 per cent more water retention and suppression of weeds to an extent of 73.52 per cent (Kumar *et al.*, 1989). Formation of terrace and crescent bund and mulching the base area with cashew leaf litter and other jungle growth available in the garden are helpful.

Soil Fertility Management

Cashew is mostly grown on marginal and sub-marginal lands with least attention of the farmers towards application of manures and or fertilizers. Loss of top fertile soil in runoff and low or no biomass recycling and imbalanced use of plant nutrients are some of the serious problems as far as soil productivity is concerned in major cashew growing regions. The depletion of organic matter in soil is responsible for declining nutrient use efficiency and widespread deficiencies of secondary and micronutrients. Soil organic matter is a key element of soil management that prevents erosion and improves water availability. Other soil physical characteristics that are linked to soil organic matter are: infiltration, water retention, bulk density and soil strength. Improving soil organic matter is essential to maintain soil health and cashew productivity under changing climate scenario. Generally, cashew grown under humid climatic zones contain higher amount of organic carbon in soils. However, maintenance of soil organic matter is difficult in arid and semiarid climatic zones, as large amount of soil carbon is lost due to high temperature. Importance of the use of organic sources of nutrients along with chemical fertilizers for maintaining soil health has been emphasized by several workers in tree crops.

Growing green manure crops like *glyricidia*, sesbania, sunhemp and cover crops between two rows of cashew have potential to improve soil moisture retention capacity and nutrient content. According to research studies at ICAR-DCR, higher soil moisture content was observed in cashew orchard with *glyricidia* (17.0 to 18.6 per cent dry basis), sunhemp (17.8 to 18.3 per cent dry basis), sesbania (15.5 to 18.2 per cent dry basis) compared to control (15.5 to 17.0 per cent dry basis). The nutrient addition to soil was about 186 kg N, 23.6 kg P_2O_5 and 126.2 kg K_2O and 141 kg N, 17.9 kg P_2O_5 and 162.3 kg K_2O/ha through *glyricidia* and sesbania, respectively.

Site Specific Nutrient Management (SSNM) is another approach with potential to mitigate effects of climate change. The SSNM is critical for GHGs mitigation so as to reduce input cost and enhance nutrient use efficiency considerably. This approach facilitates cashew grower to invest only on deficient nutrients and omit nutrient application which was in sufficient range in soils. Various benefits of SSNM practice include lowering in input cost, higher nutrient use efficiency, and reducing GHGs particularly N_2O. Application of fertilizers particularly N during water stress escalates water stress problem further as higher N improves leaf canopy which results higher evapotranspiration. Using nitrification inhibitors, and fertilizer placement practices need further consideration for GHGs mitigation. Management of soil in combination with optimum soil moisture is essential to protect the plants during weather aberrations and overall reduction of CO_2, N_2O and CH_4 from soil. For example, increased rainfall in regions that are already moist could lead to increased leaching of minerals, especially nitrates. Large increases in fertilizer applications would be necessary to restore productivity levels. Placement of fertilizer materials and split application of nutrients into soil will substantially improve both nutrient and water use efficiency. However, KNO_3 foliar sprays during intermittent droughts results in the adaptation of plants by closing stomata.

Intercropping

Cashew takes 8 to 10 years for the canopy to cover the entire area under normal spacing. In hilly regions there is a possibility of soil erosion and weed growth during initial years of. planting. Under such situations, intercropping not only minimizes erosion and conserves soil and moisture and also to realize higher returns from unit area during the early stages of cashew plantation. Growing field crops (groundnut, black gram and green gram), vegetables (cucumber and bottlegourd), tuber crops and fruit crops (pineapple), spices (turmeric, ginger and pepper) are suitable and profitable intercrops in cashew plantations. Ginger was recommended as an intercrop in the initial 3-4 years of cashew plantation in the west coast region. In east coast, crops like groundnut and cowpea are grown profitably under rainfed situations. Intercropped pineapple in cashew plantations has been adopted by farmers and resulted in 1.5 fold higher yield and net profit. Pepper vine and kokum are the other popular intercrops for enhancing the farm profitably as mixed crops grown in cashew orchards by trailing on to the stem of cashew in case of pepper and planting kokum in the middle of 4 cashew plants in west coast where rainfall is high.

Weather Related Contingencies

Cashew is sown under normal onset of monsoon during June to September. However, dry spells occur at various stages during the planting season resulting in early, mid-season and terminal droughts adversely impacting performance of the plants. The effect of drought is pronounced due to delays in onset of monsoon, prolonged breaks in monsoon and deficit rainfall. Contingent planning for rainfed cashew is important when the onset of monsoon is delayed. Small changes in climatic parameters can often be managed reasonably well by altering dates of planting. In case of normal onset followed by early season dry spell immediately after planting leading to poor crop establishment, gap filling, mulching, nutrient management etc. may be necessary. In addition, moisture conservation measures and life saving irrigation wherever possible are suggested. The irrigation through pitcher (hold pots) is recommended in dry land situations. In case of terminal drought in September to October due to dry spells or early withdrawal of the south west monsoon, crop management measures such as life saving or supplemental irrigation, if available, from harvested pond water or other sources are suggested as in the case of mid season drought wherever feasible.

Established plants survive even in adverse soil moisture conditions. The rainfall deficit or cessation of rains at early stage also adversely affects the nut yield particularly in late maturing varieties. In order to minimize the yield loss under such situations, one or two protective irrigations are highly beneficial. Supplemental irrigation of 200 L of water/plant once in 15 days during January to March from water collected in ponds through rain harvesting helps in flowering and nut development by improving the microclimate with increased humidity. It also leads to increased nut and kernel weight by reducing flower and nut drying to some extent. Drip irrigation during fruit development stages wherever water is available is helpful under drought situations to rainfed cashew crop. Normally for west coast of Dakshina Kannada, irrigation by drip at 20 L/tree/day for mature cashew plantations (10 to 15 years) is recommended. Research results indicated that yield can be doubled, if irrigated. Irrigation can be started after the commencement of flowering for better nut set, filling and yield. It has been reported that fertigation saved 50 per cent in the fertilizer requirement and doubled the cashew yield (Richards, 1993; Yadukumar and Mandal, 1994; Mishra *et al.*, 2008). To maintain the proper soil moisture regime, the harvesting of rain water and recycling them during deficit period is suggested. Moreover, adoption of soil conservation measures and installation of drip wherever water source is available will be helpful. Developing heat and drought tolerant cashew varieties by utilizing genetic resources that may be better adapted to new climatic and atmospheric conditions should be the long-term strategy.

In the recent past, continuous high rainfall in a short span leading to water logging. Heavy rainfall coupled with high speed winds in a short span are being experienced at various growth stages, of which the 2008 one was the most recent, significantly affecting the production of cashew. In addition to that, the amount and distribution of rainfall is becoming more and more erratic, causing greater incidence of drought and flood. The increase in frequency of heavy rainfall events in last 50 years over Central India points towards a significant change in climate pattern in India. Unseasonal rains received during harvesting season in Dakshina Kannada district of Karnataka caused considerable yield loss and deterioration of nut quality both in trees and on the ground. The contingency measures suggested are drying of collected matured fallen nuts immediately by sun drying/artificial drying in case there is not sufficient bright sun light.

The climatic change threats include likely increase of temperature, extreme weather conditions, increased water stress and drought, and desertification. All these may seriously increase the vulnerability of resource poor cashew farmers to global climate change. Policies that support crop insurance can provide protection to the farmers in the event of crop failure or lower production due to natural calamities. Some of the measures to cope up with climate change effects include weather based crop and an early warning system of environmental changes.

Pest and Disease Management

Tea mosquito bug (TMB) pest and inflorescence blight disease cause considerable damage to cashew. Incidence and severity of both are dependent on climate and weather factors. Need based sprays are recommended for TMB during most vulnerable periods of crops such as flushing, flowering and fruiting stage of the crop. Forewarning the outbreak of the dreaded inflorescence blight disease in west coast region of India is of utmost importance for its effective management. Proper training of cashew trees from early stages and timely pruning of side and criss-cross branches may ensure that microclimate is not congenial for build-up of sucking pests and inflorescence blight disease.

Carbon Sequestration: Cashew has substantial carbon sequestration potential for carbon sequester for mitigation of climate change and can also be grown in vast degraded/wasteland existing in cashew-growing regions. Soil carbon (C) has gained increase interest in the recent past owing to its importance in C sequestration studies and its potential impact on sustainable crop production. Carbon sequestration implies removing atmosphere carbon and storing it in natural reservoirs for extended periods (Lal, 2004; Srinivasarao *et al.*, 2013). Soil carbon sequestration is the process of transferring carbon dioxide from the atmosphere into the soil through crop residues and other organic solids, and in a form that is not immediately remitted. This transfer or "sequestering" of carbon helps off-set emissions from fossil fuel combustion and other carbon-emitting activities while enhancing soil quality and long-term agronomic productivity.

The C fixed in plants by photosynthesis and added to soil as above and below ground litter is the primary source of C in ecosystems (Warembourg and Paul, 1977). Although most carbon enters ecosystems via leaves, and carbon accumulation is most obvious when it occurs in above ground biomass, more than half of the assimilated carbon is eventually transported below ground via root growth and turnover and exudation of organic substances from roots. Numerous studies have convincingly proved that the inclusion of trees in the agricultural landscapes often improves the productivity of systems while providing opportunities to create carbon sinks.

Since C sequestration is an essential component of mitigation of green house gases, it is important to assess the sequestration potential of crops. It has been estimated that soil organic carbon sequestration potentially could offset about 15 per cent of the global CO_2 emission. Cashew has dense green leaves with good photosynthetic capacity and it can also be grown in high-density planting system. Cashew is suitable crop for C sequestration. It can be grown in vast degraded/ wasteland existing in cashew growing regions. Based on research undertaken at ICAR-DCR, it was found that, 7 years old trees of cashew genotype VTH-174 sequestered about 2.2 fold higher C under high density planting system (625 trees/

ha) as compared to normal density planting system (156 trees/ha). Carbon storage by cashew has been estimated as 32.25 and 59.22 t CO_2/ha at 5th and 7th years of growth, respectively under high density planting. The extent of C sequestered will depend on the amounts of C in standing biomass, age of the crop, tree density, variety etc.

6. Future Line of Action

- ✰ Development of cashew varieties resistant to abiotic stresses like temperature, drought, salinity, floods etc.
- ✰ Development of climate resilient agro-techniques in order to suit unfavourable abiotic stresses.
- ✰ Assessment of the impacts of elevated CO_2 and temperature on growth, development, yield and quality of cashew.
- ✰ Quantification of carbon sequestration potential of cashew.
- ✰ Development of crop simulation models for cashew.
- ✰ Early warning weather forecasts on pest incidence, rains, drought, weather extremes etc.
- ✰ Assessment of new pest and diseases in the scenario of climate change and development of cost effective, eco-friendly approaches for management of emerging pests.
- ✰ Mitigation and adaptation strategies to cope up with anticipated climate change impacts on cashew need to be strengthened.

References

Badhe, V.T. and Magar, S.S. (2004). Influence of different conservation measures on runoff, soil and nutrient loss under cashew nut in lateritic soils of south Konkan region. *Indian J. Soil Conser.*, 32(2): 143-147.

Carneiro P.T., Fernandes, P.D. and Gheyi, H.R. (2002). Germination and initial growth of precocious dwarf cashew genotypes under saline conditions. *Revista Brasileira De Engenharia Agricola E Ambiental.*, 6(2): 199-206.

Deshmukh, M.T., Sawke, D.P., Borude, S.G., Hurni, H. and Tato, K. (1992). Effects of platform bench terraces on the growth and yield of mango and cashew grafts. Erosion, conservation and small scale farming report. Department of Agricultural Engineering, Konkan Agriculture University, Ratnagiri, India, pp. 477-482.

Dhruvanarayana, V.V. and Rambabu (1983). Estimation of soil erosion in India. *Journal of Irrigation and Drainage Engineering,* 109: 419-434.

Godse, S.K., Bhosle, S.R. and Patil, B.P. (2005). Population fluctuation studies of tea mosquito bug on cashew and its relation with weather parameters. *Journal of Agrometeorology,* 7(1): 107-109.

Gupta, R.K. and Abrol, I.P. (1990). Salt affected soils - Their reclamation and management for crop production. *Advances in Soil Science,* 12: 223-275.

Haldankar, P.M., Deshpande, S.B., Chavan, V.G. and Rao, E.V.V.B. (2003). Weather associated yield variability in cashewnut. *Journal of Agrometeorology*, 5(2): 73-76.

IPCC 2007. Agriculture. In: Metz, B., Davidson, OR, Bosch PRo (Eds.), Climate Change (2007). Mitigation. Contribution of Working Group III to the Fourth Assessment Report of the Intergovernmental Panel on Climate Change. Cambridge University Press, Cambridge, United Kingdom/New York, NY, USA.

IPCC 2001. Climate Change (2001). Impacts, Adaptation, and Vulnerability, Contribution of Working Group II to the Third Intergovernmental Panel on Climate Change. In: McCarthy, J.J; Canziani, O.F; Leary, N.A; Dokken, D.J. and White, K.S. (Eds), Cambridge University Press, Cambridge, UK.

IPCC 1996. Climate Change (1995). Impacts, Adaptations and Mitigation of Climate Change: Scientific- Technical Analyses.

Krishna Kumar (2009). Impact of climate change on India's monsoon climate and development of high resolution climate change scenarios for India. Presented at MoEF, New Delhi on October 14, 2009 (http: moef.nic.in).

Kumar, D.P., Subbarayappa, A., Hiremath, I.G., Khan, M.M. and Sadashivaiah (1989). Use of coconut coir-pith: a biowaste as soil mulch in cashew plantations. *The Cashew*, 3(3): 23-24.

Lal, R. (2000). World cropland soils as a source or sink for atmospheric carbon. *Advances in Agronomy*, 71: 145-191.

Lal, R. (2004). Soil carbon sequestration impacts on global climate change and food security. *Science*, 304: 1623-1627.

Mane, M.S., Mahadkar, U.V., Ayare, B.L. and Thorat, T.N. (2009). Performance of mechanical soil conservation measures in cashew plantation grown on steep slopes of Konkan. *Indian J. of Soil Conser.*, 37(3): 181-184.

Mishra, J.N., Paul, J.C. and Pradhan, P.C. (2008). Response of cashew to drip irrigation and mulching in coastal Odisha. *J. of Soil and Water Conser.*, 7(3): 36-40.

Nawale, R.N., Sawke, D.P. and Salvi, M.J. (1985). Effect of black polyethylene mulch and supplemental irrigation on fruit retention in cashewnut. *Cashew Causerie*, 7(3): 8-9.

NOAA/ESRL. (2012). Available at http: //www.esrl.noaa.gov/gmd/ccgg/trends.

Prasada Rao, GS.L.H.V., Giridharan, M.P., Jayaprakash Naik, B., Gopakumar, C.S., Krishnakumar, K.N. and Xavier Tony (2001). Influence of weather factors on bud break, flowering and nut quality of cashew (*Anacardium occidentale* L.). *Indian Journal of Agricultural Sciences*, 71(6): 399-402.

Prasada Rao, G.S.L.H.V. (2002). Climate and Cashew. AICRP on Agro Meteorology, Department of Agricultural Meteorology, College of Horticulture, Kerala Agricultural University, Kerala, India, pp. 1-100.

Prasada Rao, G. S. L.H.V., Rao, G.G.S.N. and Rao, V.U.M. (2010). Climate Change and Agriculture over India. PHI Learning Private Limited, New Delhi, India. 328 p.

Rejani, R. and Yadukumar, N. (2010). Soil and water conservation techniques in cashew grown along steep hill slopes. *Scientia Horticulturae*, 126: 371-378.

Richards, N.K. (1993). Cashew Research in Northern Territory, Australia, 1987-1991. Tech. Bull. No. 202.

Srinivasarao, Ch., Venkateswarlu, B., Rattan Lal., Singh, A.K. and Sumantha Kundu. 2013. Sustainable management of soils of dryland ecosystems of India for enhancing agronomic productivity and sequestering carbon. *Advances in Agronomy*, 121: 253-329.

Warembourg, F.R. and Paul, E.A. (1977). Seasonal transfers of assimilated ^{14}C in grassland: plant production and turnover, soil and plant respiration. *Soil Biology and Biochemistry*, 9: 295-301.

Yadukumar, N. and Mandal, R.C. (1994). Effect of supplementary irrigation on cashewnut yield. Water Management for Plantation Crops - Problems and Prospects. Centre for Water Resource Development and Management, Calicut, Kerala, India, pp. 79-84.

Cashew: Improvement, Production and Processing (2017) *Pages* **565–593**
Editor: **Prof. P.L. Saroj**
Published by: **ASTRAL INTERNATIONAL PVT. LTD., NEW DELHI**

Impact Analysis of Production Technologies: Approaches and Outcomes

M.V. Sajeev and P.L. Saroj

1. Introduction

Impact assessment of public funded agricultural research is an important activity that helps to ensure accountability, maintain credibility, and improve internal decision-making processes and the capacity to learn from past experience. Impact assessment helps to define priorities of research and facilitate resource allocation among programs, guide researchers and those involved in technology transfer to have a better understanding of the way new technologies are assimilated and diffused into farming communities, and show evidence that farmers benefit from the various technology outputs (Manyong, *et al.*, 2001). The focus and methods of impact assessment have evolved over time in response to interest of funding agencies and organizational research mandates. In view of the need for stronger accountability in recent years, there is now a much greater emphasis on the potential impacts of agricultural research on poverty alleviation with a view to setting priorities of research (Kerr and Kolavalli, 1999; Alwang and Siegel, 2003).

Impact studies have, however, faced both conceptual and empirical challenges, partly due to the complexities of the relationships between agricultural technology and the various dimensions of poverty, with research having both direct and indirect effects on poverty alleviation (Kerr and Kolavalli, 1999; deJanvry and Sadoulet, 2002). As the goals of agricultural technology development change from increasing food production to the broader aims of reducing poverty, both technology development and studies of its impact become more complex. Qualitative and quantitative information and qualitative and quantitative methods are needed to assess the impact of research on the farmer livelihood. The context in which new technologies are released and adopted should also be examined for a better understanding of the impact of agricultural research on broader definitions of economic and social outcomes.

2. Products of Cashew Research and Impact Studies

Varietal Technologies

Improved genetic material embodied in cashew soft wood grafts is the most fundamental type of research output in cashew. "Improved" may refer to any of several desirable characteristics: higher potential nut yield, responsiveness to inputs such as fertilizer and/or irrigation, greater tolerance to stresses such as droughts, pests or diseases, length of the flowering season, better taste etc. In practice, most research on modern varieties has focused on raising yields, reducing susceptibility to various stresses, and reducing the length of the growing season (Anderson, 1997). Most impact studies have addressed commodity-specific research and, within that category, there has been an overwhelming emphasis on varietal technologies. It may be noted that more than 40 improved varieties of cashew have been brought out by research system in India. However, in cashew, studies with focus on impact of varietal technologies are found rare.

Non-varietal Technologies

Many of the products of research are not embodied in tangible inputs, such as seeds, but are provided as information, in the form of a recommendation. Some examples of these improvements are better information on the most suitable inputs, improved management techniques such as methods and levels of application of inputs, and improved cultural practices. Farmers obtain new information through explanations on field days, recommendations in extension bulletins, intermediary contacts, and fellow farmers. This can involve identifying optimal combinations and quantities of inputs or developing better management practices that do not involve material inputs, such as improved timing of operations and crop rotations. Integrated pest management practices are another category of non-varietal technologies dealing with the management of insect pests through techniques, such as crop rotations and biological control. Adoption of such technologies is often difficult to track, complicates attribution to research, and raises problems of valuation (Baidu-Forson, 1996). These are generally more complex innovations than improved varieties, and methods for impact assessment are less well developed and study results are fewer. Little impact assessment has been carried out on the impact of non varietal technologies although this type of research accounts for about one-half of all crop research in developing countries (IAEG, 1999; Pachico, 1998; 2001). However in cashew, we have enough information on impact of non-varietal technologies also which are summarized and presented later in this chapter.

Even more complex is the issue of assessing the impact of a policy recommendation. What would be the impact of the removal of subsidies on external inputs, setting new standards on the commercialization of an agricultural product, the structural adjustment program on the economies, or the application of new World Trade Organization regulations on the agricultural sector of a country? The "before and after" approach in impact assessment could be applied in such cases. As for cashew, such type of assessment of impact of policy on cashew cultivation is found extremely rare.

3. Impact Assessment Process

Impact assessment of agricultural research is a continuous process (Manyong *et al.*, 2001). Impact assessment, being a process, is better conceptualized as a cycle involving different types of impact studies at the different stages. Impact studies essentially have the same process as technology development itself. Based on the technology development process, therefore, four stages of impact assessment would constitute the impact cycle. These include impact for priority setting (*i.e., ex ante* impact), on-farm technology evaluation, adoption, and *expost* impact. The different types of impact studies are not mutually exclusive; they rather serve distinct and at the same time complementary functions in the technology development and dissemination process.

4. *Ex-Ante* Impact Assessment

Ex ante impact assessment is undertaken before the project or program is initiated as an aid in priority setting, based on the potential impacts of alternative research portfolios on aggregate net benefits or on poverty alleviation. *Ex ante* impact studies are conducted to estimate the expected returns from current alternative research efforts. Assessment of future impact includes measures of productivity impacts, distribution of economic benefits, and effects on environmental quality. Assessing expected impact is a two-stage process (Pachico, 2001): (1) scenarios are generated with the conditions expected in the future without the proposed research; and (2) the impact of potential research innovations is estimated. Considerable uncertainties exist in the generation of future scenarios as well as in the projections of expert knowledge of the potential payoffs from research and the probabilities of success (Alston *et al.*, 1995).

In practice, *ex ante* impact studies have been conducted for just a single technology development program based on information obtained from on-farm trials and thus have little or no priority setting motivation (Kristjanson *et al.*, 2002). Such studies provide valuable information on the potential impact of the technology developed and help to make the case for continued efforts and investments in technology promotion. However, their application is limited by the fact that major investments in the research have already been made and hence there is little scope for resource reallocation at that stage and hence also not found utilized in case of impact assessments in cashew.

Impact of Technology Adoption

Adoption studies are carried out to monitor the level and pathways of adoption and the impact of proven technologies on farm-level productivity during the technology promotion stage. These studies measure the extent of use of the technology, the performance of the technology (productivity changes, advantages, and disadvantages), changes in farm management induced by the new technology, and characteristics of the diffusion process. The essential information includes the following: (1) levels and speed of adoption, and reasons for non-adoption of technology; (2) farmers' perceptions of desirable traits or features of the technology options; (3) farm-level productivity and income gains due to the alleviation of biotic and abiotic constraints; (4) impact on the welfare of the farm household, for example, in terms of the intra household distribution of income, nutrition, and health; and (5) infrastructural, institutional, and policy constraints hindering technology adoption (IAEG 1999; Manyong *et al.*, 2001; Pachico 2001; Bantilan and Dar 2001).

Farmers' perceptions of important constraints, desirable cultivar traits, and management practices are very useful. First, they help in identifying binding constraints and research opportunities. Secondly, they provide an empirical basis for estimating expected ceiling levels of adoption. Technologies introduced in an environment characterized by significant constraints to adoption cannot be expected to have high levels of adoption ceilings unless these constraints are addressed. Thirdly, research options which directly address users' needs are most likely to be adopted.

Adoption studies are usually conducted as case studies, which are chosen on the basis of scientists' views on the importance and potential of different technologies, research cost, funding availability, and some balance among different research lines. Studies conducted at Directorate of Cashew Research, Puttur reveal the socio-economic determinants of farm level adoption of cashew production technologies and the technology utilization status in terms of adoption levels of recommended cashew production technologies.

5. Socio-Personal Profile of Cashew Farmers

Studies conducted by DCR shows that cashew farmers were equally distributed into young, middle and old as far as their age was concerned with mean age of 47 years (Table 34.1). Majority were high school pass (45 per cent) while 93 percent had agriculture itself as their primary occupation. Most farmers (48 per cent) had medium level of experience in farming with an average experience of 23.5 years in agriculture. These findings are in line with earlier studies done in Karnataka, Kerala and Maharashtra by Lakshmisha (2000), Shivaramu *et al.* (2004), Veerkar *et al.* (2006) and Venkattakumar (2006, 2008, 2009). Majority of the farmers (41 per cent) had low experience in cashew farming with an average experience of 10.5 years only. These findings are in line with that of Venkattakumar (2006) but in contrast with studies conducted in Maharashtra by Veerkar *et al.* (2006). Contact with extension agencies was found to be low among majority of cashew farmers (68 per cent) while participation in extension programmes was found to be medium for almost two-third of the farmers (64 per cent). These findings are in line with that of Lakshmisha

Table 34.1: Socio-Personal Profile of Cashew Farmers in Karnataka (n=75)

Independent Variables	*Mean*	*SD*	*Category*		*Respondents*	
					f	*Per cent*
Age (Years)	46.5	12.93	Young	<40	24	32
			Middle age	40-53	25	33
			Old	>53	26	35
Level of Education	3.77	1.19	Illiterate		3	4
			Primary		11	15
			Secondary		7	9
			High School		34	45
			PUC		4	5
			Degree		11	15
			PG		5	7
Primary Occupation			Agriculture		70	93
			Others		5	7
Experience in farming (Years)	23.5	13.54	Low	<17	21	28
			Medium	17-30	36	48
			High	>30	18	24
Experience in cashew farming (Years)	10.5	7.24	Low	<7	31	41
			Medium	7-14	23	31
			High	>14	21	28
Extension contact	3.03	6.29	Low		51	68
			Medium		17	23
			High		7	9
Extension participation	6.69	7.36	Low		15	20
			Medium		48	64
			High		12	16
ICT usage	10.03	5.90	Low		17	23
			Medium		42	56
			High		16	21
Cosmopoliteness	7.81	5.13	Low		27	36
			Medium		27	36
			High		21	28
Land used for cashew			Fully irrigated		2	3
			Partially irrigated		5	7
			Rain-fed		68	90
Land used for other crops			Fully irrigated		57	76
			Partially irrigated		8	11
			Rain-fed		10	13
Distance of cashew plot from home (Meters)	427	850	Less		2	3
			Moderate		60	80
			Large		13	17

(2000) and Shivaramu *et al.* (2004). More than half of the cashew farmers (56 per cent) exhibited medium levels of ICT usage while in case of Cosmopoliteness, majority were equally divided into low and medium categories (36 per cent). These findings are contrary to earlier ones by Lakshmisha (2000), Shivaramu *et al.* (2004) and Venkattakumar (2006). While three-fourth majority (76 per cent) of cashew farmers was giving irrigation for other crops grown by them, 90 percent of them cultivated cashew under rainfed system only. The average distance of cashew plots from farmers' homes were found to be around half a kilometer (427 meters) with vast majority (80 per cent) having to cover this distance to reach their plots.

6. Economic Profile of Cashew Farmers

Studies conducted in Karnataka, Kerala and Maharashtra shows that around half of the farmers (48 per cent) grew 3-4 crops on an average in their farms (Table 34.2). Almost three-fourth of them (72 per cent) gave least priority to cashew farming in Karnataka. These findings are in line with that of Venkattakumar (2008). The average farm size was found to be 1.9 acres while average area of un-used land available for cultivation was found to be 86 cents. Majority (55 per cent) had nil or negligible amount of unused land available for cultivation. The study showed that households had an average number of 173 cashew trees with a mean yield of 2.45 kg/tree. More than half of the cashew farmers (55 per cent) realized only moderate yields with an average net income of Rs. 29,664/year against an average expenditure of Rs. 9293/year. Majority (46 per cent) made low levels of yearly investment in agriculture of Rs. 90,981 with a net income to the tune of Rs. 2,40,540/year.

Table 34.2: Economic Profile of Cashew Farmers in Karnataka (n=75)

Independent Variables	*Mean*	*SD*	*Category*		*Respondents*	
					f	*Per cent*
No: of crops grown	3.28	1.62	Less	<2	20	27
			Moderate	2-4	36	48
			High	>4	19	25
Importance given to cashew	1.56	1.0	Least		54	72
			Moderate		8	11
			High		11	14
			Very High		2	3
Farm size (acres)	1.9	0.82	Low	<2.5 ha	31	41
			Medium	2.5-5 ha	23	31
			High	>5 ha	21	28
Cultivable land available (cents)	0.86	1.29	Low	<0.21	41	55
			Medium	0.21-1.51	20	27
			High	>1.51	14	18
Yielding cashew trees (Nos.)	173	220	Low	<63	26	35
			Medium	63-283	38	51
			High	>283	11	14

Contd...

Table 34.2–*Contd...*

Independent Variables	*Mean*	*SD*	*Category*		*Respondents*	
					f	*Per cent*
Yield of Cashew/tree (kg)	2.92	2.09	Low	>3.96	23	31
			Moderate	3.96-1.87	27	36
			High	<1.87	25	33
Expenditure in agriculture (Rs.)	90981	64037	Low	<52523	35	46
			Medium	52523-129258	20	27
			High	>129258	20	27
Net income from agriculture (Rs.)	240540	149649	Low	<124032	37	49
			Medium	124032-357048	20	27
			High	>357048	18	24
Expenditure in cashew farming (Rs.)	9293	11028	Low	<3780	28	37
			Medium	3780-14806	31	41
			High	>14806	16	21
Net income from cashew farming (Rs.)	29664	70426	Low	<5994	29	39
			Medium	5994-64602	40	53
			High	>64602	6	8

7. Technology Utilization Status

Production Technologies

The adoption of recommended cashew production technologies categorized to seven groups such as Planting and initial care, Soil and water conservation, Manures and fertilizers, Pruning and training, Plant protection, Intercropping and Harvesting and postharvest technologies were studied by DCR Puttur. The adoption of specific recommended practices under each production technology mentioned above was studied. The overall adoption of cashew production technologies received an index score of 40 with majority farmers (44 per cent) showing only medium level of adoption (Table 34.3). Similar findings revealing poor adoption were made by Zagade *et al.* (2000, 2003), Lakshmisha (2000), Bhairamkar *et al.* (2004), Shivaramu

Table 34.3: Classification of Farmers for Adoption of Cashew Production Technologies (n=75)

Sl.No.	*Category*	*Range*	*Respondents*	
			f	*Per cent*
1.	Low (<Mean-1/2S.D)	<36.27	18	24
2.	Medium (Mean(+/-)1/2S.D)	36.27-51.39	36	51
3.	High (>Mean+1/2S.D)	>51.39	19	25
	Mean= 43.83, S.D=15.12			

et al. (2004) and Venkattakumar (2005, 2006, 2009) from Maharashtra, Kerala and Karnataka.

Cashew farmers were found to adopt maximum practices under planting and initial care (Rank 1) including recommended varieties and planting material (grafts) (Table 34.4). This reasons the high demand for cashew grafts in government run as well as private nurseries in the locality. The findings can be read along with that of Lakshmisha (2000) and Venkattakumar *et al.* (2004). Also, these practices were easy to adopt and initial interest plays a major role in the high adoption rate of this technology. More than half of the farmers (52 per cent) exhibited medium levels of adoption for this technology. This finding is in line with earlier reports of Bhairamkar *et al.* (2004) and Shivaramu *et al.* (2004). Soil and water conservation techniques were also followed by almost half of the farmers (Rank 2). This is in contrast to findings by Shivaramu *et al.* (2004) and Venkattakumar (2009) in which adoption of soil and water conservation techniques were found to be low. However earlier studies had shown a positive perception of cashew demonstration farmers towards soil and water conservation techniques (Venkattakumar, 2005). The increased availability of heavy machineries at affordable per hour rates for land leveling, pit digging, terrace making etc. is a major reason behind this. Also, the practices under this technology are mostly adopted along with or in continuance with planting and after care thereby increasing its chance of adoption due to initial interest. Almost one-third of the farmers (35 per cent) exhibited medium levels of adoption for soil and water conservation technology. Adoption of pruning and training along with harvesting and postharvest technologies stood together at third place with adoption index of 43 while majority farmers belonged to medium level adopter category for both technologies. This finding is in line with earlier reports of Shivaramu *et al.* (2004). In case of pruning and training, 31 percent of farmers have never adopted this technique in their orchard.

Table 34.4: Adoption Levels of Recommended Cashew Production Technologies (n=75)

Cashew Production Technologies	*Adoption Index*	*Rank*	*S.D.*	*Per cent Farmers under Various Levels of Adoption*			
				High	*Medium*	*Low*	*Non adoption*
Planting and Initial Care	73	1	24.1	25	52	23	–
Soil and Water Conservation	48	2	25.0	32	35	25	8
Manures and Fertilizers	30	5	18.3	53	14	13	20
Pruning and Training	43	3	34.8	33	35	1	31
Plant Protection	20	7	19.2	29	26	28	17
Intercropping	22	6	35.7	5	20	4	71
Harvesting and Post-Harvest	43	3	12.9	37	39	24	–
Overall Adoption of CPTs	40	–	16.1	24	44	32	–

Adoption of manures and fertilizers was found to be poor among farmers with adoption index of 30 while one-fifth (20 per cent) of the farmers didn't apply any sort of manures or fertilizers to their cashew crop. Similar observations were made by Nirban and Sawant (2000) with respect to adoption of manures and fertilizers in cashew plantations. Intercropping was another technology which was poorly adopted (Rank 6) with highest rate of non adoption (71 per cent) among all the technologies. Similar observation was made by Shivaramu *et al.* (2004). Low to medium adoption with respect to most cashew production technologies could be attributed to the fact that farmers are yet to realize the importance of these technologies on the yield level and potential economic benefits that accrues from it.

Plant protection, which is one of the most important components, scored the lowest adoption rate among cashew farmers in the present study. Farmers were equally distributed among high, medium and low adopter categories for this technology while 17 percent reported complete non adoption of any crop protection technologies in their cashew crop. This finding is in line with earlier reports of Nirban and Sawant (2000) and Zagade *et al.* (2000, 2003) but in contrast with findings by Venkattakumar (2009). However, 90 percent of demonstration farmers who availed subsidies were found to have adopted plant protection measures (Venkattakumar, 2005). Non-adoption was particularly high for plant protection technologies against Cashew Stem and Root Borer (CSRB) due to the complexity of the technology while majority had adopted measures against Tea Mosquito Bug (TMB) due to less complexity, higher trialability and observability of results in comparison to measures recommended against CSRB. Dixit and Bhaskara Rao (1999) and Venkattakumar (2005) also reported farmer responses indicating that recommended control measures could not check attack of CSRB explaining poor adoption rates of plant protection technology as a whole.

8. Socio-personal Factors Contributing towards Adoption of Cashew Production Technologies

Studies at DCR, Puttur identified that five socio personal variables *viz.* level of education, years of experience in farming, years of experience in cashew farming and extension participation of cashew farmers had a significant relationship with farmers' adoption of cashew production technologies. The regression analysis revealed that three variables *viz.* the age of cashew farmer, years of experience in farming and extension participation have a significant positive contribution towards adoption while two factors; type of land used for cashew cultivation and distance of cashew plot from home had a significant negative contribution towards adoption of cashew production technologies (Table 34.5). The correlation of age with adoption is in line with findings of Venkattakumar (2006).

9. Economic Factors Contributing towards Adoption of Cashew Production Technologies

Six economic variables *viz.*, number of crops grown, importance given to cashew, cultivable land available, number of yielding cashew trees, expenditure in agriculture and net income from cashew farming were identified as having significant relationship with adoption of cashew production technologies. The regression analysis reveals that four variables *viz.* the number of crops grown,

Table 34.5: Relationship between Adoption and Socio-Personal Variables and their Contribution in Explaining the Variability in Adoption (n=75)

Sl.No.	Socio-personal Variables	'r' Value	'b' Value
1.	Age	0.134 NS	–0.0628***
2.	Level of Education	0.354**	0.073 NS
3.	Primary Occupation	–0.107 NS	–0.068 NS
4.	Experience in farming	0.380**	0.660***
5.	Experience in cashew farming	0.334**	0.143 NS
6.	Extension contact	0.205 NS	–0.159 NS
7.	Extension participation	0.292*	0.339*
8.	ICT usage	0.175 NS	0.112 NS
9.	Cosmopoliteness	0.103 NS	–0.037 NS
10.	Land used for cashew	–0.030 NS	–0.172**
11.	Land used for other crops	0.116 NS	0.066 NS
12.	Distance of cashew plot from home	–0.202 NS	–0.266***

R^2 = 0.786NS – Non-Significant, *** - Significant at 1 per cent level, ** - Significant at 5 per cent level, * - Significant at 10 per cent level.

farm size, no: of yielding cashew trees and expenditure incurred in cashew farming were the economic factors which had a significant positive contribution towards explaining the variability in adoption of cashew production technologies (Table 34.6). The socio-personal and economic variables used in the study could together explain upto 78 percent variability in adoption of cashew production technologies (R^2 = 0.786).

Table 34.6: Relationship between adoption and economic variables and their contribution in explaining the variability in adoption (n=75)

Sl.No.	Economic Variables	'r' Value	'b' Value
1.	No: of crops grown	0.450***	0.262***
2.	Importance given to cashew	0.393***	–0.075 NS
3.	Farm size	0.074 NS	0.174**
4.	Cultivable land available	0.371***	0.161 NS
5.	No: of yielding cashew trees	0.536***	0.297*
6.	Yield of Cashew/tree	0.155 NS	–0.082 NS
7.	Expenditure in agriculture	0.282**	–0.197 NS
8.	Net income from agriculture	0.176 NS	–0.120 NS
9.	Expenditure in cashew farming	0.141 NS	0.219*
10.	Net income from cashew farming	0.416***	0.280 NS

R^2 = 0.786NS – Non-Significant, *** - Significant at 1 per cent level, ** - Significant at 5 per cent level, * - Significant at 10 per cent level.

Pest Management Technologies

The adoption of pest management technologies was quantified by measuring the farmer's ability to identify the symptoms of TMB and CSRB attack, status of TMB and CSRB attack in farmer fields, application of recommended chemical in correct dose, adoption of phyto-sanitation measures and adoption of other recommended management measures. Pest management, which is one of the most important components in cashew production, scored a low adoption index (20) in the present study. Farmers were equally distributed among high, medium and low adopter categories for this technology while 17 percent of cashew farmers reported complete non adoption of any pest management measures in their cashew crop (Table 34.7) Similar findings were made by Zagade *et al.* (2000, 2003), Lakshmisha (2000), Bhairamkar *et al.* (2004) and Shivaramu *et al.* (2004).

Table 34.7: Adoption of Pest Management Measures in Cashew (n=75)

Sl.No.	*Recommended Pest Management Measures*	*Adoption by Cashew Farmers*					
		Always/ Fully		*Sometimes/ Partially*		*Never/ Non adoption*	
		f	*Per cent*	*f*	*Per cent*	*f*	*Per cent*
1.	Identifying initial symptoms of TMB attack	23	30	5	7	47	63
2.	Spraying against TMB at recommended time	19	26	7	9	49	65
3.	Spraying of recommended chemical	23	30	4	5	48	65
4.	Spraying of pesticide in recommended dose	18	24	2	3	55	73
5.	Monitoring for renewed TMB attack	24	32	4	5	47	63
6.	Identifying initial symptoms of CSRB attack	26	35	4	5	45	60
7.	Removal of grubs from CSRB affected trees	2	3	5	7	68	90
8.	Chiseling out the affected bark	6	8	10	13	59	79
9.	Application of recommended pesticide	7	9	9	12	59	79
10.	Application of pesticide in recommended dose	7	9	9	12	59	79

The poor adoption of plant protection technologies warranted a practice wise study of adoption of the recommended practices to arrive at the root causes. Studies revealed that in case of TMB, 30 percent of the farmers could identify initial attack symptoms in their orchard while majorities (63 per cent) were never able to do the same (Table 34.7). Similarly, while 26 percent of the farmers took up spraying against TMB in the flushing or flower initiation stages, majority (65 per cent) never did any spraying in their orchards. While 30 percent of the farmers followed recommendations regarding spraying the recommended chemicals (Monocrotophos/Karate), only 24 percent adhered to the recommended dosage of the chemical. It may be noted that nearly two-third of the farmers didn't spray the recommended chemicals while almost three-fourth majority overlooked the recommended dosage. Monitoring for renewed TMB attack was done by nearly one-third (32 per cent) of the farmers while 63 percent never did the activity.

In case of management of CSRB, 35 percent of the farmers could identify initial symptoms of CSRB attack in their orchards while 60 percent were unable to do so. However, 90 percent of the farmers never attempted removing grubs from CSRB affected trees. Non adoption was extremely high with respect to practices like chiseling out the affected bark, application of chlorpyriphos to chiseled portion and soil and application of chlorpyriphos in recommended dose with 79 percent of the farmers reporting to have never attempted these measures. Adoption of these practices was found to be extremely low (8 per cent, 9 per cent and 9 per cent). These findings are is in line with earlier reports of Nirban and Sawant (2000) and Zagade *et al.* (2000, 2003) but in contrast with findings by Venkattakumar (2009). However, earlier studies show that 90 percent of demonstration farmers who availed subsidies were found to have adopted pest management measures (Venkattakumar *et al.*, 2005).

It may be noted from the above findings that non-adoption was particularly high for pest management technologies against Cashew Stem and Root Borer (CSRB) due to the complexity of the technology while more farmers had adopted measures against Tea Mosquito Bug (TMB) due to less complexity, higher trialability and observability of results in comparison to measures recommended against CSRB. Dixit and Bhaskara Rao (1999) and Venkattakumar *et al.* (2005) also reported farmer responses indicating that recommended control measures could not check attack of CSRB explaining poor adoption rates of pest management technology as a whole. It is obvious from these findings that there is tremendous scope in the region for increasing adoption of recommended cashew production technologies.

Recent studies reveal that pest management component while scoring lowest adoption index also showed highly significant relation with cashew production achieved by farmers and emerged as the most significant contributor towards cashew production in Dakshina Kannada district (Sajeev. *et al.*, 2014). This clearly indicates that adoption of plant protection techniques cannot be ignored at any cost if cashew production in the district has to be improved. The findings also calls for development of plant protection measures which are user friendly (less complex), having relative advantage over existing technology and also compatible with farmer situations.

10. Socio-economic Determinants of Farm Level Adoption of Pest Management Technologies

The correlation analysis identified that two socio personal variables *viz.* years of experience in farming and extension participation of cashew farmers had a significant relationship with farmers' adoption of pest management technologies. The regression analysis revealed that two variables *viz.* the age of cashew farmer and their primary occupation have a significant positive contribution towards adoption. The study identified five economic variables viz; importance given to cashew, number of yielding cashew trees, expenditure in agriculture, net income from agriculture and net income from cashew farming as having significant relationship with adoption of pest management technologies. The regression analysis reveals that one variable *i.e.* the farm size exerts a significant positive contribution towards explaining the variability in adoption of pest management technologies. The socio-personal and economic variables used in the study could together explain upto 60 percent variability in adoption of pest management technologies ($R^2 = 0.600$).

11. *Ex-Post* Impact Assessment

Ex-post impact assessment is conducted after a technology has been widely adopted by farmers in the target areas. *Ex post* impact assessment develops the confidence of scientists, research managers, and stakeholders and makes the case for enhanced research support (Bantilan and Dar 2001). In addition, information obtained during the process of impact evaluation feeds back into research prioritization. Impact assessment begins with a priority setting task using *ex ante* impact analysis that estimates the potential impacts of alternative research portfolios on aggregate net benefits. This is based on data generated from a baseline survey, expert knowledge (*e.g.*, from biophysical scientists and research managers), and information from previous adoption and impact studies. Baseline data allow researchers to establish the current levels of adoption; information from biophysical scientists and research managers helps to predict likely changes in yields, costs, and other variables; and information from previous adoption and impact studies is used to identify alternative technologies that would address the major production constraints while at the same time taking into consideration farmers' preferences and farming conditions.

Priority setting is followed by (on-station) technology development and on-farm evaluation that are carried out to identify an appropriate technology under farmers' conditions and based on their priorities and preferences. Appropriate technologies are then promoted or scaled-up and out, starting from the initial trial-hosting villages to other villages, districts, provinces, and regions. As technology dissemination is underway, adoption studies become very important to document the process and levels of adoption and changes in productivity and cropping pattern. These are usually conducted as case studies on adoption and impact. Finally, *ex post* studies on adoption and impact are carried out following large-scale dissemination of the technology. In practice, however, case studies on adoption and impact, such as those conducted in the technology demonstration villages, are considered as *ex post* impact studies. While this does not have any practical implications for methods and results, it may be important to make a clear conceptual distinction between the two. The impact assessment process becomes complete when adoption and impact information obtained from *ex post* impact studies is fed back to *ex ante* impact studies and the process continues, as technology development itself is continuous.

12. Impact of Cashew Varieties

Studies on impact of recommended varieties on cashew area of Dakshina Kannada, Karnataka conducted at DCR, Puttur showed that major area under cashew is covered by the variety Ullal-3 (41 per cent) followed by variety Bhaskara (26.62 per cent). It may also be noted that adoption pattern also shows similar trend with variety Ullal-3 adopted by 59 per cent of farmers followed by variety Bhaskara (55 per cent). Most farmers have adopted a minimum of two cashew varieties in their field and 'Ullal-3 + Bhaskara' combination emerges to be the most popular one in the study area. Varieties Ullal-1 and Vengurla-4 (V-4) were found to cover around 8 percent each of rest of the area. But, these varieties are not popular among farmers with only 13 and 4 percent farmers respectively adopting the said varieties in the district.

Other varieties like NRCC Selection-2 (5 per cent), Madakkathara-2 (4 per cent), Vridhachalam-3 (VRI-3) (3 per cent), Ullal-4 (3 per cent) and Vengurla-7 (V-7) (1 per cent), have low impact on total cashew area. In total, improved varieties were found to be adopted by 92 percent of the farmers while 8 percent were still continuing with seedling origin plantations. However seedling origin plantations have only negligible coverage in farmer fields (0.04 per cent). Variety wise adoption and impact on area is given in Table 34.8.

Table 34.8: Variety-wise Adoption and Impact on Cashew Area (n=75)

Sl.No.	*Variety*	*Adopted by (Per cent farmers) **	*Rank*	*Area Covered (Per cent)*	*Rank*
1.	Bhaskara	55	2	26.62	2
2.	NRCC Seln-2	19	3	5.21	5
3.	Madakkathara-2	4	8	4.62	6
4.	Ullal-3	59	1	41.00	1
5.	Ullal-1	13	4	7.66	4
6.	Ullal-4	11	5	2.93	7
7.	VRI-3	7	6	2.66	8
8.	V-4	5	7	8.43	3
9.	V-7	4	8	0.75	9
10.	Other varieties	3	–	0.09	–
	Total	92**	–	99.96	–
	Seedling origin	8	–	0.04	–

* the percentages won't add upto 100 due to adoption of multiple varieties by single farmer.

** represents total percentage of farmers who have adopted released varieties

13. Impact of Cashew Varieties on Production and Productivity

Analysis of variety wise impact on cashew production in Dakshina Kannada district of Karnataka showed that farmers realized highest yield from variety Bhaskara (4.73 kg/tree) followed by Madakkathara-2 (4.45 kg/tree). This was followed by Ullal-1 (3.90 kg/tree) and Ullal-3 (3.87 kg/tree). Ullal-4, another recommended variety was at fifth position with a yield of 3.67 kg/tree while NRCC selection-2 fared low at sixth place with 3.47 kg/tree. Productivity under normal density (8x8m) as well as high density (5x5m) planting was highest for Bhaskara (737.88 and 1882.54 kg/ha) and Madakkathara-2 (694.20 and 1771.10 kg/ha). This was followed by Ullal-1 (608.40 kg/ha), Ullal-3 (603.72 kg/ha) and Ullal-4 (572.52kg/ha) under normal density. Under high density planting system, Ullal-1 (1556.10 kg/

ha), Ullal-3 (1540.26 kg/ha) and NRCC Selection-2 (1374.12kg/ha) stood respectively at third, fourth and fifth positions. High density plantations of Ullal-4, VRI-3, V-4, V-7 and other varieties were not observed during data collection for this study. Variety wise impact on production and productivity is presented in Table 34.9.

Table 34.9: Varietal Impact on Production and Productivity of Cashew (n=75)

Sl.No.	Variety	Production (kg/tree)**	Rank	Productivity (kg/ha) Under Normal (8x8m) Density	Rank	Productivity (kg/ha) Under High (5x5m) Density	Rank
1.	Bhaskara	4.73	1	737.88	1	1882.54	1
2.	NRCC Seln-2	3.47	6	541.32	6	1374.12	5
3.	Madakkathara-2	4.45	2	694.20	2	1771.10	2
4.	Ullal-3	3.87	4	603.72	4	1540.26	4
5.	Ullal-1	3.90	3	608.40	3	1556.10	3
6.	Ullal-4	3.67	5	572.52	5	–	–
7.	VRI-3	3.06	7	477.36	7	–	–
8.	V-4	1.51	9	235.56	9	–	–
9.	V-7	3.00	8	468.00	8	–	–
10.	Other varieties	2.23	–	347.88	–	–	–
11.	Seedling origin	1.23	–	147.60	–	–	

** In trees above 5 years of age.

Bhaskara variety was released during March 2006 for coastal region of Karnataka. This variety is having midseason flowering habit (Dec-Mar) with a flowering duration of 60 days and has potential to escape from the attack of the tea mosquito bug (TMB) under low to moderate outbreak situation. The average yield was reported to be 10.7 kg/tree. The nut and kernel weight are 7.4 g and 2.2 g respectively. The shelling percentage is 30.6 and kernel grade conforms to export grade W240. The potential to escape from the attack of the tea mosquito bug (TMB) along with very good yield potential have definitely favoured this variety in its high adoption among farmers of Dakshina Kannada district. Madakkathara-2 (NDR 2-1) is a selection released in 1987. The mean yield is 17 kg/tree. The nuts are bold (7.3 g nut weight) with shelling percentage of 26.2. Kernel weight is 2g having a count of W 240 export grade. However, this highly yielding variety was found to be adopted by only 4 percent of the farmers mainly due to low awareness in this district.

Ullal-3 is a selection released in 1993 from ARS, Ullal. It is early in flowering (November-January) and fruiting period is very short (50-60 days). The fruiting is from January to March and sometimes starts from last week of December. It is a high yielding variety with average yield of 14.7 kg/tree. The nut size is medium with nut weight of 7g. The shelling percentage is 30 and the kernel grade conforming to W 210 grade. Its mid season nature coupled with higher yields seems to have made it a favourite among farmers. Ullal-1 is a selection released by ARS, Ullal in 1984. The average yield is 16 kg/tree. The duration of harvest is long (about 110 days). The nut weight is 6.7g with shelling percentage of 30.7. Even though the variety recorded slightly higher yields (3.90 kg/tree) in field compared to Ullal-3 (3.87 kg/tree), the adoption by farmers (13 per cent) was found to be far below than its counterpart Ullal-3 (59 per cent) which tops in adoption among farmers in all varieties.

14. Technology Impact on Production and Productivity of Cashew

The production and productivity profile of cashew farmers were studied at DCR Puttur and the results showed that farmers achieved a mean production of 425 kg and productivity of 2.92 kg/tree. In case of production, majority fell in to medium (40 per cent) and low (43 per cent) producer categories while they were almost equally divided into high (33 per cent), medium (36 per cent) and low (31 per cent) categories with respect to productivity achieved (Table 34.10).

Table 34.10: Classification of Farmers Based on Production and Productivity of Cashew (n=75)

Categories	*Production*			*Productivity*		
	f	*Per cent*	*Range*	*f*	*Per cent*	*Range*
High	13	17	>674	25	33	>3.96
Medium	30	40	674-177	27	36	3.96-1.87
Low	32	43	<177	23	31	<1.87
Mean	425	2.92				
SD	497	2.09				

The recommended cashew production technologies starting with recommended varieties were categorized to eight groups such as Varieties, Planting and initial care, Soil and water conservation, Manures and fertilizers, Pruning and training, Plant protection, Intercropping and Harvesting and postharvest technologies. Their impact on production and productivity were studied separately and are presented here.

15. Impact of Cashew Production Technologies on Cashew Production and Productivity

Studies conducted at ICAR-DCR revealed the extent of contribution of each production technology towards variability found in levels of cashew production and productivity in the district. Plant protection component which scored the lowest adoption index and also showed highly significant relation (r=0.345**) with cashew production achieved by farmers emerged as the most significant contributor towards cashew production (b=0.339**) in the district. This clearly indicates that adoption of plant protection techniques cannot be ignored at any cost if cashew production in the district has to be improved. The findings also call for development of plant protection measures which are user friendly (less complex), having relative advantage over existing technology and also compatible with farmer situations. Soil and water conservation technology which showed a significant relationship (r=0.344**) with cashew production, is also found to have a significant contribution towards explaining the variability in cashew production (b=0.326*). Interestingly, intercropping; another poorly adopted technology was also found to have a significant contribution towards explaining the variability in cashew production (b=0.243*). Recommended varieties, manures and fertilizers and harvesting and postharvest technologies were found to have a non-significant but negative contribution towards cashew production. This is explained by the fact that in case of recommended varieties, even though the study found that highest production was given by Bhaskara, Madakkathara – 2, Ullal - 1 and Ullal - 3 varieties, the same varieties covered only 26.62, 5.21, 7.66 and 41.00 percent respectively of total area under cashew cultivation. Thus, the findings make it clear that by increasing the adoption and area coverage of above varieties we can bring a quantum increase in cashew production in the district.

Table 34.11: Contribution of Cashew Production Technologies towards Variability in Cashew Production and Productivity

Technology	*Production*	*Productivity*
	'b' value	*'b' value*
Varieties	–0.131 NS	–0.077 NS
Planting and Initial Care	0.037 NS	–0.159 NS
Soil and Water Conservation	0.326*	0.208 NS
Manures and Fertilizers	–0.195 NS	–0.184 NS
Pruning and Training	0.178 NS	0.313*
Plant Protection	0.339**	0.139 NS
Intercropping	0.243*	–0.097 NS
Harvesting and Post Harvest	–0.012 NS	0.024 NS
	R^2 = 0.406	R^2 = 0.149

NS – Non-Significant, ** - Significant at 1 per cent level, * - Significant at 5 per cent level.

Farmers in the study area were found to have poor adoption in case of manures and fertilizers for cashew. The study also found that harvesting cashew from the trees itself is a common practice to avoid theft and this coupled with improper drying practices including less number of drying days has contributed more volume to the cashew production. If proper harvesting and drying practices are followed it will decrease the total volume of cashew thus explaining the negative relation. Similarly, by increasing the adoption levels of manures and fertilizers in cashew and by adopting proper harvesting and postharvest drying techniques the cashew production can be further improved. All the recommended production technologies together could explain only up to 40 percent of variability in cashew production (R^2 = 0.406). Adoption of pruning and training in cashew orchards was found to have a significant contribution towards increasing the per unit productivity of cashew orchards. However, the adoption level of this technology is very low at present. The study reveals that increasing adoption of soil and water conservation techniques and development and popularization of user friendly plant protection measures can contribute largely to increased cashew production while increasing adoption of pruning and training in cashew orchards can significantly increase the per unit productivity of cashew orchards. The results clearly indicate that socio-economic and bio-physical factors along with policy environment have a larger contribution in explaining cashew production and productivity and technology component alone cannot be expected to bring a positive impact. Understanding the above dynamics in technology impact can help researchers and extension agencies working in cashew sector to design better innovations and effective outreach strategies.

16. Socio-personal Factors Affecting Cashew Productivity

Twelve socio-personal variables were studied for their relationship with cashew productivity. The correlation analysis revealed that extension contact and extension participation along with primary occupation registered a significant relation with cashew productivity (Table 34.12). It may be noted that contact with extension agencies was found to be low among majority of cashew farmers while participation in extension programmes was found to be medium for almost two-third of the farmers. Majority had high school education while, 15 percent of them had acquired a degree and 93 percent had agriculture itself as their primary occupation. The relationship carries importance in light of the fact that extension contact and participation is on a downward trajectory while better educated farmers were found to be shifting from cashew due to perceived high benefits from alternative crops like rubber.

Table 34.12: Relationship of Socio-personal Factors with Productivity of Cashew (n=75)

Sl.No.	*Variables*	*Productivity*
		'r' Value
1.	Age	–0.067
2.	Level of Education	0.075
3.	Primary Occupation	0.244*
4.	Experience in farming	–0.058
5.	Experience in cashew farming	0.134
6.	Extension contact	0.229*
7.	Extension participation	0.292*
8.	ICT usage	0.067
9.	Cosmopoliteness	–0.194
10.	Land used for cashew	–0.198
11.	Land used for other crops	0.065
12.	Distance of cashew plot from home	–0.001

*. Significant at the 0.05 level (2-tailed).**. Significant at the 0.01 level (2-tailed).

17. Economic Factors Affecting Cashew Productivity

Ten economic variables were studied for their relationship with cashew productivity. The correlation analysis revealed that these economic factors have a higher relationship with cashew productivity with five economic variables registering a significant relationship. These variables are importance given to cashew crop, number of yielding cashew trees, yield of cashew/tree, expenditure incurred in agriculture and net income from agriculture. It may be noted that almost three-fourth of cashew farmers had recorded least priority to cashew farming. Households had an average number of 173 cashew trees only with a mean yield of 2.92 kg/tree which was very low. More than half of the cashew farmers realized only moderate yields with an average net income of Rs. 29,664/ha/year against an average expenditure of Rs. 9293/ha/year which was on a very lower side. Majority made low levels of yearly investment in agriculture of with moderate net incomes.

Table 34.13: Relationship of Economic Factors with Productivity of Cashew (n=75)

Sl.No.	*Variables*	*Productivity* *'r' Value*
1.	No: of crops grown	–0.052
2.	Importance given to cashew	0.378**
3.	Farm size	0.077
4.	Cultivable land available	0.011
5.	No: of yielding cashew trees	0.090
6.	Yield of Cashew/tree	0.351**
7.	Expenditure in agriculture	0.310**
8.	Net income from agriculture	0.467**
9.	Expenditure in cashew farming	0.284*
10.	Net income from cashew farming	0.186

*. Significant at the 0.05 level (2-tailed).**. Significant at the 0.01 level (2-tailed).

18. Impact of Socio- personal and Economic Factors on Cashew Productivity

Studies by DCR, Puttur revealed that seven variables viz; age of the farmer, primary occupation, years of experience in farming, cosmopoliteness, and number of yielding cashew trees, yearly expenditure made in cashew farming and net income from cashew farming were found to contribute significantly to the productivity achieved by cashew farmers (Table 34.14).

It can be noted that cashew farmers were equally distributed as far as their age was concerned with mean age of 47 years while 93 percent had agriculture itself as their primary occupation. Majority farmers were having low experience of 10.5 years only in cashew farming. These above factors make them still experimenting and switching between various remunerative crops thereby increasing and decreasing their area under cashew as well as interest shown in cashew cultivation thus

Table 34.14: Contribution of Socio-economic Variables towards Variability in Cashew Productivity

Variables	Standardized Coefficients Beta	t	Sig.
(Constant)		0.343	0.733
Age	–.613**	–3.590	0.001
Level of Education	–.168	–1.334	0.188
Primary Occupation	0.370**	3.491	0.001
Experience in farming	0.192	1.210	0.232
Experience in cashew farming	0.242*	2.276	0.027
Extension contact	–.025	–.145	0.885
Extension participation	0.374	1.731	0.089
ICT usage	0.000	–.003	0.997
Cosmopoliteness	–.370**	–3.292	0.002
Land used for cashew	–.030	–.282	0.779
Land used for other crops	–.049	–.476	0.636
Distance of cashew plot from home	–.019	–.146	0.885
Number of crops grown	–.007	–.049	0.961
Importance given to cashew	–.022	–.165	0.869
Farm size	0.183	1.677	0.100
Cultivable land available	–.012	–.079	0.938
Number of yielding cashew trees	–.517*	–2.402	0.020
Expenditure in agriculture	–.148	–.853	0.397
Net income from agriculture	0.136	0.923	0.361
Expenditure in cashew farming	0.545**	3.818	0.000
Net income from cashew farming	0.684**	2.835	0.007

*. Significant at the 0.05 level (2-tailed).**. Significant at the 0.01 level (2-tailed).

Model Summary

R	R Square	Adjusted R Square	Std. Error of the Estimate
0.800[a]	0.640	0.495	1.48

explaining the negatively significant contribution towards productivity. Majority were equally divided into low and medium categories (36 per cent) with respect to their cosmopoliteness levels. Their cosmopoliteness has adversely contributed to cashew productivity with marked shifting to other remunerative crops and decreased care and interest in cashew cultivation. The number of yielding cashew trees were also found to be on lower side in most households thus negatively affecting the per unit productivity. Huge majority of cashew farmers belonged to low and medium categories with respect to average net income and average expenditure made for cashew cultivation/year. The variables selected for the study could explain upto 50 percent of the variability (R^2=0.495) in cashew productivity.

19. Socio-economic Impact of Cashew Cultivation

Ten major social and economic impact indicators were studied to arrive at the socio-economic impact of cashew farming among the respondents (Table 34.15). Study on impact on cropping pattern didn't record much of change with only 17 percent of farmers increasing area under cashew over the years (1.5 acres) while only negligible percent (3 per cent) of them purchased new lands (2 acres) for cashew cultivation. Impact on labour engagement was also low with only 20 percent farmers hiring labour for cashew and only one tenth of them opting for increased family labour engagement (11 per cent). The hiring of labour was noticed particularly for plant protection and harvesting operations with farmers mostly engaging one to two labourers during this period. While a large majority (85 per cent) reported no

Table 34.15: Socio-Economic Impact of Cashew Cultivation (n=75)

Sl.No.	*Impact Indicators*	*Increased*		*No Change*		*Mean**
		f	*Per cent*	*f*	*Per cent*	
1	**Impact on cropping pattern**					
	Area under cashew cultivation over the years	13	17	62	83	1.5 Ac
	Purchase of new land and cashew cultivation	2	3	73	97	2.0 Ac
2	**Impact on labour engagement**					
	Hired labour engagement for cashew	15	20	60	80	1-2 Nos
	Family labour engagement for cashew	8	11	67	89	1 Nos
3	**Impact on farm expenditure**					
	Cashew cultivation and farm expenditure	11	15	64	85	Rs. 2272
4	**Impact on farm income**					
	Cashew cultivation and farm income	32	43	43	57	Rs. 4188
5	**Impact on family income**					
	Cashew cultivation and family income	27	36	48	64	Rs. 4259
6	**Impact on family expenditure**					
	Profit from cashew cultivation and family expenses	15	20	60	80	Rs. 2666
7	**Impact on social participation**					
	Cashew cultivation and participation in social events	46	61	29	39	–
8	**Impact on extension contact**					
	Contacts with extension agency and research institutes	50	67	25	33	–
9	**Impact on mass media exposure**					
	Cashew cultivation and mass media exposure	42	56	33	44	–
10	**Impact on opinion leadership**					
	Cashew cultivation and opinion leadership in his/her area	47	63	28	37	–

* for preceding 10 years of cashew cultivation for only those respondents who reported an increase.

change in farm expenditure due to cashew cultivation, 43 percent of farmers reported an increase in farm income due to cashew cultivation. Farmers reported an average increase of Rs. 2272/year in farm expenditure and Rs. 4188/year in farm income due to cashew cultivation. Resultant increase in family incomes was also reported by 36 percent of the farmers to the average of Rs. 4259/year followed by an average increase of Rs. 2666/year in their family expenditure. The above indicators were measured as average for preceding 10 years of cashew cultivation. Sajeev *et al.* (2014) had reported low productivity of cashew farms in the region along with heavy price fluctuations in raw cashew nut market resulting in low economic benefits. However, earlier studies in neighbouring states of Kerala and Maharashtra revealed better economic impact among cashew farmers (Venkattakumar 2006), (Veerkar *et al.*, 2006).

Analysis of social impact presented a better picture in comparison to economic impact with majority (61 per cent) of the farmers reporting increased social participation while nearly two-third (67 per cent) majority could increase their contacts with extension agencies and research institutes due to cashew cultivation. Majority (56 per cent) reported increase in their mass media exposure while a large majority (63 per cent) reported an increase in their opinion leadership status due to cashew cultivation. Venkattakumar (2006, 2008) had earlier reported similar levels of social impact among cashew farmers of Kerala and Maharashtra. It can clearly be seen that the social benefits of cashew cultivation are far more than the economic benefits.

Classification of cashew farmers based on the social and economic benefits accrued shows that nearly half of the cashew farmers (47 per cent) belong to high social impact category while in case of economic impact large majority (80 per cent) belonged to low benefits category. Overall, nearly half of the cashew farmers (47 per cent) recorded medium levels of socio-economic impact (Table 34.16). It may be noted that the low level of economic impacts are compensated to a large extent by the high social impact due to cashew cultivation.

Table 34.16: Classification of farmers based on social and economic impact (n=75)

Categories	*Social Impact*			*Economic Impact*			*Socio-economic Impact*		
	f	*Per cent*	*Range*	*f*	*Per cent*	*Range*	*f*	*Per cent*	*Range*
Low	23	31	<0.715	60	80	<0.101	24	32	<0.493
Medium	17	22	0.715-1.445	2	3	0.101-1.329	35	47	0.493-1.299
High	35	47	>1.445	13	17	>1.329	16	21	>1.299
Mean	1.080	0.715	0.896						
SD	0.731	1.228	0.806						

20. Determinants of Socio-economic Impact

The studies conducted among cashew farmers of Karnataka identified that three personal variables *viz.* years of experience in farming, extension participation and cosmopoliteness of cashew farmers had a significant relationship with socio economic impact. The regression analysis revealed that four variables *viz.* the age of cashew farmer, years of experience in farming, cosmopoliteness and distance of cashew plot from home had a significant positive contribution towards socio economic impact. The study identified six economic variables *viz.*, importance given to cashew, cultivable land available, number of yielding cashew trees, expenditure in agriculture, net income from agriculture and expenditure in cashew farming as having significant relationship with socio economic impact. The regression analysis reveals that one variable *i.e.* importance given to cashew exerts a significant positive contribution towards explaining the variability in socio economic impact. The socio-personal and economic variables used in the study could together explain upto 60 percent variability in socio economic impact ($R^2 = 0.670$).

21. Predictors of Adoption and Impact

Stepwise regression analysis was used to identify predictors and derive selected models explaining the variation in adoption of cashew production technologies, pest management technologies and socio-economic impact of cashew cultivation. Using these models researchers can predict major impact indicators of cashew production and productivity with much accuracy with the help of only few identified predictor variables.

22. Predicting Adoption of Cashew Production Technologies

Studies were done to check the extent to which selected models could explain the variation in adoption of cashew production technologies. Eight models were tested to examine the variation in adoption among the respondents. Model 8 fitted as $CPT\ AI = 30.543 + 0.400\ X1 + 0.247\ X2 + 0.261\ X3 + 0.443\ X4 - 0.426\ X5 + 0.234\ X6 - 0.202\ X7 + 0.173\ X8$ explained up to 69.7 per cent of the variation in adoption using the predictors; no:of cashew trees (X1), other crops grown (X2), cultivable land available (X3), years of experience in farming (X4), age of the farmer (X5), extension participation (X6), distance of cashew plot from the farmers home (X7) and expenditure in cashew farming (X8). The model 8 also had the lowest standard error of the estimate (9.3513) thus making it the best model suited to predict adoption of cashew production technologies by farmers.

Future researchers and development workers in cashew can concentrate on these predictor variables to extract the maximum benefit in their technology development and dissemination programmes. The variable; number of cashew trees/unit which emerged as a predictor in the study calls for the popularization of high density orchards of cashew owing to better productivity and returns to farmers.

23. Predicting Adoption of Pest Management Technologies

Six models were tested to examine the variation in adoption of pest management technologies among the respondents. Model 6 explained up to 46.4 per cent of the variation in adoption of pest management technologies using the predictors; number of cashew trees (X1), years of experience in farming (X2), ICT usage (X3), primary occupation of farmer (X4), income from agriculture (X5) and age of the farmer (X6) (Table 34.6). The model 6 also had the lowest standard error of the estimate (14.7176) thus making it the best model suited to predict adoption of pest management technologies by farmers. The model is fitted as: *PPT AI: 40.579 + 0.028 X1 + 0.808 X2 + 0.737 X3 - 19.167 X4 + 1.916E-005 X5 - 0.412 X6* and can be used to predict adoption of pest management technologies by farmers under similar agro-ecological situations.

24. Predicting Socio-economic Impact

Seven models were tested to examine the variation in socio-economic impact among the respondents. Model 7 fitted as SEI=-0.125+0.484X1+0.301X2+0.210X3-0.277X4+0.214X5-0.168X6+0.179X7 explained up to 68 per cent of the variation in socio-economic impact using the predictors; importance given to cashew (X1), years of experience in farming (X2), cosmopoliteness (X3), distance of cashew plot from home (X4), extension participation (X5), land used for other crops (X6) and net income from cashew farming (X7) (Table 34.7). The model 7 also had the lowest standard error of the estimate (0.462) thus making it the best model suited to predict adoption of pest management technologies by farmers. The model can be used to predict socio-economic impact of cashew production technologies by farmers under similar agro-ecological situations.

25. Constraints Faced by Farmers in Cashew Cultivation

The socio-economic impact of cashew cultivation is largely influenced by the constraints faced by farmers. The present study revealed 12 constraints as reported by farmers and are classified under technical, management, economic/marketing and processing constraints (Table 34.17). Majority (83 per cent) reported poor price and the high price fluctuations in market for raw cashew nut as the major constraint (Rank 1). Lack of cashew farmer associations/groups and availability of cashew nuts from African nations allow cashew nut processors to manipulate the raw cashew nut prices. Low availability of hired labour was the second biggest constraint reported (71 per cent). Migration of workforce to urban areas, easy job availability through MNREGA scheme and respectable job avenues in many private firms for women have acted as reasons for low availability of workforce in villages. Attack of tea mosquito bug and resultant crop loss (41 per cent) and death of yielding trees due to cashew stem and root borer attack (35 per cent) were also major technical constraints (Rank 3 and 4). This is a matter of concern since cashew yields are largely influenced by the attack of Tea Mosquito Bug (TMB) while attack of Cashew Stem and Root Borer (CSRB) eliminates the crop itself. These findings derive support from Ganapathi and Akash (2013), Venkattakumar (2008), Akinwale and Ayodele (1999) and Dixit and Rao (1999).

Flower drying (Rank 5) and poor yield in some varieties (Rank 6) like NRCC selection-2 (flower drying), VRI-3, V-4 and V-7 (poor yield) were also major constraints. Problems in collection of nuts from large plantations and the resultant theft due to delay or inaccessibility was another constraint reported by farmers. Similar constraints were reported by Aravindakshan and Beevi (1992), Salam *et al.* (1992), Nirban and Sawant (2000), Mandavkar *et al.* (2003) and Veerkar *et al.* (2006). Price control and manipulation by cashew processors were also identified as a constraint by certain section of farmers. Cultivation of cashew in poor soils and wastage of cashew apple due to lack of processing avenues were the other constraints cited by farmers.

Table 17: Constraints faced by farmers in cashew cultivation (n=75)

Sl.No.	Constraints	Rank	f	Per cent
I.	**Technical Constraints**			
1.	Attack of Tea Mosquito Bug and resultant yield loss	**3**	31	41
2.	Death of yielding trees due to Cashew Stem and Root Borer attack	4	26	35
3.	Flower drying in certain varieties	5	15	20
4.	Poor yield of certain varieties	6	13	17
5.	Poor soils in cashew orchards	12	3	4
II.	**Management constraints**			
6.	Collection of nuts from large plantations/theft	7	10	13
7.	Monkey menace during fruiting	10	4	5
III.	**Economic/Marketing Constraints**			
8.	Poor price/price fluctuation	**1**	62	83
9.	Low availability of hired labour	**2**	53	71
10.	Price control/manipulation by processors	8	9	12
11.	Lack of cashew farmer associations/groups	9	9	12
IV.	**Processing constraints**			
12.	No value for cashew apple/wastage of cashew apple	11	3	4

In fact, impact assessment is a data-intensive activity. Collecting appropriate data is perhaps the most time-consuming and costly component of conducting impact assessment. Research programs usually balance data needs between the ideal (and costly) and the practical, and must draw data from various sources. If impact assessment is to become an integral part of the research process in agricultural research, it is important that an appropriate data system be institutionalized within the research system. Institutionalizing a data system also ensures that the information generated by research is available in a systematic and timely manner and is retained for future use as the staff and the Institute change and evolve.

A key issue is the role of baseline and panel surveys to provide data on benchmark-related household variables as the basis for adoption and impact. Many types of impacts can only be adequately assessed if relevant baseline data exist (especially for disembodied research products), and regular surveys, preferably of the same households, are undertaken over-time to monitor changes in farmers' practices. Having a panel of the same households allows regular monitoring of the changes in key farm practices and productivity indicators related to the most important types of research outputs. The sample should be representative of an important subset of the mandate area. The data collection system requires a clear definition of the key parameters, sample size, frequency of data collection, and benchmark sites. Collection of such data should become an integral part of agricultural research.

Impact analysis of cashew production technologies reveals that a majority of cashew farmers had medium levels of technology utilisation status. Based on the premise that a solid technology utilisation status is pre-requisite for accelerated productivity in horticultural sector, there is a reason for concern about the chances of promoting cashew cultivation in this region without higher field level adoption of recommended cashew production technologies. Plant protection, which is one of the most important technology components, scored the lowest adoption rate among cashew farmers. This is a matter of concern since cashew yields are largely influenced by the attack of Tea Mosquito Bug (TMB) while attack of Cashew Stem and Root Borer (CSRB) eliminates the crop itself. Strategic measures have to be taken to improve the utilisation of cashew production technologies in this region as one element of the foundation for technology adoption and change in cashew production sector.

There are implications for rural education and cashew outreach, given that the level of education was found to exert a positive influence on technology utilization. Cosmopoliteness makes the farmer more aware of information and its utilization and causes them to stay abreast of latest cashew production technologies. Consequently, information usage sets the stage for better technology realization in cashew fields. Extension implications might include an exploration of various mechanisms for improved education on the potential of improved technology packages, increasing adoption of cashew production technologies through research and development organizations and increased flow of information through various channels.

Even though highest production was recorded by Bhaskara, Madakkathara – 2, Ullal - 1 and Ullal - 3 varieties, these covered only minimal share of total area under cashew cultivation. By increasing the adoption and area coverage of above varieties we can bring a quantum increase in cashew production in the district. Increasing adoption of soil and water conservation techniques and development and popularization of user friendly plant protection measures can contribute largely to increased cashew production while increasing adoption of pruning and training in cashew orchards can significantly increase the per unit productivity of cashew orchards. The results clearly indicate that socio-economic and bio-physical factors along with policy environment have a larger contribution in explaining cashew production and productivity and technology component alone cannot be expected to bring a positive impact. Understanding the above dynamics in technology impact can help researchers and extension agencies working in cashew sector to design better innovations and effective outreach strategies.

Majority of the cashew farmers reported medium to low production and productivity in cashew cultivation. The analysis also revealed that a majority of cashew farmers had derived moderate socio-economic benefits with particularly high social benefits in comparison to low economic benefits. This is a matter of concern since economic impact largely influences farmer decision regarding continued adoption of cashew production technologies as well as migration to other remunerative crops. The major constraints faced by cashew farmers revealed by the study calls for government intervention in price fixation of cashew and formation of cashew farmers associations. Also, since cashew yields are largely influenced by the attack of Tea Mosquito Bug (TMB) and Cashew Stem and Root Borer (CSRB), strategic measures have to be taken to improve the utilisation of pest management technologies in this region.

There are important implications of this study in terms of the importance of technology utilization by farmers and how this relates to their ability to adopt recommended technology and ultimately to participate in the cashew value chain. The set of intrinsic motivations resulting from these factors may expand farmer aspirations towards better situations and cause farmers to set more compelling targets by adopting the latest technologies. However, one aspect that will present a barrier for development is the fact that majority of the respondents were not sure about the benefits of extension contact and participation. The state run public extension system has demonstrated poor performance in recent times, and this has obviously contributed to poor participation in extension programmes among farmers in the study area. Understanding some of the dynamics in technology utilization process can help researchers working in cashew sector to design better innovations. The variables that were identified as key indicators towards explaining impact of cashew production technologies can be utilized in this context.

References

Akinwale, T. O. and Ayodele, E. A. (1999). Production Constraints. *Quarterly Nigeria's first Magazine Journal*, 2(9): 47-88.

Alston, J., Norton, G. and Pardey, P. (19950. Science under scarcity: Principles and practice for agricultural research evaluation and priority setting. Cornell University Press, Ithaca, NY, USA.

Alwang, J. and Siegel, P. B. (2003). Measuring the impacts of agricultural research on poverty reduction. *Agricultural Economics*, 29: 1–14.

Anderson, J. (1997). On grappling with the impact of agricultural research. Background paper for presentation for the CGIAR Centers' Week, October 1997, The World Bank, Washington, DC, USA.

Aravindakshan, M. and Pathummal Beevi. (1992). Cashew production technologies relevant to Kerala. *Proceedings of National Workshop on Cashew, Kannur.* Balasubramanian pp. (Eds). Directorate of Cashew and Cocoa Development, Kochi, Kerala.

Baidu-Forson, J. (1996). Methodology for evaluating crop and resource management technologies. Page 40 *in* Partners in impact assessment. Summary proceedings of the ICRISAT/NARS workshop on methods and joint impact targets in Western and Central Africa, (Eds. J. Baidu-Forson, M.C.S. Bantilan, S.K. Debrah, and D.D. Rohrbach), ICRISAT, Patancheru, India.

Bantilan, M.C.S. and Dar, W.D. (2001). Impact assessment linked with research priority setting: experience at the International Crops Research Institute for the Semi-Arid Tropics (ICRISAT). Pages 57–59 (Annex) In: The future of impact assessment in the CGIAR: needs, constraints and options. Proceedings of a workshop organized by the Standing Panel on Impact Assessment of the Technical Advisory Committee, 3–5 May 2000. FAO, Rome, Italy.

Bhairamkar, M. S., Mehta, P. G., Sawant, P. A. and Tawade, N. D. (2004). Adoption of recommended cashew cultivation technologies by the cashew growers. *The Cashew*, 18(2): 7-12.

De Janvry, A. and Sadoulet, E. (2002). World poverty and the role of agricultural technology: direct and indirect effects. *Journal of Development Studies*, 38(4): 1–26.

Dixit, S. and Bhaskara Rao, E.V.V. (1999). Problems in cashew cultivation. *The Cashew*, 13(3): 3-16.

Dixit, S., Raviprasad, T.N. and Yadukumar, N. (1998). Cashew plant protection campaign. *The Cashew*, 12(1): 21-31.

Ganapathi, H.A. and Akash, S.B. (2013). Constraints of farmers in cashew production – A case study of Udupi district of Karnataka State. *Cashew Bulletin*, 51(7): 7-11.

IAEG (Impact Assessment and Evaluation Group of the CGIAR). (1999). Impact assessment of agricultural research: context and the state of the art. Paper presented at the ASARECA/ECART/CTA workshop on impact assessment of agricultural research in Eastern and Central Africa, Entebbe, Uganda, 16–19 November 1999.

Kerr, J. and Kolavalli, S. (1999). Impact of agricultural research on poverty alleviation: a conceptual framework with illustration from the literature.?

Kristjanson, P., Tarawali, S., Okike, I., Singh, B. B., Thornton, P. K., Manyong, V. M., Kruska, R. L. and G. Hoogenboom. (2002). Genetically improved dual-purpose cowpea: assessment of adoption and impact in the dry savanna region of West Africa. ILRI Impact Assessment Series No. 9. International Livestock Research Institute (ILRI), Nairobi, Kenya. p. 68.

Lakshmisha, R. (2000). Impact of cashew demonstrations on knowledge, adoption and yield levels of farmers in Dakshina Kannada district. M.Sc. (Agri.) thesis, University of Agricultural Sciences, Bangalore.

Mandavkar P M, Kokate K D and Bairamkar M S. 2003. Adoption of improved varieties of cashew nut. *Maharashtra Journal of Extension Education*, 22(1): 113-116.

Manyong, V.M., Douthwaite, B., Coulibaly, O. and Keatinge, J. D. H. (2001). Participatory impact assessment at the International Institute of Tropical Agriculture: functions and mechanisms. Pages 69–74 (Annex) In: The future of impact assessment in the CGIAR: needs, constraints and options. Proceedings of a workshop organized by the Standing Panel on Impact Assessment of the Technical Advisory Committee, 3–5 May 2000. FAO, Rome, Italy.

Nirban, A.J. and Sawant, P.A. (2000). Constraint analysis of cashewnut growers. *The Cashew*, 14(1): 45-49.

Pachico, D. (1998). Conceptual framework for natural resource management research and basic methodological issues in impact assessment. Paper presented at the international workshop on assessing the impact of research on natural resource management, 27–29 April 1998, Nairobi, Kenya.

Pachico, D. (2001). Approaches and challenges in impact assessment of agricultural and natural resources management research. Pages 3–10 (Annex). In: The future of impact assessment in the CGIAR: needs, constraints and options.

Sajeev, M. V., Saroj, P. L.and Lakshmisha, R. (2014). Technology impacts on area, production and productivity of cashew in Dakshina Kannada district, Karnataka. *Journal of Plantation Crops,* 42(1): 62-69.

Salam, M. A., Satheesh Babu and Balasubramanian, P. P. (1992). Trends in cashew production in Kerala - An Analysis. *The Cashew,* 5(3): 3-7.

Shivaramu, K., Manjunatha, D., Mohan, E. and Guruprasad, T. R. (2004). Impact of cashew demonstration on adoption of recommended cultivation practices by farmers. *The Cashew,* 18(2): 16-21.

Veerkar, P. D., Wadkar, S. S., Bhosale, S. S., and Bagade, S.R. (2006). Constraint analysis of cashew production in South Konkan region of Maharashtra. *The Cashew,* 20(3): 11-22.

Venkattakumar, R. (2006). Socio-economic impact of cashew cultivation in Kannur District of Kerala, India. *Journal of Plantation Crops,* 34 (3): 709-714.

Venkattakumar, R. (2008). Socio-economic impact of cashew cultivation in Sindhudurg district of Maharshtra. *Journal of Plantation Crops,* 36(2): 117-122.

Venkattakumar, R. (2009). Socio-economic factors for cashew production and implicative strategies: An overview. *Indian Research Journal of Extension Education,* 9 (3): 55-62.

Venkattakumar, R. and Bhat, M.G. (2003). Widening cashew growers' perspectives. *The Cashew* 17(2): 5-30.

Venkattakumar, R., Ravi Prasad, T. N and Nayak, M. G. (2005). Perception of cashew demonstration farmers towards recommended soil and water conservation and plant protection measures. *Journal of Plantation Crops,* 33 (2): 135-137.

Venkattakumar, R., Sreekanth, P. D. and Muralikrishna, H. (2004). Analysis of state-wise availability of cashew grafts. *The Cashew,* 18 (4): 9-17.

Zagade, P. M., Tawade, N. D. and Kokate, K. D. (2000). Knowledge and adoption of recommended crop protection measures by the cashew growers in Sindudurg District of Maharashtra. *The Cashew,* 14(4): 22-25.

Zagade, P. M., Tawade, N. D. and Kokate, K. D. (2003). Problems of cashew nut growers in adoption of recommended plant protection measures. *Maharashtra Journal of Extension Education,* 22(1): 44-46.

Cashew: Improvement, Production and Processing (2017) *Pages* **595–623**
Editor: **Prof. P.L. Saroj**
Published by: **ASTRAL INTERNATIONAL PVT. LTD., NEW DELHI**

Climate Change and Cashew Pests

K. K. Srikumar, P.S. Bhat, T.N. Raviprasad and P.L. Saroj

1. Introduction

Awareness about global climatic changes has rapidly increased in the past two decades which has triggered widespread apprehension among scientists and governments about their long-term implications (Gadgil, 1996). Climate change induced global warming has become an issue of serious concern worldwide for existence of life on the earth (Abrol *et al.*, 1996; IPCC, 2007). Over the past hundred years, global temperature has increased by 0.8°C and is expected to register an increase of 1.1-5.4°C by the end of the next century. On the other hand, CO_2 concentration in the atmosphere has increased drastically from 280 to 370 ppm and is likely to double by 2100 (IPCC, 2007). This change is attributed mainly to the overexploitation and improper usage of natural resources for various anthropogenic developmental activities such as, increased urbanization, deforestation and industrialization resulting in aberrant weather changes in rainfall pattern, frequent droughts and floods, increased intensity and frequency of heat and cold waves, outbreaks of insect-pests and diseases, *etc.* which profoundly affect many biological systems and ultimately the human society (IPCC, 2007; Fand *et al.*, 2012).

Agro-ecosystem environment is largely governed by diverse interactions between abiotic (temperature, humidity, rainfall, soil factors, pollutants) and biotic (crop-plants, weeds, insect-pests, pathogens, nematodes) components. The abiotic stress factors strongly modulate the effects of biotic stresses and are most harmful when they occur in combination (Mittler, 2006), greatly influencing crop growth and productivity to an extent of 80 per cent (Oerke *et al.*, 1994; Theilert, 2006).

The India Meteorological Department was established as a National agency in 1875, amalgamating various provincial meteorological services which existed during the 19th century (Kelkar, 2000). However, instrumental data and meteorological records for a few stations in India existed since 18th century. Some of the oldest observatories include Madras (presently Chennai) (September 1793), Bombay (presently Mumbai) (1823) and Calcutta (present Kolkata) (December 1829). The first seismological observatory was set up in Alipore (Calcutta) in 1898. Basically, the annual climate of India is divided into four seasons: (i) winter (January and February) (ii) pre-monsoon or hot weather season (March-May) (iii) southwest or summer monsoon season (June - September) and (iv) post monsoon season (October-December).

The all-India mean annual surface temperature derived from 73 meteorological stations across India showed a significant warming of 0.4°C over the past 100 years, which is comparable to global mean trend of 0.3°C increase per 100 years (Hingane *et al.*, 1985). Wigley (1989) reported a warming of 0.5°C by 1995-2005, 1.5°C by 2015-2050 and 3.0°C by 2050-2100. Pant *et al.* (1996) analyzed mean rainfall over South Asian region for the period 1813-1980 and reported a statistically increasing long-term trend. However, an analysis of a representative rainfall series over the past 176 years for India as a whole by Sontakke (1990) did not suggest any significant trend. The monthly, seasonal and annual rainfall time series of 36 meteorological sub-divisions of India were constructed using a fixed but a large network of about 1476 rain-gauge stations which revealed that the contribution of June, July and September rainfall to annual rainfall is decreasing for few sub-divisions while contribution of August rainfall is increasing in few other sub-divisions (Guhathakurta and Rajeevan, 2008).

Being a sub-tropical country, India is more challenged with impacts of climate change (Chahal *et al.*, 2008). The productivity of Indian agriculture is limited by its high dependency on monsoon rainfall which is erratic in its distribution and inadequate in its distribution (Chand and Raju, 2009). The country is experiencing declining trend of agricultural productivity due to fluctuating temperatures (Aggarwal, 2008; Joshi and Viraktamath, 2004), frequently occurring droughts and floods (Samra, 2003), problematic soils, and increased outbreaks of insect-pests (Joshi and Viraktamath, 2004; Srikanth, 2007; IARI News, 2008; IRRI News, 2009) and due to a variety of plant diseases. These problems are likely to aggravate further due to the changing climate which puts forth a major challenge to achieving the goal of food security.

2. Effect of Climate Change on Insect Pests

Insects being poikilotherms, temperature is probably the single most important environmental factor influencing their behaviour, distribution, development, survival, and reproduction (Yamamura and Kiritani, 1998; Bale *et al.*, 2002; Petzoldt and Seaman, 2010). Therefore, it is highly expected that, the major drivers of climate change *i.e.*, elevated CO_2 levels increased temperature and depleted soil moisture can impact population dynamics of insect-pests (Figure 35.1) and the extent of crop losses, significantly (Caulifield and Bunce, 1994; Petzoldt and Seaman, 2010).

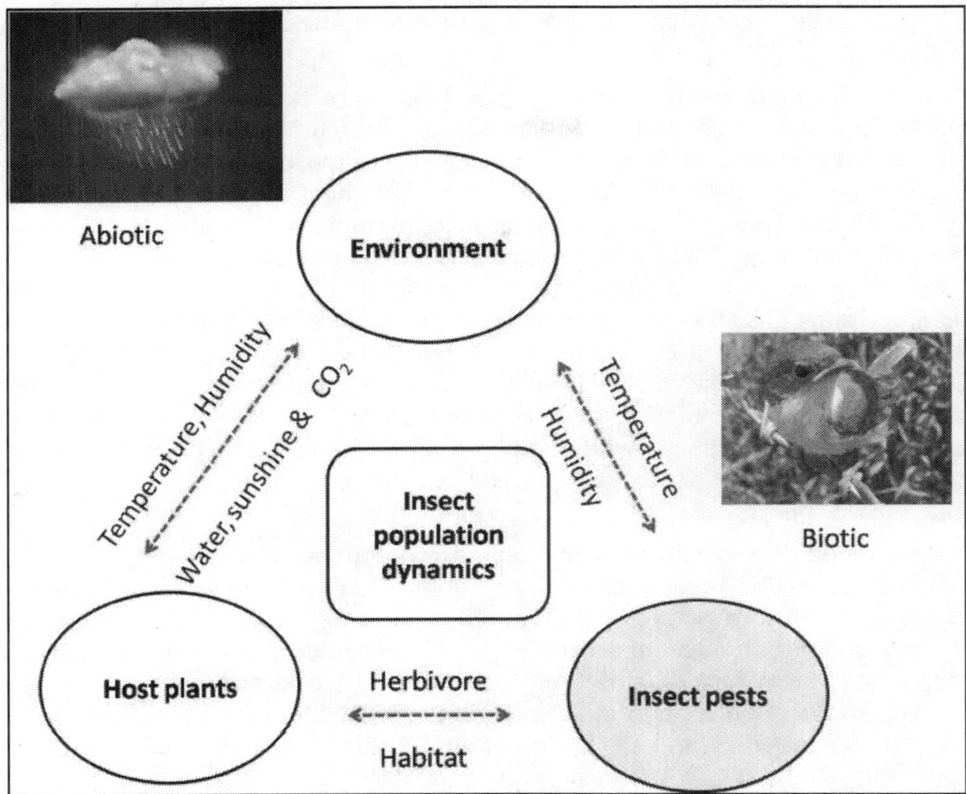

Figure 35.1: Crop-pest Triangle Showing Interactions between Abiotic and Biotic Factors.

Impact of climate change on interactions between crop plants and insect-pests has been extensively reviewed (Yamamura and Kiritani, 1998; Rao *et al.*, 2009; Deka *et al.*, 2008; Petzoldt and Seaman, 2010).

The geographic distribution and abundance of plants and animals in nature is determined by species-specific climatic requirementic essential for their growth, survival and reproduction. Altered temperature and rainfall regimes with the predictable changes in climate will determine the future distribution, survival and reproduction of a species (NACCAP, 2008). The differential rates of range adjustments between annual and perennial plant species along with local extinctions will definitely affect distribution and survival of insect fauna associated with them (Thomas *et al.*, 2004).

Temperature being the single most important regulating factor for insect development, global increase in temperature within favourable range may accelerate the rates of development, reproduction and survival in tropical and subtropical insects. Consequently, insects will be capable of completing more number of generations per year and ultimately resulting in more crop damage (Yamamura and Kiritani, 1998; Bale *et al.*, 2002; Petzoldt and Seaman, 2010). The impacts of climate change on seasonality of insects have been studied by many workers (Bale *et al.*, 2002).

With a rise in temperature, the insect-pests are expected to extend their geographic range from tropics and sub-tropics to temperate regions at higher altitudes along with simultaneously shifts in cultivation areas of their host plants (Hill and Dymock, 1989; Parry and Carter, 1989; Sharma *et al.*, 2005; 2010; Fand *et al.*, 2012). This may lead to an increased abundance of tropical insect species (Patterson *et al.*, 1999; Bale *et al.*, 2002; Diffenbaugh *et al.*, 2008) and sudden outbreaks of insect-pests can wipe out certain crop species (Kannan and James, 2009). At the same time; warming in temperate regions may lead to decrease in relative abundance of temperature sensitive insect populations (Petzoldt and Seaman, 2010: Sharma *et al.*, 2005; 2010). Mostly, the polar regions are generally without insect outbreaks due to low temperature and frequently occurring frosts (Volney and Fleming, 2000). In future, projected climate warming (Carroll *et al.*, 2004) and increased drought incidence (Logan *et al.*, 2003) are expected to cause more frequent insect outbreaks in temperate regions also.

Climate change resultant abiotic environment (increased temperature, elevated CO_2 and depleted soil moisture) will significantly affect the diversity and abundance of insect-pests through geographic range expansion, increased overwintering survival and more number of generations per year, thereby increasing the extent of crop losses. It may result in upsetting ecological balance because of unpredictable changes in the population of insect-pests along with their existing and potential natural enemies which are existing in a natural equilibrium (Rao *et al.*, 2009; IPCC, 2007).

Global warming resulting in altitude wise range expansion and increased overwintering survival of corn earworms, *Heliothis zea* (Boddie) and *Helicoverpa armigera* (Hubner) may cause heavy yield loss and put forth major challenge for pest management in maize, a staple food crop of USA (Diffenbaugh *et al.*, 2008). Range extension in a migratory polyphagous species like *Helicoverpa armigera* (Hubner), a major pest of cotton, pulses and vegetables in north India is predicted with global climate warming (Sharma *et al.*, 2005; 2010). Subsequently, these ongoing shifts in insect-pest distribution and range due to changing climate may alter regional structure, diversity and functioning of locally existing ecosystems (Timoney, 2003; UN-HABITAT, 2004; IPCC, 2007).

The temperature in India is expected to increase by 1.0-5.°C within next 100 years (De and Mukhopadhyay, 1998; IPCC, 2007). Thus, insects undergoing a winter diapause are likely to experience the most significant changes in their thermal environment (Bale and Hayward, 2010).

Accelerated metabolic rates at higher temperatures shorten the duration of insect diapause due to faster depletion of stored nutrient resources (Hahn and Denlinger, 2007). Warming in winter may cause delay in onset of diapause and early summer may lead to faster termination of diapause in insects, which can then resume their active growth and development. This implies that an increase in temperature in the range of 1.0 to 5.0°C would increase insect survival due to low winter mortality, increased population built-up, early infestation and resultant crop damage by insect-pests under global warming scenario (Harrington *et al.*, 2001; Sharma *et al.*, 2005; 2010). Very few studies have been concentrated on the direct effects of higher winter temperatures on rate of development and reproduction in insects (Konestabo *et al.*, 2007; Bale and Hayward, 2010).

Changes in climatic variables have been recorded to increase frequency and intensity of outbreaks of insect-pests. Outbreak of sugarcane woolly aphid *Ceratovacuna lanigera* Zehntner, in sugarcane belt of Karnataka and Maharashtra during 2002-03 resulted in 30 per cent yield losses. These situations of increased and frequent pest damage to the crops have added to the woes of the already distressed farmers, by increasing the cost of plant protection and reducing the profit margins.

Expression of the host plant resistance is considerably influenced by environmental factors like temperature, sunlight, soil moisture and air pollution. Under stressful environment, plants become more susceptible to attack by insect-pests resulting in pest outbreaks and excessive crop damage (Rhoades, 1985). Breakdown of plant resistance which is associated with thermal and drought induced stress have been widely reported (Rhoades 1985; Sharma *et al.*, 2005; Volney and Fleming, 2000; Logan *et al.*, 2003). With a rise in global temperature and increased water stress, sub-tropical countries like India may face the problem of severe yield loss in sorghum due to breakdown of resistance against sorghum midge, *Stenodiplosis sorghicola* (Coq.) and stem borer, *Chilo partellus* Swinhoe (Sharma *et al.*, 2005).

3. Cashew Yields and Climate Change

Cashew (*Ancardium occidentale* L.) is a native of Brazil which was introduced into India by the Portuguese travelers about 400 years ago. Cashew requires relatively dry atmosphere and mild winter (15 to 20°C minimum temperature) coupled with moderate dew during night for profuse flowering. Suitability studies for cashew using GIS at Directorate of Cashew Research (DCR), Puttur showed that cashew grows at elevations up to 1000 m above MSL. However, its productivity is the highest at altitudes of 750 m above MSL. The average annual rainfall distribution in cashew areas ranged from low rainfall (300-600 mm in Gujarat) to high rainfall (2700-3000 mm in west coast and NEH region) but the productivity is highest in regions with a mean annual rainfall distribution of 600-1500 mm. The productivity of cashew is higher in regions with mean annual temperature range from 22.5 to 27.5°C, and the minimum temperature ranges from 10 to 22°C, however it is lower in regions where the minimum temperature drops below 10°C (Rupa *et al.*, 2013).

Reproductive phase of cashew is highly sensitive to weather aberrations. In all major cashew growing areas, pest complex of cashew is another constraint which affects cashew production to a great extent, as the peak pest incidence coincides with the reproductive phase of the crop. The meteorological factors, which are conductive for better flowering, generally trigger the pest build up. Both higher maximum temperature and lower minimum temperature during the reproductive phase, influences the cashew production and its quality.

Unseasonal rains and heavy dew during flowering and fruiting period aggravated the incidence of pests. Total nut yield losses ranged from 50 to 65 per cent during peak harvest period with poor quality in March, 2008. Further, the rains received during harvesting period resulted in soaking of nuts and affected shelf life of nuts and quality of processed kernels. Continuous and unseasonal rains received during March, 2008 did not permit drying of nuts immediately thereby resulted in poor quality of cashew kernels. Huge quantity of nuts which could not be harvested in time and germinated in the field itself may cause of high postharvest losses. Unseasonal rains at ripening stage led to blackening of nuts as well as rotting of apples on trees itself. Flowers exposed to incessant rain accompanied with wind led to severe in flower drop and occurrence of fungal diseases.

The weather parameters recorded at DCR for Puttur, Dakshina Kannada during past 23 years are presented in Figure 35.2. The annual maximum temperature increased significantly over the past two decades. On the other hand, no significant trend can be observed in case of minimum temperature. The mean temperature for past two decades showed significant increase, while sunshine hours showed slight decreasing trend. There was also a decreasing trend for evaporation and annual rainfall. The forenoon humidity had an increasing trend whereas, afternoon humidity decreased over the period.

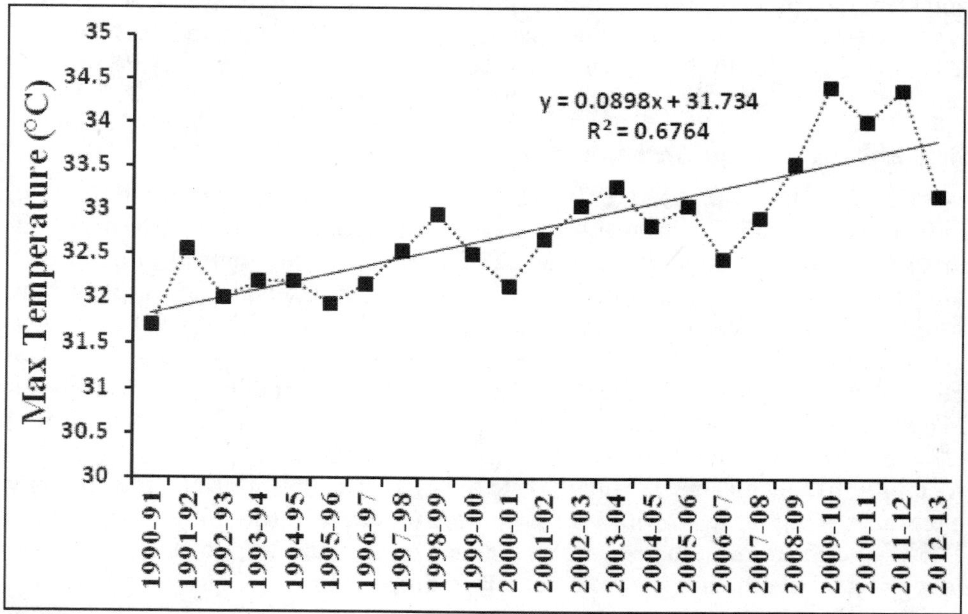

Figure 35.2: Max. Temperature Recorded at DCR, Puttur during 1990-2013.

4. Effect of Climate Change on Cashew Pests

Pest problem is one among the factors responsible for low productivity in cashew. The details of various pests infesting cashew have been described from India (Pillai, 1976; Rai, 1984 and Sundararaju, 1993). The cashew stem and root borer (CSRB) and tea mosquito bug (TMB) *Helopeltis* spp. are the major pests of cashew. The intensity of infestation and extent of damage by different insect pests in

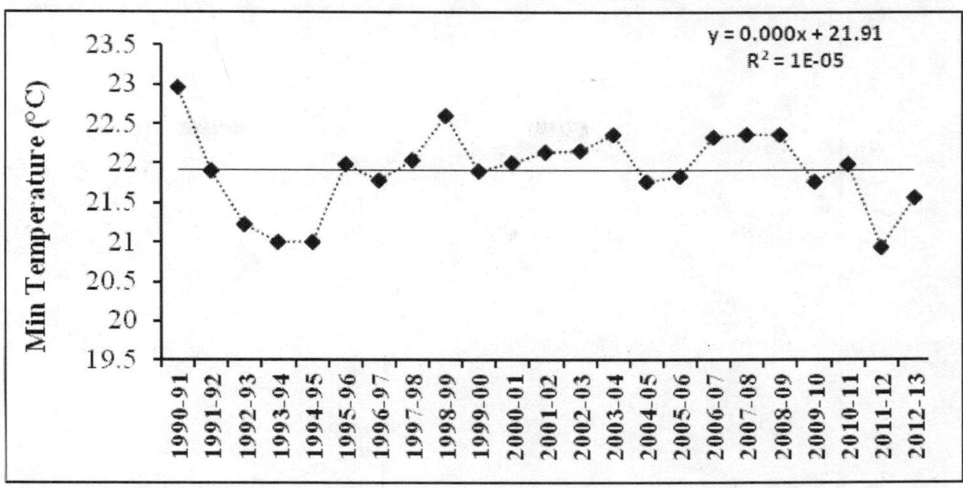

Figure 35.3: Min. Temperature Recorded at DCR, Puttur during 1990-2013.

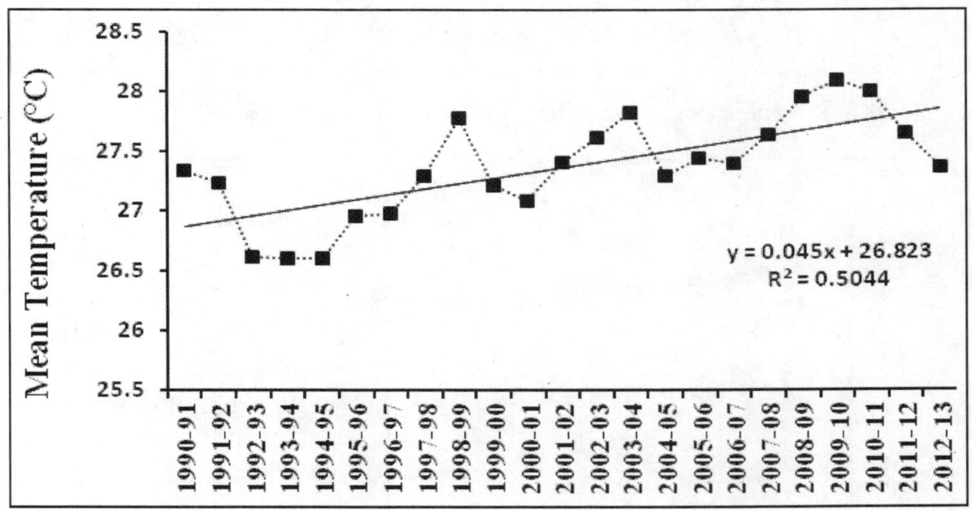

Figure 35.4: Mean Temperature Recorded at DCR, Puttur during 1990-2013.

different climatic regions are not alike and abiotic factors play an important role in the insect population build up and subsequent damage on cashew crop. The studies done at Andhra Pradesh Agricultural University as given in Table 35.1 revealed the seasonality in occurrence of some cashew pests (Ramadevi and Murthy, 1983).

In coastal tracts of West Bengal, the interrelationship of insect pests incidence and damage with meteorological parameters has been systematically documented by Chakraborti and Majumder (2007). TMB mostly appeared in three months during winter (December, January and February). The damage caused by leaf and blossom webber (*Lamida moncusalis* Walker) was observed throughout the year except the

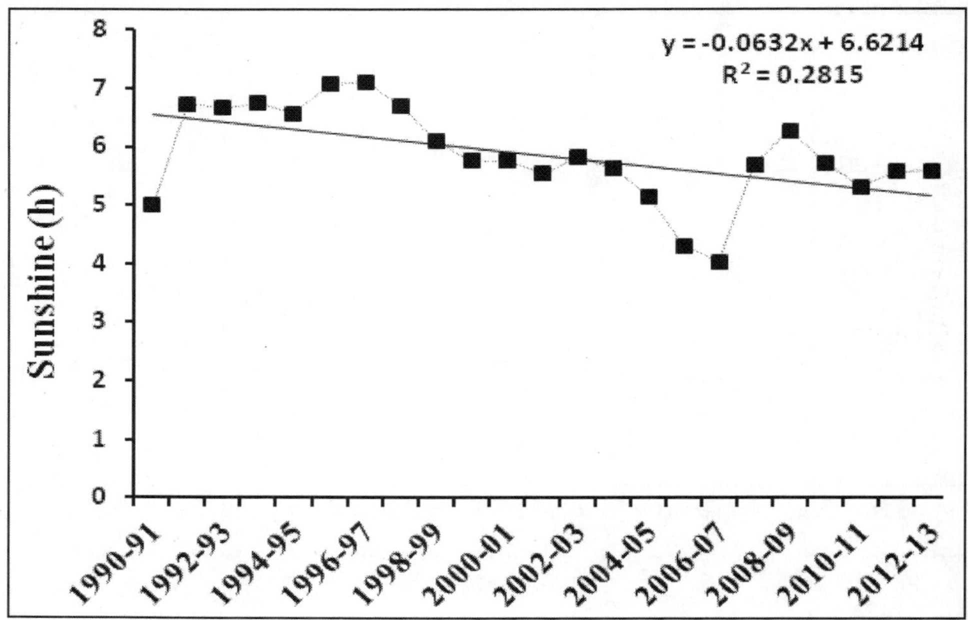

Figure 35.5: Sunshine Hours Recorded at DCR, Puttur during 1990-2013.

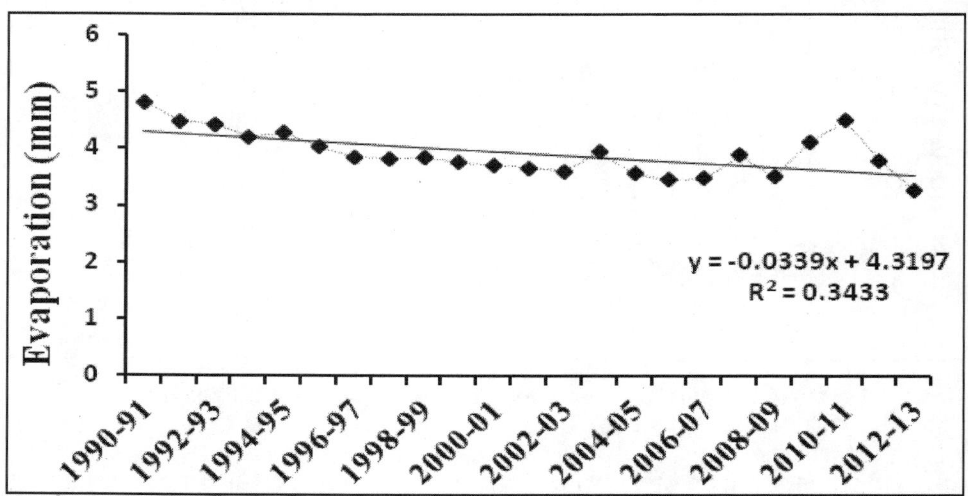

Figure 35.6: Evaporation Recorded at DCR, Puttur during 1990-2013.

summer months (April, May, June and July) (damage: 0.0-22.1 per cent). The damage caused by shoot tip caterpillar (*Hypotima haligramma* M.) and leaf miner (*Acrocercops* (=*Conopomorpha*) *syngramma* M.) were observed during January, February, March, November, and December in all the years (damage: 0.0-24.8 and 0.0-23.6 per cent, respectively). The population of leaf thrips (*Ripiphorothrips cruentatus* Hood)

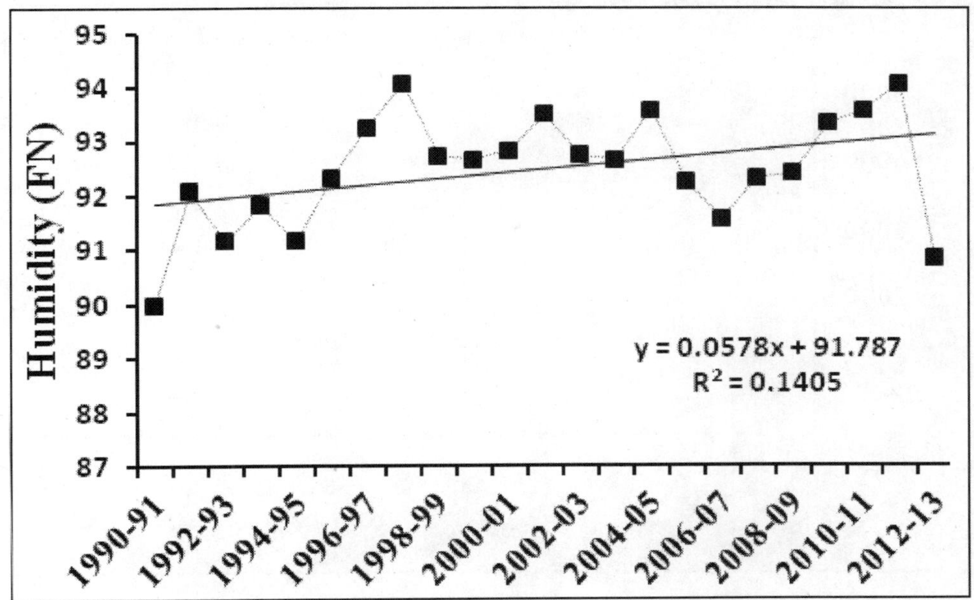

Figure 35.7: Humidity (FN) Recorded at DCR, Puttur during 1990-2013.

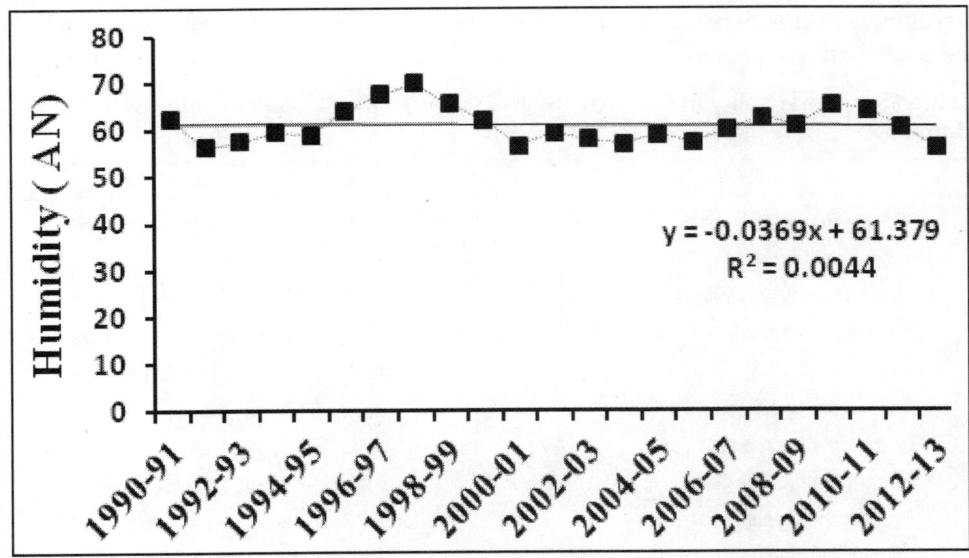

Figure 35.8: Humidity (AN) Recorded at DCR, Puttur during 1990-2013.

was encountered for 6 months (January, February, March, April, November and December) (Mean population: 0-180.40) in each year. Flower thrips (*Frankliniella schultzei* Trybom.) and leaf folder infestation occurred during 5 months (January, February, March, April and December) (Mean population: 0-190.40 and damage

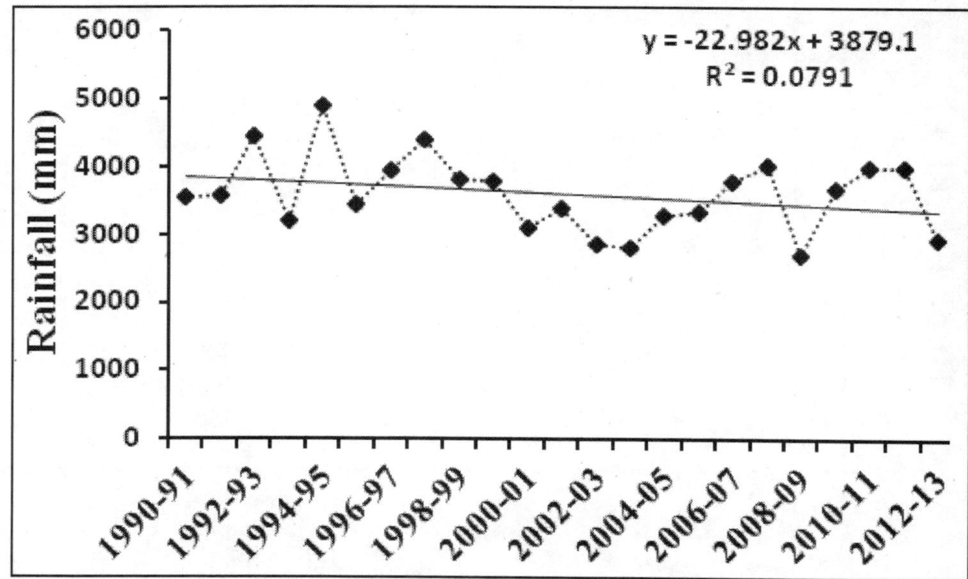

Figure 35.9: Rainfall Recorded at DCR, Puttur during 1990-2013.

0-9.6 per cent, respectively) in each year. Cashew stem and root borer (*Plocaederus ferrugineus* L.) incidence on the trees was observed during the whole year (4.0-30.0 per cent) (Table 35.2).

Table 35.1: Seasonality in Occurrence of Cashew Pests at Cashew Research Station, Bapatla

Cashew Pest	*Seasonal Occurrence*
Plocaederus ferrugineus L.	May-July, Aug-Sept, Oct- April
Indarbela tetraonis M.	June-Sept, Oct-May
Macalla moncusalis Wlk.	June-Sept, Oct-Jan, Feb-May
Orthaga exvinacea M.	Aug.-Oct.
Holotricha sp.	June-July, Aug-Oct, Nov-May
Thallssodes guadraris Guen.	Oct-Dec, Jan.-Sept.
Bombetellia reacosatrix Guen.	Oct-Dec, Jan-Sept
Metanastria hyrtaca Cram.	Jan-Aug, Sept-Doc
Acrocercops Syngramma M.	Jan -April, May-Aug, Sept-Dec
Rhipiphorothrips cruentatus H	March-May, June- Feb
Rhipiphorothrips rapensis G.	Feb-April, May-Jan

The seasonal incidence of TMB is similar across the various cashew tracts of India as reported by several cashew workers *viz.*, Rai (1984), Jeevaratnam and Rajapakse (1981), Pillai and Abraham (1975), Pillai *et al.* (1984), Satapathy (1993), Prasad Rao (1994), Sundararaju and Sundarababu (1999) and Sundararaju (2005).

Table 35.2: Effect of Meteorological Parameters on Incidence of Eight Cashew Insect Pests at Two Locations (Contai and Digha Midnapur, West Bengal) for the Years 2002-2004

Insect Pest	*Regression Coefficient*						
	Temp. Max. (°C)	*Temp. Min. (°C)*	*RH Max. (per cent)*	*RH Min. (per cent)*	*Rainfall (mm)*	*Rainy Days*	*MSH*
TMB	–0.04	–0.35**	0.06	0.02	0.02	–0.18	0.42
LT	–0.56	–7.27**	2.61*	–0.32	0.63**	–6.23**	0.66
LBW	–1.38**	0.54**	0.07	0.10	–0.11	0.28	7.28
LF	–0.03	–0.42**	0.007	0.01	0.03	–0.18	0.06
FT	–1.40	–7.66**	3.16	–0.13	0.75	–7.88	0.15
CSRB	–0.07	0.02	–0.21	–0.10	–0.04	–0.12	–2.57**
LM	–0.21	–1.00**	0.09	0.03	0.03	–0.08	0.49
STC	0.52**	–0.63**	0.11	0.06	0.01	0.26	1.46

*: Significant at 5 per cent; **: Significant at 1 per cent (Chakraborti and Majumder, 2007).

MSH: Mean sunshine hour/day; TMB: Tea mosquito bug; LBW: Leaf and blossom webber; STC: Shoot tip caterpillar; LM: Leaf miner; LT: Leaf thrips; FT: Flower thrips; LF: Leaf folder; CSRB: Cashew stem and root borer.

The study at DCR, Puttur also revealed that the peak population of TMB varied between December and February during the cropping season depending upon biology of the pest influenced by favourable weather, which triggers the pest population. Inter-annual variation in pest population was also noticed (Figure 35.10).

Seasonal fluctuations in the population density of *Helopeltis* spp. (TMB) was studied in Kasaragod during June 1972 to May 1977 at weekly intervals (Table 35.3). The buildup of pest population commenced in October-November synchronizing with the emergence of new flushes in cashew after the cessation of monsoon showers.

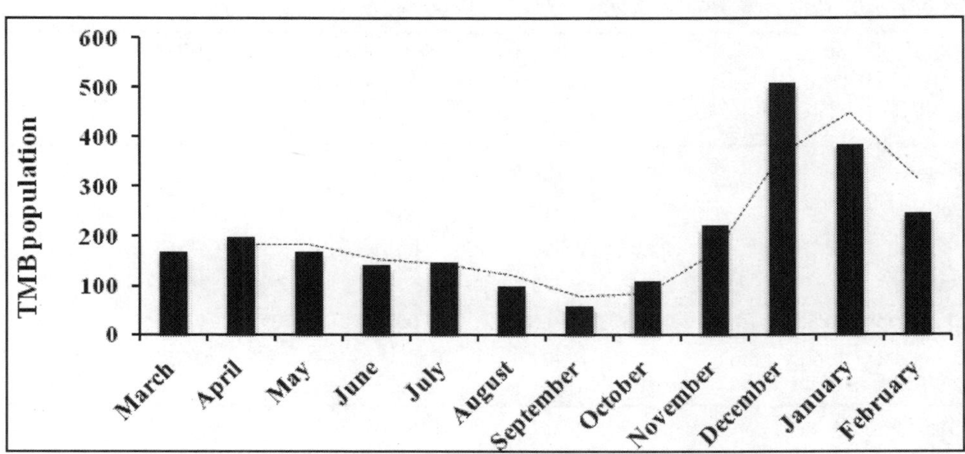

Figure 35.10: Seasonality of TMB Population at DCR, Puttur (2009-2013).

The population buildup of the pest was negatively correlated with minimum temperature, minimum relative humidity and rainfall and positively with sunshine. The most favourable period for the rapid multiplication and population build up of the pest was December to February. During the monsoon season, tea mosquito population was totally absent (Pillai *et al.*, 1984).

Table 35.3: Mean Meteorological Parameters at CPCRI Farm, Kasaragod and Correlation Coefficients with Population of TMB (1972-1977)

Season	*Minimum Temp (°C)*	*Relative Humidity (RH per cent)*	*Rainfall (mm)*	*Bright Sunshine Hours (BSH)*
June-September	23.0	80.4	2539.9	4.4
October-November	22.4	67.6	253.6	7.2
December- February	20.3	53.3	140.9	9.4
March-May	24.2	64.7	498.8	8.7
Correlation coefficient (r)	–0.782*	–0.875*	–0.719*	0.783*

*: Significant at 1 per cent.

TMB causes yield losses ranging from 30-90 per cent depending on the severity of pest incidence. The pest population starts building up from September to October and gradually reaches a peak during February. The pest population is synchronized with the phenological stages of the crop, which is also dependant on prevalent weather parameters. Various weather parameters influence the pest buildup of which temperature and relative humidity have been reported to play a significant role. During 1998-99, the flowering in most of the cashew plantations on the West Coast indicated a bumper crop; however, this was devastated due to occurrence of the pest complex resulted in less than half the expected nut yield. Variation of pest population according to weather was also noticed in different years (Yadukumar *et al.*, 2008).

Table 35.4: Mean Meteorological Parameters at ARS, Chintamani and Correlation Coefficient with Pest Population of TMB

Season	*Temperature (°C)*		*Relative Humidity (RH per cent)*		*Rainfall (mm)*	*Number of Rainy Days*
	Max	*Min*	*Forenoon*	*Afternoon*		
June-Sept	29.7	20.5	78.4	59.2	413.0	24.3
Oct-Nov	27.8	18.3	78.9	63.3	180.2	11.9
Dec-Feb	27.8	14.7	74.6	47.8	25.2	1.9
Mar-May	34.0	20.7	68.3	39.7	103.3	7.1
Correlation coefficient (r)	–0.385*	–0.664*	–0.095	–0.173*	–0.305*	–0.397*
Direct effect through path coefficient analysis	–0.364	–0.351	0.077	–0.182	0.016	0.262

*: Significant at 1 per cent.

A field experiment was conducted to monitor pest surveillance on TMB of cashew orchards in relation to weather parameters at fortnightly intervals for ten years (at the Agricultural Research, Chintamani, Karnataka) (Table 35.4). Path coefficient analysis was carried out to understand the direct and indirect effects of the individual weather elements on tea mosquito incidence. The pest population was low under cold (< 12°C of minimum temperature) and hot (> 35°C of maximum temperature) weather conditions (Prasad Rao *et al.*, 2001).

Data given in Table 35.5 showed the influence of temperature on population fluctuations of TMB in Bastar cashew region of Chhattisgarh on population (Chaudhary and Sahu, 2009).

Table 35.5: Correlation of Weather Parameters with the TMB at Jagdalpur

Weather Parameters	*Correlation Coefficients Values (r) of TMB Damages*	
	2004-05	*2005-06*
Max Temp. (°C)	0.592*	0.095
Min Temp. (°C)	–0.055	0.033
R.H (FN) (per cent)	–0.459	–0.084
R.H (AN) (per cent)	–0.506	–0.178
Rainfall (mm)	–0.358	–0.094
Rainy days	–0.428	–0.081

*: Significant at 1 per cent.

Table 35.6: Correlations among Mean TMB Population by Absolute Estimate, its Growth and Weather Parameters during Cropping Season of 1992-93 (September to April)

Weather Parameters	*Correlation Coefficient 'r' Values for Average TMB Population vs.*			*Growth of TMB Population vs.*
	FPW	*SPW*	*TPW*	*CWP*
Max Temp. (°C)	0.04	–0.21	–0.28	0.14
Min Temp. (°C)	–0.40*	–0.28	–0.01	0.10
Mean Temp. (°C)	0.34	0.11	0.13	0.17
Humidity FN (per cent)	–0.38*	–0.21	–0.09	–0.08
Humidity AN (per cent)	–0.40*	–0.20	–0.04	–0.14
Mean Humidity (per cent)	0.32	0.16	0.12	0.13
Rainfall (mm)	–0.29	–0.21	–0.26	0.02
Rainy days(No.)	–0.43*	–0.21	0.02	–0.37*
Bright sunshine (h)	0.45*	0.26	0.11	0.02
Wind speed (Km/h)	0.18	0.01	–0.06	0.11

*: Significant at 5 per cent level.

Studies conducted by Sundararaju (2005) in cashew plantations of DCR, Shanthigodu, Dakshina Kannada Karnataka during 1992-93, when the trees were three years old and in 1999-2000 to 2002-2003 when they were 10-13 years old indicated that only minimum temperature had a consistent negative relationship negative relationship with the pest population. When the current week average of TMB population of absolute estimate was correlated with weather parameters of first, second and third preceding week, it was noticed that maximum and minimum temperatures, morning and evening relative humidity and number of rainy days of first preceding week showed significant relationship, and bright sunshine hours exhibited positive relationship (Tables 35.6 and 35.7).

Table 35.7: Correlations among Mean TMB Population by Absolute Estimate, its Growth and Weather Parameters during Cropping Season from 1999-2000 to 2002-2003 (October-May)

Weather Parameters	*Correlation Coefficient 'r' Values for Average TMB Population vs.*			*Growth of TMB Population vs.*
	FPW	*SPW*	*TPW*	*CWP*
Max Temp.	0.04	–0.02	–0.08	0.03
Min Temp.	–0.20*	–0.29**	–0.38**	0.02
Mean Temp.	0.21*	0.25**	0.32**	0.00
Humidity FN	0.06	0.01	–0.09	0.12
Humidity AN	–0.16	–0.17	–0.24**	–0.07
Mean Humidity	0.19*	0.18*	0.24**	0.10
Rainfall	0.09	–0.09	–0.09	–0.01
Rainy days(No.)	–0.13	–0.13	–0.13	–0.02
Bright sunshine h	0.12	0.00	0.11	0.02
Evaporation	–0.05	–0.08	–0.05	0.03
Evening soil temp.	0.06	–0.02	–0.01	0.00

*: Significant at 5 per cent level.

FPW: First previous week weather parameters; SPW: Second previous week weather parameters; TPW: Third previous week weather parameters and CPW: Current week weather parameters.

The pest incidences in different climatic conditions along east coast (Vridhachalam, Baptla, Bhubaneswar and Jhargram) and west coast (Madakkathara and Vengurle) cashew belt were studied for a period of 5 years (AICRP, 2009; AICRP, 2010; AICRP, 2011; AICRP, 2012 and AICRP, 2013).

The simple correlation studies conducted at Vridhachalam revealed that maximum temperature, relative humidity and sunshine had a positive relation with the activity of *H. antonii*, whereas negative correlation was established with rainfall (Figures 35.11 and 35.12). Aphid population had positive correlation with relative humidity and minimum temperature. Similarly, blossom webber, leaf miner, leaf roller and shoot tip caterpillar have negative correlation with maximum temperature.

The studies conducted at Bapatla showed that the maximum temperature and minimum temperature exercised a non-significant but positive influence on the

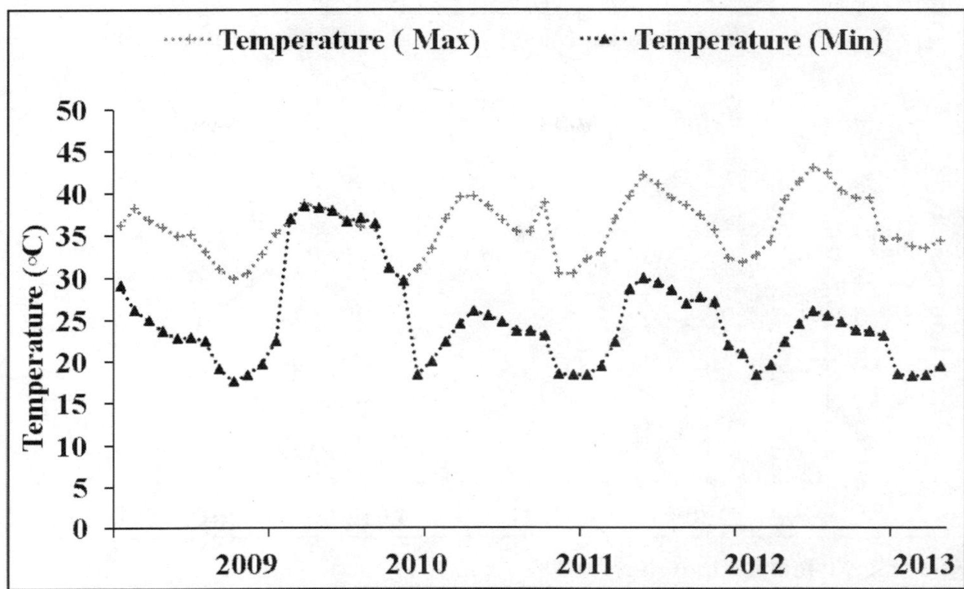

Figure 35.11: Temperature Recorded at Vridhachalam.

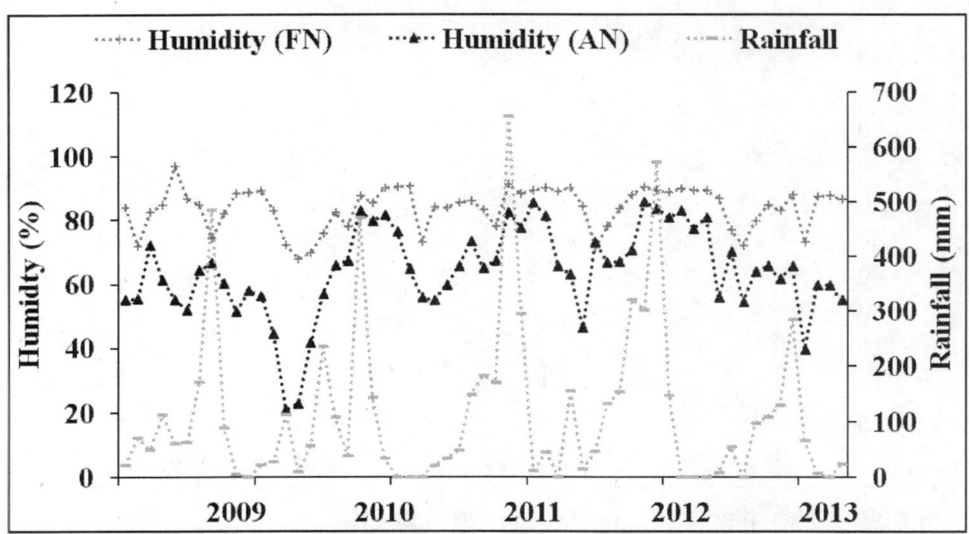

Figure 35.12: Humidity and Rainfall Recorded at Vridhachalam.

activity of the leaf and blossom webber, whereas the relative humidity showed a negative influence (Figures 35.13 and 35.14). The activities of leaf miner, leaf folder and shoot tip caterpillar were found to be influenced by both maximum temperature and minimum temperature.

At Bhubaneswar, TMB was observed from December to April with maximum intensity during February (9.5 adult and nymph/52 leader shoots). Evening RH had

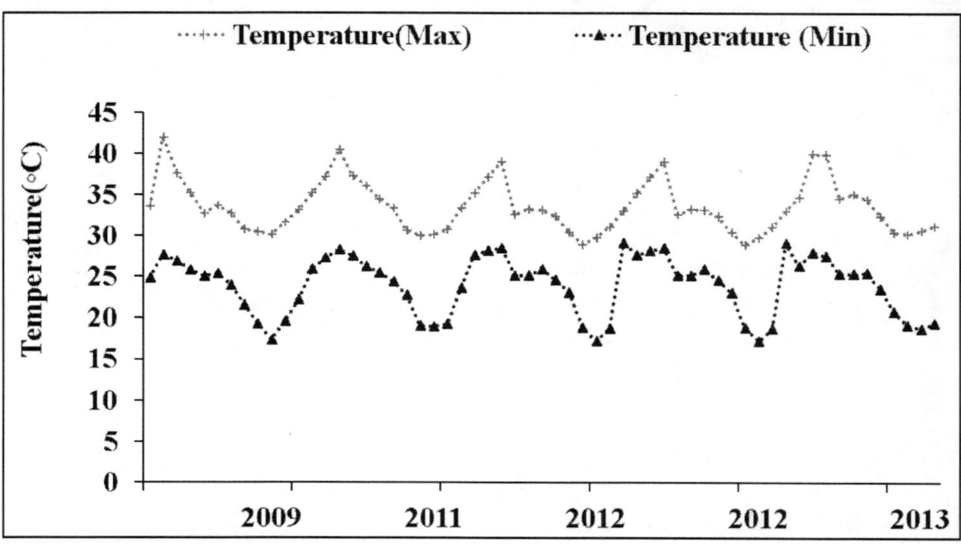

Figure 35.13: Temperature Recorded at Bapatla.

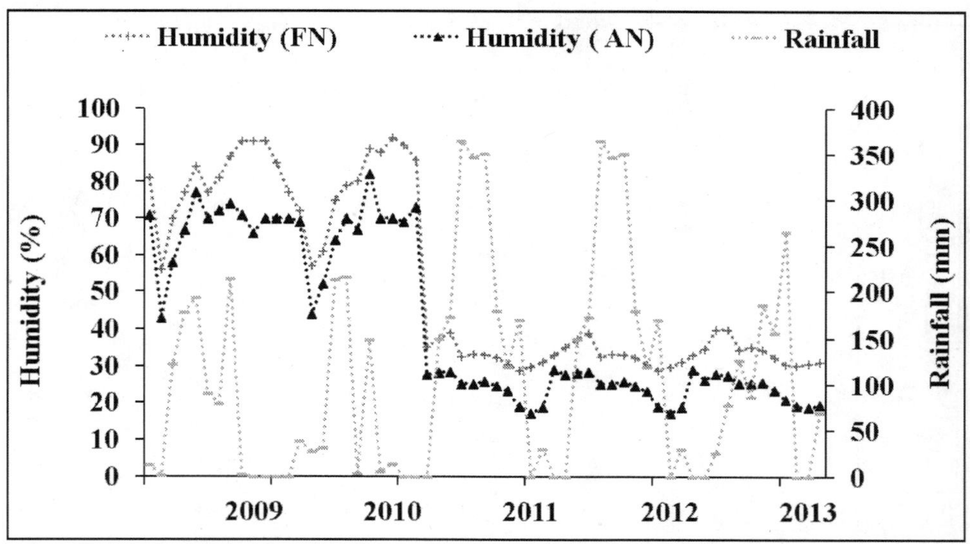

Figure 35.14: Humidity and Rainfall Recorded at Bapatla.

negative correlation with the incidence of the *H. antonii*. Shoot tip caterpillar, leaf miner and thrips were also recorded during the period (Figures 35.15 and 35.16). The population of TMB was negligible at Jhargram during the period (Figures 35.17 and 35.18).

Correlation studies at Madakkathara showed significant negative correlations between TMB infestations with maximum and minimum temperature, bright sunshine hours and wind speed. The incidence of leaf miner showed two population peaks first during May-June and second during September-December. Highest

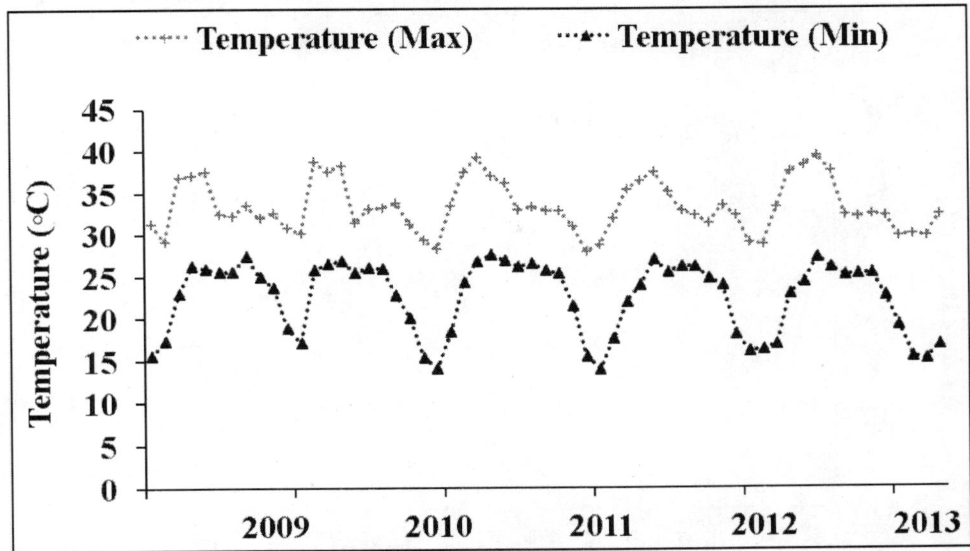

Figure 35.15: Temperature Recorded at Bhubaneshwar.

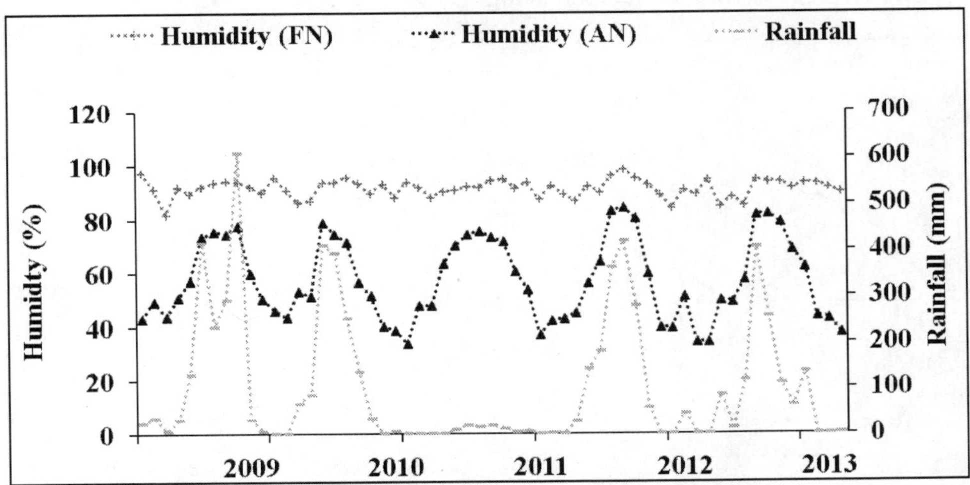

Figure 35.16: Humidity and Rainfall Recorded at Bhubaneshwar.

incidence of leaf miner was noticed in the months of October to November coinciding with the flushing (Figures 35.19 and 35.20).

Studies at Vengurle showed TMB infestation showed positive correlation with maximum temperature and negative significant correlation with minimum temperature, rainfall, number of rainy days and evening humidity (Figures 35.21 and 35.22). The infestation of thrips showed positive relationship with maximum temperature and significant negative correlation with minimum temperature and rainfall. The leaf miner showed significant positive correlation with maximum

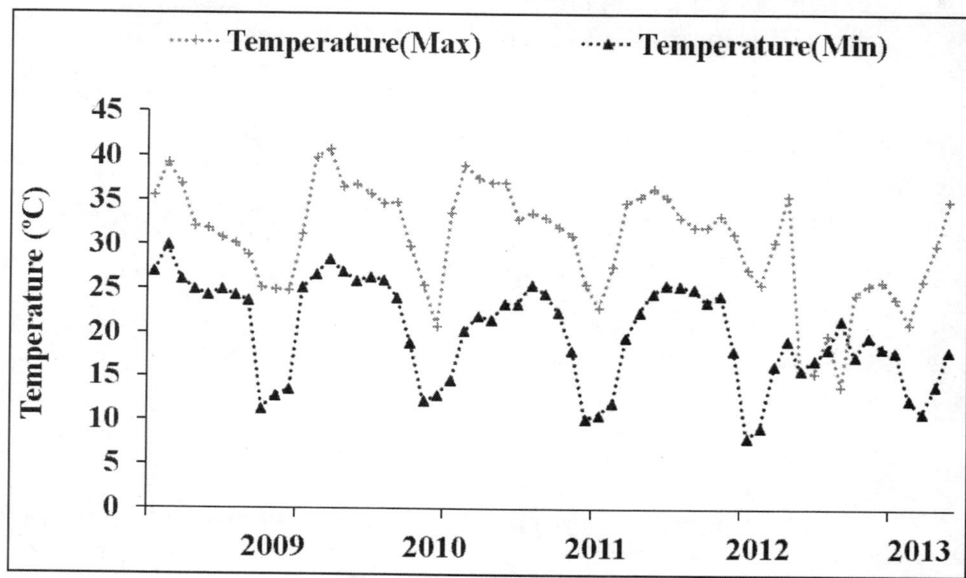

Figure 35.17: Temperature Recorded at Jhargram.

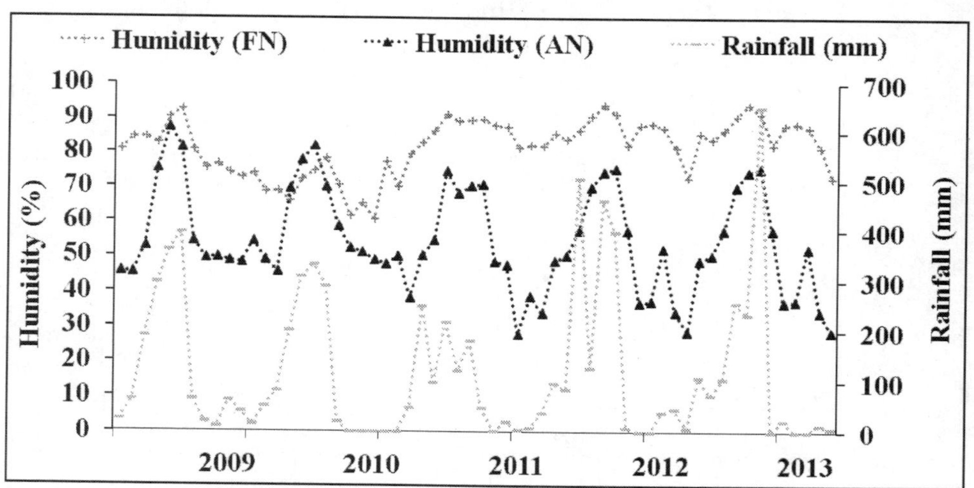

Figure 35.18: Humidity and Rainfall Recorded at Jhargram.

temperature and negative correlation with rainfall. The incidence of apple and nut borer showed negative correlation with rainfall.

The studies at DCR, Puttur (2009-2013) revealed that minimum temperature (r = -0.622) and bright sunshine hours (r = 0.481) had high correlation indices in increasing phase of *H. antonii* population (Table 35.5). Further rainfall (r = -0.382), maximum temperature (r = 0.334) and afternoon humidity (r = -0.352) also had significant effect on *H. antonii* incidence. The pest build up was observed during winter wherein, minimum temperature and sunshine hours had significant

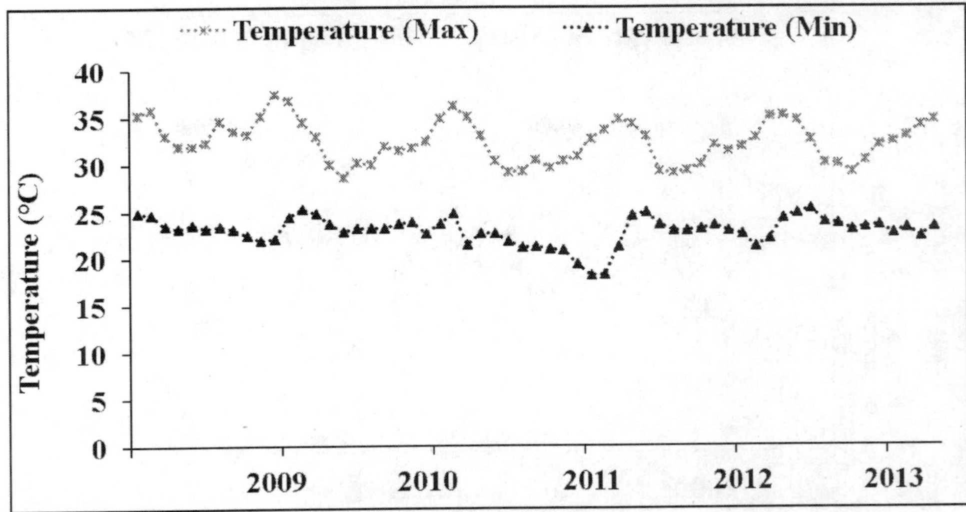

Figure 35.19: Temperature Recorded at Madakkathara.

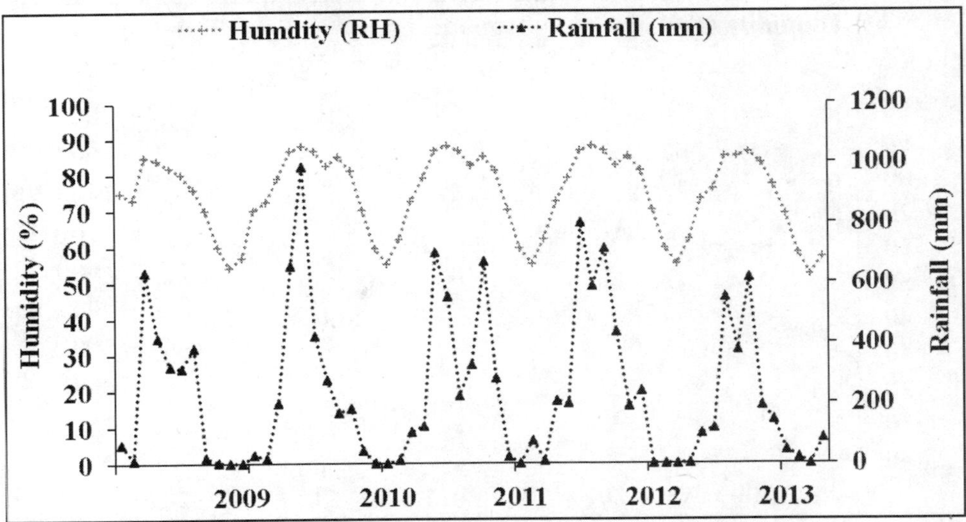

Figure 35.20: Humidity and Rainfall Recorded at Madakkathara.

correlation indices. The highest peak of TMB population occurred in December and January, of each year observed (Table 35.8).

A multiple regression prediction model was developed based on weather data and pest population with 88 per cent accuracy of prediction of incidence (Prasad Rao, 2002). The predicted pest population is a function of sunshine hours, rainfall, relative humidity (min and max), temperature (min and max) and the number of rainy days.

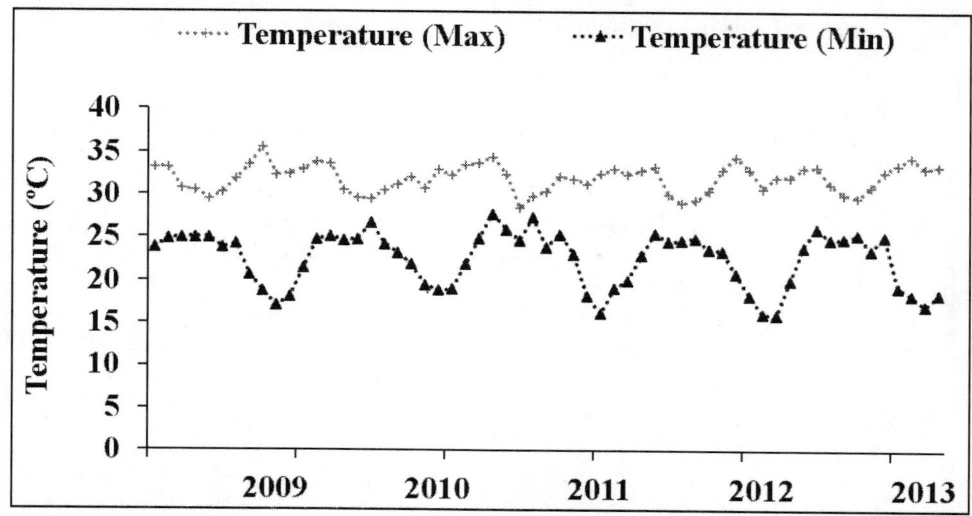

Figure 35.21: Temperature Recorded at Vengurle.

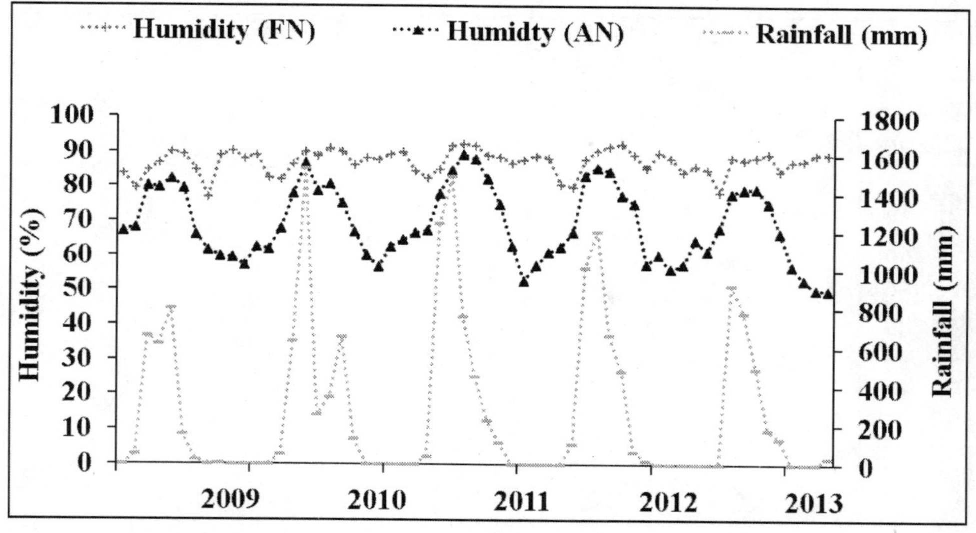

Figure 35.22: Humidity and Rainfall Recorded at Vengurle.

$Y_1 = 0.648x_1 + 0.002x_2 + 0.048x_3 - 0.073\ x4 - 0.597x6 - 0.272\ x7 + 30.473$ ($R^2 = 0.54$) – in case of pest population

$Y_2 = 1.738x_1 + 0.013x_2 + 0.138x_3 - 0.230\ x_4 - 1.762\ x_5 - 1.589\ x_6 - 0.791\ x_7 + 83.629$ ($R^2 = 0.54$) - in case of percentage damage due to pest

Where,

Y_1 = Predicted pest population;

Y_2 = percentage damage due to pest population

x_1 = Sunshine hours (h/day);

x_2 = Rainfall (mm);

x_3 = Morning relative humidity (per cent)

x_4 = Evening relative humidity (per cent);

x_5 = Maximum temperature (°C);

x_6 = Minimum temperature (°C)

x_7 = Number of rainy days.

Both the regression equations could account for more than 50 per cent occurrence in pest population as well percent damage due to pest. However, it accounted up to 90 per cent when individual years were considered. The prediction models developed by Satapathy (1993) also revealed that they explained between 80.0 and 90.0 per cent in pest population and per cent damage, respectively based on weather variables. The prediction equations will be useful in understanding the effects of weather elements on pest populations.

Table 35.8: Correlation Coefficients (r) of *Helopeltis* Population with Temperature, Humidity, Rainfall and Sunshine

TMB Population	*Temperature (%C)*		*Humidity (per cent)*		*Rainfall (mm)*	*Sunshine (h)*
	Max	*Min*	*Forenoon*	*Afternoon*		
H.antonii	0.334*	–0.622*	–0.262	-0.352*	–0.382*	0.481**
H.bradyi	0.102	–0.439**	–0.068	–0.169	–0.236	0.280

*: Significant at 5 per cent level; **: Significant at 1 per cent level.

Population fluctuation of *H. antonii* was also reported in cashew plantations at Wonogiri, Indonesia for two years (2004-2006). The *H. antonii* population fluctuated in a cyclical pattern with the peak population in July. The TMB population did not occur during high rainfall but, began to increase at the end of the rainy season and was high during periods of low and intermittent rainfall. Number of shoots and inflorescences of cashew significantly influenced the number of *H. antonii* population. This trend of population abundance was not directly associated with rainfall, but rainfall influenced the physiology of shoot flushes and inflorescence production. Results of correlation and regression analysis showed that rainfall was not significantly correlated to *H. antonii* population and does not significantly contribute to the number of *H. antonii* population on cashew. However, the availability of food in the form of number of shoots and inflorescence positively correlated with the abundance of *H. antonii* population. The analysis between rainfall and number of shoots and inflorescence revealed that these parameters were negatively correlated. This indicated that rainfall did not directly influence the number of *H. antonii* population, but appeared to influence the number of shoots and inflorescence (Table 35.9). The population increased just after the end of rains and reached a peak when rainfall was intermittently low. The population dropped when the rainfall was persistently low, and absent with increased rainfall until peak rainfall (Siswanto *et al.*, 2008).

Table 35.9: Coefficient of correlation (r) between *H. antonii* population and rainfall, temperature and relative humidity during two consecutive flushing-flowering stages

Stages	*Month*	*Rainfall*	*Temp.*	*Relative Humidity*	*Shoots + Inflore-scence*	*Predator*
Flushing-flowering	March-Sept. 2004	(ns)	–0.61	(ns)	0.73**	(ns)
Post flowering	Sept-March 2004-2005	(ns)	(ns)	(ns)	(ns)	(ns)
Flushing-flowering	March-Nov. 2005	(ns)	(ns)	(ns)	0.58*	(ns)
Post flowering	Nov-April 2005-2006	(ns)	(ns)	(ns)	0.78**	(ns)
Pooled stages	March-May 2004-2006	–0.37**	–0.41*	(ns)	0.63**	(ns)

ns: Not significant; *: Significant at 5 per cent; **: Significant at 1 per cent.

The hot spot areas of TMB across the cashew tract of India have been demarcated by taking into account the optimum night temperature during the flushing and flowering stages. The criteria adopted for intensity of pest based on night surface air temperature as the pest population was moderate to severe across the west coast, commencing from northern districts of Kerala to southern districts of Maharashtra were night surface air temperature prevails between 14-18°C. The pest population was low to moderate across the east coast and is the high ranges and relatively absent in West Bengal and north eastern states with night surface air temperature between 12-14°C. In the areas were the surface temperature was in between 10-12°C the population was low and nil when temperature was below 10°C (Prasad Rao, 2002).

The surveys conducted during 2010 to 2013 in west coast and east coast belts of India suggested the existence of four species of TMB (*H. antonii, H. bradyi, H. theivora* and *Pachypeltis maesarum* Kirk.) in Karnataka and Goa while, *H. antonii, H. bradyi* and *H. theivora* occurred in Tamil Nadu; *H. antonii, H. theivora* and *P. maesarum* were observed in Kerala and *H. bradyi* and *H. theivora* in Meghalaya. *H. antonii* was extensively prevalent in Gujarat, Maharashtra, Chhattisgarh, Andaman and Nicobar Islands and Odisha. In Darjeeling tea plantations of West Bengal, the only species encountered was *H. theivora* (Figure 35.23) (Srikumar *et al.*, unpublished).

Understanding the impact of climate change on cashew pests helps the crop protection scientist to orient their research on various futuristic possibilities that can help in mitigating and adapting to menace of anticipated climate change. Although the effect of climate change on cashew pests has been studied by biologists, thus far, large-scale assessments of climate change and cashew have not included the impact of pests. It is evident that weather plays a major role in triggering pest population of foliage pests of cashew. The endemic areas could be well identified with systematic collection of data on pest surveillance and weather variables can be concluded that, there is a need for systematic documentation of major and minor pests of cahew, through investigation of metabolic alteration in insects to changing environment. Developing prediction models and studying evolutionary changes under modified environment would be useful to face the challenge in near future.

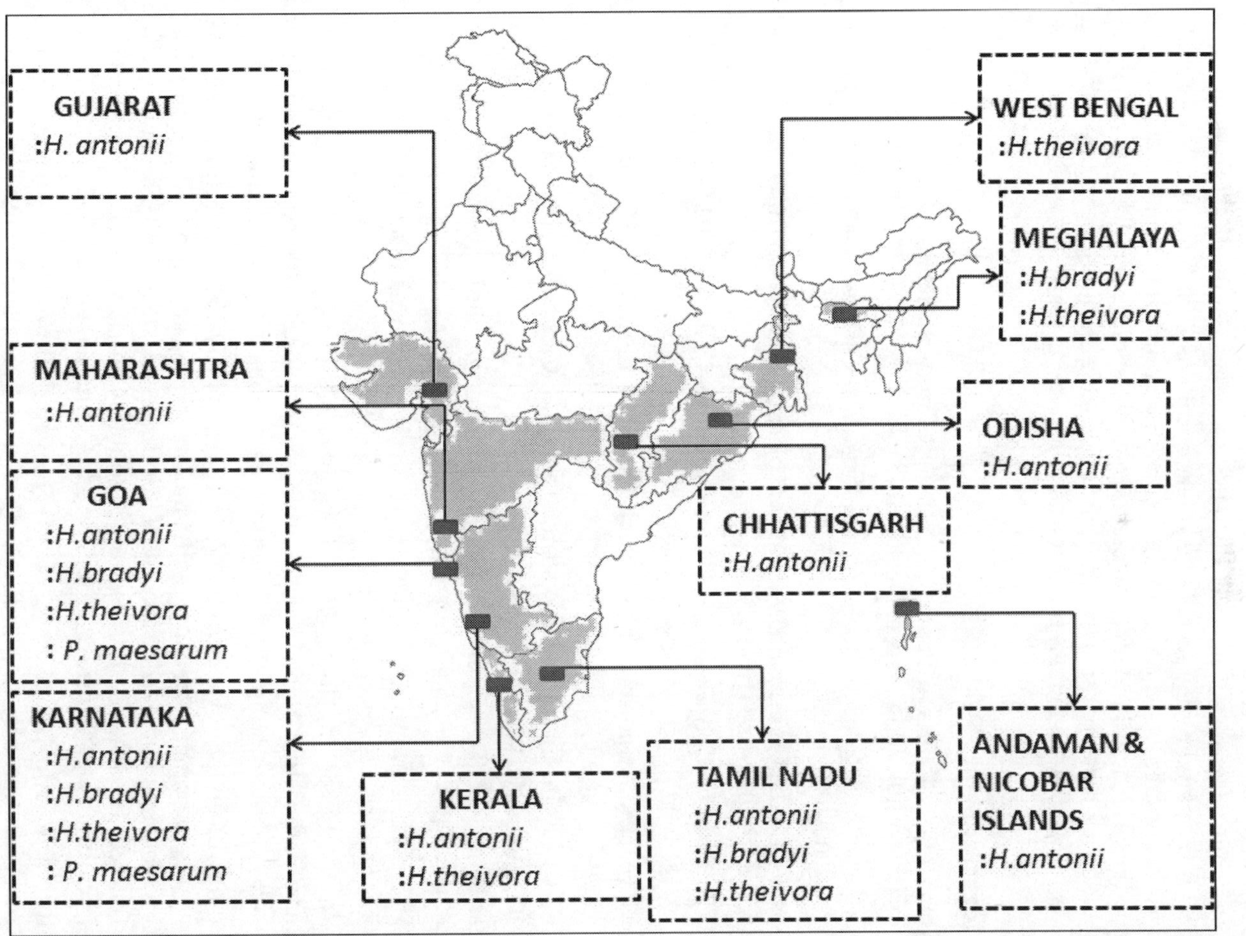

Figure 35.23: Distribution of TMB Population in India.

Population dynamics of cashew pests were influenced by the seasonal weather conditions. A better understanding of population dynamics and seasonal weather variation mechanism can fine-tune crop response to pest attacks and pest management strategies. Close monitoring of weather data alarms the need for proper timing for control of pests since the biological parameters and population of pest is triggered by favourable weather conditions. It is evident that minimum and maximum temperature, humidity, sunshine hours and rainfall influence pest population. The development of prediction equation will be useful in understanding the effects of these parameters. It is indispensable in hot spot regions and the successful pest management programme relies on the knowledge of these abiotic factors.

References

Abrol, Y., Gadgil, S. and Pant, G. B. (1996). Climate variability and Agriculture. Narosa Publishing House, New Delhi.

AICRP (All India Coordinated Research Project on Cashew). 2009. Annual Report for 2008-2009, pp. 111.

AICRP (All India Coordinated Research Project on Cashew). 2010. Annual Report for 2009-2010. pp. 110.

AICRP (All India Coordinated Research Project on Cashew). 2011. Annual Report for 2010-2011. pp. 124.

AICRP (All India Coordinated Research Project on Cashew). 2012. Annual Report for 2011-2012. pp. 112.

AICRP (All India Coordinated Research Project on Cashew). 2013. Annual Report for 2012-2013. pp. 106.

Aggarwal, P. K. (2008). Climate change and Indian Agriculture: impacts, adaptation and mitigation, *Indian Journal of Agricultural Sciences*, 78: 911-919.

Bale, J. S. and Hayward, S. A. L. (2010). Insect overwintering in a changing climate. *The Journal of Experimental Biology*, 213: 980-994.

Bale. J., Masters, G. J., Hodkins, I. D., Awmack, C., Bezemer, T. M., Brown, V. K. and Buterfield, J. (2002). Herbivory in global climate change research: direct effects of rising temperature on insect herbivores. *Global Change Biology*, 8: 1-16.

Carroll, A. L, Taylor, S. W., Regniere, J. and Safranyik, D. L. (2004). Effects of climate change on range expansion by the mountain pine beetle in British Columbia. In: *Mountain Pine Beetle Symposium: Challenges and Solutions* (Eds.: Shore, T.L., Brooks, J.E. and Stone, J.E.), Natural Resources Canada, Canadian Forest Service, Pacific Forestry Centre, Victoria, British Columbia. pp. 223-232.

Caulifield, F. and Bunce, J. A. (1994). Elevated atmospheric carbon dioxide concentration affects interactions between *Spodoptera exigua* (Lepidoptera: Noctuidae) larvae and two host plant species outdoors. *Environmental Entomology*, 23: 999-1005.

Chahal, S. K., Bains, G. S. and Dhaliwal, L. K. (2008). Climate change: Mitigation and Adaptation. Proceedings of International Conference on Climate Change, Biodiversity and Food Security in the South Asian Region, 3-4 November, 2008, Punjab State Council for Science and Technology, Chandigarh and United Nations Educational, Scientific and Cultural Organization, New Delhi. pp. 12.

Chakraborti, S. and Majumder. 2007. Meteorological determinants of the incidence of insect pests and their damage on cashew (*Anacardium occidentale* L.) in coastal region of West Bengal. *Indian Journal of Agricultural Sciences,* 77(4): 264-266.

Chand, R. and Raju, S. S. (2009). Instability in Indian Agriculture during different phases of technology and policy, *Indian Journal of Agricultural Economics,* 64: 187-207.

Chaudhary, J.L. and Sahu, K.R. (2009). Association of weather parameters with population fluctuations of tea mosquito bug of cashew in Bastar region of Chhattisgarh. *Journal of Agrometeorology,* 11: 153-155.

Deka, S., Byjesh, K., Kumar, U. and Choudhary, R. (2008). Climate change and impacts on crop pests- a critique. ISPRS Archives XXXVIII-8/W3 Workshop Proceedings: Impact of climate change on Agriculture, pp. 147-149.

De, U. S. and Mukhopadhyay, R. K. 1999. Severe heat wave over the Indian subcontinent in 1998 in perspective of global climate, *Current Science,* 75: 1308-1315.

Diffenbaugh, N. S., Krupke, C. H., White, M. A. and Alexander, C. E. (2008). Global warming presents new challenges for maize pest management. *Environmental Research Letter,* 3: 1-9.

Fand, B. B., Ankush, L. K. and Kumar, M. (2012). Will climate change pose serious threat to crop pest management: A critical review? *International Journal of Scientific and Research Publications,* 2 (11): 1-14.

Gadgil, S. (1996). Climate change and agriculture-an indian perspective. In : "Climatic Variability and Agriculture". Abrol, Y.P. *et al.* (Eds), Narosa Publishing House, New Delhi, India, pp. 1-18.

Guhathakurta, P. and Rajeevan, M. (2008). Trends in the rainfall pattern over India, International *Journal of Climatology,* 28(11): 1453-1469.

Hahn, D. A. and Denlinger, D. L. (2007). Meeting the energetic demands of insect diapause: nutrient storage and utilization. *Journal of Insect Physiology,* 53: 760-773.

Harrington, R., Fleming, R. and Woiwood, I. P. (2001). Climate change impacts on insect management and conservation in temperate regions: can they be predicted? *Agriculture and Forest Entomology,* 3: 233-240.

Hill, M. G. and Dymock, J. J. (1989). Impact of Climate Change: Agricultural/ Horticultural Systems. DSIR Entomology Division Submission to the New Zealand Climate Change Program. Department of Scientific and Industrial Research, Auckland, New Zealand, pp. 16.

Hingane, L. S., Rupakumar. and Ramana, M. V. (1985). Long-term trends of surface air temperature in India. *Journal of Climatology*, 5: 521-528.

IARI (Indian Agricultural Research Institute) News. (2008). Brown plant hopper outbreak in rice, 24, pp. 1-2.

IPCC (Intergovernmental Panel on Climate Change) (2007). Climate Change-Impacts, Adaptation and Vulnerability. In: (Eds.: Parry, M.L., Canziani, O.F., Palutikof, J.P., van der Linden, P.J., Hanson, and C.E.) Cambridge University Press, Cambridge, UK, pp. 976.

IRRI (International Rice Research Institute) News (2009). Pest outbreaks in India. *Rice Today*, 6. www.ricenews.irri.org.

Jeevaratnam, K. and Rajapakse, R. H. S. (1981). Biology of *Helopeltis antonii* Sign. (Heteroptera: Miridae) in Sri Lanka. *Entomon*, 6: 247-251.

Joshi, S. and Viraktamath, C. A. (2004). The sugarcane woolly aphid, *Ceratovacuna lanigera* Zehntner (Hemiptera: Aphididae): its biology, pest status and control. *Current Science*, 87: 307-316.

Kannan, R. and James, D. A. (2009). Effects of climate change on global diversity: a review of key literature, *Tropical Ecology*, 50: 31-39.

Kelkar, R. R. (2000). The National Weather Service in India, 125 years, WMO Bulletin, 49, pp. 375-382.

Konestabo, H. S., Michelsen, A. and Holmstrup, M. (2007). Responses of springtail and mite populations to prolonged periods of soil freeze-thaw cycles in a sub-arctic ecosystem. *Applied Soil Ecology*, 36: 136-146.

Kuchlein, J. H. and Ellis, W. N. (1997). Climate-induced changes in the microlepidoptera fauna of the Netherlands and the implications for nature conservation. *Journal of Insect Conservation*, 1: 73-80.

Logan, J. A., Regniere, J. and Powell, J. A. 2003. Assessing the impacts of global warming on forest pest dynamics. *Frontiers in Ecology and Environment*, 1: 130–137.

Mittler, R. (2006). Abiotic stress, the field environment and stress combination, *Trends in Plant Science*, 11: 15-19.

NACCAP. (2008). Climate change impacts on pest animals and weeds. Communicating Climate Change. National Agriculture and Climate Change Action Plan (NACCAP), Bureau of Meteorology, Department of Agriculture, Fisheries and Forestry, Australian Government, pp. 1-6.

Oerke, E. C., Debne, H. W., Schonbeck, F. and Weber, A. (1994). Crop Production and Crop Protection. Elsevier, Amsterdam.

Pant, G. B., Kumar, K. R. and Kumar, K. K. (1996). Climate change and variability over south asia. In: "Climate variability and agriculture", Abrol Y.P. *et al.* (Eds.), Narosa Publishing House, New Delhi, India, pp. 51-68.

Parry, M. L. and Carter, T. R. (1989). An assessment of the effects of climatic change on agriculture, *Climatic Change* 15: 95-116.

Patterson, D. T., Westbrook, J. K., Joyce, R. J. V., Lingren, P. D. and Rogasik, J. (1999). Weeds, Insects and Disease, *Climatic Change*, 43: 711-727.

Petzoldt, C. and Seaman, A. (2010. Climate Change Effects on Insects and Pathogens. Climate Change and Agriculture: Promoting Practical and Profitable Responses, 6. 16. http: //www.climateandfarming.org/pdfs/FactSheets/III.2Insects. Pathogens.pdf.

Pillai, G. B. and Abraham, V. A. (1975). Tea mosquito-a serious menace to cashew. *Indian Cashew Journal*, 10: 5-7.

Pillai, G. B., Dubey, O. P. and Singh, V. (1976). Pests of cashew and their control in India. A review of current status. *Journal of Plantation Crops*, 4: 37-50.

Pillai, G. B., Singh, V., Dubey, O. P. and Abraham, V. A. (1984). Seasonal abundance of tea mosquito bug, *Helopeltis antonii* on cashew in relation to meteorological factors. In: Cashew research and development (Eds: E.V.V.B. Roa and H.H Khan) Indian Society for Plantation Crops, CPCRI, Kasaragod, India, pp.103-110.

Prasada Rao, G. S. L. H. V. and Gopakumar, C. S. (1994). Climate and cashew. *The Cashew*, 8(4): 3-9.

Prasada Rao, G. S. L. H. V. (2002). Climate and cashew. AICRP on Agro meteorology, Department of Agricultural Meteorology, College of Horticulture, Kerala Agricultural University, 100 pp.

Prasad Rao, G. S. L. H. V., Giridharan, M. P. and Jayaprakash Naik, B. (2001). Influence of weather factors on bud break, flowering and nut quality of cashew. *The Indian Journal of Agricultural Sciences*, 71 (6): 199-402.

Rai, P. S. (1984). *Handbook on Cashew Pests.* Researchco Publications, Delhi, pp. 124.

Ramadevi, M. and Krishna Murthy, P. R. (1983). Schedule of pest occurrence on cashew (*Anacardium occidentale* L.) *Cashew Causerie*, 5(1): 14-16.

Rao, G.G.S.N., Rao, A.V.M.S., Rao, V. U. M. (2009). Trends in rainfall and temperature in rainfed India in previous century. In: *Global climate change and Indian Agriculture case studies from ICAR network project*, (Ed.: PK Aggarwal), ICAR Publication, New Delhi, pp. 71-73.

Rhoades, D. F. (1985). Offensive-defensive interactions between herbivores and plants: their relevance in herbivore population dynamics and ecological theory. *American Naturalist*, 125: 205-238.

Rupa, T. R., Rejani, R. and Bhat, M. G. (2013). Climate Change Effects on Cashew and Adaptation Strategies. 2013. Climate-Resilient Horticulture: Adaptation and Mitigation Strategies, pp. 189-198.

Samra, J. S. (2003). Impact of Climate and Weather on Indian Agriculture. *Indian Society of Soil Science*, 51: 418-430.

Satapathy, C. R. (1993). Bioecology of major insect pests of cashew (*Anacardium occidentale* Linn.) and evaluation of certain pest management practices. *Ph.D. Thesis*, U.A.S., Bangalore, pp. 224.

Sharma, H. C., Dhillon, M. K., Kibuka, J. and Mukuru, S. Z. (2005). Plant defense responses to sorghum spotted stem borer, *Chilo partellus* under irrigated and drought conditions. *International Sorghum and Millets Newsletter*, 46: 49-52.

Sharma, H. C., Srivastava, C. P., Durairaj, C. and Gowda, C. L. L. (2010). Pest management in grain legumes and climate change. In *Climate Change and Management of Cool Season Grain Legume Crops* (eds.Yadav, S.S., McNeil, D.L. Redden, R, and Patil, S.A,), Business Media, Springer Science, Dordrecht, The Netherlands, pp. 115-140.

Siswanto, R. M., Dzolkhifli, O. and Karmawati, E. (2008). Population Fluctuation of *Helopeltis antonii* Signoret on Cashew *Anacardium occidentale* L., in Java, Indonesia. *Pertanika Journal of Tropical Agricultural Science*, 31(2): 191-196.

Sontakke, N. A. (1990). Indian summer monsoon rainfall variability during the longest period 1813-1988. M.Sc. thesis, Dept. of Physics, University of Poona, pp. 146.

Srikanth, J. (2007). World and Indian scenario of sugarcane woolly aphid. In Woolly Aphid Management in Sugarcane (eds. Mukunthan, N., Srikanth, J., Singaravelu, B., Rajula Shanthy, T., Thiagarajan, R. and Puthira Prathap, D.), Extension Publication, Sugarcane Breeding Institute, Coimbatore, 154, pp. 1-12.

Sundararaju, D. (1993). Compilation of recently recorded and some new pests of cashew in India. *The Cashew*, **7**: 15-19.

Sundararaju, D. (2005). Seasonal abundance and extent of damage of tea mosquito bug on cashew. *Journal of Plantation Crops*, 33: 53-58.

Sundararaju, D. and Sundarababu, P. C. (1999). *Helopeltis* spp. (Heteroptera: Miridae) and their management in plantation and horticultural crops of India. *Journal of Plantation Crops*, 27: 155-174.

Theilert, W. (2006). A unique product: The story of the imidacloprid stress shield. *Pflanzenschutz-Nachrichten Science Forum Bayer*, 59: 73-86.

Thomas, C. D, Cameron A and Green R E. (2004). Extinction risk from climate change. *Nature*, 427: 145-148.

Timoney, K. P. (2003). The changing disturbance regime of the boreal forest of the Canadian Prairie Provinces. *Forest Chronicle*, 79: 502-516.

UN-HABITAT (United Nations Human Settlements Programme). (2004). The State of the World's Cities 2004/2005- Globalization and Urban Culture, Earthscan, London, pp. 198.

Volney, W. J. A. and Fleming, R. A. (2000). Climate change and impacts of boreal forest insects. *Agriculture Ecosystem and Environment*, 82: 283-294.

Wigley, T. M. L. and Jones, P. D. (1989). Detecting CO_2, induced climatic change. *Nature*, 292: 205.

Yamamura, K. and Kiritani, K. (1998). A simple method to estimate the potential increase in the number of generations under global warming in temperate zones. *Applied Entomology and Zoology*, 33: 289-298.

Yadukumar, N., Raviprasad, T. N. and Bhat, M.G. (2008). Status of effect of climatic change on yield and insect pest incidence in cashew. In: Impact Assessment of Climatic Change for Research Priority Planning in Horticultural Crops. (Eds. Lal, S.S., Govindakrishnan P.M., Dua V. K., Singh, J. P.and Pandey, S.K.) CPRI, Shimla. pp. 100-104.

Index

C

D

E

F

G

H

N

O

P

U

V

W

X

Y

Z

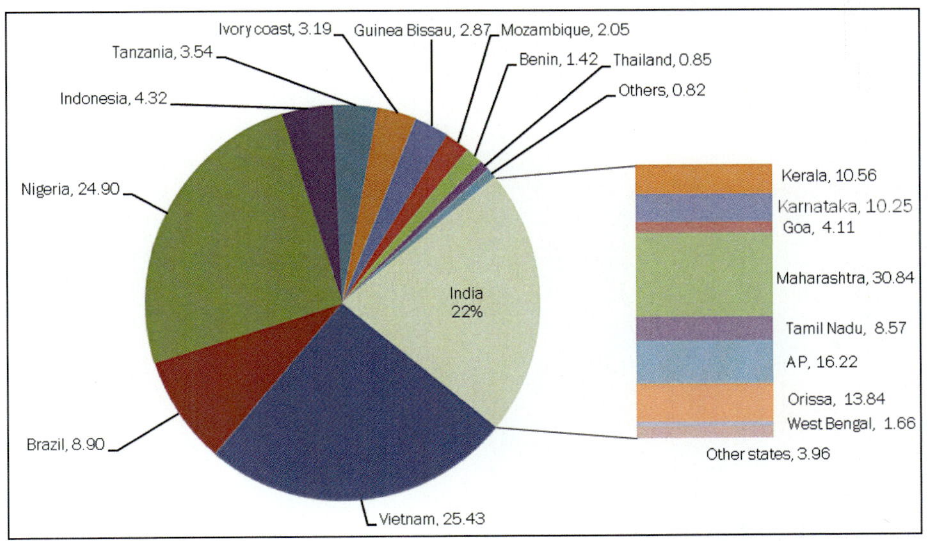

Figure 3.1: Share of Cashewnut Production in the Domestic and International Level. (Pg. 50)

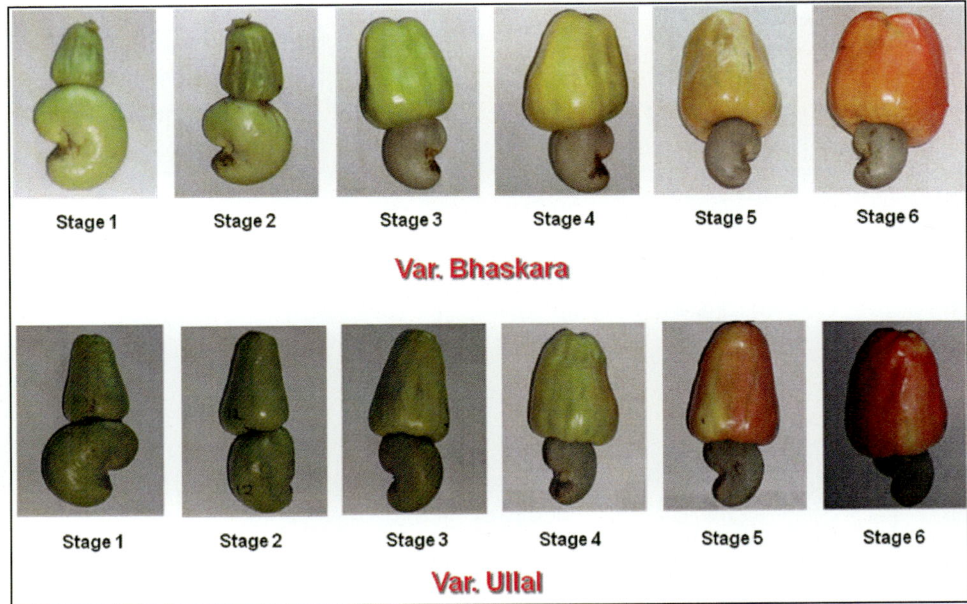

Figure 3.2: Changes in the Size and Surface Colour of Raw Cashewnuts at Various Stages of Maturity. (Pg. 51)

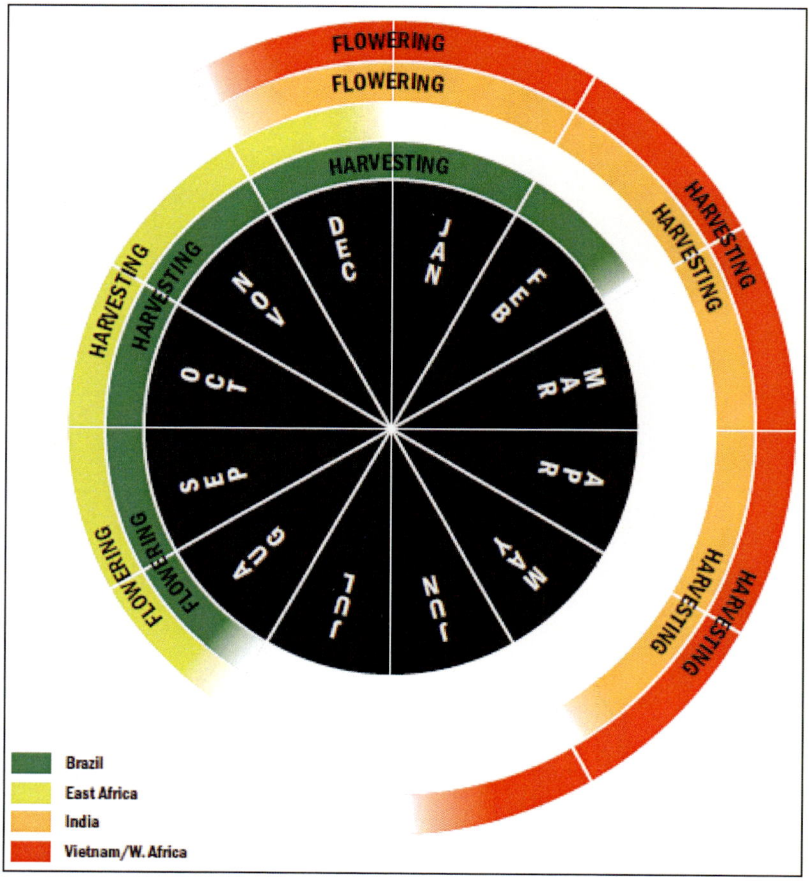

FLOWERING
FLOWERING
HARVESTING
HARVESTING
HARVESTING
HARVESTING
HARVESTING
FLOWERING
FLOWERING
HARVESTING
HARVESTING
DEC
JAN
FEB
MAR
APR
MAY
JUN
JUL
AUG
SEP
OCT
NOV
Brazil
East Africa
India
Vietnam/W. Africa

Figure 3.3: Harvesting Period of Cashew in Major Cashew Growing Countries. (Pg. 53)

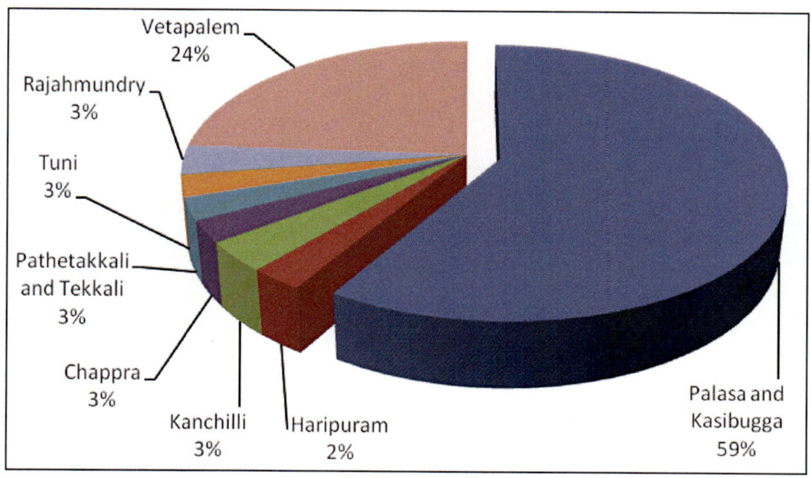

Figure 3.20: Distribution of Cashew Processing Units in Srikakulam District, Andhra Pradesh. (Pg. 72)

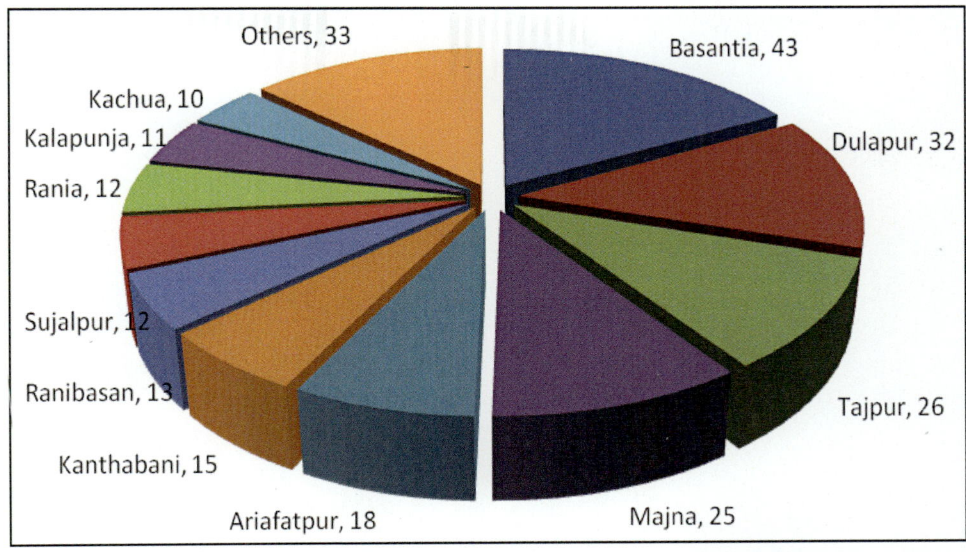

Figure 3.21: Distribution of Cashewnut Processing Units in West Bengal. (Pg. 75)

Figure 4.1: A Cashew Plant in Flowering and Fruiting. (Pg. 96)

Figure 4.3: Male Flower. (Pg. 98)

Figure 4.4: Hermaphrodite Flower. (Pg. 98)

Figure 4.5: Cluster of Nuts. (Pg. 98)

Figure 4.6: Cashew Apple with Nut. (Pg. 101)

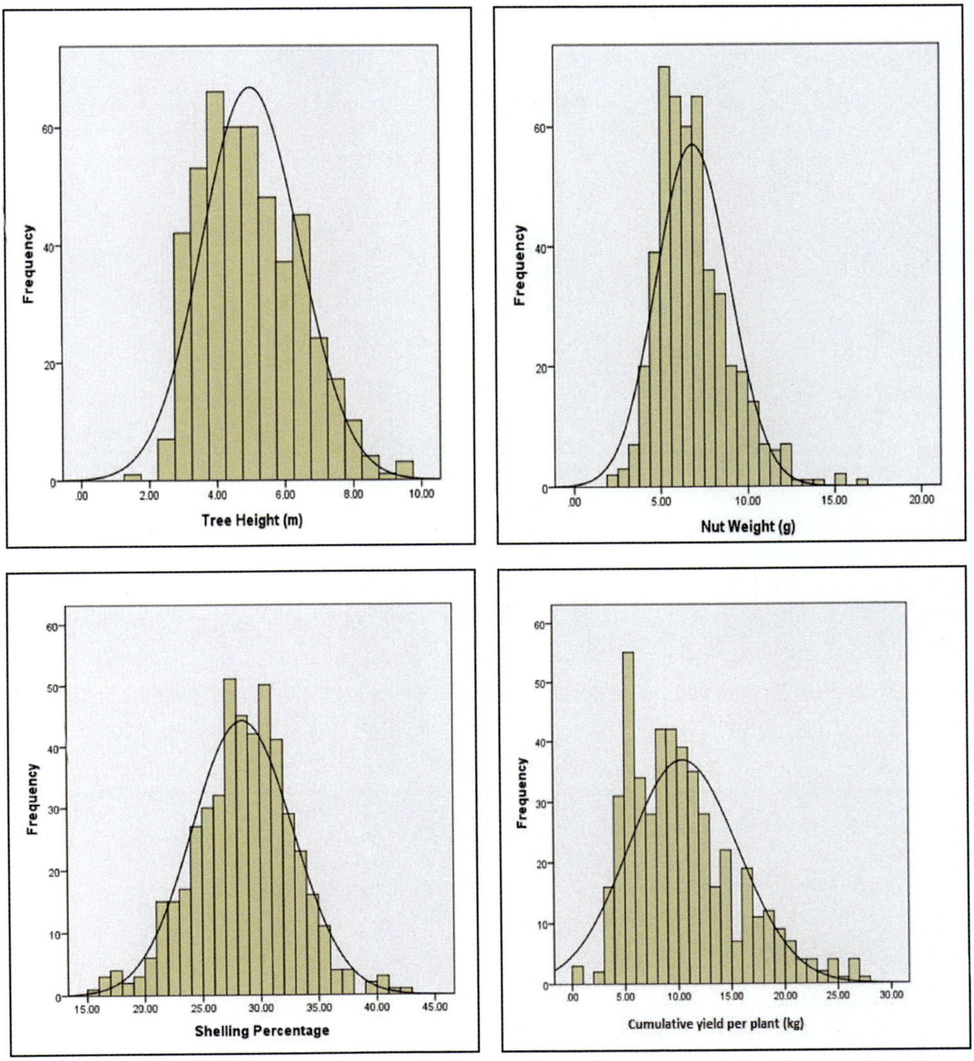

Figure 5.2: Frequency Distribution of some Quantitative Characters. (Pg. 112)

Figure 5.3: Compactness of Inflorescence. (Pg. 112)

Figure 5.1: Purple Accession of Cashew. (Pg. 111)

Figure 5.4: Different Shapes of Cashew Apple. (Pg. 115)

Figure 5.5: Different Tree Habits. (Pg. 115)

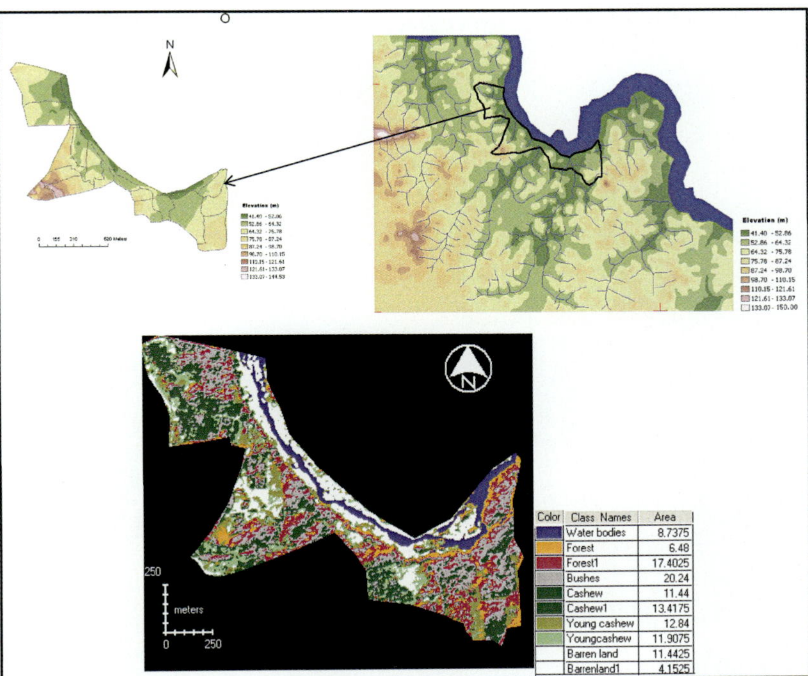

Figure 7.2: LISS IV Image Classification at Farm Level (Area in ha). (Pg. 151)

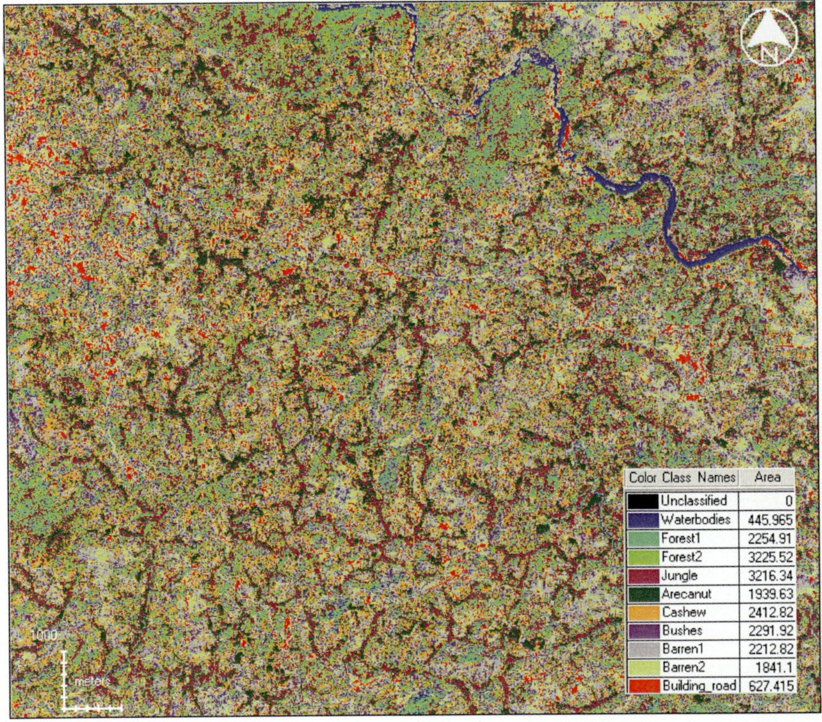

Figure 7.3: LISS IV Image Classification for Larger Area (Area in ha). (Pg. 152)

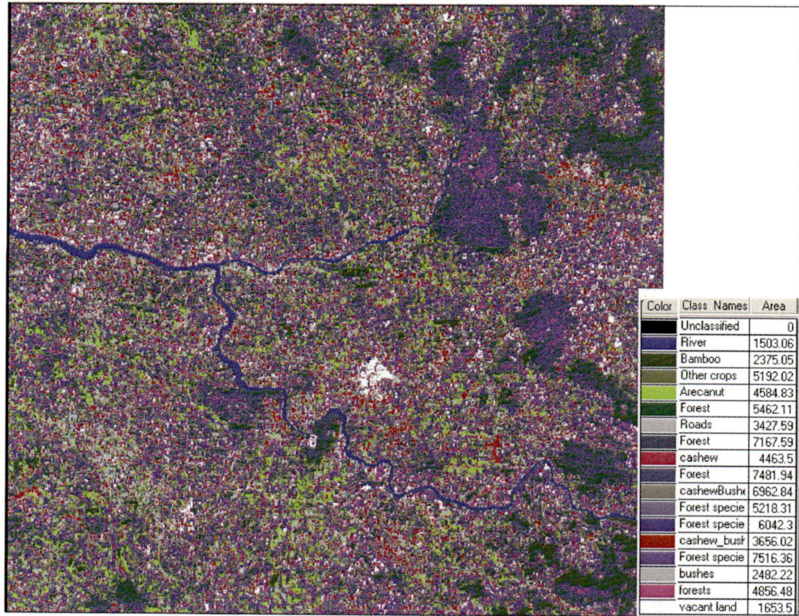

Figure 7.4: LISS III Image Classification for Larger Area (Area in ha). (Pg. 153)

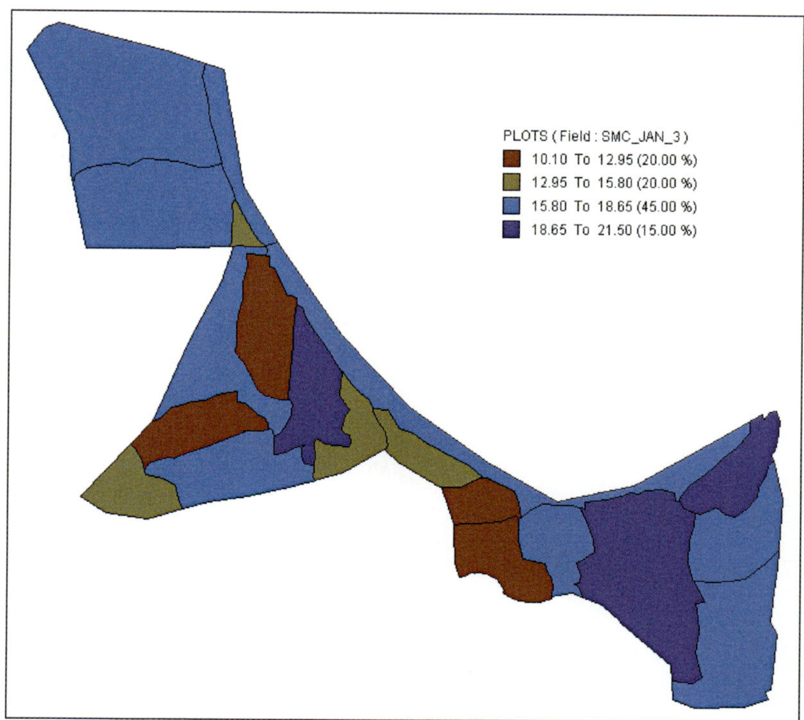

Figure 7.5: Soil Moisture Content (Per cent dry basis) at DCR Farm (0-30 and 31-60 cm depth) (January 2008). (Pg. 154)

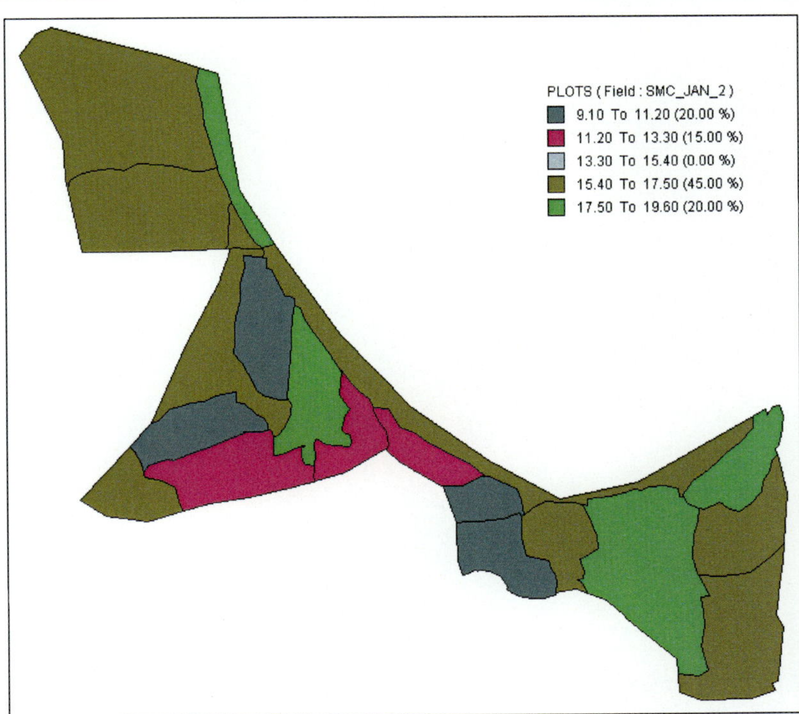

Figure 7.6: Soil Moisture Content (Per cent dry basis) at DCR Farm (0-30 and 31-60 cm depth) (January 2008). (Pg. 154)

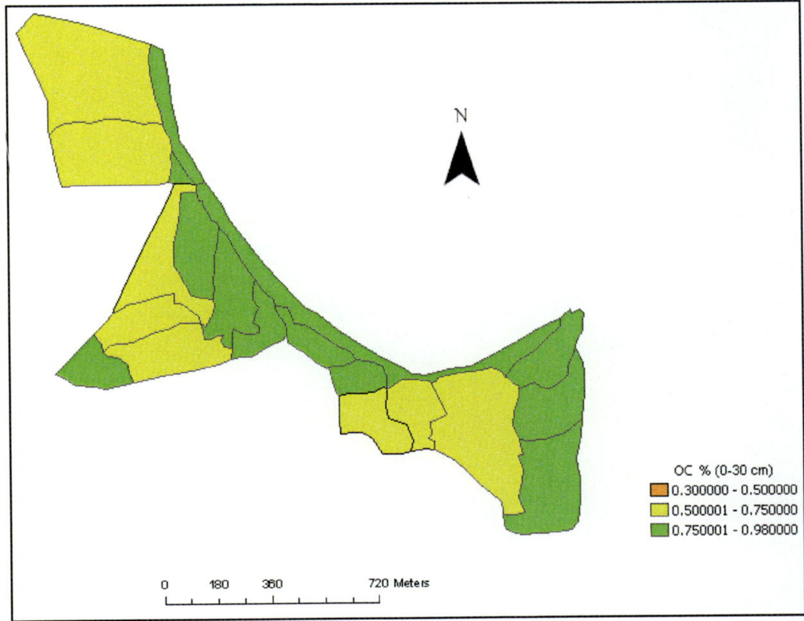

Figure 7.7: Organic Carbon Content (Per cent) and Available P_2O_5 Content (kg/ha) of the Soil (0-30 cm depth). (Pg. 155)

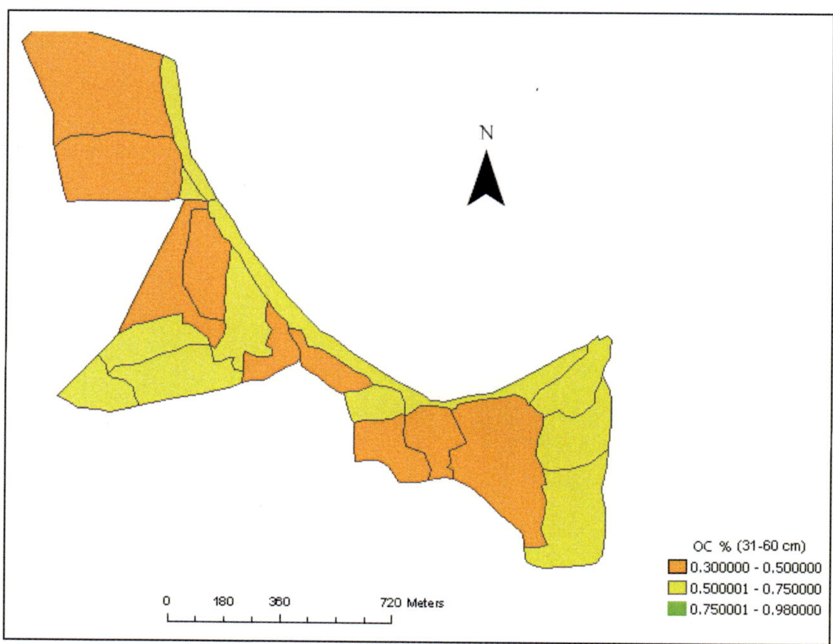

Figure 7.8: Organic Carbon Content (Per cent) and Available P_2O_5 Content (kg/ha) of the Soil (0-30 cm depth). (Pg. 155)

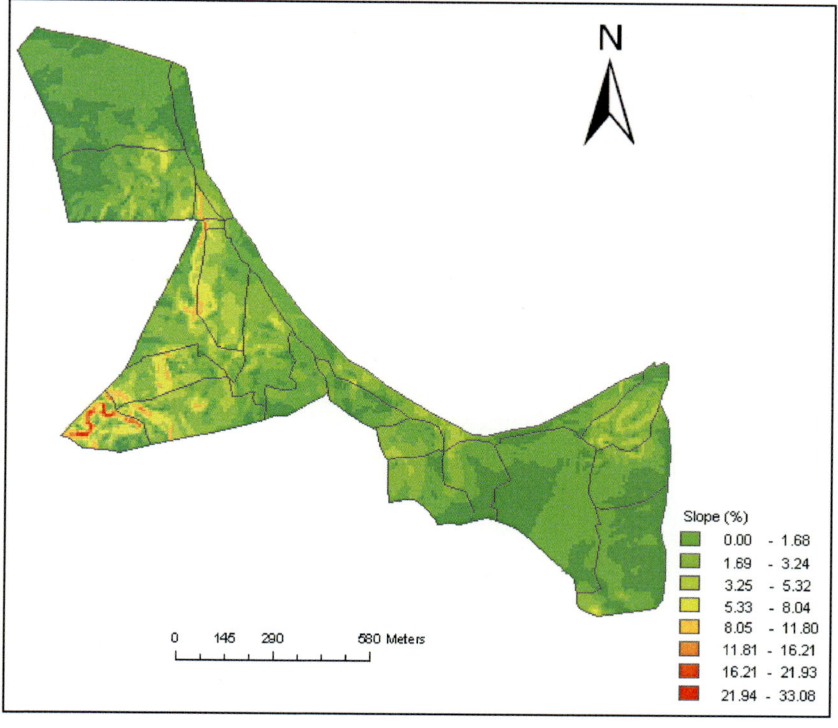

Figure 7.9: Land Slope of DCR Farm. (Pg. 157)

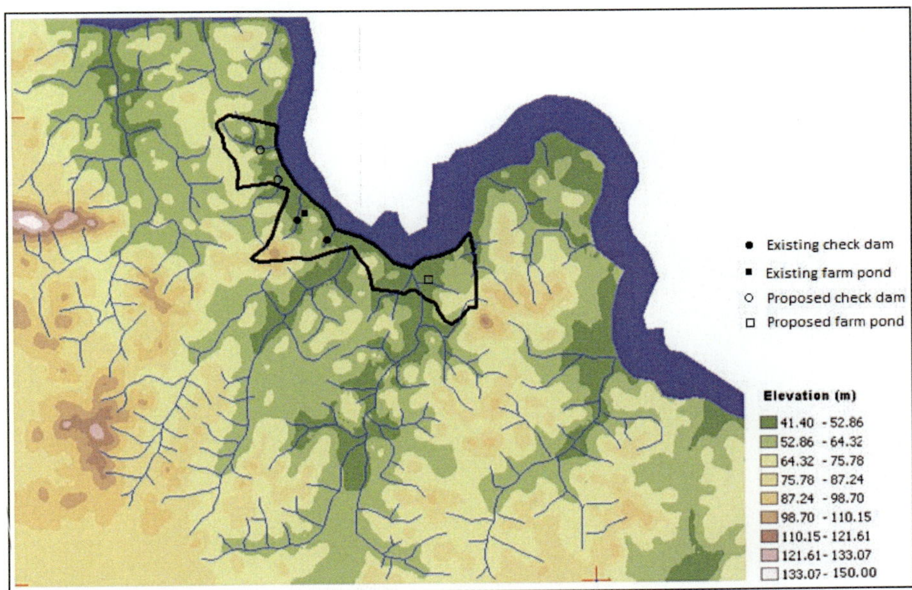

Figure 7.10: Structures Proposed for the DCR Farm. (Pg. 157)

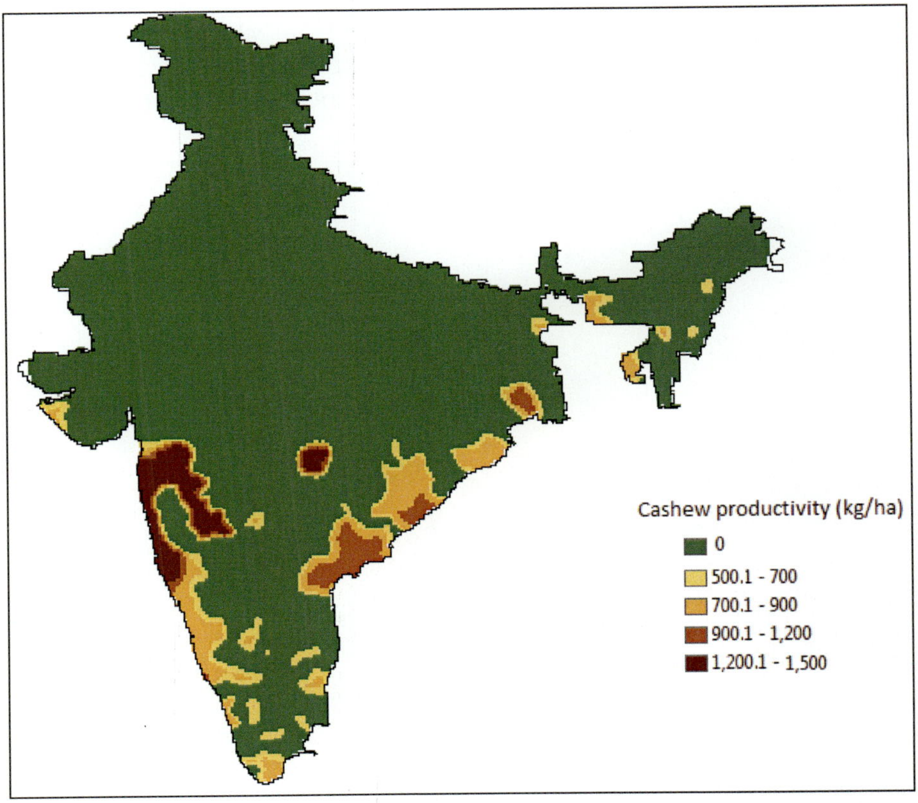

Figure 7.12: Distribution of Cashew Areas in India. (Pg. 160)

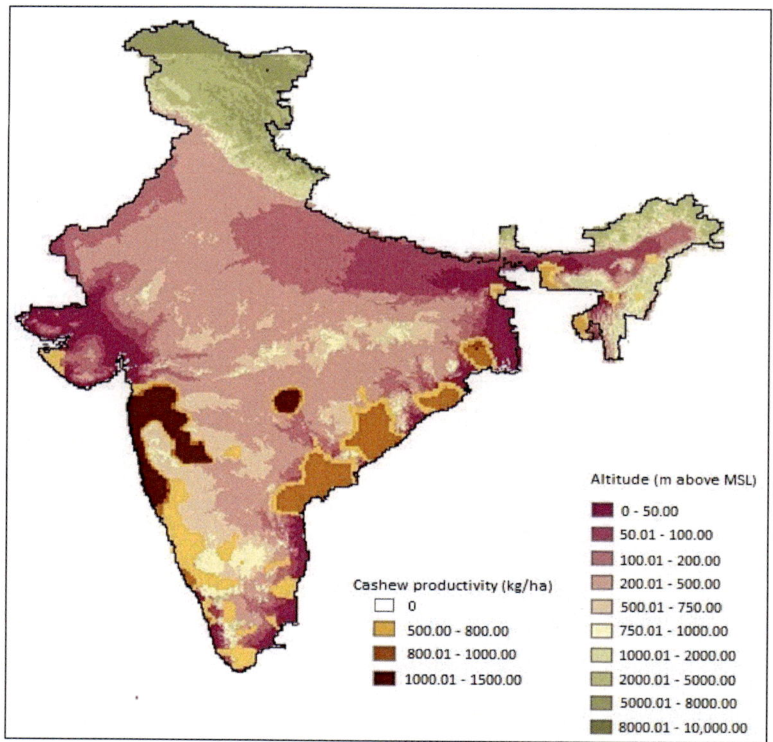

Figure 7.13: Cashew Area and Productivity (kg/ha) Overlay with Altitude (m above MSL). (Pg. 162)

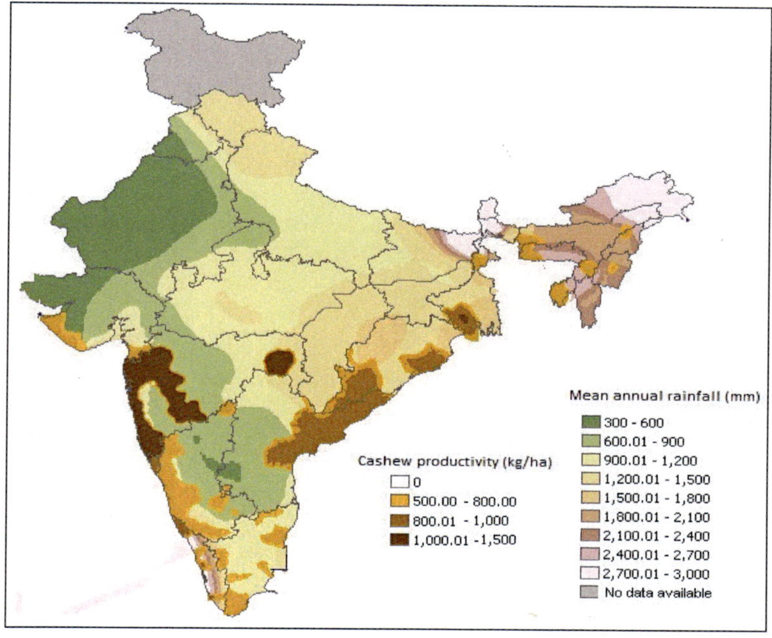

Figure 7.14: Cashew Area and Productivity (kg/ha) Overlay with Mean Annual Rainfall (mm). (Pg. 163)

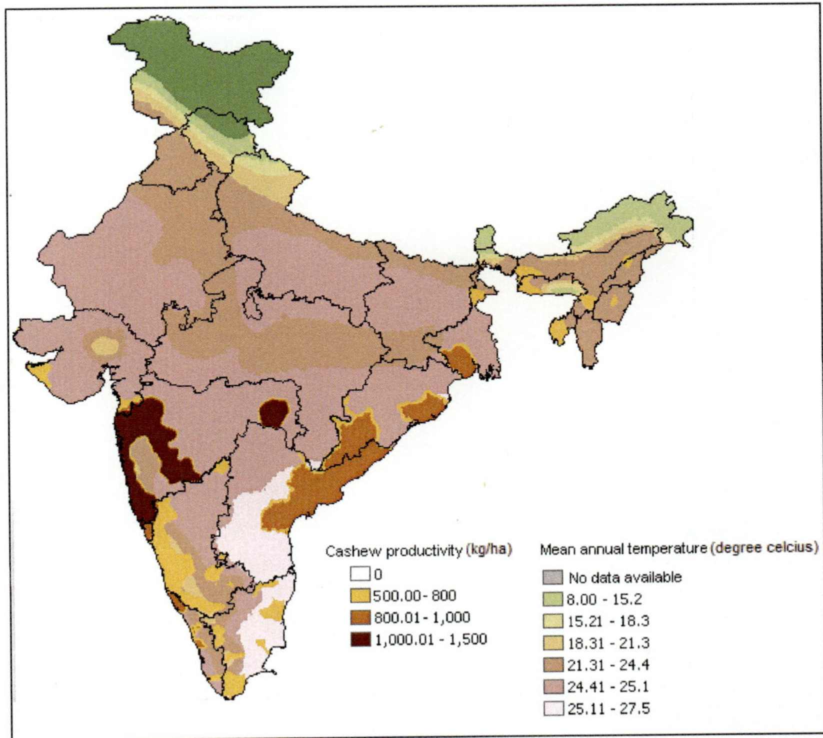

Figure 7.15: Cashew Area and Productivity (ka/ha) Overlay with Mean Annual Temperature (ºC). (Pg. 165)

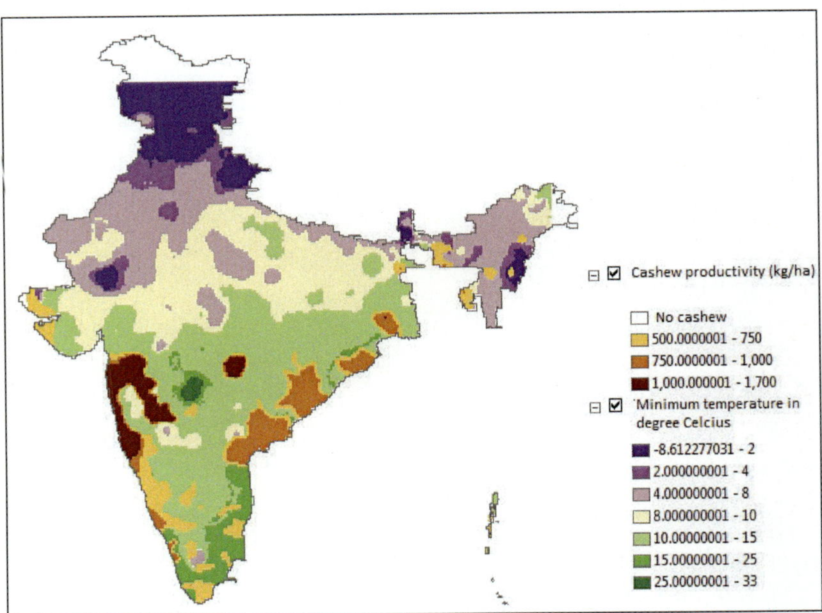

Figure 7.16: Cashew Area and Productivity Overlay with Decrease in Minimum Temperature (-2 ºC). (Pg. 166)

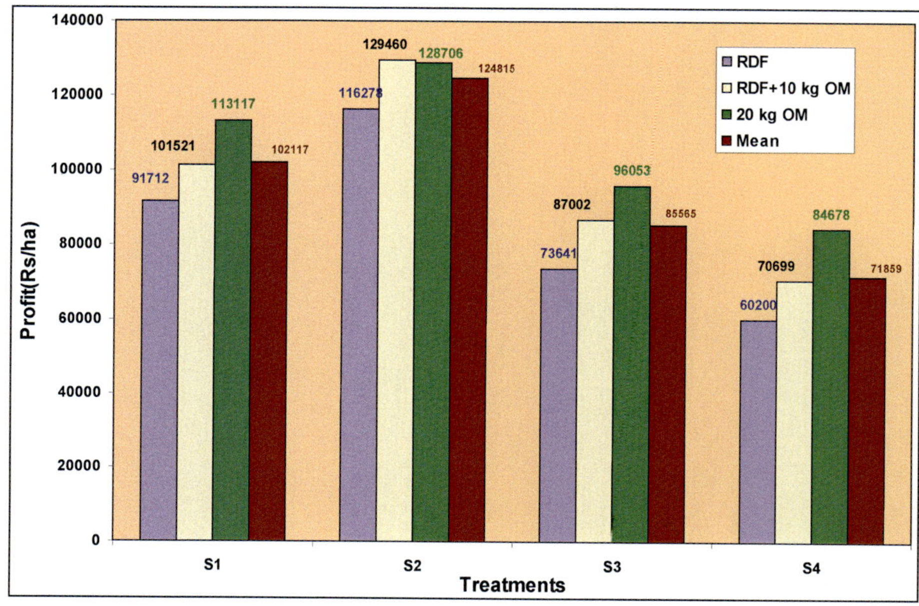

Figure 8.7: Net Profit for Five Years (2002-06). (Pg. 183)

Figure 12.3: High Density Planting at Regional Research Station, Vridhachalam. (Pg. 222)

Cashew under UHDP – At planting. (Pg. 224)

Cashew under UHDP – One year after planting. (Pg. 224)

Cashew under UHDP – 1 ½ years after planting. (Pg. 224)

Cashew under UHDP – 2 years after planting. (Pg. 224)

Cashew under UHDP – At flowering.

Cashew under UHDP – At fruiting.

Figure 12.4: UHDP at RRS, Vridhachalam, Tamil Nadu. (Pg. 224)

Figure 12.5: UHDP trial by Jain Irrigation, Udumalaipet, Tamil Nadu. (Pg. 225)

Figure 15.1: ***Chromolaena odorata*** **(Young plant). (Pg. 266)**

Figure 15.2: ***Chromolaena odorata*** **(with flower). (Pg. 266)**

Figure 15.3: ***Pennisetum polystachyon.*** **(Pg. 267)**

Figure 15.4: ***Lantana camara.*** **(Pg. 268)**

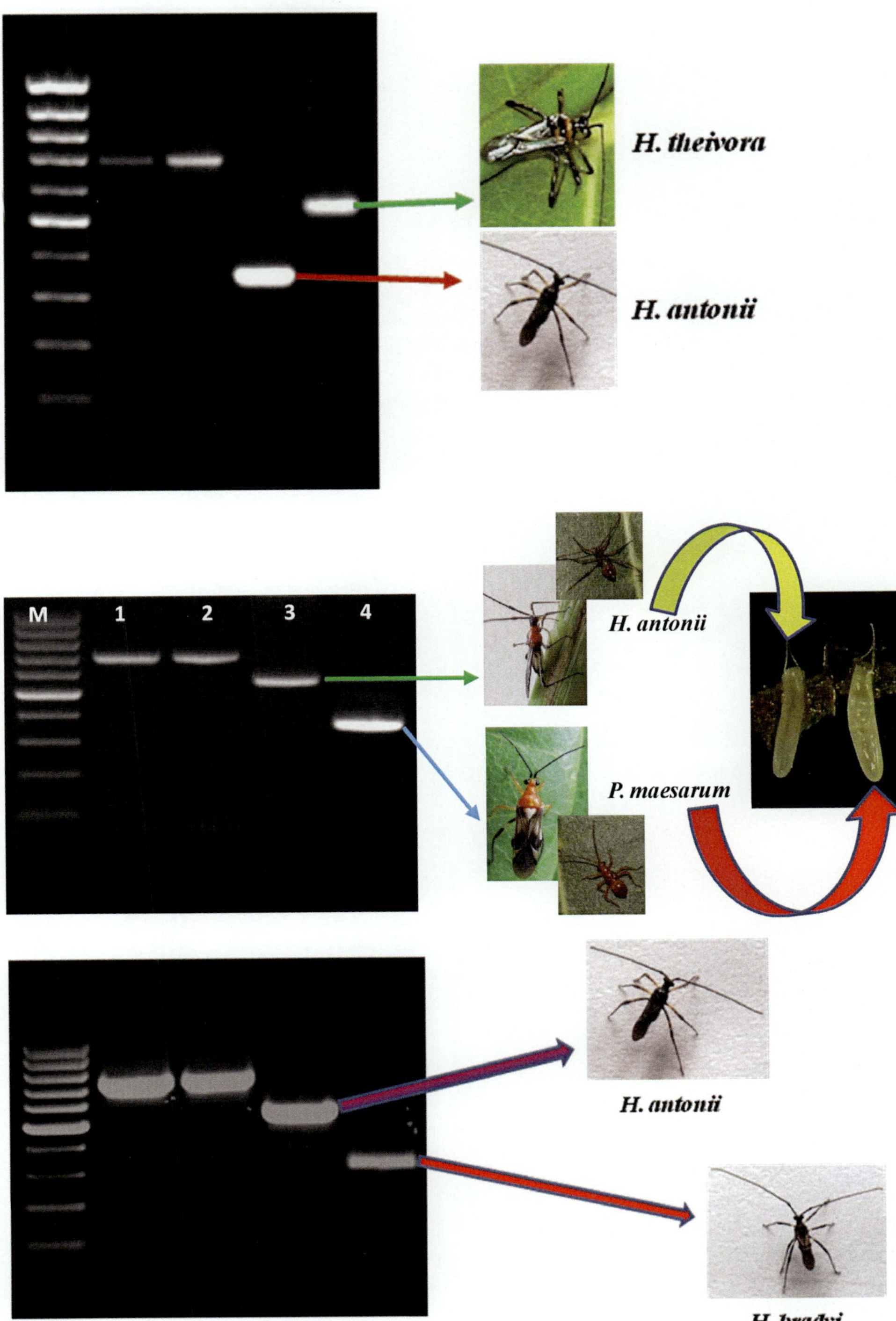

Figure 19.1: Species-Specific Markers Developed for Various Mirids Species in India. (Asokan *et al.,* 2012, Rebijith *et al.,* 2012). (Pg. 337)

Figure 20.1: ***H. antonii.***

Figure 20.2: ***H. bradyi.***

Figure 20.3: ***H. theivora.***

(Pg. 348)

Figure 20.4: Damaged Shoot.

Figure 20.5: Damaged Panicle.

Figure 20.6: Damaged Immature Nut and Apple.

(Pg. 250)

Figure 20.7: Oviposition Behaviour of *H. antonii.* (Pg. 353)

Figure 20.8: Eggs Laid in Groups by *H. antonii* (Extra-chorionic process exposed). (Pg. 353)

Figure 20.9: *Telenomus* sp. Emerging from *Helopeltis* Egg. (Pg. 356)

Figure 20.10: *Telamonia dimidiata* Predating on *H. antonii.* (Pg. 358)

Figure 20.11: Reduviid Predating on *H. antonii.* (Pg. 358)

Figure 21.1: Trunk Damage by Cashew Stem and Root Borer. (Pg. 368)

Figure 21.2: *Plocaederous* Adult. (Pg. 369)

Figure 21.3: *Plocaederous* Grub. (Pg. 369)

Figure 21.4: Cashew Stem Borer Grub and Pupae. (Pg. 369)

Figure 21.5: Chipping off the Damaged Portion of the Trunk to Remove the Grubs. (Pg. 370)

Figure 21.6: Treating the Damaged Portion of Trunk with Sealer-cum-Healer. (Pg. 370)